SEDIMENTARY GEOLOGY OF MARS

EDITED BY:

JOHN P. GROTZINGER
CALIFORNIA INSTITUTE OF TECHNOLOGY

RALPH E. MILLIKEN
UNIVERSITY OF NOTRE DAME

SEPM SPECIAL PUBLICATION NO. 102

SEPM and the authors are grateful to the following for their generous contributions to the cost of publishing and an online open access version at www.sepmonline.org at a six-month post-publication date

California Institute of Technology
Central Michigan University
LZ Technology/Jacobs-ESCG
Johns Hopkins University—Applied Physics Laboratory
Lakehead University
Michigan State University
NASA Ames Research Center
Purdue University
University of Notre Dame
University of Texas–Austin
University of Utah

This book includes a CD/DVD which contains a complete digital copy of this volume, with many color figures substituted for printed black and white versions.

SEPM (Society for Sedimentary Geology) is an international not-for-profit Society based in Tulsa, Oklahoma. Through its network of international members, the Society is dedicated to the dissemination of scientific information on sedimentology, stratigraphy, paleontology, environmental sciences, marine geology, hydrogeology, and many additional related specialties.

The Society supports members in their professional objectives by publication of two major scientific journals, the Journal of Sedimentary Research (JSR) and PALAIOS, in addition to producing technical conferences, short courses, and Special Publications. Through SEPM's Continuing Education, Publications, Meetings, and other programs, members can both gain and exchange information pertinent to their geologic specialties.

For more information about SEPM, please visit **www.sepm.org**.

Gary J. Nichols and Brian Ricketts, SEPM Special Publication Editors
SEPM Special Publication 102

ISBN 978-1-56576-312-8
CD/DVD version 978-1-56576-313-5

4111 S Darlington, Suite 100
Tulsa, Oklahoma 74135-6373, USA

Sedimentary Geology of Mars

John P. Grotzinger and Ralph E. Milliken, Editors

CONTENTS

THE SEDIMENTARY ROCK RECORD OF MARS: DISTRIBUTION, ORIGINS, AND GLOBAL STRATIGRAPHY

JOHN P. GROTZINGER
Division of Geological and Planetary Sciences, California Institute of Technology, Pasadena, California 91108 USA
e-mail: grotz@gps.caltech.edu

AND

RALPH E. MILLIKEN
Department of Civil Engineering and Geological Sciences, University of Notre Dame, Notre Dame, Indiana 46556 USA

ABSTRACT: The last decade of Mars exploration produced a series of discoveries that provide compelling evidence for the existence of sedimentary rocks on Mars. Previously, Mars was regarded principally as a volcanic planet, the dominant surface processes of which were eruption of lavas and pyroclastic deposits, although early studies did recognize valley networks, enormous outflow channels, and the required transport of sedimentary materials to the northern plains of Mars. In contrast, our new view of Mars shows a rich history of interactions between water and the surface, with weathering, transport, and deposition of sediments by water as well as eolian processes. Surprisingly thick accumulations of stratified rocks extend back into the Noachian Era—the oldest of which were likely formed over 4 billion years ago, making these rocks much older than any sedimentary rocks preserved on Earth.

Some sedimentary rocks were formed and deposited locally, whereas others accumulated as vast sheets that can be correlated for hundreds of kilometers or farther. Local deposits were formed in alluvial fan, deltaic, sublacustrine fan, and lacustrine environments in addition to deposits that fill canyons and valleys possibly carved during catastrophic floods. These former deposits indicate more gradual erosion and sedimentation, perhaps even involving meteoric precipitation, and they provide support for the notion of clement conditions on early Mars. In contrast, rapid erosion and sedimentation may have occurred within large, regional outflow channels thought to have resulted from outbursts of groundwater. Regionally extensive sedimentary deposits have less obvious origins, but the presence of hydrated sulfate minerals indicates that some of these deposits may have formed as lacustrine evaporites, particularly in the Valles Marineris network of open and closed basins. Others may have involved eolian reworking of previously deposited sulfates, or perhaps aqueous (groundwater) alteration of previously deposited basaltic sediments. Another major type of regionally extensive sedimentary deposit occurs as meter-scale stratification with highly rhythmic organization. These deposits occur in several places in the Arabia Terra region of Mars and are also observed at the top of a 5-km-thick stratigraphic section in Gale Crater. The significant lateral continuity of relatively thin beds, their distribution over broadly defined highs as well as lows, and the lack of strong spectral absorption features indicate that these rocks may be duststones, formed by weak lithification of fine particles that settled from the Martian atmosphere. The most ancient sedimentary deposits on Mars may be dominated by stacked, impact-generated debris sheets, similar to those seen on the Moon, and may include impact melts.

In the absence of plate tectonics, it appears that the flux of sediment on Mars has declined over time. Early on, primary sediments may have consisted mainly of impact- and volcanic-generated particles that would have been transported by fluvial and eolian processes. Chemical weathering of fragmented bedrock in the presence of circum-neutral pH fluids would have generated clay minerals and carbonates, though the latter are surprisingly rare; weathering under more acidic conditions generated dissolved salts that precipitated as sulfates, halides, and oxides. With time, Mars is regarded to have evolved from a rather wet planet, in which chemical weathering by circum-neutral pH fluids was common, to a regime in which more acidic chemical weathering took place and, eventually, to a cold, dry environment dominated by physical weathering. As the flux of impactors and volcanism declined, and as the planet's hydrologic cycle decreased in vigor, the formation of sedimentary rocks also declined. Today the Martian highlands appear to be in a net state of erosion, and outcrops of sedimentary rocks are exposed as a result of wind-driven denudation. This erosion is likely balanced by deposition of sediments in the Martian lowlands.

Orbiter observations of depositional framework, bed-scale textural/morphologic attributes, and mineralogy provide the basis for an "orbital facies" classification scheme. Orbital facies include Massive Breccia (MBR); Complexly Stratified Clay (CSC); Laterally Continuous Sulfate (LCS); Laterally Continuous Heterolithic (LCH); Distributary Network (DNW); and Rhythmite (RHY). These orbital facies are observed in several key reference sections, and their succession allows for correlation between widely separated regions of Mars, leading to a more refined understanding of environmental history. The oldest terrains on Mars are dominated by MBR and CSC facies, whereas younger terrains are characterized by LCS, DNW, and RHY facies. However, some occurrences of clay-bearing DNW and LCH facies may be contemporaneous with large sulfate deposits of the LCS facies, which are typically regarded as Hesperian in age. This indicates that the climatic evolution of Mars may be more complex than a simple global alkaline–acidic transition and that important regional variations in aqueous geochemistry and the relative roles of surface waters and groundwaters may be preserved in the Martian sedimentary record.

KEY WORDS: Mars, sedimentology, stratigraphy, facies

Sedimentary Geology of Mars
SEPM Special Publication No. 102, Copyright © 2012
SEPM (Society for Sedimentary Geology), Print ISBN 978-1-56576-312-8, CD/DVD ISBN 978-1-56576-313-5, p. 1–48.

INTRODUCTION

The previous decade witnessed several milestones in the history of sedimentary geology. In the year 2000, the first high-resolution image-based evidence for the occurrence of sedimentary rocks on Mars was published (Malin and Edgett 2000), and in 2004 the *Opportunity* rover provided in situ confirmation of their existence (Squyres 2004). The discipline of extraterrestrial sedimentary geology was revitalized[1] and is now undergoing significant growth, not just as a result of discoveries on Mars, but also as a result of the discovery of Saturn's extraordinary moon Titan (Hayes et al. 2008, Rubin and Hesp 2009).

The uninterrupted exploration of Mars over the past ~15 years has dramatically unveiled the rich record of sedimentary rocks on Mars at increasing levels of resolution (Fig. 1). Beginning with the Mars Orbiter Camera (MOC) on the Mars Global Surveyor (MGS) orbiter, sedimentary rocks were discovered to be widespread on the red planet. The MOC camera provided images with a resolution of about 1.5 to 12 m/pixel and showed clear evidence for sand dunes, alluvial fans, deltas, and vast "layered terrains"[2] with strata covering hundreds to thousands of kilometers (Malin et al. 2010). Following MOC, the High Resolution Stereo Camera (HRSC) on the European Space Agency's Mars Express mission acquired images of the Martian surface at resolutions of ~10 m/pixel at multiple wavelengths, providing both color and stereo views of Mars. This instrument was revolutionary in that for the first time much of the planet was imaged at a rather high spatial resolution, and the multiple viewing angles for each image allowed ~50 m/pixel digital terrain models to be constructed for detailed topographic studies (see Jaumann et al. [2007] for a review of the HRSC instrument and science results). In addition, HRSC contains a "super resolution" channel that images the surface in grayscale at ~2.3 m/pixel. More recently, the High Resolution Imaging Science Experiment (HiRISE) camera on the Mars Reconnaissance Orbiter (MRO) spacecraft has provided a dramatic increase in spatial resolution and currently images Mars at a stunning scale of about 25 cm/pixel, including three-channel false color (McEwen et al. 2010). The HiRISE camera provides detail of the surface of Mars that is unparalleled even for parts of Earth, and the data are good enough to show, for example, that some sand dunes—previously thought to be relict and immobile—exhibit evidence of movement when imaged sequentially from year to year (Bridges et al. 2010). The MRO spacecraft also hosts the Context Camera (CTX), which images the surface at a lower spatial resolution of ~6 m/pixel, but at the benefit of covering thousands of square kilometers per image (Malin et al. 2007). The combination of extremely high-resolution HiRISE images with the moderate-resolution (but greater coverage) CTX images has been revolutionary in allowing detailed, local observations to be properly placed in their regional context.

The MOC, HRSC, CTX, and HiRISE cameras provide a wealth of data of interest to sedimentary geologists, and it would be possible to fill a volume with descriptions of just the surficial deposits formed during the most recent phases of Martian history. Indeed, there has been much attention given to these younger deposits, particularly those of eolian origin. Therefore, it is the goal of this special publication to more specifically examine the record of the ancient sedimentary rocks of Mars. Orbiter-based visible, near-infrared, and thermal infrared spectral data and ground-based rover data provide information on the composition and petrogenesis of these ancient sedimentary rocks as well as information on their physical textures and spatial distribution. Integration of all available data sets indicates a new view of the environmental evolution of Mars that goes beyond the limited paradigm of old = warm/wet vs. young = cold/dry. Mineralogical inferences based on visible–near-infrared reflectance data hypothesize a time-dependent evolution from an early environment conducive to formation and deposition of clay-rich deposits, to a younger environment conducive to formation of sulfate-rich sediments, to even-younger strata that generally lack hydrous minerals and that may be dominated by finely ground unaltered basalt and anhydrous Fe-oxides (Bibring et al. 2006). For the first time, we have evidence for a diverse sedimentary record that likely spans billions of years of history and evidence that proxies of environmental change are embedded within this record. The view that global records of stratigraphy and mineralogy combine to inform evaluation of local processes is now a major driver in selection of landing sites for next-generation rovers (Grotzinger 2009, Grant et al. 2011). Sedimentary geology has rapidly become an important tool for understanding Mars, particularly where understanding past surface processes, habitability, and potential for preservation of organics is important.

In many respects, these alien sedimentary rocks seem very Earth-like: they show a variety of styles and thicknesses of bedding; they exhibit simple to complex stratal geometries; they show patterns of variability that relate to depositional position within basins, in which source-to-sink transport pathways can be traced; in some places they form thicknesses that exceed 5 km; they have expressions that indicate deposition from mass-flow, traction, and suspension transport; and they bear evidence of complex diagenetic histories. Even cyclic or quasi-periodic meter-scale deposits are present. But for all of these visual similarities, there are also aspects of the Martian sedimentary record that are very different from those of Earth. These differences include the lack of evidence for deposition within subsiding basins; the abundance of sulfate deposits not associated with carbonates; deposits of likely wind-blown dust that vastly exceed in scale anything known from Earth; the potential for preservation of very old strata formed primarily by impact-related processes; and the possible dominance of strata formed of eolian rather than subaqueous origin, opposite to what we see on Earth. One remarkable observation is that the oldest strata on Mars are likely older than the oldest rocks on Earth. This last point is important, as it serves as a constant reminder that Mars is not Earth, its history has been different, and there may be sedimentary rock types, facies, and processes on Mars for which the Earth provides no suitable analog. Though we should be guided by our knowledge gained from the study of Earth, we must take care to not impose our terrestrial biases too strongly on the interpretation of Mars, lest we obscure a novel perspective.

In the middle of the last century, the Precambrian record was viewed as a frontier of sedimentary geology. Compared to Phanerozoic Earth history, our understanding of Precambrian Earth history was greatly limited as a result of our access to only the coarsest age constraints, the absence of fossils for use in correlation, poor preservation of strata due to deformation, and restriction of outcrops to outliers lacking regional context. It is hard to not see the sedimentary rock record of Mars in a similar light today, although admittedly through an even darker lens. The original locations of the few meteorite samples we have from Mars are unknown, and it is not possible for us to walk the outcrops; but the planet is mapped at a scale of which we can be envious even on Earth, and the absence of plate tectonics has resulted in minimal deformation and metamorphism of strata. The obvious differences between

[1]The discipline of extraterrestrial sedimentology was born with speculations based on remote observations of dust storms and dust deposition on Mars suggested by McLaughlin (1954) and Sagan and Pollack (1969). Early petrogenetic studies of lunar sedimentary materials were conducted by Lindsay (1972).

[2]In this article we prefer use of the term "stratified" over "layered" to draw specific attention to the increase in understanding of these deposits over the past decade. In earlier decades the possibility that "layered terrains" could have had igneous, metamorphic, or shock-induced origins could not be so easily discounted. Today it is clear that many of these rocks are of sedimentary origin, and it is possible to even apply principles of sequence stratigraphy. However, we revert to use of "layered" in two instances: where historical accounts of discovery refer to the original use of term and where previous classification has ingrained this term (e.g., "Interior Layered Deposits" or "Light-toned Layered Deposits").

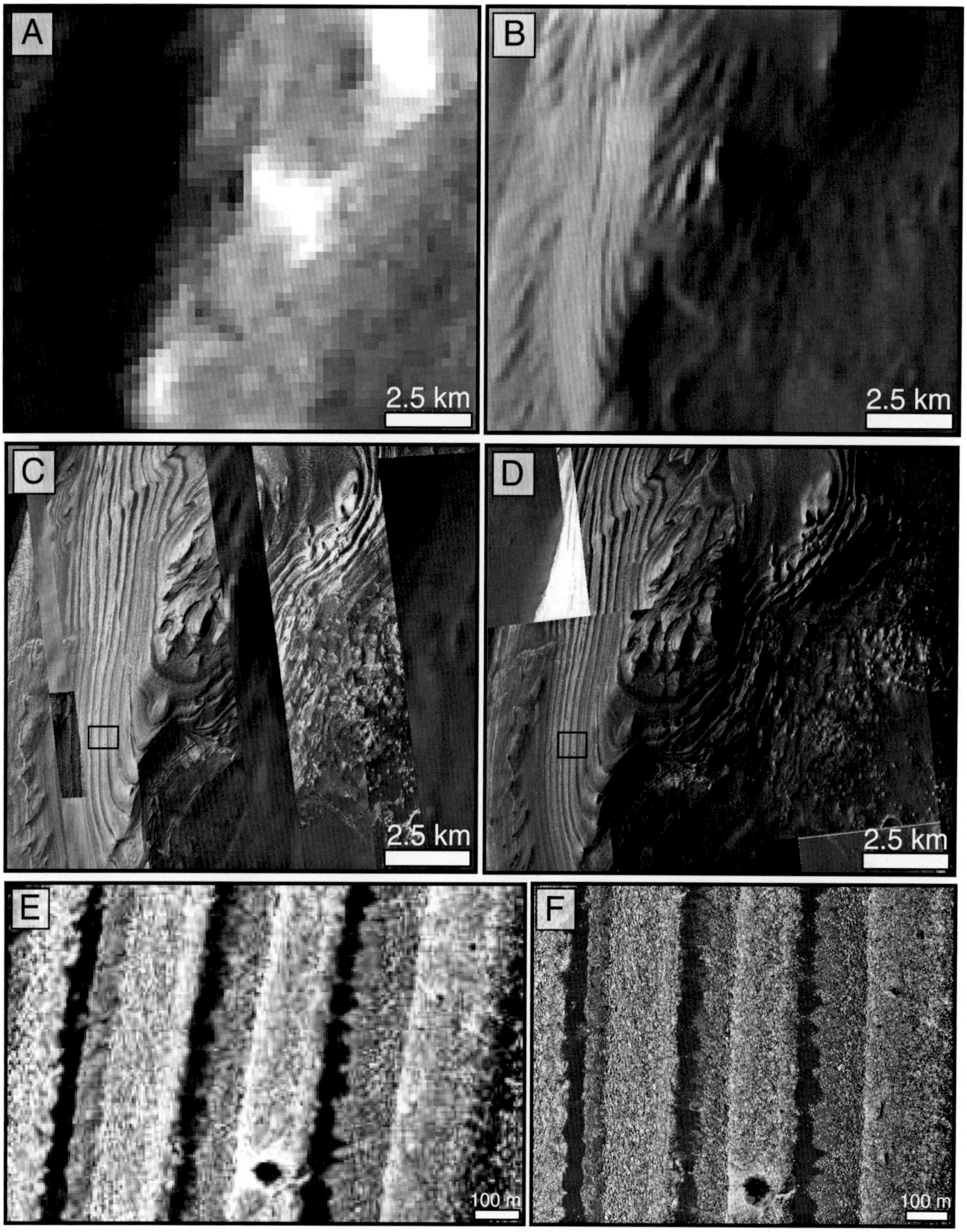

FIG. 1.—Images of layers in Juventae Chasma showing changes in orbiter camera resolution from the Viking orbiters (mid-1970s) to the HiRISE camera (mid-2000s). Development of the MOC represents a milestone in that its resolution was fine enough that sedimentary deposits were finally obvious. Most stratification on Mars is revealed at this scale of 3 to 6 m, although HiRISE images reveal further detail at scales of 25 cm/pixel. (A) Viking MDIM (~200 m/pixel); (B) THEMIS daytime infrared mosaic (100 m/pixel); (C) MOC Narrow Angle (~3 m/pixel); (D) HiRISE (25 cm/pixel); (E) MOC NA E0202546 (4.33 m/pixel). Subset box in (C) corresponds to this image. (F) HiRISE ESP_020470_1755 (25 cm/pixel). Subset box in (D) corresponds to this image.

exploring Mars and the Precambrian Earth notwithstanding, it is hard to not feel a great sense of excitement in studying the data returned from Mars orbiters and rovers. Intriguingly, we may have much to learn about geologic processes during the most ancient times on Earth by studying our neighbor, Mars. Each new set of observations promises fresh insight into an extraordinary time in the evolution of our solar system, when Earth and Mars set on different courses in the evolution of surface environments of terrestrial planets.

This volume comprises a set of articles that chronicle some of these events in the evolution of Mars. The articles presented here illustrate Martian sedimentary processes, including weathering, sediment transport, and deposition. Other articles describe potential terrestrial analogs, including consideration of microbial habitats. A third group of articles develop the case for unique Martian processes that lack clear terrestrial analogs—modern or ancient. Finally, an atlas of HiRISE images has been prepared that provides the reader with a sense of what can be observed from orbit—a representative subset of potential study areas, very few of which have been examined in detail.

HISTORY OF DISCOVERY

The first suggestion that Mars may have a sedimentary record was provided by Carpenter (1948). He acknowledged the role of wind on Mars and suggested that the smoother surface of Mars (as compared to that of the Earth's moon) might well result from abrasion of topography by the sediment-laden wind, acting over millions of years. Interestingly, he did not expect that such a process would directly contribute to the accumulation of compacted sediments to form rocks because of the suggested absence of water.

In the 1970s, images of the Valles Marineris canyon system from the Mariner 9 and Viking orbiters provided the first strong indication that layered rocks, possibly sedimentary, could be present on Mars. Furthermore, it was recognized that such materials could provide windows into past surface environments (Blasius et al. 1977, Malin 1979, Lucchitta and Ferguson 1983, Nedell et al. 1987, McKay and Nedell 1988, Goldspiel and Squyres 1991, Komatsu et al. 1993, Williams and Zimbelman 1994, Forsythe and Zimbelman 1995), including those that might be habitable for microbes (McKay and Stoker 1989). What were less clear were the texture and composition of sediments that might compose these layers. Different workers interpreted the layered terrains of Mars—in and beyond Valles Marineris—in various ways, with ideas spanning from the more common suggestion of volcanic ignimbrites (Scott and Tanaka 1982) or volcanic tephra (e.g., Chapman and Tanaka 2002) offset by more provocative ideas, including lacustrine deposits (Lucchitta and Ferguson 1983, Williams and Zimbelman 1994, Cabrol and Grin 2001) or even carbonates (McKay and Nedell 1988). More recently, astrobiologists pointed out the potential for sedimentary rocks to exist on Mars, though no unambiguous examples were identified (Farmer and Des Marais 1999). Despite the limited data and clear indications for extensive volcanism on Mars, members of the terrestrial sedimentary geology community held the hope that sedimentary rocks were present on Mars and emphasized the need to search for Mars-derived sandstone meteorites (Ashley and Delaney 1999).

Before and following the arrival of the MGS orbiter, an important discussion regarding the Martian layered terrains centered on interpretation of hardness, and therefore their degree of lithification. In many cases, cliffs of layered materials can be seen weathering without producing talus, an observation first made by Malin (1976, 1979). This tended to discount a volcanic origin of the layers as stacked lavas or ignimbrites, though poorly consolidated tephra remained a possibility. This was further supported by MGS Thermal Emission Spectrometer (TES) data, which indicated that certain types of layered terrains were characterized by relatively low thermal inertia, suggestive of poorly lithified materials. One particular outcrop, known as "White Rock" and located in Pollack Crater, has attracted much attention as a result of this property (compare McCauley [1974], Thomas [1984], and Ruff et al. [2001]). In contrast, basalts and most other volcanic or pervasively lithified materials would be expected to exhibit relatively high thermal inertia (Christensen et al. 2001, Putzig et al. 2005). Finally, in many cases these layered deposits were observed to be light-toned, often being described as "white" as a result of their appearance in grayscale images (Williams and Zimbelman 1994), further reinforcing the possibility of sediments and/or sedimentary rocks on Mars. In some instances these light-toned layered deposits were interpreted as lacustrine deposits (Forsythe and Zimbelman 1995). It was clear at that time that Mars had a distinctive record of layered materials that was not easily interpreted as lava flows, as had been the case on the Moon (Head 1982). Estimates of the physical properties of these deposits, such as thermal inertia, were important in making the case that these layered materials were likely sedimentary rocks (e.g., Christensen et al. 2003), and this point is examined further below.

The Mariner 9 and Viking orbiters are credited with the observation of deep channels incised into bedrock, accepted by many to have formed during catastrophic discharge of subsurface groundwater. In addition to these large "outflow channels" (e.g., Sharp and Malin 1975), other more highly branched networks with greater saturation densities, referred to as "valley networks," were also observed in those early images (e.g., Carr 1996). The apparent densities of these valley networks are even more pronounced in later MOC and recent HiRISE data, and some have been interpreted to represent periods of meteoric precipitation (e.g., Hynek and Phillips 2003, Mangold et al. 2004, Barnhart et al. 2009). During the Mariner 9–Viking era, much emphasis was placed on outflow channels and valley networks, though some authors did provide evidence for sediment-bearing polar ice deposits (Cutts 1973). Over the past decade, strong evidence has been provided for the existence of ancient "source-to-sink" networks, in which drainage networks that cut into bedrock can be traced down topographic gradients to fewer numbered or even single channels, and then farther downslope into sedimentary deposits represented by terminal distributary networks, including alluvial fans (Moore and Howard 2005), deltas (Wood 2006), and even sublacustrine fans (Metz et al. 2009a).

The observation of distributary networks from orbit, and the logical inference that they form sediments and/or sedimentary rocks of fluvial origin, has been relatively straightforward, and there is strong consensus for this process. This is also true for liquid water as the transport medium of choice (but see Hoffman [2000] for a different view). In contrast, the case for a sedimentary origin of the vast layered terrains that are not obviously related to any eolian, fluvial, or other aqueous processes has been more difficult. Initial studies (Malin and Edgett 2000, Edgett and Malin 2002) emphasized regional stratigraphy more than petrogenesis, and a compelling case was made that these layered deposits were younger than the rocks of the older, more extensively cratered southern highlands.[3] Evidence was found in key locations that showed the physical onlap of these younger strata on older bedrock. Furthermore, the recognition that these sediments accumulated over a substantial interval of time was provided through the identification of craters that were once buried but that are now being exhumed and exposed at the surface. The significance of this discovery cannot be underestimated in that it indicates that Mars, in the absence of plate tectonics, still has a rock cycle: older rocks are exposed to weathering, erosion, and transport as sediments that accumulate to some considerable thickness (kilometers in places), but which are then subjected to erosion themselves and reworked to form a second generation of sediments. However, whereas the extent of these processes distinguishes Mars from the Moon, they still fall short of

[3]Studies by Malin and Edgett also indicate the possibility that the ancient cratered terrains, or older bedrock, may be composed of stratified rocks in some cases.

providing an easy comparison to the tectonically driven rock cycle of Earth.

The composition of inferred sedimentary rocks on Mars was often assumed to be basaltic, likely involving little chemical weathering. This interpretation, which prevailed until the middle of the past decade, arose primarily because of the absence of clear spectroscopic data indicative of weathering by-products, such as carbonates and clay minerals (e.g., Christensen et al. 2001). Some evidence for chemical weathering was provided by the observation of sulfur enrichment in the soils observed by the Viking landers, the Sojourner rover, and the few locations of hematite enrichment observed by the TES instrument. In addition, studies of the Martian meteorites indicate a relatively low degree of weathering in subsurface environments on Mars that has produced clays, carbonates, sulfates, and halides (Velbel, this volume).

The arrival of the *Opportunity* rover at the surface of Meridiani Planum in 2004 dramatically changed this view. After 7 years of observations it is now clear that on the order of 10 to 20 m of rocks composed of windblown sulfate–silicate sands existed in this region, reworked rarely by sulfate-rich stream flows, and that they have been subjected to salt-rich diagenetic fluids during burial. Contemporaneous with these discoveries, the Mars Express orbiter was returning a wealth of data in 2004. In addition to high-resolution images acquired by HRSC, the visible–near-infrared Observatoire pour la Minéralogie, l'Eau, les Glaces et l'Activité (OMEGA) spectrometer began to map the Martian surface at high spectral resolution and revealed the widespread presence of clay minerals and sulfate salts in the ancient crust (Bibring et al. 2005, Gendrin et al. 2005, Poulet et al. 2005). Together, these orbital and in situ measurements have revolutionized the way we view water–rock interactions on Mars. The initial findings of the OMEGA instrument have since been confirmed by the Compact Reconnaissance Imaging Spectrometer for Mars (CRISM) instrument on the MRO, which has provided a significant improvement in spatial resolution (Murchie et al. 2007). Just as the number of hydrous minerals detected on Mars continues to increase (e.g., Mustard et al. 2008), so too does the evidence for the formation and preservation of sedimentary rocks formed in a variety of possible depositional environments.

DISTRIBUTION OF SEDIMENTARY ROCKS ON MARS

Geologic Timescale

In considering the distribution of sedimentary rocks on Mars it is important to review how age constraints are achieved. The geologic timescale for Mars is based entirely on the density distribution of impact craters, which in very broad terms constrains the age of geomorphic surfaces (e.g., Hartmann and Neukum 2001). With the exception of the Martian meteorites, rocks, per se, cannot be directly dated; the ages of Martian rocks are inferred from the ages of surfaces that expose rocks of all types. The older the surface, the greater the frequency of craters in general and the frequency of larger craters in particular.

A relative timescale based on crater density distribution is converted into an absolute timescale via comparison to the surface of the Moon (Binder 1966, Hartmann 1966, Öpik 1966, Ivanov 2001). Lunar crater density distributions for different terrains were known to vary in relative age (e.g., lowlands vs. highlands), and some of these different surfaces were sampled by Apollo astronauts who returned rocks to Earth for geochronometric calibration. Once dated, the rocks provided a constraint on the maximum age of the surface from which they were sampled. A first-order approach would be to apply the impact density distributions calibrated for the Moon to Mars, accounting for the flux of bolides to Mars and the influence of higher gravity on crater excavation diameters (smaller bolides produce larger impact craters on the Moon; Ivanov 2001). In addition, Mars differs significantly from the Moon in that it has an atmosphere, and in the very ancient past this atmosphere may have been more substantial. Driven by the dynamics of the atmosphere, surface processes that cause erosion and sedimentation (e.g., eolian, fluvial) will all tend to degrade craters through erosion of their uplifted margins and infilling of their interiors. In this manner, the distribution of craters on surfaces will become modified, and relative to the Moon, some surfaces on Mars may appear younger than their true age (Hartmann and Neukum 2001). In some cases sediment accumulation rates may have been high enough that craters became fully buried, giving rise to the important concept of a "cratered volume" rather than a "cratered surface" (Malin and Edgett 2000, Edgett and Malin 2002). To obtain the correct age for the corresponding surface one would need to sum all the craters in the sedimentary succession for a given area. Therefore, crater counts of surfaces may instead reflect exposure ages or crater retention ages rather than absolute depositional ages.

With these caveats in mind, a geologic timescale for Mars has been erected and is used on a routine basis for discussion of the ages of rocks and terrains on Mars. The three principal subdivisions are known as the Noachian (4.5–3.6 Ga), Hesperian (3.6–2.6 Ga), and Amazonian (2.6 Ga–present) eras, for which the Noachian–Hesperian boundary has uncertainties of ± 0.1 Gyr and the Hesperian–Amazonian boundary has uncertainties of ± 0.6 Gyr (Tanaka and Hartmann 2008). In recent years, global mineralogical mapping from orbit has been interpreted to indicate an evolution in weathering and production of hydrated minerals with stages that roughly correspond to the three principal subdivisions mentioned above (Fig. 2) (Bibring et al. 2006). The Noachian Era appears to have been marked by circum-neutral pH aqueous alteration that generated clay minerals; the Hesperian Era witnessed generation of sulfates, some under apparently acidic conditions; and the Amazonian Era was dominated by formation of anhydrous ferric oxides in a slow superficial weathering regime without liquid water (Bibring et al. 2006, McLennan and Grotzinger 2008, Murchie et al. 2009b).

Global Distribution

As a first pass it is useful to examine the global distribution of inferred sedimentary rocks on Mars. In general, most strata inferred to be of sedimentary origin (i.e., not including polar ice or lava plains) are restricted to more equatorial latitudes, residing between 50°N and 50°S. The global distribution of Martian sedimentary strata is shown in Figure 3. All of the localities discussed in this article are shown in Figure 4.

Figure 3 was constructed by compiling observations from MOC,

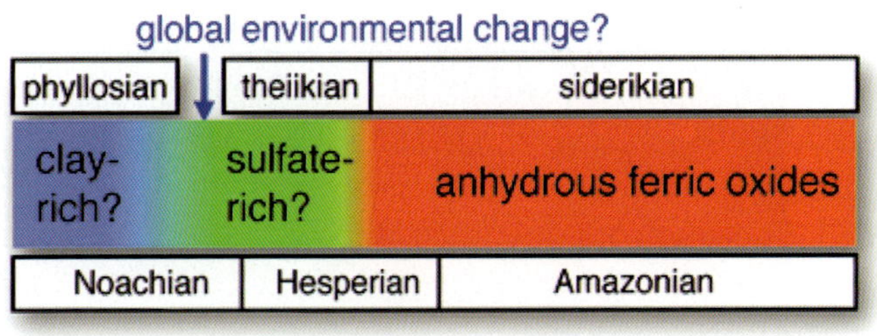

FIG. 2.—One current scenario for the environmental evolution of Mars surface environments through time (Bibring et al. 2006). In some sites, such as Gale Crater, the stratigraphic record of hydrated minerals supports this model (Milliken et al. 2010). Figure modified after figure 5 in Bibring et al. (2006).

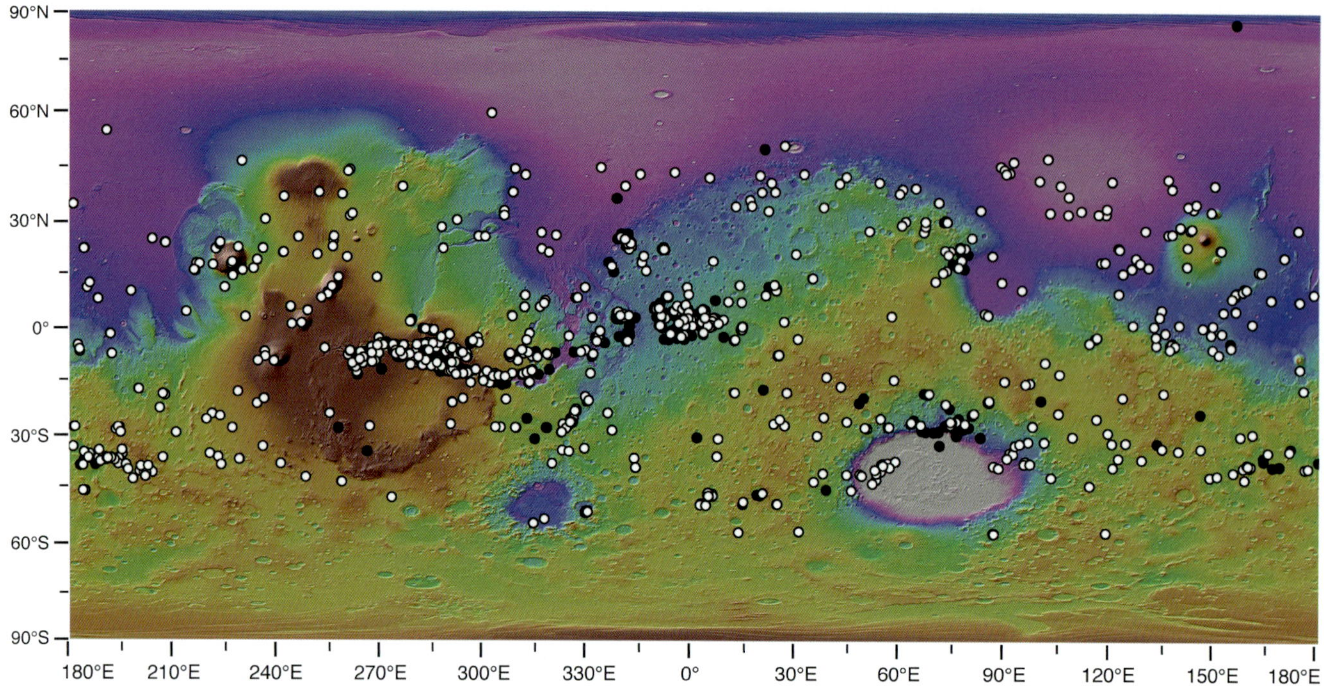

FIG. 3.—Distribution of inferred sedimentary rocks on Mars plotted on a Mars Orbiter Laser Altimeter (MOLA) elevation map. White dots correspond to stratified deposits observed in HiRISE images. Black dots correspond to MOC images with words such as "light-toned" or "layered" deposits in the formal image description tag. Note the concentration of stratified rocks in the walls of major canyon systems such as Valles Marineris (high concentration of dots in center–left part of image) as well as in Sinus Meridiani (~0°E, 0°N).

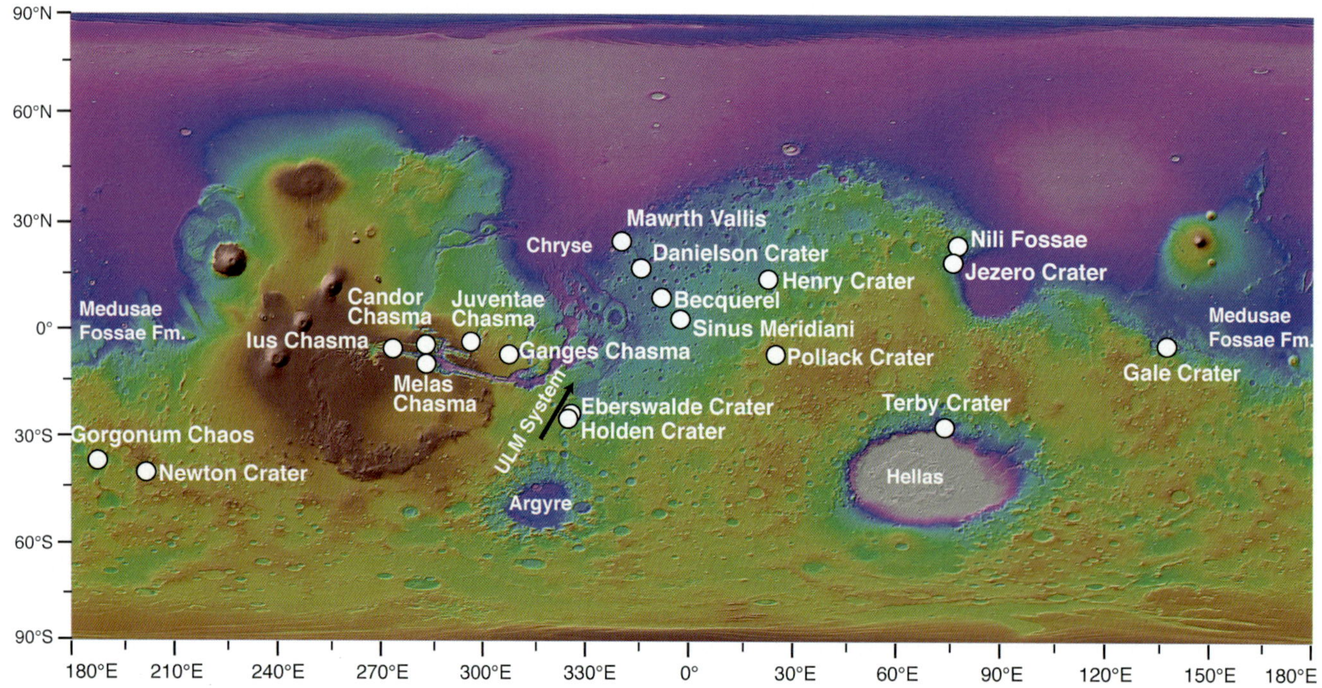

FIG. 4.—Reference map showing important localities discussed in this article. Base map is MOLA topography. Note the strong contrast in elevation between the northern lowlands and the more ancient southern highlands. Black arrow shows downslope transport direction within ULM fluvial system.

CTX, and HiRISE data sets, building on earlier efforts based only on MOC data. However, even this current effort is limited to a small fraction of the available data, consisting of examination of 7016 HiRISE images, of which 2045 images exhibit possible sedimentary strata (white circles in Fig. 3).[4] A first pass was made to search for bedrock; all images that contained bedrock were then examined for stratification; then, if stratified, an attempt was made to distinguish if the rocks are volcanic lava flows or sedimentary, including possible pyroclastic volcanic deposits. In addition to the examination of HiRISE data, the MOC image database was searched for images whose description tag contained words such as "layered terrain," "layers," "strata," or "light-toned deposits," with the thought that many of these images may contain sedimentary strata (black circles in Fig. 3), although visual inspection of each image is still required for confirmation. Here, 1388 potential sedimentary rock targets were identified.

Our initial work (and that of Malin et al. [2010], their fig. 16) indicates that the stratified deposits of Mars are located globally, although many are found in mid-equatorial regions (Fig. 3). We note that the apparent paucity of clearly stratified deposits in the mid- to high-latitude regions of Mars may be a result of overprinting and obscuration by ice-related processes known to occur in these regions under current climate conditions. Airfall deposition of ice–dust mixtures that mantle the surface (Mustard et al. 2001), relaxation of topography due to possible creep of ground ice (Squyres and Carr 1986), and widespread deposition of ice during high-obliquity "ice ages" (Head et al. 2003) are all believed to occur at these latitudes, and all could act to cover or erase stratified outcrops. Therefore, the apparent increase in stratified deposits for near-equatorial regions may represent the effect of ice-related processes at higher latitudes on a cold and dry Mars over the past several billion years rather than an inherent distribution in depositional environments conducive to formation of sedimentary strata on ancient Mars. However, recent studies have shown that some of these geologically young latitude-dependent ice–dust deposits (Mustard et al. 2001) also exhibit layering (Schon et al. 2009), raising the question of whether or not such processes may ultimately contribute to the formation of stratified rocks at these latitudes.

Many of the interpreted sedimentary deposits depicted in Figure 3 are not in close enough proximity to currently identified volcanoes to explain their deposition exclusively via volcanism. Nevertheless, it should be possible to recognize likely volcanic lava flows if they are present (see, for example, Milazzo et al. [2009]). In some cases, the nature of the stratified deposit is such that a definitive case for a sedimentary deposit may be made (Malin and Edgett 2003). In other cases the evidence is suggestive but not compelling (Malin and Edgett 2000). However, in many cases it is not possible to distinguish one from the other with a high degree of confidence based on initial study. Therefore, we have developed additional criteria that, with further testing and evaluation, may become helpful in distinguishing classes of sedimentary rocks in addition to distinguishing lava flows from sedimentary rocks. These criteria are also used in helping to define the categories of sedimentary rocks illustrated in the atlas of Beyer et al. (this volume).

Based on preliminary work, we recognize the following broadly defined attributes that usefully characterize particular types of strata. These include tone/albedo (MOC, HRSC, CTX, HiRISE grayscale and false color), apparent thickness of stratification and presence of apparent rhythmic beds, weathering character (rough, blocky vs. smooth), larger-scale textures and patterns (e.g., polygonal or reticulated), and spectral signature, as seen in visible–near-infrared (VNIR) and thermal infrared data. The spectral signature attribute includes the distinct lack of absorption features associated with mafic or hydrous minerals, most often attributed to modern dust cover, but possibly also associated with a distinct class of sedimentary rocks suggested to be composed of ancient lithified dust (Bridges and Muhs, this volume). These attributes are shared between sequences of strata across widely separated regions of Mars. For example, compare the light-toned, apparently thickly bedded, rough to blocky weathering deposits of the lower part of the ~5-km-thick mound in Gale Crater (Fig. 5A) with the light-toned, rough to blocky weathering deposits of an ~2-km-thick mound in Juventae Chasma (Fig. 5B). Both of these stratigraphic successions are also known to contain hydrated sulfates based on mapping using OMEGA and CRISM reflectance spectra (Gendrin et al. 2005, Milliken et al. 2010). This type of stratified rock can be contrasted with the medium-toned, smooth-weathering, apparently thinly bedded and rhythmic deposits in the upper part of the mound in Gale Crater (Fig. 5C) and with similar deposits in Becquerel Crater (Fig. 5D). In both cases these latter deposits have been shown to be thin-bedded (beds are at least as thin as a few meters) and have a statistically significant rhythmic thickness distribution (Lewis et al. 2008a, Lewis 2009).

It is possible to subdivide the Martian layered terrains with probable sedimentary origins based on these characteristics. However, classification reaches its fullest potential if regional topographic/structural context is considered as well. To be clear, the analysis of sedimentary rocks on Mars is still in a very immature stage, and therefore these generalizations must be approached with appropriate caution. As such, this represents the beginning of an effort to build an objectively defined stratigraphy for parts of Mars, with the hope that interregional correlations might eventually be possible. The regional- to local-scale sedimentary terrain types we define here include (1) underfilled basins; (2) overfilled craters; (3) chasm and canyon systems; (4) plains-covering deposits; and (5) very ancient strata. In addition, Beyer et al. (this volume) also recognize the polar ice caps as examples of sedimentary strata, including well-developed stratal geometries (Kocurek and Ewing, this volume); however, these are not discussed further here. The following section briefly describes a few members in each of these five categories. These examples are by no means exhaustive, but rather they provide an overview of the diversity of sedimentary rocks and depositional environments that exist on Mars.

Underfilled Basins

Topographic depressions created by tectonic extension, subterranean collapse, or impact events have acted as significant sediment traps on Mars. Specifically, it has been known for decades that many craters in the Martian southern highlands exhibit clear evidence for partial infilling, presumably the result of volcanic or sedimentary processes (see Craddock et al. 1997, Malin and Edgett 2000, Craddock and Howard 2002 [and references therein]). These craters, as well as other topographic basins on Mars, are quite variable in terms of the amount of material they preserve. One class of these features, which we refer to as "underfilled basins," exhibits limited sediment accumulation in the sense that accommodation space in the crater or basin still remains significant. In other words, these basins, most of which are impact craters, contain sediments or sedimentary rocks but lack evidence for having been filled close to or beyond their rims. Many craters within this category exhibit flat floors (e.g., Chapman et al. 1968, McGill and Wise 1972, Craddock et al. 1997), and the interior deposits lack clear geomorphic evidence for a unique depositional mode, making a sedimentary origin ambiguous. However, in a number of rather spectacular cases the geomorphology of these deposits indicates that fluvial systems were an important agent for sediment transport and that at least a fraction of the sediment filling many of these basins is externally derived.

Sedimentary deposits at the termini of some of these fluvial

[4]Data analysis to generate Figure 3 was undertaken by Jennifer Griffes and Kathryn Stack from the California Institute of Technology.

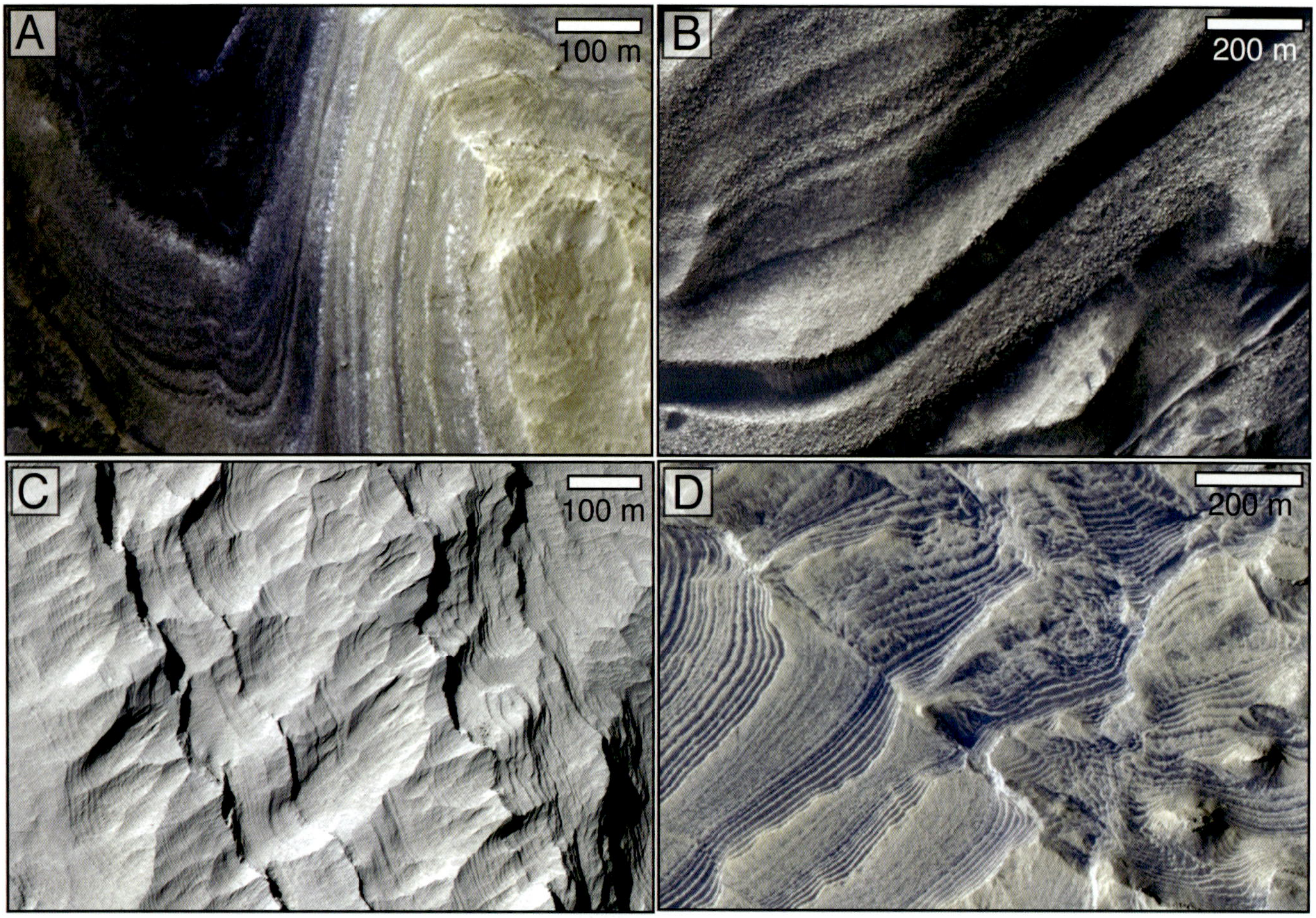

FIG. 5.—Examples of inferred sedimentary rocks that exhibit distinct textures in orbital images. (A) Strata in Lower formation of mound in Gale Crater (PSP_009294_1750). (B) Interior Layered Deposit in Juventae Chasma (PSP_007126_1755). (C) Strata in Upper formation of mound in Gale Crater (PSP_009927_1750). (D) Strata in Becquerel Crater (ESP_019268_2015). See text for discussion.

networks include alluvial fans, deltas, and stratified units for which the mode of emplacement is uncertain (Fig. 6). In the ancient and topographically high southern hemisphere of Mars, well-developed drainage networks often lead to single-strand channels that route to the margins of craters and then incise through the crater rim, sometimes grading laterally into wedge-shaped alluvial deposits (Howard et al. 2005). Many examples of these alluvial fans have been documented (Howard et al. 2005), but there are fewer examples of well-defined deltas. Fan-shaped features, possibly deltas, have been identified in a number of locations (e.g., Di Achille et al. 2006a, 2006b; Hauber et al. 2009), and those lying outside of craters that face the northern plains of Mars and along the topographic dichotomy boundary have been inferred to constitute evidence of a northern ocean (Di Achille and Hynek 2010). However, these latter features do not show clearly defined channel networks indicative of distributary networks (Kraal et al. 2008), and they may represent alternative modes of deposition, possibly including debris flows. As evidence for a northern ocean, these possible fans or deltas must be regarded as highly conjectural.

Holden Crater: The ~150-km-diameter Holden Crater (26°S, 326°E; Fig. 4) hosts well-developed alluvial fans along the interior margins of the crater wall. As with other alluvial fans on Mars, these examples typically exhibit a cone-shaped morphology, are tens of kilometers in length, and have gradients of a few degrees (Howard et al. 2005). Holden Crater provides a representative example, with numerous fans coalescing to form long bajadas (Grant et al. 2008). Fans typically radiate from deep, extensively dissected alcoves in the crater wall and are best expressed along the western and southwestern walls. Differential eolian abrasion of these alluvial systems has produced inverted topography of distributary channel networks (Fig. 6A). The thickness of these alluvial deposits is at least tens if not hundreds of meters, as projected above the flat-lying, possibly lacustrine strata that they apparently overlie (Grant et al. 2008). Clay minerals have been detected in some of these deposits, with the strongest clay signatures occurring in the lowermost (possibly lacustrine) strata. The presence of clay minerals in the surrounding crater wall indicates that at least some of these clays are detrital, having been transported from the crater wall to their current location by fluvial systems (Milliken and Bish 2010). The strata in Holden Crater also record a late-stage, high-energy flooding event, caused when the southwestern portion of the crater rim was breached by water that had ponded in the adjacent Uzboi Vallis (Grant et al. 2008). This event, possibly akin to the Lake Bonneville flood on Earth (Jarrett and Malde 1987), formed large subaqueous dunes that contain meter-sized blocks and whose paleocurrent direction is aligned with Uzboi Vallis (Grant et al. 2008).

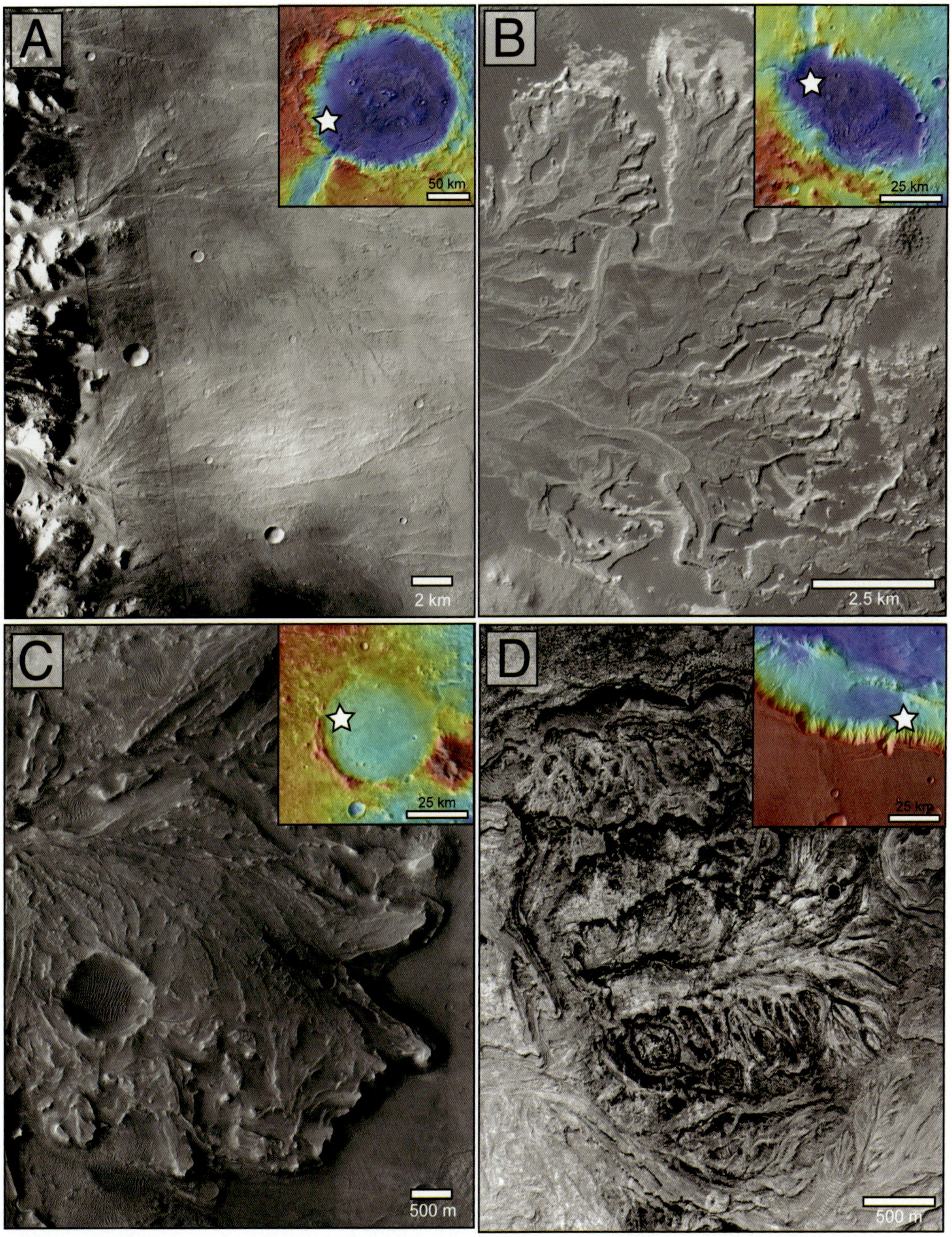

FIG. 6.—Sedimentary deposits in several underfilled basins. (A) Holden Crater, showing alluvial fan deposits within crater-fringing bajada. (B) Eberswalde Crater, showing a well-preserved delta. (C) Jezero Crater, with only partially preserved delta or alluvial fan. (D) Southwest Melas Basin within the larger Melas Chasma, where a topographic depression (not a crater) is filled with possible sublacustrine fan sediments. See text for discussion. Inset maps show location of images.

Eberswalde Crater: Deltaic deposits are very well developed in Eberswalde Crater (24°N, 327°E; Fig. 4), showing a remarkable degree of preservation (Malin and Edgett 2003). In Eberswalde Crater (Fig. 6B), which sits along the northern rim of Holden Crater, the deltaic feature is approximately 100 m thick and may largely preserve its original extent (Lewis and Aharonson 2006, Pondrelli et al. 2008). The delta comprises six distinct lobes that have prograded on the order of 17 km from their apex (Bhattacharya et al. 2005, Wood 2006) and represents an estimated 6 km^3 of sediment (Jerolmack et al. 2004). The geometry of the deposit features flat topset strata, comprising distinct meandering channels that are fronted by more steeply dipping strata interpreted as clinoforms (Lewis and Aharonson 2006, Pondrelli et al. 2008). Clay-bearing sediments apparently comprise the bottomset strata (Milliken and Bish 2010), and though their abundance is unknown, their presence is consistent with the inferred increase in the relative proportion of fine-grained sediment that a deltaic facies model would predict. Though the deltaic origin of these strata is generally agreed upon, estimates for the duration of deltaic deposition and associated surface runoff range widely, from ~10 years (Jerolmack et al. 2004) to >100,000 years (Bhattacharya et al. 2005). Eberswalde Crater pre-dates the formation of Holden Crater, as evidenced by the presence of ejecta from Holden Crater that resides in Eberswalde. It has recently been suggested (Grant and Wilson 2011) that the fan deposits in both of these craters are much younger than previously recognized, possibly Hesperian or younger. If true, the Eberswalde Delta may be somewhat anomalous in that it would thus postdate the major period of valley network formation, which had largely ceased by the Early Hesperian (Fassett and Head 2008a).

Jezero Crater: On the opposite side of the planet, the inferred deltaic deposits at Jezero Crater (18°N, 78°E; Fig. 4) have received significantly less attention. There are two distinct deposits, but each is significantly eroded, and their original maximum extent is not known. Accordingly, it is difficult to distinguish between a deltaic vs. alluvial fan origin. In contrast to Eberswalde, Jezero Crater has a significant outlet channel on the opposite side of the crater, from which the delta is developed, indicating a significant history of discharge and throughput; estimates of duration are similar to that calculated for Eberswalde (Fassett and Head 2005). The present upper surface of the deltas/alluvial fans is below the elevation of the outlet breach, indicating that the crater was underfilled at the time at which deposition ceased. However, postdepositional eolian deflation could have reduced the lateral extent as well as the thickness of strata associated with the fluvial deposits. Both deposits are estimated to collectively represent 5 km^3 of sediment. The western deposit is shown in Figure 6C. It has a thickness of 50 to 100 m and extends outward for ~10 km from its apex. Distributary channels are clearly truncated, however, indicating a greater primary extent (Fassett and Head 2005). This erosional truncation reveals the internal layering of the deposit, and spectroscopic data indicate clay minerals are concentrated in the delta relative to the surrounding region (Ehlmann et al. 2008a).

Melas Chasma: Distributary networks representing terminal sediment sinks also are present in topographic depressions other than craters. One well-studied region is an enclosed basin, hereafter referred to as the "southwest Melas Basin" (10°S, 283°E), which resides within the larger Melas Chasma in the Vallis Marineris canyon system (Figs. 4, 11A). Similar to the underfilled craters, there is no clear evidence that the entire basin was ever filled with sediment. Furthermore, in contrast to the deposits that are exposed in the broader Valles Marineris system (see below), the morphologic characteristics of the Melas deposits place tighter constraints on their origin. Specifically, on the floor of this basin exists a well-preserved deposit (Fig. 6D) that has been interpreted to represent a set of sublacustrine fans (Metz et al. 2009a).

The highlands surrounding the basin and fans are marked by dense, highly organized valley networks (Mangold et al. 2004). Because the heads of the valleys occur at different elevations, including near the tops of wall rock ridgelines, they have been interpreted to have been fed by meteoric precipitation (Mangold et al. 2004, Quantin et al. 2005). These valleys drain into a subcircular closed depression within the Melas Basin that is interpreted to have been filled by a body of standing water (Quantin et al. 2005). The basin contains numerous light-toned, flat-lying layers that can be traced over several kilometers. Four fans have been identified in the southwest Melas Basin, two in the western part of the basin and two in the eastern part of the basin (Metz et al. 2009a). One of these, shown in Figure 6D, is particularly well preserved and appears to preserve most of its primary extent and finest scale distributaries. This fan has unique morphologic features, including multiple lobes with dendritic finger-like terminations that branch off at high angles in the downstream direction (up to 90° to the overall transport direction). The mean branching angles for the fingers are 75°, with a median of 80° (Metz et al. 2009a).

The Melas fans appear morphologically similar to deltas and submarine fans; however, the details of the morphology and their position in the basin indicate that they likely formed in a sublacustrine environment (Metz et al. 2009a). The combination of low slope, channel branching geometry, presence of distinctive small-scale lobes, and position near the center (rather than near the margin of a topographic basin) show that the Melas fans are distinct from previously identified fluvial features on Mars. Estimates of fan formation timescales indicate that they formed in a minimum of 10^2 to 10^4 years, and, thus, a stable body of water must have been present for at least this long.

Overfilled Craters

One of the most fascinating types of stratified deposits on Mars involves sequences of rocks, often hundreds and sometimes thousands of meters thick, whose upper parts approach or even exceed the elevation of the margins of the crater in which they are preserved. Therefore, the entirety of these deposits could not simply have accumulated in local topographic lows. As depocenters, these craters are probably best regarded as random sediment traps, and the more significant question has to do with how sediment accumulated up to and possibly beyond the rims of these older craters over such large regions.

These deposits are often preserved as mounds within large (>100-km-diameter) craters, sometimes with a "moat" occurring between the mound and crater walls. Their origin is uncertain, and some workers have argued that because the deposits form the shape of a mound today they may have been deposited as a mound (e.g., Rossi et al. 2008). However, inspection of layers in digital terrain models shows that thicknesses remain constant toward the margins of these mounds, indicating that the mounds are erosional remnants of deposits that were once more laterally extensive; i.e., there is no evidence for thinning and downlap toward the margins of the mound. In this sense they may be similar to outliers of terrestrial Proterozoic cratonic cover rocks that overlie older Archean basement and have been eroded to small remnants. These outliers were once thought to be isolated minibasins, but subsequent work (e.g., Ware and Hiscott 1985, Rainbird et al. 2003) revealed that they are erosional remnants of once–laterally continuous strata. By analogy, it is startling to consider that parts of Mars may have once been covered with regionally extensive successions of sedimentary rocks that have been stripped away over the course of billions of years by eolian abrasion, only to leave isolated remnants behind as evidence of their existence (Malin and Edgett 2000).

Gale Crater: There are several examples that illustrate this curious relationship of intracrater strata that exceed the height of portions of

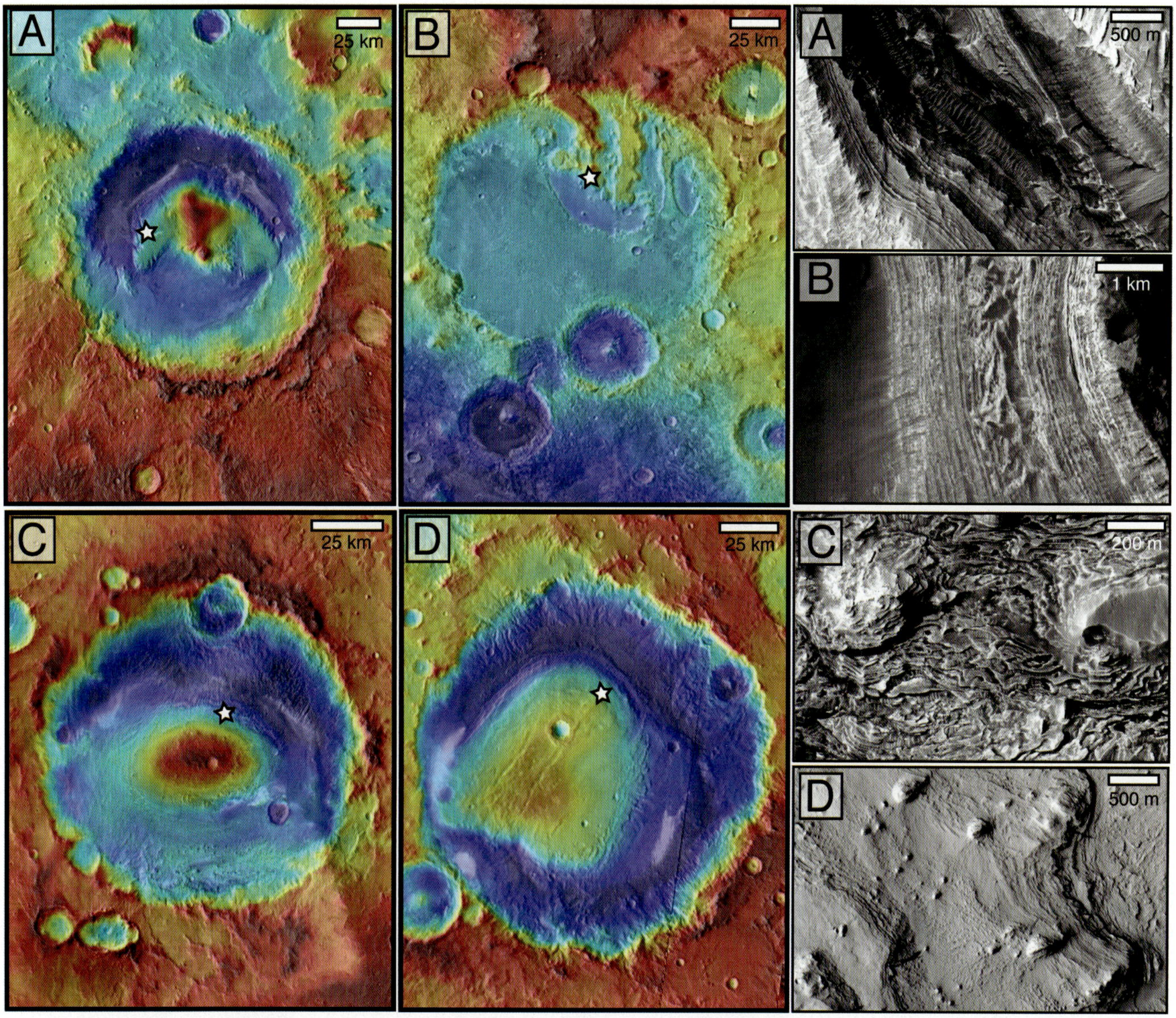

FIG. 7.—Sedimentary deposits in overfilled craters. Subsets are from HiRISE images. (A) Gale Crater, in which strata form a mound over 5 km in thickness that exceeds the elevation of the northern crater rim. Subset is PSP_008002_1750. (B) Terby Crater; note large mesas formed of stratified deposits. Subset is PSP_002572_1520. (C) Crommelin Crater, showing central mound formed of stratified deposits. Subset is PSP_001876_1850. (D) Henry Crater, in which strata form a significant mound, the peak elevation of which coincides with elevation of the crater rim. Subset is PSP_006569_1915. See text for discussion.

their bounding crater rim. One of the most interesting is Gale Crater (5°S, 137°E; Fig. 4), an ~155-km-diameter Noachian Age crater located along the geomorphic and topographic boundary that separates the ancient heavily cratered southern highlands from the younger and topographically smoother northern lowlands. The crater contains a remarkably thick, ~5-km succession of stratified rocks that form a mound, the maximum elevation of which is similar to that of the southern crater rim but that is several kilometers higher than the degraded northern rim (Fig. 7A) (Anderson and Bell 2010, Thomson et al. 2011). The mound exhibits no volcanic landforms (e.g., lava flows, vents, cones), and, thus, the rocks are considered to be sedimentary in origin (Malin and Edgett 2000, Anderson and Bell 2010, Milliken et al. 2010, Thomson et al. 2011). Other researchers have interpreted the strata as being volcanic ash, lacustrine (Cabrol et al. 1999), eolian, spring mound (Rossi et al. 2008), or ancient polar deposits (Schultz and Lutz 1988).

The mound in Gale Crater can be divided into two formations based on visible images (Milliken et al. 2010). The Lower formation is composed of parallel beds that dip gently to the northwest at 2 to 4° and that vary in thickness, albedo, and surface texture, whereas strata in the Upper formation appear relatively homogeneous, dip more steeply toward the north–northeast, and have fewer impact craters. In addition, the Lower formation exhibits a net transition from clay–sulfate to sulfate–oxide mineral assemblages and is separated from the overlying Upper formation by an erosional unconformity. Superposition and crater statistics indicate that strata in the Lower formation lie along the

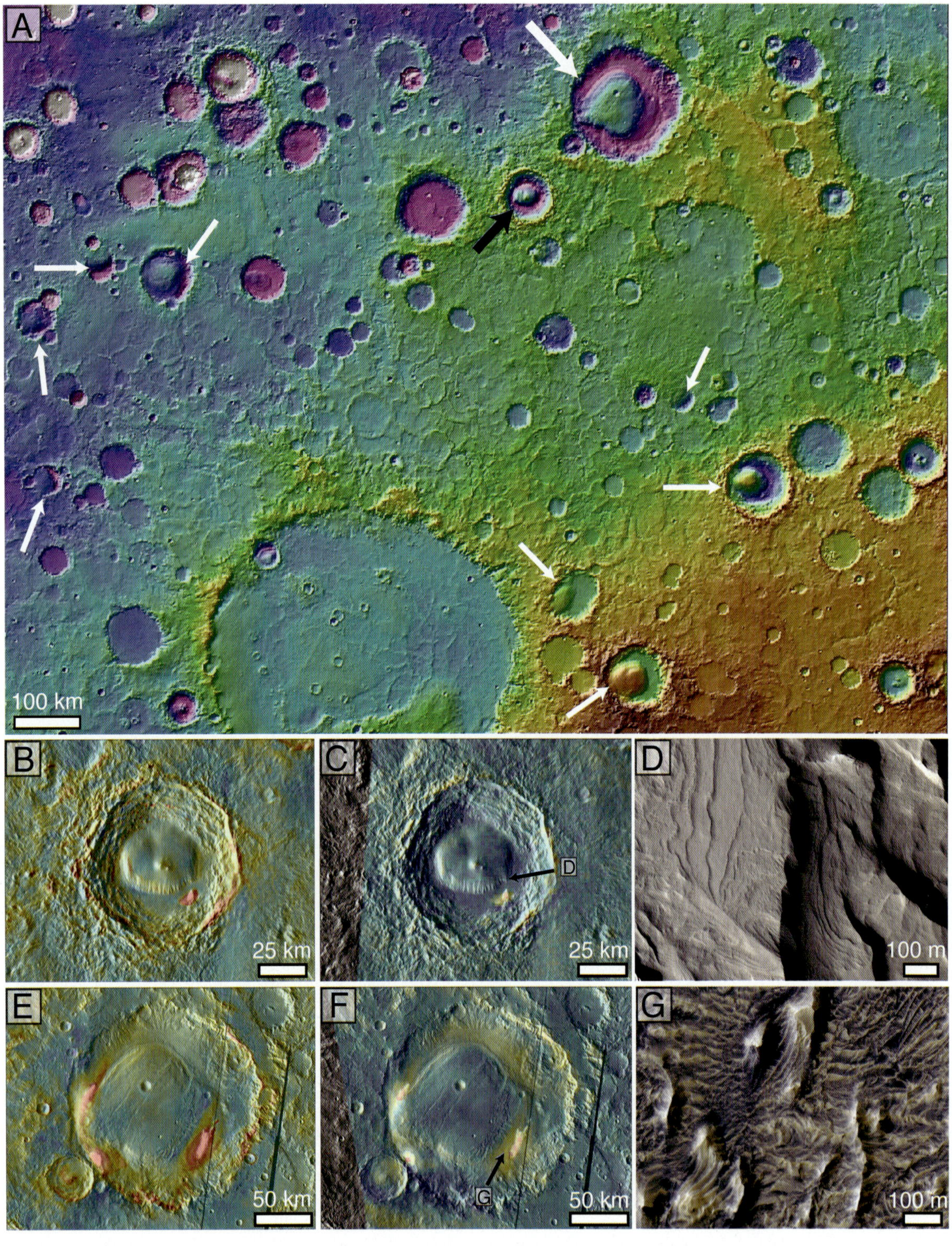

FIG. 8.—Overfilled to partially filled craters of Arabia Terra, including THEMIS nighttime infrared and OMEGA hydration responses. (A) Regional topographic map showing locations of craters that exhibit various degrees of infilling. Small white arrows point to partially filled/exhumed craters. Large white arrow points to an unnamed crater, shown in B, C, and D. Large black arrow points to Henry Crater, shown in E, F, and G. (B) THEMIS nighttime infrared image; yellow and red tones indicate higher thermal inertia relative to surroundings, consistent with large particles or bedrock. (C) H_2O content map derived from OMEGA data; yellow and red tones indicate $\geq$4 wt. % H_2O; note that this crater does not exhibit an increase in hydration associated with the stratified mound; black arrow marks location of D. (D) Close-up of strata in central mound of crater (HiRISE PSP_003655_1885). (E) THEMIS nighttime infrared image; color scale same as B. (F) H_2O content map

Noachian–Hesperian time-stratigraphic boundary (Thomson et al. 2011), whereas beds in the Upper formation, which lack signatures indicative of hydrous minerals, are thinner, more regularly spaced, and are clearly younger. In addition to the hydrated mineralogy, a broad array of geomorphic features indicates the activity of water within Gale Crater (Anderson and Bell 2010).

The observed stratigraphic trends are consistent with the rocks at Gale Crater recording a global transition from a climate favorable for clay formation to one more favorable to forming sulfates, succeeded by younger strata that lack evidence for hydrated minerals (Milliken et al. 2010). These transitions may represent the progressive "drying out" of Mars from early clement conditions to water-limited acidic and oxidizing conditions, and ultimately to the cold, dry climate of today in a single stratigraphic sequence. As such, the Gale mound constitutes an important "reference section" for establishing the chronostratigraphic history of Mars.

Terby Crater: Terby Crater (28°S, 74°E; Fig. 4), located along the northern rim of the giant Hellas Impact Basin, forms a basin 175 km in diameter. The crater is flat-floored, indicating significant infill, and is marked by three significant linear mesas—or ridges—of stratified deposits (Wilson et al. 2007). These ridges are equivalent in elevation to the crater's northern rim but are several kilometers higher than the degraded southern rim (Fig. 7B). Layers are generally flat in all three ridges and form successions that are several kilometers thick. As is the case at Gale Crater, many of the layers are of uniform thickness when traced laterally (Ansan et al. 2011), indicating that they are erosional remnants of once–more extensive deposits.

Strata in Terby are suggested to be either fine grained or lacking in pervasive cementation as a result of the lack of talus at the base of slopes. Blocks produced by erosion and collapse of outcrop faces via mass wasting are generally not apparent in orbital images. This is particularly interesting given Terby's latitude and proximity to Hellas, which is in a region that has been subject to ice-related processes for the past several hundred million years or longer (e.g., Head et al. 2003, Crown et al. 2005). Freeze–thaw cycles might be expected to increase physical weathering at Terby and produce large boulders along the steep walls of the mounds, yet such boulders are oddly absent.

In general, the modes of deposition for the strata in Terby are not known, although both lacustrine and eolian loess-like processes have been invoked (Wilson et al. 2007, Ansan et al. 2011). Most recently, recognition of locally steeper dips has indicated the presence of clinoforms, equated with deltaic progradation (Ansan et al. 2011). However, the presence of large-scale trough cross-bedding indicates the possibility that these sets of clinoforms, on the order of 6 to 10°, could also be of eolian origin, similar to the draa-scale stratification observed in the Burns formation, Meridiani (Edgar et al., this volume). Finally, hydrous minerals, including clay minerals, are present in the Terby strata, and the spectral signatures are often strongest in the upper part of the succession, indicating that at least part of the succession was formed or modified by aqueous activity. As with all Martian clay minerals that occur in sedimentary rocks, these hydrous phases could have either a detrital or authigenic origin (Ansan et al. 2011).

Crommelin Crater/Arabia Terra: A broad region of Mars known as Arabia Terra (Fig. 4) is marked by a number of craters that are partly filled with strata of likely sedimentary origin (Malin and Edgett 2000). The intracrater stratified deposits of Arabia Terra contain successions on the order of at least several hundred meters in thickness that are distributed widely from 350°E to 30°E and from the equator up to 25°N. The region is on the order of 500 km by 1000 km, and although widely separated by large distances, occurrences of sedimentary strata are very similar in their gross morphologic properties (Lewis et al. 2008a, Lewis 2009). Part of this region is shown in Figure 8. Unlike the examples described above, many of these deposits have not been demonstrated to be associated with fluvial features, and their origin(s) remains enigmatic. Though fluvial features incise the rims of many craters in Arabia Terra, the sedimentary deposits in these craters are commonly disconnected from these features and do not exhibit fan morphologies, raising the possibility that they could be eolian in origin. Recently, Fergason and Christensen (2008) presented a detailed study of the thermal inertia properties of deposits in this region and concluded that the crater mounds were composed of weakly indurated sediments, possibly formed by cementation of airfall dust, the accumulation of which was linked to orbital variations. Thermal inertia data also reveal that the region as a whole is covered by a centimeters- to meter-thick blanket of dust, which commonly obscures the underlying bedrock (Fergason and Christensen 2008).

The Arabia Terra region hosts numerous examples of overfilled craters. One of the most dramatic is Crommelin Crater (5°N, 350°E; Fig. 4). Similar to the Gale Crater, the Crommelin Crater preserves a large mound localized in the center of the crater and surrounded by a moat (Fig. 7C). The bedding in the Crommelin Crater mound appears rhythmic (Lewis 2009), and the occurrence of rhythmic bedding in a number of craters within Arabia Terra indicates that this climate-based control was regional in extent (Lewis et al. 2008a, Lewis 2009). However, it is also possible that climatic forcing may have been pervasive across Mars, as rhythmic bedding is also observed at the top of the mound at Gale Crater (Lewis 2009, Milliken et al. 2010) and in the Medusae Fossae Formation (Lewis 2009; see below).

Henry Crater/Arabia Terra: One of the larger examples of these mound-forming intracrater deposits is found in Henry Crater (11°N, 337°E; Figs. 4, 8A), which is also ~170 km in diameter (Figs. 7D, 8A). Similar to Crommelin Crater, the mound in Henry Crater also reaches an elevation equivalent to its rim, again indicating a process of deposition that exceeds simple filling of a topographic depression (Malin and Edgett 2000). The mound is estimated to be on the order of 1500 to 2000 m thick (Andrews-Hanna et al. 2010), and, as with other crater-filling mounds in Arabia, the Henry mound is well-bedded with strata on the order of a few meters to tens of meters thick. As with other examples of mounds that fill craters, the deposit has been deeply eroded to form a "moat" around its base.

Much of the mound in Henry Crater (and much of Arabia Terra) is covered with dust, making compositional identification via orbital spectroscopy difficult. However, analysis of OMEGA data has shown that small exposures of bedrock in Arabia Terra have increased water contents, consistent with the increased H^+ signal seen throughout the region by the Mars Odyssey Gamma Ray Spectrometer (Feldman et al. 2004). This indicates that hydrous minerals may be present in some of the strata in Arabia Terra. Though not all overfilled craters exhibit this property (Fig. 8B–D), small and relatively dust-free outcrops in Henry Crater do exhibit an apparent increase in hydration, as observed in the OMEGA data (Fig. 8E–G). Visual inspection indicates that the bedding in the Henry Crater mound may also be rhythmic, consistent with a climate-based control on sedimentary accumulation. However,

←

derived from OMEGA data; color scale same as C; note that this crater does exhibit an increase in hydration associated with some portions of the stratified mound; black arrow marks location of G. (G) Close-up of strata in central mound (HiRISE PSP_008072_1905). For reference, typical dusty regions on Mars exhibit 2 to 4 wt. % H_2O, as seen by OMEGA (Milliken et al. 2007).

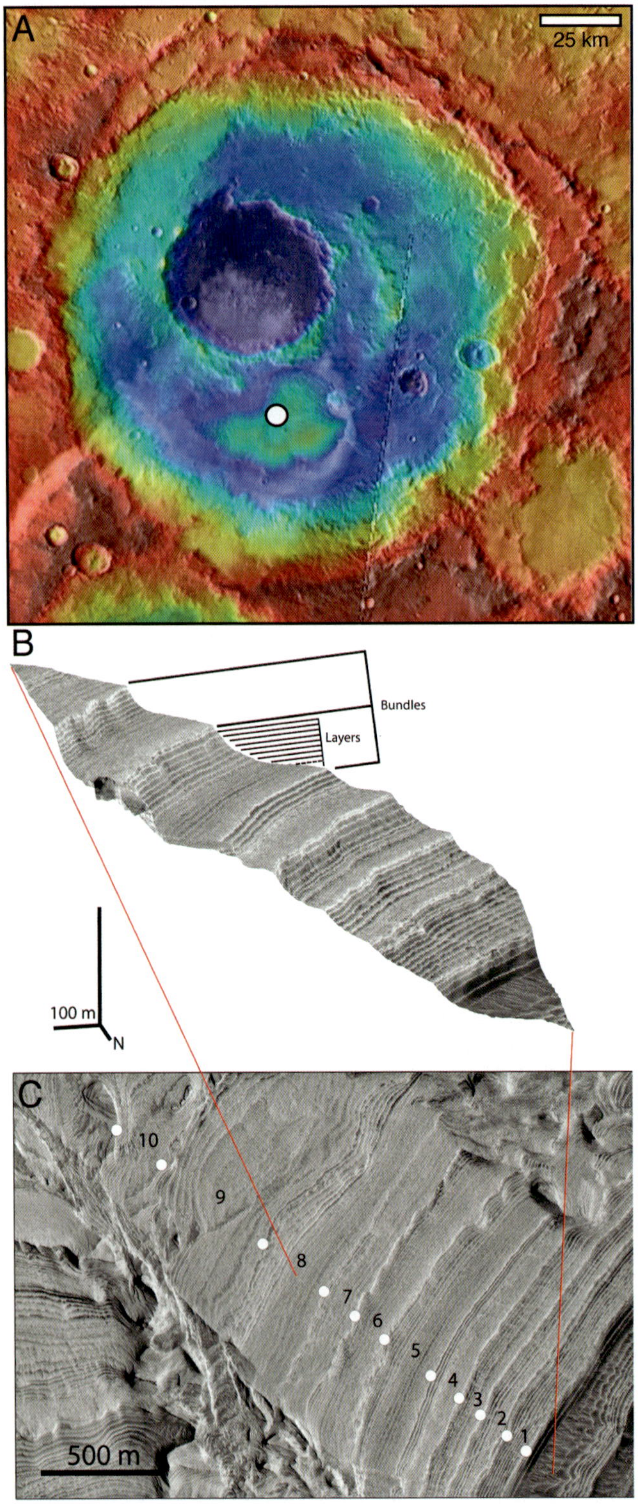

FIG. 9.—Strata in Becquerel Crater. (A) MOLA topography reveals that deposits infill a significant portion of crater. White filled circle shows location of strata analyzed in B and C. (B) HiRISE Digital Elevation Models (DEMs) showing a three-dimensional view of typical Becquerel strata. Note that there are two scales of bedding, marked as beds and bundles. The ratio of these thicknesses indicates forcing of the cyclicity by obliquity-driven climate variations (Lewis et al. 2008a). HiRISE image PSP_001546_2015 is shown draped over digital stereo topography. Scale bars, 100 m (both horizontal and vertical). (C) Plan view of HiRISE image PSP_001546_2015, showing context for B; north is down. Numbers mark the boundaries between successive bundles, defined by variations in topography. Figure constructed by Kevin Lewis using DEM made by K. Lewis and T. Suer.

this has yet to be confirmed using time-series analysis (cf. Lewis et al. 2008a, Lewis 2009).

Becquerel Crater/Arabia Terra: Becquerel Crater (22°N, 353°E; Fig. 4) also preserves a mound (Fig. 9A), but its elevation is considerably below the rim of the crater; the highest part of the stratigraphic succession lies over a kilometer beneath the highest elevation of the crater rim. Becquerel Crater has a width of ~170 km in diameter, similar to the Henry Crater. Although the mound is much lower in elevation than the crater rim, the similarity of the deposits that form the mound, combined with the range of erosion elevations for mounds in other craters of Arabia Terra—from overfilled to underfilled—warrant its inclusion in the overfilled crater category. Importantly, the analysis of bed thickness distributions makes it a kind of "type section" for this style of deposit.

The stratification at Becquerel shows striking differential resistance to erosion that has produced a distinctive "staircase" weathering pattern (Fig. 9B, C), likely reflecting changes in facies or cementation patterns within the rocks (Lewis et al. 2008a). Furthermore, these strata are characterized by a distinctive cyclicity in bedding thicknesses. Lewis et al. (2008a) analyzed the distribution of bed thicknesses at a number of locations, including Becquerel, and found a roughly 10:1 ratio of frequencies over several hundred meters of section, for a total of at least 10 bundles (Fig. 9B, C). Individual beds here have a mean thickness of 3.6 ± 1 m, and the bundles are 36 ± 9 m thick. This ~10:1 thickness ratio indicates cyclicity in environmental conditions, possibly as a result of astronomical forcing due to obliquity variations. If deposition were forced by orbital variation, the stratigraphic succession at Becquerel may have been deposited over tens of millions of years, although the true age is poorly constrained.

Danielson Crater/Arabia Terra: Another spectacular example of rhythmic bedding in a mound-forming deposit occurs in Danielson Crater, also located in Arabia Terra (8°N, 353°E; Fig. 4). This crater has received little study, although much of the crater floor is covered by these intriguing rhythmic strata (Fig. 10). Bedding in the center of the crater seems to be relatively flat-lying, in contrast to bedding at the margin of the crater, which steepens significantly, dipping inward toward the center of the crater (Fig. 10B). It could be that this change in dip is a result of simple compaction of thicker deposits toward the center of the crater.

Chasm and Canyon Systems

One of the more amazing discoveries from the Mariner 9 mission was the presence of enormous east–west trending canyons east of the Tharsis volcanic complex and just south of the equator (McCauley et al. 1972, Sharp 1973). Now known as the Valles Marineris, these canyons were created primarily by tectonic forces associated with the formation of the Tharsis rise, with later modification by collapse, fluvial, and eolian processes. Studies of the regional tectonic setting indicate multiple periods of extension, uplift, and collapse to form the graben and canyon systems found throughout the region, with

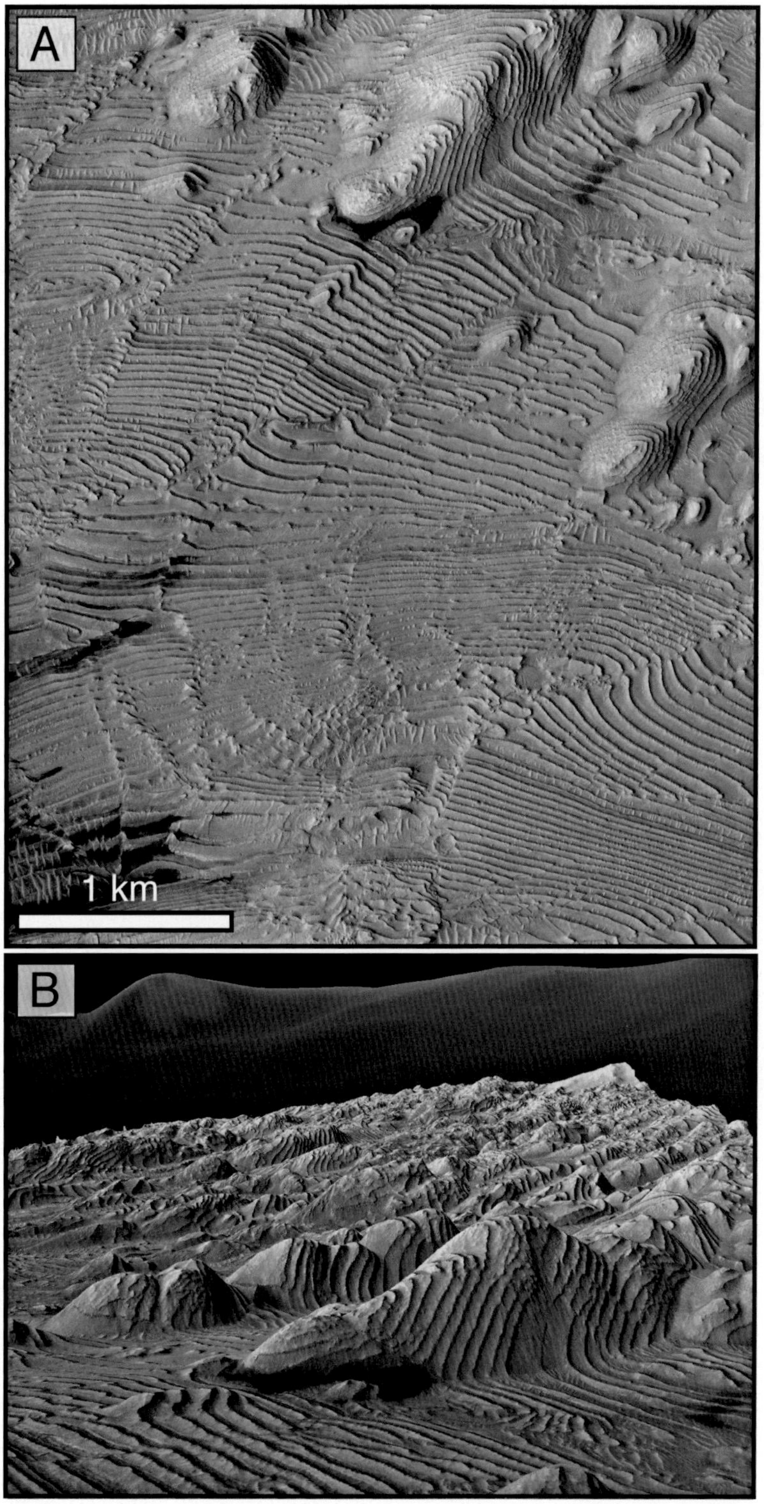

FIG. 10.—Rhythmic strata in Danielson Crater (8°N, 353°E; Fig. 4) in the Arabia Terra region. (A) In addition to clear effects of structural deformation, note the remarkable uniformity of bed thickness and the lateral persistence of bedding. This indicates a possible climate control on sediment accumulation (Lewis et al. 2008a). HiRISE image PSP_002733_1800. (B) HiRISE DEM showing orientation of deformed strata and rhythmic bedding. The nose of the main ridge is shown at the top of the image in A and is about 500 m wide. Figure based on DEM made by K. Lewis and T. Suer.

formation extending from the Early Noachian until the Early Amazonian (Tanaka and Davis 1988, Schultz 1998). Today the Valles Marineris system consists of both interconnected, open canyons and isolated, closed basins (Fig. 11A). In many instances the interiors of these chasms contain thick sequences of layered deposits, more commonly referred to as "Interior Layered Deposits" (e.g., Lucchitta et al. 1994), or ILD.[5] The ILD are generally light-toned compared to surrounding terrains and reach thicknesses of up to several kilometers. In some cases, the maximum elevations of the ILD approach the same elevations as the surrounding plains, raising the possibility that their lateral (and possibly vertical) extent was once much greater.

Stratification is visible at the subkilometer to meter scale in many of the deposits, and the prevailing view has been that the ILD postdate canyon formation. In some locations the ILD show evidence of onlapping the surrounding wallrock (e.g., Metz et al. 2009a), which is often characterized by spur-and-gully morphology, but in many cases the stratal relationships are unclear as a result of debris cover along the contacts. In other cases, it has been shown that light-toned deposits are exposed within the walls of some chasms in the Valles Marineris (Malin and Edgett 2000, Montgomery and Gillespie 2005), which indicates that light-toned deposits may underlie the surrounding plains traditionally interpreted as thick stacks of lava flows (e.g., McEwen et al. 1999). Other evidence indicates that some ILD were formed in older ancestral basins, the topographic boundaries of which are no longer visible today (Lucchitta et al. 1994, Schultz 1998), and although not the predominant hypothesis, the ambiguity in stratigraphic relationships between some ILD and the canyon walls leaves open the possibility that some of these deposits may pre-date canyon formation. The origin of the ILD is a major unresolved question in Martian geology, and volcanic ash; subaqueous volcanism; and fluvial, eolian, and lacustrine processes have all been invoked (e.g., Peterson 1981, Nedell et al. 1987, McKay and Nedell 1988, Lucchitta et al. 1992). Spectral data have revealed the presence of sulfate salts (Gendrin et al. 2005, Mangold et al. 2008, Bishop et al. 2009, Murchie et al. 2009a, Roach et al. 2010) and Fe-oxides (Christensen et al. 2001, Bibring et al. 2007, Le Deit et al. 2008) associated with the ILD, and while this mineralogical similarity to deposits in Sinus Meridiani (described below) could reflect similar depositional environments (Bibring et al. 2007, Murchie et al. 2009a), their exact origin remains unknown.

Candor Chasma: Candor Chasma (6°S, 289°E; Fig. 4) lies in the north–central portion of the Valles Marineris and hosts several kilometers–thick ILD (Fig. 11A–C). Crude layers were visible in these deposits even from Mariner 9 and Viking images, but MOC and HiRISE data have revealed that they are also stratified at the meter to decameter scale (e.g., Malin and Edgett 2000, Okubo 2010). OMEGA and CRISM data indicate that these strata are composed at least in part of polyhydrated sulfates (likely Mg-varieties such as hexahydrite, $MgSO_4{\cdot}6H_2O$), kieserite ($MgSO_4{\cdot}1H_2O$), and crystalline ferric oxides (Gendrin et al. 2005, Bibring et al. 2007, Mangold et al. 2008, Murchie et al. 2009a, Roach et al. 2010). At longer wavelengths, TES data have revealed the presence of gray (specular) hematite (Christensen et al. 2001), similar to the signatures observed for Sinus Meridiani. Despite this knowledge of the composition of the ILD, whether these minerals are detrital or authigenic is not known, and the depositional environment of the sediments is also ambiguous. However, similarities in composition to strata examined by the *Opportunity* rover in Sinus Meridiani have led to the interpretation that they are eolian sediments cemented by sulfates during interaction with groundwater (Bibring et al. 2007, Murchie et al. 2009a). It is also possible that the Candor ILD formed as primary lacustrine evaporites, accumulating in one of the planet's great topographic depressions. In either scenario, these thick ILD would postdate the formation of Candor Chasma, which is supported by the apparent onlap of ILD onto the surrounding wallrock.

In contrast to most ILD and other stratified deposits within Valles Marineris, the western region of Candor contains strata that have experienced extensive deformation (Fig. 11C, D). Deposits surrounding the westernmost ILD exhibit significant folding and faulting (Okubo et al. 2008, Metz et al. 2010, Okubo 2010), and the structural geometries are consistent with thin-skinned deformation (Metz et al. 2010). Although the abundance of sulfates and other salts within these deposits is unknown, their mere presence may have facilitated deformation of these units. Indeed, it has even been suggested (Montgomery et al. 2009) that vast salt deposits might underlie the entire region east of Tharsis (the Thaumasia Plateau) and were responsible for the opening of Valles Marineris via gravity gliding along a regionally extensive detachment surface localized within the salt. Locally within western Candor, strike-and-dip measurements derived from 1 m/pixel digital elevation models indicate that undeformed strata in this region dip toward the basin interior and that folds in the deformed zones have a gentle plunge of $\sim$5° (Metz et al. 2010, Okubo 2010). The deformed rocks exhibit meter- to decameter-scale stratification, and bedding is often accentuated by the presence of wind-blown, dark-toned sediment that has accumulated in troughs formed by eroded layers. In some locations this dark material exhibits strong sulfate and Fe-oxide signatures (Murchie et al. 2009a), indicating that at least some fraction of it is sourced from the surrounding light-toned ILD. The original extent of the ILD in Candor and the terminal sink for the sediment eroded from them are not known.

Juventae Chasma: Juventae Chasma is a large basin that lies to the northeast of Candor Chasma, at 4°S, 299°E; Juventae Chasma is the source for the large Maja Valles outflow channel that empties into Chyrse Planitia. A small outlet channel along the northern margin debouches into a separate smaller basin, and together these two basins are lower in elevation than the surrounding regions. Therefore, although Juventae may not technically be a closed basin today, this appears to have been the case in the past. Juventae contains at least four distinct ILD (Catling et al. 2006) that are known to contain hydrated sulfates (Gendrin et al. 2005, Bishop et al. 2009). Layering is apparent in several of these ILD, even in low-resolution images, as a result of differential erosion that has produced stair-step topography (see Fig. 1). However, meter- to decameter-scale stratification (Fig. 11E) is also apparent in high-resolution MOC and HiRISE images (e.g., Chapman et al. 2003, Catling et al. 2006, Bishop et al. 2009). As with other ILD, the depositional environment(s) of these sediments is poorly constrained. Detailed studies indicate that these light-toned deposits, which are possibly $\sim$3 to 6 km thick, were once much more laterally extensive and that they may be eroding at a rate of $\sim$1 km/Gyr (Catling et al. 2006). Origins consistent with these observations have included, but are certainly not limited to, precipitation of evaporites in bodies of standing water, accumulation of wind-blown sulfates (McLennan and Grotzinger 2008), and volcanic sulfate aerosols deposited with ice during high-obliquity cycles (Catling et al. 2006).

Original OMEGA-based interpretations of the composition of the ILD indicated that the lower strata in one of the mounds contained kieserite ($MgSO_4{\cdot}H_2O$) and were overlain by gypsum-bearing strata (Gendrin et al. 2005). If formed by direct precipitation via evaporative concentration, this mineral sequence is hard to explain because of the

[5]For the unfamiliar, use of the term "Interior Layered Deposit" is full of genetic implications. Many in the Mars research community accept that these are sedimentary deposits and that they are layered (although the term "stratified" might be more appropriate if the term "layered" had not been in use for over 15 years). However, the designation "interior" indicates that these deposits are restricted to the interior of the Valles Marineris (VM) system. On this point there is lack of agreement, with many workers suggesting they formed after the VM system developed, but with a few studies indicating that they predate opening of the VM system, in some cases extending into the subsurface beneath the adjacent plains.

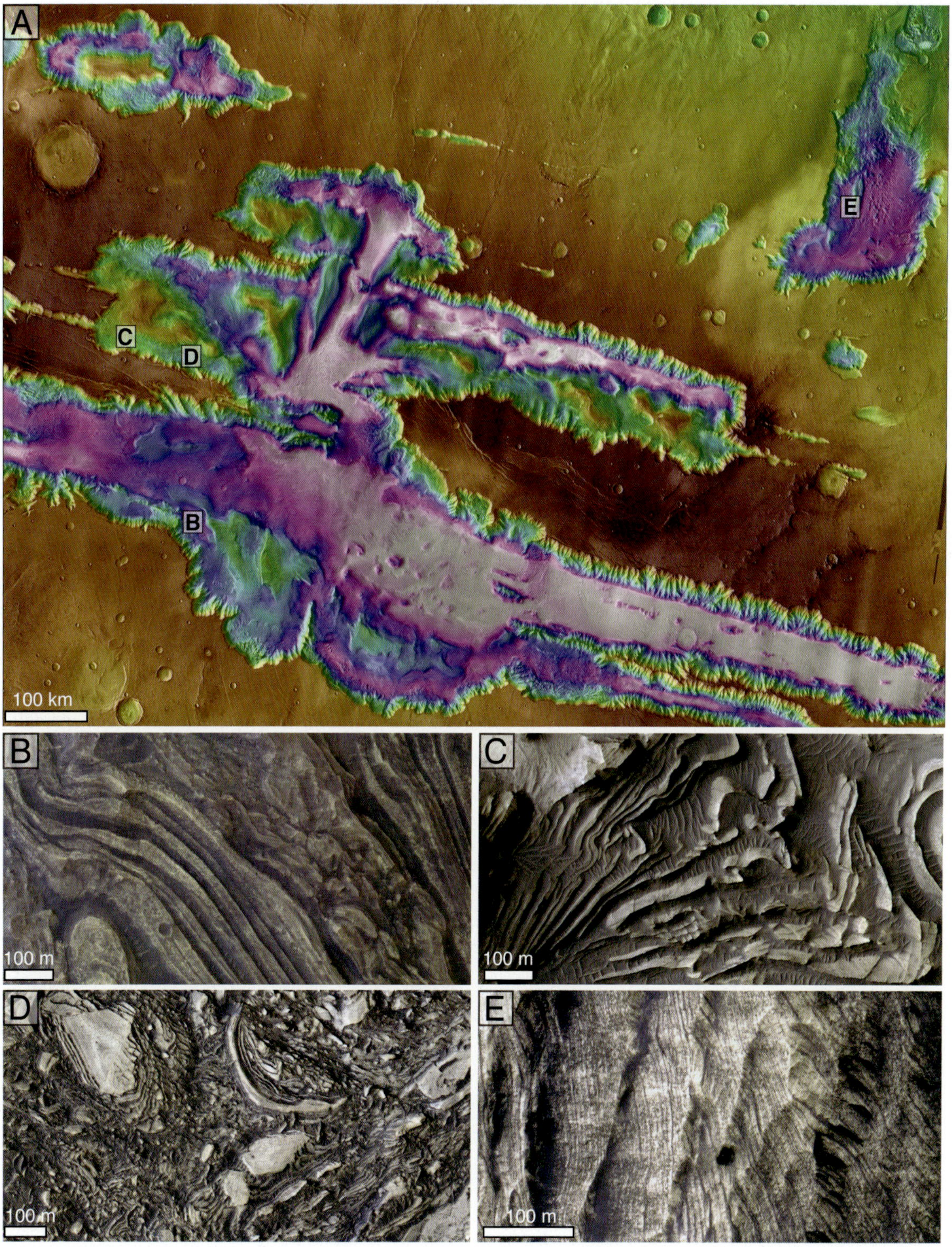

FIG. 11.—Sedimentary deposits in Valles Marineris. (A) Overview of central Valles Marineris; MOLA elevation on THEMIS daytime infrared mosaic. (B) Close-up of strata in Melas Chasma, HiRISE image ESP_019442_1700. (C) Close-up of deformed (folded) strata in west Candor, HiRISE image PSP_001918_1735. (D) Close-up of brecciated and deformed strata in west Candor. HiRISE image ESP_011372_1730. (E) Close-up of finely stratified ILD in Juventae Chasma, HiRISE image PSP_006915_1760. CRISM data indicate that sulfates are present in all locations.

high solubility of kieserite and the low solubility of gypsum: the sequence should be the opposite of what is observed if given a simple, single-stage evaporation process. However, more recent analysis using higher-resolution CRISM data has shown that the identification of gypsum was likely an artifact during processing of the OMEGA data in this region. Monohydrated sulfates have been confirmed in the CRISM data (kieserite and possibly the Fe-sulfate szomolnokite, $FeSO_4 \cdot H_2O$), and polyhydrated sulfates consistent with Mg and Fe varieties (but not gypsum) have also been identified (Bishop et al. 2009). The variation in hydration states of sulfates may reflect hydration–dehydration reactions that occur as these phases are exposed to the atmosphere by continual erosion. The hydrated Mg- and Fe-sulfate minerals are particularly sensitive to changes in relative humidity, and thermodynamic calculations and laboratory studies have predicted that changes in the hydration states of these sulfates (e.g., between monohydrated and polyhydrated states) should occur on Mars (Vaniman et al. 2004, Hasenmueller and Bish 2005, Bish and Scanlan 2006, Vaniman and Chipera 2006, Chipera and Vaniman 2007). However, such changes have not yet been observed for a given sulfate deposit, even after several years of repeat observations by OMEGA and CRISM, implying that the rates of these reactions are too slow to be observed on these timescales (Roach et al. 2010). Regardless, the diversity of hydrated sulfates in the sedimentary rocks of Juventae provides an interesting location for further testing of this hypothesis. As with other ILD, additional measurements of the strike, dip, and stratal geometries of layers within these units would help to place further constraints on their mode(s) of emplacement.

Plains Covering Deposits

At several locations on Mars, outcrops of regional extent show clear evidence of stratification, are laterally continuous on a scale of hundreds of kilometers, and form the bedrock beneath broad plains. These strata lack signs of accumulation in an obvious topographic depression. Exposures occur along relatively high plateaus, such as the rim of Valles Marineris, as well as in lower regions, such as Arabia Terra and Sinus Meridiani. What is important about these regions is that these strata seem to represent sequences that blanket large regions of the planet and that in some cases (e.g., Sinus Meridiani) achieve substantial thickness, on the order of many hundreds of meters. Though the paleotopography of these regions on Mars is unknown, it is clear that they were sinks for sediment accumulation.

Plains Surrounding Valles Marineris: The plains immediately adjacent to the Valles Marineris are predominantly Hesperian in age, though some Noachian terrains are exposed along the far eastern portions (e.g., south of Eos Chasma) and north and south of eastern Candor Chasma (Scott and Tanaka 1982). Historically, these plains were interpreted as sequences of lava flows, and many exhibit compression ridges (Scott and Tanaka 1982). However, examination of high-resolution images has revealed the presence of very thinly bedded deposits of alternating light- and dark-toned strata in some locations along the margins of these plains. These deposits consist of submeter- to meter-thick beds, and the well-stratified portions of these sequences are typically <60 m thick (Weitz et al. 2010). Analysis of CRISM data has shown that these deposits contain opaline silica and hydrated Fe-sulfates, consistent with acidic alteration of basalt or deposition in acidic fluids (Milliken et al. 2008). In addition, some of these deposits exhibit inverted channels and drainage networks, indicating that fluvial processes were related to their emplacement (Milliken et al. 2008; Weitz et al. 2008, 2010). Similar fluvial networks observed in other plains regions adjacent to Valles Marineris have been interpreted as evidence for precipitation-driven runoff (Mangold et al. 2004), and in all cases these deposits and features are interpreted to postdate the primary period of valley network formation, which was prior to the Early Hesperian (Fassett and Head 2008a).

The best examples of this category of sedimentary plains deposits occur along the western rim of Juventae Chasma (~4.6°S, 296.5°E; Fig. 4) and the southern rim of Ius Chasma (~8.3°S, 275.1°E; Fig. 12A–D). In some locations these deposits occur between ridges, although whether they onlap the ridges or instead represent erosional windows into the subsurface is not always clear. Furthermore, although these deposits are typically the locally highest stratigraphic unit, they often recede beneath a thin, dark capping unit when traced farther away from the canyon rim. The origin of this capping unit (eolian?, volcanic?, pedogenic?), its thickness, and whether or not it has a similar origin to the regionally extensive volcanic Hesperian plains are currently unclear. Therefore, these opal- and sulfate-bearing strata may be restricted to the edges of the canyon system, where they sit atop Hesperian-age lava flows. Alternatively, they may be regionally extensive but commonly obscured by the thin dark capping unit, or they may represent a sedimentary sequence that sits within a stack of Hesperian lava flows. If either of the latter two scenarios are correct, then the full extent of these sediments may be much greater than currently recognized, exposed only near the edges of the canyons, where erosion has been more extensive. Regardless of their age and origin, the finely stratified nature of these deposits, the presence of mineral phases that require water for their formation, and the association with fluvial features imply aqueous activity and alteration much more recently in Mars' history than previously thought (Milliken et al. 2008, Weitz et al. 2010). This supports growing geomorphic evidence for localized valley incision during the Late Hesperian or even Amazonian, as is also observed in the Margaritifer Terra region (Grant and Wilson 2011) and in the Newton and Gorgonum basins (Howard and Moore 2011).

Sinus Meridiani: Sinus Meridiani is a large low-albedo region centered roughly near 0° longitude, 0° latitude (Fig. 4). Outcrops are represented by cliff-forming, light-toned stratified deposits covering a region somewhat larger than the Colorado Plateau (300,000 km^2), and the northern part of Sinus Meridiani is particularly well exposed and free of mantling dust, silt, and sand (Edgett and Malin 2002). A variety of origins for these strata have been proposed, including lacustrine, paleopolar deposits (dust), eolian dust/dune, volcaniclastic, and other subaqueous environments (summarized in Edgett and Malin [2002]).

Edgett (2005) estimated the thickness of the strata at Sinus Meridiani to be >800 m, composed of four distinct units. Differentiation of these units is based on geomorphic style (plains vs. ridge forming), tone/albedo, and the presence of buried/inverted channels. The section shows extensive exhumation, exposing inverted channels as well as numerous buried craters. The presence of craters, formed and buried at the time of sediment accumulation and subsequently exposed during erosion and exhumation (Fig. 12E, F), is strong evidence for a Martian cycle of sedimentation followed by erosion (Malin and Edgett 2000, Edgett and Malin 2002). Toward the upper part of this succession, in a region called Meridiani Planum, are the outcrops encountered by the *Opportunity* rover (Edgett 2005). On the ~7-km traverse from Eagle to Endurance to Erebus to Victoria craters, outcrops were observed and studied systematically using the rover payload (Squyres et al. 2004; Grotzinger et al. 2005, 2006; Metz et al. 2009b; Squyres et al. 2009; Hayes et al. 2011; Edgar et al. [this volume]); the bulk of these sediments are of eolian origin.

The Burns formation is the informal name given to the portion of the succession examined in outcrop by *Opportunity*, which revealed a succession of well-sorted, moderately indurated sandstones (Fig. 13A) preserved immediately beneath the surface of the Meridiani plains. Detailed stratigraphic measurements (Fig. 14) and sedimentological observations (Fig. 13) indicate that eolian and local subaqueous processes deposited these sedimentary rocks as part of an interstratified

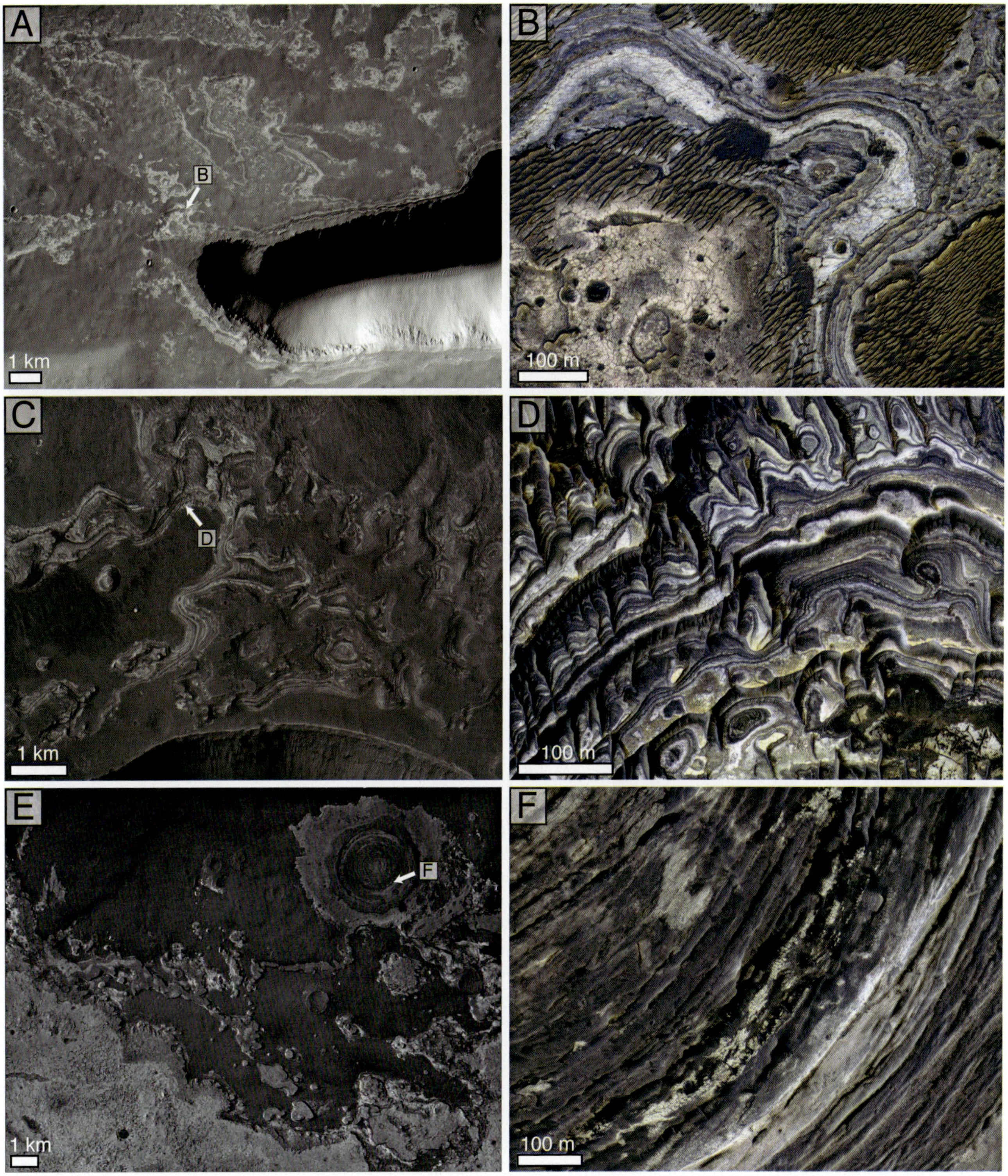

FIG. 12.—Sedimentary deposits formed on the plains surrounding the Valles Marineris and other broad open tablelands of Mars. (A) CTX mosaic showing light-toned outcrops along the plains south of Ius Chasma; some light-toned deposits contain opaline silica. (B) Close-up of opal-bearing units in plains adjacent to Ius Chasma; arrow in A marks location; HiRISE image PSP_002459_1715. (C) CTX mosaic showing light-toned outcrops along the plains west of Juventae Chasma; some light-toned deposits contain opaline silica and/or Fe-sulfates. (D) Close-up of opal- and sulfate-bearing units; location marked by arrow in C; HiRISE image PSP_003434_1755. (E) CTX mosaic showing light-toned outcrops in eastern Sinus Meridiani including large exhumed crater. (F) Close-up of finely stratified light-toned units in a filled and exhumed crater; location marked by arrow in E; HiRISE image PSP_001981_1825.

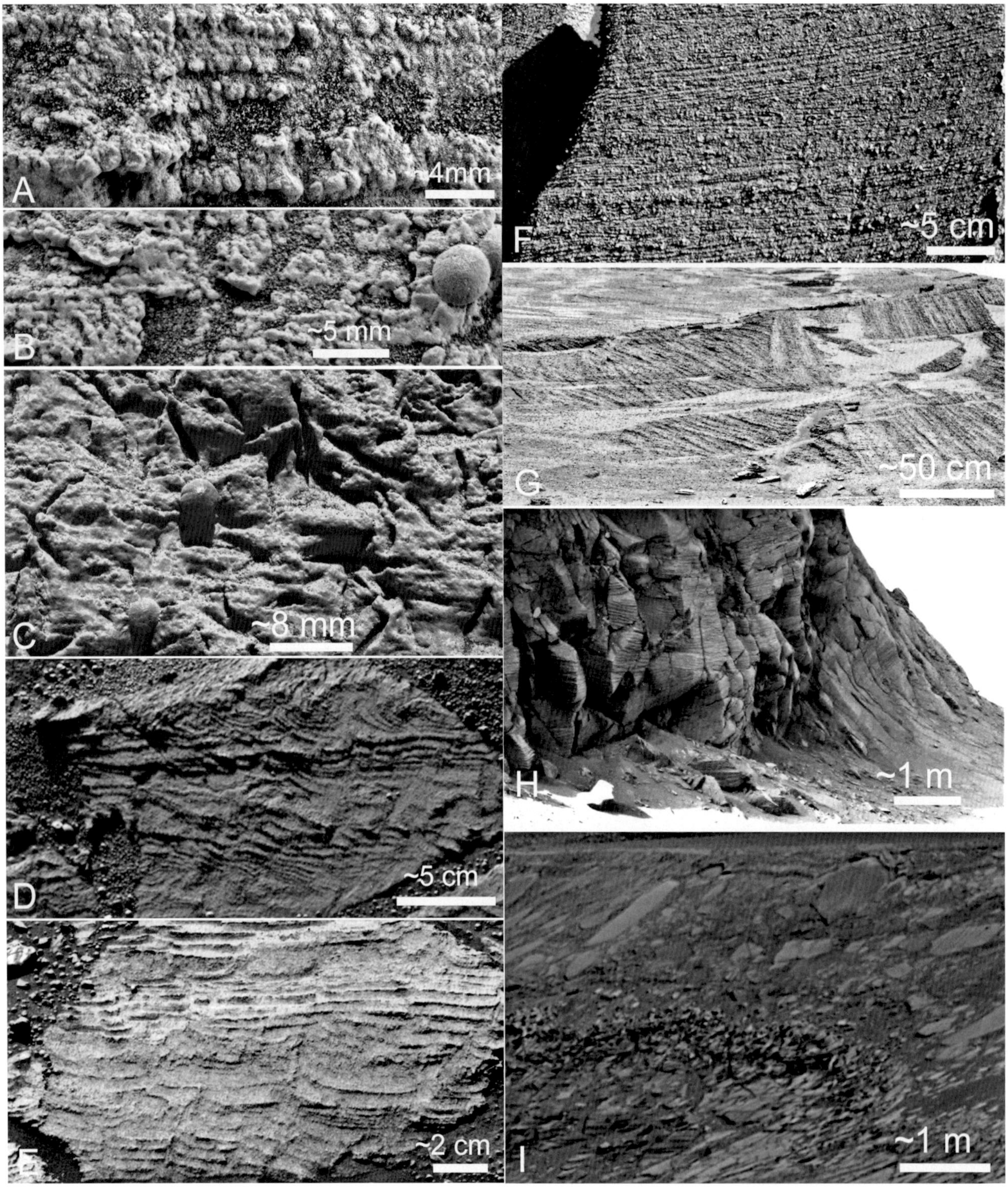

FIG. 13.—Sedimentary facies and textures in the Burns formation, Meridiani Planum. (A) Microscopic Imager (MI) image showing distinct grains that form lamination ("Flatrock," Eagle Crater), indicating that most of the Burns formation was composed of sandstones, the grains of which were derived from reworked evaporites. Grains range in size from 0.1 to 1.0 mm, are moderately well rounded, and are well sorted into discrete laminae. The excellent size sorting on the scale of individual laminae is strongly indicative of "translatent" strata, formed by sediment transport within migrating eolian impact ripples (see Grotzinger et al. [2005] for more details). Fabric preservation, which reveals discrete grains, is rare as a result of cementation recrystallization; preservation is best where early cementation was minimal, permitting grains to weather in positive relief. Compare with Figure 13B, which shows the effects of increasing cementation. The image (1M131912465) was

dune—sand sheet—interdune depression sequence (Fig. 14). Primary textures are overprinted by a variety of diagenetic features (Fig. 13B, C).

Eolian dune facies are characterized by the occurrence of large-scale, cross-bedded, well-sorted sandstones (Grotzinger et al. 2005, Metz et al. 2009b, Hayes et al. 2011, Edgar et al. [this volume]). This facies is interpreted to represent a migrating dune system of unknown extent that was deposited under dry conditions, such that the sediment was noncohesive and thus transported in migrating dune fields. Bedset thicknesses indicate moderately large dunes in excess of 4 m (Fig. 13H). Eolian sand sheet facies are characterized by planar-laminated to low-angle cross-stratified, well-sorted sandstones (Grotzinger et al. 2005, Metz et al. 2009b). This type of stratification (Fig. 13F) results from sediment limitation, forming low-relief bedforms produced by migrating impact ripples. In Endurance Crater, the contact between the dune and overlying sand sheet facies marks the boundary between the lower and middle stratigraphic units (Fig. 13G). This important stratigraphic boundary, termed the "Wellington contact," is interpreted to be a deflation surface formed where the groundwater capillary fringe limited erosion of previously deposited dune sediments (Grotzinger et al. 2005).

Wet to evaporitic interdune facies are characterized by centimeter-scale trough cross-lamination, diagnostic of subaqueous current transport in the lower flow regime (Grotzinger et al. 2005, 2006). At Endurance, Eagle, and Erebus craters, this facies marks where the groundwater table breached the surface and wind-driven subaqueous currents transported the sediment. Centimeter-scale trough cross-lamination (Fig. 13D) is particularly well developed at Erebus Crater, where additional features indicative of desiccation also are present (Grotzinger et al. 2006). The sediments that form these ripples were likely transported by brines with potentially high viscosities (Lamb et al., this volume). Prism cracks (Fig. 13E) are interpreted to have formed during multiple wetting and drying events, and soft sediment deformation features are consistent with sediment liquefaction. Stratigraphic sections studied by *Opportunity* are exposed in the walls of craters, and, therefore, intact bedrock is usually capped by a unit of impact breccia (Fig. 13I). The presence of ferric sulfates (i.e., jarosite) in the rocks of Meridiani Planum indicates interaction with or deposition by acidic fluids, where the acidity may have been generated by the oxidation of Fe(II)-rich water as it emerged from the subsurface (Hurowitz et al. 2010). Indeed, global hydrologic models predict that groundwater would emerge in the Sinus Meridiani region (Andrews-Hanna et al. 2007), and terrestrial studies of acid–saline lakes may prove useful for constraining the origin of acidic brines on Mars (see Bowen et al., this volume).

Medusae Fossae Formation: Perhaps one of the most enigmatic sedimentary deposits on Mars is the Medusae Fossae Formation (MFF). The MFF is a large discontinuous deposit that straddles the equator between 130 and 230°E longitude and that is flanked by Noachian- and Hesperian-aged plains units. The deposit can be as thick as several kilometers, although locally it can be significantly thinner, and it has an estimated volume of 1.4×10^6 km^3 (Bradley et al. 2002). An Amazonian age has traditionally been assigned to the MFF (Scott and Tanaka 1986), though recent analysis of superposition by Hesperian lavas, as seen in higher resolution images, has indicated that it may have been emplaced in the Hesperian (Kerber and Head 2010). The deposit is extremely variable in its morphology, but major textural characteristics indicate that it is poorly cemented, easily eroded, and wind scoured. The friable nature of the deposit is likely a contributing factor in the interpreted young age, as determined by crater counting; many craters have been eroded, and the Amazonian age may reflect an exposure age that is much younger than its depositional age (Kerber and Head 2010). Large portions of the surface are characterized by yardangs, and stratification is developed at a various scales; <10-m beds are visible only at MOC or HiRISE resolution, and these occurrences are locally restricted (Bradley et al. 2002). However, this stratification is not visible in SHARAD shallow sounding radar data, indicating that the upper several hundred meters of the deposit have a high porosity and that this region is rather

←

acquired on sol 042. (B) MI image showing effects of early cementation and recrystallization, visible in the center left ("Algerita," Eagle Crater). Note set of three or four laminae, expressed by fused grains. The lack of expression of individual grains, as compared to Figure 13A, is interpreted to reflect the greater extent of early cementation. At center right, the effects of recrystallization on destruction of primary fabrics are well illustrated by a halo, which envelops a hematitic concretion. Note complete loss of primary fabric and development of blocky, interlocking crystals. See McLennan et al. (2005) for more details. The image was acquired on sol 028. (C) MI mosaic showing both diagenetic concretions (positive relief) and molds after diagenetic minerals (negative relief). Note distinct geometric shapes of crystal molds, which cross-cut primary lamination. The images were acquired on sol 029. (D) Pancam image of trough (or "festoon") cross-lamination (upper "Overgaard," Erebus Crater). Small-scale trough cross-lamination is interpreted to have formed in low-velocity subaqueous flows. See Grotzinger et al. (2006) for further discussion. This image was acquired on Sol 716. (E) Pancam image of probable desiccation cracks and soft-sediment deformation features ("Skull Valley," Erebus Crater). Cracks cross-cut lamination, some oblique to bedding. Note characteristic upward-deflected laminae along crack margins; termination of prominent crack in center of rock at a discrete bedding plane; and truncation of upward-deflected laminae along discrete bedding planes in center and upper parts of rock. This image was acquired on sol 713. See Grotzinger et al. (2006) for further details. (F) Pancam image of low-angle cross-stratification featuring well-developed pinstripe texture, which has likely formed as a result of differential cementation of eolian translatent strata ("Tipuna," Endurance Crater). The Tipuna outcrop represents the uppermost 30 to 40 cm of the Burns middle unit. The image was obtained on sol 307. (G) Pancam image of the contact between the lower and upper units of the Burns formation. Below the contact, large-scale cross-bedding forms a bedset at least 2 m thick that was likely formed by migrating eolian dunes. The contact—a truncation surface—is interpreted to represent deflation of the dune field down to the level of the groundwater table. See Grotzinger et al. (2005) for further details. The image was acquired on sol 288. (H) Pancam image of large-scale cross-bedding in the cliffs of Victoria crater, along the west-facing wall of Cape St. Mary. This cross-stratification is also diagnostic of migrating eolian dunes, as indicated by the presence of meter-scale scours and wedge-shaped sets. The image is part of a mosaic acquired during sols 970–991. See also Edgar et al. (this volume) for additional images. (I) Pancam image of ejecta blanket deposited on Burns formation bedrock at Victoria Crater, along the east face of Cabo Frio. The blanket consists of coarse block breccia and is capped by a stratified upper unit, which drapes protruding breccia blocks. The image is part of a mosaic acquired on sol 952.

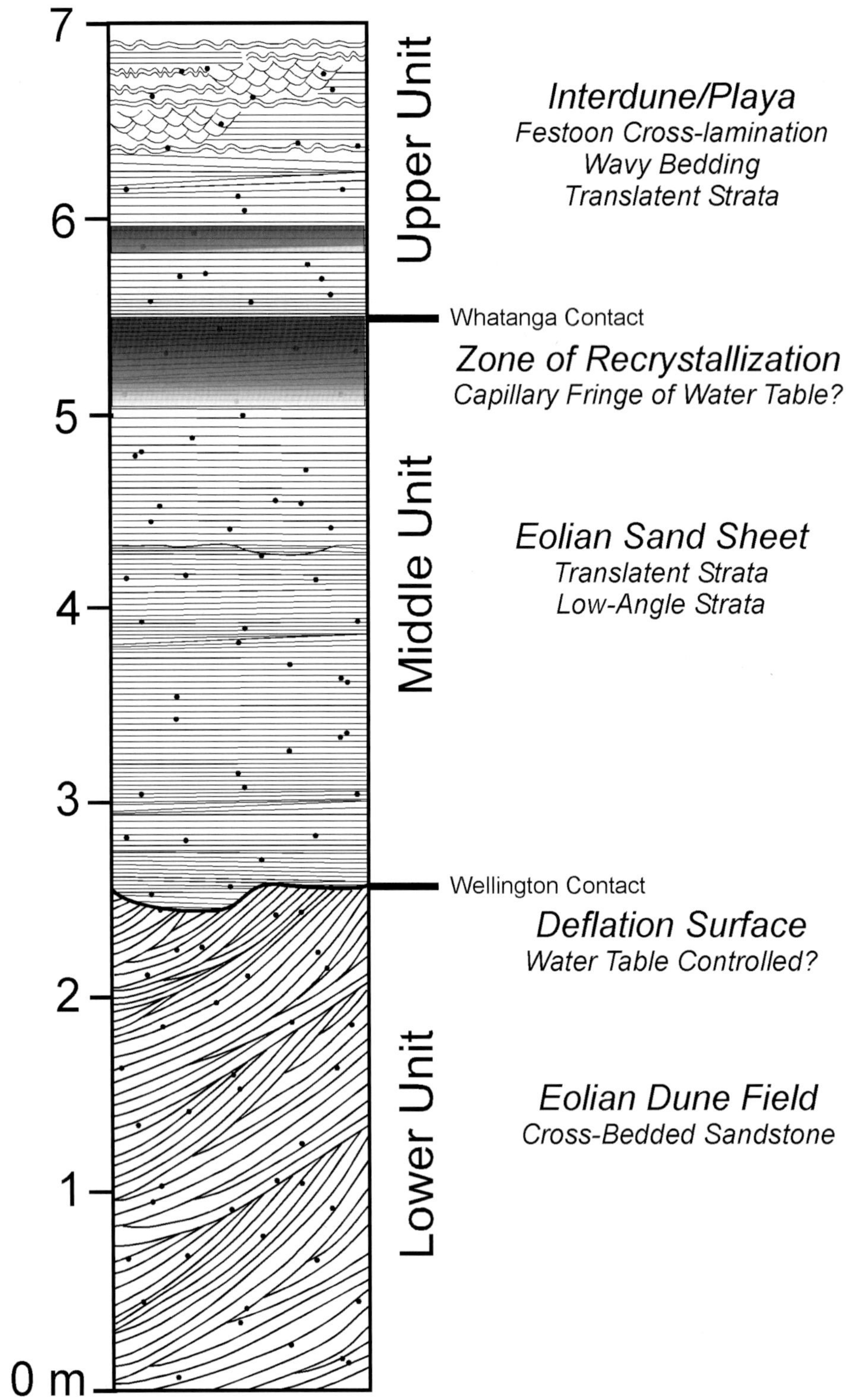

FIG. 14.—Stratigraphy of the Burns formation exposed at Burns cliff and within Eagle Crater, Meridiani Planum. Names and letters on the left side denote the approximate stratigraphic locations of key rocks and targets. Sedimentological interpretations of the three stratigraphic units are shown on the right-hand side of the column. See text for further discussion. From Grotzinger et al. (2005).

homogeneous (low permittivity contrast) or has discontinuous bedding.

The ultimate origin of the deposit is still unknown, but volcanic airfall, ignimbrites, or eolian dust are leading hypotheses (Bradley et al. 2002, Mandt et al. 2008, Kerber and Head 2010). Alternatively, radar data indicate that the dielectric properties of the deposit are also consistent with the presence of significant amounts of water ice (Watters et al. 2007), possibly indicating a formation mechanism similar to the polar layered deposits (Schultz 1988). To date, spectral data have revealed no unique or diagnostic mineral signatures associated with the MFF, adding further mystery to its origin. Of the sedimentary deposits discussed here, the MFF may be the leading candidate for a thick sequence of volcanic ash on Mars. The general absence of well-defined, meter-scale stratification developed across a significant fraction of the aerial extent of each deposit helps differentiate the MFF from the spectrally bland but well-bedded deposits that fill and sometimes overtop the craters of Arabia Terra (see preceding discussion, "Overfilled Craters"). As discussed above, these latter deposits are regarded as possible eolian dust deposits as a result of their highly rhythmic bedding (Lewis et al. 2008a, Lewis 2009). However, Lewis (2009) examined the bed thicknesses of several stratified deposits within the MFF and found that these regions did in fact exhibit a periodic thickness of 3 m. Given that the majority of the MFF does not exhibit this style of bedding, this raises the possibility of facies variability in time or space, such that massive deposits may substitute for rhythmic deposits.

Another intriguing attribute of the MFF is the numerous fluvial features that mark its surface, many of which are inverted (Burr et al. 2009). These features indicate that erosion, transport, and redeposition of the MFF sediments by fluvial systems have played an important role in its history, and although the primary depositional mechanism(s) of the MFF remains uncertain, it is clear that subsequent reworking has formed fluvial deposits (Burr et al. 2009, Kerber and Head 2010).

Very Ancient Strata

Some of the stratified deposits on Mars that have captured significant attention over the past several years are those preserved as part of the most ancient Noachian crust. It is widely believed that Mars, like the Moon, was subject to a period of intense bombardment and impact crater formation from 4.1 to 3.8 Ga. The size and number of impact events during the Late Heavy Bombardment imply that much of the preexisting crust would have been destroyed and that vast regions of the planet would have been covered by impact ejecta. Therefore, it may be somewhat surprising that any sedimentary sequences of appreciable thickness formed prior to this period would remain visible from orbit today. Indeed, much of the ancient Noachian crust in the southern highlands of Mars lacks clear stratification at the scale of tens and hundreds of meters. However, Noachian stratified rocks, possibly of sedimentary origin, have been identified in a handful of instances. These regions typically contain some of the strongest spectral signatures of clay minerals observed on Mars, and their mineralogical diversity is striking. On Earth, deciphering the original depositional environments of such ancient rocks, which have been subjected to billions of years of postdepositional geologic processes, can be daunting. The same is true for Mars, although the volume of stratified crust preserved from the earliest days of the planet is much greater than for Earth. Furthermore, the fraction of volcanic vs. sedimentary rocks, the possible depositional environments of any sedimentary rocks, and alteration conditions of these strata continue to be a subject of significant debate.

Mawrth Vallis: The Mawrth Vallis outflow channel, located near ~23°N, 341.5°E (Fig. 4), is a very old flood channel that empties into the northern lowlands of Mars. The channel incises bedrock formed of stratified rocks up to several hundred meters in thickness (Loizeau et al. 2007, Michalski and Noe Dobrea 2007). Orbital images show that these rocks are stratified at the meter to decameter scale, typically light-toned, significantly eroded, and heavily cratered (Fig. 15A–C). They span a region measuring >80,000 km^2 and are assigned an Early to Middle Noachian age (4.1–3.8 Ga) (Michalski and Noe Dobrea 2007), implying they were emplaced during or prior to the Late Heavy Bombardment.

In addition to being some of the oldest possible sedimentary rocks yet observed, these rocks also exhibit some of the strongest clay mineral signatures detected on Mars (Poulet et al. 2005, Loizeau et al. 2007, Michalski et al. 2010). The rocks at Mawrth Vallis exhibit a broad-scale bipartite mineralogic stratigraphy in which Fe/Mg-rich smectitic clays persist for >100 m in thickness and ultimately transition upsection to more Al-rich clays (e.g., kaolinite, montmorillonite), possibly mixed with opaline silica (Poulet et al. 2005, Bishop et al. 2008, Michalski et al. 2010, Noe Dobrea et al. 2010). The thickness of the Al-rich zone is variable as a result of erosion, but locally it can measure up to 50 m (Loizeau et al. 2007). It is not clear whether this mineralogical transition (Fig. 16A) is linked to variations in primary depositional environments or whether it cuts across primary bedding, which would indicate a diagenetic origin for at least the Al-rich rocks. However, this mineralogical sequence is observed elsewhere in the surrounding region, and studies have indicated that the Al-bearing phases at the top of the succession are due to diagenetic overprinting, leaching, or pedogenesis (e.g., Noe Dobrea et al. 2010). The entire succession is unconformably overlain by an erosionally resistant, dark-toned capping unit of unknown origin (Michalski and Noe Dobrea 2007, Noe Dobrea et al. 2010).

As with all clay minerals detected on Mars, it is also unclear if the Fe/Mg-smectites at Mawrth Vallis are detrital, authigenic, or diagenetic; if the latter is true, then the clay minerals could be much younger than the rocks in which they reside. Regardless of these complexities, it is clear that these strata are very ancient, and their compositional diversity indicates changes in fluid chemistry, parent rock composition, and/or depositional environment. The potential for the Al- to Mg/Fe-clay transition to record the surface weathering process on ancient Mars is particularly intriguing as it would imply the persistent stability of water in the near-surface.

Despite their strong mineral signatures and great areal extent, the modes of emplacement for these rocks are also poorly constrained. Though interpreted as sedimentary in origin (Michalski and Noe Dobrea 2007), the rocks may represent fluvial, lacustrine, altered volcanic ash, eolian, or impact deposits (Loizeau et al. 2007; Michalski and Noe Dobrea 2007; Bishop et al. 2008; Michalski et al. 2010; Noe Dobrea et al. 2010; D Sumner, personal communication, 2011). An origin for some strata, through accretion of impact deposits, is indicated by the occurrence of buried impacts and their adjacent ejecta (Fig. 17). In addition, examination of stratal geometries indicates that impact breccias may form a significant part of the section (Sumner 2010; D Sumner, personal communication, 2011), similar to the stratification present in the Fra Mauro Formation at Silver Spur in the lunar highlands (Lindsay 1972, Cadogan 1981). Primary bedding truncations form depressions that are overlain by draping (not onlapping) layers (Fig. 16B). Such features may have formed by scour and drape related to impact surge deposition, and stratigraphically bound breccias may represent impact deposits. In addition, lenticular pods of breccia are also present, nearly parallel to bedding, and may represent injectites (Fig. 16C). However, rover-based observations of small-scale features such as accretionary lapilli will likely be required to confirm an impact surge origin for these deposits (see Fralick et al., this volume). Postdepositional tectonic overprinting includes extensive low-angle faulting (Fig. 16D) and small block rotation/imbrication (Fig. 16E); in other cases, trains of disharmonic waveforms are present (Fig. 16F). Finally, these already-complex relationships may be

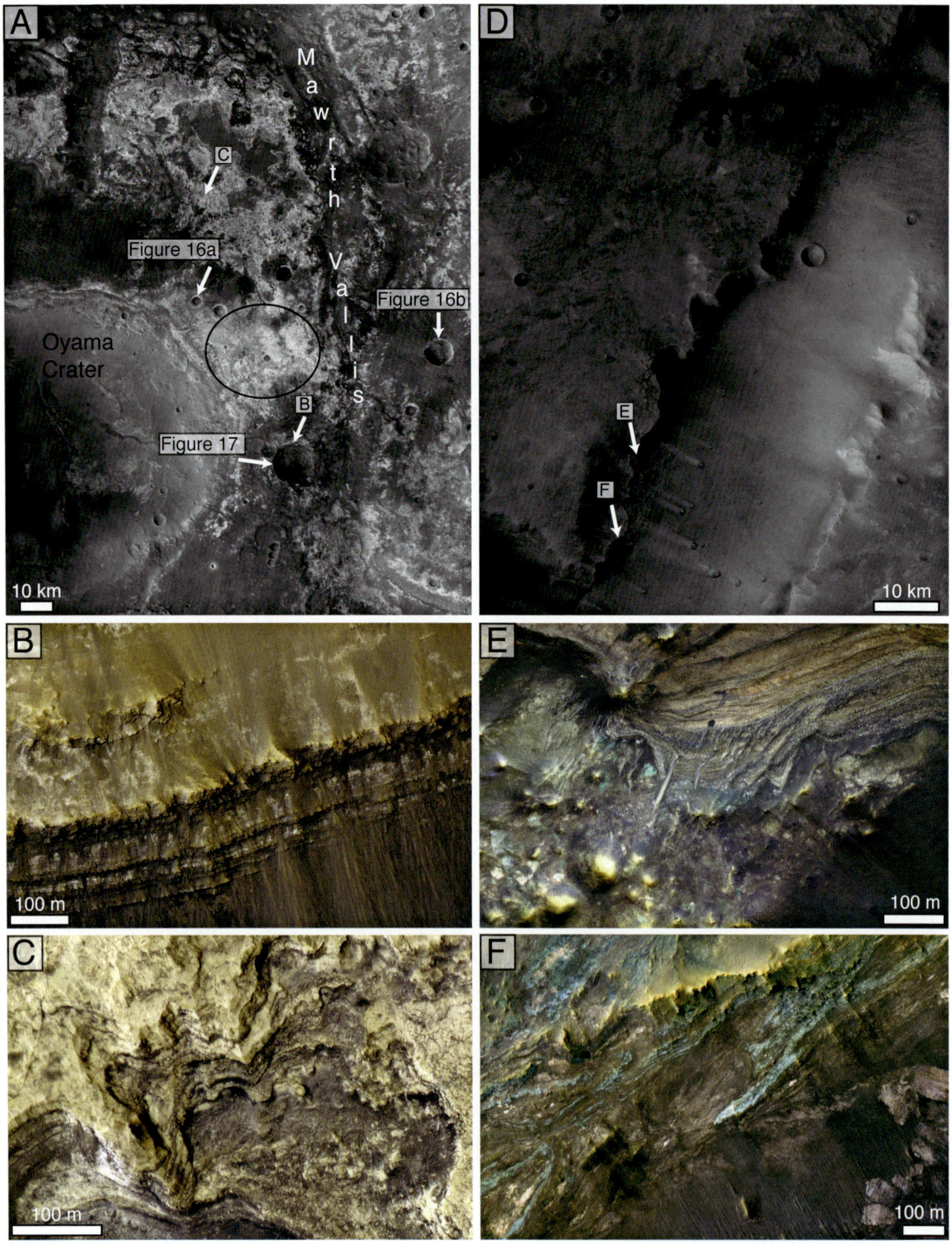

FIG. 15.—Very ancient (Noachian) rocks of Mars. (A) CTX mosaic of Noachian-aged light-toned strata surrounding Mawrth Vallis. Note light-toned outcrops within large crater (Oyama Crater) in lower left of image. (B) False-color close-up of strata exposed in a crater wall in Mawrth Vallis; strata contain clay minerals and other hydrous (sulfate?) minerals; location marked by arrow in A; HiRISE image ESP_016829_2040. (C) False-color close-up of light-toned strata exposed near Mawrth Vallis; rocks contain Al-, Mg-, and Fe-bearing clay minerals; location marked by arrow in A; HiRISE image PSP_001929_2050. (D) CTX mosaic of Noachian-aged units in the Nili Fossae region; the region of lower topography (trough) is a large graben on the western edge of the Isidis Impact Basin. (E) False-color close-up of stratified rocks and breccia exposed in the Nili Fossae trough; rocks in this region contain a variety of clay minerals; location marked by arrow in D; HiRISE image

overprinted by pervasive high-angle conjugate joint sets (Fig. 16G) and brecciation that obscure primary stratal geometries and earlier low-angle deformation features.

It is important to note that many of these complex stratal geometries are observed within a one-crater radius of the edge of Oyama Crater, though similarly complex geometries are also observed farther away in the floor of Mawrth Vallis itself (Fig. 15A). However, strata even farther away from the rim of Oyama, north of Mawrth Vallis, show evidence of noticeably lower strain and may better preserve primary stratification.

Nili Fossae: In addition to strata surrounding Mawrth Vallis, rocks exposed within and adjacent to the Nili Fossae region (22°N, 75°E; Fig. 4) are interpreted to represent ancient, aqueously altered crust (Mangold et al. 2007; Mustard et al. 2007, 2009; Ehlmann et al. 2009). Nili Fossae consists of several large concentric grabens that lie to the northwest of the large Isidis Impact Basin, which is centered near 13°N, 88°E. The fossae (grabens) are believed to have formed as a result of tectonic readjustment after the Isidis Impact event at approximately 3.9 Ga, and they cut through and expose several hundred meters of ancient Noachian crust. Unlike the rather uniform and relatively flat stratigraphy found near Mawrth Vallis, the rocks in and around Nili Fossae often exhibit evidence of extreme brecciation and ductile deformation in addition to decameter-scale stratification (Fig. 15D–F). This region of Mars also hosts the widest range of hydrous minerals yet observed on Mars, including Mg/Fe-smectite, chlorite, illite/muscovite, kaolinite, zeolite (analcime), sulfates, carbonate, serpentine, prehnite, and opal (Ehlmann et al. 2009). Brecciated zones have been interpreted as impact breccia deposits, and the large clasts within these units can exhibit layering and often exhibit mafic spectral signatures (primarily pyroxene) or absorptions consistent with Mg/Fe-clay minerals, as determined from CRISM data (Mustard et al. 2009). In addition to these breccias, some of the strata have experienced ductile deformation, possibly as a result of the Isidis Impact (Mustard et al. 2009) (Fig. 15F).

Despite these complexities, certain outcrops within this region exhibit clear stratification and are consistent with a sedimentary origin. The lowermost strata observed in vertical section in this region exhibit very strong visible–near-infrared absorptions attributed to Mg/Fe-clay minerals (e.g., Mangold et al. 2007, Mustard et al. 2007). These clay-rich units are often succeeded by either olivine-rich deposits, which appear to drape preexisting topography (Mustard et al. 2009), or thin deposits containing kaolinite (Ehlmann et al. 2009). The olivine unit has been interpreted as possible impact melt from the Isidis Basin, which would imply that the strata beneath this unit pre-date that event (Mustard et al. 2007). In some locations the olivine is mixed with Mg/Fe-carbonate, indicating that a period of aqueous alteration occurred after the Isidis Basin had formed (Ehlmann et al. 2009). Such aqueous activity is also supported by the presence of the Jezero delta/alluvial fan complex, which is found to the east of the fossae and also postdates the Isidis Impact event. The stratigraphy of Nili Fossae is rather complex, which is not unexpected for ancient crust that has experienced Late Heavy Bombardment. Nevertheless, this region is intriguing in that it appears to contain aqueously altered Noachian-age igneous, sedimentary, and possibly metamorphic rocks, many of which are covered by younger, unaltered Hesperian lava flows. The preponderance and diversity of hydrous minerals at Nili Fossae, the spatial distribution of those phases, the presence of high-temperature phases (e.g., prehnite), and the proximity to the Isidis Impact Basin have led to the hypothesis that the ancient crust in this region may have been overprinted by impact-induced hydrothermal alteration (e.g., Ehlmann et al. 2009, Mustard et al. 2009). Mg/Fe-clay minerals are found in smaller deposits throughout the ancient southern highlands (Bibring et al. 2006) and they, like those at Nili, may be the result of deep crustal alteration (Murchie et al. 2009b). However, the environments of formation of the stratified clay-bearing rocks at Nili remain unknown, and further research is needed to determine which, if any, of the sections at Nili still preserve information related to ancient sedimentary environments.

ORIGINS OF MARTIAN STRATA

The forgoing descriptions make clear that there should be several modes of sedimentation on Mars. For this discussion we exclude pyroclastic deposits, although eruptions on Mars are likely capable of transporting material over much greater distances than on Earth (Wilson and Head 1994), and pyroclastic deposits are most certainly present in the Martian stratigraphic record. Based on surface geomorphology, as observed from orbit, the most obvious transport mechanism is via fluvial networks, often as alluvial fans but occurring locally as meandering rivers in deltas or across plains. In outcrop (e.g., the Burns formation) we see clear evidence for wind-driven bedload transport of sand, and planet-wide dust storms indicate that eolian transport of sediment occurs on a global scale. It is also possible that the upper part of the ~2-m-thick "Home Plate" deposit in Gusev Crater, studied by the *Spirit* rover, involved eolian reworking of sand-sized volcaniclastic sediments (Lewis et al. 2008b). However, the amount of outcrop collectively studied by the *Opportunity* and *Spirit* rovers is trivial. *Opportunity*'s investigation of the Burns formation established the environmental significance of the uppermost part of the sequence at Meridiani Planum, but it represents only ~2% of the estimated thickness of the sequence (~20 m out of >800 m). This limitation of in situ investigations by rovers highlights the importance and necessity of interpretation based on orbital data in order to understand sedimentary rocks and processes on Mars in a regional and global context. The following section describes the primary modes by which sediment may be transported, concentrated, and converted to rock on Mars, informed largely by orbital observations.

The Role of Surface Water: Fluvial Networks and Source to Sink

As described above, many regions in the ancient southern highlands of Mars exhibit extensive fluvial networks. Crater counts indicate that the majority of the fluvial or "valley" networks on Mars were carved during the Noachian period and that valley network formation had largely ceased by the Early Hesperian (Fassett and Head 2008a). It also has been estimated that their formation potentially stripped several hundreds of meters of material from the surface of the southern highlands, further evidence for extensive aqueous sediment transport on ancient Mars (Hynek and Phillips 2001). This apparently widespread fluvial activity declined over time, possibly associated with the loss of atmospheric volatiles and associated drop in atmospheric pressure and temperature; only localized fluvial bedrock incision is observed on Late Hesperian and younger surfaces (e.g., Carr 1996, Mangold et al. 2004, Fassett and Head 2008a, Grant and Wilson 2011, Howard and Moore 2011). However, recent crater-counting studies based on new high-resolution images have shown that smaller,

←

ESP_019476_2005. (F) False-color close-up of deformed, stratified rocks exposed in the Nili Fossae trough; rocks are highly deformed and often contain clay and mafic minerals; location is just south of E; HiRISE image ESP_0016153_2005.

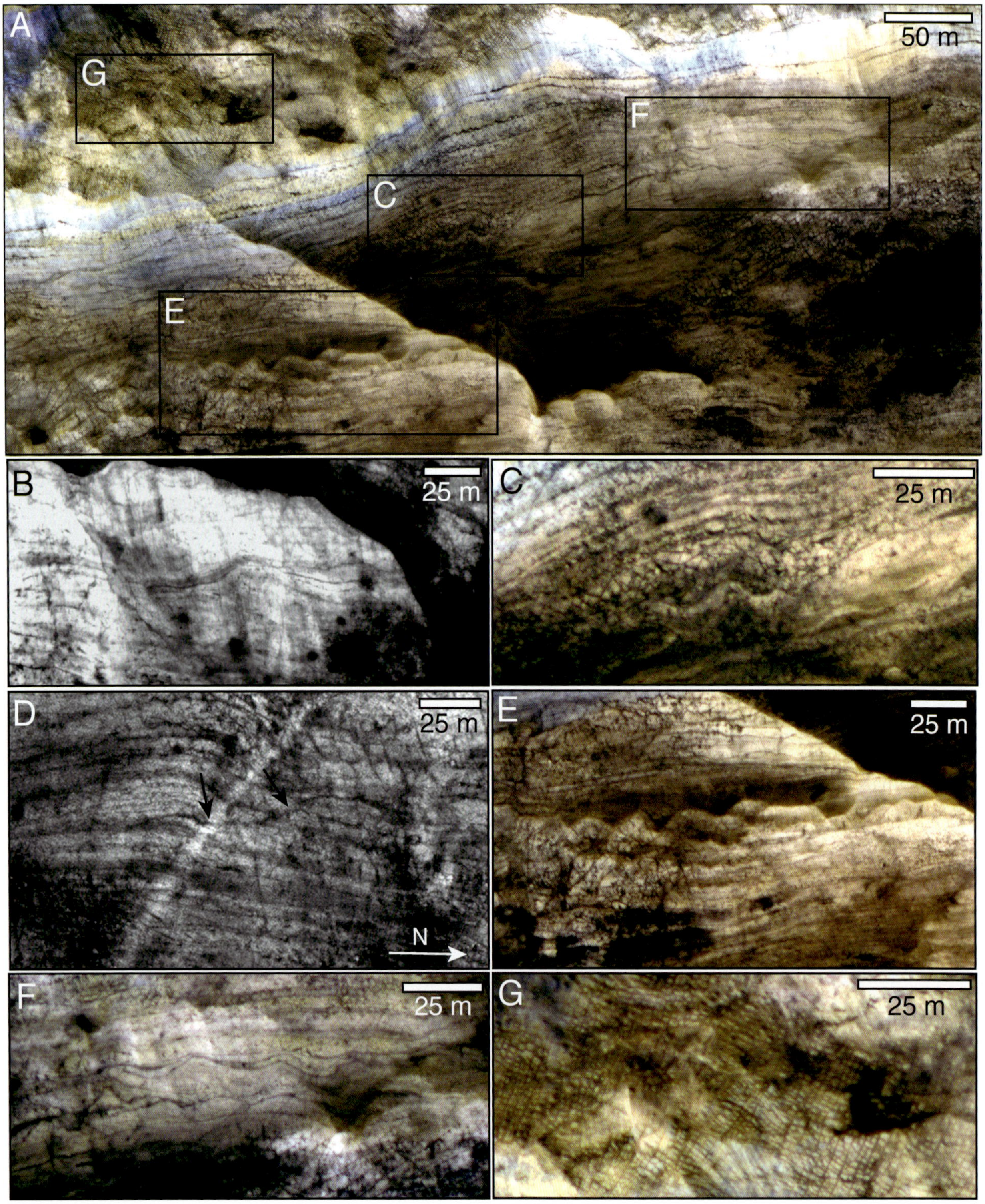

FIG. 16.—Complex stratal geometries at Mawrth Vallis, in the vicinity of the proposed landing ellipse for the Mars Science Laboratory. Location shown in Figure 15A. (A) HiRISE false-color image showing uneven, laterally discontinuous stratification. Transition from beige color in lower part of image to blues and yellows shown in upper part of image corresponds approximately to the Fe–Mg phyllosilicate to Al–phyllosilicate transition. Black boxes show regions highlighted in Figure 16C–G. (B) Stratal truncation surface is overlain by draping strata that contour truncation surface. Location is from axis of Mawrth Vallis, shown in Figure 15A. (C) Pod of breccia appears stratigraphically bound within layered sequence. Note how upper layers bend around pod, indicating a possible origin through injection. (D) Possible low-angle fault, marked by black arrows, penetrates succession, causing offset of strata. (E) Deformation by rotation and/or imbrication of blocks of

but still substantial, regions of Mars have experienced overland flow, fluvial incision, and channel network formation during Late Hesperian to Amazonian times, indicating a late pulse of fluvial activity (Grant and Wilson 2011, Howard and Moore 2011).

Extensive fluvial erosion of the southern highlands generated significant amounts of sediment. In some cases there was enough to fill >10-km-diameter craters with several hundred meters of material (Craddock and Howard 2002). Eroded landscapes, including the inner walls of craters, achieved channel densities approaching terrestrial values (Irwin and Howard 2002, Mangold et al. 2004). The environment that created such significant erosion and sediment transport is still debated, but many agree that widespread, although episodic, precipitation likely occurred in association with runoff and groundwater seepage (Hynek and Phillips 2003, Howard et al. 2005, Irwin et al. 2008).

One of the longest and largest source-to-sink systems on Mars has its source along the northern rim of the Argyre Impact Basin (50°S, 318°E; Fig. 4) and extends all the way to the northern lowlands, inferred to be the terminal sink. The total length of the outflow system, from where Uzboi Vallis begins near the rim of Argyre to where the outflow system discharges onto the northern plains at Chryse Planitia, is over 4000 km (see Fig. 18). It is hypothesized that fine-grained sediments transported through this and other outflow systems accumulated along the fringe of the northern lowlands, where their burial triggered subsurface mud diapirism (Oehler and Allen, this volume).

A series of valleys and basins, known as the Uzboi–Ladon–Morova network (ULM), link Argyre with the northern lowlands (Fig. 18A). Several ancient craters along this network are interpreted to once have been lakes, and younger craters within the ULM system may also have hosted lakes when their formation blocked the large drainage network (Grant et al. 2008, Grant and Wilson 2011). These impact basins are regarded as significant local sinks for sediments, including clay minerals (Milliken and Bish 2010). The ancient, heavily eroded Ladon and Holden basins are the largest potential lake basins in the ULM system (Fig. 18B), although younger craters such as the Holden and Eberswalde craters also likely hosted lakes (Fig. 18C). Sedimentary, clay-bearing strata of possible lacustrine origin are preserved throughout Ladon Basin, in terraces in Ladon Vallis (Fig. 18D), and in Eberswalde and Holden craters (Milliken and Bish 2010).

For many locations on Mars the sedimentary record captured within craters and other depressions defines complete erosional–depositional systems, albeit at much smaller scales than the hemispheric-scale system described above. In the Melas Chasma Basin we see a diverse set of geomorphic elements that ranges from the fluvially incised source region in the surrounding highlands to sediment sinks at progressively lower elevations formed by the alluvial fans, shoreline/delta, and sublacustrine fans in the topographically lowest part of the basin (Fig. 19). On Earth, sediments are moved from their source in the surrounding mountainous areas to their sink in depositional areas by the sediment transport system (Allen 2008). On Mars, fluvial incision of bedrock (Fig. 19A), interpreted by some to be caused by runoff from precipitation, drains the ridges and plains bordering topographic depressions that form sedimentary basins (e.g., Mangold et al. 2004, Howard et al. 2005, Quantin et al. 2005, Grant et al. 2008). Sediments generated during erosion of the upland areas were transported by fluvial drainage systems and formed classic cone-shaped alluvial fans (Fig. 19B) along basin margins, where confined channel flow emerged onto the fan surface (Howard et al. 2005, Quantin et al. 2005).

The preservation of sediments in these fringing alluvial fans depends on whether there is accommodation space available to store the sediment over the long term. Accommodation space can be generated when the graded profile of these streams moves upward in response to a rise in base level or to uplift of the source area (Viseras et al. 2003). The amount of sediment permanently stored in the alluvial fans is likely small compared to the total flux of sediment in the system (Allen 2008). Sediments that passed through the alluvial fans could then be deposited as shoreline deposits if bodies of standing water were present. Clinoforms also developed in some Martian sedimentary basins, which could record a potential shoreline or the upslope channel levee part of a sublacustrine fan (Dromart et al. 2007). In other cases clinoforms are associated with delta progradation (e.g., Eberswalde Delta; Lewis and Aharonson 2008). At Melas Chasma, observed clinoforms (Fig. 19C) give way farther down the topographic profile to a sublacustrine fan (Fig. 19D), very similar in morphology to the Mississippi submarine fan. This ultimate topographic low in the system then provides the terminal sink for the sediments.

In all cases, such as Melas Chasma, these interpretations invoke a significant assumption—that the various elements of the geomorphic system are all the same age. If these features have different ages, then the linkages between elements in the sediment transport system would not necessarily hold. However, even if some of these elements are not the same age, they still represent pieces of the sediment transport system. For example, cross-cutting relationships indicate that there are several generations of valley networks preserved in the ridges surrounding the small basin in Melas Chasma (Quantin et al. 2005). It is not possible to determine which generation of the drainages may have fed the depositional fans currently preserved in the bottom of the basin. Similar to the valley networks, perhaps the currently exposed clinoforms overlie a set of older buried clinoforms that are the same age as the depositional fan. As sediments were transported into the basin, the depositional system could have responded with retrogradation, aggradation, or progradation of the sediment depocenter farther into the basin; these dynamic responses depend on how the incoming sediment flux balances with basin subsidence and lake level (Flemings and Grotzinger 1996, Hodgson et al. 2006).

The Role of Groundwater

Widespread accumulation of sedimentary rocks in regions such as Arabia Terra and Sinus Meridiani may have been facilitated by regional groundwater flow, which would have driven oscillations in the water table (Grotzinger et al. 2005, Andrews-Hanna et al. 2010). In turn, such a process could have promoted accumulation of largely wind-blown strata, either as saltation-driven sand sheets or suspended dust that settled during intervals of decreased wind velocities (Fig. 20A). Evaporation of emerging groundwater, possibly brines, would have resulted in the precipitation of evaporites and cementation of eolian sediments. This cementation mechanism could have driven the accumulation of strata, perhaps even above the rims of large craters, if the regional groundwater table rose to sufficiently high elevations.

Arabia Terra is a unique region of Mars wherein elevation values define a broadly depressed surface relative to the southern highlands of Mars (Hynek and Phillips 2001). Groundwater could have sourced from the highlands and would have emerged downgradient, trapping sediments and inducing evaporite sedimentation in Arabia Terra and similar regions (Andrews-Hanna et al. 2010). Such a process could

←

strata. (F) Waveforms that represent either primary bedforms or disharmonic folding. Bedforms seem unlikely as they have little laterally continuity. (G) Penetrative high-angle conjugate joint sets are superimposed over all strata in this location.

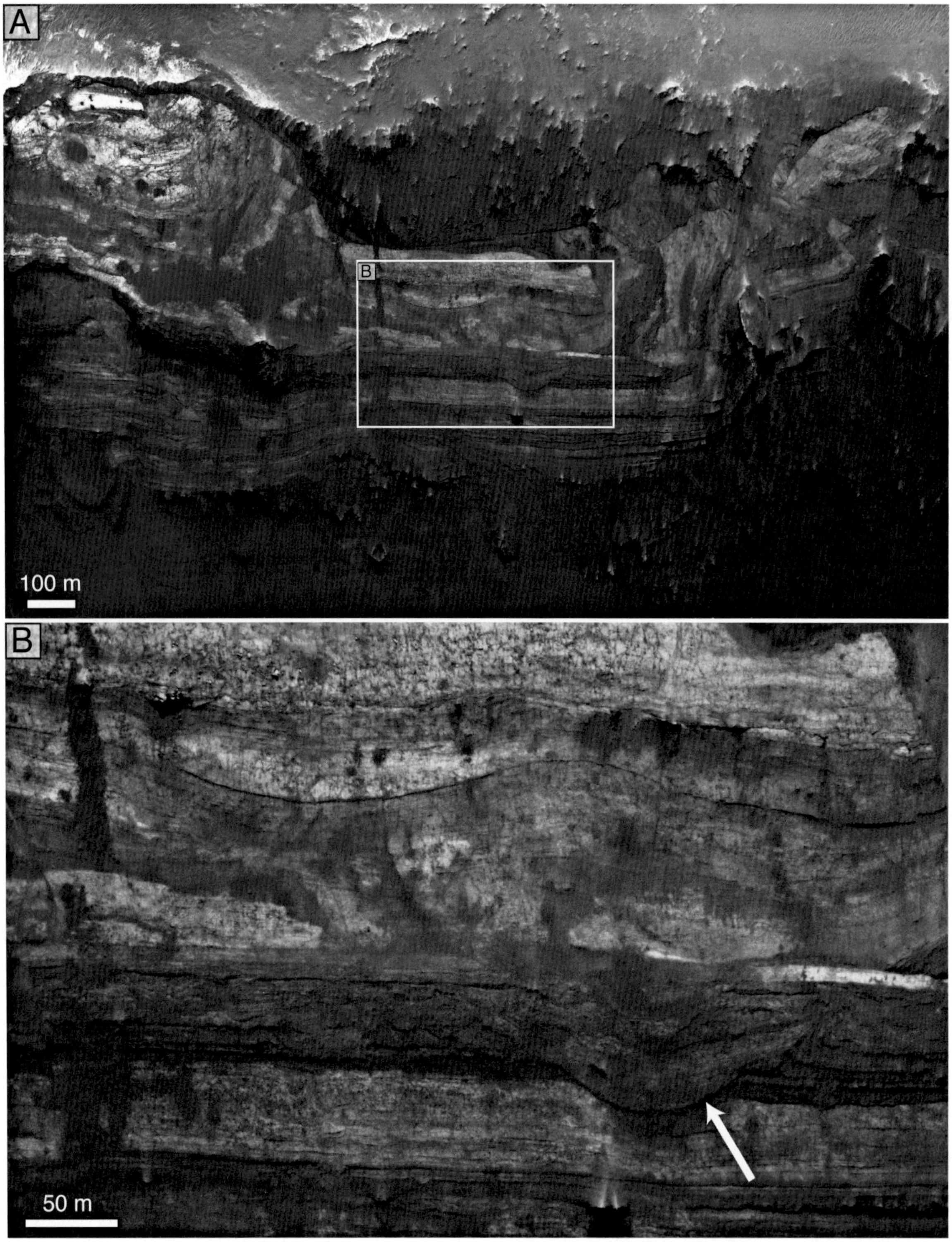

FIG. 17.—Strata exposed on the gently sloping wall of a crater in the Mawrth Vallis region. Location is the west wall of the impact crater marked in Figure 15A. North is to the right in A and B. (A) Strata in this location have been faulted, and some regions contain localized breccia, likely generated by impacts. (B) An important feature of the strata is an inferred buried impact crater (white arrow). Note dark stratigraphic unit pierced by impact depression that is also marked by overturned strata along right margin (to the upper right of white arrow). Younger strata onlap the margins of this paleo-crater, and bounding strata may represent reworking of impact-derived detritus. HiRISE image ESP_018820_2035.

lead to significant infilling or even overfilling of craters to form extensive stratigraphic sheets (Fig. 20B). Subsequent lowering or drying up of the groundwater system would result in stripping of the section by eolian abrasion. Over very long timescales, this transition from a wet period driven by changes in regional or global groundwater levels to a dry period of eolian denudation could reflect the transition from wet to dry Mars during the Hesperian.

Evidence for groundwater at a local scale is provided by stratigraphic relationships and textural and geochemical data in the Burns formation (Grotzinger et al. 2005, McLennan and Grotzinger 2008). At a more regional scale, evidence for groundwater is provided by fractures filled with possible cements in a number of locations (Okubo and McEwen 2007, Okubo et al. 2009, Anderson and Bell 2010). In addition, rounded, amphitheater-headed canyons that incise the plains along the rim of Valles Marineris are morphologically similar to classic groundwater-sapping features observed on Earth, leading to the hypothesis that similar processes operated on Mars (Pieri 1980). However, recent studies have indicated that such features may instead be carved by catastrophic floods (Lamb et al. 2008). Other evidence for regional groundwater includes the "chaos" terrains of Mars, which are large regions of brecciation and collapse, similar to karst topography and believed to have formed by melting of ground ice (Sharp 1973) or by dissolution of soluble rocks by emerging groundwaters (Carr 1979, Spencer and Croft 1986, Spencer and Fanale 1990). Finally, recent studies of inferred lakes developed in "open" basins—craters or topographic depressions with both an inlet and outlet channel—have indicated that the volume of water in some of these inferred lakes was much greater than can be explained by surface runoff from the surrounding watershed, implying that groundwater played a significant role (Fassett and Head 2008b). It is interesting to note that many of these inferred groundwater-fed paleolakes are located in Arabia Terra, the same region that hosts many of the overfilled craters discussed above (Fig. 8) (Fassett and Head 2008b). Therefore, an active groundwater system may be a crucial component for lithification and net accumulation of thick, crater-filling sedimentary deposits, such as those observed in Henry Crater and elsewhere in this region (e.g., Fig. 20).

Light-Toned Layered Deposits and Their Origins

In addition to the presence of stratification, many of the interpreted sedimentary rocks on Mars are apparent in orbital images because they are light-toned. However, caution must be used with such descriptions. Many images of the Martian surface are acquired as single-band, grayscale images. When processed, these images are often stretched to highlight features over the full albedo range within each individual image. Therefore, deposits often only appear light-toned or dark-toned relative to their surroundings. In addition, the true albedo values at visible wavelengths are rarely reported for features observed in MOC, HiRISE, or CTX images. Therefore, deposits that appear "light-toned" in one image may have a true albedo that is higher, equal, or lower than "light-toned" deposits from another image. Indeed, nearly all of Mars is relatively dark in that albedo values are typically <0.35, and most regions are <0.25 (where the highest possible albedo value is 1, meaning 100% of incident light is reflected from the surface). It is also useful to keep in mind that Mars is the "Red Planet," and deposits that appear light-toned or white in grayscale images are often light-orange, reddish, or brownish tones in true color.

Despite these complications, many of the sedimentary deposits on Mars have the intriguing attribute that they are brighter or "light-toned" at visible wavelengths when compared to their surroundings. The juxtaposition of such deposits against the darker basaltic compositions that typify the Martian crust can be striking. Indeed, this was one of the primary reasons that the enigmatic "White Rock" feature in Pollack Crater (discussed above) was so surprising when first imaged by Mariner 9. Most of the crust of Mars is basaltic (e.g., Taylor and McLennan 2009), and many of the bedrock exposures, sand dunes, and other sediments are relatively dark, as would be expected for basaltic compositions. Therefore, the identification of light-toned rocks, and light-toned stratified bedrock in particular, immediately raises the possibility that nonbasaltic compositions may be present. On a planet with a CO_2-rich atmosphere and a clear history of aqueous processes, identification of bright sedimentary deposits sustains hope that hypothesized large carbonate deposits may one day be recognized (compare Kahn [1985] with Milliken et al. [2009] and Michalski and Niles [2010]). Although local occurrences of carbonates have been detected both from orbit (Ehlmann et al. 2008b) and in situ (Boynton et al. 2009, Morris et al. 2010), large carbonate deposits are conspicuously absent; rather, many large, light-toned deposits are known to contain sulfates (Gendrin et al. 2005). This has raised the possibility that Mars and its sedimentary record may have been dominated by a sulfur cycle rather than a carbon cycle (McLennan and Grotzinger 2008; McLennan, this volume).

Although the origin of White Rock is still ambiguous, we now know that it is merely one of many examples of light-toned sedimentary deposits on Mars (Malin and Edgett 2000). Orbital spectroscopy provides a means by which to characterize the composition of these and other deposits, and an increase in the spatial resolution and wavelength range of instruments on Mars missions over the past decade has fundamentally changed our knowledge of their possible origins. Many light-toned deposits identified in high-resolution visible images (MOC, HiRISE, CTX) were below the kilometers/pixel spatial resolution of the TES instrument on MGS. The Thermal Emission Imaging Spectrometer (THEMIS) instrument on Mars Odyssey acquires images at much greater spatial resolutions (~100 m/pixel), but its limited spectral resolution makes unique identification of compositions difficult. However, by combining data from these two thermal infrared systems it has been shown that some small light-toned deposits contain chloride salt (Osterloo et al. 2008). As discussed above, the advent of even higher spatial and spectral visible–near-infrared spectrometers in orbit around Mars has provided greater insight into the compositions, and thus the processes, that may have formed light-toned deposits.

As with TES, the rather coarse spatial resolution of data acquired by the visible–near-infrared OMEGA spectrometer (often several kilometers/pixel) precluded mineralogical identification in many of the smaller or thinly bedded light-toned deposits. However, the OMEGA data were revolutionary in that they revealed the widespread presence of hydrous minerals in regionally extensive intermediate- and light-toned deposits, including clay minerals in very ancient strata (e.g., Mawrth Vallis; Poulet et al. 2005) and sulfate salts in plains covering deposits (e.g., Sinus Meridiani; Gendrin et al. 2005). The large and often light-toned ILD in Valles Marineris were also found to contain hydrated sulfates and Fe-oxides (Gendrin et al. 2005, Le Deit et al. 2008, Mangold et al. 2008). In addition to these areally extensive deposits of light-toned strata, higher spatial resolution OMEGA data (~250 m/pixel) revealed the presence of hydrous minerals in some smaller light-toned deposits, hinting at the rewards that higher spatial resolution spectral data would offer. These rewards were realized in 2006 when the visible–near-infrared CRISM instrument on the MRO spacecraft began to return images in 544 channels over the 0.35 to 4-μm wavelength region at a maximum resolution of ~18 m/pixel.

Studies utilizing CRISM data have built on those based on OMEGA data to show that many of the light-toned deposits identified on Mars contain hydrous minerals and that the diversity of hydrous minerals is much greater than previously recognized (Mustard et al. 2008, Murchie et al. 2009b). Specifically, the high-resolution CRISM data have revealed previously unrecognized occurrences of clay minerals in spatially restricted light-toned outcrops of the Eberswalde and Jezero deposits (Ehlmann et al. 2008a, Milliken and Bish 2010), opaline silica in thinly bedded strata adjacent to Valles Marineris (Milliken et al.

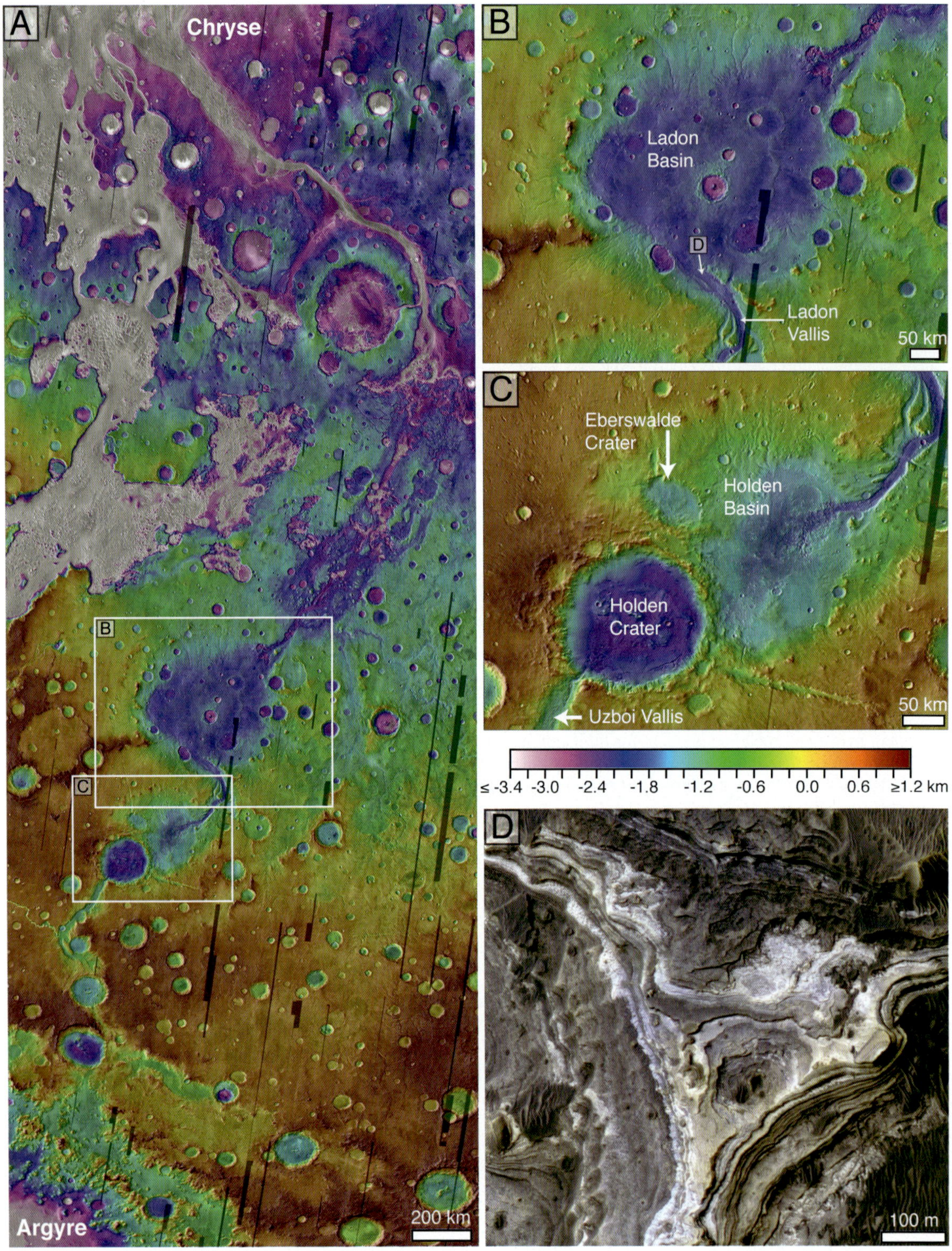

FIG. 18.—The Uzboi–Ladon–Morava (ULM) fluvial–lacustrine transport system. (A) MOLA elevation on THEMIS day infrared mosaic, showing the low elevation of the ULM system relative to the surrounding ancient highlands. The ULM system was once through-going from the northern rim of the Argyre Impact Basin to Chryse Planitia at the edge of the northern lowlands, but it was blocked at one point by the formation of Holden Crater (see C). Much of the channel system exhibits evidence for sedimentary fill and multiple episodes of incision. (B) Close-up of the ancient Ladon Basin, which has been filled with sediment of an unknown thickness that is known to contain clay minerals (Milliken and Bish 2010); numerous fluvial features are found along the heavily degraded rim and wall. (C) Close-up of Eberswalde Crater, Holden Crater, and the more ancient Holden Basin; breach of Uzboi Vallis into Holden Crater is visible in the southwest portion of the image;

2008, Weitz et al. 2010), carbonates in ancient, light-toned rocks in the Nili Fossae region (Ehlmann et al. 2009), and interbedded clays and sulfates in intermediate- and light-toned deposits in Gale and Columbus craters (Milliken et al. 2010, Wray et al. 2011).

The occurrence of hydrous minerals in these and other light-toned deposits implies that these strata record a much more complex geologic history than do the typical lava flows associated with Martian volcanism. However, in many (perhaps most) cases the origin of these minerals is unclear. Are they secondary minerals produced by in situ surface alteration of primary mafic minerals? Were they formed deep in the crust, only to be brought up to the surface by impacts? Do they occur as pore-filling cements formed by fluctuating groundwater levels, or are they precipitated directly out of bodies of standing water? Answering questions such as these is critical in order to determine the origin of the light-toned deposits in which some of these minerals reside, but such questions are exceedingly difficult to answer when limited to orbital data. In the case of the light-toned sulfate and hematite-bearing rocks detected from orbit in Sinus Meridiani, the *Opportunity* rover was able to perform in situ investigations to place tighter constraints on their origin, as discussed above. But what, if anything, does this imply about light-toned, sulfate-bearing rocks identified elsewhere on Mars? The transition between light- and dark-toned rocks in Sinus Meridiani examined by *Opportunity* was shown to result in part from the presence of a diagenetic contact that cuts across bedding (see Whatanga contact in Fig. 14; Grotzinger et al. 2005). The true relationship between the dark–light tone transition and actual bedding is not apparent from orbital images. Elsewhere on Mars, many light-toned strata are overlain, underlain, or interbedded with dark-toned strata, and although the sharpness of these contacts often appears to represent bedding, it cannot be ruled out that these transitions are not bedding contacts at all, but rather the result of overprinting by diagenesis.

A major step forward in determining the origin of light-toned deposits would be to determine the mineral abundances in these rocks and whether the alteration and secondary minerals in each deposit are detrital, authigenic, or diagenetic. Using the sulfate-bearing ILD in Valles Marineris as an example, the determination that sulfates comprise 30% of the rocks would be consistent with pore-filling cements, possibly indicating an origin similar to the sulfate-cemented sandstones in Meridiani Planum. However, if the rocks are determined to contain more sulfate than can be accommodated in pore space (i.e., $>30\%$), then direct precipitation from a body of standing water may better explain their origin. Unfortunately, the process of deriving accurate mineral abundances from VNIR data can be highly nonlinear, and the uncertainties based on current models are rather large and poorly constrained for fine-grained sediments (though recent attempts at modeling OMEGA data have been made by Poulet et al. [2008]). Additional laboratory experiments and modeling are needed for this possibility to reach its full potential.

Similarly, determining whether hydrous minerals, especially clay minerals, in a given rock are detrital or authigenic can be challenging when examining the terrestrial rock record, and the problem could be intractable for Mars when limited to orbital data. Clay minerals detected in the bottomsets of the Eberswalde Delta are spectrally similar to clay mineral signatures in the surrounding watershed, indicating that at least some of the clays in the delta are detrital (Milliken and Bish 2010). However, the presence of neoformed clays in the deltaic strata cannot be ruled out. Similarly, whether the strong clay mineral signatures at Mawrth Vallis represent concentration of transported, fine-grained sediment or in situ alteration of primary mafic deposits such as volcanic ash remains unknown (Loizeau et al. 2007, Michalski and Noe Dobrea 2007, Bishop et al. 2008). Despite these ambiguities, it is clear that light-toned stratified deposits on Mars are diverse in their mineralogy and were likely formed in a wide range of depositional environments, even though the context and evidence for a specific environment are not always clear from orbit.

We conclude our discussion on light-toned deposits by noting that there is not a one-to-one correlation between detections of hydrous minerals and light-toned strata. In fact, many occurrences of clays and sulfates on Mars are not associated with light-toned materials at all, and some occur in rocks that lack clear stratification altogether. The converse is also true, in that not all light-toned rocks on Mars exhibit spectral signatures diagnostic of hydrous or mafic minerals. Even the enigmatic White Rock deposit itself, which is intermediate- to light-toned when compared to its immediate surroundings, appears to be spectrally bland (Ruff et al. 2001). The origin and composition of such deposits remain unknown, but one possibility is that they represent deposits of poorly lithified dust, or "duststones" (Bridges and Muhs, this volume).

Martian Duststones

The striking red color of Mars is due primarily to the ubiquity of fine-grained, Fe-oxide–bearing dust that coats much of the surface in a layer that is optically thick at visible wavelengths, though possibly only several hundreds of micrometers thick by measurement (Fig. 21). Significant amounts of dust are mobilized by typical local and regional winds and also by global dust storms. Several decades of spacecraft observations have shown that bright and dark regions on Mars are not necessarily fixed; some dark regions become coated with dust over time, and other regions are blown free of dust (Christensen 1988, Geissler 2005, Szwast et al. 2006, Fenton et al. 2007). Indeed, it is safe to say that eolian transport of dust has played a significant role in the Martian sediment cycle for hundreds of millions and possibly billions of years (Bridges and Muhs, this volume). Dust-sized particles (2–5 μm; Kahn et al. 1992, Lemmon et al. 2004) are constantly transported across the surface, and it is estimated that 2.9×10^{12} kg/yr is exchanged between the surface and atmosphere (Pollack et al. 1977). Rover measurements have shown that the major constituents of this material are consistent with pulverized, unaltered basalt (Goetz 2005, Hamilton and Christensen 2005). Perhaps this is not unexpected on a planet where liquid water has apparently been unstable or meta-stable in the near-surface for several billion years, but the lack of abundant liquid water does not preclude the possibility that this dust can make a significant contribution to the rock record. One final point about these dust particles is that they may become aggregated to form larger silt- or sand-sized particles that may then become transported by saltation as well as suspension (Herkenhoff et al. 2008a, Sullivan 2008). Aggregation of these particles may be stimulated by electrostatic forces or weak cementation during gas exchange and condensation in soils.

Accumulation and compaction of dust, perhaps aided by precipitation of salts from thin films of water along grain boundaries, could form weakly lithified strata. If composed primarily of dust, these

←

note that the formation of Holden Crater blocked water flow through the larger ULM system. (D) Close-up of finely stratified and clay-bearing outcrops exposed in a terrace in Ladon Vallis where it enters Ladon Basin along the southern rim; location marked by arrow in B; HiRISE image PSP_006637_1590.

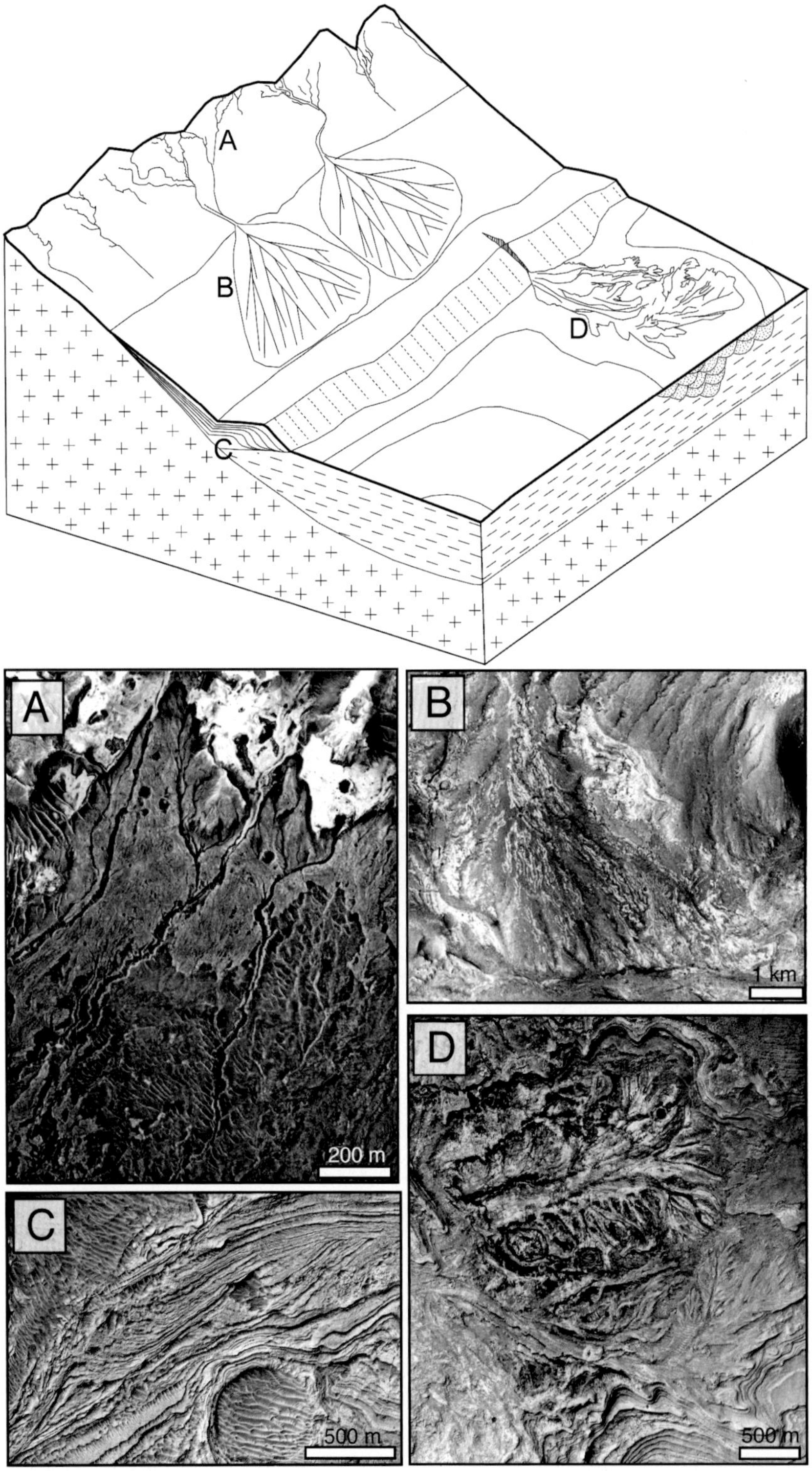

FIG. 19.—Cartoon of the Melas Basin source-to-sink system (after Metz et al. 2009a). (A) Upper reaches of the system show a well-developed tributary network cut into bedrock (HiRISE image PSP_005452_1700) that passes downslope into (B) an alluvial fan complex (CTX image B10_013561_1704_XN_09S076W). This passes farther downslope into a set of (C) clinoforms, the origin of which—although uncertain—could represent a deltaic or shoreline deposit (HiRISE image PSP_008735_1700). The terminal sediment sink (D) is represented by a set of inferred sublacustrine fans (HiRISE image PSP_007667_1700). The accompanying images come from successively lower geomorphic/sedimentologic positions within the Melas Basin.

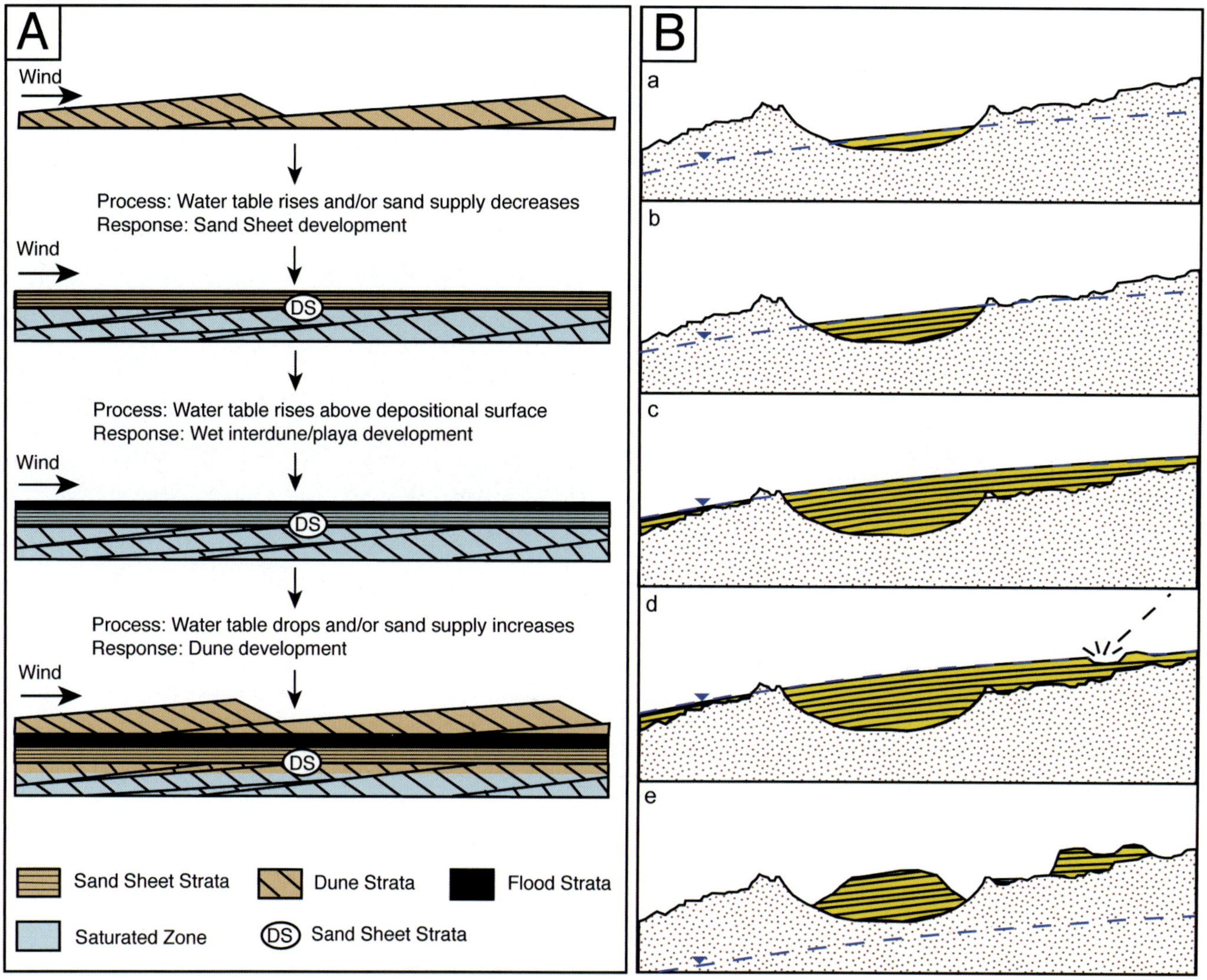

FIG. 20.—Accumulation of strata under the influence of a regional groundwater system. (A) Depositional model for the Burns formation in the region of Eagle and Endurance craters (from Grotzinger et al. 2005). Initially, accumulation is driven by migration of eolian dunes composed of sulfate-cemented altered basalt sand grains. Elevation of the water table (or decrease in sediment flux) results in the formation of a deflation surface (DS) down to the level of the capillary fringe of the water table. Sand sheet strata are deposited above the deflation surface. At some point, the groundwater table breaches the surface, resulting in deposition of subaqueous interdune sandstones ("Flood Strata"). Such a model predicts the cycle would be repeated as the level of the groundwater table fluctuated. (B) Application of this process can be extended to a more regional scale, whereby a rising groundwater table traps sediments. Initially, sediments are trapped just within the largest craters and other topographic depressions. With time and a rising groundwater table, sediment accumulates, fills, and overtops craters, spreading across the intercrater plains. Impact structures will be buried in the stratigraphy as a function of accumulation rate. If the water table drops over the long term then the strata are exposed to eolian deflation, creating erosional remnants as buttes and mesas. Modified after Grotzinger et al. (2005) and Andrews-Hanna et al. (2010).

"duststones" may be akin to terrestrial mudrocks with regard to their particle size, but they would be inherently different in that they would be dominated by physically, and not chemically, weathered particles (Bridges and Muhs, this volume). Unlike on Earth, the clay size fraction of dust on Mars may be lacking in actual clay minerals, if it has any at all. Therefore, Martian duststones may be hard to identify on Mars based solely on mineralogy, as derived from spectral observations. In fact, the apparent lack of unique spectral signatures may be one of their defining characteristics. Strata such as those at the top of the Gale Crater mound and in some craters in Arabia Terra are rhythmically bedded (Lewis et al. 2008a, Lewis 2009) and are often spectrally "bland." Their VNIR spectra are often equivalent to that of typical Martian dust, and they lack absorption features diagnostic of hydrous minerals. This could be caused by the presence of a thin dust cover, but it could also indicate that these rocks are composed of material that is compositionally equivalent to Martian dust.

Rocks such as these could be duststones that capture information about the dry climate of the past several billion years in the Martian rock record. The amount, source regions, lofting, and transport distance of dust can be directly related to atmospheric properties. Thus, if some Martian strata are composed largely of this material, then these rocks could act as ancient climatically sensitive deposits (Lewis et al. 2008a). As such, this type of stratified sedimentary rock, with its distinct set of characteristics, may also have a distinct origin. The nature and origins of these duststones are discussed in greater detail by Bridges and Muhs (this volume). The essential point is that Mars may have produced significant quantities of fine-grained sediments early in its history as a result of impact fracturing of bedrock, in addition to volcanism. Over time, the flux of both sediment sources would have been reduced. Remobilization by winds and subsequent deposition would have created duststone deposits wherever circulation patterns preferred net accumulation over geologically significant timescales. Today, this process may still be in effect, albeit at much lower rates and with a depocenter represented by the northern plains of Mars.

Lithification, Burial, and Diagenesis

An important question in sedimentary geology is how sediments undergo conversion to rock. On Earth, the circulation of pore fluids upon burial creates the geochemical environment in which most lithification occurs (in addition to diagenetic transformations). The pore space in sedimentary rocks of Precambrian age is often entirely occluded. In contrast, the absence of both an atmosphere and subsurface water on the Moon insures that 4.4 billion–year-old impact-generated deposits have not been cemented by aqueous processes (Carrier et al. 1991). Mars, it seems, is somewhere in between.

There is evidence that some cementation and lithification have occurred on Mars, perhaps pervasively over broad regions, but there is little evidence that this has ever occurred as completely as on Earth. Diverse observations constitute evidence for cementation and lithification of Martian sediments. These include preservation of resistant, cliff-forming outcrops of stratified light-toned materials (Malin and Edgett 2000); occurrence of hydrous minerals in outcrops of stratified light-toned materials (Gendrin et al. 2005); thermal inertia of stratified materials (Christensen et al. 2001, Putzig et al. 2005, Fergason et al. 2006, Edwards et al. 2009); and observation of probable cements and diagenetic fronts in outcrop (Grotzinger et al. 2005; McLennan and Grotzinger 2005; Squyres et al. 2007; Metz et al. 2009b; Edgar et al., this volume). Of these, the thermal inertia values for globally distributed layered deposits may best capture the degree of regional or global lithification. Interestingly, the majority of thermal inertia values for the Martian surface, including many light-toned layered deposits, are lower than would be expected for completely lithified rock, indicating relatively low rock densities (Fergason et al. 2006, Edwards et al. 2009) and possibly higher preserved porosities. TES-derived thermal inertia values for White Rock, for instance, were shown to be inconsistent with either loose dust or well-cemented bedrock (Ruff et al. 2001). Similarly, thermal inertia values for stratified deposits in Hebes Chasma are too low to be consistent with well-lithified rock and are more consistent with poorly cemented or highly porous eolian or volcanic sediments (Fergason et al. 2006). The general paucity of high thermal inertia surfaces on Mars (Edwards et al. 2009) is supported by the absence of rock fall talus adjacent to cliff-forming outcrops, which is observed for many light-toned deposits on Mars and which indicates that these rocks are weakly indurated materials that are easily abraded by wind-blown particles (Malin and Edgett 2000, Fergason and Christensen 2008).

Direct observations by *Opportunity* and *Spirit* indicate that stratified outcrops formed of granular materials contain pore-filling materials, interpreted to represent cements or cemented matrix sediment. However, deposits at both Meridiani Planum and Home Plate (an outcrop in Gusev Crater) exhibit textural evidence for the presence of significant porosity (McLennan et al. 2005, McLennan and Grotzinger 2008); possible volcaniclastic sediments in Gusev Crater may be better lithified (Squyres et al. 2007). Finally, measurements by the Viking landers were consistent with surface soils having porosities on the order of 50% (Arvidson et al. 1989).

FIG. 21.—Dust on Mars can form irregularly shaped aggregates of silt- to sand-sized particles. These aggregates are easily crushed and molded by the ~1-N contact force generated by the Mössbauer plate. The image spans approximately 31 mm in width. Microscopic Imager image 2M147677362IFF8800P2976M2F1 taken on sol 240 by the *Opportunity* rover (see Herkenhoff et al. 2008b).

More recently, radar data from the MARSIS instrument on Mars Express has shown that the dielectric properties of the extensive MFF are consistent with either an ice-rich deposit or very low-density materials. The latter case would require that the deposit be composed of poorly consolidated material with a very high porosity (average bulk density of <1.9 g/cm^3 over an ~2.5-km depth; Watters et al. 2007). At a minimum, these examples show that at least some sedimentary rocks on Mars are likely poorly lithified, even if they have been buried to several kilometers in depth. If accumulation of eolian sediment and fine-grained dust has been the dominant process in the Martian rock cycle for the past several billion years, then the high porosities of these sediments, the lower planetary gravity, and the cold, dry environment indicate that conversion of sediment to rock is, on average, less complete on Mars than it is on Earth.

Although the conversion of sediment to rock may be incomplete for many sedimentary terrains of Mars, there is significant evidence that diagenesis occurred, perhaps associated with significant burial. On a regional scale, the transition from Al-phyllosilicates such as kaolinite to Mg/Fe-rich phyllosilicates such as nontronite with depth at Mawrth Vallis could represent diagenesis and leaching, either at the surface or in the subsurface (Michalski and Noe Dobrea 2007, Bishop et al. 2008, Noe Dobrea et al. 2010). Kilometers-thick sequences of sedimentary rocks, such as the ILD in Valles Marineris or the mound in Gale Crater, are believed to be older than the vast majority of rocks on Earth. Strata at the base of these sections would have likely experienced groundwater interaction simply because they formed in significant topographic depressions and were deeply buried. As discussed above, spectral data indicate that the ILD contain highly soluble Mg-sulfates (Gendrin et al. 2005, Mangold et al. 2008, Murchie et al. 2009a), raising the possibility that some of these rocks may have been dissolved and recrystallized after their initial formation. In addition, even a low, early geothermal gradient of 12° C/km (McSween et al. 2011) would imply temperatures conducive to diagenesis and mineral transitions at the great depths experienced by strata at the base of kilometers-thick ILD in Valles Marineris or clays at the base of the Gale Crater mound.

The detection of clay minerals on Mars, and specifically smectites (Poulet et al. 2005), is of particular interest in this context. On Earth, the vast majority of smectites are ultimately converted to illite or chlorite during burial diagenesis. The extent and rate of reaction depend on factors such as time, temperature, and fluid chemistry, and intermediate products include mixed-layered illite/smectite or chlorite/smectite (see Srodon [1999] for a review on terrestrial mixed-layered clays). The oldest smectites on Earth may be on the order of 600 million years old (Bristow et al. 2009), yet the vast majority of clays on Mars are believed to have formed over 3.5 billion years ago (Bibring et al. 2006). This raises the questions of whether or not mixed-layered clays are present on Mars and to what extent burial diagenesis may have occurred in clay-bearing sedimentary rocks (McSween et al. 2011, Milliken et al. 2011). Given the interpreted age of the clay minerals, time is not a limiting factor, and it is unlikely that temperature is a limiting factor for clays at the base of thick sequences, such as in Gale Crater. Therefore, a lack of mineralogic evidence for diagenesis may indicate that chemical sediments on Mars are juvenile and have had little if any interaction with water since their formation (Tosca and Knoll 2009). However, recent analyses of CRISM data indicate that mixed-layered chlorite/smectite, and thus diagenesis, may indeed be more common on Mars than was previously recognized (Milliken et al. 2011). Ultimately, the role and extent of diagenesis on Mars remains an open question, yet it is an extremely important one when assessing the preservation potential of organic compounds in Martian sediments, especially given the often destructive effects of diagenesis.

MINERALOGY AND STRATIGRAPHY: TOWARD A GEOLOGIC TIMESCALE

Martian Mineral Stratigraphy

On Earth, the great majority of stratified rocks are of sedimentary origin and have formed as a result of accumulation of loose particles, including precipitated minerals. Stratigraphic studies traditionally have focused on the correlation of spatially separated successions of sedimentary rocks for the purpose of establishing a temporally comprehensive record of Earth history—the geologic timescale.

The geologic timescale has two components: a *relative* timescale based on correlation of stratigraphic successions and an *absolute*; timescale based on geochronologic calibration of key tie points that link those successions (Gradstein et al. 2004). Correlation commonly is achieved by delineating temporal trends in the successions of fossils, chemical markers, and magnetic properties. Radiometric age determination of volcanic rocks interspersed with sedimentary rocks provides absolute age control for even the oldest successions of terrestrial strata. Advances in U-Pb zircon dating have revolutionized calibration of the Precambrian timescale so that events may now be constrained to $\pm 10^5$ years (Bowring et al. 2007).

For Mars, the same principles apply: a relative sequence of superimposed and cross-cutting rocks (terrains with distinct properties) is established (Scott and Tanaka 1986, Greeley and Guest 1987, Nimmo and Tanaka 2005), and this record is then calibrated in "absolute" time using crater counting (Hartmann and Neukum 2001). In comparison to Earth, the geologic timescale for Mars is far less well resolved. This stems from a lack of precision with both the ordering of events (relative component) and the absence of direct geochronologic calibration of those events (absolute component). In contrast to lunar rocks returned by the Apollo program, age-dating of Martian meteorites provides little help because the locations on Mars from which these rocks come are unknown. For the near future, there is little prospect for an improved measurement scheme for absolute dating of strata on Mars. In contrast, the past decade of global mapping and the ongoing MRO mission offer the chance for a significant improvement in the relative component of the Martian geologic timescale. This will likely improve precision in understanding the relative order of regional geomorphic[6] and compositional terrains on Mars. New high-resolution data (e.g., HiRISE) provide the capability to produce a high-fidelity relative timescale for individual sedimentary deposits that was previously not possible. In turn, this could enable correlation within more highly deformed terrains, which remains problematic.

Most recently, our understanding of the geologic history of Mars has taken a fresh turn as a result of the discoveries by ongoing missions such as Mars Express and MRO. Mars may have a distinct history of aqueous events that left distinct time-dependent patterns embedded within the rock record (Fig. 22), and recent studies have attempted to classify deposits based on distinct mineral assemblages, as observed from orbit (e.g., Murchie et al. 2009b). A rising paradigm is that the long-term environmental evolution of Mars is captured in its history of mineral assemblages (Bibring et al. 2006). In this hypothesis, Mars began as a planet characterized by aqueous alteration of impact-brecciated, primary ancient crust under pH conditions that were conducive to the formation of clay minerals. Subsequently, it transitioned to a drier and more acidic environment that resulted in the formation of vastly extensive bedded sulfate minerals, and ultimately it transitioned to a dry and cold planet dominated by the accumulation of anhydrous iron oxides, as is observed today (Bibring et al. 2006, Murchie et al. 2009b). The foundation for this hypothesis is the apparent temporal separation of these mineral assemblages. Phyllosilicates are most commonly found in Noachian-age terrains, sulfates are primarily associated with younger Hesperian terrains, and anhydrous Fe-oxides are dominant in Amazonian terrains; Bibring et al. (2006) refer to these "mineralogical" time periods as the phyllosian, theiikian, and siderikian, respectively (Figs. 2, 22).

In order for this model for hydrous environmental evolution to be correct, the age of the observed alteration minerals must be closely related in age to the strata in which they now reside, and the observations themselves must stand up to independent testing. This is a delicate proposition given the difficulty in assessing the age of alteration minerals relative to their parent rocks, but it strongly motivates research aimed at evaluating the evolutionary record and testing the ordering and environmental history of stratified deposits. Though new for Mars, the importance of unique mineral assemblages in defining environmental breakpoints in the geologic timescale is not without precedent. It is in exactly this mode that the geologic history of Earth was established, for example, based on recognition of the temporal restriction of Precambrian banded iron formation; distinct mineral deposits with profound environmental significance, in that case signaling progressive oxygenation of Earth's atmosphere and oceans (Knoll et al. 2004). In the course of construction of the Phanerozoic geologic timescale the mineralogic name "Cretaceous" (from the Latin "creta," for chalk) was applied to that period of time in which chalks were noted to be particularly abundant.

The determination of similar patterns in the geologic history of Mars offers the same promise of insight into early environmental evolution, and one of the goals of future research will be to continue to work

[6]As used by the Mars community of scientists, the term "geomorphic" connotes all surface textural properties of rocks on Mars, including stratal geometries and scales of bedding of sedimentary rocks, in addition to its more traditional use on Earth, where emphasis is placed mostly on terrain surface morphology.

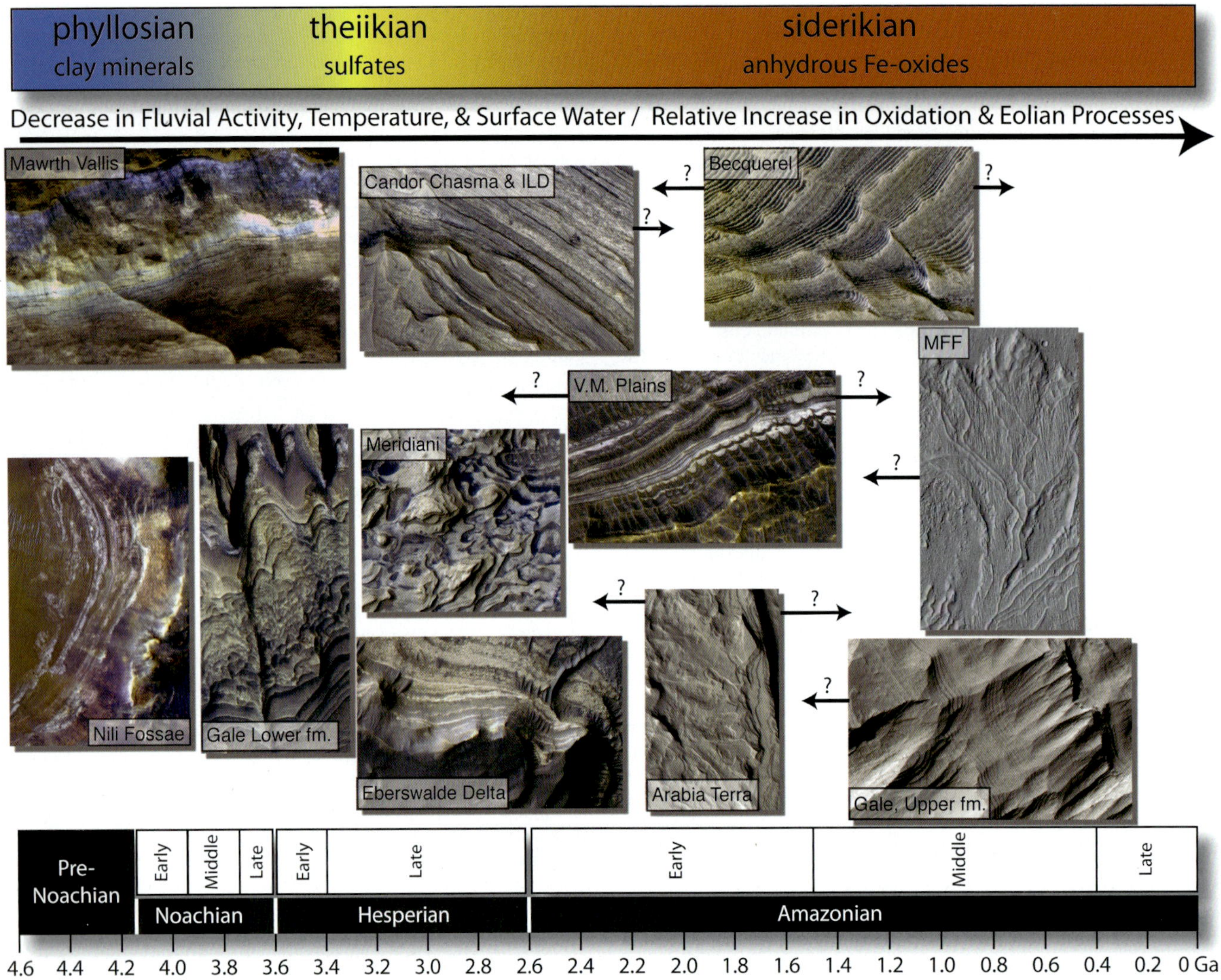

FIG. 22.—Mars chronostratigraphy illustrating representative outcrops of stratified rocks formed during different geologic eras. Lower scale shows the traditional Martian timeline and eras based on crater size frequency distributions (Tanaka and Hartmann 2008). Upper scale and nomenclature are based on the recent hypothesis of Bibring et al. (2006), which relates observed variations in mineralogy to geologic time. Images have been placed to be more in line with the upper (mineralogical) scale based on their inferred compositions. However, we note that the ages of stratified rocks depicted on the right half of the figure are poorly constrained, and though many of these examples are likely Hesperian, some may in fact be Amazonian in age.

toward establishing the relative chronology of different types of stratified deposits. In this context, we now highlight several localities on Mars that could provide key reference sections for refining the Martian geologic timescale.

Orbital Facies and Environmental History

The categories of Martian sedimentary rocks described above (underfilled basins, overfilled craters, chasma/canyon deposits, plains covering deposits, very ancient strata) were defined based on their large-scale geomorphic context. Similarly, Murchie et al. (2009b) recently discussed and defined 10 classes of deposits on Mars based primarily on mineral assemblages indicative of aqueous alteration. However, the preceding discussion makes it clear that both the mineral assemblages in the Martian sedimentary rock record as well as their bedding-scale textural attributes are important distinguishing attributes for classification. Together, mineralogy and texture/morphology provide critical information about petrogenesis; in turn, this informs our understanding of depositional environments and paleoclimate, the water–rock ratios involved in diagenesis, aqueous geochemistry, and how these factors have changed throughout the history of the planet. We now integrate the textural/morphologic and mineralogical aspects of Martian sedimentary strata to define several distinct "orbital facies." Similar to terrestrial facies defined on the basis of outcrop observations (Walker and James 1992) or Martian facies defined on the basis of rover-based in situ observations (Grotzinger et al. 2005), these "orbital facies" are defined and characterized by distinctive attributes of deposits observed from orbit. However, in contrast to terrestrial and Martian facies schemes derived from outcrop-scale observations, these orbital facies cannot offer the same resolution of insight into past depositional or diagenetic processes.

As a result of the inherent limitations of orbital data, the specific depositional environment(s) or formation conditions of the rocks in question are often unknown; thus, these conditions need not be unique to each orbital facies. Rather, each orbital facies represents a rock type or group of rock types that exhibits similar attributes in visible images (texture/morphology) and/or VNIR data (mineralogy). We discuss below the major orbital facies that we have identified on Mars and their relationships to what we believe are key stratigraphic sections ("reference" sections). Because the characteristics we use to define these orbital facies are primarily objective, we do not assume a priori that certain orbital facies are restricted in either time or space, and key sections may contain more than one orbital facies. Indeed, our goal is to determine which orbital facies are present in the different reference sections. Coupled with the inferred age of these reference sections, we can then attempt to determine whether or not certain orbital facies are restricted to specific time intervals or whether there are spatial transitions in dominant orbital facies for a constant time interval. We now briefly describe each orbital facies (see Table 1); this is followed by a discussion of several stratigraphic reference sections (or regions) on Mars that could act as key tie points for refining the Martian geologic timescale.

Massive Breccia (MBR): This orbital facies consists of rock units that contain breccia blocks with diameters commonly in the range of <1 to ~100 m that are present in a finer-grained matrix. Units are massive and marked by poorly expressed layering that occurs at a scale of tens of meters in thickness; ductile deformation, including recumbent folding, may also be present. These units are variable in their tonality, laterally discontinuous, and are known to contain mafic and/or clay minerals. Though the definition of this facies does not assume an origin, many such deposits have been interpreted as impact breccias. The reference section for this orbital facies can be found in the walls of the Nili Fossae (~73.6°E, 20.0°N; see Mustard et al. [2009] for examples).

Complexly Stratified Clay (CSC): The defining characteristics of this orbital facies are units that exhibit generally thin (often <10-m), light-toned layers that are known to contain clay minerals. The composition of clay minerals can be variable (e.g., Mg-, Fe-, or Al-rich varieties). The reference section for this orbital facies is within and adjacent to the landing ellipse that was proposed for Mars Science Laboratory (MSL) near Mawrth Vallis (~29°N, 341°E), although other examples can be found within Mawrth Vallis itself as well as in the surrounding plains (e.g., Poulet et al. 2005, Michalski and Noe Dobrea 2007) (Fig. 16A).

Though the areal extent of the clay-bearing rock body is significant (>80,000 km^2), the strata themselves are generally discontinuous when traced over hundreds of meters to kilometer-length scales. Stratal geometries can be very complex as a result of primary erosional truncation as well as postdepositional deformation (Fig. 16B–G). In the case of Mawrth Vallis, and as discussed above, many of the observed complex stratal geometries are proximal to Oyama Crater, whereas deposits farther from Oyama exhibit less strain and may better preserve primary stratification. We propose that the latter examples constitute a less-deformed subfacies of the CSC orbital facies. Future mapping could determine which is more dominant. We cannot tell at this time if the inferred lower strain of the CSC subfacies is simply due to distance from Oyama Crater or rather relates to variability within a facies mosaic created by Noachian impact–induced erosional, depositional, and subsequently overprinting deformational events. The riot of impacts predicted during Late Heavy Bombardment implies that all such processes may have been happening more or less simultaneously on Mars at that time.

Laterally Continuous Sulfate (LCS): Highly variable bed thickness (~1 to ~100 m), presence of sulfate minerals often associated with Fe-oxides, and great lateral and/or vertical extent of layering are the primary characteristics of this facies. Planar bedding is the dominant primary stratal geometry observed (this excludes postdepositional deformation). Units are commonly light-toned, etched or wind-scoured, and friable. Units formed by this orbital facies can be exposed over hundreds of kilometers and can approach several kilometers in thickness. Apparent massive beds may be the result of amalgamation of thinner beds or may be due to the resolution limitations of orbital images. Type examples can be found in the ILD of Valles Marineris (e.g., Gendrin et al. 2005) and in Sinus Meridiani (e.g., Edgett 2005, Griffes et al. 2007).

Laterally Continuous Heterolithic (LCH): This facies consists of units that exhibit thin, cliff-forming beds (~1 to <10 m) and that can be traced laterally for tens to hundreds of kilometers. Planar bedding is the dominant primary stratal geometry observed. In contrast to the laterally continuous clay and sulfate strata, units in this facies can contain opal, sulfates, or clay minerals or can lack spectral signatures indicative of hydrous minerals. In addition to the thin bedding, another defining characteristic of layers in this facies is their variegated tonality. Strata vary from light-toned to very dark-toned vertically through a section, sometimes bed by bed. Type examples include opal-/sulfate-bearing strata along the plains west of Juventae Chasma (Milliken et al. 2008, Weitz et al. 2010) and clay-bearing strata in Ladon Vallis (Milliken and Bish 2010) (compare Fig. 12B, D and Fig. 18D).

Distributary Network (DNW): Units in the "distributary network" orbital facies are characterized by moderately thick layers (commonly 5–20 m), the tonality of which ranges from light to moderately dark. These layers can contain clay minerals, carbonate, or other hydrous minerals, but many examples also lack spectral evidence for hydrous phases. The defining aspect of this facies is the occurrence of depositional bodies in clear association with distributary channel systems. In addition, nonplanar stratal geometries are present and may include well-developed clinoforms with slopes in excess of several degrees. This facies includes alluvial fan, deltaic, and sublacustrine fan deposits. Type examples include Holden Crater (alluvial fans; Grant et al. 2008), Eberswalde Crater (delta; Malin and Edgett 2003), Jezero Crater (delta; Fassett and Head 2005, Ehlmann et al. 2008), and the southwest Melas Basin (sublacustrine fan; Metz et al. 2009a).

Rhythmite (RHY): Primary attributes of the "rhythmite" facies are very thin (~1–5 m) beds that exhibit a repeatable thickness (are rhythmic) within a vertical sequence and that often lack spectral signatures indicative of sulfate, clay, or other hydrous phases. Planar bedded stratal geometries dominate, perhaps uniquely so. Visible–near-infrared spectra are similar to typical Martian dust, indicating that these units are either composed of dust or covered by an optically thick veneer of dust. Type localities include craters within Arabia Terra (e.g., Becquerel Crater; Lewis et al. 2008a) and the Upper formation of the Gale Crater mound (Lewis 2009, Milliken et al. 2010).

Key Reference Sections

Mawrth Vallis and Nili Fossae: The ancient Noachian crust of Mars contains a multitude of outcrops of clay mineral-bearing rocks (Bibring et al. 2006, Mustard et al. 2008, Murchie et al. 2009b), and the two largest expanses of clay-bearing terrains on Mars are found in the regions surrounding Mawrth Vallis and Nili Fossae (Poulet et al. 2005). The section at Mawrth Vallis is dominated by the CSC facies, whereas rocks at Nili are best characterized as MBR orbital facies and local occurrences of the CSC orbital facies (Fig. 23). At Mawrth, much of

TABLE 1.—Description of orbital facies and their mineralogical and morphological attributes.

Orbital facies name:	Massive Breccia	Complexly Stratified Clay	Laterally Continuous Sulfate strata	Laterally Continuous Heterolithic strata	Distributary Network	Rhythmite
Abbreviation	MBR	CSC	LCS	LCH	DNW	RHY
Stratification	Discontinuous stratification on hundreds of meters– to kilometer-length scale; large breccia blocks in massive matrix; ductile deformation including recumbent folding; massive layering	Discontinuous stratification on hundreds of meters– to kilometer-length scale; forms successions up to a few hundred meters in thickness; stratified breccia; thin bedding; irregular bedding associated with buried impact craters	Beds extend laterally for >10 km; forms successions that can exceed hundreds of meters in thickness; bed thickness highly variable; easily eroded	Laterally extensive for tens to hundreds of kilometers; forms successions up to ~100 m in thickness; thin and variegated (alternating light/dark) bedding; cliff-forming	Close association with bedrock channels and upstream contributory networks; deposits traceable over 1 km to tens of kilometers; beds locally discontinuous	Extreme lateral continuity of bedding, > tens of kilometers; forms successions in excess of hundreds of meters; uniform bed thickness distributions; apparent lack of hydrous minerals
Dominant mineralogy	Mafic, clay minerals	Clay minerals	Sulfates, Fe-oxides	Opal, sulfates, clay minerals	Clay minerals, anhydrous phases	Anhydrous phases
Representative bed thickness	>10 m to massive	<10 m	<10 m to tens of meters	<10 m	<20 m	1 to ~5 m typically
Tonality	Variable	Generally light	Generally light	Light to dark	Light to moderately dark	Light to intermediate
Reference locality	Nili Fossae	Mawrth Vallis	Sinus Meridiani; ILD; Gale Lower formation	Ladon Vallis; plains surrounding Valles Marineris	Eberswalde Crater; Holden Crater	Becquerel; Gale Upper formation
Example figures	Figure 15E, F	Figure 15B, C	Figures 5A, B; 11; 12F; 13	Figures 12A–D; 18D	Figure 6	Figures 5C, D; 9; 10

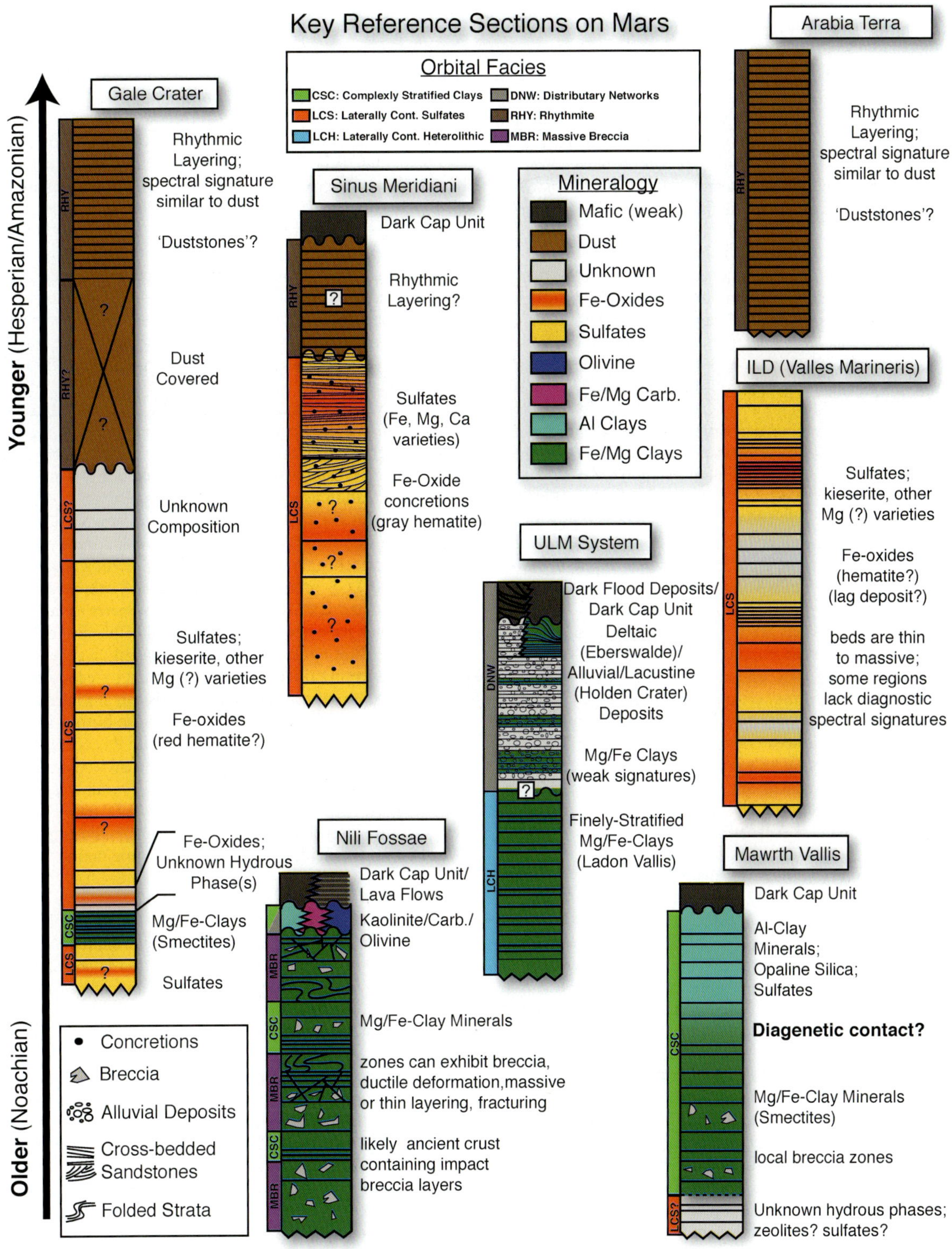

FIG. 23.—Correlation of stratigraphic reference sections on Mars at a regional to global scale. Orbital facies identified in each section are also noted. Oldest (Noachian) sections are dominated by MBR and CSC facies, whereas younger (Hesperian–Amazonian) sections are dominated by LCS and RHY orbital facies. However, important outcrops of clay-bearing DNW facies are also present in post-Noachian terrains and may be contemporaneous with LCS and RHY facies (e.g., stratigraphic section for ULM system). See text for detailed description and discussion of orbital facies and key sections. Angled lines indicate that one or more units may be present at a given stratigraphic position (e.g., dark flood deposits and dark capping unit at the top of the ULM section). This figure is purely schematic; individual sections and units are not to scale.

the section is heavily fractured and affected by low-angle to high-angle faulting, possibly as a result of postdepositional impact events. In addition, depositional breccias related to impact processes may be present, but they appear to be uniform in mineralogical composition. In contrast, MBR units at Nili Fossae include polymict breccias with both mafic and clay-rich clasts (some of which are themselves stratified) and highly deformed zones, all capped by younger (Hesperian) lava flows (e.g., Mustard et al. 2009). At Mawrth Vallis, parallel stratal geometries with significant lateral continuity indicate that a CSC subfacies is present, but it is not yet clear if this subfacies also exists in the Nili Fossae region.

It is intriguing that these rocks, which are among the oldest on the planet, also host the strongest clay–mineral signatures observed from orbit, indicating that alteration by circum-neutral pH fluids was widespread on ancient Mars. The mineralogy and age of both locations indicates early conditions that were favorable to the formation of smectitic clay, but overlying mineral assemblages indicate that later environmental conditions may have differed between these two regions.

As discussed above, the original depositional environment(s) of any possible sedimentary rocks, other than impact deposits, in these regions are currently unconstrained, although much of the stratigraphy is believed to have formed over ~3.6 Gyr ago in the Noachian Era. The ambiguity in the origins of these deposits is not unexpected given their antiquity, the effects of the Late Heavy Bombardment, and the limitations of orbital data. Furthermore, it seems that any future work on the stratigraphy of these sites will have to wrestle with palinspastic reconstruction of overprinting deformation events in order to reconstruct primary depositional geometries. However, it is clear that these regions are important stratigraphic tie points because they likely record global processes that were occurring on early Mars: large and small impact events, deformation, erosion, and aqueous alteration of primary crust. Sedimentation by processes other than impacts may have occurred but has not yet been documented. Indeed, the Noachian rock record of Mars may largely be a product of aqueous alteration of primary basaltic crust punctuated and significantly overprinted by impact events. The orbital facies in the stratigraphic reference sections at both Mawrth Vallis and Nili Fossae likely record these events.

Sinus Meridiani: Strata at Sinus Meridiani, the upper part of which includes the 10 to 20 m of the Burns formation examined by *Opportunity*, are dominated by the LCS facies (Fig. 23). Orbital data indicate that these rocks contain sulfate and Fe-oxide minerals, and rover observations have shown that at least part of this section occurs as sulfate-cemented sandstones with hematite concretions. The strata examined by *Opportunity* also indicate formation and/or alteration under acidic conditions, although the extent to which this is true for other rocks in Sinus Meridiani and other regions exhibiting the LCS orbital facies remains unknown. We note that the dominance of the LCS orbital facies in this region does not require or imply that all of the rocks have a similar origin (i.e., reworked sulfate-cemented sandstones). The LCS orbital facies within this region may also include hypothesized evaporites from playa lakes (Grotzinger et al. 2005, McLennan et al. 2005) or sulfate-bearing rocks formed in other environments, such as the alternative (hydrothermal) explanation for the Burns formation provided by McCollom and Hynek (2005).

We also note that several craters in northern Sinus Meridiani exhibit strata that appear to be rhythmically bedded (intracrater deposits unit of Hynek et al. [2001]), although their power spectrum has not yet been computed, as was done for beds in Becquerel, other Arabia Terra craters, and the top of the mound at Gale Crater (Lewis et al. 2008a, Lewis 2009). Nevertheless, it is possible that the sulfate- and hematite-bearing plains in Meridiani Planum are overlain by a section of the RHY orbital facies, similar to the RHY deposits in Arabia Terra (Fig. 23). However, Hynek et al. (2001) noted that the stratigraphic relationships between these intracrater deposits in northern Sinus Meridiani and the hematite-bearing plains is unclear and that the former may be erosional remnants of the latter. If true, deposits formed of the LCS and RHY orbital facies at Sinus Meridiani could indicate time-equivalent, lateral changes in orbital facies.

Interior Layered Deposits: Similar to what is observed at Sinus Meridiani, the ILD in Valles Marineris are dominated by the LCS orbital facies, and rocks are known to contain sulfates and Fe-oxides (Fig. 23). Though we classify strata in these two regions as the same orbital facies, the definitions of our orbital facies (Table 1) and the limits of orbital data preclude us from determining their origin. However, the similarity in mineralogy to rocks at Meridiani Planum has led others to hypothesize that these deposits have a common origin (e.g., Bibring et al. 2007, Murchie et al. 2009a). Specular (gray) hematite and other Fe-oxides (e.g., red hematite) are associated with some ILD (Christensen et al. 2001, Bibring et al. 2007), and it is possible that these phases occur as lag deposits of concretions produced by groundwater interaction, similar to the Meridiani Planum deposits. Future work that helps prove they are indeed lag deposits like those at Meridiani Planum (e.g., Jerolmack et al. 2006) or that shows the presence of diagenetic contacts within the ILD would help support their interpretation as sulfate-cemented sandstones.

We note that the ILD are distinct from Sinus Meridiani in that the deposits are significantly thicker (up to several kilometers) and occur in well-defined basins; thus, their depositional setting is quite different. In addition, the LCS orbital facies in the ILD exhibit a wide range of morphologic and textural characteristics at meter to decameter scales, many of which have not yet been observed in this orbital facies at Sinus Meridiani, indicating that a single depositional/diagenetic environment may be an oversimplification. It is important to recognize that sulfates formed at the top of a sedimentary succession at Sinus Meridiani have the opposite stratigraphic relationship to those formed in the ILD at the base of a stratigraphic succession—and with this comes the possibility that the latter may have formed in lacustrine environments, whereas the former are known to have accumulated as a result of eolian processes. Despite these possible differences, the ILD and Sinus Meridiani deposits are likely both Hesperian in age, indicating that the LCS orbital facies was widespread during this time period.

Uzboi–Ladon–Morava System: This system is complex and involves at least intermittent events of channel incision and sedimentation beginning in Noachian time and possibly extending through to early Amazonian time (Grant and Parker 2002, Grant and Wilson 2011). Therefore, strata deposited at various positions along the network, and at various points in time, have the potential to overlap in age with very different orbital facies exposed at other reference sections.

Bedrock incision and depositional infilling of the oldest parts of the ULM system, including Holden Basin, Ladon Basin, and Ladon Vallis, likely began during late Noachian time or earlier (Grant and Parker 2002). Outcrops in Ladon Basin and exposed terraces at the northern end of Ladon Vallis contain Mg/Fe-phyllosilicates and vary widely in their tonality; thus, they are classified as the LCH orbital facies. The specific mode of emplacement for these valley and basin-filling deposits remains somewhat ambiguous, but their mineralogy is consistent with weathering, transport, deposition, and/or alteration by circum-neutral pH fluids at this time. This through-going transport network was later blocked by the impact event that formed Holden Crater, but the initial formation of the ULM system and its early activity as a possible sediment conveyor to the northern lowlands may have been contemporaneous with the uppermost parts of the reference sections at Mawrth Vallis and Nili Fossae, and all three sections are dominated by clay minerals rather than sulfates.

Recent studies have shown that alluvial fan/lacustrine deposits in

Holden Crater and deltaic deposits in Eberswalde Crater, which we classify as DNW orbital facies, may be much younger than previously recognized, having formed in Hesperian or even younger times (Grant and Wilson 2011). These younger ages would place these deposits in the era proposed to represent the time of peak sulfate formation on Mars (Bibring et al. 2006). However, these ULM strata are known to contain clay minerals (Grant et al. 2008, Milliken and Bish 2010), whereas sulfates and other salts are conspicuously absent (Milliken et al. 2009). The presence of such phases (clays) and the absence of others (sulfates) in aqueous deposits with ages possibly equivalent to the LCS orbital facies in the ILD and Meridiani deposits (Fig. 23) poses a significant challenge to the hypothesis invoking "global" acidic environments that arose during the Late Noachian–Early Hesperian (e.g., Bibring et al. 2006, Murchie et al. 2009b).

Considering their broader context, the sulfate deposits in the ILD and at Meridiani Planum likely formed in the presence of groundwater, or locally emergent groundwater (playa lakes). In contrast, the deposits characterized by the DNW orbital facies in the ULM system formed largely as a result of surface runoff and sediment transport, although groundwater may have been important in the upper reaches of contributory networks. Therefore, it is possible to invoke a difference in pH between locally emerging groundwater in the lowlands of Mars and surface water (especially if meteoric precipitation occurred) in the highlands of Mars. This may also help explain why Hesperian alluvial fan and deltaic systems (e.g., Holden and Eberswalde craters) were able to transport and deposit clay minerals without destroying them, as would be expected for smectitic clays interacting with acidic fluids. If correct, this would also account for the absence of observed sulfates in these strata.

Regardless of their origin, the rocks within the ULM system provide an important reference section on Mars because they may contain the longest temporal record of clay mineral formation, transport, and deposition on the planet, stretching from the Noachian (LCH orbital facies in Holden Basin, Ladon Basin, Ladon Vallis) to the Hesperian–Early Amazonian (DNW orbital facies in Holden and Eberswalde craters). If the inferred young ages of the latter deposits are correct, then this occurrence of clay-bearing DNW and LCH orbital facies would stand in marked contrast to the more commonly observed LCS or RHY orbital facies characteristic of post-Noachian terrains (Fig. 23).

Arabia Terra: The Arabia Terra region of Mars hosts many overfilled and underfilled crater deposits that, unfortunately, are commonly covered with veneers of dust that can obscure spectroscopic determination of their composition (Fergason and Christensen 2008). However, as discussed earlier, in some cases dust is absent or limited enough to allow the hypothesis that these deposits may be formed of dust and not simply covered by dust. Specific locations of "key sections" in the vast Arabia Terra region include the deposits in Danielson, Crommelin, and Becquerel craters, discussed above. The distinct rhythmic layering (Lewis et al. 2008a) and often "bland" spectral signatures of these deposits lead us to classify them as the RHY orbital facies.

Many of these RHY deposits have been heavily eroded as a result of their weak lithification, which has undoubtedly affected the number of impact craters visible on their surfaces. However, these deposits are clearly younger than the craters in which they reside, indicating a likely Hesperian or possibly Early Amazonian age. Although some of these RHY deposits exhibit an increase in H_2O absorption features at near-infrared wavelengths, most of them appear to lack clay, sulfate, or other common hydrous minerals (Fig. 8). Therefore, the composition of these rocks is consistent with deposition in a dry environment.

Based on their mineralogical and physical stratigraphy we assign a younger chronostratigraphic age to the RHY-dominated deposits relative to the LCS-dominated rocks of the ILD and Sinus Meridiani deposits (Fig. 23). In support of this hypothesis we note that examples of the RHY orbital facies are restricted to relatively young terrains. Furthermore, in at least one place the RHY orbital facies occurs—in a single continuous section—stratigraphically above LCS orbital facies (Gale Crater); possibly another place is Sinus Meridiani, where RHY orbital facies may fill a crater formed in LCS orbital facies. Finally, to date no examples of the RHY orbital facies have been observed in Noachian-age deposits. Therefore, the RHY orbital facies may be limited in geologic time, possibly to the last several billion years, when the planet had dried and was dominated by transport and accumulation of anhydrous dust. However, its formation does not appear to be limited in space, as examples of the RHY orbital facies are found in widely separated regions of the planet (Lewis 2009).

The Section at Gale Crater: We conclude our list of key reference sections on Mars by examining the ~5-km-thick mound in Gale Crater. This deposit is likely the thickest continuously exposed single succession of sedimentary rocks on the planet (Milliken et al. 2010). The succession cannot, however, record continuous accumulation as it contains at least one significant erosional unconformity (Malin and Edgett 2000). Although the absolute ages of the deposit are not well constrained, the lower part of the mound may date to the Noachian–Hesperian chronostratigraphic boundary, and the upper part of the mound may be as young as Amazonian (Thomson et al. 2011). Albeit discontinuously, the mound in Gale may therefore record significant events in the environmental evolution of Mars (Anderson and Bell 2010, Milliken et al. 2010, Thomson et al. 2011).

The Lower formation, located beneath the significant unconformity (see Milliken et al. 2010), is best classified as LCS orbital facies, with a thin interval of CSC orbital facies near the base. In contrast, the Upper formation is composed at least in part of the RHY orbital facies (Fig. 22). The CSC facies at the base of Gale, however, is most similar to the minimally deformed subfacies observed at Mawrth as a result of its lateral continuity. This interval is on the order of tens of meters thick, and although the abundance of clay minerals are unknown, they are spectrally the dominant hydrous phase in these beds, possibly mixed with lesser amounts of sulfates (Milliken et al. 2010). Through recognition of CSC orbital facies (minimally deformed subfacies) it is proposed that the lower-most strata in the Lower formation at Gale may be contemporaneous with the youngest rocks at the Nili Fossae and Mawrth Vallis reference sections and the lower-most parts of the section at the Sinus Meridiani reference section. Indeed, most of the Lower formation at Gale is the same orbital facies (LCS) as the deposits in Sinus Meridiani, and these rocks may have formed contemporaneously. That said, the outcrop facies at Gale may differ significantly from the outcrop facies of the Burns formation at Meridiani, and we reiterate that not all occurrences of an orbital facies need to have the same petrogenetic origin. Testing this hypothesis will be one of the principal objectives of the MSL science investigation at Gale.

In a similar fashion, the RHY orbital facies at the top of the Gale mound may be equivalent in time to the RHY-dominated deposits observed in Arabia Terra, though more robust crater counts are needed to test this hypothesis. The mound at Gale is intriguing in that it captures several different orbital facies and, remarkably, may in a single section record the global transitions from clay to sulfate to anhydrous Fe-oxide stages of Mars' environmental evolution, effectively chronicling the "drying out" of Mars (Milliken et al. 2010).

Synthesis: The extent to which any of the key reference sections described above record local vs. regional or global-scale environmental conditions is currently poorly understood. Many reference sections, independently constrained in time to avoid circular reasoning, would be required to build a more confident history of the environmental evolution of Mars. But there must always be a beginning, and as a group the reference sections discussed above provide a well-defined starting point. The most ancient rocks on Mars, dating to the Noachian Era, are dominated by the CSC and MBR orbital facies and record extensive aqueous activity on a young planet that was still experiencing a significant flux of impactors. Younger Hesperian rocks in many places are dominated by the LCS orbital facies, and the presence of Fe-sulfates in some of these locations indicates a very acidic environment. However, local occurrences of clay-bearing DNW orbital facies also are observed in rocks deposited at this time, indicating that regional differences in aqueous geochemistry were important and are preserved in the Martian rock record. The youngest sedimentary rocks on Mars may consist dominantly of the RHY orbital facies and may support the traditional hypothesis that eolian processes have dominated the rock cycle on a cold, dry Mars for the past several billion years. Fortunately, possible duststones in the RHY facies may contain much information about the Martian atmosphere and climate over this time. It is our hope that future work will continue to identify important mineralogical and morphological key reference sections on Mars to build on what is presented here, providing a framework with which to understand the evolution of climate and geologic processes through time.

SUMMARY

1. Mars has a surprising diversity of sedimentary rocks. From orbit, these rocks are distinctive in their stratification, depositional morphology, mineral composition, tone/albedo variations, and thermal inertia. These attributes combine to form distinct classes of inferred sedimentary rocks. In many cases it is impossible to confirm their mode(s) of origin because of non-uniqueness; origins for those with distinctive morphologies are best known (deltas, alluvial fans, sublacustrine fans); followed by eolian deposits, the large-scale cross-bedding of which can in some cases be observed even from orbit—and which are confirmed by in situ rover observations. Lacustrine deposits are largely inferred based on context, most commonly expressed as flat, continuous layers in topographic lows. Regionally extensive sulfate deposits, some of which are kilometers thick, indicate precipitation as evaporites; reworking by wind and precipitation of cements from groundwater brines may have helped them attain their great thickness. Regionally extensive sheets of flat-lying strata that apparently lack hydrated mineral signatures may represent "duststones" formed by prolonged settling of dust from the atmosphere, although regionally extensive volcanic ash deposits could have similar stratigraphic expression.

2. The oldest sedimentary rocks on Mars are likely older than the oldest rocks (gneisses) on Earth. The absence of plate tectonics on Mars has resulted in impressive preservation of the earliest records of aqueous processes on a terrestrial planet. However, the sedimentary record of Mars is fundamentally different from that of Earth in several ways. Mars is a basaltic planet, and so the provenance of sedimentary materials, including particulate detritus and the geochemistry of fluids from which chemical constituents precipitate, reflects mafic compositions rather than intermediate to felsic compositions. Furthermore, Mars seems to have progressed from a very early (primarily Noachian) stage, in which aqueous alteration was dominated by circum-neutral pH reactions that formed clay minerals, to a later (primarily Hesperian) stage, in which alteration occurred at lower pH. This stage of alteration—hypothesized as acidic and global in extent, and therefore very different from terrestrial weathering—saw high mobility of Al and Fe, resulting in the formation of exotic sulfate minerals deposited in volumes unknown on the Earth. Mg-, Ca-, Fe-, and Al-sulfates likely dominate Martian evaporite deposits. During late Hesperian–Early Amazonian time Mars may have seen a last pulse of significant water, expressed as bedrock-incising fluvial channel networks, before arriving at its current dry state, which has been dominated by eolian and ice-related processes for the past ~2 billion years.

3. The absence of plate tectonics on Mars leads to critical differences between Earth and Mars with regard to the character of the stratigraphic record. Most (all?) sedimentary deposits on Mars formed in preexisting topographic depressions created either by impact events or by faulting associated with the development of the Valles Marineris rift system. To date, there is no evidence for the syndepositional subsidence that is so common on Earth. Consequently, most stratal geometries are very flat or otherwise exhibit simple downlap and onlap associated with the infilling of crater basins. Rare examples of more complex stratal geometries have been observed in Melas Chasma and Eberswalde Crater, likely associated with progradation of shoreline/deltaic sediment wedges. Again, the lack of tectonics results in only rare exposures of stratigraphic cross sections that can be observed at only slight off-nadir angles by orbiting cameras; this bias tends to emphasize flatter stratal geometries.

4. Source-to-sink systems can be well defined on Mars. For aqueous transport systems, the source regions can exhibit very densely spaced fluvial networks, analogous to those observed on Earth, and consistent with meteoric precipitation and overland flow. In other cases, particularly for the largest systems on Mars, the source regions are more consistent with emerging groundwater, particularly where networks and outflow channels originate in "chaos" regions (breccia, collapse features). Distributary networks are represented by alluvial fans, possible debris flow fans, deltas, and sublacustrine fans. In contrast, eolian systems are poorly defined. Source regions can only be imagined, but as such are regarded to be the cratered highlands of Mars, where impact-generated sediments would have been available for transport. In the early history of Mars, both chemical and physical weathering also would have been important in the generation of fine-grained materials. Later in the history of Mars, physical weathering processes became dominant, and erosion of basalt yielded quantities of unaltered olivine and pyroxene that form the eolian bedforms and soils observed across the surface of Mars today. The sinks for fine-grained eolian materials might have been vast duststone deposits, the abundance of which decreased through time as impact, volcanic, and weathering byproducts all decreased in importance. Today these deposits are in a state of net erosion. Eolian deposits formed by traction processes are less evident, though both *Opportunity* and *Spirit* have observed cross-bedded eolianites, and some stratified outcrops observed from orbit exhibit large-scale cross-stratification and have preserved large-scale bedforms most simply interpreted as eolian traction deposits. These tend to occur at high stratigraphic positions (paleoelevations) in their respective sections.

5. Integrating regional setting, bed-scale physical properties, and mineral assemblages as observed from orbital data leads to identification of distinct "orbital facies" that can be identified in different key reference sections on Mars. These orbital facies may not be restricted in time, and none appear to be restricted in space—beyond the apparent general limitation of most sedimentary rocks occurring at lower latitudes. Several key reference sections exhibit multiple orbital facies developed in a stratigraphic context that allows potential correlation to other locations. The most ancient strata preserved on Mars are dominated by MBR and CSC orbital facies. In contrast, younger strata in key reference sections are dominated by LCS orbital facies and RHY orbital facies. This transition is largely consistent with an apparent widespread, perhaps global, transition from clay-forming to sulfate/Fe-oxide–forming conditions followed by the desiccation of Mars. However, important occurrences of post-Noachian clay-bearing orbital facies (e.g.,

LCH) have been identified in the southern highlands of Mars in strata clearly linked to distributary networks, indicating that environmental shifts throughout Mars' history may be more complex than a simple clay–sulfate–oxide transition. Further refinement of a relative timescale based on rock types will depend on identification of additional key reference sections and their orbital facies, improved estimates of relative ages based on crater counts using new high-resolution images, detailed geologic mapping, and correlation of strata. In implementing this integrated approach we can hope to decipher the geologic and environmental history of Mars.

ACKNOWLEDGMENTS

This article has benefited from many discussions with our colleagues on Mars missions led by Ken Edgett (MOC, CTX), Mike Malin (MOC, CTX), Alfred McEwen (HiRISE), Scott Murchie (CRISM), Steve Squyres (MER), and also the MSL team. Interactions and presentations given at MSL Landing Site workshops and the First International Conference on Mars Sedimentology and Stratigraphy were key sources of knowledge and inspiration. Individually, we would like to thank Ken Edgett and Mike Malin for sharing their knowledge of sedimentary rocks on Mars, including early articles that speculate on the presence of possible sedimentary rocks; John Grant and Tim Parker for their insights into the evolution of crater lakes and the ULM fluvial network; Gary Kocurek, Dave Rubin, and Rob Sullivan for discussions of eolian systems; Kevin Lewis and Katie Stack for insights into Mars bed thickness distributions; Dawn Sumner and Horton Newsom for their work on impact processes and stratigraphy; David Bish for fruitful discussions relating to clay minerals and diagenesis on Mars; and Joel Hurowitz and Scott McLennan for insights into Martian geochemistry. Support for writing this work was provided the National Aeronatics and Space Administration Astrobiology Institute and the Mars Exploration Rover and Mars Science Laboratory projects. We thank Ken Edgett, Sangeev Gupta, Dawn Sumner, and Brad Thomson for their helpful reviews of the manuscript. Jennifer Griffes is thanked for help in figure construction and compiling all of the observations on which Figure 3 is based.

REFERENCES

Allen PA. 2008. From landscapes into geological history. *Nature* 451:274–276.

Anderson RB, Bell JF III. 2010. Geologic mapping and characterization of Gale Crater and implications for its potential as a Mars Science Laboratory landing site. *Mars* 5:76–128.

Andrews-Hanna JC, Phillips RJ, Zuber MT. 2007. Meridiani Planum and the global hydrology of Mars. *Nature* 446:163–166.

Andrews-Hanna JC, Zuber MT, Arvidson RE, Wiseman SM. 2010. Early Mars hydrology: Meridiani playa deposits and the sedimentary record of Arabia Terra. *Journal of Geophysical Research* 115:E06002.

Ansan V, Loizeau D, Mangold N, Le Mouelic S, Carter J, Poulet F, Dromart G, Lucas A, Bibring J-P, Gendrin A, Gondet B, Langevin Y, Masson P, Murchie S, Mustard JF, Neukum G. 2011. Stratigraphy, mineralogy, and origin of layered deposits inside Terby crater, Mars. *Icarus* 211:273–304.

Arvidson RE, Gooding JL, Moore HJ. 1989. The Martian surface as imaged, sampled, and analyzed by the Viking landers. *Review of Geophysics* 27:39–60.

Ashley GM, Delaney JS. 1999. If a meteorite of Martian sandstone hit you on the head would you recognize it? *In* 20th Annual Lunar and Planetary Science Conference; Houston, Texas. Abstract 1273.

Barnhart CJ, Howard AD, Moore JM. 2009. Long-term precipitation and late-stage valley network formation: Landform simulations on Parana Basin, Mars. *Journal of Geophysical Research* 114:E01003.

Bhattacharya JP, Payenberg THD, Lang SC, Bourke M. 2005. Dynamic river channels suggest a long-lived Noachian crater lake on Mars. *Geophysical Research Letters* 32:L10201.

Bibring J-P, Arvidson RE, Gendrin A, Gondet B, Langevin Y, Le Mouelic S, Mangold N, Morris RV, Mustard JF, Poulet F, Quantin C, Sotin C. 2007. Coupled ferric oxides and sulfates on the Martian surface. *Science* 317:1206–1210.

Bibring J-P, Langevin Y, Gendrin A, Gondet B, Poulet FO, Berthe M, Soufflot A, Arvidson R, Mangold N, Mustard J, Drossart P, The Omega Team. 2005. Mars surface diversity as revealed by the OMEGA/Mars Express observations. *Science* 307:1576–1581.

Bibring J-P, Langevin Y, Mustard JF, Poulet FO, Arvidson R, Gendrin A, Gondet B, Mangold N, Pinet P, Forget F, The OMEGA team, Berthe M, Gomez CC, Jouglet D, Soufflot A, Vincendon M, Combes M, Drossart P, Encrenaz TRS, Fouchet T, Merchiorri R, Belluci G, Altieri F, Formisano V, Capaccioni F, Cerroni P, Coradini A, Fonti S, Korablev O, Kottsov V, Ignatiev N, Moroz V, Titov D, Zasova L, Loiseau D, Doute S, Schmitt B, Sotin C, Hauber E, Hoffmann H, Jaumann R, Keller U, Arvidson R, Duxbury T, Neukum G. 2006. Global mineralogical and aqueous Mars history derived from OMEGA/Mars Express data. *Science* 312:400–404.

Binder AB. 1966. Mariner IV: Analysis of preliminary photographs. *Science* 152:1053–1055.

Bish DL, Scanlan MK. 2006. The hydration and dehydration of hydrous mixed-cation sulfates. *In* 37th Annual Lunar and Planetary Science Conference; League City, Texas; Abstract 1011.

Bishop JL, Dobrea EZN, McKeown NK, Parente M, Ehlmann BL, Michalski JR, Milliken RE, Poulet F, Swayze GA, Mustard JF, Murchie SL, Bibring J-P. 2008. Phyllosilicate diversity and past aqueous activity revealed at Mawrth Vallis, Mars. *Science* 321:830–833.

Bishop JL, Parente M, Weitz CM, Noe Dobrea EZ, Roach LH, Murchie S, McGuire PC, McKeown NK, Rossi CM, Brown AJ, Calvin WM, Milliken RE, Mustard JF. 2009. Mineralogy of Juventae Chasma: Sulfates in the light-toned mounds, mafic minerals in the bedrock, and hydrated silica and hydroxylated ferric sulfate on the plateau. *Journal of Geophysical Research* 114:E00D09.

Blasius KR, Cutts JA, Guest JE, Masursky H. 1977. Geology of the Valles Marineris: First analysis of imaging from the Viking 1 Orbiter Primary Mission. *Journal of Geophysical Research* 82:4067–4091.

Bowring SA, Grotzinger JP, Condon DJ, Ramezani J, Newall M, Allen P. 2007. Geochronologic constraints on the chronostratigraphic framework of the Neoproterozoic Huqf Supergroup, sultanate of Oman. *American Journal of Science* 307:1097–1145.

Boynton WV, Ming DW, Kounaves SP, Young SMM, Arvidson RE, Hecht MH, Hoffman J, Niles PB, Hamara DK, Quinn RC, Smith PH, Sutter B, Catling DC, Morris RV. 2009. Evidence for calcium carbonate at the Mars Phoenix landing site. *Science* 325:61–64.

Bradley BA, Sakimoto SEH, Frey H, Zimbelman JR. 2002. Medusae Fossae Formation: New perspectives from Mars Global Surveyor. *Journal of Geophysical Research* 107(E8):5058.

Bridges NT, Ayoub F, Leprince S, Avouac J, Necsoiu M, Fenton LK, Kirk RL, Colon C. 2010. Detecting bedform migration on Mars: A review of current results and plans for sub-pixel detection techniques. *In* 2010 American Geophysical Union Fall Meeting; San Francisco, California; Abstract EP51A-0531.

Bristow TF, Kennedy MJ, Derkowski A, Droser ML, Jiang G, Creaser RA. 2009. Mineralogical constraints on the paleoenvironments of the Ediacaran Doushantuo Formation. *Proceedings of the National Academy of Sciences of the United States of America* 106:13190–13195.

Burr DM, Enga M-T, Williams RME, Zimbelman JR, Howard AD, Brennand TA. 2009. Pervasive aqueous paleoflow features in the Aeolis/Zephyria Plana region, Mars. *Icarus* 200:52–76.

Cabrol NA, Grin EA. 2001. The evolution of lacustrine environments on Mars: Is Mars only hydrologically dormant? *Icarus* 149:291–328.

Cabrol NA, Grin EA, Newson HE, Landheim R, McKay CP. 1999. Hydrogeologic evolution of Gale Crater and its relevance to the exobiological exploration of Mars. *Icarus* 139:235–245.

Cadogan PH. 1981. *The Moon—Our Sister Planet*: Cambridge University Press, UK. 393 p.

Carpenter AH. 1948. Principles of historical geology applied to neighboring planets and life on Mars. *Popular Astronomy* 56:233–246.

Carr MH. 1979. Formation of Martian flood features by release of water from confined aquifers. *Journal of Geophysical Research* 84(B6):2995–3007.

Carr MH. 1996. *Water on Mars*. Oxford University Press, Mississippi. 229 p.

Carrier WD III, Olhoeft GR, Mendell W. 1991. Physical properties of the lunar surface. *In* Heiken GH, Vaniman DT, French BM (Editors). *The Lunar Source Book*: Cambridge University Press, Massachusetts. p. 475–594.

Catling DC, Wood SE, Leovy C, Montgomery DR, Greenberg HM, Glein CR, Moore JM. 2006. Light-toned layered deposits in Juventae Chasma, Mars. *Icarus* 181:26–51.

Chapman CR, Pollack JB, Sagan C. 1968. *An Analysis of the Mariner 4 Photography of Mars*: Smithsonian Astrophysical Observatory, Washington, DC. Special Report 268:1–65.

Chapman MG, Gudmundsson MT, Russell AJ, Hare TM. 2003. Possible Juventae Chasma subice volcanic eruptions and Maja Valles ice outburst floods on Mars: Implications of Mars Global Surveyor crater densities, geomorphology, and topography. *Journal of Geophysical Research* 108(E10):5113.

Chapman MG, Tanaka KL. 2002. Related magma–ice interactions: Possible origins of chasmata, chaos, and surface materials in Xanthe, Margaritifer, and Meridiani Terrae, Mars. *Icarus* 155:324–339.

Chipera SJ, Vaniman DT. 2007. Experimental stability of magnesium sulfate hydrates that may be present on Mars. *Geochimica et Cosmochimica Acta* 71:241–250.

Christensen PR. 1988. Global albedo variations on Mars: Implications for active aeolian transport, deposition, and erosion. *Journal of Geophysical Research* 93(B7):7611–7624.

Christensen PR, Bandfield JL, Bell JF III, Gorelick N, Hamilton VE, Ivanov A, Jakosky BM, Kieffer HH, Lane MD, Malin MC, McConnochie T, McEwen AS, McSween HY, Mehall GL, Moersch JE, Nealson KH, Rice JW, Richardson MI, Ruff S, Smith MD, Titus TN, Wyatt MB. 2003. Morphology and composition of the surface of Mars: Mars Odyssey THEMIS results. *Science* 300:2056–2061.

Christensen PR, Bandfield JL, Hamilton VE, Ruff SW, Kieffer HH, Titus TN, Malin MC, Morris RV, Lane MD, Clark RL, Jakosky BM, Mellon MT, Pearl JC, Conrath BJ, Smith MD, Clancy RT, Kuzmin RO, Roush T, Mehall GL, Gorelick N, Bender K, Murray K, Dason S, Greene E, Silverman S, Greenfield M. 2001. Mars Global Surveyor Thermal Emission Spectrometer experiment: Investigation description and surface science results. *Journal of Geophysical Research* 106(E10):23823–23871.

Craddock RA, Howard AD. 2002. The case for rainfall on a warm, wet early Mars. *Journal of Geophysical Research–Planets* 107(E11):5111.

Craddock RA, Maxwell TA, Howard AD. 1997. Crater morphometry and modification in the Sinus Sabaeus and Margaritifer Sinus region of Mars. *Journal of Geophysical Research* 102(E6):13321–13340.

Crown DA, Bleamaster LF III, Mest SC. 2005. Styles and timing of volatile-driven activity in the eastern Hellas region of Mars. *Journal of Geophysical Research* 110:E12S22.

Cutts JA. 1973. Nature and origin of layered deposits of the Martian polar regions. *Journal of Geophysical Research* 78:4231–4249.

Di Achille G, Hynek BM. 2010. Ancient ocean on Mars supported by global distribution of deltas and valleys. *Nature Geoscience* 3:459–463.

Di Achille G, Marinangeli L, Ori GG, Hauber E, Gwinner K, Reiss D, Neukum G. 2006a. Geological evolution of the Tyras Vallis paleolacustrine system, Mars. *Journal of Geophysical Research* 111:E04003.

Di Achille G, Ori GG, Reiss D, Hauber E, Gwinner K, Michael G, Neukum G. 2006b. A steep fan in Coprates Catena, Valles Marineris, Mars, as seen by HRSC data. *Geophysical Research Letters* 33:L07204.

Dromart G, Quantin C, Broucke O. 2007. Stratigraphic architectures spotted in southern Melas Chasma, Valles Marineris, Mars. *Geology* 35:343–366.

Edgett KS. 2005. The sedimentary rocks of Sinus Meridiani: Five key observations from data acquired by the Mars Global Surveyor and Mars Odyssey orbiters: Mars. *The International Journal of Mars Science and Exploration* 1:5–58.

Edgett KS, Malin MC. 2002. Martian sedimentary rock stratigraphy: Outcrops and interbedded craters of northwest Sinus Meridiani and southwest Arabia Terra. *Geophysical Research Letters* 29(24):2179.

Edwards CS, Bandfield JL, Christensen PR, Fergason R. 2009. Global distribution of bedrock exposures on Mars using THEMIS high-resolution thermal inertia. *Journal of Geophysical Research* 114:E1101.

Ehlmann BL, Mustard JF, Fassett CI, Schon SC, Head JW III, Des Marais DJ, Grant JA, Murchie SL. 2008a. Clay minerals in delta deposits and organic preservation potential on Mars. *Nature Geoscience* 1:355–358.

Ehlmann BL, Mustard JF, Murchie SL, Poulet F, Bishop J, Brown A, Calvin W, Clark R, Des Marais D, Milliken RE, Roach L, Roush T, Swayze G, Wray J. 2008b. Orbital identification of carbonate-bearing rocks on Mars. *Science* 322:1828–1832.

Ehlmann BL, Mustard JF, Swayze G, Clark RN, Bishop JL, Poulet F, Des Marais DJ, Roach LH, Milliken RE, Wray JJ, Barnouin-Jha O, Murchie S. 2009. Identification of hydrated silicate minerals on Mars using MRO-CRISM: Geologic context near Nili Fossae and implications for aqueous alteration. *Journal of Geophysical Research* 114:E00D08.

Farmer JD, Des Marais DJ. 1999. Exploring for a record of ancient Martian life. *Journal of Geophysical Research* 104(E11):26977–26995.

Fassett CI, Head JW III. 2005. Fluvial sedimentary deposits on Mars: Ancient deltas in a crater lake in the Nili Fossae region. *Geophysical Research Letters* 32:L14201.

Fassett CI, Head JW III. 2008a. The timing of Martian valley network activity: Constraints from buffered crater counting. *Icarus* 195:61–89.

Fassett CI, Head JW III. 2008b. Valley network-fed, open-basin lakes on Mars: Distribution and implications for Noachian surface and subsurface hydrology. *Icarus* 198:37–56.

Feldman WC, Prettyman TH, Maurice S, Plaut JJ, Bish DL, Vaniman DT, Mellon MT, Metzger AE, Squyres SW, Karunatillake S, Boynton WV, Elphic RC, Funsten HO, Lawrence DJ, Tokar RL. 2004. Global distribution of near-surface hydrogen on Mars. *Journal of Geophysical Research–Planets* 109:E09006.

Fenton LK, Geissler PE, Haberle RM. 2007. Global warming and climate forcing by recent albedo changes on Mars. *Nature* 446:646–649.

Fergason RL, Christensen PR. 2008. Formation and erosion of layered materials: Geologic and dust cycle history of eastern Arabia Terra, Mars. *Journal of Geophysical Research* 113:E12001.

Fergason RL, Christensen PR, Kieffer HH. 2006. High-resolution thermal inertia derived from the Thermal Emission Imaging System (THEMIS): Thermal model and application. *Journal of Geophysical Research* 111:E12004.

Flemings PB, Grotzinger JP. 1996. STRATA: Freeware for analyzing classic stratigraphy problems. *Geological Society of America Today* 6:1–7.

Forsythe RD, Zimbelman JR. 1995. A case for ancient evaporite basins on Mars. *Journal of Geophysical Research* 100(E3):5553–5563.

Geissler PE. 2005. Three decades of Martian surface changes. *Journal of Geophysical Research* 110:E02001.

Gendrin A, Mangold N, Bibring J-P, Langevin Y, Gondet B, Poulet F, Bonello G, Quantin C, Mustard JF, Arvidson R, Lemouelic S. 2005. Sulfates in Martian layered terrains: The OMEGA/Mars Express view. *Science* 307:1587–1591.

Goetz WEA. 2005. Indication of drier periods on Mars from the chemistry and mineralogy of atmospheric dust. *Nature* 436:62–65.

Goldspiel JM, Squyres SW. 1991. Ancient aqueous sedimentation on Mars. *Icarus* 89:392–410.

Gradstein FM, Ogg JG, Smith AG. 2004. *A Geologic Time Scale 2004*: Cambridge University Press, UK. 610 p.

Grant JA, Golombek MP, Grotzinger JP, Wilson SA, Watkins MM, Vasavada AR, Griffes JL, Parker TJ. 2011. The science process for selecting the landing site for the 2011 Mars Science Laboratory. *Planetary and Space Science* 59:1114–1127.

Grant JA, Irwin RP III, Grotzinger JP, Milliken RE, Tornabene LL, McEwen AS, Weitz CM, Squyres SW, Glotch TD, Thomson BJ. 2008. HiRISE imaging of impact megabreccia and sub-meter aqueous strata in Holden Crater, Mars. *Geology* 36:195–198.

Grant JA, Parker TJ. 2002. Drainage evolution in the Margaritifer Sinus region, Mars. *Journal of Geophysical Research* 107(E9):5066.

Grant JA, Wilson SA. 2011. Late alluvial fan formation in southern Margaritifer Terra, Mars. *Geophysical Research Letters* 38:L08201.

Greeley R, Guest JE. 1987. *Geologic Map of the Eastern Equatorial Region of Mars*: US Geological Survey Map I-1802-B.

Griffes JL, Arvidson RE, Poulet F, Gendrin A. 2007. Geologic and spectral mapping of etched terrain deposits in northern Meridiani Planum. *Journal of Geophysical Research* 112:E08S09.

Grotzinger J, Bell J, Herkenhoff K, Johnson J, Knoll A, McCartney E, McLennan S, Metz J, Moore J, Squyres S, Sullivan R, Ahronson O, Arvidson R, Joliff B, Golombek M, Lewis K, Parker T, Soderblom J. 2006. Sedimentary textures formed by aqueous processes, Erebus crater, Meridiani Planum, Mars. *Geology* 34:1085–1088.

Grotzinger JP. 2009. Beyond water on Mars. *Nature Geoscience* 2:231–233.

Grotzinger JP, Arvidson RE, Bell JF III, Calvin W, Clark BC, Fike DA, Golombek M, Greeley R, Haldemann A, Herkenhoff KE, Jolliff BL, Knoll AH, Malin M, McLennan SM, Parker T, Soderblom L, Sohl-Dickstein JN, Squyres SW, Tosca NJ, Watters WA. 2005. Stratigraphy and sedimentology of a dry to wet eolian depositional system, Burns formation, Meridiani Planum, Mars. *Earth and Planetary Science Letters* 240:11–72.

Hamilton VE, Christensen PR. 2005. Evidence for extensive, olivine-rich bedrock on Mars. *Geology* 33:433–436.

Hartmann WK. 1966. Martian cratering. *Icarus* 5:565–576.

Hartmann WK, Neukum G. 2001. Cratering chronology and the evolution of Mars. *Space Science Reviews* 96:165–194.

Hasenmueller EA, Bish DL. 2005. The hydration and dehydration of hydrous ferric iron sulfates. *In Lunar and Planetary Science Conference XXXVI*; Houston, Texas; Abstract 1164.

Hauber E, Gwinner K, Kleinhans M, Reiss D, Di Achille G, Ori G-G, Scholten F, Marinangeli L, Jaumann R, Neukum G. 2009. Sedimentary deposits in Xanthe Terra: Implications for the ancient climate on Mars. *Planetary and Space Science* 57:944–957.

Hayes A, Aharonson O, Callahan P, Elachi C, Gim Y, Kirk R, Lewis K, Lopes R, Lorenz R, Lunine J, Mitchell K, Mitri G, Stofan E, Wall S. 2008. Hydrocarbon lakes on Titan: Distribution and interaction with a porous regolith. *Geophysical Research Letters* 35:L09204.

Hayes AG, Grotzinger JP, Edgar LA, Squyres SW, Watters WA, Sohl-Dickstein J. 2011. Reconstruction of eolian bed forms and paleocurrents from cross-bedded strata at Victoria Crater, Meridiani Planum, Mars. *Journal of Geophysical Research–Planets* 116:E00F21.

Head JW. 1982. Lava flooding of ancient planetary crusts: Geometry, thickness, and volumes of flooded lunar impact basins. *The Moon and Planets* 26:61–88.

Head JW, Mustard JF, Kreslavsky MA, Milliken RE, Marchant DR. 2003. Recent ice ages on Mars. *Nature* 426:797–802.

Herkenhoff KE, Grotzinger J, Knoll AH, McLennan SM, Weitz C, Yingst A, Anderson R, Archinal BA, Arvidson RE, Barrett JM, Becker KJ, Bell JF, Budney C, Chapman MG, Cook D, Ehlmann B, Franklin B, Gaddis LR, Galuszka DM, Garcia PA, Geissler P, Hare TM, Howington-Kraus E, Johnson JR, Keszthelyi L, Kirk RL, Lanagan P, Lee EM, Leff C, Maki JN, Mullins KF, Parker TJ, Redding BL, Rosiek MR, Sims MH, Soderblom LA, Spanovich N, Springer R, Squyres SW, Stolper D, Sucharski RM, Sucharski T, Sullivan R, Torson JM. 2008a. Surface processes recorded by rocks and soils on Meridiani Planum, Mars: Microscopic Imager observations during Opportunity's first three extended missions. *Journal of Geophysical Research–Planets* 113:E12S32.

Herkenhoff KE, Golombek MP, Guinness RA, Johnson JB, Kusack AK, Richter L, Sullivan RJ, Gorevan S. 2008b. *In situ* observations of the physical properties of the Martian surface. *In* Bell J (Editor). *The Martian Surface*: Cambridge University Press, UK. p. 451–467.

Hodgson DM, Flint SS, Hodgetts D, Drinkwater NJ, Johannessen EP, Luthi SM. 2006. Stratigraphic evolution of fine-grained submarine fan systems, Tanqua Depocenter, Karoo Basin, South Africa. *Journal of Sedimentary Research* 76:20–40.

Hoffman N. 2000. White Mars: A new model for Mars' surface and atmosphere based on CO_2. *Icarus* 146:326–342.

Howard AD, Moore JM. 2011. Late Hesperian to early Amazonian mid-latitude Martian valleys: Evidence from Newton and Gorgonum basins. *Journal of Geophysical Research* 116:E05003.

Howard AD, Moore JM, Irwin RP. 2005. An intense terminal epoch of widespread fluvial activity on early Mars: 1. Valley network incision and associated deposits. *Journal of Geophysical Research–Planets* 110:E12S04.

Hurowitz JA, Fischer WW, Tosca NJ, Milliken R. 2010. Origin of acidic surface waters and the evolution of atmospheric chemistry on early Mars. *Nature Geoscience* 3:323–326.

Hynek BM, Phillips RJ. 2001. Evidence for extensive denudation of the Martian highlands. *Geology* 29:407–410.

Hynek BM, Phillips RJ. 2003. New data reveal mature, integrated drainage systems on Mars indicative of past precipitation. *Geology* 31:757–760.

Irwin RP III, Howard AD. 2002. Drainage basin evolution in Noachian Terra Cimmeria, Mars. *Journal of Geophysical Research–Planets* 107(E7):5056.

Irwin RP III, Howard AD, Craddock RA. 2008. Fluvial valley networks on Mars. *In* Rice SP, Roy AG, Rhoads BL (Editors). *River Confluences, Tributaries and the Fluvial Network*: John Wiley & Sons, Ltd, Chichester, UK. p. 419–451.

Ivanov BA. 2001. Mars/moon cratering rate ratio estimates. *Space Science Reviews* 96:87–104.

Jarrett RD, Malde HE. 1987. Paleodischarge of the late Pleistocene Bonneville flood, Snake River, Idaho, computed from new evidence. *Geological Society of America Bulletin* 99:127–134.

Jaumann R, Neukum G, Behnke T, Duxbury TC, Eichentopf K, Flohrer J, Gasselt SV, Giese B, Gwinner K, Hauber E, Hoffmann H, Hoffmeister A, Köhler U, Matz K-D, McCord TB, Mertens V, Oberst J, Pischel R, Reiss D, Ress E, Roatsch T, Saiger P, Scholten F, Schwarz G, Stephan K, Wählisch M, The HRSC Co-Investigator Team. 2007. The high-resolution stereo camera (HRSC) experiment on Mars Express: Instrument aspects and experiment conduct from interplanetary cruise through the nominal mission. *Planetary and Space Science* 55:928–952.

Jerolmack DJ, Mohrig D, Grotzinger JP, Fike DA, Watters WA. 2006. Spatial grain size sorting in eolian ripples and estimation of wind conditions on planetary surfaces: Application to Meridiani Planum, Mars. *Journal of Geophysical Research* 111:E12S02.

Jerolmack DJ, Mohrig D, Zuber MT, Byrne S. 2004. A minimum time for the formation of Holden Northeast Fan, Mars. *Geophysical Research Letters* 31:L21701.

Kahn R. 1985. The evolution of CO_2 on Mars. *Icarus* 62:175–190.

Kahn RA, Martin TZ, Zurek RW, Lee SW. 1992. The Martian dust cycle. *In* Kieffer HH, Jakosky BM, Snyder CW, Matthews MS (Editors). *Mars*: University of Arizona Press, Tucson. p. 1017–1053.

Kerber L, Head JA III. 2010. The age of the Medusae Fossae Formation: Evidence of Hesperian emplacement from crater morphology, stratigraphy, and ancient lava contacts. *Icarus* 206:669–684.

Knoll AH, Walter MR, Narbonne GM, Christie-Blick N. 2004. A new period for the geologic time scale. *Science* 305:621–622.

Komatsu G, Geissler PE, Strom RG, Singer RB. 1993. Stratigraphy and erosional landforms of layered deposits in Valles Marineris, Mars. *Journal of Geophysical Research* 98(E6):11105–11121.

Kraal ER, Van Dijk M, Postma G, Kleinhans MG. 2008. Martian stepped-delta formation by rapid water release. *Nature* 451:973–976.

Lamb MP, Dietrich WE, Aciego SM, Depaolo DJ, Manga M. 2008. Formation of Box Canyon, Idaho, by Megaflood: Implications for seepage erosion on Earth and Mars. *Science* 320:1067–1070.

Le Deit L, Le Mouelic S, Bourgeois O, Combe J-P, Mege D, Sotin C, Gendrin A, Hauber E, Mangold N, Bibring J-P. 2008. Ferric oxides in East Candor Chasma, Valles Marineris (Mars) inferred from analysis of OMEGA/Mars Express data: Identification and geological interpretation. *Journal of Geophysical Research* 113:E07001.

Lemmon MT, Wolff MJ, Smith MD, Clancy RT, Banfield D, Landis GA, Ghosh A, Smith PH, Spanovich N, Whitney B, Whelley P, Greeley R, Thompson S, Bell JF, Squyres SW. 2004. Atmospheric imaging results from the Mars Exploration Rovers: Spirit and Opportunity. *Science* 306:1753–1756.

Lewis K, Aharonson O. 2008. Geomorphic Aspects of the Eberswalde Delta and Potential MSL Traverses. http://marsoweb.nas.nasa.gov/landingsites. Accessed on September 2011.

Lewis KW. 2009. The rock record of Mars: Structure, sedimentology and stratigraphy [PhD dissertation]: California Institute of Technology, Pasadena, 136 p.

Lewis KW, Aharonson O. 2006. Stratigraphic analysis of the distributary fan in Eberswalde Crater using stereo imagery. *Journal of Geophysical Research–Planets* 111:E06001.

Lewis KW, Aharonson O, Grotzinger JP, Kirk RL, McEwen AS, Suer T-A.

2008a. Quasi-periodic bedding in the sedimentary rock record of Mars. *Science* 322:1532–1535.

Lewis KW, Aharonson O, Grotzinger JP, Squyres SW, Bell JF, Crumpler LS, Schmidt ME. 2008b. Structure and stratigraphy of Home Plate from the Spirit Mars Exploration Rover. *Journal of Geophysical Research–Planets* 113:E12S36.

Lindsay JF. 1972. Sedimentology of clastic rocks from the Fra Mauro region of the moon. *Journal of Sedimentary Petrology* 42:19–32.

Loizeau D, Mangold N, Poulet F, Bibring JP, Gendrin A, Ansan V, Gomez C, Gondet B, Langevin Y, Masson P, Neukum G. 2007. Phyllosilicates in the Mawrth Vallis region of Mars. *Journal of Geophysical Research–Planets* 112:E08S08.

Lucchitta BK, Ferguson HM. 1983. Chryse Basin channels: Low gradients and ponded flows: Proceedings of the 13th Lunar and Planetary Science Conference, Part 2. *Journal of Geophysical Research* 88(Supplement): A553–A568.

Lucchitta BK, Isbell NK, Howington-Kraus A. 1994. Topography of Valles Marineris: Implications for erosional and structural history. *Journal of Geophysical Research* 99(E2):3783–3798.

Lucchitta BK, McEwen AS, Clow GD, Geissler PE, Singer RB, Schutz RA, Squyres SW. 1992. The canyon system on Mars. *In* Kieffer HH, Jakosky BM, Snyder CW, Mathews MS (Editors). *Mars*: University of Arizona Press, Tucson. p. 453–492.

Malin MC. 1976. Nature and origin of intercrater plains on Mars. *In* Studies of the surface morphology of Mars [PhD dissertation]: California Institute of Technology, Pasadena. p. 101–176.

Malin MC. 1979. Mars: Evidence of indurated deposits of fine materials. *In* Second International Colloquium of Mars: National Aeronatics and Space Administration (NASA), Pasadena, California. NASA Conference Publication 2072.

Malin MC, Bell JF III, Cantor BA, Caplinger MA, Calvin WM, Clancy RT, Edgett KS, Edwards L, Haberle RM, James P, Lee SW, Ravine MA, Thomas PC, Wolff MJ. 2007. The Context Camera investigation onboard the Mars Reconnaissance Orbiter. *Journal of Geophysical Research* 112:E05S04.

Malin MC, Edgett KS. 2000. Sedimentary rocks of early Mars. *Science* 290:1927–1937.

Malin MC, Edgett KS. 2003. Evidence for persistent flow and aqueous sedimentation on early Mars. *Science* 302:1931–1934.

Malin MC, Edgett KS, Cantor BA, Caplinger MA, Danielson GE, Jensen EH, Ravine MA, Sandoval JL, Supulver KD. 2010. An overview of the 1985–2006 Mars Orbiter Camera. *Mars* 5:1–60.

Mandt KE, De Silva SL, Zimbelman JR, Crown DA. 2008. Origin of the Medusae Fossae Formation, Mars: Insights from a synoptic approach. *Journal of Geophysical Research–Planets* 113:E12001.

Mangold N, Gendrin A, Gondet B, Lemouelic S, Quantin C, Ansan V, Bibring J-P, Langevin Y, Masson P, Neukum G. 2008. Spectral and geological study of the sulfate-rich region of West Candor Chasma, Mars. *Icarus* 194:519–543.

Mangold N, Poulet F, Mustard JF, Bibring J-P, Gondet B, Langevin Y, Ansan V, Masson P, Fassett C, Head JW, Hoffmann J, Neukum G. 2007. Mineralogy of the Nili Fossae region with OMEGA/Mars Express data: 2. Aqueous alteration of the crust. *Journal of Geophysical Research* 112:E08S04.

Mangold N, Quantin C, Ansan VR, Delacourt C, Allemand P. 2004. Evidence for precipitation on Mars from dendritic valleys in the Valles Marineris area. *Science* 305:78–81.

McCauley JF. 1974. White rock: A Martian enigma: Mars as viewed by Mariner 9. National Aeronautics and Space Administration (NASA) special publication 329. p. 170–171.

McCauley JF, Carr MH, Cutts JA, Hartmann WK, Masurskey H, Milton DJ, Sharp RP, Wilhelms DE. 1972. Preliminary Mariner 9 report on the geology of Mars. *Icarus* 17:289–327.

McCollom TM, Hynek BM. 2005. A volcanic environment for bedrock diagenesis at Meridiani Planum on Mars. *Nature* 438:1129–1131.

McEwen AS, Malin MC, Carr MH, Hartmann WK. 1999. Voluminous volcanism on early Mars revealed in Valles Marineris. *Nature* 397:584–586.

McEwen AS, and 68 others. 2010. The High Resolution Imaging Science Experiment (HiRISE) during MRO's Primary Science Phase (PSP). *Icarus* 205:2–37.

McGill GE, Wise DU. 1972. Regional variations in degradation and density of Martian craters. *Journal of Geophysical Research* 77:2433–2441.

McKay CP, Nedell SS. 1988. Are there carbonate deposits in the Valles Marineris, Mars? *Icarus* 73:142–148.

McKay CP, Stoker CR. 1989. The early environment and its evolution on Mars: Implication for life. *Reviews of Geophysics* 27:189–214.

McLaughlin DB. 1954. Volcanism and aeolian deposition on Mars. *Geological Society of America Bulletin* 65:715–718.

McLennan S, Grotzinger JP. 2008. The sedimentary rock cycle of Mars. *In* Bell J (Editor). *The Martian Surface*: Cambridge University Press, UK. p. 541–577.

McLennan SM, Bell JF III, Calvin WM, Christensen PR, Clarke BC, De Souza PA, Farmer J, Farrand WH, Fike DA, Gellert R, Ghosh A, Glotch TD, Grotzinger JP, Hahn BC, Herkenhoff KE, Hurowitz JA, Johnson JR, Johnson SS, Joliff B, Klingelhofer G, Knoll AH, Learner Z, Malin MC, McSween HY, Pocock J, Ruff SW, Soderblom LA, Squyres SW, Tosca NJ, Watters WA, Wyatt MB, Yen A. 2005. Provenance and diagenesis of the evaporite-bearing Burns formation, Meridiani Planum, Mars. *Earth and Planetary Science Letters* 240:95–121.

McSween HY, Hahn BC, Viviano CE, Moersch J. 2011. Onset of metamorphism in the Martian crust. *In* 42nd Lunar and Planetary Science Conference; Houston, Texas; Abstract 1064.

Metz J, Grotzinger JP, Okubo CH, Milliken RE. 2010. Thin-skinned deformation of sedimentary rocks in Valles Marineris, Mars. *Journal of Geophysical Research* 115:E11004.

Metz JM, Grotzinger JP, Mohrig D, Milliken R, Prather B, Pirmez C, McEwen AS, Weitz CM. 2009a. Sublacustrine depositional fans in southwest Melas Chasma. *Journal of Geophysical Research–Planets* 114:E10002.

Metz JM, Grotzinger J, Rubin DM, Lewis KW, Squyres SW, Bell JF III. 2009b. Sulfate-rich eolian and wet interdune deposits, Erebus Crater, Meridiani Planum, Mars. *Journal of Sedimentary Research* 79:247–264.

Michalski J, Bibring J-P, Poulet F, Loizeau D, Mangold N, Noe Dobrea EZ, Bishop JL, Wray JJ, McKeown NK, Parente M, Hauber E, Altieri F, Carrozzo FG, Niles PB. 2010. The Mawrth Vallis Region of Mars: A potential landing site for the Mars Science Laboratory (MSL) Mission. *Astrobiology* 10:687–703.

Michalski J, Noe Dobrea EZ. 2007. Evidence for a sedimentary origin of clay minerals in the Mawrth Vallis region, Mars. *Geology* 35:951–954.

Michalski JR, Niles PB. 2010. Deep crustal carbonate rocks exposed by meteor impact on Mars. *Nature Geoscience* 3:751–755.

Milazzo MP, Keszthelyi LP, Jaeger WL, Rosiek M, Mattson S, Verba C, Beyer RA, Geissler PE, McEwen AS; the HiRISE Team. 2009. Discovery of columnar jointing on Mars. *Geology* 37:171–174.

Milliken RE, Bish DL. 2010. Sources and sinks of clay minerals on Mars. *Philosophical Magazine* 90:2293–2308.

Milliken RE, Bristow TF, Bish DL. 2011. Diagenesis of clay minerals on Mars and implications for the Mars Science Laboratory rover. *In* 42nd Lunar and Planetary Science Conference; Houston, Texas; Abstract 2230.

Milliken RE, Fischer WW, Hurowitz JA. 2009. Missing salts on early Mars. *Geophysical Research Letters* 36:L11202.

Milliken RE, Grotzinger JP, Thomson BJ. 2010. Paleoclimate of Mars as captured by the stratigraphic record in Gale Crater. *Geophysical Research Letters* 37:L04201.

Milliken RE, Mustard JF, Poulet F, Jouglet D, Bibring J-P, Gondet B, Langevin Y. 2007. Hydration state of the Martian surface as seen by Mars Express OMEGA: 2. H_2O content of the surface. *Journal of Geophysical Research* 112:E08S07.

Milliken RE, Swayze GA, Arvidson RE, Bishop JL, Clark RN, Ehlmann BL, Green RO, Grotzinger JP, Morris RV, Murchie SL, Mustard JF, Weitz C. 2008. Opaline silica in young deposits on Mars. *Geology* 36:847–850.

Montgomery DR, Gillespie A. 2005. Formation of Martian outflow channels by catastrophic dewatering of evaporite deposits. *Geology* 33:625–628.

Montgomery DR, Som SM, Jackson MPA, Schreiber BC, Gillespie A, Adams JB. 2009. Continental-scale salt tectonics on Mars and the origin of Valles Marineris and associated outflow channels. *Geological Society of America Bulletin* 121:117–133.

Moore JM, Howard AD. 2005. Large alluvial fans on Mars. *Journal of Geophysical Research* 110:E04005.

Morris RV, Ruff SW, Gellert R, Ming DW, Arvidson RE, Clark BC, Golden DC,

Siebach K, Klingelhofer G, Schroder C, Fleischer I, Yen AS, Squyres SW. 2010. Identification of carbonate-rich outcrops on Mars by the Spirit Rover. *Science* 329:421–424.

Murchie S, Arvidson RE, Bedini P, Beisser K, Bibring J-P, Bishop J, Boldt J, Cavender P, Choo T, Clancy RT, Darlington EH, Des Marais D, Espiritu R, Fort D, Green R, Guinness E, Hayes J, Hash C, Heffernan K, Hemmler J, Heyler G, Humm D, Hutcheson J, Izenberg N, Lee R, Lees J, Lohr D, Malaret E, Martin T, McGovern JA, McGuire P, Morris RV, Mustard J, Pelkey S, Rhodes E, Robinson M, Roush T, Schaefer E, Seagrave G, Seelos FP, Silverglate P, Slavney S, Smith M, Shyong W-J, Strohbehn K, Taylor H, Thompson P, Tossman B, Wirzburger M, Wolff M. 2007. Compact Reconnaissance Imaging Spectrometer for Mars (CRISM) on Mars Reconnaissance Orbiter (MRO). *Journal of Geophysical Research* 112:E05503.

Murchie S, Roach L, Seelos F, Milliken RE, Mustard J, Arvidson R, Wiseman S, Lichtenberg K, Andrews-Hanna J, Bishop J, Bibring J-P, Parente M, Morris R. 2009a. Evidence for the origin of layered deposits in Candor Chasma, Mars, from mineral composition and hydrologic modeling. *Journal of Geophysical Research* 114:E00D05.

Murchie SL, Mustard JF, Ehlmann BL, Milliken RE, Bishop JL, McKeown NK, Dobrea EZN, Seelos FP, Buczkowski DL, Wiseman SM, Arvidson RE, Wray JJ, Swayze G, Clark RN, Marais DJD, McEwen AS, Bibring JP. 2009b. A synthesis of Martian aqueous mineralogy after 1 Mars year of observations from the Mars Reconnaissance Orbiter. *Journal of Geophysical Research–Planets* 114:E00D06.

Mustard JF, Cooper CD, Rifkin MK. 2001. Evidence for recent climate change on Mars from the identification of youthful near-surface ground ice. *Nature* 412:411–414.

Mustard JF, Ehlmann BL, Murchie S, Poulet F, Mangold N, Head JW, Bibring J-P, Roach LH. 2009. Composition, morphology, and stratigraphy of Noachian crust around the Isidis Basin. *Journal of Geophysical Research–Planets* 114:E00D12.

Mustard JF, Murchie SL, Pelkey SM, Ehlmann BL, Milliken RE, Grant JA, Bibring J-P, Poulet F, Bishop J, Dobrea EN, Roach L, Seelos F, Arvidson RE, Wiseman S, Green R, Hash C, Humm D, Malaret E, McGovern JA, Seelos K, Clancy T, Clark R, Des Marais D, Izenberg N, Knudson A, Langevin Y, Martin T, McGuire P, Morris R, Robinson M, Roush T, Smith M, Swayze G, Taylor H, Titus T, Wolff M. 2008. Hydrated silicate minerals on Mars observed by the Mars Reconnaissance Orbiter CRISM instrument. *Nature* 454:305–309.

Mustard JF, Poulet F, Head JW, Mangold N, Bibring J-P, Pelkey S, Fassett C, Langevin Y, Neukum G. 2007. Mineralogy of the Nili Fossae region with OMEGA/Mars Express data: 1. Ancient impact melt in the Isidis Basin and implications for the transition from the Noachian to Hesperian. *Journal of Geophysical Research–Planets* 112:E08S03.

Nedell SS, Squyres SW, Andersen DW. 1987. Origin and evolution of the layered deposits in the Valles Marineris, Mars. *Icarus* 70:409–441.

Nimmo F, Tanaka K. 2005. Early crustal evolution of Mars. *Annual Review of Earth and Planetary Sciences* 33:133–161.

Noe Dobrea EZ, Bishop JL, McKeown NK, Fu R, Rossi CM, Michalski JR, Heinlein C, Hanus V, Poulet F, Mustard JF, Murchie S, McEwen AS, Swayze G, Bibring J-P, Malaret E, Hash C. 2010. Mineralogy and stratigraphy of phyllosilicate-bearing and dark mantling units in the greater Mawrth Vallis/ west Arabia Terra area: Constraints on geological origin. *Journal of Geophysical Research* 115:E00D19.

Okubo CH. 2010. Structural geology of Amazonian-aged layered sedimentary deposits in southwest Candor Chasma, Mars. *Icarus* 207:210–225.

Okubo CH, Lewis KW, McEwen AS, Kirk RL. 2008. Relative age of interior layered deposits in southwest Candor Chasma based on high-resolution structural mapping. *Journal of Geophysical Research–Planets* 113:E12002.

Okubo CH, McEwen AS. 2007. Fracture-controlled paleo-fluid flow in Candor Chasma, Mars. *Science* 315:983–985.

Okubo CH, Schultz RA, Chan MA, Komatsu G, Team H. 2009. Deformation band clusters on Mars and implications for subsurface fluid flow. *Geological Society of America Bulletin* 121:474–482.

Öpik EJ. 1966. The Martian surface. *Science* 153:255–265.

Osterloo MM, Hamilton VE, Bandfield JL, Glotch TD, Baldridge AM, Christensen PR, Tornabene LL, Anderson FS. 2008. Chloride-bearing materials in the Southern Highlands of Mars. *Science* 319:1651–1654.

Peterson C. 1981. A secondary origin for the central plateau of Hebes Chasma. *In* Proceedings of Lunar and Planetary Science 12B: Pergamon, New York. p. 1459–1471.

Pieri DC. 1980. Martian valleys: Morphology, distribution, age, and origin. *Science* 210:895–897.

Pollack J, Colburn D, Kahn RA, Hunter J, Van Camp W, Carlston CE, Wolf MR. 1977. Properties of aerosols in the Martian atmosphere, as inferred from Viking Landing imaging data. *Journal of Geophysical Research* 82:4479–4496.

Pondrelli M, Rossi AP, Marinangeli L, Hauber E, Gwinner K, Baliva A, Di Lorenzo S. 2008. Evolution and depositional environments of the Eberswalde Fan Delta, Mars. *Icarus* 197:429–451.

Poulet F, Bibring J-P, Mustard JF, Gendrin A, Mangold N, Langevin Y, Arvidson RE, Gondet B, Gomez C; the OMEGA Team. 2005. Phyllosilicates on Mars and implications for early Martian climate. *Nature* 438:623–627.

Poulet F, Mangold N, Loizeau D, Bibring J-P, Langevin Y, Michalski J, Gondet B. 2008. Abundance of minerals in the phyllosilicate-rich units on Mars. *Astronomy and Astrophysics* 487:L41–L44.

Putzig NE, Mellon MT, Kretke KA, Arvidson RE. 2005. Global thermal inertia and surface properties of Mars from the MGS mapping mission. *Icarus* 173:325–341.

Quantin C, Allemand P, Mangold N, Dromart G, Delacourt C. 2005. Fluvial and lacustrine activity on layered deposits in Melas Chasma, Valles Marineris, Mars. *Journal of Geophysical Research* 110:E12S19.

Rainbird RH, Hadlari T, Aspler LB, Donaldson JA, Lecheminant AN, Peterson TD. 2003. Sequence stratigraphy and evolution of the Paleoproterozoic intracontinental Baker Lake and Thelon basins, western Churchill Province, Nunavut, Canada. *Precambrian Research* 125:21–53.

Roach LH, Mustard JF, Swayze GA, Milliken RE, Bishop JL, Murchie S, Lichtenberg K. 2010. Hydrated mineral stratigraphy of Ius Chasma, Valles Marineris. *Icarus* 206:253–268.

Rossi AP, Neukum G, Pondrelli M, Van Gasselt S, Zegers T, Hauber E, Chicarro A, Foing B. 2008. Large-scale spring deposits on Mars? *Journal of Geophysical Research* 113:E08016.

Rubin DM, Hesp PA. 2009. Multiple origins of linear dunes on Earth and Titan. *Nature Geoscience* 2:653–658.

Ruff SW, Christensen PR, Clark RN, Kieffer HH, Malin MC, Bandfield JL, Jakosky BM, Lane MD, Mellon MT, Presley MA. 2001. Mars' "White Rock" feature lacks evidence of an aqueous origin: Results from Mars Global Surveyor. *Journal of Geophysical Research* 106(E10):23921–23927.

Sagan C, Pollack JB. 1969. Windblown dust on Mars. *Nature* 223:791–794.

Schon S, Head JW III, Milliken RE. 2009. A recent ice age on Mars: Evidence for climate oscillations from regional layering in mid-latitude mantling deposits. *Geophysical Research Letters* 36:L15202.

Schultz PH, Lutz AB. 1988. Polar wandering of Mars. *Icarus* 73:91–141.

Schultz RA. 1998. Multiple-process origin of Valles Marineris basins and troughs, Mars. *Planetary and Space Science* 46:827–829.

Scott DH, Tanaka KL. 1982. Ignimbrites of Amazonis Planitia Region of Mars. *Journal of Geophysical Research* 87(B2):1179–1190.

Scott DH, Tanaka KL. 1986. *Geologic Map of the Western Equatorial Region of Mars*: U.S. Geological Survey, Flagstaff, AZ.

Sharp RP. 1973. Mars: Troughed terrain. *Journal of Geophysical Research* 78:4063–4072.

Sharp RP, Malin MC. 1975. Channels on Mars. *Geological Society of America Bulletin* 86:593–609.

Spencer JR, Croft SK. 1986. Valles Marineris as karst: Feasibility and implications for Martian atmospheric evolution. *In* Carr M, James P, Leovy C, Pepin R, Pollack J (Editors). *Workshop on the Evolution of the Martian Atmosphere*: Honolulu, Hawaii, Lunar and Planetary Institute, p. 40–41.

Spencer JR, Fanale FP. 1990. New models for the origin of Valles Marineris closed depressions. *Journal of Geophysical Research* 95(B9):14301–14313.

Squyres SW, Aharonson O, Clark BC, Cohen BA, Crumpler L, De Souza PA, Farrand WH, Gellert R, Grant J, Grotzinger JP, Haldemann AFC, Johnson JR, Klingelhofer G, Lewis KW, Li R, McCoy T, McEwen AS, McSween HY, Ming DW, Moore JM, Morris RV, Parker TJ, Rice JW, Ruff S, Schmidt M,

Schroder C, Soderblom LA, Yen A. 2007. Pyroclastic activity at Home Plate in Gusev Crater, Mars. *Science* 316:738–742.

Squyres SW, Carr MH. 1986. Geomorphic evidence for the distribution of ground ice on Mars. *Science* 231:249–252.

Squyres SW, Grotzinger JP, Arvidson RE, Bell JF, Calvin W, Christensen PR, Clark BC, Crisp JA, Farrand WH, Herkenhoff KE, Johnson JR, Klingelhofer G, Knoll AH, McLennan SM, McSween HY, Morris RV, Rice JW, Rieder R, Soderblom LA. 2004. In situ evidence for an ancient aqueous environment at Meridiani Planum, Mars. *Science* 306:1709–1714.

Squyres SW, Knoll AH, Arvidson RE, Ashley JW, Bell JF, Calvin WM, Christensen PR, Clark BC, Cohen BA, De Souza PA, Edgar L, Farrand WH, Fleischer I, Gellert R, Golombek MP, Grant J, Grotzinger J, Hayes A, Herkenhoff KE, Johnson JR, Jolliff B, Klingelhofer G, Knudson A, Li R, McCoy TJ, McLennan SM, Ming DW, Mittlefehldt DW, Morris RV, Rice JW, Schroder C, Sullivan RJ, Yen A, Yingst RA. 2009. Exploration of Victoria Crater by the Mars Rover Opportunity. *Science* 324:1058–1061.

Srodon J. 1999. Nature of mixed-layer clays and mechanisms of their formation and alteration. *Annual Review of Earth and Planetary Sciences* 27:19–53.

Sullivan REA. 2008. Wind-driven particle mobility on Mars: Insight from Mars Exploration Rover observations at "El Dorado" and surroundings in Gusev Crater. *Journal of Geophysical Research* 113:E06S07.

Sumner D. 2010. Physical outcrop characteristics of the Mawrth candidate landing site and the potential role of impacts in shaping stratigraphy. http://marsoweb.nas.nasa.gov/landingsites. Accessed September 2011.

Szwast MA, Richardson MI, Vasavada AR. 2006. Surface dust redistribution on Mars as observed by the Mars Global Surveyor and Viking orbiters. *Journal of Geophysical Research* 111:E11008.

Tanaka KL, Davis PA. 1988. Tectonic history of the Syria Planum province of Mars. *Journal of Geophysical Research* 93(B12):14893–14917.

Tanaka KL, Hartmann WK. 2008. The Planetary Time Scale. *In* Ogg JG, Ogg G, Gradstein FM (Editors). *The Concise Geologic Time Scale*: Cambridge University Press, Cambridge University Press, UK. p. 177.

Taylor SR, McLennan SM. 2009. Mars: Crustal composition and evolution. *In* Planetary Crusts: Their Composition, Origin and Evolution: Cambridge University Press, UK. p. 141–180.

Thomas P. 1984. Martian intracrater splotches: Occurrence, morphology, and colors. *Icarus* 57:205–227.

Thomson BJ, Bridges NT, Milliken RE, Baldridge AM, Hook SJ, Crowley JK, Marion GM, De Souza Filho CR, Brown AJ, Weitz CM. 2011. Constraints on the origin and evolution of the layered mound in Gale Crater, Mars using Mars Reconnaissance Orbiter data. *Icarus*, 214:413–432.

Tosca NJ, Knoll AH. 2009. Juvenile chemical sediments and the long term persistence of water at the surface of Mars. *Earth and Planetary Science Letters* 286:379–386.

Vaniman DT, Bish DL, Chipera SJ, Fialips CI, Carey JW, Feldman WC. 2004. Magnesium sulphate salts and the history of water on Mars. *Nature* 431:663–665.

Vaniman DT, Chipera SJ. 2006. Transformations of Mg- and Ca-sulfate hydrates in Mars regolith. *American Mineralogist* 91:1628–1642.

Viseras C, Calvache ML, Soria JM, Fernandez J. 2003. Differential features of alluvial fans controlled by tectonic or eustatic accommodation space. Examples from the Beltic Cordillera, Spain. *Geomorphology* 50:181–202.

Walker R, James NP. 1992. *Facies Models*: Geological Association of Canada, St. Johns. 276 p.

Ware MJ, Hiscott RN. 1985. Sedimentology of Proterozoic cratonic sheet sandstones of the eastern Canadian Shield: Sims Formation, Labrador, Canada. *Precambrian Research* 30:1–26.

Watters TR, Campbell B, Carter L, Leuschen CJ, Plaut JJ, Picardi G, Orosei R, Safaeinili A, Clifford SM, Farrell WM, Ivanov AB, Phillips RJ, Stofan ER. 2007. Radar sounding of the Medusae Fossae Formation Mars: Equatorial ice or dry, low-density deposits? *Science* 318:1125–1128.

Weitz CM, Milliken RE, Grant JA, McEwen AS, Williams RME, Bishop JL. 2008. Light-toned strata and inverted channels adjacent to Juventae and Ganges chasmata, Mars. *Geophysical Research Letters* 35:L19202.

Weitz CM, Milliken RE, Grant JA, McEwen AS, Williams RME, Bishop JL, Thomson BJ. 2010. Mars Reconnaissance Orbiter observations of light-toned layered deposits and associated fluvial landforms on the plateaus adjacent to Valles Marineris. *Icarus* 205:73–102.

Williams SH, Zimbelman JR. 1994. 'White Rock': An eroded Martian lacustrine deposit(?). *Geology* 22:107–110.

Wilson L, Head JW. 1994. Mars: Review and analysis of volcanic eruption theory and relationships to observed landforms. *Review of Geophysics* 32:221–263.

Wilson SA, Howard AD, Moore JM, Grant JA. 2007. Geomorphic and stratigraphic analysis of Crater Terby and layered deposits north of Hellas Basin, Mars. *Journal of Geophysical Research* 112:E08009.

Wood LJ. 2006. Quantitative geomorphology of the Mars Eberswalde Delta. *Geological Society of America Bulletin* 118:557–566.

Wray JJ, Milliken RE, Dundas CM, Swayze GA, Andrews-Hanna JC, Baldridge AM, Chojnacki M, Bishop JL, Ehlmann BL, Murchie SL, Clark RN, Seelos FP, Tornabene LL, Squyres SW. 2011. Columbus crater and other possible groundwater-fed paleolakes of Terra Sirenum, Mars. *Journal of Geophysical Research* 116:E01001.

AN ATLAS OF MARS SEDIMENTARY ROCKS AS SEEN BY HIRISE

ROSS A. BEYER
Sagan Center at the SETI Institute and NASA Ames Research Center
e-mail: Ross.A.Beyer@nasa.gov

KATHRYN M. STACK AND JENNIFER L. GRIFFES
Department of Geologic and Planetary Sciences, California Institute of Technology

RALPH E. MILLIKEN
University of Notre Dame

KEN E. HERKENHOFF
United States Geological Survey

SHANE BYRNE
Lunar and Planetary Laboratory, The University of Arizona

JOHN W. HOLT
University of Texas at Austin

AND

JOHN P. GROTZINGER
Department of Geologic and Planetary Sciences, California Institute of Technology

"*There are no foreign lands. It is only the traveler who is foreign*"
—Robert Louis Stevenson

INTRODUCTION

Images of distant and unknown places have long stimulated the imaginations of both explorers and scientists. The atlas of photographs collected during the Hayden (1872) expedition to the Yellowstone region was essential to its successful advocacy and selection in 1872 as America's first national park. Photographer William Henry Jackson of the Hayden expedition captured the public's imagination and support, returning home with a treasure of images that confirmed the existence of western landmarks previously regarded as glorified myths: the Grand Tetons, Old Faithful, and strange pools of boiling hot mud. Fifty years later, photographer Ansel Adams began his long legacy of providing the public with compilations of iconic images of natural wonders that many only see in prints.

Photography in space has provided its own bounty. Who can forget the first image of Earthrise taken by astronaut William Anders in 1968 from *Apollo 8*; the solemnity of the first photos of the surface of the Moon from the *Apollo 11* astronauts; and the startling discovery of the tallest mountain in the solar system (Olympus Mons) on the surface of Mars in images sent from *Mariner 9*? The images from *Mariner 9* also allowed for a game-changing discovery. Earlier, based on very limited *Mariner 4* data that covered less than 10% of the planet's surface, Chapman et al. (1968) speculated that "If substantial aqueous erosion features—such as river valleys—were produced during earlier epochs of Mars, we should not expect any trace of them to be visible on the *Mariner IV* photographs unless they were of greater extent than typical features on Earth." Mapping a much greater part of the planet, *Mariner 9* provided the first evidence of such spectacular features and ushered in the modern era of understanding Mars in the context of its aqueous history. Our understanding of the ways in which terrestrial planetary surfaces evolve through time was transformed by the notion that Mars may have been warmer or wetter at some point in the past.

Almost three decades later, this first evidence for aqueous erosion of bedrock on Mars was finally matched by evidence for deposition of aqueous sedimentary materials (Malin and Edgett 2000). The identification of "layered terrains" thought to be of sedimentary origin on Mars is regarded as a major discovery. Observations of erosion in one place require, naturally, that deposition of the eroded materials occurs elsewhere, so this discovery was not so much surprising as exciting. Unlike erosion, deposition records a time series of past events and can thus be analyzed as a record of environmental history. Furthermore, the state of preservation of these outcrops suggests a past cycle of erosion and deposition, followed by erosion of the deposited sediments. Evaluation and understanding of the Martian "rock cycle," which is very different from Earth's, are some of the major objectives of the next decade of Mars research, and the study of sedimentary rocks is central to this theme (McLennan and Grotzinger 2008, Grotzinger et al. 2011).

The most recent image data of the surface of Mars show that there are at least several types of distinct layered deposits that are consistent with a sedimentary origin. In some cases, they form vast, laterally extensive sheets that cover the ancient cratered terrains and fill topographic depressions such as impact craters. The spectacular images in this atlas present just a tiny fraction of the material that exists for further study and examination. This atlas is not intended to capture all of this diversity, but rather to illustrate some of the most important

Sedimentary Geology of Mars
SEPM Special Publication No. 102, Copyright © 2012
SEPM (Society for Sedimentary Geology), Print ISBN 978-1-56576-312-8, CD/DVD ISBN 978-1-56576-313-5 p. 49–95.

TABLE 1.—Classification of sedimentary deposits on Mars.

Category	Subcategory	Example locations
Underfilled crater interior	Preserved depositional system	Eberswalde crater delta Holden crater alluvial deposits Xanthe Terra terraced fans Melas Chasma sublacustrine fans
	No observed depositional system	Columbus crater deposits
Overfilled crater interior		Gale crater layered mound Terby crater layered mesas Galle crater layers Becquerel crater layered mound
Chasm/canyon	Undeformed layered deposits	Ophir Chasma Juventae Chasma Ganges Mensa Melas Chasma
	Fractured and faulted layered deposits	West Candor Chasma
Plains-covering deposits	Plains surrounding Valles Marineris	West of Juventae Chasma Near Ganges Chasma South of Ius and Melas Chasmata
	Stratified plains	Meridiani Planum layers
Ancient terrain		Mawrth Vallis Nili Fossae
	Central uplift	Oudemans crater central uplift
Polar deposits		North polar layered deposits and basal unit South polar layered deposits

end members defined by representative exposures, some of which might be viewed as key "reference" sections (see Grotzinger and Milliken this volume). Our selections of sedimentary deposits fall into several very broad categories based on the environment in which these outcrops are found today (Table 1). These categories include: (1) underfilled crater interiors, (2) overfilled crater interiors, (3) chasm/canyon units, (4) plains-covering deposits, (5) very ancient terrain, and (6) polar deposits.

Determining the ages of rocks and units on Mars is problematic due to our inability to age-date samples. However, Scott and Carr (1978) were the first to establish three general time-stratigraphic periods, Noachian, Hesperian, and Amazonian, which were further refined into eight epochs and assigned relative ages based on the traditional principles of superposition and intersection, as well as the size-frequency distribution of impact craters (Tanaka 1986, Tanaka et al. 1992). The connection to absolute ages was further refined by Hartmann and Neukum (2001): the Noachian Period stretches from the beginning of Martian time to 3.6 ± 0.1 Ga, the Hesperian Period from 3.6 to 3.1 ± 0.6 Ga, and the Amazonian from 3.1 Ga to the present (Hartmann 2005, Fassett and Head 2008, Tanaka and Hartmann 2008). More details on the Martian time scale can be found in Grotzinger and Milliken (this volume).

Underfilled crater interiors contain sediments that are interpreted not to have exceeded the volume of their host crater. This category is further subdivided into deposits that exhibit distinct fan morphologies (e.g., Eberswalde crater delta, Holden crater alluvial deposits, and the Xanthe Terra terraced fans), and those that do not (e.g., Columbus crater deposits). Although the putative sublacustrine fans in Melas Chasma (Metz et al. 2009) are within the Valles Marineris and not a crater, they are better grouped in this category, which includes other fan morphologies, because the fluvial system that created the Melas deposits is well expressed.

Overfilled crater interiors are those with stratified deposits that reside primarily within a crater but rise close to or above the current elevation of the crater's rim. While there is no straightforward explanation for these deposits, they may be remnants of once-larger deposits that both filled the craters and extended beyond to blanket the surrounding plateaus. Many of these deposits occur as tall mounds near the center of the crater, surrounded by a lower elevation "moat." The specific erosional processes that would have scoured away the surrounding deposits and denuded nearly down to the crater floor while allowing an isolated mound to be preserved are currently unknown. The Gale crater mound, the layered deposits in Terby crater, the Henry crater layered mound, and the stratified deposits in Galle crater all exhibit this remnant high-standing topography and may provide evidence for formerly extensive deposits that had greater lateral extent than that currently preserved.

Chasmata and canyons on Mars provide topographic lows and often closed basins in which sediments can accumulate, and their deposits are often similar to those found in craters. The Martian chasmata of Valles Marineris are very large and extensive, providing a vast setting to test a variety of sedimentary hypotheses. Tectonics have played a role in the geologic history here, from the activity that may have formed the chasmata themselves, to other tectonic forces that caused deformation of the sedimentary deposits observed within those chasmata. We divide the sedimentary deposits in the Valles Marineris into those that have not been tectonically disrupted (deposits in Ophir, Melas, Juventae, and Ganges Chasmata) and those that have, like the west Candor Chasma deformed layered deposits (Metz et al. 2010, Okubo 2010).

Plains-covering deposits include complex sequences of stratified units that vary widely from place to place over the surface of Mars. Volcanism, sedimentation, and fluvial action have all played a role in the formation of these terrains. Unique features among these terrain units are the plateaus surrounding the large canyons of Valles Marineris, which exhibit layered exposures not seen elsewhere (Milliken et al. 2008, Weitz et al. 2010). These sediments might be the evidence of fluvial systems that either existed prior to the chasma openings or those that acted as overland flows for systems (e.g., Mangold et al. 2004) that ultimately drained into these giant cavities.

Other very ancient terrains consist of the oldest identifiable portions of the Martian crust (from the Martian Noachian Period), which, in many cases, have been altered by a variety of chemical and physical processes. This ancient crust contains numerous deposits with clay minerals (e.g., Bibring et al. 2006, Murchie et al. 2009b), the presence of which indicates water–rock interaction either on the surface or in the subsurface. Although the specific timing and duration of this alteration are unknown, our knowledge of the diversity of clay minerals on Mars has increased dramatically over the past several years, and it is beginning to rival that of Earth. The ancient terrain consists of layered rocks, many of which are presumably sedimentary in origin, but it also hosts many deposits that lack clear stratification. These deposits may represent primary crust, thick deposits of crater ejecta, or regions that have been heavily altered by hydrothermal processes. The ancient crust on Mars is complex in both its morphology and mineralogy, but it is clear that sedimentary processes are recorded in the Martian rock record to at least 4 billion years before present, and likely longer. This makes Mars a particularly interesting and possibly unique place for understanding the role of aqueous and sedimentary processes in the earliest history of the Solar System.

In addition to areally extensive sequences of stratified deposits, very ancient terrains on Mars are also characterized by craters in which central uplifts occasionally exhibit large, layered structural blocks brought up from depth during the crater formation process. These deep layers may be a way to sample Martian geologic history from the distant past that would otherwise remain buried. In the Southern Highlands, many of these central uplifts contain clay minerals that are indicative of aqueous alteration of primary crust, possibly as a result of deep alteration within the ancient crust (Murchie et al. 2009a).

Polar deposits at both the north and south Martian poles exhibit a variety of ice and dust layering sequences that likely encode the recent variability of Martian climate. Polar layered deposits were first recognized in *Mariner 9* images (Murray et al. 1972) and motivated theoretical studies of climate changes on Mars (Ward 1974, 1979; Bills 1990; Touma and Wisdom 1993). The tilt of the rotational axis and eccentricity of the orbits of both Earth and Mars have undergone periodic variations due to gravitational interactions with the other planets, but the amplitude of these variations is much greater on Mars. Because these orbital/axial changes are known to cause climate changes (ice ages) on Earth, they are expected to cause even greater climate variability on Mars (e.g., Head et al. 2003). More recent observations have shown that the polar layered deposits are ice-rich, with less sediment than previously thought (Byrne 2009).

This atlas is largely based on image data obtained by the High Resolution Imaging Science Experiment (HiRISE; McEwen et al. 2007) on board the *Mars Reconnaissance Orbiter* (MRO). This camera has allowed for a more detailed examination of the Martian surface than previous imagers, providing images at 25 cm/pixel scale, with a swath width of about 6 km. HiRISE images and digital terrain models (DTMs) can be obtained from http://hirise.lpl.arizona.edu. In addition, other image data for Mars can be found at National Aeronautics and Space Administration's (NASA) Planetary Data System (PDS) at http://pds.jpl.nasa.gov. This atlas includes some images from the Context Camera (CTX; Malin et al. 2007) on MRO. CTX acquires grayscale images of Mars with a spatial scale of approximately 6 m/pixel, and a swath width of about 30 km. There are also data from the Shallow Radar (SHARAD; Seu et al. 2007) instrument on MRO. This atlas discusses mineralogical data from hyperspectral mapping instruments such as the Compact Reconnaissance Imaging Spectrometer for Mars (CRISM; Murchie et al. 2007) and the Observatoire pour la Minéralogie, l'Eau, les Glaces et l'Activité (OMEGA; Bibring et al. 2004). These hyperspectral imaging systems measure the sunlight reflected off of the surface over visible and near-infrared wavelengths, providing important information about the types and distributions of primary and secondary minerals across the Martian surface.

The goal of this atlas is to provide a pictorial sampling of the diversity of Martian "layered terrains," which are increasingly regarded to have formed through the action of sedimentary processes. Our hope is that these images will motivate the reader to further pursue their own investigations.

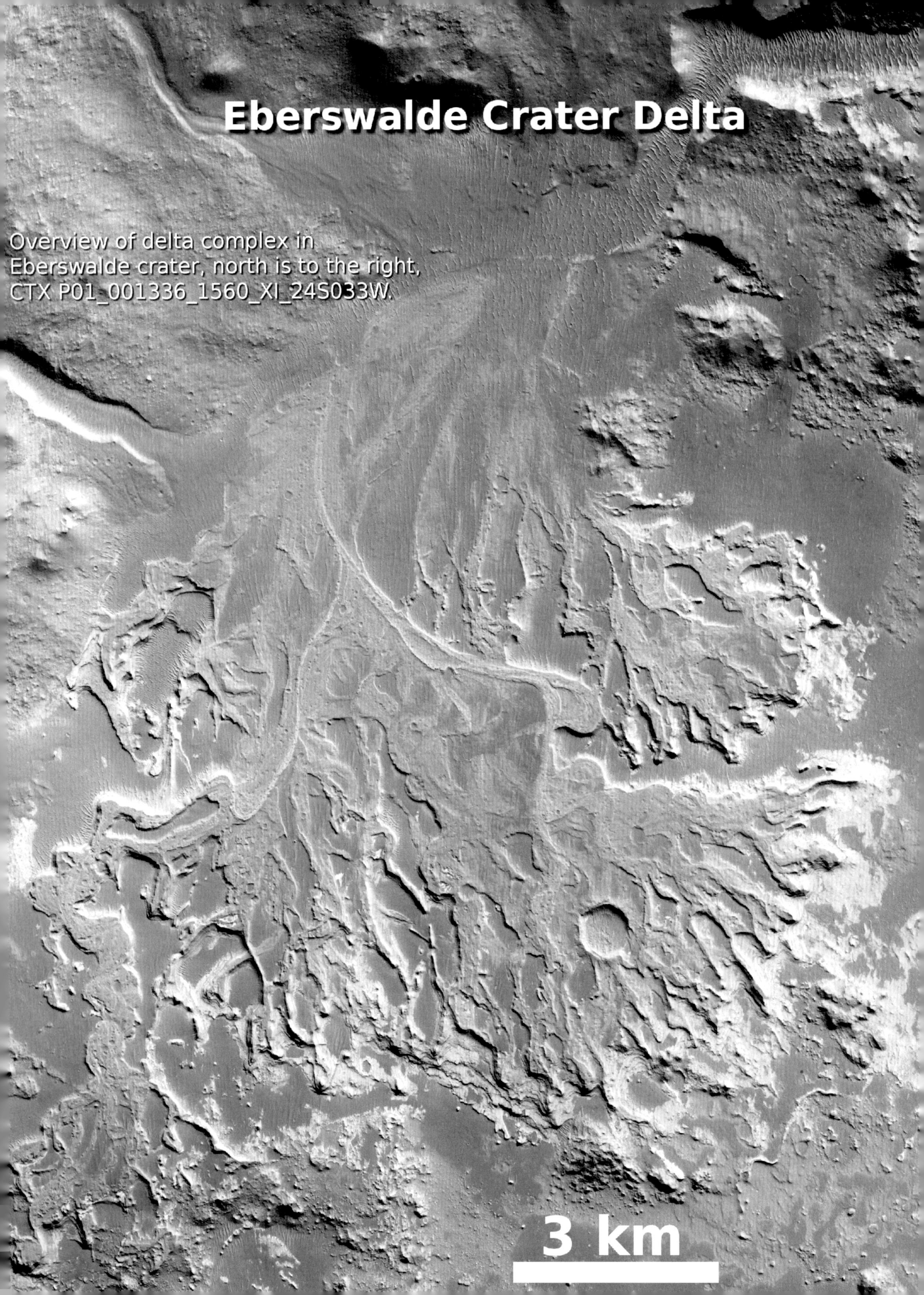
Eberswalde Crater Delta
Overview of delta complex in Eberswalde crater, north is to the right, CTX P01_001336_1560_XI_24S033W.
3 km

Potential down-lapping of layered deposits at toe of Eberswalde delta, perspective view of terrain model created from HiRISE PSP_001534_1560 and PSP_001336_1560.

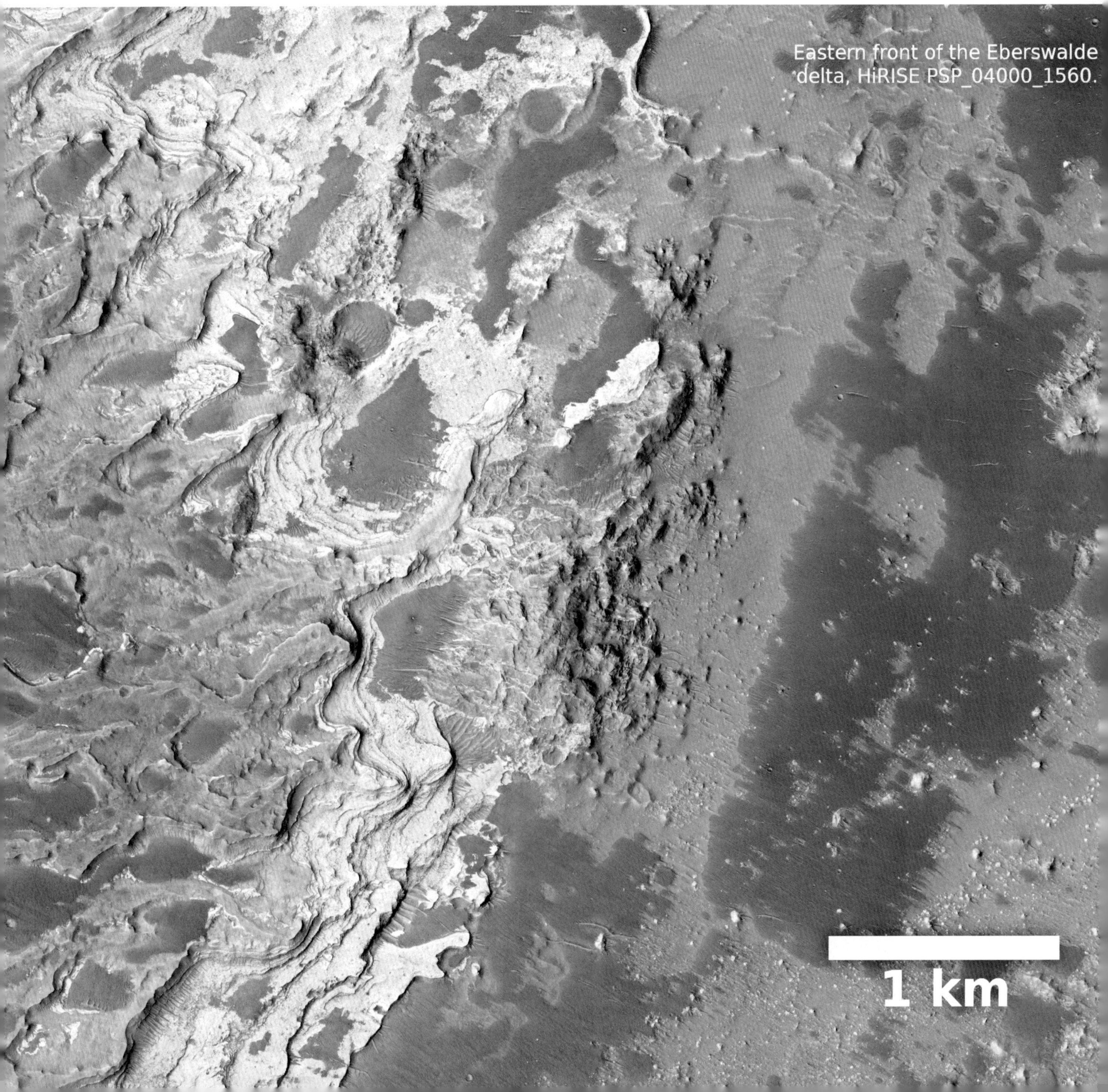

Eastern front of the Eberswalde delta, HiRISE PSP_04000_1560.

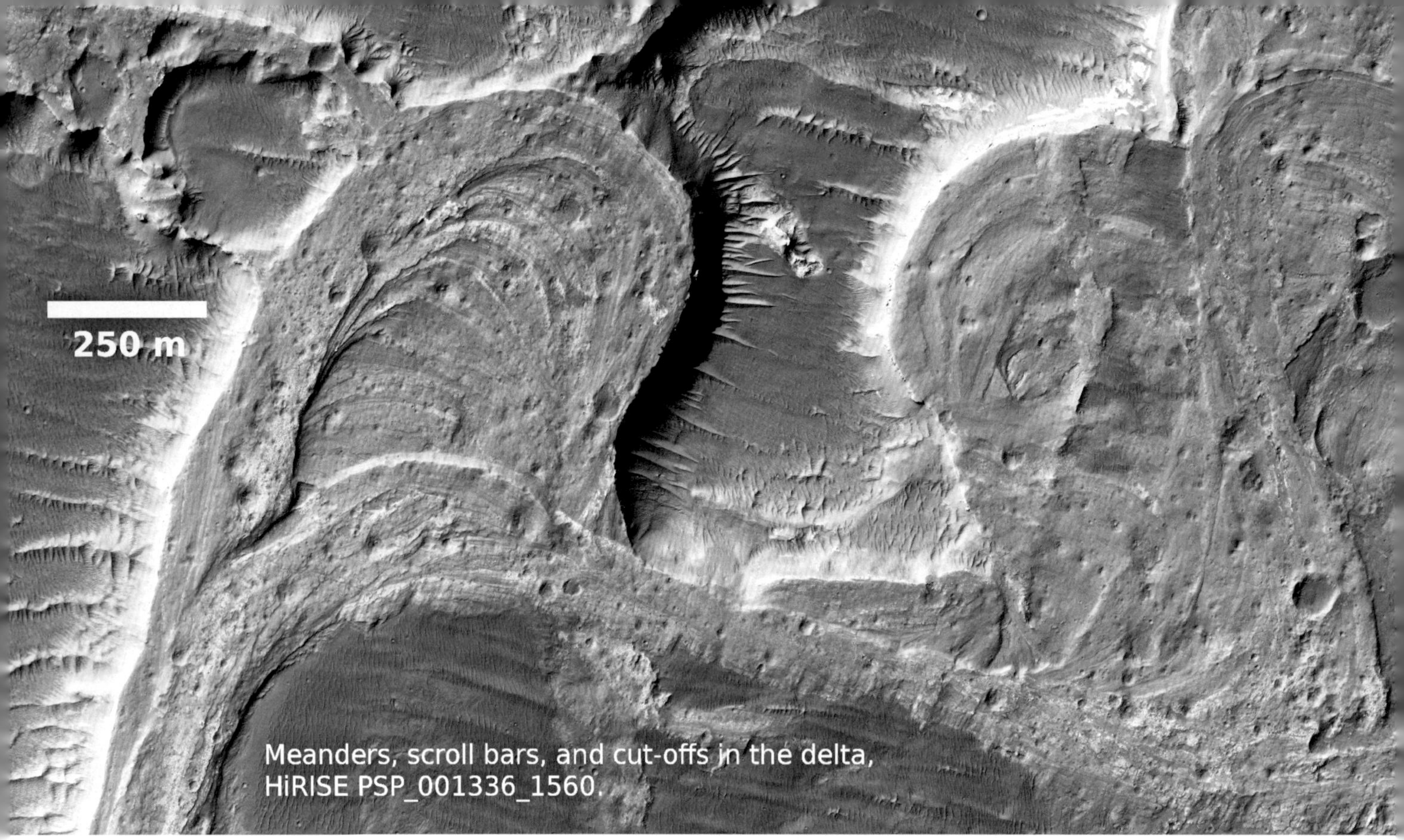

Meanders, scroll bars, and cut-offs in the delta, HiRISE PSP_001336_1560.

Eberswalde crater contains ~100 m of strata exposed at the terminus of a well-preserved delta form (Malin and Edgett 2003, Moore et al. 2003), most likely dating from the Hesperian Period (Irwin 2011), although it may be as old as Late Noachian (Moore et al. 2003) or as young as Early Amazonian (Grant and Wilson 2011). This sedimentary deposit contains dozens of shallowly dipping (~2°), alternating bright and dark layers of varying thickness (~1–10 m). HiRISE terrain models reveal topset, foreset, and bottomset strata, and these bottomset beds are interpreted as lake-floor deposits (Lewis and Aharonson 2006, Pondrelli et al. 2008). This delta is distinguished from other fan-shaped deposits on Mars by the presence of a preserved distributary network including lobes, inverted channels, and meander cutoffs. Another example of a fan with a distributary network can be found in Jezero crater, which may represent a more degraded version of the same kind of system.

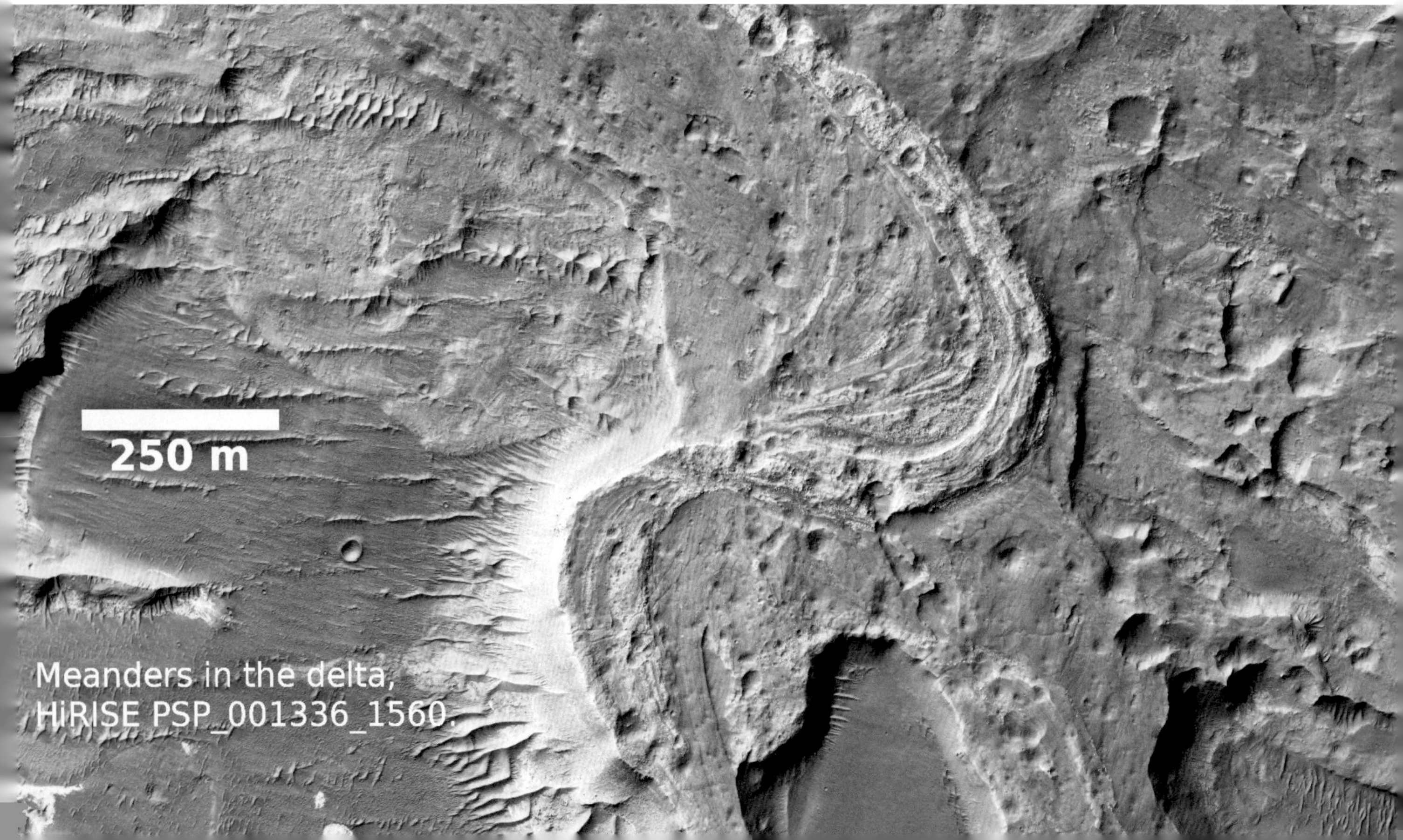

Meanders in the delta, HiRISE PSP_001336_1560.

Holden Crater Alluvial Deposits

Flat-lying bright layered units in upper left overlain by steeply dipping darker fluvial beds in Holden crater, HiRISE PSP_003077_1530.

The brighter layers in Holden crater are exposed as rounded slopes and buttes that underlie alluvial-fan deposits distributed along the interior rim of the crater. These relatively flat-lying strata are interpreted as distal alluvial or lacustrine deposits, although no obvious time-equivalent fan delta is preserved. The well-stratified deposits are traceable for hundreds of meters and define the lower unit of Grant et al.'s (2008) Holden lithostratigraphy emplaced during Hesperian or Early Amazonian wet phases (Grant et al. 2010, Grant and Wilson 2011). The lower unit overlies a basal megabreccia composed of poorly sorted blocks tens of meters in diameter, and it underlies a dark-toned, crudely layered upper unit. Fe/Mg smectite clay minerals and possibly mixed-layer smectite–chlorite phyllosilicates have been observed within the Holden crater stratigraphy (Milliken and Bish 2010). Moore and Howard (2005) discuss a number of other alluvial-fan locations on Mars.

Stratigraphic relationship of H
sedimentary units, HiRISE PSP_002088_

Layered units, HiRISE PSP_009841_1530.

Well stratified layers, HiRISE ESP_019256_1530.

Fractured blocks of stratified units, HiRISE ESP_019256_1530.

Xanthe Terra Terraced Fans

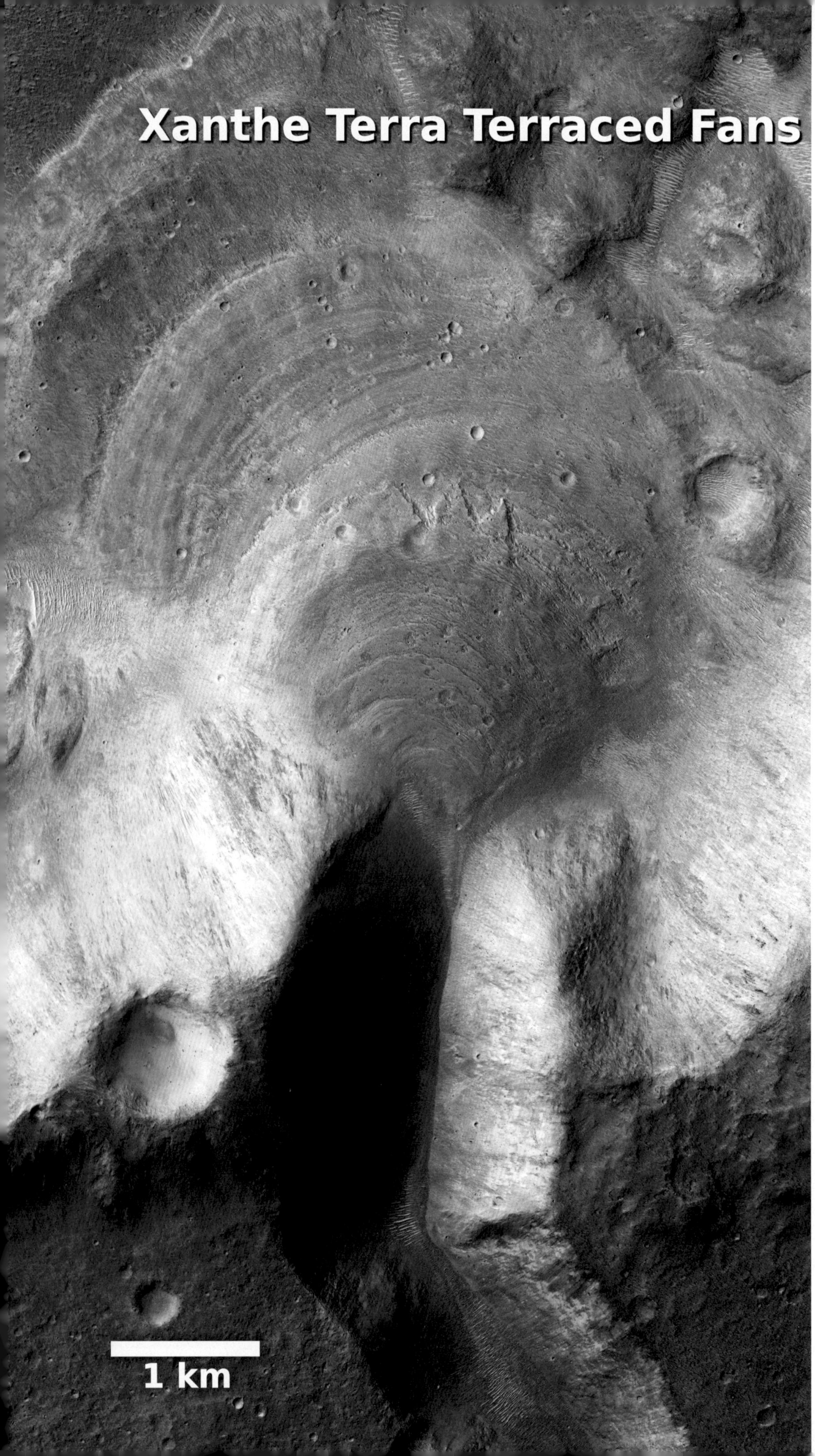

The Xanthe Terra region is host to several fan-shaped deposits that occur where deeply entrenched channels intersect crater walls. These fans are distinguished from alluvial fans or deltas on Mars because they are composed of distinctive concentric terraces, they lack lobes or evidence for distributary channels, and little to no incision has occurred on the surfaces of the fans. The upper part of these fans usually consist of a flat plain and a steep front (Hauber et al. 2009), and they are assumed to be Late Noachian in age. The terraces are present at distal parts of the deposit and are hypothesized to be erosional scarps from wave action or resedimentation processes (Hauber et al. 2009). There are many hypotheses for the origin of these fans in Xanthe Terra and similar features in the Memnonia region, including formation from a single outflow event (Kraal et al. 2008), or interpretation as Gilbert-type deltas (Ori et al. 2000, Hauber et al. 2009). Alternately, it seems possible that these fans could be viscous debris-flow deposits that spread laterally at the mouth of the canyon and crater walls. Similar terraced-fan features are also found in Aeolis Mensae and Coprates Catena.

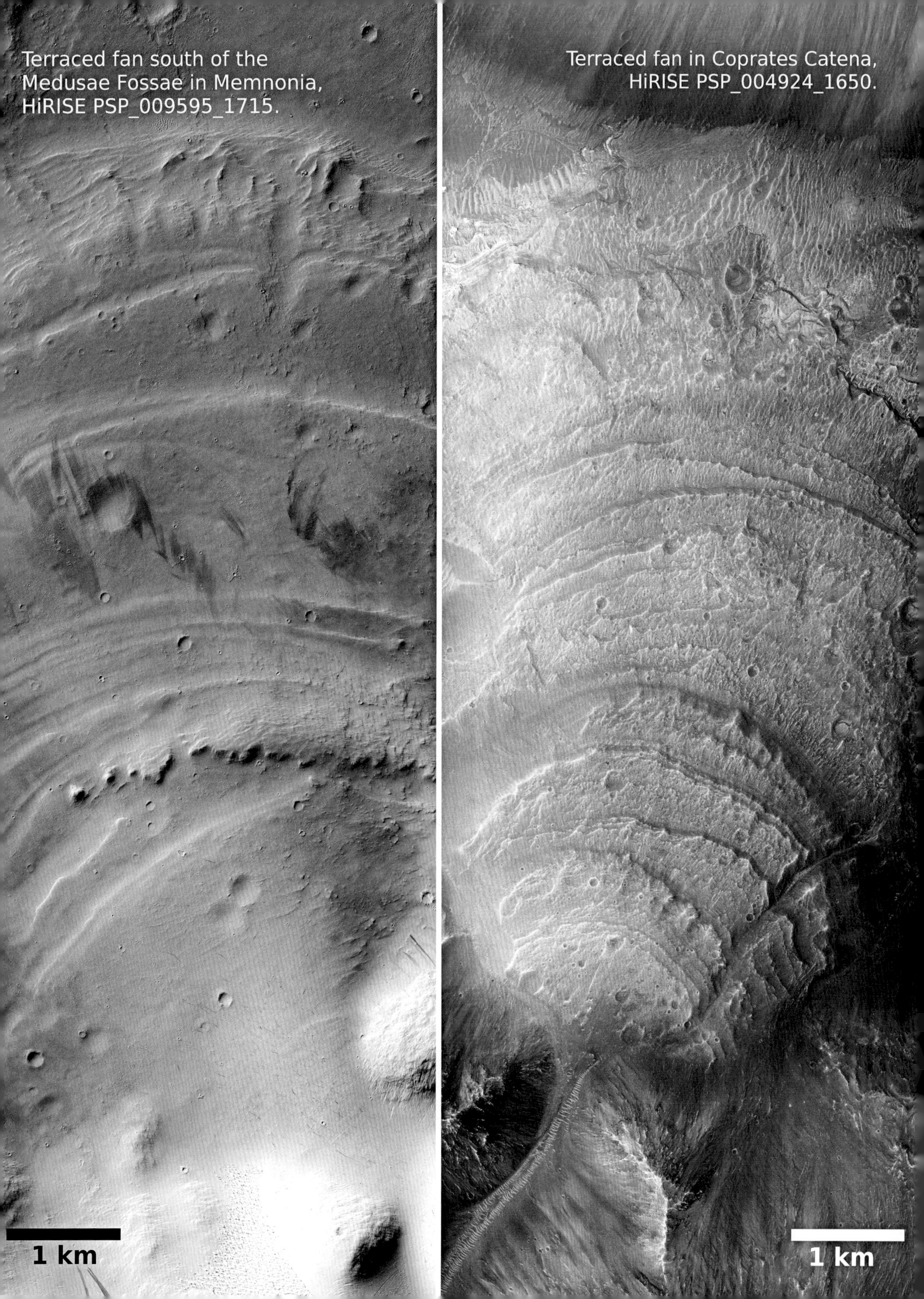
Terraced fan south of the
Medusae Fossae in Memnonia,
HiRISE PSP_009595_1715.
1 km
Terraced fan in Coprates Catena,
HiRISE PSP_004924_1650.
1 km

These Noachian- or Hesperian-aged fans, located at the center of the Melas Chasma topographic basin, are unique amongst the undeformed layered deposits in Valles Marineris because they suggest a sublacustrine origin (Metz et al. 2009). Layers observed in the fans are characterized by their low slope, and they preserve evidence of channel-branching geometry and dendritic small-scale lobes. Mono- and polyhydrated sulfates, Fe-oxides, opaline silica, and jarosite have been detected in the vicinity of the Melas fan (Metz et al. 2009). There are no other confirmed sublacustrine fan deposits outside of Melas Chasma.

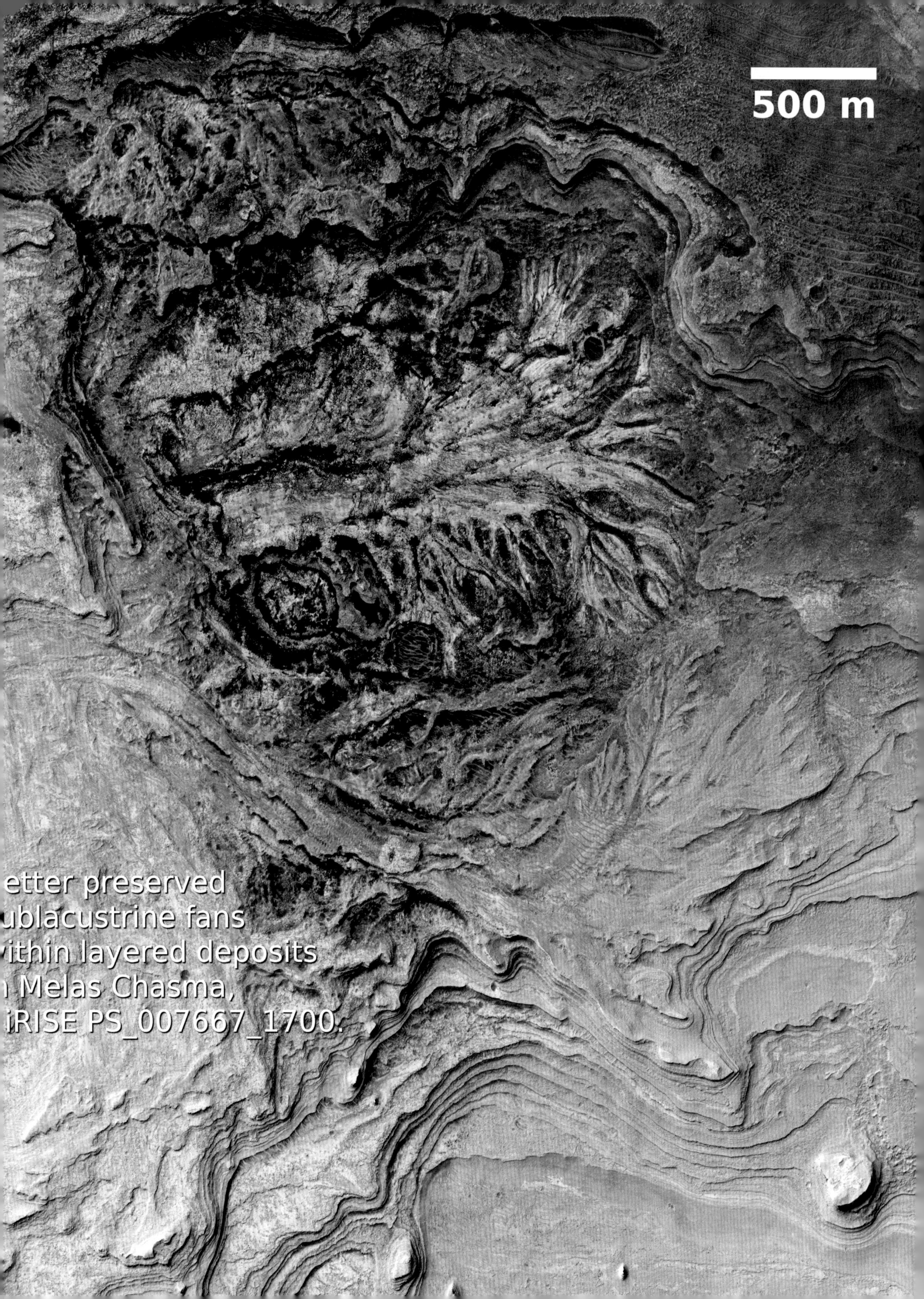

etter preserved
ublacustrine fans
vithin layered deposits
n Melas Chasma,
iRISE PS_007667_1700.

Layered deposits are found on the floor and inner walls of Columbus crater. Because no fluvial networks have been identified in the region, a groundwater-fed playa lake environment has been hypothesized to explain the layered deposits and observed mineralogy within the crater (Wray et al. 2011). Polyhydrated sulfates (gypsum and Mg-sulfate) interbedded with kaolinite have been detected in these layers, and jarosite and Fe/Mg smectites have also been observed and are assumed to have formed in the Late Noachian (Wray et al. 2011). The layered material consists of polygonally fractured bright layers that can be traced around the circumference of the crater. Other locations of similar deposits can be found in Terra Sirenum, and the *Opportunity* rover site in Meridiani Planum.

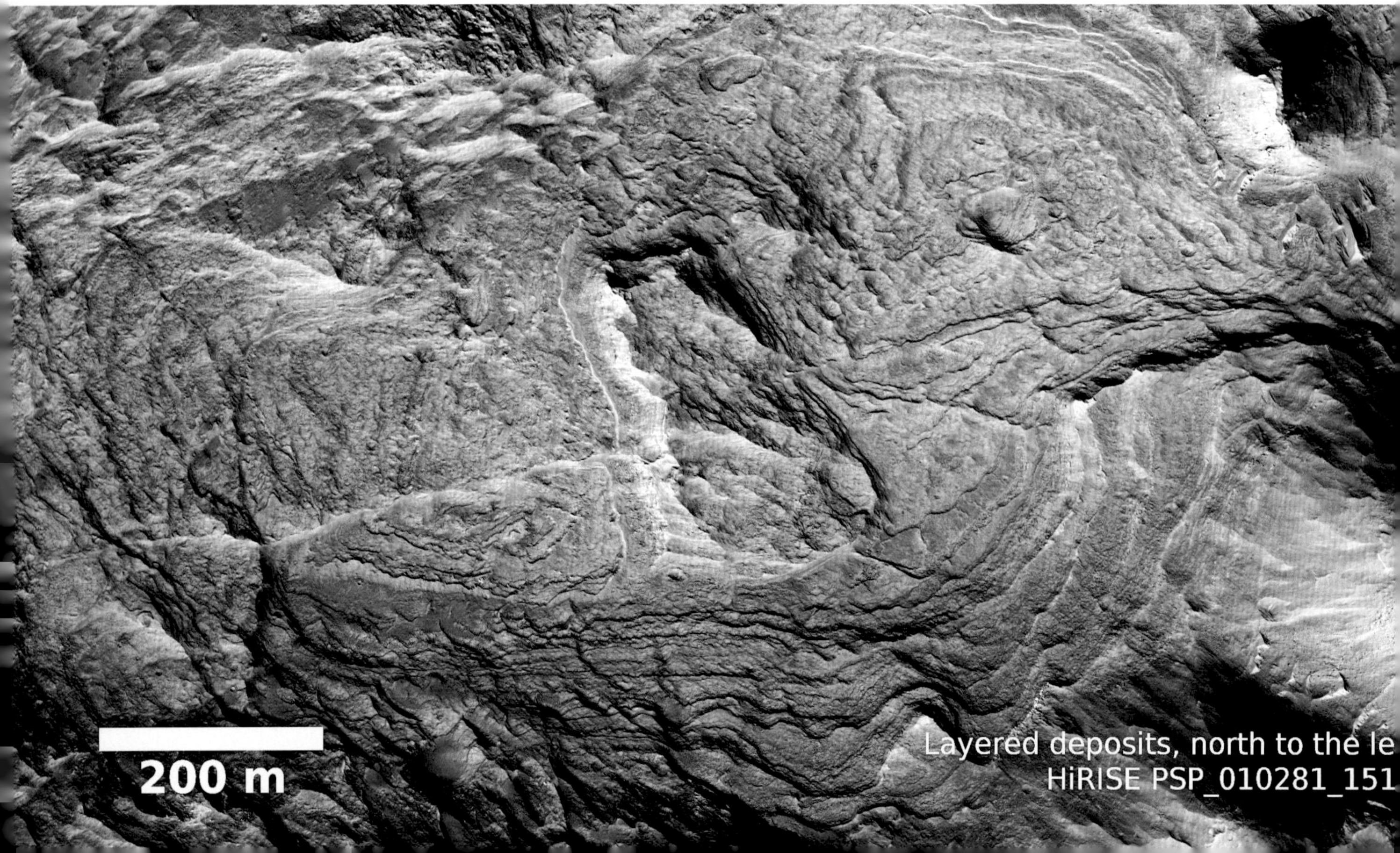

Becquerel Crater Layered Mound

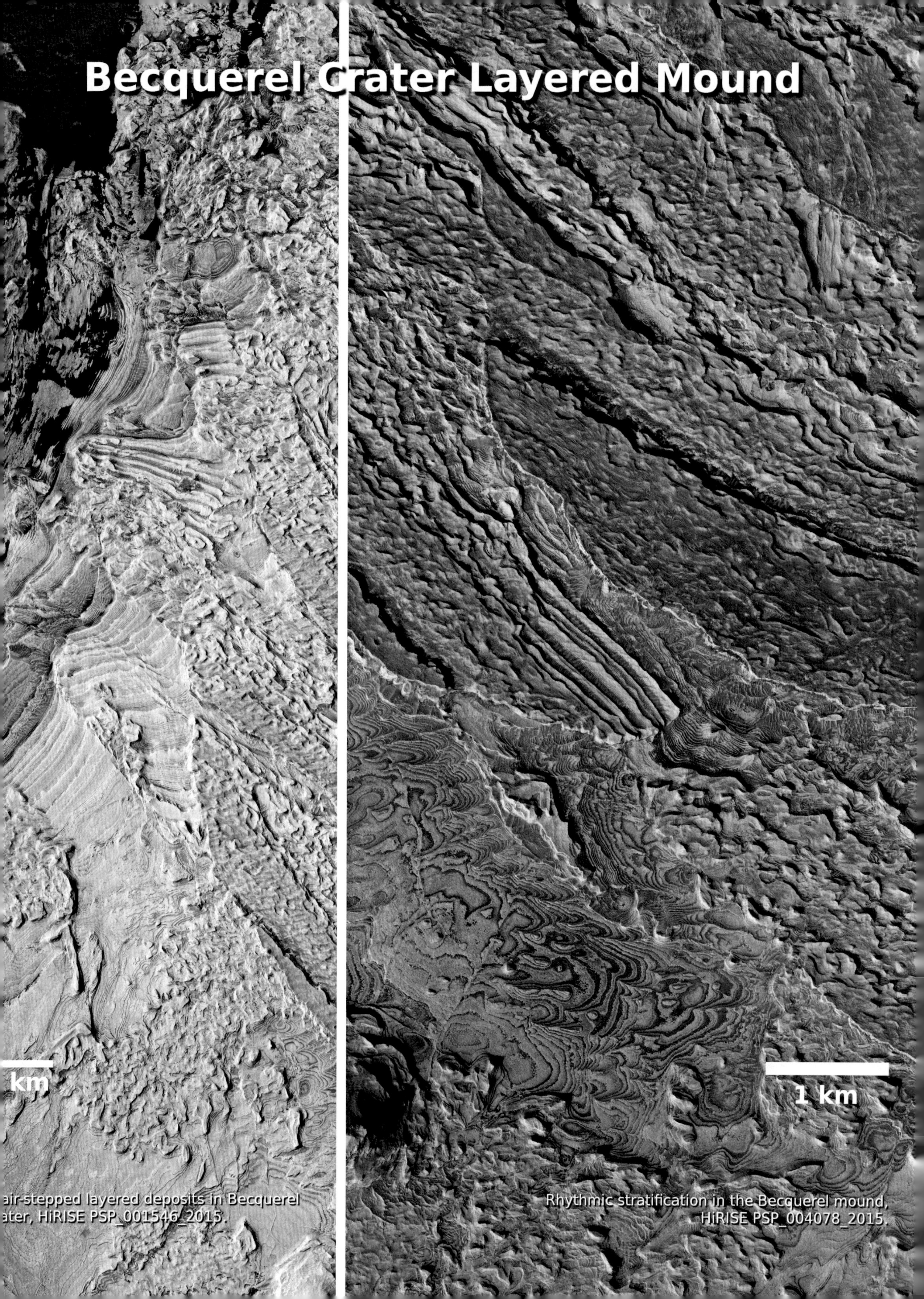

air-stepped layered deposits in Becquerel ater, HiRISE PSP_001546_2015.

Rhythmic stratification in the Becquerel mound, HiRISE PSP_004078_2015.

The layered mound in Becquerel crater contains repetitive indurated beds exhibiting nearly uniform thickness that appear to be expressed in rhythmic bund (Lewis et al. 2008). The strata are generally of uniform tone and are often fractured and faulted, resulting in the structural offset of layers throughout the secti Several interpretations have been put forward to explain the origin of the Becquerel mound, including air-fall dust, lacustrine deposition, or polar layered depos (Bridges et al. 2008, Lewis et al. 2008). Similar exposures can be found elsewhere in the Arabia Terra region, for example, in Henry crater, Schiaparelli cra Vernal crater, and Crommelin crater, and they are thought to be Noachian in age (Malin and Edgett 2000). At Henry crater, the layered mound is at a hig elevation than the rim of the crater, suggesting deposition in an overfilled basin. At Becquerel crater, the highest point on the mound is below the crater rim, maki it difficult to determine whether these strata represent deposition in an underfilled basin, or are the degraded remnant of a more extensive overfilled basin depo

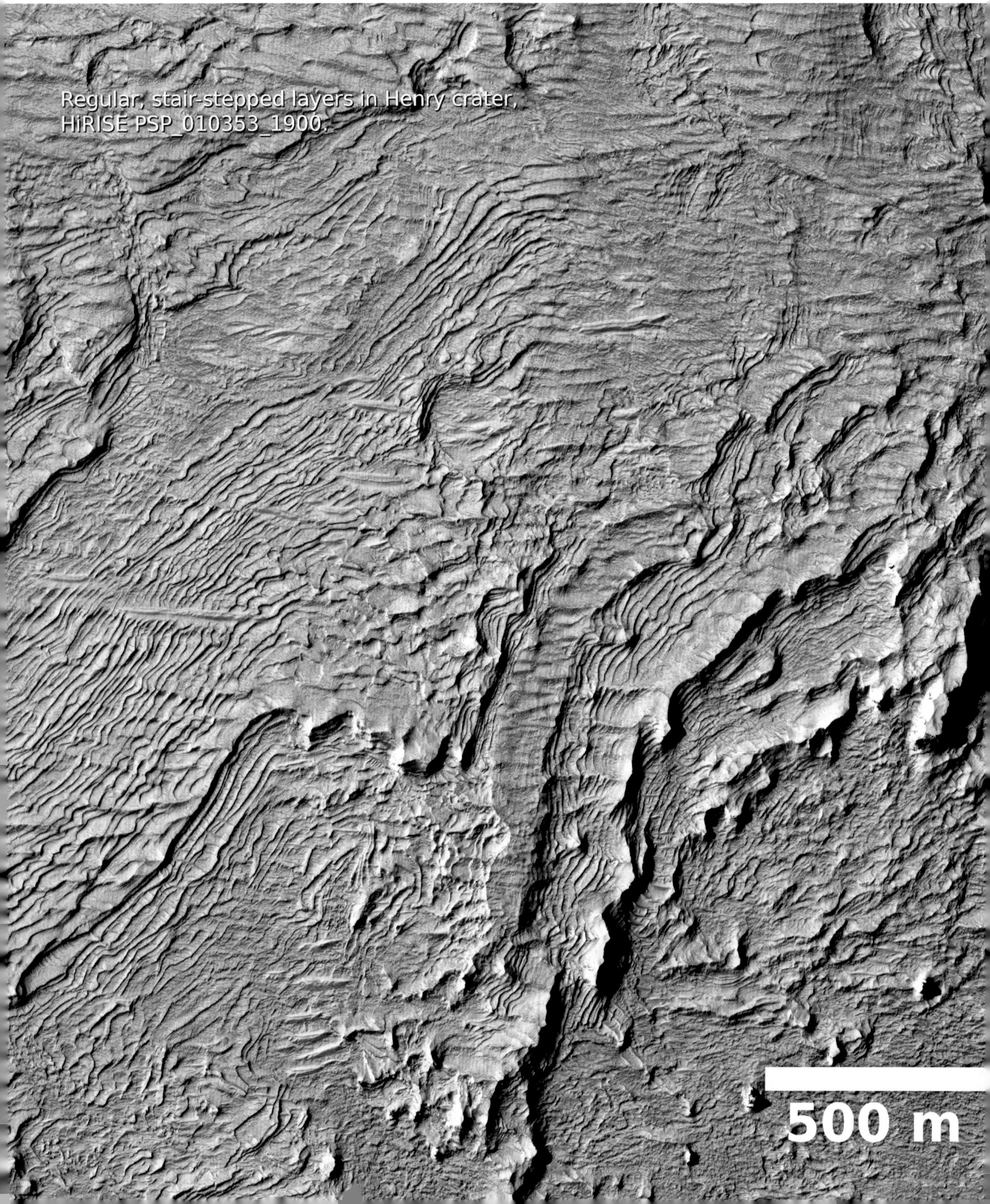

Regular, stair-stepped layers in Henry crater, HiRISE PSP_010353_1900.

ayered mound in Schiaparelli crater,
HiRISE PSP_005897_1790.
00 m

Layered mound in Schiaparelli crater,
HiRISE PSP_002930_1800.
200 m

Galle Crater Layers

Finely-bedded layers in Galle crater mound, HiRISE PSP_003934_1275.

The southern part of Galle crater hosts a 600-m-thick stack of layered deposits. The lower portion of the deposit consists of thin layers with unconformable contacts and numerous truncation surfaces. This layered mound has been interpreted as lacustrine (Reiss et al. 2006), but because the style of layering in Galle is reminiscent of stratigraphy observed in polar layered deposits, this mound and similar outcrops in Spallanzani crater have been interpreted as ancient polar layered deposits (Ansan and Mangold 2003). The ages of these deposits are unknown.

Truncated layers in the Galle crater mound, HiRISE PSP_003934_1275.

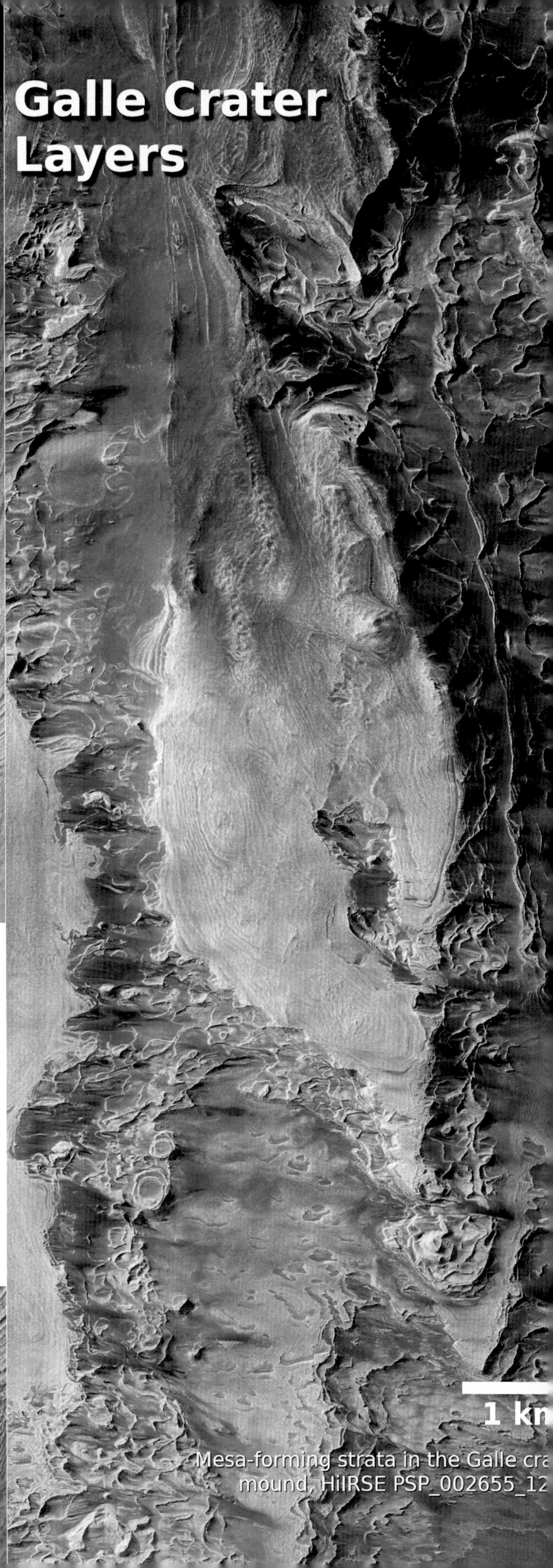

Mesa-forming strata in the Galle crater mound, HiRISE PSP_002655_12

Eroded strata in Galle crater,
HiRISE PSP_002866_1280.

Eroded strata in Spallanzani crater,
HiRISE PSP_006778_1220 (above)
and PSP_03442_1215.

Rhythmic stratification in the upper part of the Gale mound, HiRISE PSP_008002_1750.

Portion of the Gale crater mound, higher elevations to the bottom right, and a large channel that goes from the lower right to the upper left, HiRISE PSP_009149_1750.

1 km

Gale crater contains a layered mound exhibiting diverse mineral signatures and layer morphologies. At ~5 km in thickness, the Gale mound is one of the most substantial continuous sections on Mars. A clearly expressed mineral stratigraphy is observed throughout the mound. The layered deposits at the base of the mound are sulfate- and clay-bearing, whereas the middle section is clay-poor but rich in mono- and polyhydrated sulfates. The upper portion of the mound appears to be spectrally neutral (Milliken et al. 2010), consistent with the composition of Martian dust. The Gale crater mound may record a large section of Martian time, from the Noachian to Hesperian age of its lower units to a Hesperian to Amazonian age of its upper formation (Milliken et al. 2010). These mineralogical changes are coupled with morphological changes. The clay- and sulfate-rich layers appear to vary in thickness, albedo, and texture, whereas the spectrally neutral layers of the upper mound exhibit a stair-stepped morphology with uniform layer thickness, homogeneity in tone, and a lack of impact craters. These rhythmic strata near the top of the mound are comparable to those observed in Arabia at Becquerel crater. Many hypotheses have been invoked to explain the origin of this stack of sedimentary layers, including volcanic ash, lacustrine (Cabrol et al. 1999), eolian, spring mound (Rossi et al. 2008), or ancient polar deposits (Schultz and Lutz 1988).

spective view of a HiRISE ain model looking theast, up the channel.

Eolian erosional scalloped texture near the base of the Gale mound, HiRISE PSP_008437_1750.

Layers of alternating tone and thickness in the lower part of the Gale mound, HiRISE PSP_009149_1750.

Strata exposed in lower portion of Gale mound, HiF PSP_006855_17

Terby Crater Layered Mesas

Westernmost mesa in Terby crater, HiRISE PSP_002216_1525.

Possible truncation surface, HiRISE ESP_013160_1530.

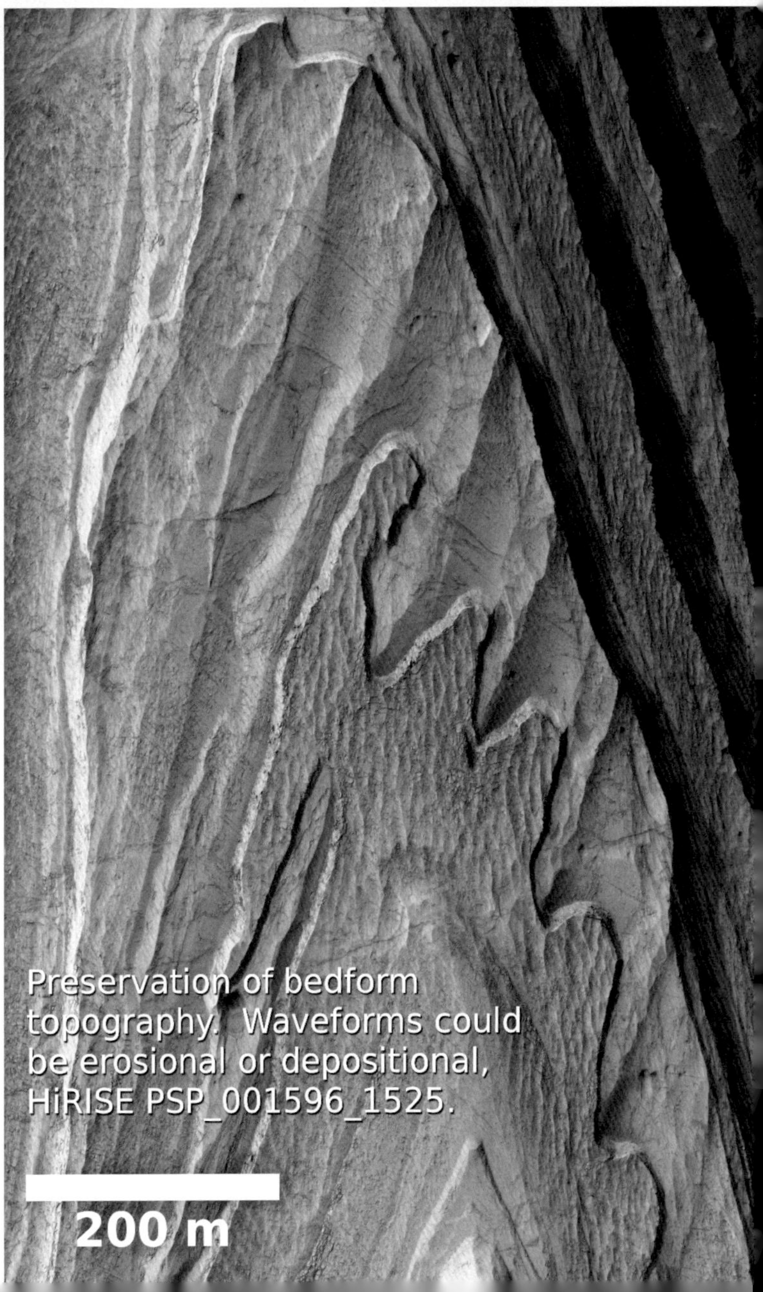

Preservation of bedform topography. Waveforms could be erosional or depositional, HiRISE PSP_001596_1525.

Southernmost tip of the central mesa, HiRISE ESP_014017_1520.

Similar layers 150 km to the east of Terby crater, HiRISE PSP_010839_1525.

There is a 2.5-km-thick sequence of meter-scale strata exposed in three mesas within Terby crater, located along the north rim of Hellas Basin. Layers in the northern part of the crater are higher in elevation than the degraded southern rim. This deposit has been variously interpreted to be the result of eolian (Lewis et al. 2010), lacustrine (Wilson et al. 2007), volcanic air-fall, and deltaic processes (Ansan et al. 2011). The layered mesas display Fe/Mg smectite clay mineral signatures, and hydrated Mg-sulfates; zeolites or hydrated silicates may be present as well (Ansan et al. 2011). These strata vary in albedo from bright to dark and are laterally continuous for kilometers. Locally, these strata are observed to truncate unconformably. Layers are polygonally fractured in places and exhibit a grooved texture thought to be the result of eolian deflation (Wilson et al. 2007), and they are assumed to have been emplaced during the Noachian. Similar deposits can be observed along the northern rim of Hellas Basin.

Thermal Emission Imaging System (THEMIS) day time infrared mosaic showing context for the layered mesas in Terby crater.

Oudemans Crater Central Uplift

Steeply dipping layered rocks are exposed in the central uplift of Oudemans crater, located at the western end of Valles Marineris. The strata exposed in Oudemans were likely excavated from a depth of several kilometers below the surface, as a result of the impact-forming process. It is possible that these steeply dipping strata are possibly correlative with strata exposed in the walls of nearby Valles Marineris. Several other craters in the general Tharsis region (e.g., Martin and Mazamba craters) also exhibit well-preserved strata in their central uplifts, suggesting a possible link between these layered deposits and ash fallout from the Tharsis volcanic field. Caudill et al. (2011) identified a number of other layered units in the central peaks of craters. Therefore, these strata may be basaltic pyroclastic deposits rather than deposits of sedimentary origin. These kinds of exposures are found in craters on Hesperian and Amazonian surfaces, but they may well be bringing up rocks formed during the Noachian.

Valles Marineris Undeformed Layered Deposits

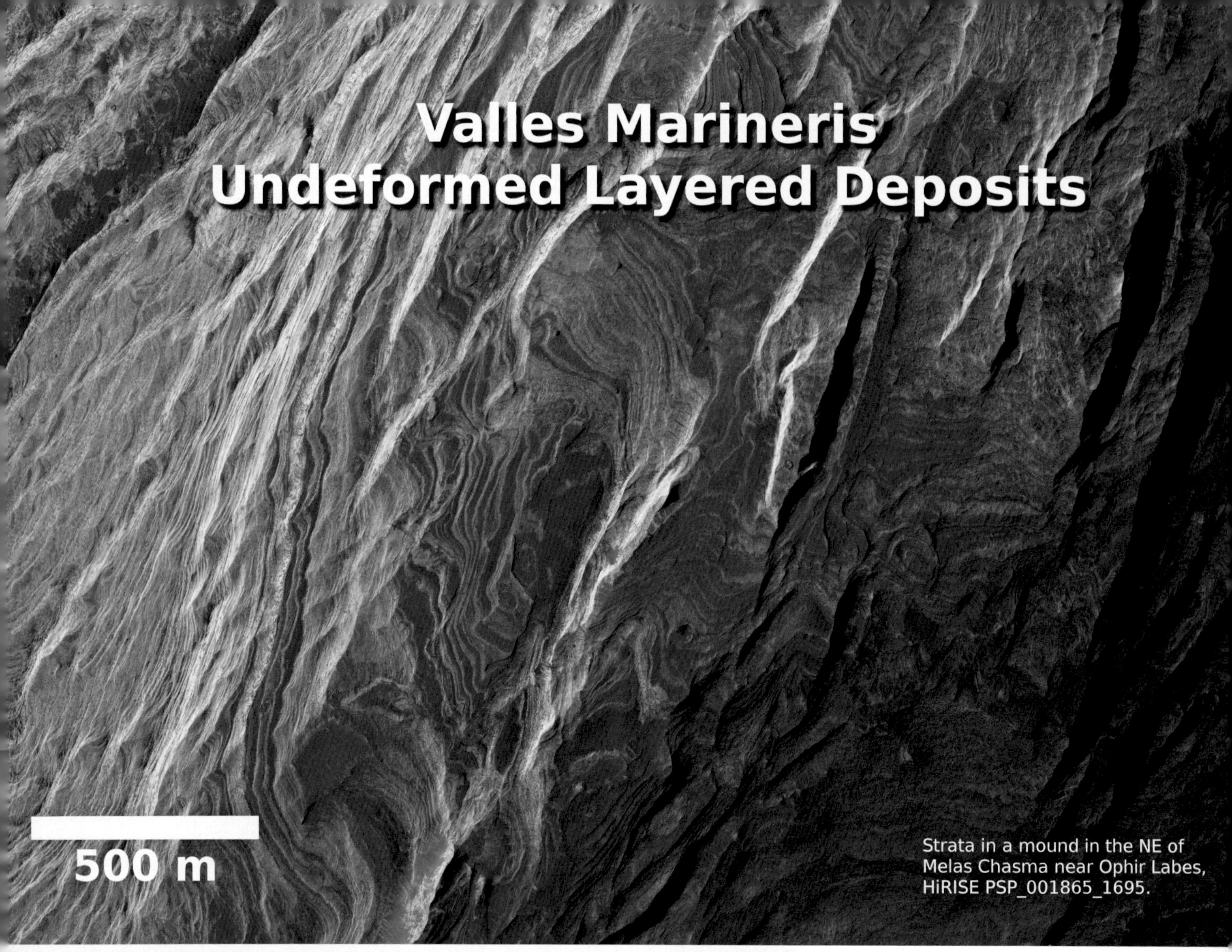

Strata in a mound in the NE of Melas Chasma near Ophir Labes, HiRISE PSP_001865_1695.

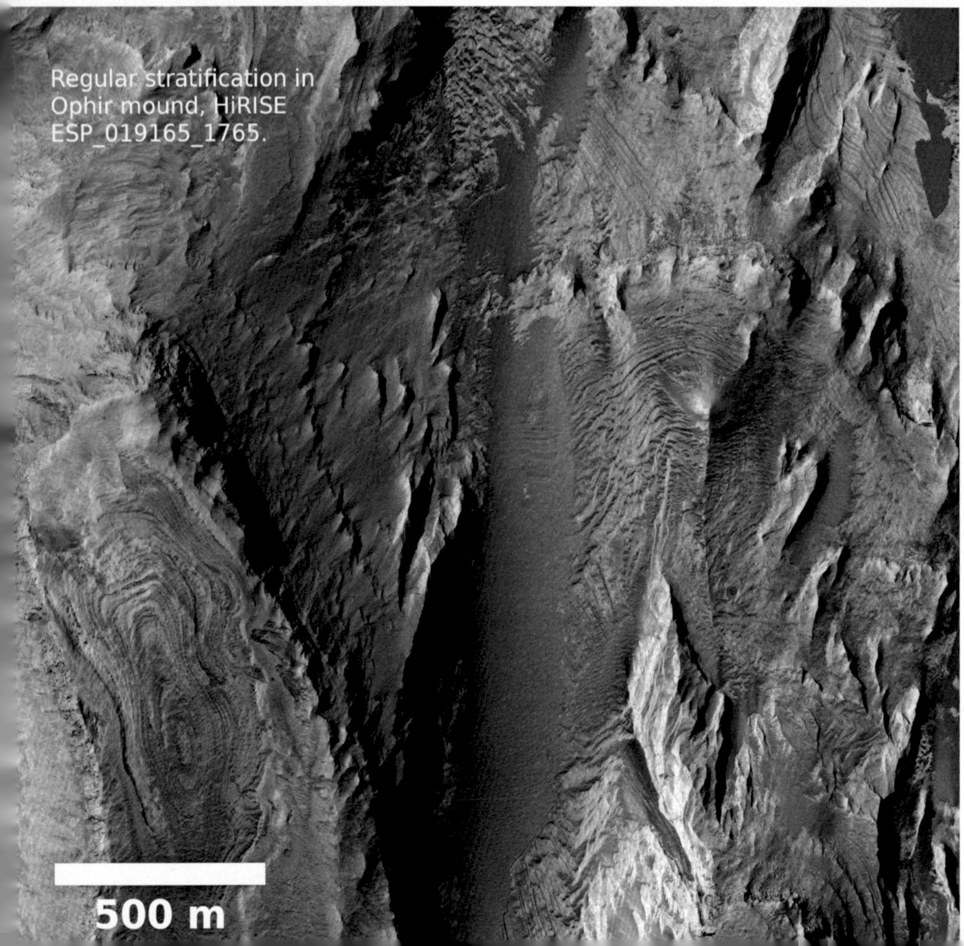

Regular stratification in Ophir mound, HiRISE ESP_019165_1765.

A variety of layered deposits are exposed as isolated plateaus, on the chasm floor, and adjacent to the walls of Valles Marineris. No single origin for these strata has been identified, and various explanations are likely needed to explain the diversity of deposits within the Valles Marineris (Lucchitta 2010), which were likely emplaced in Late Noachian to Hesperian times. In general, these deposits are hundreds-of-meters- to kilometers-thick stacks of undeformed strata that vary in thickness and albedo. Mono- and polyhydrated sulfates, Fe-oxide, and phyllosilicates have been detected in many of these outcrops (Gendrin et al. 2005, Bibring et al. 2007, Le Deit et al. 2007). Layered deposits in Ophir Chasma, Juventae Chasma, Ganges Mensa, and Melas Chasma fall under this broad category.

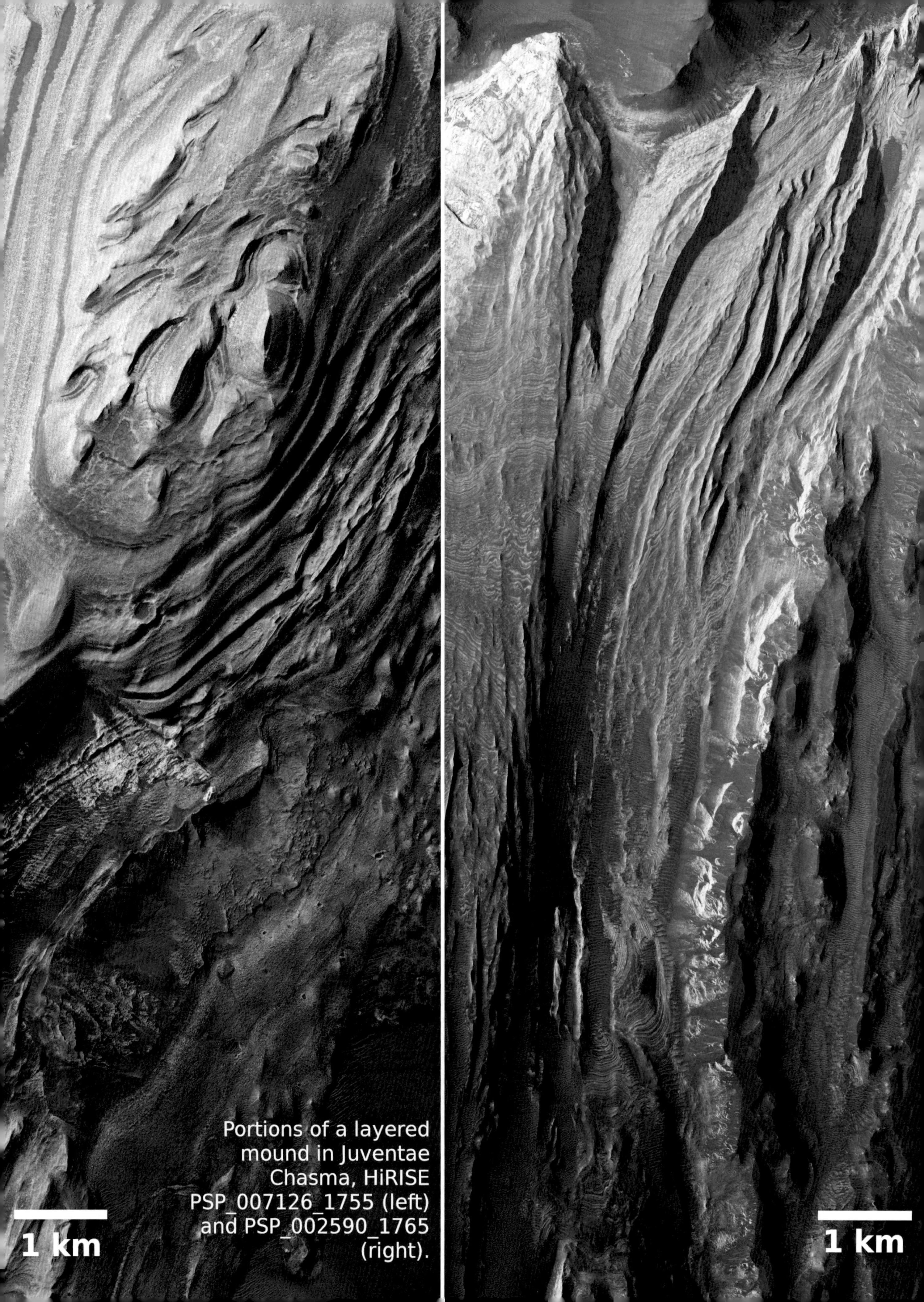

Portions of a layered mound in Juventae Chasma, HiRISE PSP_007126_1755 (left) and PSP_002590_1765 (right).

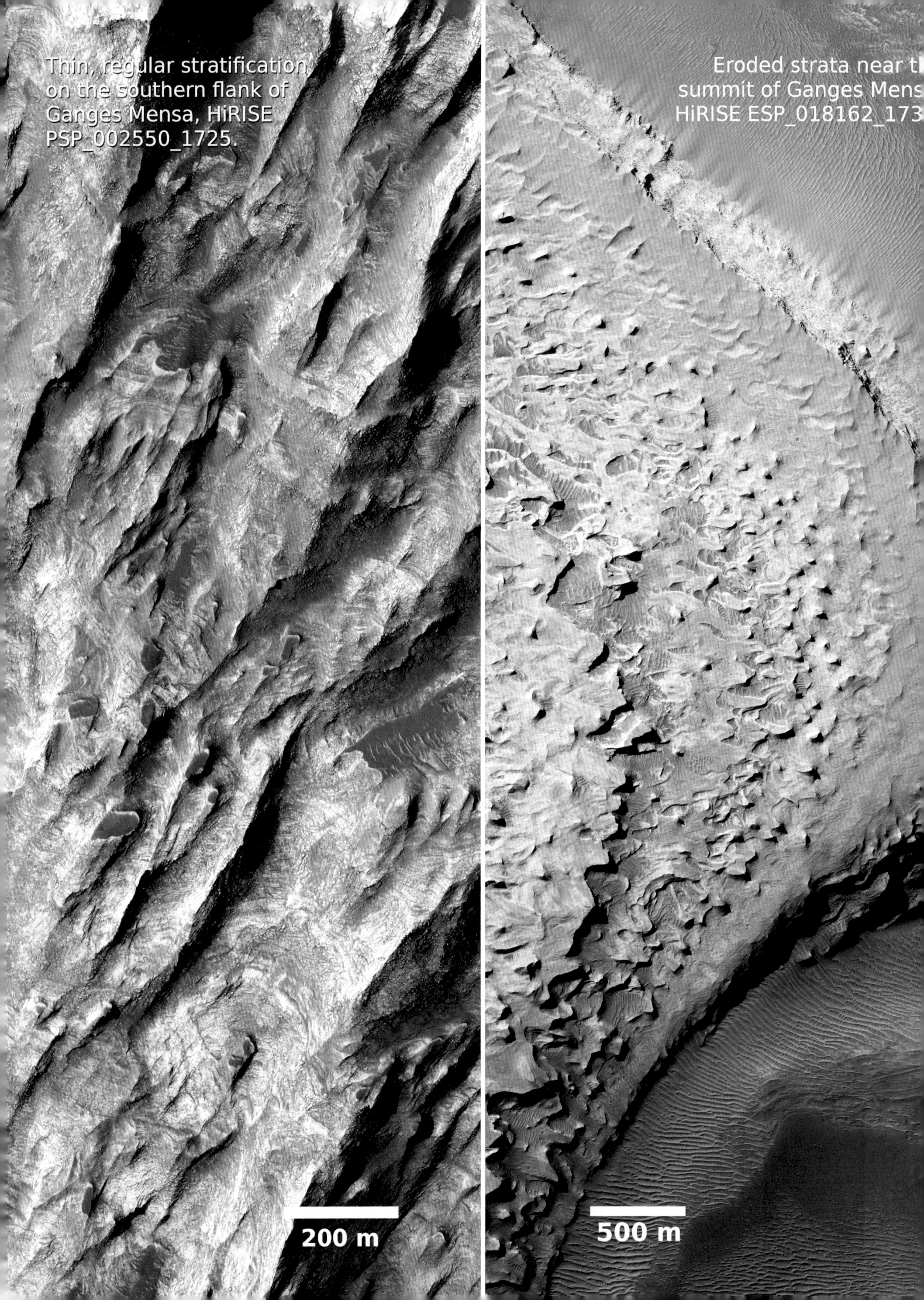
Thin, regular stratification
on the southern flank of
Ganges Mensa, HiRISE
PSP_002550_1725.
200 m
Eroded strata near th
summit of Ganges Mens
HiRISE ESP_018162_173
500 m

West Candor Chasma Deformed Layered Deposits

Faulted and folded layered bright strata, HiRISE PSP_003540_1735.

Faulted bright strata in western Candor Chasma, HiRISE PSP_003540_1735.

Although the Valles Marineris system contains many stratified units, there are some locations where those units show evidence of extensive brittle and ductile deformation, making them unique amongst these extensive deposits. This kind of deformation was first detailed in west Candor Chasma (Okubo et al. 2008, Okubo 2010), but it is also observed in Melas and Ius Chasmata (Metz et al. 2010). Kieserite, polyhydrated sulfates, and nanophase ferric oxides and oxyhydroxides have been detected in west Candor Chasma's layered mounds (Mangold et al. 2008, Murchie et al. 2009b). Mangold et al. (2008) observed that strata containing polyhydrated-sulfates have a higher thermal inertia than those containing kieserite, suggesting that these polyhydrated-sulfate-bearing layers may be coarser or better indurated than the kieserite layers (Mangold et al. 2008).

200 m
Folded and fau
bright strata, Hi
PSP_003540_17

Stratified Plains Surrounding Valles Marineris

Much of the dark material in the scene is loose dark fines, but there are some actual dark layers alternating with the brighter layers here on the plains outside of Juventae Chasma, HiRISE PSP_003434_1755.

Well-stratified layered deposits have been observed on the plains surrounding Valles Marineris. One of the best examples is located on the plains west of Juventae Chasma, where stratified rocks (<1-m-thick beds) occur beneath preserved crater ejecta and inside inverted channels (Milliken et al. 2008, Weitz et al. 2010). Layers form narrow, sinuous laminations that are continuous for up to tens of kilometers. Hydroxylated ferric sulfate and hydrated silica (opaline silica) have been detected in the deposits (Milliken et al. 2008, Weitz et al. 2010). Several locations on the plains surrounding Valles Marineris also expose layers exhibiting a similar composition and style of stratification to those near Juventae Chasma, including plains around the Louros Valles south of Ius Chasma, Melas Chasma, Candor Chasma, and Ganges Chasma (see Le Deit et al. 2008, Milliken et al. 2008, Weitz et al. 2010). These units are thought to be Late Hesperian in age or younger.

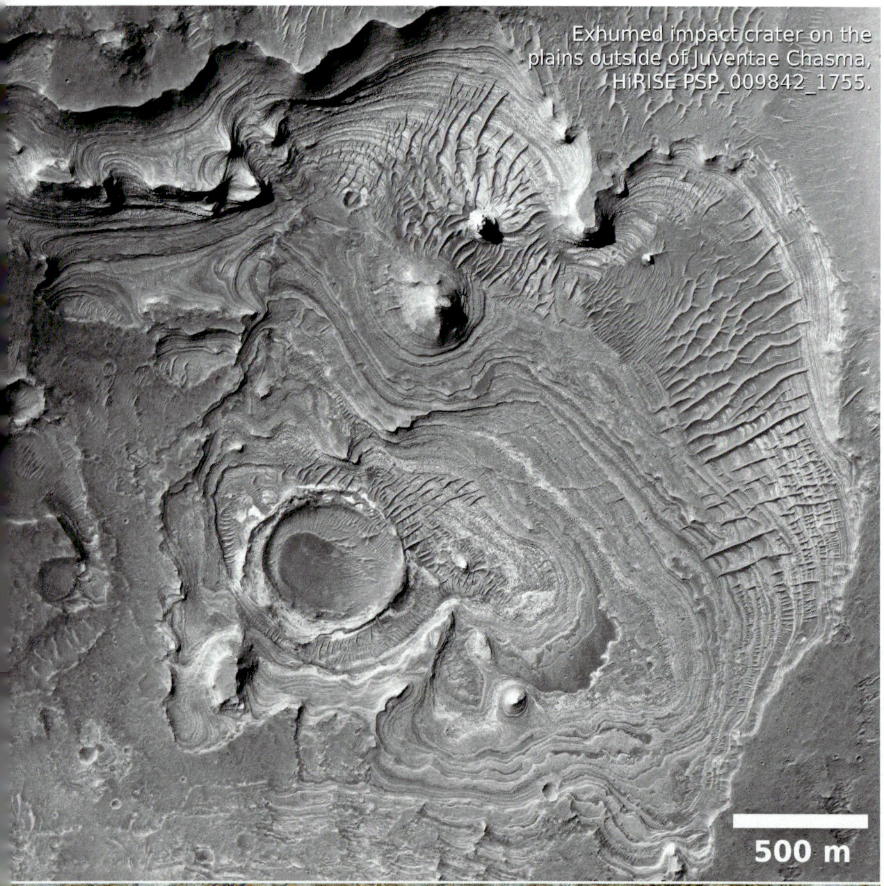

Exhumed impact crater on the plains outside of Juventae Chasma, HiRISE PSP_009842_1755.

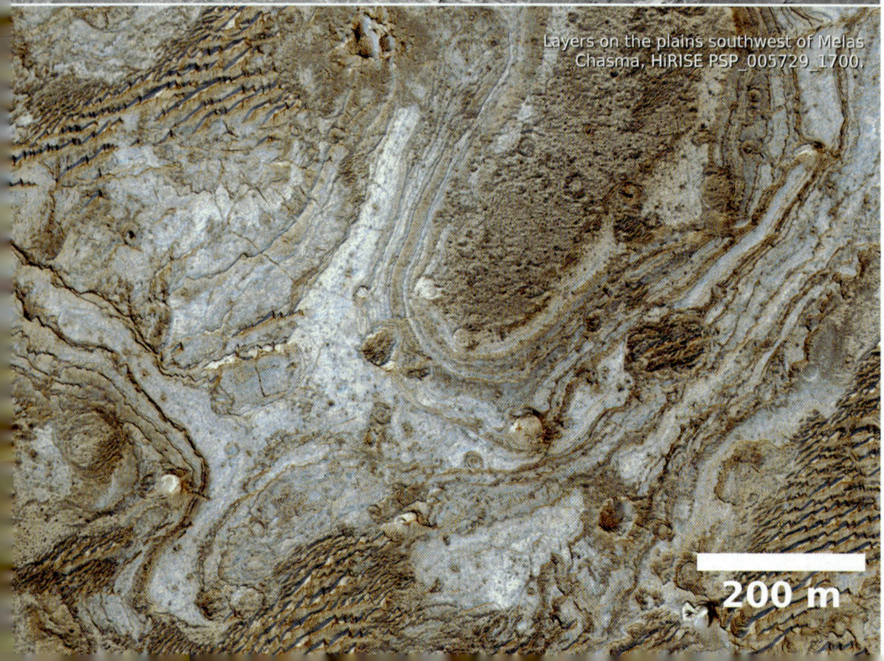

Layers on the plains southwest of Melas Chasma, HiRISE PSP_005729_1700.

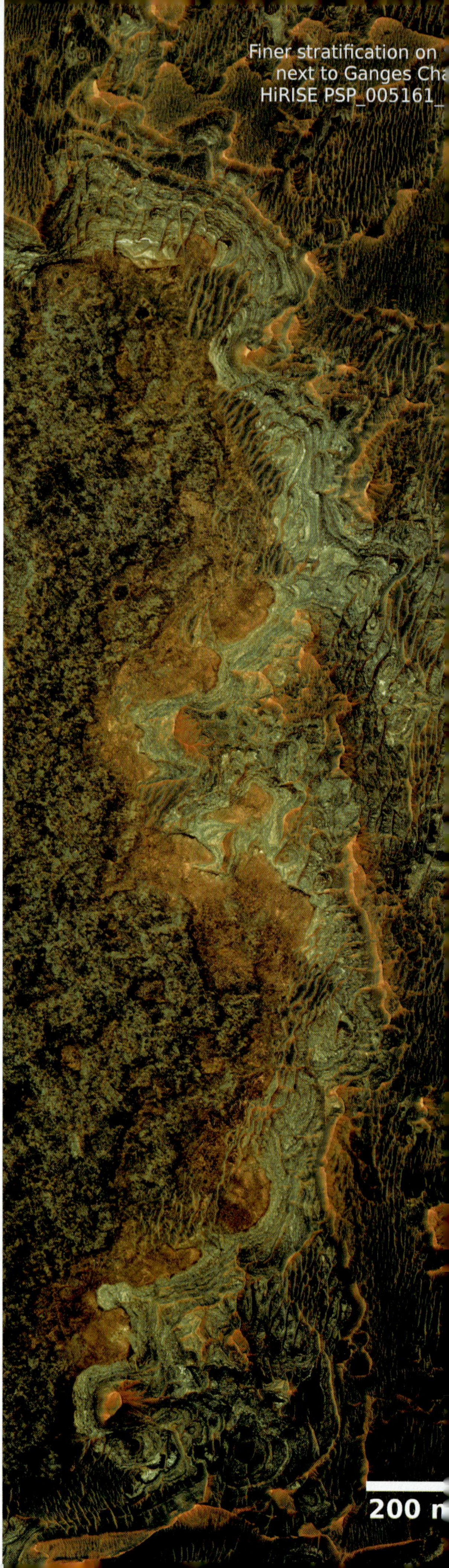

Finer stratification on
next to Ganges Cha
HiRISE PSP_005161_

Meridiani Planum Layers

These deposits are only the remnants of more extensive units which once filled the craters and blanketed the region, HiRISE PSP_008930_1880.

Meridiani Planum is an equatorial region with abundant sedimentary rocks that blanket several hundred thousand square kilometers of the Martian surface. It has a rich diversity of distinct stratigraphic units and mineralogy formed during the Late Noachian to Early Hesperian as mapped by CRISM and OMEGA (Wiseman et al. 2010). The sediments here show evidence for an extensive aqueous history that evolved over time, including deposition, alteration, and repeated erosion. The *Opportunity* rover (Squyres et al. 2004) is currently studying the top of a stratigraphic section (Burns formation) that is composed of dominantly eolian sulfate-rich sandstones, locally reworked by fluvial processes, that were diagenetically altered by groundwater brines (Grotzinger et al. 2005, Edgar et al. this volume, Lamb et al. this volume).

These rhythmic stair-stepped layers fill a crater in Meridiani Planum and show fractures and faults. HiRISE PSP_002878_1880.

Mawrth Vallis, an ancient outflow channel located at the transition between the Southern Highlands and the Northern Lowlands, is localized within some of the oldest Noachian-aged layered deposits on the planet. Flat plains adjacent to Mawrth Vallis expose Fe, Mg, and Al clay-rich strata covering over 100,000 square kilometers (Poulet et al. 2005, Michalski and Noe Dobrea 2007, Loizeau et al. 2007, Bishop et al. 2008). These rocks are exposed as irregularly shaped and eroded mounds and scarps. While there are a number of depositional interpretations for the layered deposits, including altered volcaniclastic sediments or a lacustrine origin, impact processes likely played an important role in the formation, transport, and distribution of sediments in this location.

Large-scale undulating stratification within clay-rich deposits in a Mawrth Vallis area crater wall, north is to the left, HiRISE PSP_004052_2045.

Nili Fossae and Syrtis Major Terrains
200 m
Complexly deformed
layered deposits in the Nili
Fossae region, HiRISE
ESP_019898_2000.

Thicker layers and a massive blocky deposit in a Nili Fossae crater wall, HiRISE PSP_002176_2025.

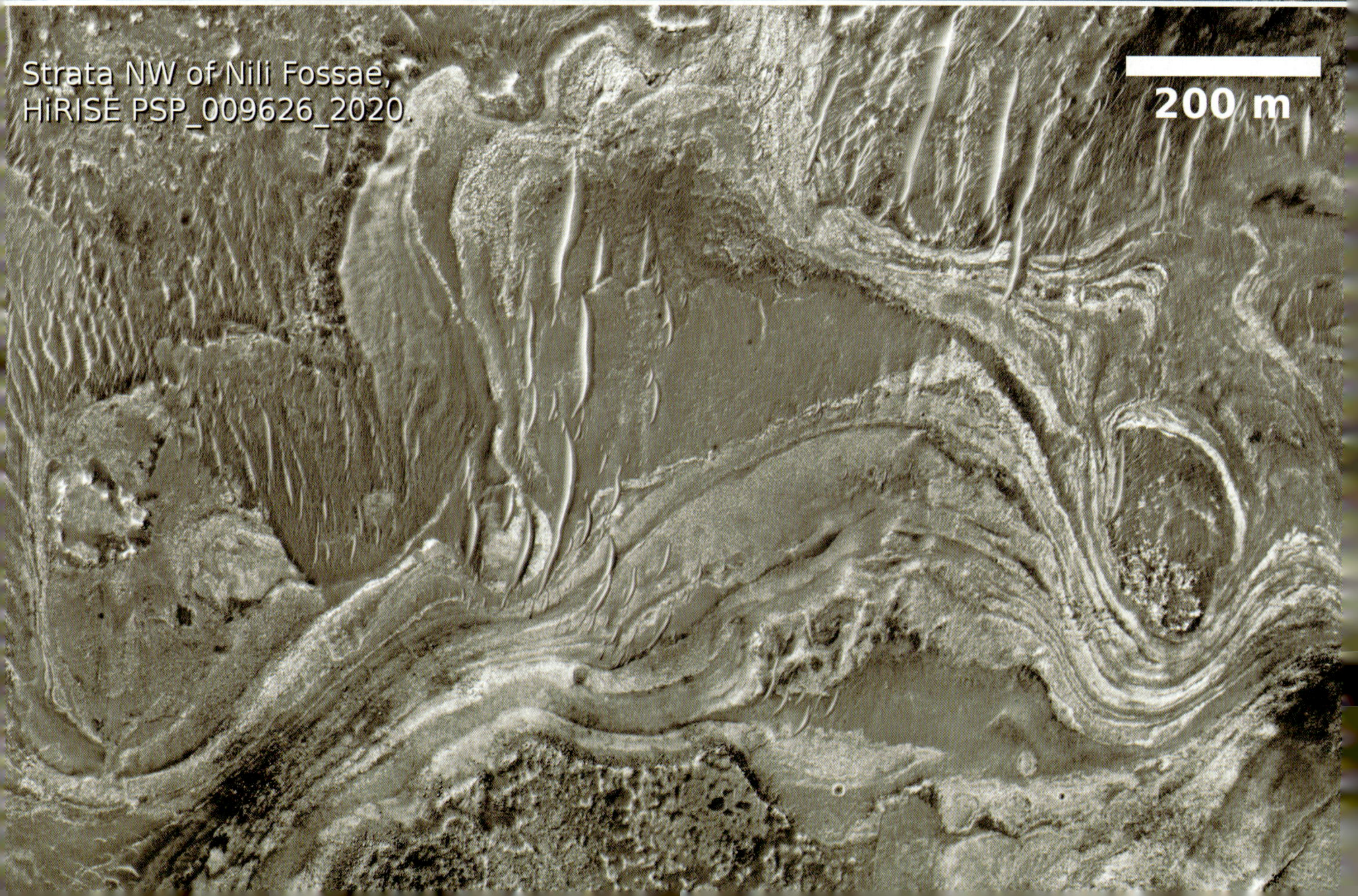

Strata NW of Nili Fossae, HiRISE PSP_009626_2020.

The Nili Fossae and Syrtis Major regions of Mars are on the northwest edge of the large Isidis impact basin. The former are large, concentric grabens that likely formed by tectonic readjustment after the Isidis impact event, whereas the latter are ancient Noachian terrains capped by younger Hesperian lava flows from the Syrtis Major volcanic system (Mangold et al. 2007, Mustard et al. 2007). The layered rocks in this region are among the oldest on Mars, likely dating back to >3.7 Ga. These rocks also host the greatest diversity of minerals observed on Mars, including Fe/Mg smectite, kaolinite, opaline silica, sulfates, carbonates, serpentine, prehnite, and zeolite (analcime) (Ehlmann et al. 2009). Layered deposits in this region have been interpreted as hydrothermally or fluvially altered mafic/ultramafic lava flows and impact breccia deposits, both of which likely comprise much of the ancient Noachian basement. In contrast, the overlying Hesperian-age lava flows likely represent low-temperature, near-surface aqueous alteration (Mustard and Ehlmann 2010). The origin and depositional environments of stratified rocks exposed within Nili Fossae are largely unknown, but their antiquity and mineralogical diversity indicate that aqueous processes were important and widespread on ancient Mars. Indeed, these rocks may provide some of the best examples of altered primary crust that has survived the Late Heavy Bombardment.

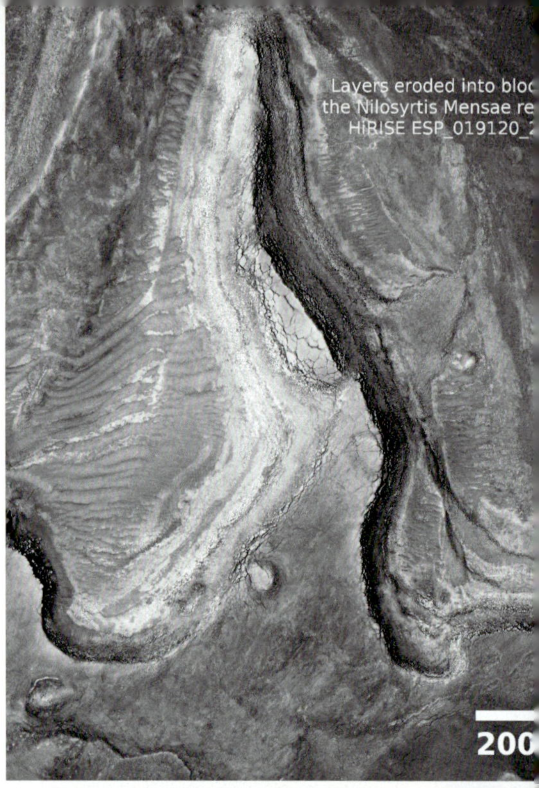

Layers eroded into bloc the Nilosyrtis Mensae re HiRISE ESP_019120_

Possibly thrust-faulted phyllosilicate-bearing layers exposed along a scarp, HiRISE ESP_016153_2005.

North Polar Layered Deposits and Basal Unit

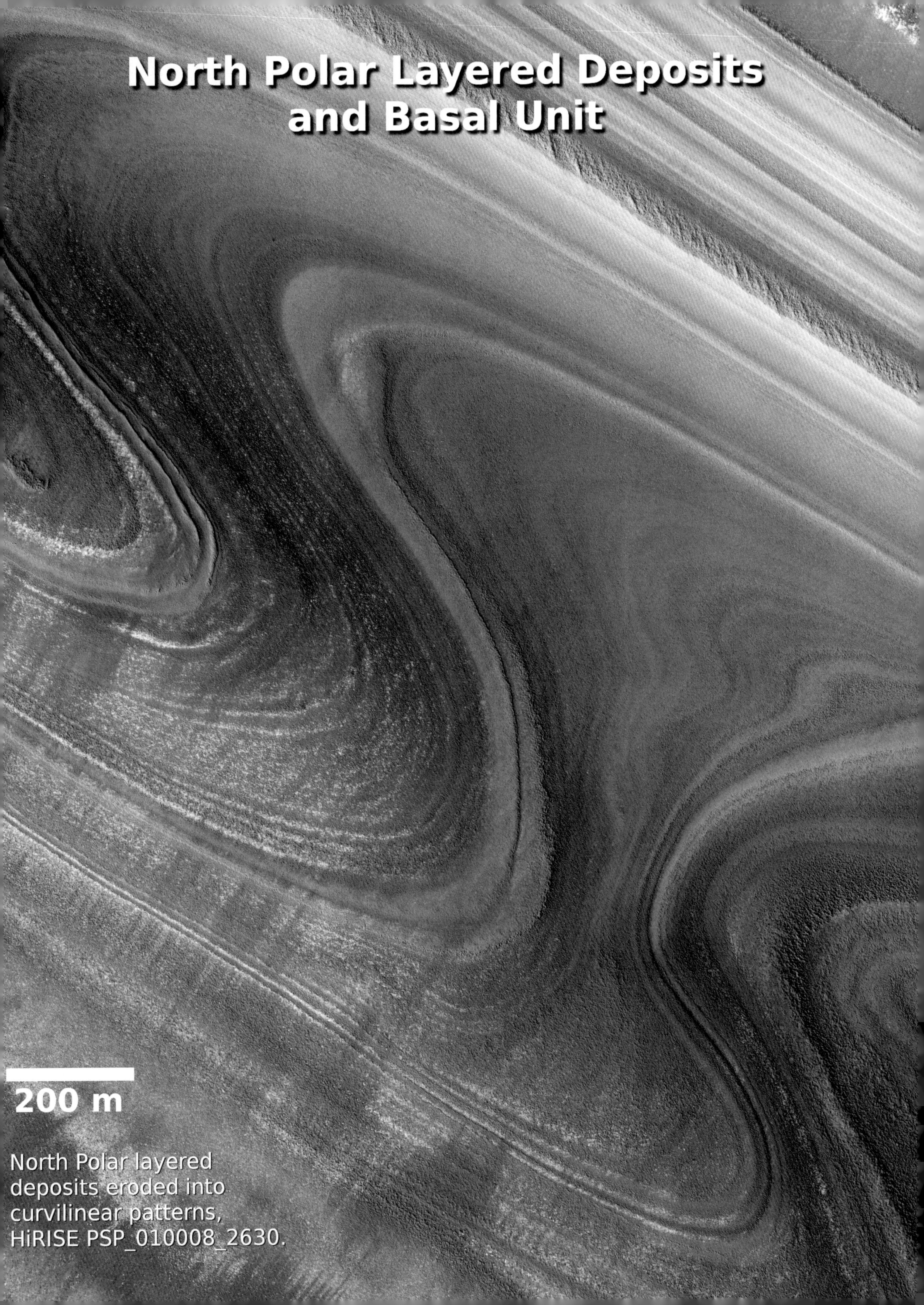

North Polar layered deposits eroded into curvilinear patterns, HiRISE PSP_010008_2630.

Dune-scale cross-stratification or clinoforms within the north polar basal cavi unit, HiRISE PSP_001334_2645.

Undulating dune-forms in north polar basal unit, HiRISE ESP_019047_2640.

The north polar layered deposits (NPLDs) form an ~3-km-thick stack of young strata exposed in the walls of troughs and scarps in the north polar regions of Mars (Byrne 2009). The NPLDs are composed primarily of water-ice, based on radar penetration to their base, and are thought to record recent climate variations. Craters are sparse, but their population statistics suggest ongoing deposition over the past few thousands of years. Layers of the NPLD are laterally continuous over scales of hundreds of kilometers, and they exhibit variable brightness, roughness, and slope where exposed. The thinnest layers detected by HiRISE are ~10 cm, but most observable layers are of meter-scale thickness (Fishbaugh et al. 2010). Morphological differences between the lower and upper layers within the NPLDs have been identified; the lower layers are steeply exposed and exhibit extensive polygonal fracturing, while upper layers are often exposed in troughs with gentle slopes, and lack polygonal fracturing (Herkenhoff et al. 2007). Radar reflectors within the NPLDs identified from orbit are generally subhorizontal and exhibit vertical variations in brightness, resulting in distinct packets (Philips et al. 2008). Radar-layer unconformities and lateral thickness changes have been observed and indicate accumulation rates that vary in both space and time, including the early formation of two accumulation centers that later partially merged (Holt et al. 2010). Radar stratigraphy also indicates the more recent, constructional formation of large-scale spiral troughs and ridges within the NPLDs via deposition concurrent with eolian transport (Smith and Holt 2010). One of two basal units identified beneath the NPLDs (Byrne and Murray 2002, Fishbaugh and Head 2005, Tanaka et al. 2008) appears to have alternating layers of ice and sand-sized material organized into successions dominated by low-angle clinoforms. Erosion of this unit supplies sand-sized material to the current circumpolar sand sea. Langevin et al. (2005) found that calcium-rich sulfates (gypsum) comprised a minor constituent of the dune material. The gypsum is concentrated at 240°E, decreases slowly to the west, and locally appears most abundant at dune crests (Roach et al. 2007). Hydrous alteration of material comprising the dunes (Fishbaugh et al. 2007) or within dikes radiating from Alba Patera (Tanaka 2006) has been suggested as a possible source; however, CRISM data show that trace amounts of gypsum also exist within the NPLDs and are concentrated as a sublimation lag in places before being transported to the dunes (Massé et al. 2010). Thus, where, when, and how this mineral formed remain unknown.

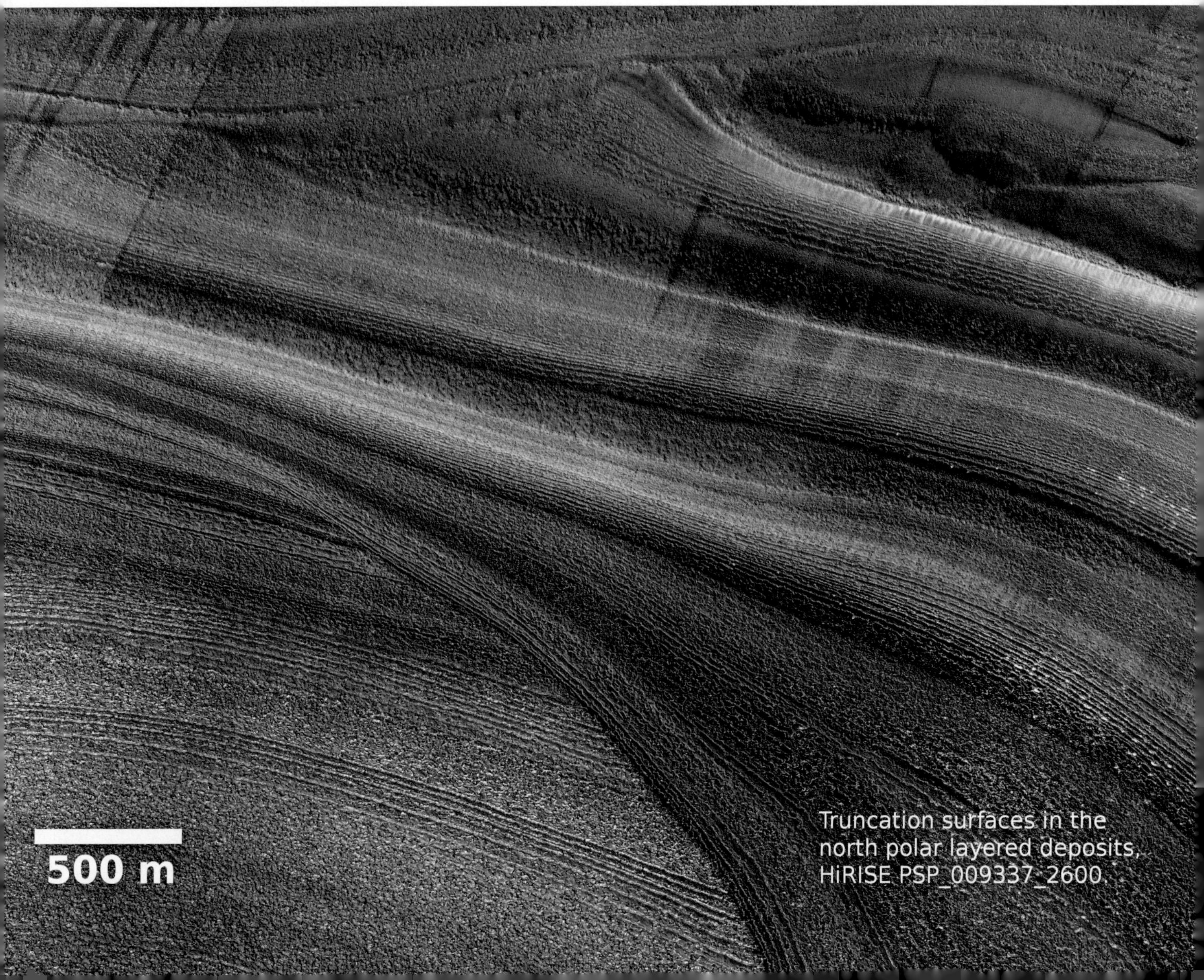

Truncation surfaces in the north polar layered deposits, HiRISE PSP_009337_2600.

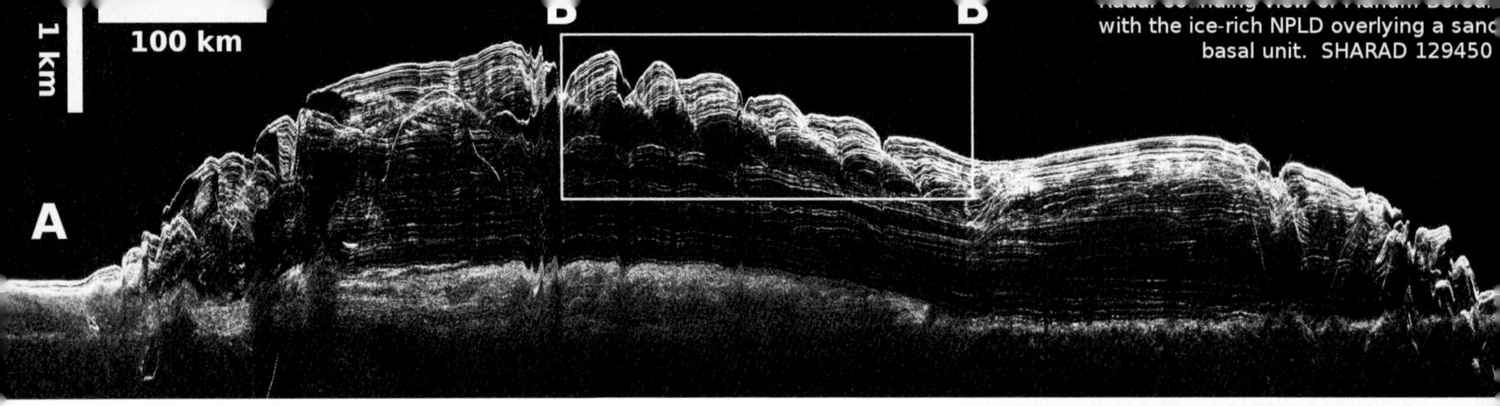

with the ice-rich NPLD overlying a sanc basal unit. SHARAD 129450

Stratal geometries associated with the northward and upward migration of spiral troughs, resulting from aeolian transport concurrent with new deposition. SHARAD 1294501.

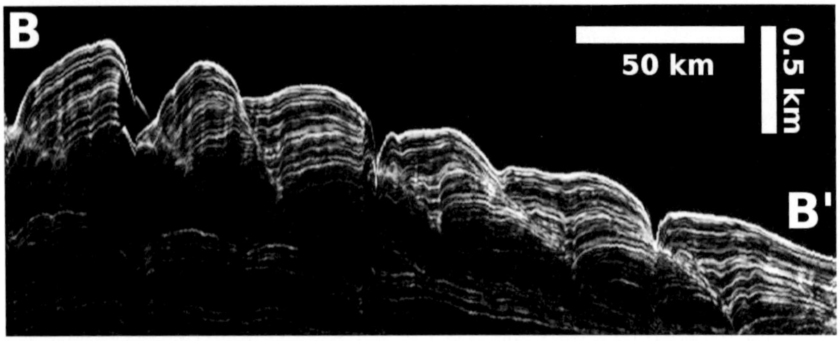

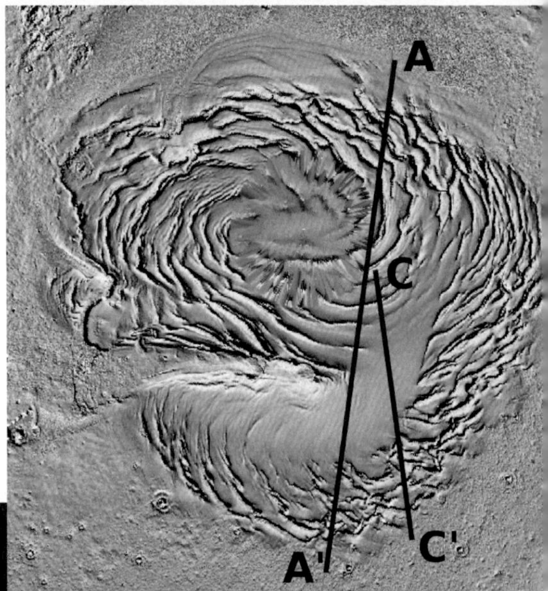

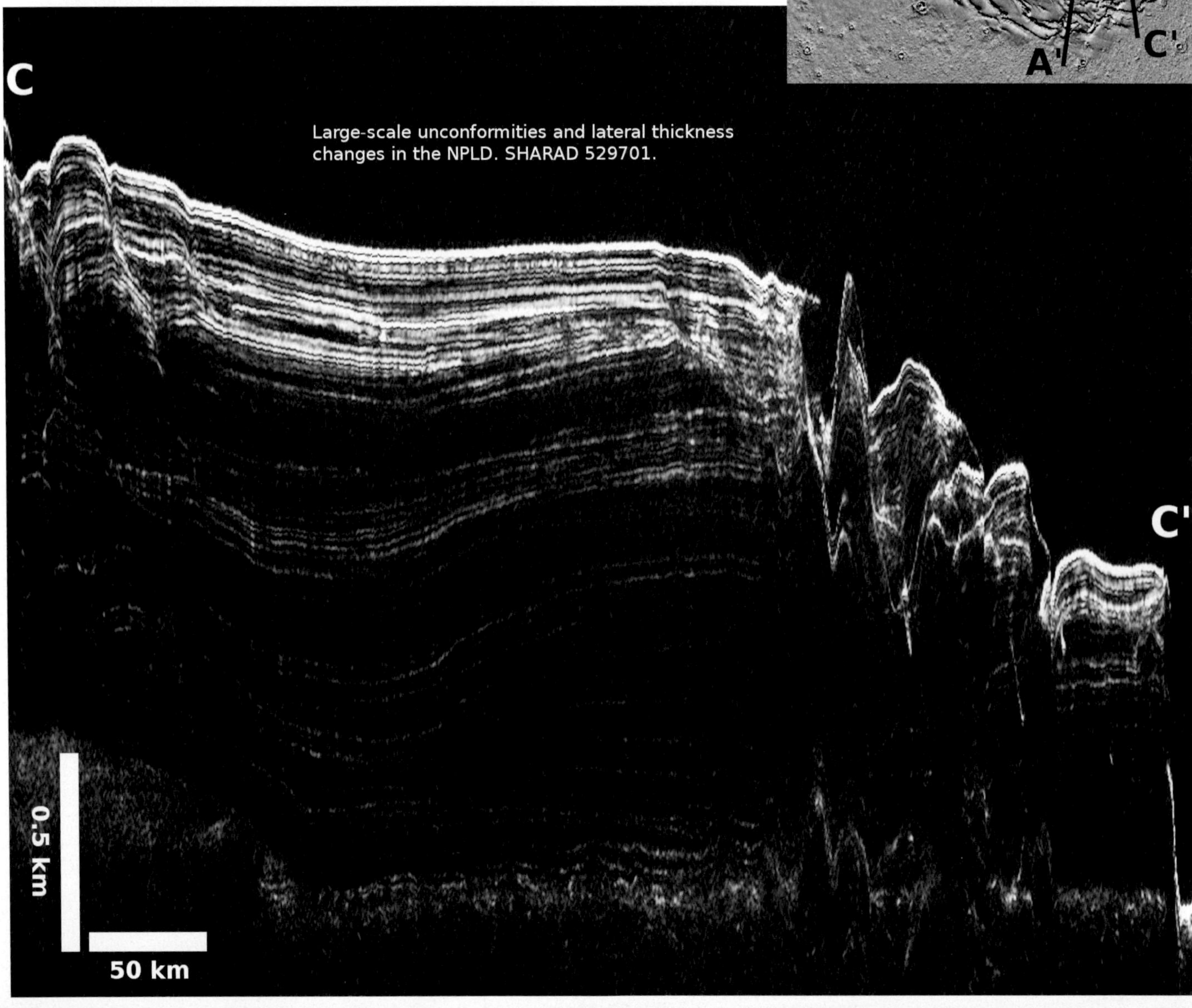

Large-scale unconformities and lateral thickness changes in the NPLD. SHARAD 529701.

South Polar Layered Deposits

Folded and faulted layers in the South Polar layered deposits, HiRISE PSP_004708_1000.

The south polar layered deposits (SPLDs) form a stack of strata up to 3 km thick observed in the walls of troughs and scarps in the south polar regions (e.g., Chasma Australe, Promethei Chasma, Ultimum Chasma; Byrne 2009). The SPLDs are composed of varying amounts of water-ice and dust, with gravity data indicating concentrations of the latter at ~15%. Impact craters are more common on the SPLDs, indicating that their surface is older than the surface of the NPLDs, with age estimates ranging from 30–100 Ma. The SPLDs have been broken out into three main units, although each of these units may be subdivided into rhythmic strata formed of smaller-scale bedding. The smallest-scale bedding observed in the SPLDs is typically thicker as compared to the NPLDs, and subsurface relationships are less evident in orbiter-based radar observations of the SPLDs. However, it is likely that the SPLDs also record climate variations on Mars (Byrne 2009). Orbiter-based surface-penetrating radar data reveal pervasive volume scattering that obscures layering in many locations and that may be due to fracturing at the scale of the radar wavelength. The surface of the SPLDs appears to be a low-thermal-inertia material consistent with a dust lag.

yered Deposits (center d lower right) and "Swiss eese terrain" (upper left) the South Polar terrain, RISE PSP_002856_0875.

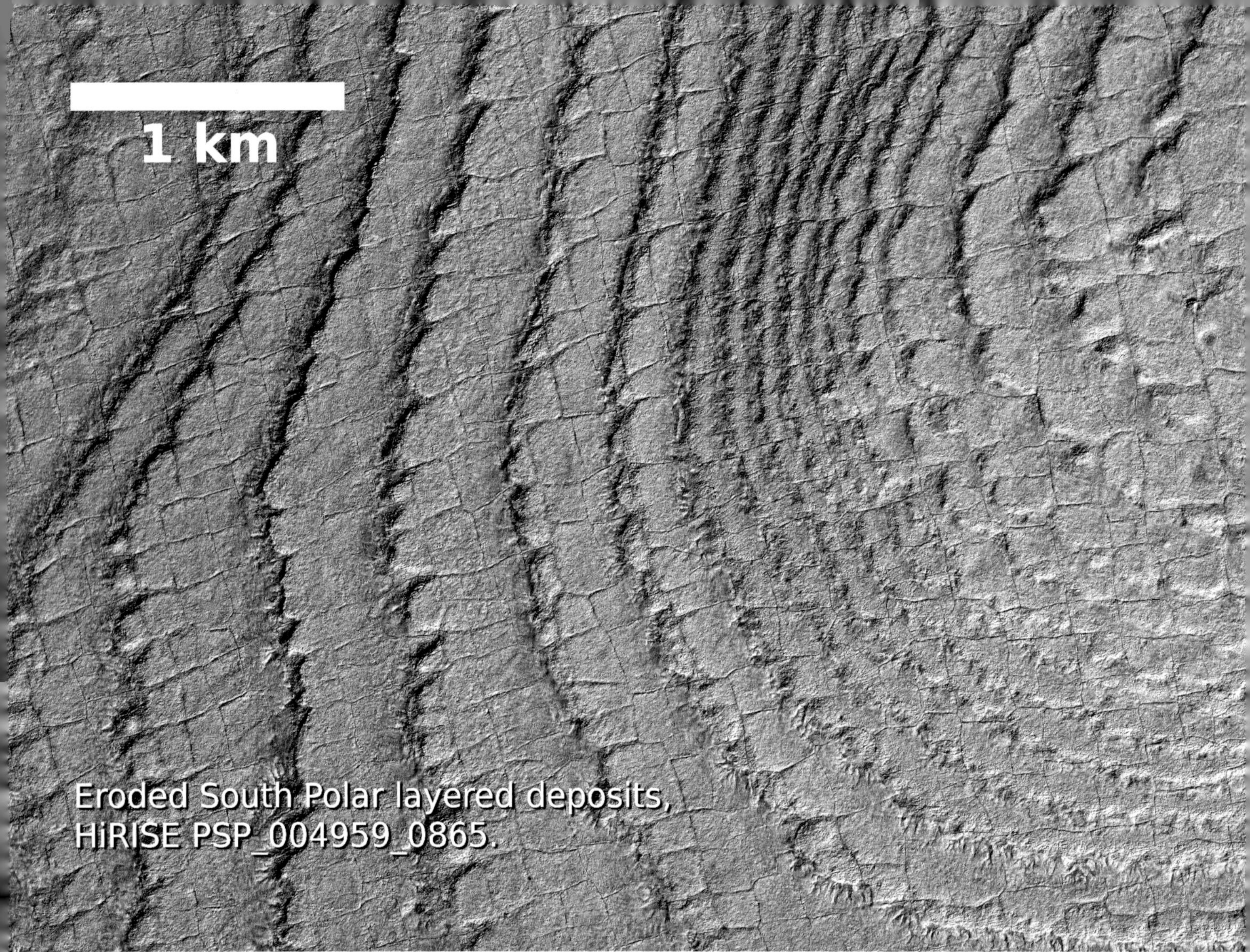

Eroded South Polar layered deposits, HiRISE PSP_004959_0865.

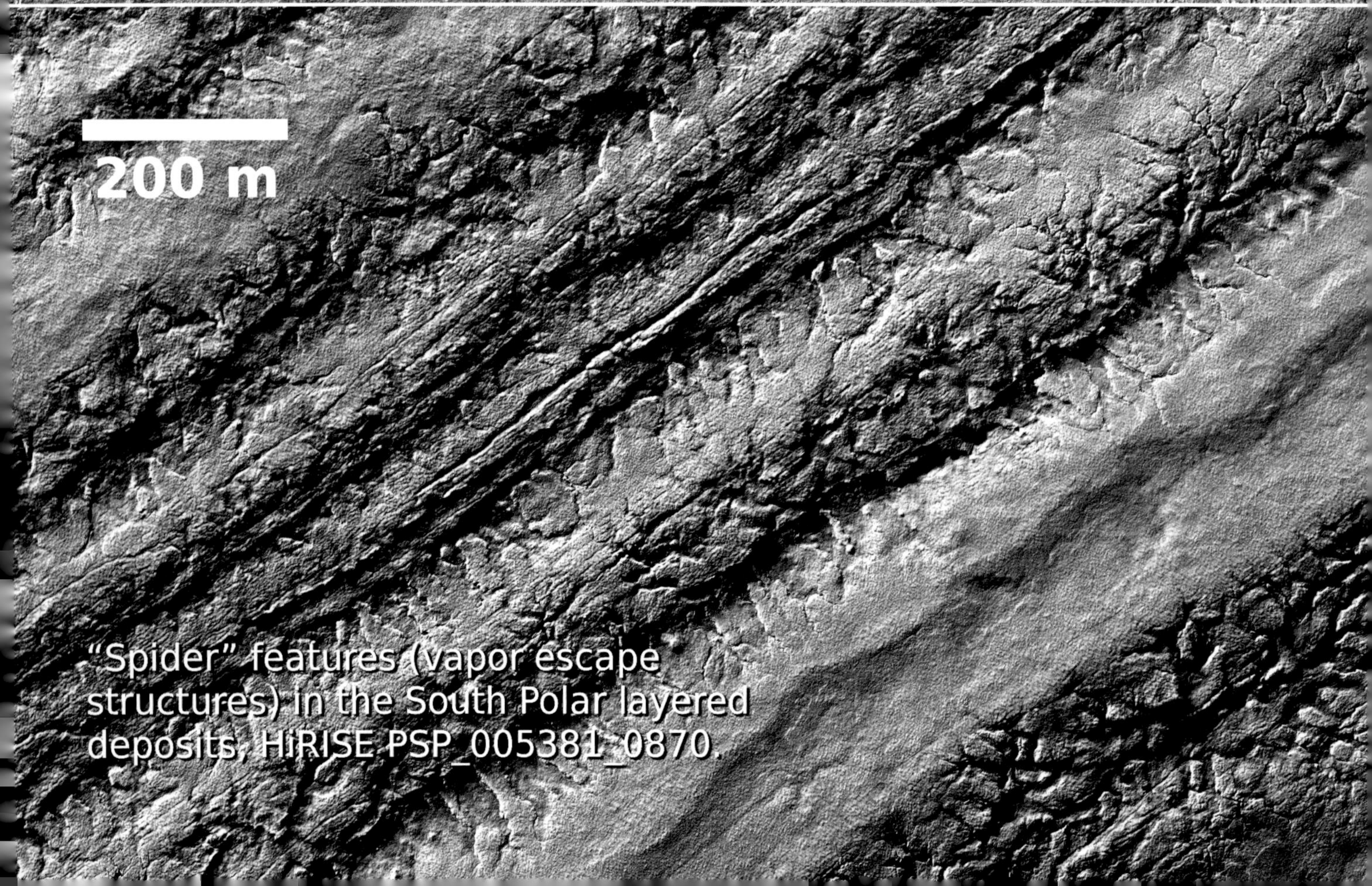

“Spider” features (vapor escape structures) in the South Polar layered deposits, HiRISE PSP_005381_0870.

ACKNOWLEDGMENTS

This work has made use of National Aeronautics and Space Administration's Astrophysics Data System and the US Geological Survey's Integrated Software for Imagers and Spectrometers (ISIS). This material is based upon work supported by the National Aeronautics and Space Administration under awards issued through the Mars Reconnaissance Orbiter program. The authors would like to thank Ken Tanaka and Jenny Blue for their early review of the draft manuscript. The authors thank Paul Harris and Joe Michalski for their careful reviews.

REFERENCES

Ansan V, Loizeau D, Mangold N, Le Mouélic S, Carter J, Poulet F, Dromart G, Lucas A, Bibring J-P, Gendrin A, Gondet B, Langevin Y, Masson Ph, Murchie S, Mustard JF, Neukum G. 2011. Stratigraphy, mineralogy, and origin of layered deposits inside Terby crater, Mars. *Icarus* 211(1):273–304. DOI:10.1016/j.icarus.2010.09.011

Ansan V, Mangold N. 2003. Identification of past polar deposits among layered terrains on Mars: preliminary results. *In* Third International Conference on Mars Polar Science and Exploration; October 13–17, 2003; Alberta, Canada, abstract #8071.

Bibring J-P, Arvidson RE, Gendrin A, Gondet B, Langevin Y, Le Mouelic S, Mangold N, Morris RV, Mustard JF, Poulet F, Quantin C, Sotin C. 2007. Coupled ferric oxides and sulfates on the Martian surface. *Science* 317:1206–1210. DOI:10.1126/science.1144174

Bibring J-P, Langevin Y, Gendrin A, Gondet B, Poulet F, Berthé M, Soufflot A, Arvidson R, Mangold N, Mustard J, Drossart P, and the OMEGA Team. 2005. Mars surface diversity as revealed by the OMEGA/Mars Express observations. *Science* 307:1576–1581. DOI:10.1126/science.1108806

Bibring J-P, Langevin Y, Mustard JF, Poulet F, Arvidson R, Gendrin A, Gondet B, Mangold N, Pinet P, Forget F, and the OMEGA Team. 2006. Global mineralogical and aqueous Mars history derived from OMEGA/Mars Express data. *Science* 312:400–404. DOI:10.1126/science.1122659

Bibring J-P, Soufflot A, Berthé M, Langevin Y, Gondet B, Drossart P, Bouyé M, Combes M, Puget P, Semery A, Bellucci G, Formisano V, Moroz V, Kottsov V, Bonello G, Erard S, Forni O, Gendrin A, Manaud N, Poulet F, Poulleau G, Encrenaz T, Fouchet T, Melchiori R, Altieri F, Ignatiev N, Titov D, Zasova L, Coradini A, Capacionni F, Cerroni P, Fonti S, Mangold N, Pinet P, Schmitt B, Sotin C, Hauber E, Hoffmann H, Jaumann R, Keller U, Arvidson R, Mustard J, Forget F. 2004. OMEGA: Observatoire pour la Minéralogie, l'Eau, les Glaces et l'Activité. *In* Wilson A, Chicarro A (Editors). *Mars Express: The Scientific Payload*: ESA Publications Division, Noordwijk, Netherlands. ESA SP-1240, p. 37–49.

Bills BG. 1990. The rigid body obliquity history of Mars. *Journal of Geophysical Research* 95:14137–14153.

Bishop J, Noe Dobrea EZ, McKeown NK, Parente M, Ehlmann BL, Michalski JR, Milliken RE, Poulet F, Swayze GA, Mustard JF, Murchie SL, Bibring J-P. 2008. Phyllosilicate diversity and past aqueous activity revealed at Mawrth Vallis, Mars. *Science* 321:830–833. DOI:10.1126/science.1159699

Bridges JC, Kim JR, Tragheim DG, Muller J-P, Balme MR, Pullan D. 2008. Sedimentary rocks in Becquerel crater: origin as polar layered deposits during high obliquity. *In* Lunar and Planetary Science Conference 39, abstract #1913.

Byrne S. 2009. The polar deposits of Mars. *Annual Review of Earth and Planetary Sciences* 37(1):535–560. DOI:10.1146/annurev.earth.031208.100101

Byrne S, Murray BC. 2002. North polar stratigraphy and the paleo-erg of Mars. *Journal of Geophysical Research* 107(E6):11-1, CiteID 5044. DOI:10.1029/2001JE001615

Cabrol NA, Grin EA, Newsom HE, Landheim R, McKay CP. 1999. Hydrogeologic evolution of Gale crater and its relevance to the exobiological exploration of Mars. *Icarus* 139:235–245. DOI:10.1006/icar.1999.6099

Caudill C, Tornabene L, McEwen AS, Wray J. 2011. Crater-exposed intact stratigraphy blocks and volcanogenic origin. *In* Lunar and Planetary Science Conference 42, abstract #2393.

Chapman CR, Pollack JB, Sagan C. 1968. An Analysis of the Mariner 4 Photography of Mars. SAO Special Report 268.

Ehlmann BL, Mustard JF, Swayze G, Clark RN, Bishop JL, Poulet F, Des Marais DJ, Roach LH, Milliken RE, Wray JJ, Barnouin-Jha O, Murchie S. 2009. Identification of hydrated silicate minerals on Mars using MRO-CRISM: geologic context near Nili Fossae and implications for aqueous alteration. *Journal of Geophysical Research* 114. DOI:10.1029/2009JE003339

Fassett CI, Head JW III. 2008. The timing of Martian valley network activity: constraints from buffered crater counting. *Icarus* 95:61–89. DOI:10.1016/j.icarus.2007.12.009

Fishbaugh KE, Head JW. 2005. Origin and characteristics of the Mars north polar basal unit and implications for polar geologic history. *Icarus* 174:444–474.

Fishbaugh KE, Hvidberg CS, Byrne S, Russell PS, Herkenhoff KE, Winstrup M, Kirk R. 2010. The first high-resolution stratigraphic column of the Martian north polar layered deposits. *Geophysical Research Letters* 37:L07201. DOI:10.1029/2009GL041642

Fishbaugh KE, Poulet F, Chevrier V, Langevin Y, Bibring JP. 2007. On the origin of gypsum in the Mars north polar region. *Journal of Geophysical Research* 112:7002.

Gendrin A, Mangold N, Bibring J-P, Langevin Y, Gondet B, Poulet F, Bonello G, Quantin C, Mustard J, Arvidson R, Le Mouélic S. 2005. Sulfates in Martian layered terrains: the OMEGA/Mars Express view. *Science* 207(5715):1587–1591. DOI:10.1126/science.1109087

Grant JA, Irwin RP, Grotzinger JP, Milliken RE, Tornabene LL, McEwen AS, Weitz CM, Squyres SW, Glotch TD, Thomson BJ. 2008. HiRISE imaging of impact megabreccia and sub-meter aqueous strata in Holden crater, Mars. *Geology* 36(3):195–198. DOI:10.1130/G24340A.1

Grant JA, Irwin RP, Wilson SA, Buczkowski D, Siebach K. 2010. A lake in Uzboi Vallis and implications for Late Noachian–Early Hesperian climate on Mars. *Icarus* 212:110–122. DOI:10.1016/j.icarus.2010.11.024

Grant JA, Wilson SA. 2011. Late alluvial fan formation in Margaritifer Terra. *In* 5th MSL Landing Site Workshop; Monrovia, California.

Grotzinger JP, Arvidson RE, Bell JF, Calvin W, Clark BC, Fike DA, Golombek M, Greeley R, Haldemann A, Herkenhoff KE, Jolliff BL, Knoll AH, Malin M, McLennan SM, Parker T, Soderblom L, Sohl-Dickstein JN, Squyres SW, Tosca NJ, Watters WA. 2005. Stratigraphy and sedimentology of a dry to wet eolian depositional system, Burns formation, Meridiani Planum, Mars. *Earth and Planetary Science Letters* 240(1):11–72. DOI:10.1016/j.epsl.2005.09.039

Grotzinger JP, Beaty D, Dromart G, Gupta S, Harris M, Hurowitz J, Kocurek G, McLennan S, Milliken R, Ori GG, Sumner D. 2011. Mars sedimentary geology: key concepts and outstanding questions. *Astrobiology* 11(1):77–87. DOI:10.1089/ast.2010.0571

Hartmann WK. 2005. Martian cratering 8: Isochron refinement and the chronology of Mars. *Icarus* 174:294–320. DOI:10.1016/j.icarus.2004.11.023

Hartmann WK, Neukum G. 2001. Cratering chronology and the evolution of Mars. *Space Science Reviews* 96(1/4):165–194. DOI:10.1023/A:1011945222010

Hauber E, Gwinner K, Kleinhans M, Reiss D, di Achille G, Ori GG, Scholten F, Marinangeli L, Jaumann R, Neukum G. 2009. Sedimentary deposits in Xanthe Terra: implications for the ancient climate on Mars. *Planetary and Space Science* 57(8–9):944–957. DOI:10.1016/j.pss.2008.06.009

Hayden FV. 1872. Preliminary Report of the United States Geological Survey of Montana and Portions of Adjacent Territories being a Fifth Annual Report of Progress: US Department of Interior, Washington, D.C.

Head JW, Mustard JF, Kreslavsky MA, Milliken RE, Marchant DR. 2003. Recent ice ages on Mars. *Nature* 426:797–802.

Herkenhoff KE, Byrne S, Russell PS, Fishbaugh KE, McEwen AS. 2007. Meter-scale morphology of the north polar region of Mars. *Science* 317(5845):1711. DOI:10.1126/science.1143544

Holt JW, Fishbaugh KE, Byrne S, Christian S, Tanaka K, Russell P, Herkenhoff K, Safaeinili A, Putzig N, Phillips R. 2010. The construction of Chasma Boreale on Mars. *Nature* 465:446–449. DOI:10.1038/nature09050

Irwin RP III. 2011. Timing, duration, and hydrology of the Eberswalde crater paleolake, Mars. *In* Lunar and Planetary Science Conference 42, abstract #2748.

Kraal ER, van Dijk M, Postma G, Kleinhans MG. 2008. Martian stepped-delta formation by rapid water release. *Nature* 451:973–976. DOI:10.1038/nature06615

Langevin Y, Poulet F, Bibring JP, Gondet B. 2005. Sulfates in the north polar region of Mars detected by OMEGA/Mars Express. *Science* 307:1584–1586.

Le Deit L, Le Mouélic S, Bourgeois O, Combe J-P, Mège D, Sotin C, Gendrin A, Hauber E, Mangold N, Bibring J-P. 2007. Ferric oxides in East Candor Chasma, Valles Marineris (Mars), inferred from analysis of OMEGA/Mars Express data: identification and geological interpretation. *Journal of Geophysical Research* 113:E07001. DOI:10.1029/2007JE002950

Le Deit L, Le Mouélic S, Bourgeois O, Mège D, Massé M, Quantin-Nataf C, Sotin C, Bibring J-P, Gondet B, Langevin Y. 2008. Composition and morphology of hydrated layered deposits on the plains around Valles Marineris (Mars). *In* Workshop on Martian Phyllosilicates: Recorders of Aqueous Processes; Paris, France: Lunar and Planetary Institute Contribution No. 1441, p. 39–40.

Lewis KW, Aharonson O. 2006. Stratigraphic analysis of the distributary fan in Eberswalde crater using stereo imagery. *Journal of Geophysical Research* 111(E6). CiteID E06001. DOI:10.1029/2005JE002558

Lewis KW, Aharonson O, Grotzinger JP, Kirk RL, McEwen AS, Suer T-A. 2008. Quasi-periodic bedding in the sedimentary rock record of Mars. *Science* 322(5907):1532–1535. DOI:10.1126/science.1161870

Lewis KW, Aharonson O, Grotzinger JP, McEwen AS, Kirk RL. 2010. Global significance of cyclic sedimentary deposits on Mars. *In* Lunar and Planetary Science Conference 41, abstract #2648.

Loizeau D, Mangold N, Poulet F, Bibring J-P, Gendrin A, Ansan V, Gomez C, Gondet B, Langevin Y, Masson P, Neukum G. 2007. Phyllosilicates in the Mawrth Vallis region of Mars. *Journal of Geophysical Research* 112:E08S08. DOI:10.1029/2006JE002877

Lucchitta B. 2010. Lakes in Valles Marineris. *In* Cabrol NA, Grin EA (Editors). *Lakes on Mars*: Elsevier, p. 111.

Malin MC, Bell JF, Cantor BA, Caplinger MA, Calvin WM, Clancy RT, Edgett KS, Edwards L, Haberle RM, James PB, Lee SW, Ravine MA, Thomas PC, Wolff M. 2007. Context Camera investigation on board the Mars Reconnaissance Orbiter. *Journal of Geophysical Research* 112(E5). CiteID E05S04. DOI:10.1029/2006JE002808

Malin MC, Edgett KS. 2000. Sedimentary rocks of early Mars. *Science* 290:1927–1937.

Malin MC, Edgett KS. 2003. Evidence for persistent flow and aqueous sedimentation on early Mars. *Science* 302(5652):1931–1934. DOI:10.1126/science.1090544

Mangold N, Gendrin A, Gondet B, Le Mouelic S, Quantin C, Ansan V, Bibring J-P, Langevin Y, Masson P, Neukum G. 2008. Spectral and geological study of the sulfate-rich region of West Candor Chasma, Mars. *Icarus* 194(2):519–543. DOI:10.1016/j.icarus.2007.10.021

Mangold N, Poulet F, Mustard JF, Bibring J-P, Gondet F, Langevin Y, Ansan V, Masson P, Fassett C, Head JW, Hoffmann J, Neukum G. 2007. Mineralogy of the Nili Fossae region with OMEGA/Mars Express data: 2. Aqueous alteration of the crust. *Journal of Geophysical Research* 112:E08S04.

Mangold N, Quantin C, Ansan V, Delacourt C, Allemand P. 2004. Evidence for precipitation on Mars from Dendritic Valleys in the Valles Marineris area. *Science* 305:78–81. DOI:10.1126/science.1097549

Massé M, Bourgeois O, Le Mouélic S, Verpoorter C, Le Deit L, Bibring JP. 2010. Martian polar and circum-polar sulfate-bearing deposits: sublimation tills derived from the north polar cap. *Icarus* 209:434–451.

McEwen AS, Eliason EM, Bergstrom JW, Bridges NT, Hansen CJ, Delamere WA, Grant JA, Gulick VC, Herkenhoff KE, Keszthelyi L, Kirk RL, Mellon MT, Squyres SW, Thomas N, Weitz CM. 2007. Mars Reconnaissance Orbiter's High Resolution Imaging Science Experiment (HiRISE). *Journal of Geophysical Research* 112(E5). CiteID E05S02. DOI:10.1029/2005JE002605

McLennan SM, Grotzinger JP. 2008. The sedimentary rock cycle of Mars. *In* Bell J III (Editor). *The Martian Surface—Composition, Mineralogy, and Physical Properties*: Cambridge, UK, Cambridge University Press. 652 p.

Metz JM, Grotzinger JP, Mohrig D, Milliken R, Prather B, Pirmez C, McEwen AS, Weitz CM. 2009. Sublacustrine depositional fans in southwest Melas Chasma. *Journal of Geophysical Research* 114:E10002. DOI:10.1029/2009JE003365

Metz J, Grotzinger J, Okubo C, Milliken R. 2010. Thin-skinned deformation of sedimentary rocks in Valles Marineris, Mars. *Journal of Geophysical Research* 115:E11004. DOI:10.1029/2010JE003593

Michalski J, Noe Dobrea E. 2007. Evidence for a sedimentary origin of clay minerals in the Mawrth Vallis region, Mars. *Geology* 35:951–954.

Milliken RE, Bish D. 2010. Sources and sinks of clay minerals on Mars. *Philosophical Magazine* 90(17):2293–2308.

Milliken RE, Grotzinger JP, Thomson BJ. 2010. Paleoclimate of Mars as captured by the stratigraphic record in Gale crater. *Geophysical Research Letters* 37(4). CiteID L04201. DOI:10.1029/2009GL041870

Milliken RE, Swayze GA, Arvidson RE, Bishop JL, Clark RN, Ehlmann BL, Green RO, Grotzinger JP, Morris RV, Murchie SL, Mustard JF, Weitz C. 2008. Opaline silica in young deposits on Mars. *Geology* 26(11):847–850. DOI:10.1130/G24967A.1

Moore JM, Howard AD. 2005. Large alluvial fans on Mars. *Journal of Geophysical Research* 110:E04005. DOI:10.1029/2004JE002352

Moore JM, Howard AD, Dietrich WE, Schenk PM. 2003. Martian layered fluvial deposits: implications for Noachian climate scenarios. *Geophysical Research Letters* 30(24):PLA 6–1. CiteID 2292. DOI:10.1029/2003GL019002

Murchie S, Arvidson R, Bedini P, Beisser K, Bibring J-P, Bishop J, Boldt J, Cavender P, Choo T, Clancy RT, Darlington EH, Des Marais D, Espiritu R, Fort D, Green R, Guinness E, Hayes J, Hash C, Heffernan K, Hemmler J, Heyler G, Humm D, Hutcheson J, Izenberg N, Lee R, Lees J, Lohr D, Malaret E, Martin T, McGovern JA, McGuire P, Morris R, Mustard J, Pelkey S, Rhodes E, Robinson M, Roush T, Schaefer E, Seagrave G, Seelos F, Silverglate P, Slavney S, Smith M, Shyong W-J, Strohbehn K, Taylor H, Thompson P, Tossman B, Wirzburger M, Wolff M. 2007. Compact Reconnaissance Imaging Spectrometer for Mars (CRISM) on Mars Reconnaissance Orbiter (MRO). *Journal of Geophysical Research* 112:E05S03. DOI:10.1029/2006JE002682

Murchie SL, Mustard JF, Ehlmann BL, Milliken RE, Bishop JL, McKeown NK, Dobrea EZN, Seelos FP, Buczkowski DL, Wiseman SM, Arvidson RE, Wray JJ, Swayze G, Clark RN, Des Marais DJ, McEwen AS, Bibring J-P. 2009a. A synthesis of Martian aqueous mineralogy after 1 Mars year of observations from the Mars Reconnaissance Orbiter. *Journal of Geophysical Research–Planets* 114. DOI:10.1029/2009JE003342

Murchie S, Roach L, Seelos F, Milliken R, Mustard J, Arvidson R, Wiseman S, Lichtenberg K, Andrews-Hanna J, Bishop J, Bibring J-P, Parente M, Morris R. 2009b. Evidence for the origin of layered deposits in Candor Chasma, Mars, from mineral composition and hydrologic modeling. *Journal of Geophysical Research* 114. CiteID E00D05. DOI:10.1029/2009JE003343

Murray BC, Soderblom LA, Cutts JA, Sharp RP, Milton DJ, Leighton RB. 1972. Geological framework of the south polar region of Mars. *Icarus* 17:328–345.

Mustard JF, Ehlmann BL. 2010. Three Distinct Habitable Environments Defined by Aqueous Alteration Traversing the Alkaline–Acidic Transition. Memoranda to the MSL Landing Site Selection Process Describing the NE Syrtis Site. http://marsoweb.nas.nasa.gov/landingsites/msl/memoranda/sites_jan10/Aqueous_enviroments_MSL-Mustard.pdf. Accessed.

Mustard JF, Poulet F, Head JW, Mangold N, Bibring J-P, Pelkey S, Fassett C, Langevin Y, Neukum G. 2007. Mineralogy of the Nili Fossae region with OMEGA/Mars Express data: 1. Ancient impact melt in the Isidis Basin and implications for the transition from the Noachian to Hesperian. *Journal of Geophysical Research–Planets* 112:E08S03.

Okubo C. 2010. Structural geology of Amazonian-aged layered sedimentary deposits in southwest Candor Chasma, Mars. *Icarus* 207(1):210–225. DOI:10.1016/j.icarus.2009.11.012

Okubo CH, Lewis KW, McEwen AS, Kirk RL. 2008. Relative age of interior layered deposits in southwest Candor Chasma based on high-resolution structural mapping. *Journal of Geophysical Research* 113:E12002. DOI:10.1029/2008JE003181

Ori GG, Marinangeli L, Baliva A. 2000. Terraces and Gilbert-type deltas in crater lakes in Ismenius Lacus and Memnonia (Mars). *Journal of Geophysical Research* 105(E7):17629–17642. DOI:10.1029/1999JE001219

Phillips RJ, Zuber MT, Smrekar SE, Mellon MT, Head JW, Tanaka KL, Putzig NE, Milkovich SM, Campbell BA, Plaut JJ, Safaeinili A, Seu R, Biccari D, Carter LM, Picardi G, Orosei R, Mohit PS, Heggy E, Zurek RW, Egan AF, Giacomoni E, Russo F, Cutigni M, Pettinelli E, Holt JW, Leuschen CJ, Marinangeli L. 2008. Mars north polar deposits: stratigraphy, age, and geodynamical response. *Science* 320(5880):1182. DOI:10.1126/science.1157546

Pondrelli M, Rossi AP, Marinangeli L, Hauber E, Gwinner K, Baliva A, di Lorenzo S. 2008. Evolution and depositional environments of the Eberswalde fan delta, Mars. *Icarus* 197(2):429–451. DOI:10.1016/j.icarus.2008.05.018

Poulet F, Bibring J-P, Mustard JF, Gendrin A, Mangold N, Langevin Y, Arvidson RE, Gondet B, Gomez C. 2005. Phyllosilicates on Mars and implications for early Martian climate. *Nature* 438:623–627. DOI:10.1038/nature04274

Reiss D, Hauber E, Gwinner K, Scholten F, Jaumann R, di Achille G, Marinangeli L, Ori GG, Neukum G. 2006. Geologic evolution of the Galle crater, Mars. *In* European Planetary Science Congress; September 18–22, 2006; Berlin, Germany: p. 529.

Roach LH, Mustard JF, Murchie S, Langevin Y, Bibring J-P, Bishop J, Bridges N, Brown A, Byrne S, Ehlmann BL, Herkenhoff K, McGuire PC, Milliken RE, Pelkey S, Poulet F, Seelos FP, Seelos K, and the CRISM Team. 2007. CRISM spectral signatures of the North Polar Gypsum Dunes. *In* Lunar and Planetary Science Conference 38, abstract #1970.

Rossi AP, Neukum G, Pondrelli M, van Gasselt S, Zegers T, Hauber E, Chicarro A, Foing B. 2008. Large-scale spring deposits on Mars? *Journal of Geophysical Research* 113(E8). CiteID E08016. DOI:10.1029/2007JE003062

Schultz P, Lutz AB. 1988. Polar wandering of Mars. *Icarus* 73:91–141. DOI:10.1016/0019–1035(88)90087–5

Scott DH, Carr MH. 1978. *Geologic Map of Mars*: US Geological Survey. Miscellaneous Investigations Series Map I-1083. scale 1:25,000,000.

Seu R, Phillips RJ, Biccari D, Orosei R, Masdea A, Picardi G, Safaeinili A, Campbell BA, Plaut JJ, Marinangeli L, Smrekar SE, Nunes DC. 2007. SHARAD sounding radar on the Mars Reconnaissance Orbiter. *Journal of Geophysical Research* 112(E5). CiteID E05S05. DOI:10.1029/2006JE002745

Smith IB, Holt JW. 2010. Onset and migration of spiral troughs on Mars revealed by orbital radar. *Nature* 465:450–453. DOI:10.1038/nature09049

Squyres SW, Arvidson RE, Bell JF, Brückner J, Cabrol NA, Calvin W, Carr MH, Christensen PR, Clark BC, Crumpler L, Des Marais DJ, d'Uston C, Economou T, Farmer J, Farrand W, Folkner W, Golombek M, Gorevan S, Grant JA, Greeley R, Grotzinger J, Haskin L, Herkenhoff KE, Hviid S, Johnson J, Klingelhöfer G, Knoll AH, Landis G, Lemmon M, Li R, Madsen MB, Malin MC, McLennan SM, McSween HY, Ming DW, Moersch J, Morris RV, Parker T, Rice JW, Richter L, Rieder R, Sims M, Smith M, Smith P, Soderblom LA, Sullivan R, Wänke H, Wdowiak T, Wolff M, Yen A. 2004. The *Opportunity* rover's Athena science investigation at Meridiani Planum, Mars. *Science* 306:1698–1703. DOI:10.1126/science.1106171

Tanaka KL. 1986. The stratigraphy of Mars. *Journal of Geophysical Research* 91:E139–E158. DOI:10.1029/JB091iB13p0E139

Tanaka KL. 2006. Mars' North Polar Gypsum: possible origin related to early Amazonian magmatism at Alba Patera and Aeolian Mining. *In* 4th International Conference on Mars Polar Sci. Expl., Davos, Switzerland, p. 8024.

Tanaka KL, Hartmann WK. 2008. Planetary time scale. *In* Ogg JG, Ogg G, Gradstein FM (Editors). *The Concise Geologic Time Scale*: Cambridge, UK, Cambridge University Press.

Tanaka KL, Rodriguez AP, Skinner JA Jr, Bourke MC, Fortezzo CM, Herkenhoff KE, Kolb EJ, Okubo CH. 2008. North polar region of Mars: advances in stratigraphy, structure, and erosional modification. *Icarus* 196(2):318–358. DOI:10.1016/j.icarus.2008.01.021

Tanaka KL, Scott DH, Greeley R. 1992. Global stratigraphy. *In* Kieffer HH, Jakosky BM, Conway CW, Matthews MS (Editors). *Mars*: The University of Arizona Press, Tucson, Arizona.

Touma J, Wisdom J. 1993. The chaotic obliquity of Mars. *Science* 259:1294–1297.

Ward WR. 1974. Climatic variations on Mars. 1. Astronomical theory of insolation. *Journal of Geophysical Research* 79:3375–3386.

Ward WR. 1979. Present obliquity oscillations of Mars: fourth-order accuracy in orbital e and I. *Journal of Geophysical Research* 84:237–241.

Weitz CM, Milliken RE, Grant JA, McEwen AS, Williams RME, Bishop JL, Thomson BJ. 2010. Mars Reconnaissance Orbiter observations of light-toned layered deposits and associated fluvial landforms on the plateaus adjacent to Valles Marineris. *Icarus* 205(1):73–102. DOI:10.1016/j.icarus.2009.04.017

Wilson SA, Howard AD, Moore JM, Grant JA. 2007. Geomorphic and stratigraphic analysis of crater Terby and layered deposits north of Hellas Basin, Mars. *Journal of Geophysical Research* 112(E8). CiteID E08009. DOI:10.1029/2006JE002830

Wiseman SM, Arvidson RE, Morris RV, Poulet F, Andrews-Hanna JC, Bishop JL, Murchie SL, Seelos FP, Des Marais D, Griffes JL. 2010. Spectral and stratigraphic mapping of hydrated sulfate and phyllosilicate-bearing deposits in northern Sinus Meridiani, Mars. *Journal of Geophysical Research* 115:E00D18. DOI:10.1029/2009JE003354

Wray JJ, Milliken RE, Dundas CM, Swayze GA, Andrews-Hanna JC, Baldridge AM, Chojnacki M, Bishop JL, Ehlman BL, Murchie SL, Clark RN, Seelos FP, Tornabene LL, Squyres SW. 2011. Columbus crater and other possible groundwater-fed paleolakes of Terra Sirenum, Mars. *Journal of Geophysical Research* 116:E01001. DOI:10.1029/2010JE003694

AQUEOUS ALTERATION IN MARTIAN METEORITES: COMPARING MINERAL RELATIONS IN IGNEOUS-ROCK WEATHERING OF MARTIAN METEORITES AND IN THE SEDIMENTARY CYCLE OF MARS

MICHAEL A. VELBEL

Department of Geological Sciences, 206 Natural Science Building, Michigan State University, East Lansing, Michigan 48824-1115 USA

e-mail: velbel@msu.edu

Abstract: Many of the minerals observed or inferred to occur in the sediments and sedimentary rocks of Mars, from a variety of Mars-mission spacecraft data, also occur in Martian meteorites. Even Martian meteorites recovered after some exposure to terrestrial weathering can preserve preterrestrial evaporite minerals and useful information about aqueous alteration on Mars, but the textures and textural contexts of such minerals must be examined carefully to distinguish preterrestrial evaporite minerals from occurrences of similar minerals redistributed or formed by terrestrial processes. Textural analysis using terrestrial microscopy provides strong and compelling evidence for preterrestrial aqueous alteration products in a number of Martian meteorites. Occurrences of corroded primary rock-forming minerals and alteration products in meteorites from Mars cover a range of ages of mineral–water interaction, from ca. 3.9 Ga (approximately mid-Noachian), through one or more episodes after ca. 1.3 Ga (approximately mid–late Amazonian), through the last half billion years (late Amazonian alteration in young shergottites), to quite recent. These occurrences record broadly similar aqueous corrosion processes and formation of soluble weathering products over a broad range of times in the paleoenvironmental history of the surface of Mars.

Many of the same minerals (smectite-group clay minerals, Ca-sulfates, Mg-sulfates, and the K-Fe–sulfate jarosite) have been identified both in the Martian meteorites and from remote sensing of the Martian surface. This suggests that both kinds of samples—Martian meteorites and Mars' surface rocks, regolith, and soils—were altered under broadly similar conditions. Temporarily and locally occurring but likely stagnant aqueous solutions reacted quickly with basaltic/mafic/ultramafic minerals at low water–rock ratios. Solutes released by primary mineral weathering precipitated locally on Mars as cation-rich clays and evaporite minerals, rather than being leached away, as on Earth.

The main secondary host minerals for Fe differ between Martian meteorites and Mars' surface materials. In Martian meteorites, sideritic–ankeritic carbonate is the predominant secondary host mineral for Fe, whereas in Mars' surface materials, ferric oxides and ferric sulfates are the predominant secondary host minerals for Fe. Differences in the initial compositions of the altering solutions are implied, with carbonate/bicarbonate dominating in the solutions that altered Martian meteorites, and sulfate dominating the solutions that altered most Mars' surface materials. During impact on and ejection from Mars, Martian meteorites may have been exhumed from depths sufficient to have isolated them from large quantities of Mars' surface solutions. Pre-ejection weathering of the basaltic rocks occurred in grain-boundary fracture microenvironments at high pH values in aqueous solutions buffered by reactions with basalt minerals.

INTRODUCTION

On Earth, coarse clastic detritus is dominated by primary rock-forming minerals and rock fragments that persisted against the weathering to which the source rocks were subjected; fine clastic particles are products of weathering of primary minerals in source rocks; and chemical sediments formed by precipitation of dissolved products of weathering. Most introductory treatments of weathering on Earth begin with the weathering of Earth's dominant continental igneous rock type, granite (e.g., Grotzinger and Jordan 2010). Martian meteorites are mafic and ultramafic rocks from the predominantly basaltic planet Mars, and they are presently the only samples available for study in terrestrial laboratories. The purpose of this paper is to examine the weathering of primary rock-forming minerals and the formation of secondary minerals on Mars from the perspective of mineral and textural evidence in Martian meteorites.

Mineral–water interactions during weathering at the interface between a solid rocky planet and its fluid envelope (hydrosphere and/or atmosphere) modify primary minerals, produce new minerals as alteration products, and modify the solute loads of the solutions. The relationships among these phenomena can be summarized in the following generic mineral weathering reaction:

$$\text{Primary mineral} + \text{water} + \text{solutes} \rightarrow \text{secondary mineral(s)} + \text{modified solutes} \quad (1)$$

A usually mobile solvent (water) brings mobile solutes (e.g., environmental acidity) to the surface of primary minerals in a soil, regolith, or rock undergoing weathering. If the solution is sufficiently undersaturated with respect to the primary mineral, the primary mineral and the solution react. The primary mineral may be completely destroyed if geochemical reaction kinetics are fast, but partially weathered primary mineral grains survive for a variety of reasons. Factors favoring persistence of primary minerals in weathered residues include:

- slow reaction kinetics caused by intrinsically sluggish reactions due to strong bonding in the reactant mineral (Velbel 1999) or attributes of the reactant solution, such as low degree of undersaturation of solution with respect to the primary mineral (Velbel 1989a, White and Brantley 2003);
- restricted abundance of reactive solution (Nahon 1991, Taylor and Eggleton 2001); or
- physical removal of grains from the soil or regolith by erosion before weathering completely destroys them (Grantham and Velbel 1988; Johnsson et al. 1988; Johnsson and Stallard 1989; Johnsson 1990, 1992; Velbel 2007; Velbel et al. 2007).

Secondary minerals are weathering products (e.g., clay minerals and oxides/oxyhydroxides of major trivalent cations, Al^{3+}, Fe^{3+}) that form in the soil or weathered regolith. Other secondary minerals (carbonates, sulfates) can form from dissolved products if leaching is restricted, for example, due to the paucity of percolating water. The anions (e.g.,

Sedimentary Geology of Mars
SEPM Special Publication No. 102, Copyright © 2012
SEPM (Society for Sedimentary Geology), Print ISBN 978-1-56576-312-8, CD/DVD ISBN 978-1-56576-313-5, p. 97–117.

carbonate, sulfate, chloride) are often the anions of dissociated and consumed environmental acidity (e.g., carbonic acid, sulfuric acid, hydrochloric acid, respectively) that drove hydrolysis reactions. If abundant mobile solvent is present, dissolved products may be removed as solutes and transported elsewhere.

Reaction Eq. 1 also has applications to the sedimentary cycle as a whole. Residual primary minerals, which are unreacted either because of slow geochemical reaction kinetics (persistence) or because of environmental factors (e.g., limited water or reactive solutes), often occur in soils and regoliths as coarse (sand-size) grains. When eroded, transported, and deposited, these become sands, and with burial diagenesis, sands become sandstones, the volumetrically dominant terrestrial coarse clastic sedimentary rock (Garrels and Mackenzie 1971, Railsback 1993; conglomerates make up only a small volume fraction of Earth's clastic sedimentary rocks). Secondary minerals include clay minerals and oxides/oxyhydroxides of major trivalent cations (Al^{3+}, Fe^{3+}) that commonly form during the weathering of weatherable (nonpersistent) Al- and Fe-bearing silicate minerals. Clay minerals are generally fine-grained (clay-size). When eroded, transported, and deposited, these become muds, and with burial diagenesis, muds become mudstones and shales, the volumetrically dominant terrestrial clastic sedimentary rocks (Garrels and Mackenzie 1971, Railsback 1993; shales make up the largest volume fraction of all Earth's sedimentary rocks).

Soluble ions can form other secondary minerals (carbonates, sulfates) at the weathering site if leaching is restricted by, for example, a paucity of mobile water. However, on Earth, with an active and long-lived hydrologic cycle operating over much of the planet's surface much of the time, soluble ions are more typically leached from a weathering profile, moving with the abundant mobile solvent (water). Eventually, many dissolved products of weathering become the constituents of chemical sediments, recombined either by abiotic processes (e.g., evaporation) to form evaporite minerals and rocks or by biologically mediated processes (e.g., biochemical production of shells, skeleta, etc.) to form carbonate minerals, sediments, and sedimentary rocks (limestones, dolostones). Carbonates and other chemical sedimentary rocks make up a smallest fraction of the total volume of Earth's sedimentary rock (Garrels and Mackenzie 1971, Railsback 1993). Thus, several of the various reactants and products in reaction Eq. 1 correspond to categories of sediments and sedimentary rocks.

The reaction for the congruent dissolution of olivine in the presence of natural acidity illustrates many attributes embodied in reaction Eq. 1 and elaborated upon in the preceding discussion.

$$Mg_2SiO_4 + 4H^+ \rightarrow 2Mg^{2+} + H_4SiO_{4(aq)} \qquad (2)$$

In this example, a primary mineral, olivine (represented for simplicity by end-member forsteritic olivine), reacts with reactant solutes (an acid, represented by H^+), producing a modified solute load in the resulting aqueous solution. If the acid (e.g., carbonic or sulfuric acid) is known and specified, the undissociated acid appears in place of the H^+ on the reactant side of reaction Eq. 2, and the anion group resulting from the dissociation of the acid (HCO_3^-, CO_3^{2-}, or SO_4^{2-}) appears on the product side. Solubility considerations then influence whether some of the dissolved products precipitate as products (e.g., silica, Mg-hydroxide, Mg-carbonate, Mg-sulfate, hydrous Mg-silicate), and where. Depending on the concentration of solutes in the solution, and the mobility of the solution itself, some or all of the secondary minerals (solid products) may form near the olivine-dissolution site; alternatively, some or all solutes may be leached away to precipitate as chemical sediments elsewhere. Reactions involving more realistic olivine compositions (e.g., some fayalite in solid solution, releasing Fe), and reactions involving incongruent dissolution of rock-forming silicates (e.g., plagioclase feldspar plus acid reacting to produce kaolinite plus solutes) become slightly more complicated, but they can be interpreted in the same framework provided by reaction Eq. 1.

Mineral–water interactions modify primary minerals, produce new minerals as alteration products, and modify the solute loads of the solutions. Reactants (minerals and mobile species) react by processes governed by interactions among structure, composition, surface properties, and solute composition (the latter itself influenced by the history of the solution, including its origin, and other reactions in which the solution participated prior to arriving at the current reaction site). Product minerals form, and the solution composition is altered. Except for situations in which the solutions still coexist with reactant and product minerals, properties of the solutions (e.g., the abundances of various solutes; the duration of their contact with the minerals) must be inferred from the mineral assemblages left behind. In many Earth and planetary materials, the solutions left the system long ago, and only the solids survive; it is from the surviving solids that we infer the former processes through a process-product framework (Nahon 1991, Berner and Berner 1996, Delvigne 1998, Lasaga 1998, Taylor and Eggleton 2001).

This paper is about minerals and reactions. Abundant literature exists on identifying and interpreting the paleoenvironmental significance of preserved primary minerals and secondary minerals observed on Mars from orbital and surface missions (e.g., clay minerals and other phyllosilicates; sulfates; carbonates) and their implications for understanding past aqueous solution–mediated reactions. The Martian meteorites are the only available samples of Mars, so the smectite-group clays, carbonates, sulfates, and halite they contain are the only samples of Martian secondary minerals available for study on Earth. Although many Martian meteorites have an overprint of terrestrial aqueous alteration, this can be avoided with careful petrographic work and can under some circumstances be informative in its own right. Gooding (1992) reviewed aqueous alteration phenomena and products observed in Martian meteorites, but many more Martian meteorites have been recovered and recognized since then, and much research has been published in the ensuing decades. This paper examines alteration phenomena of Martian origin in meteorites from Mars, and explores implications for Martian mineral–water interactions.

METEORITES: A PRIMER

Meteorites and other rock samples from solar-system bodies preserve in their chemical compositions, minerals, and textures direct evidence of the processes by which our solar system's planets and small bodies originated, were modified, and evolved to their present state. Returned samples of solar-system materials are of the highest scientific value because they are known to come from bodies for which other kinds of information are available to complement studies of the samples in terrestrial laboratories. However, only a few solar-system bodies other than Earth have been sampled by human or robotic missions. Lunar rocks returned by *Apollo* and *Luna* missions (Mason and Melson 1970, Taylor 1975, Papike et al. 1998, Warren 2005), comet dust returned from comet 81P/Wild 2 by the National Aeronautics and Space Administration's (NASA) *Stardust* mission (Brownlee et al. 2006, Zolensky et al. 2006), and mineral grains from the regolith of asteroid 25143 Itokawa recently returned by the Institute of Space and Astronautical Science, Japan Aerospace Exploration Agency (JAXA) *Hayabusa* mission (Yano et al. 2006) sample only a few specific solar-system bodies (the Moon, one comet, and one asteroid).

Most meteorites are naturally delivered asteroidal and possibly cometary samples; the remainder are samples from larger, differentiated solar-system bodies. Other than samples returned from the Moon, comet 81P/Wild 2, and asteroid 25143 Itokawa, and interplanetary dust particles recovered in Earth's upper atmosphere and generally not demonstrably associated with specific comets (Rietmeijer 1998, Bradley 2005), meteorites are our only direct samples of a large variety of materials formed and modified on numerous parent bodies during the earliest stages of solar-system evolution. Studies of meteorites also contribute to the knowledge base for interpreting asteroidal and cometary composition and history (including aqueous alteration) by both direct (sample return; e.g., NASA *Stardust*; ISAS/JAXA *Hayabusa*) and remote (fly-by) asteroidal and/or cometary missions (e.g., NASA Deep Impact; NASA Stardust NeXT; European Space Agency Rosetta).

A summary of the entire series of processes involving ejection of a meteoroid from its parent body, interplanetary transit, dramatic passage through Earth's atmosphere, and arrival at Earth's surface as a meteorite is beyond the scope of this paper. Interested readers are referred to McSween (1999), Bevan and de Laeter (2002), and Hutchison (2004) for up-to-date introductions to these phenomena. For brief descriptions of two case studies describing a broad and representative range of phenomena associated with the recent fall and recovery of two stony meteorites, see Osborn et al. (1997) and Velbel et al. (2002). Several specific aspects are discussed as needed elsewhere in this paper.

Meteorites are named for major geographic features near their recovery site. Where natural processes result in the accumulation of large numbers of meteorites, the recovery site name is followed by a number identifying each individual object within the accumulated population. For the abundant meteorites recovered in Antarctica, the recovery location name is abbreviated to one letter for meteorites recovered by Japanese research teams, and three letters for research teams from the United States of America. For example, the full name of the oldest known Martian meteorite, Allan Hills (Antarctica) 84001 (Fig. 1), is condensed to ALH 84001.

At the broadest level of compositional classification, meteorites include objects that consist mainly of metal (irons), predominantly of silicate minerals (stones or stony meteorites), and subequal abundances of metal and silicates (stony irons) (Dodd 1981, Wasson 1985, McSween 1999, Hutchison 2004). Stony meteorites are subdivided into those with textural evidence of familiar (to geologists) igneous differentiation processing on their parent bodies, and those containing textural evidence of cooling and solidification of individual grains from vapor and dust followed by assembly of the solids into rocks and parent bodies that did not differentiate to produce magmas and igneous rocks (Dodd 1981, Wasson 1985, McSween 1999, Hutchison 2004). The defining features of the latter group are "chondrules," which are generally spherical, millimeter- to submillimeter-sized, silicate-dominated solids that are solidified droplets formed by flash heating, melting, and rapid cooling of precursor solids in the early solar nebula (Lofgren and Russell 1986; Lofgren 1989, 1996; Lofgren and Lanier 1990; McSween 1999; Hutchison 2004). Stony meteorites containing chondrules, and closely related meteorites in which evidence of chondrules may have been destroyed during the meteorite's history, are called "chondrites." Stony meteorites with igneous textures lacking chondrules are called "achondrites." There are many more specific taxonomic classes of both chondrites and achondrites (Dodd 1981, Wasson 1985, McSween 1999, Hutchison 2004).

Most achondrites have igneous crystallization ages almost as old as the age of the solar system and likely formed on small (asteroid-size) parent bodies shortly after the accumulation of solids into planetesimals in the early solar system (Dodd 1981, Wasson 1985, McSween 1999, Hutchison 2004). However, one clan of compositionally and isotopically related achondrites consists of mafic–ultramafic igneous rocks with crystallization ages of ca. 1.3 Ga or younger. This group and more recently identified relatives are the subject of this paper. Before introducing this clan, some more general recovery-related phenomena are discussed.

FIG. 1.—Martian meteorite Allan Hills, Antarctica (ALH) 84001. This orthopyroxenite meteorite is a find, recovered long after it fell and after prolonged exposure to the Antarctic weathering environment. Weathering removed some of the black fusion crust, most of which remains. This meteorite was the first (and so far only) to expand the definition of Martian meteorites beyond SNCs (Mittlefehldt 1994). Within a few hundred million years of the igneous crystallization of this meteorite ca. 4 Ga (Lapen et al. 2010), concentrically zoned disks or rosettes of carbonate minerals (Fig. 8) formed by reactions with aqueous solutions. See text for further details. NASA Johnson Space Center image S85–39565.

Meteorites from Mars, and How Scientists Know Martian Meteorites Are from Mars

Most of the >50 meteorites now attributed to Mars belong to one of three compositional classes. The longest-known members of these three classes, Shergotty (fell 1865; Grady 2000), Nakhla (fell 1911; Grady 2000), and Chassigny (fell 1815; Grady 2000), each gave their name to a group of similar meteorites. The most abundant are the "shergottites," which are basalts and harzburgites with radiometric ages indicating crystallization in the ca. 150–475 Ma range (McSween 1994, 2002, 2008; Meyer 2005). There are subclasses of shergottites (McSween and Treiman 1998; Hutchison 2004; Meyer 2005; McSween 2002, 2008): basaltic shergottites, lherzolitic shergottites, and olivine–phyric shergottites. Intermediate in abundance, there are "nakhlites" (eight examples presently known), which are clinopyroxenites with radiometric crystallization ages ca. 1.3 Ga (Treiman 2005). The smallest class (presently known from two examples) is the "chassignites," which are dunites with radiometric crystallization ages ca. 1.3 Ga (McSween 1994, 2002; Meyer 2005). The ensemble of these three classes became known as the SNCs (pronounced "snicks"). Oxygen isotopes suggest that the SNCs are related to each other and are not related to the Earth–Moon system or any other meteorite group (Clayton et al. 1976; Wasson and Wetherill 1979; Wood and Ashwal 1981; Clayton and Mayeda 1983, their Fig. 1; Wasson 1985; McSween 1999, his Fig. 4.4; Meyer 2005, his Fig. I-3). The unique orthopyroxenite ALH 84001 (Fig. 1) has oxygen-isotopic affinities with the nakhlites (Clayton 1993, Mittlefehldt 1994) and a crystallization age of ca. 4.1 Ga (Lapen et al. 2010). Being a different rock type than any of the SNCs, ALH 84001 has expanded the definition of "Martian meteorite" beyond the SNCs proper (Mittlefehldt 1994). However, all pre–ALH 84001 literature equates "SNCs" with "Martian meteorites"; where previous inferences were drawn from SNCs alone

FIG. 2.—Martian meteorite Elephant Moraine, Antarctica (EET) A79001. This shergottite (basaltic) meteorite is a find, recovered long after it fell and after prolonged exposure to the Antarctic weathering environment. Weathering removed much of the black fusion crust, patches of which remain. This meteorite was the first SNC meteorite analyzed for inert and noble gases that matched those of Mars' atmosphere (Bogard and Johnson 1983) and was found to contain low-temperature aqueous alteration products, including clay mineraloids, carbonates, and gypsum, many of which are preserved in textures indicating their preterrestrial origin on Mars (Gooding 1986a, Gooding et al. 1988). See text for further details. NASA Johnson Space Center image S80–37480.

before recovery and recognition of ALH 84001, then-known Martian meteorites will be referred to here as "SNCs."

Even before a specific parent body was identified, it was recognized that magma genesis on the "SNC" parent body persisted longer than on the parent bodies of any other then-known meteorite group (McSween et al. 1979, Wood and Ashwal 1981), and longer than on the Moon (McSween and Treiman 1998, McSween 1999, Hutchison 2004). This in turn suggested that the "SNC" parent body retained internal heat longer, and was thus larger, than any asteroid (McSween et al. 1979, Wood and Ashwal 1981) or the Moon. Mars volcanoes, the largest volcanoes in the solar system (Carr et al. 1977, Carr 2006), were first observed in data from the *Mariner* spacecraft (Mutch et al. 1976), and data from the *Viking* orbiters on impact-crater populations indicated that the volcanic landforms were relatively young Martian geomorphic features (Carr et al. 1977, McSween 1999, Carr 2006). By the early 1980s, the case for Mars as the parent body of SNCs was strong (Wasson and Wetherill 1979, Wood and Ashwal 1981), if still circumstantial.

The case for a Martian origin of shergottite EET 79001 (Fig. 2), and by extension other shergottites and all the SNCs to which they are related on the basis of the evidence described previously, was strengthened in the early 1980s. Several research groups simultaneously and independently measured abundances of inert gases and noble gas isotopes implanted in preterrestrial veins and pockets of shock glass in EETA79001. The gases implanted in those shock glasses were essentially identical to the Martian atmosphere (elemental and isotopic abundances) as analyzed by the *Viking* landers, and different from those of Earth's atmosphere or other known gas reservoirs in the solar system (Bogard and Johnson 1983, Becker and Pepin 1984, Wiens et al. 1986, Treiman et al. 2000). Similar gases were soon found in several other SNC meteorites, confirming their consanguinity and their connection to Mars (Carr et al. 1985, Marti et al. 1995, Garrison and Bogard 1998, Grady et al. 1998, Mathew et al. 1998, Terribilini et al. 1998, Treiman et al. 2000). By the mid-1980s, a Martian origin of SNC meteorites was becoming widely accepted (McSween 1984, 1985), and a decade later, the Martian origin of what had finally become known as "Martian meteorites" (including SNCs and the related ALH 84001; Mittlefehldt 1994) was firmly established (Treiman 1996, Treiman et al. 2000).

Absolute crystallization ages of Martian meteorites, all of which are igneous rocks, are known from geochronology using common radioactive-decay series (e.g., U-Pb, Rb-Sr, K-Ar, Ar-Ar). Relating igneous crystallization to other aspects of Martian geologic history requires relating absolute crystallization ages of Martian meteorites to the Martian geologic timescale. Relative ages of the landscapes and surface features on Mars are based on morphostratigraphy. The three Martian relative-age intervals are, from oldest to youngest, Noachian, Hesperian, and Amazonian. In the absence of radiometrically datable samples from known localities, absolute-age calibration of surface ages inferred from size-frequency distributions of impact craters on Martian surface units (Hartmann 2005, Soderblom and Bell 2008) is not yet possible, and errors for the absolute ages of boundaries, especially younger boundaries, are large (Nyquist et al. 2009). At present, the best estimate for the age of the Noachian–Hesperian boundary is ca. 3.7–3.5 Ga, and the best age for the Hesperian–Amazonian boundary is ca. 3.2–2.0 Ga (Hartmann 2005, Soderblom and Bell 2008). Igneous crystallization ages (see previous) of all SNCs all fall within the Amazonian; only one Martian meteorite, ALH 84001, samples ancient Noachian material (McSween 2002). The question, "Specifically where on Mars does any specific Martian meteorite come from?" has not yet been answered for any Martian meteorite (McSween 2002, 2008).

While specific Martian meteorites cannot be matched to specific source craters on Mars, meteorites with similar ejection ages can be identified using cosmic-ray–induced radionuclides. During transit from their parent body to Earth, all meteoroids accumulate cosmic-ray–induced radionuclides (Herzog 2005, Eugster et al. 2006, McSween 2008). From measurements of such nuclides, it is possible to infer a cosmic-ray exposure (CRE) age, the time at which such radionuclide accumulation began, which can in turn be interpreted as the time when the meteoroid was ejected from its parent body. For Martian meteorites, it appears that, despite the large number of individuals in many groups, most members of each group have a small number of ejection ages. For example, ejection ages of Chassigny and all of the longer-known, most thoroughly studied nakhlites have been determined to be similar (Fritz et al. 2005, Herzog 2005, Eugster et al. 2006, McSween 2008; ejection ages have not yet been determined for the most recently recovered nakhlites). This is interpreted to indicate that all 10 NC meteorites (eight known nakhlites, two known chassignites) not only have essentially identical igneous crystallization ages (ca. 1.3 Ga; McSween 1994, 2008; Treiman 2005), but they were also all ejected together with one another, at a different time than the ejection of other Martian meteorites (Fritz et al. 2005, Herzog 2005, Eugster et al. 2006, McSween 2008). It is unlikely that 10 different clinopyroxenites and dunites with identical crystallization ages were all ejected from different places on Mars with identical geologic histories, and all at the same time. It is much more likely that one impact ejected these 10 meteorites from a single geologic locale. This insight has led Martian meteorite petrologists to interpret the seven longest-known nakhlites as different samples of the same thick flow (Friedman Lentz et al. 1999) or shallow sill; systematic variations between nakhlites in compositional zoning or homogeneity of olivines and pyroxenes, and in the degree of crystallization and devitrification of interstitial igneous glass, are interpreted in terms of different nakhlites sampling different distances from the cooling surfaces of the flow or sill (Mikouchi et al. 2003, 2005, 2006; McKay et al. 2007). Nakhlites with interstitial glass and compositionally zoned olivine and pyroxene are interpreted to have cooled quickly, near the upper and possibly lower margins of the magma body, while nakhlites with devitrified interstitial glass and compositionally homogenous olivines and pyroxenes cooled slowly in the interior of the magma body (Mikouchi et al. 2003, 2005, 2006; Treiman 2005; McKay et al. 2007). The inferred solidification products of this magma body are referred to as the "nakhlite stack" (Friedman Lentz et al. 1999; Mikouchi et al. 2003, 2005, 2006; McKay et al. 2007). Thus, the eight known nakhlites and two known chassignites may all represent samples of different parts of the same rock unit on Mars, all ejected at the same time during the same impact.

Similarly, all lherzolitic shergottites seem to have the same CRE age (Herzog 2005, Eugster et al. 2006, McSween 2008). In contrast, the CRE ages of basaltic shergottites cluster into four groups (up to six meteorites per cluster), suggesting four different ejection events (Herzog 2005, Eugster et al. 2006, McSween 2008).

The overall sense of current understanding suggests that the >50 presently known Martian meteorites may have been ejected by a total of some seven (some worker suggest as few as five, some as many as eight) impact events on Mars (Herzog 2005, Fritz et al. 2005, Eugster et al. 2006, McSween 2008) and thus sample the geology of a comparably small number of geologically distinct but locally lithologically homogeneous source areas on Mars. The localities of the half-dozen or so impact craters from which the currently known Martian meteorites were ejected have not been identified. However, the Martian meteorite source localities they represent are comparable in number to the six surface locations on Mars examined to date by Mars landers (*Viking 1* and *2*, and *Phoenix*) and rovers (Mars Pathfinder rover *Sojourner* and Mars Exploration Rovers [MER] *Spirit* and *Opportunity*) (McSween 2008) and thus provide detailed information from a number of (not specifically known) locations on the surface of Mars other than those visited by robotic missions.

Meteorite Recovery and Associated Effects

Among the many phenomena of meteorite arrival at Earth (McSween 1999, Bevan and de Laeter 2002, Hutchison 2004), one, the formation of fusion crust, bears on the interpretation of hydrous minerals and thus warrants discussion here. When the orbit of a meteoroid intersects Earth's orbit and both objects are at the same point in space at the same time, the meteoroid enters Earth's atmosphere at high relative velocity (kilometers per second). At these velocities, heating caused by friction between the fast-moving meteoroid and Earth's atmosphere melts and ablates the outer surface of the meteoroid, without significant heating of the meteoroid's interior (except immediately beneath the ablated surface—a few millimeters in stony meteorites; up to a centimeter in irons). Understanding of this natural phenomenon inspired the design of the heat shields on the space capsules of the previous century, which allowed the heat of spacecraft re-entry to be dissipated without thermal harm to the *Mercury*, *Gemini*, and *Apollo* astronauts.

Often, the meteorite that is eventually recovered represents only a small fraction of the pre-atmospheric mass, the remainder of the meteoroid's pre-atmospheric mass having been lost by ablation and related processes and/or because of loss of fragments that survive fall but are not recovered. For example, cosmogenic-nuclide systematics indicate that the 469 g Coleman (Michigan) ordinary chondrite was the only recovered fragment of a meteoroid with a preatmospheric mass in the 10–100 kg range (Osborn et al. 1997), implying a recovery fraction of ~0.5–5%. Before arrival at Earth's surface, meteoroids smaller than ~1 m diameter decelerate through the velocity below which too little frictional heat is produced to melt rock or metal (skydivers at terminal fall velocity do not burn up). On meteorites, the last remnant of the material melted at high velocity solidifies when the meteoroid decelerates, and it remains on the meteoroid's surface, forming a thin (almost always <1 mm thick on stony meteorites) usually black glassy "fusion crust" (McSween 1999, Bevan and de Laeter 2002, Hutchison 2004). Figures 1–4 show fusion crust at several different scales. The fusion crust is a relative age datum for distinguishing preterrestrial (including parent-body) from terrestrial (postarrival) phenomena (e.g., Gooding 1986a).

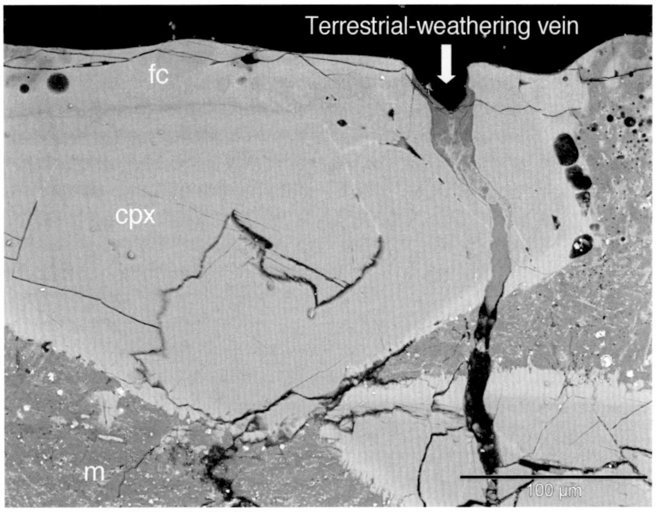

FIG. 3.—Backscattered-electron (BSE) image of a polished thin section of Antarctic nakhlite find MIL 03346. In BSE imagery, contrast is determined by atomic number (Z). Large areas of uniform medium gray are clinopyroxene (cpx). Darker-gray material with high-Z-contrast speckles is slightly devitrified mesostasis (volcanic glass; m). The black area across the top is outside the sample. Between the primary meteorite phases and the sample exterior, there is the vesicular fusion crust (fc; see text for discussion). At right (below arrow), a vein crosscuts the fusion crust and penetrates into the meteorite's interior. Preliminary EDS analyses of vein material indicate Ca-S, K-Al-S, and K-Al-Si associations consistent with a Ca-sulfate (probably gypsum), an alunite–jarosite group K-sulfate, and K-rich Si-bearing alteration product. Since this vein crosscuts the fusion crust, the precipitation of the various vein phases postdates the arrival of this meteorite at Earth's surface, and the vein materials were redistributed and/or formed by terrestrial weathering. Similar phases elsewhere in this meteorite must be examined for textural indications of preterrestrial origin; occurrences with textures indicative of preterrestrial origin are included in inventories of Mars' minerals in Martian meteorites. Occurrences that lack such textures may have formed by terrestrial weathering as did the vein material shown here; alteration minerals formed by terrestrial weathering should not be included in inventories of Mars' minerals in Martian meteorites.

When interpreting aqueous alteration phenomena in meteorites, it is necessary to distinguish between preterrestrial (parent body) and terrestrial alteration. Chemical and mineralogical properties of meteorites, including indicators of preterrestrial aqueous alteration, may be affected by their journey from their parent body to the laboratory especially if it involved a long hiatus at Earth's surface prior to recovery. In freshly fallen meteorites ("falls") recovered promptly after their witnessed arrival, aqueous alteration minerals crosscut by or

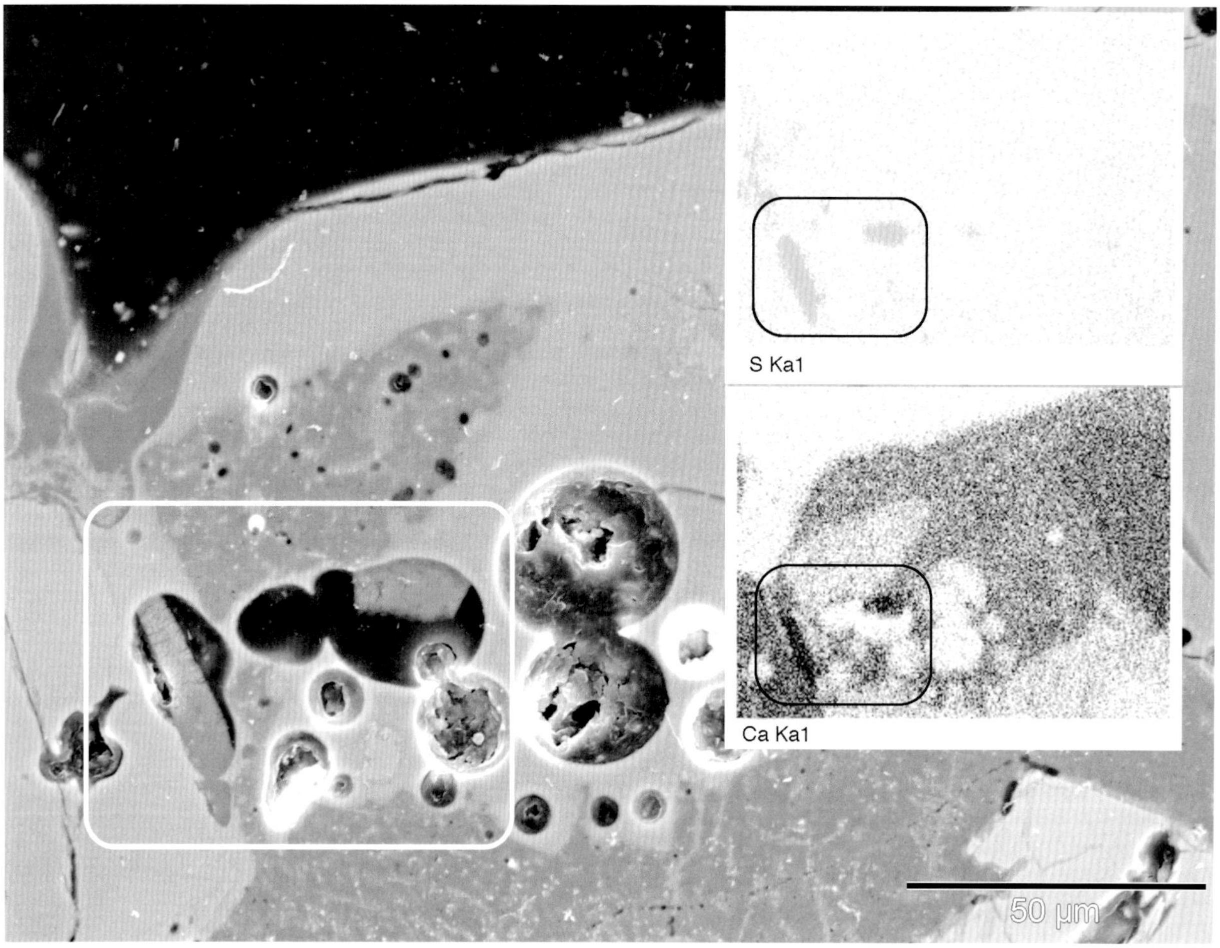

FIG. 4.—Backscattered-electron (BSE) image of a polished thin section of Antarctic nakhlite find MIL 03346. In BSE imagery, contrast is determined by atomic number (Z). Dark-gray material with high-Z-contrast speckles across the bottom of this image is slightly devitrified mesostasis (volcanic glass). The black area across the top is outside the sample. Between the primary meteorite phases and the sample exterior, there is the vesicular fusion crust (see text for discussion). Several vesicles contain mineral matter. The inset preliminary EDS maps for Ca and S show a Ca-S association (consistent with a Ca-sulfate, most likely gypsum) in several of the vesicles. Since these vesicles occur in the fusion crust, the precipitation of the various vesicle-fill phases postdates the arrival of this meteorite at Earth's surface, and the vesicle-fill minerals were redistributed and/or formed by terrestrial weathering. Similar phases elsewhere in this meteorite must be examined for textural indications of preterrestrial origin; occurrences with textures indicative of preterrestrial origin are included in inventories of Mars' minerals in Martian meteorites. Occurrences that lack such textures may have formed by terrestrial weathering as did the material shown here; alteration minerals formed by terrestrial weathering should not be included in inventories of Mars' minerals in Martian meteorites.

metamorphosed immediately beneath the fusion crust are preterrestrial and record aqueous alteration on their parent body (Reid and Bunch 1975, Gooding et al. 1991). Most Martian meteorites available for scientific study (e.g., Figs. 1–4) are "finds," recovered after unwitnessed arrival and some exposure to the terrestrial surface environment, often over millennial or longer timescales. In finds, effects of terrestrial aqueous weathering and contamination (Figs. 3, 4) are superimposed upon the aqueous alteration mineral assemblages of preterrestrial origin. Such phenomena are broadly understood for stony meteorites in general (Gooding 1986a, 1986b; Jull et al. 1988; Velbel 1988; Velbel and Gooding 1990; Velbel et al. 1991; Losiak and Velbel 2011), achondritic meteorites (Gooding 1986a), and Martian meteorites in particular (Gooding and Muenow 1986, Gooding et al. 1988, Schwandt 2005, Treiman 2005).

AQUEOUS ALTERATION IN MARTIAN METEORITES

Minerals and mineral groups understood to occur in Martian meteorites and surface materials on Mars are summarized in McSween (2008) and Ming et al. (2008). Pyroxene and olivine are the most abundant minerals in Martian meteorites and constitute, by volume, >70% of the shergottites, >85% of the nakhlites, and >90% of the chassignites and ALH 84001 (Meyer 2005). This section summarizes terrestrial-analog case studies, reviews existing concepts from terrestrial analogs, and then follows with existing applications to Mars samples (meteorites) and mission images. Many Martian meteorites contain small quantities of aqueous alteration minerals (Reid and Bunch 1975; Gooding, 1986a, 1986b; Gooding and Muenow 1986; Gooding et al. 1988, 1991; Treiman et al. 1993; Wentworth and

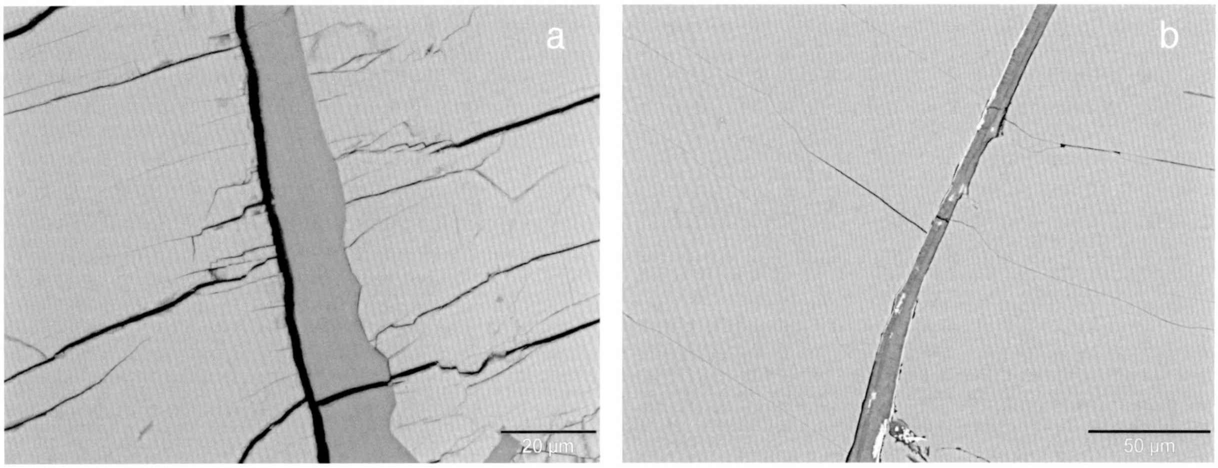

FIG. 5.—Smooth, nonserrated interfaces between olivine and phyllosilicate products of olivine alteration. (a) Iddingsite vein in cumulate olivine from the interior of nakhlite MIL 03346. (b) Serpentine vein in olivine of slightly weathered terrestrial dunite from the Webster–Addie ultramafic complex near Sylva, North Carolina, one of the southern Appalachian Blue Ridge sample localities reported on by Velbel (2009) (see also his Fig. 1a).

Gooding 1993, 1994; Bridges and Grady 1999, 2000; Bridges et al. 2001; Meyer 2005; Treiman 2005; Wentworth et al. 2005; Fisk et al. 2006; McSween 2008). Secondary minerals are discussed after primary minerals. Data and images acquired by optical petrography, scanning electron microscopy (SEM) in backscattered-electron and secondary-electron imaging modes, energy-dispersive (EDS), electron-probe microanalysis (EPMA), X-ray diffraction (XRD), transmission electron microscopy (TEM), and selected-area electron diffraction (SAED) are reviewed.

Weathering of Primary Rock-Forming Minerals in Martian Meteorites

Olivine: Olivine is a major constituent of Martian meteorites; the chassignites are nearly pure olivine, the lherzolitic shergottites contain up to 60% olivine, and the nakhlites contain >10% olivine (Meyer 2005, Treiman 2005). Serrated or sawtooth interfaces separating olivine from iddingsite veins are widely reported (Treiman et al. 1993, Bridges and Grady 2000, Bridges et al. 2001, Treiman 2005, Leshin and Vicenzi 2006), but less visually attractive smooth interfaces also occur in nakhlites (Fig. 5a), which resemble the smooth featureless interfaces commonly observed in terrestrial serpentinization (Fig. 5b; see also Velbel, 2009, and references therein). Several paragenetic sequences are possible; iddingsites and associated minerals may have formed

- after (a) smooth or (b) conchoidal, jagged-edge–forming mechanical fracturing of olivine, as vein fillings, where the first-precipitated product precipitated directly upon the olivine surface and later precipitates were superposed upon earlier ones, as proposed by Changela and Bridges (2010),
- after jagged-edge–forming chemical corrosion (etching) of olivine, as documented in terrestrially weathered olivines by Velbel (2009), followed by precipitation of vein-filling minerals in the sequence suggested by Changela and Bridges (2010), or
- simultaneously with shaping of the olivine surface at an isovolumetric replacement front that originated at a fracture around or across the primary mineral grain and replaced the primary mineral from the fracture into the remnant primary mineral, a texture known as peripheral or centripetal replacement (Velbel 1984, 1993; Delvigne 1998). The replacement front may be either smooth as in terrestrial serpentinization of olivine (Fig. 5b; see also Velbel, 2009, and references therein) or serrated by corrosion along fractures or dislocations, forming interfaces identical in geometry to etch pits known to form by terrestrial weathering of terrestrial olivine (Velbel, 2009). Olivine–iddingsite interfaces in altered terrestrial volcanic rocks imaged at TEM scales by Banfield et al. (1990) have jagged sawtooth features that are identical in orientation and geometry to arrays of etch pits formed by corrosion rather than replacement of olivine during weathering (Velbel, 2009). In the case of peripheral/centripetal replacement, the paragenetic sequence of secondary minerals would be the opposite of that inferred assuming mineral precipitation in a preexisting fracture—in the replacement case, the oldest, first-formed product mineral would be in the center of the "vein," and successively later generations of replacement product would form the outer margins, as the interface between the replaced and replacing minerals progressed away from the initial fracture and toward the interior of the primary mineral remnant.

Some carbonates in nakhlites are associated with olivine surfaces showing etch pits typical in form for weathered olivine (Fig. 6), suggesting that the olivine was actively corroded prior to or during carbonate precipitation. Distinguishing whether veins form by precipitation into a fracture or by replacement originating from a fracture is important to correctly interpreting the sequence of environmental changes recorded by the different vein minerals and their textures. It is not clear whether all reported interfaces between olivine and iddingsite and/or carbonate in nakhlites formed by the same process; it is possible that different assemblages and/or occurrences formed by different specific processes among those hypothesized here. Much detailed work on the assemblages and textures of olivine's alteration products in nakhlites is required.

Iddingsite is dated at ca. 670 Ma in the nakhlite Lafayette (Swindle et al. 2000), but from whole-rock geochronology of the three longest-known and best-studied nakhlites, it appears more likely that aqueous alteration occurred in multiple episodes since the igneous crystallization of the nakhlites rather than occurring as a single discrete datable aqueous alteration event (Swindle and Olson 2004). Regardless of the

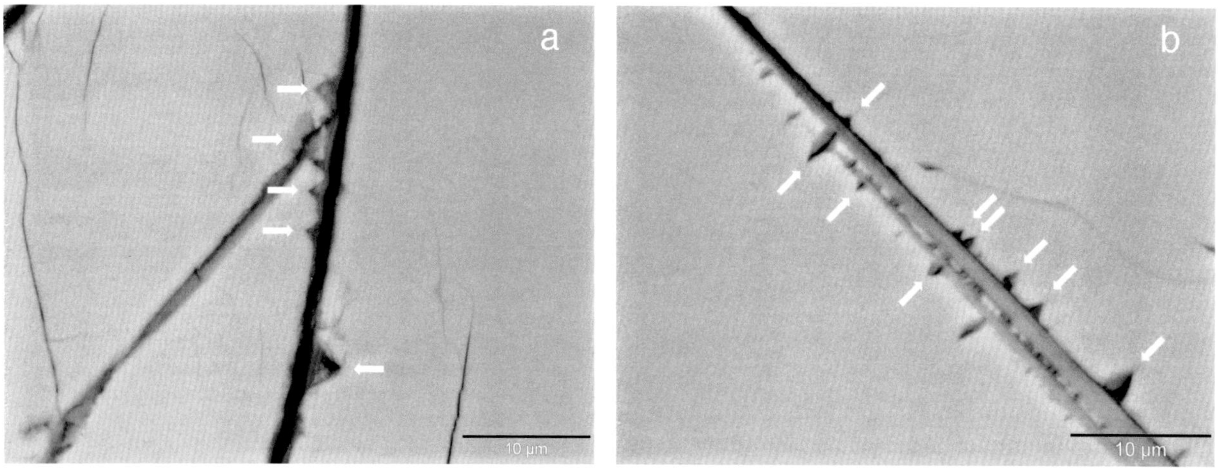

FIG. 6.—Olivine etch pits as seen in backscattered-electron image of a polished thin sections. Wedge-shaped features (shown by arrows) are cross sections through conical (funnel-shaped) etch pits along fractures in olivine. The overall geometry of typical olivine etch pits, the various geometric expressions of different sections through such pits, and their shared orientation in individual olivine crystals were described by Velbel (2009). The conical surfaces of naturally weathered olivine etch pits do not conform to individual crystal planes and faces of the dissolving mineral's structure (Velbel 2009). (a) Olivine etch pits in the Martian meteorite Nakhla. In this example, the etch pits are filled, and the fractures are lined by carbonates similar to those previously reported in Nakhla by Bridges and Grady (2000). (b) Olivine etch pits along grain boundary fractures between olivine and a micaceous inclusion in a weathered (terrestrial) dunite from the Webster–Addie ultramafic complex near Sylva, North Carolina, one of the southern Appalachian Blue Ridge sample localities reported on by Velbel (2009).

specific number and timing of aqueous alteration events or episodes to affect the ca. 1.3 Ga nakhlites, their aqueous alteration indicates that aqueous solutions with solute abundances rendering them capable of altering olivine and forming various alteration products were at least episodically present as recently as the otherwise arid mid–late Amazonian.

Pyroxene: Pyroxene is a major constituent of Martian meteorites, including nakhlites (clinopyroxenites, dominated by clinopyroxene [cpx]; Meyer 2005, Treiman 2005), many shergottites (basalts; Meyer 2005), and the orthopyroxenite (dominated by orthopyroxene [opx]); Allan Hills (ALH) 84001 (Meyer 2005). Martian meteorites contain multiple examples of textures consistent with low-temperature aqueous corrosion of pyroxene, at a range of times in the paleoenvironmental history of the surface of Mars.

"Straight-pointed alteration" features, identical in size and geometry to denticles on naturally weathered terrestrial pyroxene (e.g., Berner et al. 1980, Berner and Schott 1982, Delvigne 1998, Velbel 2007, Velbel and Barker 2008, Velbel and Losiak 2010) and related chain-silicate minerals (e.g., amphiboles; Velbel 1989b, 2007; Velbel et al. 2009), have been reported from transmitted-light images of pyroxene associated with iddingsite formed by preterrestrial aqueous alteration (Treiman 2005) in petrographic thin sections of the Martian clinopyroxenite meteorite Nakhla (Fisk et al. 2006). Smaller-scale versions of denticles occur on pyroxene in the nakhlite MIL 03346 (Antarctic find), immediately beneath fusion crust, where they may have been formed by terrestrial weathering (Fig. 7).

Thomas-Keprta et al. (2009) reported a variety of possible aqueous alteration textures in ALH 84001, including a number associated with this meteorite's carbonate rosettes (Fig. 8a). Among them, there is a disk-shaped carbonate patch on pyroxene (e.g., Thomas-Keprta et al. 2009, their Fig. 1C). Fracturing during SEM sample preparation exposed both the top of disk and a section through it. The close-up view (their Fig. 1F) shows (1) the nonporous nature of the carbonate patch; (2) possible twinning striae, suggesting recrystallization of the carbonate into a single crystal occupying the disk-shaped space along the fracture; and (3) possible fine-scale corrosion textures on the pyroxene surface exposed from areas formerly beneath the carbonate. The fine-scale corrosion textures on pyroxene beneath carbonate in ALH 84001 (Fig. 8b) strongly resemble microdenticles on terrestrially weathered terrestrial chain-silicate minerals (Fig. 8c) (Velbel 2011). A lower-magnification view shows that the fine-scale corrosion texture is confined to an area beneath the former extent of the carbonate patch, whereas the pyroxene away from the formerly covered area lacks the corrosion texture. This suggests a long and complex history of the carbonate (not at all surprising in light of the abundant literature on carbonates in ALH 84001, stimulated by the initial report of McKay et al. 1996; see carbonates section later herein), including reaction with the pyroxene beneath. Such reaction might either have accompanied or followed precipitation of the carbonate and, as such, is different from instances in which mineral surfaces were inert substrates for evaporites.

Microdenticles (with lengths in the micron–submicron range rather than tens of microns; Velbel 2011; Fig. 8b, c) are a small-scale variant of the denticles (a.k.a. sawteeth, or hacksaw terminations, or cockscomb terminations) commonly formed during natural weathering of pyroxenes and amphiboles as described in a previous paragraph. Microdenticles share the shape and orientation of the larger more typical denticles, producing arrays of microdenticles that give the larger host denticle the appearance of a surface covered with imbricate pointed or rounded scales (Velbel 2011), as expressed in both the terrestrial analog example (Fig. 8c; Velbel 2011) and ALH 84001 (Thomas-Keprta et al. 2009; Fig. 8b). In the terrestrial case, microdenticles are superposed upon and modify the lateral surfaces of larger "classic" denticles formed by natural weathering of the parent chain silicate and are therefore either a concurrent modification of low-temperature denticulation or postdate it as a later episode of small-scale modification after production of the larger denticles by weathering (Velbel 2011). In either case, the microdenticles are weathering features formed under low-temperature surface conditions (Velbel 2011). In the case examined to date, microdenticles are associated with weathering solutions that have compositions suggesting possible

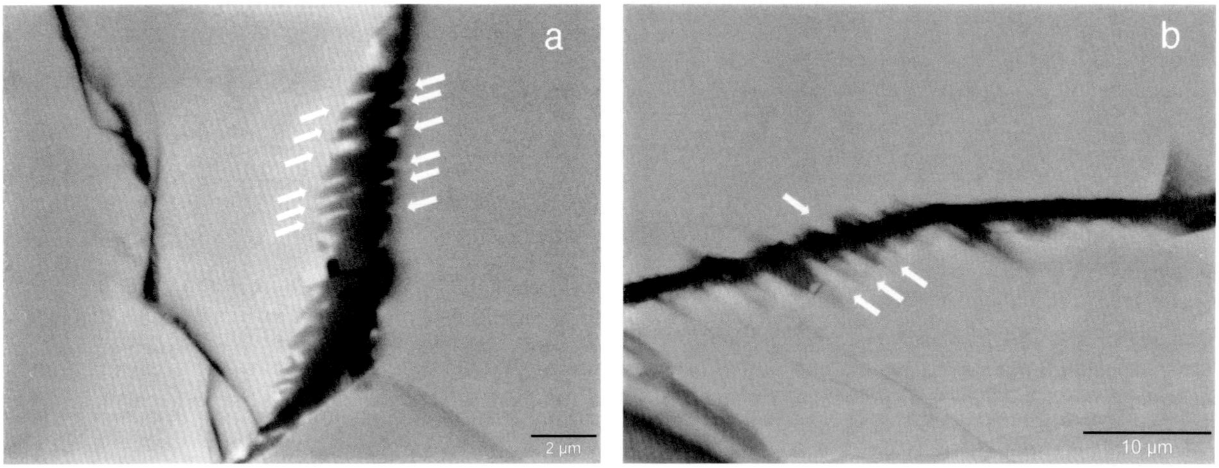

FIG. 7.—Pyroxene denticles (shown by arrows) as seen in backscattered-electron images of polished thin sections. This corrosion form, larger-scale versions of which are easily visible in thin sections and grain mounts in transmitted light, is also known as sawtooth, hacksaw, or cockscomb termination. Denticles are the remnants of the walls of elongate, almond-shaped etch pits that are the typical corrosion form produced during weathering of chain-structure silicates of the pyroxene and amphibole groups. See Velbel (2007) for additional background on this corrosion form, and Velbel and Losiak (2010) for detailed discussion of occurrences of denticles in Martian meteorites and Mars' surface mission imagery. (a) Denticles on clinopyroxene in Antarctic nakhlite find MIL 03346. With lengths of approximately one micron, small-scale sawtooth (denticulated) margins along fractures in this example are smaller than but otherwise typical of aqueous corrosion features formed by terrestrial weathering of terrestrial pyroxenes and amphiboles (Berner et al. 1980; Velbel 1989b, 2007). This occurrence is from immediately beneath fusion crust, where the pyroxene in this find may have been subjected to terrestrial weathering in Antarctica before recovery. Denticles of similar size have been described from orthopyroxene in ALH 84001 (Thomas-Keprta et al. 2009; see Fig. 8b); larger, more typical denticles have been described from clinopyroxene in Nakhla (Fisk et al. 2006). (b) Denticles several microns in length on etched clinopyroxene in weathered dacite from the Qtapg unit north of Volcan Tecuamburro, Guatemala.

control by extreme undersaturation of solutions with respect to the primary chain silicate (Velbel 2011). The occurrence of microdenticles on pyroxene beneath carbonate in ALH 84001 (Thomas-Keprta et al. 2009; Fig. 8b) may indicate a locally extreme microenvironment for pyroxene corrosion associated only with the microenvironment of carbonate formation, relative to the conditions to which the remainder of the pyroxene surface was exposed.

Alteration Product Minerals in Martian Meteorites

Many Martian meteorites contain small quantities of aqueous alteration minerals (Gooding 1986a, 1986b; Gooding and Muenow 1986; Gooding et al. 1988, 1991; Treiman et al. 1993; Wentworth and Gooding 1993, 1994; Bridges and Grady 1999, 2000; Bridges et al. 2001; Meyer 2005; Treiman 2005; Wentworth et al. 2005; Fisk et al. 2006; McSween 2008). In finds, it is important to distinguish products of terrestrial aqueous alteration from products of preterrestrial aqueous alteration indigenous to the meteorite, present when it arrived at Earth, and therefore likely formed on Mars (e.g., Gooding and Muenow 1986, Gooding et al. 1988). In falls, it is widely assumed that any aqueous alteration products are indigenous to the meteorite and were formed on its parent body. Where such minerals are crosscut by or thermally modified near fusion crust, preterrestrial origin is proven (Reid and Bunch 1975, Gooding et al. 1991). However, some of the inventory of highly soluble minerals (e.g., halite) is known to occur in fractures and vugs in the fusion crust of Nakhla (fall) and therefore postdates Earth arrival of Nakhla (Gooding et al. 1991), and elemental redistribution and formation of halite are known to occur even during curatorial exposure of finds to laboratory atmosphere (Schwandt 2005). Consequently, careful textural study is required to distinguish preterrestrial (Martian) aqueous alteration products from products of terrestrial weathering in Martian meteorites.

Aqueous alteration products are present in only small quantities in Martian meteorites, so their characterization is often only partial. Identification is commonly by a combination of morphological crystallography (by SEM) and qualitative abundances of some elements (by EDS or EPMA). Usually, the volume of product minerals available is insufficient for determination of crystal structure (e.g., by XRD) or a complete structural formula (including light elements, e.g., H in OH or structural H_2O).

Minerals are defined by composition and structure. Much of what is identified as clay minerals in Martian meteorites is classified on the basis of composition (by EPMA or EDS). Some structural data from TEM SAED exist for the material called iddingsite in nakhlites (Gooding et al. 1991, Treiman et al. 1993, Gillet et al. 2002, Noguchi et al. 2009, Changela and Bridges 2010). However, in general, structural data requiring larger or more abundant sample material (e.g., XRD, SAED) are uncommon, and most occurrences of such materials would be more accurately described as clay mineraloids (following Gooding 1986a, 1986b). Changela and Bridges (2010) have shown that most iddingsite in nakhlites lacks crystal structure at TEM scale.

Similarly, carbonate and sulfate minerals are usually identified by the major cations present using EPMA or EDS (e.g., Harvey and McSween 1996; Bridges and Grady 1999, 2000; Bridges et al. 2001; Thomas-Keprta et al. 2009). Only sometimes is compositional information supplemented by crystal morphology (from SEM) and XRD (e.g., Gooding 1986a; Gooding et al. 1988, 1991; Wentworth and Gooding 1994). It is generally possible to discern different solid-solution carbonates and end members from one another by composition, and to confidently identify a mineral by composition as, for example, a calcium carbonate. However, insufficient crystallo-

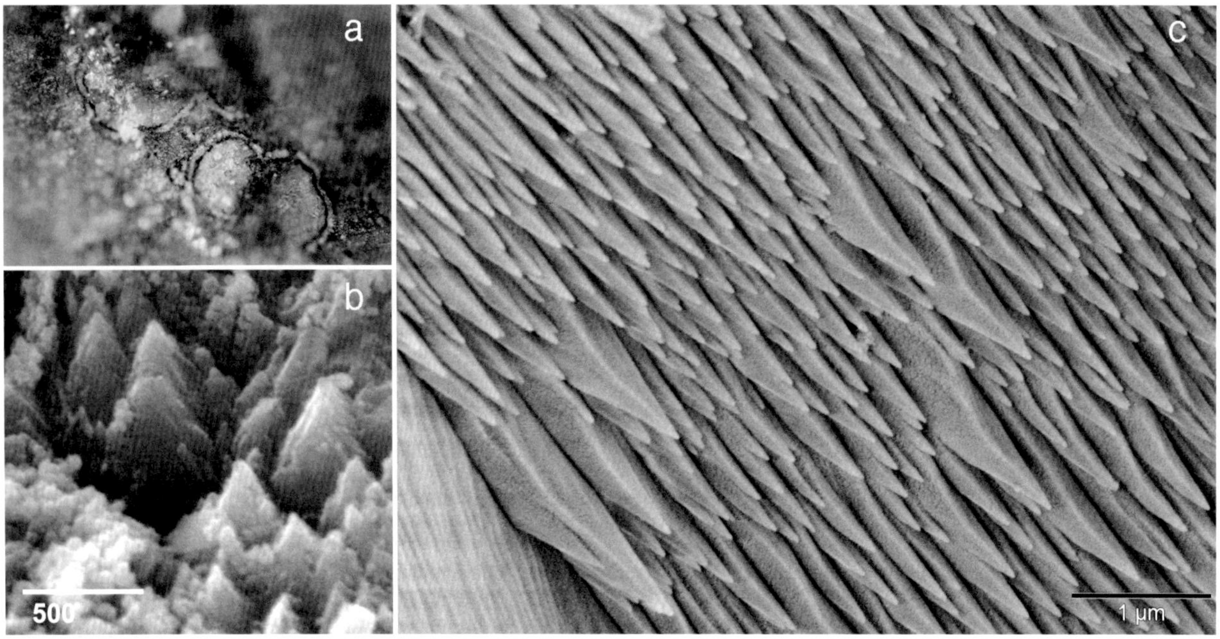

FIG. 8.—(a) Optical microscope image of concentrically zoned carbonate rosettes on orthopyroxene in Martian meteorite Allan Hills (ALH) 84001. The large central portions consist of magnesite–siderite ($Sd_{40}Mgs_{40}Cal_{20}$) carbonate; dark bands formed next at the perimeter of the earlier-formed carbonate and consist of more Fe-rich magnesite; white bands formed last and are more Mg-rich (Romanek et al. 1994, Harvey and McSween 1996). These carbonates formed only a few hundred million years after the igneous crystallization of this meteorite ca. 4 Ga (Borg et al. 1999, Lapen et al. 2010). Compositionally diverse sideritic–magnesitic and dolomitic–ankeritic carbonates also occur in Martian meteorites of younger igneous crystallization ages, but usually without the concentrically zoned growth sequences seen in ALH 84001. See text for further details. NASA Johnson Space Center image S95–00690. (b) An array of identically oriented microdenticles of submicron length on pyroxene beneath carbonate in ALH 84001. See text for further details. Image courtesy of Kathie Thomas-Keprta. (c) An array of identically oriented microdenticles of submicron length on terrestrially weathered terrestrial hornblende. Previously unpublished field-emission SEM image of the same sample examined and reported on by Velbel (2011). Microdenticles on terrestrially weathered chain-silicates are a corrosion feature of low-temperature weathering origin (Velbel 2011). See text for further details.

graphic data generally exist to identify a specific polymorph (e.g., calcite or aragonite), although the TEM-SAED characterization of siderite in Lafayette by Bridges et al. (2004) and Changela and Bridges (2010) is a welcome improvement. Similarly, Ca-sulfates, Mg-sulfates, and mixed-cation sulfates can be distinguished from one another on the basis of compositional evidence, but it is usually difficult to distinguish between different hydration states of sulfates of the same cations (e.g., gypsum from anhydrite; or the many known hydration states of Mg-sulfates). The recent TEM-SAED identification of gypsum as the specific Ca-sulfate in Nakhla by Changela and Bridges (2010) is an important step forward. Intergrowths of multiple phases at scales smaller than the EDS or EPMA probe spot size can give the same compositions as many solid solutions, so it is not always possible to distinguish among several possible mineralogical interpretations of the same analyses (Gooding et al. 1988). Interestingly, this is also true for mission data from the surface of Mars. Partial (major element) compositional data, some spectroscopic data for some specific ions, species, bonds, and sites (e.g., monohydrated sulfates can be distinguished spectroscopically from polyhydrated sulfates; Gendrin et al. 2005, Bibring and Langevin 2008), and rarely morphological information are available to arrive at plausible identifications of minerals likely to be present in any analyzed area. However, complete data sets consisting of both chemical analyses and crystallographic data that would definitively characterize mineral structure (e.g., XRD or SAED), the combination of which would thereby permit the writing of mineral structural formulae for individual phases, are lacking.

Phyllosilicates: Among Martian meteorites, only nakhlites contain noteworthy abundances of phyllosilicate minerals. TEM lattice-fringe images from the phyllosilicate component of nakhlite "iddingsite" indicate d(001) = 1.1 ± 0.2 nm (consistent with collapsed smectite-group clay minerals) in Nakhla (Gooding et al. 1991), Lafayette (Treiman et al. 1993, Changela and Bridges 2010), NWA 817 (Gillet et al. 2002), and Yamato 000593 (Noguchi et al. 2009). A phyllosilicate with d(001) = 0.7 nm (consistent with serpentine-group minerals) has recently been reported to be part of the phyllosilicate assemblage in Yamato 000593 (Noguchi et al. 2009) and Lafayette (Changela and Bridges 2010). The occurrence of smectite-group phyllosilicates has also been inferred from Fourier transform infrared (FTIR) spectroscopy in the Yamato 000593 nakhlite (Imae et al. 2003) and from electron probe microanalysis in MIL 03346 (Imae and Ikeda 2007). Noncrystalline Fe-Mg-Al-Si material with smectite- and serpentine-like compositions occurs as a volumetrically dominant part of the "iddingsite" vein assemblage in all five of the nakhlites examined by Changela and Bridges (2010).

The material called iddingsite in Martian meteorites resembles its terrestrial counterpart in being a mixture of the same constituent minerals (a smectite-hydroxide assemblage) and in being associated exclusively with olivine, but it differs in being volumetrically dominated by amorphous material without the systematic crystallographic relationships with the reactant olivine common among terrestrial iddingsites examined by optical petrography (Delvigne et al. 1979) and TEM (Eggleton 1984, 1986; Smith et al. 1987).

Iddingsite in the Lafayette nakhlite contains alkalis and Br well in excess of amounts attributable to olivine, and the chemical compositions of Lafayette iddingsites are effectively modeled as local olivine + local siliceous mesostasis + water containing magnesium, sulfate, and chloride (Treiman et al. 1993, Treiman and Lindstrom 1997). Changela and Bridges (2010) interpreted measured compositional variations in the predominantly amorphous material of nakhlite iddingsite veins in terms of impact-induced hydrothermal flow upward through the vertical arrangement of nakhlites in their source igneous body suggested by the "nakhlite stack" hypothesis introduced previously (Friedman Lentz et al. 1999; Mikouchi et al. 2003, 2005, 2006; McKay et al. 2007).

While other more soluble minerals (carbonates, sulfates, halite) might form readily during natural (Velbel et al. 1991) or curatorial/laboratory (Schwandt 2005) weathering of stony meteorites, including achondrite falls (Gooding et al. 1991, Schwandt 2005), the occurrence of demonstrably preterrestrial iddingsite in the fall Nakhla (Gooding et al. 1991, Wentworth et al. 2005) permits the interpretation of all similar occurrences in other nakhlites as also being preterrestrial. Preterrestrial aqueous alteration of olivine to iddingsite in the nakhlite Lafayette occurred ca. 670 Ma (Swindle et al. 2000). Phyllosilicates are unknown from either of the two chassignites that crystallized at the same time as the nakhlites.

Unspecified phyllosilicates have also been noted by SEM-EDS in the shergottite finds EETA 79001 (Gooding 1986a, 1986b; Gooding and Muenow 1986) and ALHA 77005 (Wentworth and Gooding 1993). These phyllosilicates are no older than the shergottites in which they occur, but they cannot be unequivocally attributed to preterrestrial origin from finds alone.

Laihunite: Laihunite, a mineral with a structure similar to olivine but with oxidized Fe and vacancies, has been reported from Yamato 000593 (Noguchi et al. 2009). Laihunite forms during subsolidus oxidation of olivine (Banfield et al. 1990). Noguchi et al. (2009) inferred an alteration temperature of 400–800° C.

Carbonates: Carbonate minerals occur in all classes of Martian meteorites, and in both falls and finds. Carbonates (Fig. 8) in ALH 84001 (Fig. 1) have a broad range of compositions, including calcite, dolomite–ankerite solid-solution, siderite–magnesite solid-solution, and intermediate compositions (Mittlefehldt 1994; Romanek et al. 1994; Harvey and McSween 1996; McKay et al. 1996; Valley et al. 1997; Scott et al. 1997, 1998; Treiman and Romanek 1998; Scott 1999; Barber and Scott 2003, 2006; Treiman 2003; Corrigan and Harvey 2004; Thomas-Keprta et al. 2009). Compositionally diverse carbonate minerals often occur adjacent to one another in concentrically zoned disks or rosettes (Fig. 8) (Mittlefehldt 1994; Romanek et al. 1994; Harvey and McSween 1996; McKay et al. 1996; Kirschvink et al. 1997; Valley et al. 1997; Scott et al. 1998; Scott 1999; Brearley 2003; Treiman 2003; Barber and Scott 2003, 2006; Corrigan and Harvey 2004; described recently and in great geometric detail by Thomas-Keprta et al. 2009). Carbonates occurring as veins and fracture and void fillings are also noted (Harvey and McSween 1996; Barber and Scott 2003; Corrigan and Harvey 2004), but many are compositionally zoned like the disks (Scott et al. 1997, 1998; Scott 1999) and are probably just disks exposed as apparent slabs in cross section (Corrigan and Harvey 2004). ALH 84001 carbonates are associated with corroded opx, but this is only seen using high-magnification/high-resolution microscopic methods (FEG-SEM, TEM) employed by only a few published studies (McKay et al. 1996; Barber and Scott 2003, 2006; Thomas-Keprta et al. 2009). If the textures imaged at the opx–carbonate interface by high-magnification methods are typical of carbonate-disk–opx interfaces, this association may be quite widespread. The genesis of carbonates in ALH 84001 (especially the temperature of their formation) as inferred from mineral-composition/phase-equilibrium relationships, stable isotopes, and paleomagnetism, has been the subject of considerable debate and disagreement (Mittlefehldt 1994, Harvey and McSween 1996, McKay et al. 1996, Hutchins and Jakosky 1997, Kirschvink et al. 1997, Valley et al. 1997, Farquhar et al. 1998, McSween and Harvey 1998, Scott et al. 1998, Treiman and Romanek 1998, Warren 1998, Scott 1999, Golden et al. 2000, Brearley 2003, Treiman 2003, Corrigan and Harvey 2004, Romanek et al. 2004, Bell 2007, Steele et al. 2007, Niles et al. 2009, Thomas-Keprta et al. 2009), as has their possible subsequent shock-metamorphic modification (Ash et al. 1996; Scott et al. 1997; Treiman 1998, 2003; Brearley 2003; Barber and Scott 2003, 2006; Bell 2007; Steele et al. 2007). Carbonate (associated with corroded opx) in ALH 84001 formed ca. 3.6–4.0 Ga (Knott et al. 1995, Borg et al. 1999). Interested readers should acquaint themselves with the primary papers cited in this paragraph.

A Ca-bearing calcite-group mineral has been inferred to occur in Nakhla from morphological crystallography using SEM-EDS (Gooding et al. 1991). Subsequent EPMA analyses of carbonates in multiple nakhlites revealed a broad range of compositions, including near-end-member siderite (structure confirmed by ATEM-SAED; Bridges et al. 2004, Changela and Bridges 2010), Mg- and Mn-rich siderite, and compositions intermediate between the dolomite–ankerite and siderite–magnesite solid-solutions (Bridges and Grady 1999, 2000; Bridges et al. 2001, 2004; Changela and Bridges 2010). Unlike ALH 84001 carbonates, compositionally distinct carbonates in nakhlites are spatially discrete or related as bands or laminae rather than juxtaposed in concentrically zoned disks or rosettes, and Ca-rich carbonates (e.g., more Ca-rich than the dolomite–ankerite solid-solution series) have not been reported from nakhlites.

Carbonates are reported from both chassignites. Calcite, magnesite, and hydromagnesite were identified in the fall Chassigny by morphological crystallography in SEM, with EDS (Wentworth and Gooding 1994). Wentworth et al. (2005) reported additional occurrences of Ca-carbonate on exposed olivine grain-boundary fracture surfaces in Chassigny. Occurrences of carbonates in the fall Chassigny are consistent with a preterrestrial origin, but formation of hydromagnesite after the recovery of olivine-bearing chondritic stony meteorites has been documented (Velbel et al. 1991). In addition to occurrences of individual evaporite minerals, some disk-shaped masses occur with nonporous Ca-sulfate in the center, surrounded by microgranular or porous (dehydrated? devolatilized?) Ca-carbonate (Wentworth et al. 2005). Calcite and aragonite in NWA 2737 (chassignite find) were distinguished from one another by Raman spectroscopy, and some carbonates appear to have experienced shock metamorphism, indicating preterrestrial origin for at least those occurrences (Beck et al. 2006).

Calcite (confirmed by XRD), and unspecified Ca-carbonate and low-Ca–carbonate (identified by SEM-EDS) have all been reported in the shergottite EETA 79001 (Fig. 2; Gooding 1986a, Gooding et al. 1988). Some Ca-carbonate in the shergottite EETA 79001 occurs in veins and pockets of shock melt formed during an impact on Mars (Gooding et al. 1988; the same shock melt that trapped Martian atmospheric gases; Bogard and Johnson 1983). This suggests that at least these Ca-carbonate grains were already present in the rock when it experienced impact and shock on Mars, and they are carbonate grains of Martian origin (Gooding et al. 1988). Evidence for the pre-ejection origin of some Ca-carbonates in EETA 79001 is strong (Gooding et al. 1988), although this may not extend to other carbonate occurrences in this meteorite; as a find, it may have been affected by terrestrial aqueous alteration during its Antarctic exposure prior to recovery.

In stony meteorites, carbonate minerals are known to form by terrestrial weathering of the meteorite before recovery. Witnessed falls can be similarly affected; samples collected many decades after the fall of the diogenite Tatahouine contain carbonates that are not present in samples collected immediately after the fall (Barrat et al. 1998, 1999;

Gillet et al. 2000; Benzerara et al. 2003, 2005a, 2005b, 2006). In one well-studied ordinary chondrite find from Antarctica (LEW 85320), carbonate minerals (the hydrous Mg-carbonates nesquehonite and hydromagnesite) formed both naturally before recovery and in the laboratory after recovery during sample processing and storage (e.g., Jull et al. 1988, Velbel et al. 1991). Consequently, given that 46 of the 50 presently known Martian meteorites are finds, considerable care must be taken to avoid overinterpreting occurrences of carbonates in Martian meteorite finds as indicators of carbonate-forming processes on Mars.

Sulfates: The nakhlites Nakhla (fall), Governador Valadares (find), NWA 998 (find), Yamato 000749 (find), and MIL 033346 (find; Figs. 3, 4), and the chassignite Chassigny (fall) all contain Ca-sulfates. Morphological crystallography indicates gypsum or bassanite in Nakhla (Gooding et al. 1991) and Chassigny (Wentworth and Gooding 1994). Gypsum and anhydrite are distinguished from one another by EPMA in Nakhla (fall) and Governador Valadares (find); analyses with totals near 100% are interpreted as anhydrite, whereas those with significantly lower totals are interpreted as gypsum (Bridges and Grady 1999, 2000). Gypsum in Yamato 000749 has been identified by Raman spectroscopy (Noguchi et al. 2009), and in Nakhla, it has been confirmed by TEM-SAED (Changela and Bridges 2010). In Nakhla, striated (corroded? dehydrated?) Ca-sulfate occurs on visibly unmodified silicate surfaces (Wentworth et al. 2005).

Multiple hydration states of Mg-sulfate are known to occur on Earth, and many of the known hydration states are stable or metastable under the range of Mars' surface conditions (Vaniman et al. 2004, Vaniman and Chipera 2006, Peterson et al. 2007). Unspecified shrunken and cracked Mg-sulfate occurs immediately beneath fusion crust in Nakhla (Gooding et al. 1991, Wentworth et al. 2005). One interpretation is that the hydrous sulfate was metamorphosed (dehydrated?) by heating associated with fusion crust formation; this would indicate the sulfate was preterrestrial (Wentworth et al. 2005). However, it is less clear is the Mg-sulfates are intact away from fusion crust. Consequently, an alternate interpretation is that hydrous sulfates throughout Nakhla might have been dehydrated regardless of proximity to fusion crust; this would render the timing of sulfate formation ambiguous. However, there are published images of Ca-sulfates (reported to be anhydrite) in samples away from fusion crust, in the interior of Nakhla, that lack internal cracks or shrinkage textures (Bridges et al. 2001, Leshin and Vicenzi 2006). Further study of Nakhla is required to ascertain the state of Mg-sulfate away from the thermal influence of fusion crust.

Jarosite occurs as part of a jarosite–hematite assemblage in the Antarctic nakhlite find MIL 03346 (McCubbin et al. 2009). Identified by EPMA and Raman spectroscopy and situated in a pyroxene-hosted melt inclusion with no obvious indications of terrestrial weathering or contamination, this mineral assemblage in the melt inclusion leads to the interpretation that the jarosite formed from magma-associated hydrothermal fluids (McCubbin et al. 2009). Hydrothermal sulfate would not be closely related to the widespread primary sedimentary and diagenetic sulfates (and jarosite in particular; Christensen et al. 2004, Klingelhöfer et al. 2004) reported by MER *Opportunity* from Meridiani Planum on Mars (Christensen et al. 2004, Herkenhoff et al. 2004, Rieder et al. 2004, McLennan et al. 2005) and interpreted to reflect acidic diagenetic pore fluids in Martian sedimentary rocks (Klingelhöfer et al. 2004; Morris et al. 2004, 2006; Squyres et al. 2004, 2006; McLennan et al. 2005; Tosca et al. 2005; Hurowitz and McLennan 2007). Furthermore, jarosite or jarosite-like K-sulfate of possible terrestrial-weathering origin has been reported in several non-Mars achondrite finds from anhydrous parent bodies (e.g., asteroidal achondrites of the howardite–eucrite–diogenite group; eucrite EETA79005 and diogenite ALHA77256; Gooding 1986a). Additionally, Noguchi et al. (2009) and Changela and Bridges (2010) documented authoritatively that jarosite occurs in the more-weathered Yamato nakhlite Y-000749 rather than the less-weathered but otherwise similar Y-000593, and it is a product of terrestrial weathering. However, the jarosite in MIL 03346 (McCubbin et al. 2009) demonstrates that preterrestrial jarosite occurs in at least one Martian meteorite, and that preterrestrial (Martian) jarosite can survive even in a find. The search for Mars-surface–related preterrestrial (as distinct from hydrothermal preterrestrial or terrestrial-weathering–produced) jarosite in Martian meteorites has recently been and remains an area of active research at the time of this writing.

Wentworth et al. (2005) observed Ca-sulfate phases on exposed olivine grain-boundary fracture surfaces in the Martian dunite Chassigny. In addition to occurrences of individual evaporite minerals, some disk-shaped masses occur with nonporous Ca-sulfate in the center, surrounded by microgranular or porous (dehydrated? devolatilized?) Ca-carbonate (Wentworth et al. 2005).

More than three dozen shergottites are presently known, but only two (Shergotty and Zagami) are falls, and almost all observations of aqueous alteration products in shergottites are from a single Antarctic find, EETA 79001 (Fig. 2). Gypsum has been identified by crystal morphology using SEM in EETA 79001 (Gooding et al. 1988). Some Ca-sulfate in the shergottite EETA 79001 occurs in veins and pockets of shock melt formed during an impact on Mars (Gooding et al. 1988; the same shock melt that trapped Martian atmospheric gases; Bogard and Johnson 1983). Many of these sulfates are euhedral; some are rounded and cracked, suggesting thermal decrepitation (Gooding et al. 1988). This suggests that at least these Ca-sulfate grains were already present in the rock when it experienced impact and shock on Mars, and that these sulfate grains are of Martian origin (Gooding et al. 1988). However, this may not extend to other sulfate occurrences in this meteorite; as a find, it may have been affected by terrestrial aqueous alteration during its Antarctic exposure prior to recovery.

Sulfates filling fractures that crosscut fusion crust (e.g., Fig. 3) and filling vesicles in fusion crust (Fig. 4) (in other words, of terrestrial origin) are known to occur in various achondrite finds (e.g., asteroidal achondrites of the howardite–eucrite–diogenite group; gypsum in eucrite EETA79004, K-sulfate in eucrite EETA79005 and diogenite ALHA77256; Gooding 1986a). Mobilization of (possibly indigenous) Ca-sulfate during exposure to moisture in laboratory atmosphere has been documented in a carbonaceous chondrite fall by Gounelle and Zolensky (2001). It is clear that specific stony-meteorite finds have been affected by terrestrial aqueous alteration, raising the legitimate concern that soluble minerals in any find may have been redistributed or formed during terrestrial (rather than Martian) weathering. Consequently, given that most of the presently known Martian meteorites are finds, considerable care must be taken to avoid overinterpreting occurrences of sulfates in Martian meteorite finds as indicators of sulfate-forming processes on Mars. However, even Martian meteorite finds (e.g., EETA 79001, MIL 03346) can preserve preterrestrial evaporite minerals in textures for which evidence of a preterrestrial (Martian) origin is compelling (Gooding et al. 1988, McCubbin et al. 2009), allowing retrieval of useful information about aqueous alteration on Mars. The textures and textural contexts of such minerals must be examined carefully to distinguish preterrestrial evaporite minerals from occurrences of similar minerals formed or redistributed by terrestrial processes.

Halite: Halite has been reported in several allocations of Nakhla (fall) (Gooding et al. 1991; Bridges and Grady 1999, 2000; Wentworth et al. 2005). Where secondary-electron imagery enables study of grain shape, rounded halite grains rest on a pyroxene surface in the meteorite's interior (Wentworth et al. 2005). Rounding of halite (and other highly soluble minerals) is characteristic of (diffusion-limited) dissolution in undersaturated solutions (Velbel 2004). Dissolution experiments commonly proceed to complete destruction of the halite (Simon 1981, Alkattan et al. 1997), and although grain shapes and textures have been described at intermediate stages in such experiments (Walker and Kiefer 1985, Kerr 1995), it is not known how partially dissolved halite grains would look if their dissolution were interrupted (by, for example, a very small amount of water reaching saturation with respect to halite before complete dissolution of the halite grain).

It is possible that the halite in Nakhla was euhedral before sample preparation. Before the meteoritics community became aware of the possibility of halite occurring in meteorites, meteorites were cut and polished for thin sectioning using water, as have been most terrestrial rocks not expected to contain easily soluble minerals. Halite dissolves readily in dilute aqueous solutions (Simon 1981, Walker and Kiefer 1985, Kerr 1995, Alkattan et al. 1997; sand-size grains of halite dissolve completely in tap water or distilled water in tens of seconds at Earth-surface temperatures—Walker and Kiefer 1985, Velbel 2004). Some halite occurs in Nakhla (fall) fractures and vugs in the fusion crust and therefore postdates Earth arrival of Nakhla (Gooding et al. 1991). Elemental redistribution and formation of halite are known to occur even during curatorial exposure of finds to laboratory-atmosphere fluctuations in relative humidity (Schwandt 2005). Thus, the preservation of halite, and especially its distribution and growth and dissolution textures, is highly unlikely, even if the halite is preterrestrial, as is likely in Nakhla. Given the ease with which even minimal exposure to terrestrial moisture can mobilize halite in falls (Nakhla; Gooding et al. 1991) and during laboratory curation of finds (ALH 84001; Schwandt 2005), the absence of halite from other nakhlites (Lafayette, Governador Valadares; Bridges and Grady 2000) cannot be safely inferred to indicate that halite was absent from nakhlite finds. It is just as likely that halite has been leached from Martian meteorites during prerecovery terrestrial weathering, and by postrecovery humidity-related redistribution in the laboratory (even in falls), as it is that halite was naturally absent from the evaporite-mineral inventory of any given Martian meteorite.

Synthesis: Primary and Secondary Minerals in Martian Meteorites

Textural features of aqueous corrosion origin on olivine and pyroxene, and various mineral products of aqueous alteration, occur in a variety of Martian meteorites, in a variety of assemblages, and over a range of ages. Orthopyroxene corrosion and associated carbonate formation dated to ca. 3.9 Ga (Borg et al. 1999, Thomas-Keprta et al. 2009; approximately mid-Noachian) occur in orthopyroxenite ALH 84001 that crystallized ca. 4.1 Ga. Clinopyroxene and olivine corrosion and diverse alteration minerals including phyllosilicates, carbonates, sulfates, and halite, some dated as ca. 670 Ma (Swindle et al. 2000, Treiman 2005, Fisk et al. 2006; approximately mid–late Amazonian), occur in nakhlites that crystallized ca. 1.3 Ga. Carbonates and sulfates occur in, and therefore must be younger than, their shergottite host rocks with igneous crystallization ages of ca. 150–475 Ma (Meyer 2005; late Amazonian). The restricted range of halite occurrence (only in Nakhla) is likely a preservational bias that places no limits on how long before or after halite formation in Nakhla halite might have been precipitated in other meteorites. However, carbonates occur in Martian meteorites of all crystallization and alteration ages, and sulfates occur in all Martian meteorites crystallized and altered within the past 1.3 Ga. Martian aqueous alteration responsible for forming corrosion textures and/or alteration products in Martian meteorites has varied little over time, and aqueous activity has persisted in at least small amounts relatively recently on Mars.

The aqueous alteration record inferred from Mars' surface mineralogy (McSween et al. 2004, Bibring et al. 2006, Herkenhoff et al. 2006, Bibring and Langevin 2008, McLennan and Grotzinger 2008, Ming et al. 2008) indicates that the parent minerals pyroxene and olivine were and are present, available for weathering, and detectably modified by alteration, and that phyllosilicates, carbonates, and sulfates formed during aqueous alteration. However, different alteration products are associated with surface units of different ages. Orbital spectroscopic observations of the distribution of hydrous minerals among Mars' surface geomorphic units indicate that the main interval of phyllosilicate formation was in the early–mid-Noachian, followed by the main interval of sulfate formation in the latest Noachian to late Hesperian (Bibring et al. 2006, Bibring and Langevin 2008, Ming et al. 2008). The widespread persistence of primary silicates indicates that water was generally minor in abundance, limited in movement, and present for too short a time to consume primary minerals (McLennan and Grotzinger 2008, Ming et al. 2008). Data sets from landed surface missions show primary igneous minerals, cation-rich phyllosilicates, and salt minerals juxtaposed with one another, indicating that soluble elements have not been leached away from the site of dissolution of primary mineral phases/lithologies, apparently due to limited water abundances (McLennan and Grotzinger 2008, Ming et al. 2008). However, for many of the ancient altered-bedrock and ancient sedimentary assemblages observed from orbit, phyllosilicates do not coexist with the soluble salt–mineral phases that should have formed as a result of igneous mineral alteration and subsequent phyllosilicate mineral formation (Milliken et al. 2009). At least in isolated localities on the ancient Noachian surface of Mars, leaching and removal of soluble components may have been more prevalent, perhaps implying a more water-rich environment than has been inferred from landed-mission data sets and the alteration-mineral assemblages of Martian meteorites (Milliken et al. 2009).

The aqueous alteration record in Martian meteorites resembles that inferred from Mars' surface mineralogy (McSween et al. 2004, Bibring et al. 2006, Herkenhoff et al. 2006, Bibring and Langevin 2008, McLennan and Grotzinger 2008, Ming et al. 2008, Velbel and Losiak 2010) in that the parent minerals pyroxene and olivine are present, available for weathering, and detectably modified by alteration, and that phyllosilicates, carbonates, and sulfates formed during aqueous alteration without leaching of soluble products. The persistence of both primary silicates and many soluble products indicates that water was minor in abundance, limited in movement, and present for too short a time to consume primary minerals and/or leach away soluble products (McLennan and Grotzinger 2008; Ming et al. 2008).

The aqueous alteration record in Martian meteorites differs from that inferred from Mars' surface mineralogy (McSween et al. 2004, Bibring et al. 2006, Herkenhoff et al. 2006, Bibring and Langevin 2008, McLennan and Grotzinger 2008, Ming et al. 2008, Velbel and Losiak 2010) in that meteorites record phyllosilicate and sulfate formation long after the main phyllosilicate- and sulfate-forming intervals inferred from the distribution hydrous minerals among Mars' surface geomorphic units (early–mid-Noachian and latest Noachian to late Hesperian, respectively; Bibring et al. 2006, Bibring and Langevin 2008, Ming et al. 2008). Phyllosilicates and sulfates in 1.3-Ga-old nakhlites formed after the nakhlites crystallized, indicating aqueous alteration during the Amazonian, which appears from orbit to have been a period of oxide, rather than phyllosilicate or sulfate, formation (Bibring et al. 2006, Bibring and Langevin 2008). Meteorites may sample rock-dominated aqueous microenvironments within which alteration-mineral assemblages formed in near isolation from surface acidity (Newsom 2005, Tosca and McLennan 2006, McLennan and Grotzinger 2008). Such relatively benign geochemical environments are potentially promising subsurface targets for future exploration of potentially habitable environments on Mars (Grotzinger 2009).

Evaporation Sequences in Martian Meteorites and on Mars

Abundant S and Cl have been observed in chemical analyses of Mars' surface materials since the *Viking* landers arrived at Mars in 1976, and they have been interpreted to indicate mediation of Mars' surface chemistry by mobilization of sulfate and chloride by small amounts of water with sufficiently high solute loads to readily precipitate sulfate and chloride minerals (Clark et al. 1977, 1982, 2005; Clark and van Hart 1981; Banin et al. 1992; Gooding et al. 1992; Clark 1993; McLennan and Grotzinger 2008; Ming et al. 2008). Much remains to be resolved. The recent recognition at the *Phoenix* mission's landing site of perchlorate compounds (Hecht et al. 2009) has motivated reexamination of oxidation–reduction phenomena in *Viking* lander data (Navarro-Gonzáles et al. 2010), and will stimulate new understanding of chlorine dynamics and oxidation pathways in weathering reactions well beyond common terrestrial experience. The sulfate that eventually precipitates as sulfate minerals in Martian materials may have originated by localized volcanic hydrothermal or fumarolic processes, widespread dispersal of volcanogenic volatiles in acid fog, oxidation of primary igneous or sedimentary diagenetic sulfides, and/or redistribution of any such sulfur by hydrothermal alteration after impacts (see reviews by McLennan and Grotzinger [2008] and Ming et al. [2008]). Furthermore, multiple processes can produce brines; evaporation is most commonly considered (McLennan and Grotzinger 2008, Ming et al. 2008), but freezing or ablation of even initially dilute aqueous solutions produces brines as water (solvent) is removed, leaving the same quantity of solutes in a decreasing amount of solvent until the solution is supersaturated with respect to "evaporite" minerals (Herut et al. 1990, Marian et al. 1999, Starinksy and Katz 2003, McLennan and Grotzinger 2008, Morin et al. 2008, Niles and Michalski 2009, Frank et al. 2010). Nevertheless, ambiguity about sulfate and chloride sources and detailed concentration mechanisms notwithstanding, the ubiquity and importance of sulfate and chloride in Mars materials have long been and remain well established (Clark et al. 1977, 1982, 2005; Clark and van Hart 1981; Banin et al. 1992; Gooding et al. 1992; Clark 1993; McLennan and Grotzinger 2008; Ming et al. 2008). Consequently, meteoriticists were prepared to consider the processes by which sulfate and chloride are concentrated in Mars' surface materials (e.g., evaporation) even before the MERs and *Mars Express OMEGA* supplied evidence of abundant discrete sulfate minerals on Mars. One line of argument for the low-temperature origin of the carbonates in ALH 84001 invoked precipitation from evaporating brines (McSween and Harvey 1998, Warren 1998). Mission data indicate less carbonate at the surface of Mars than expected, but recently carbonates have been detected by both orbital and surface-mission spectroscopy at several locations on Mars (Ehlmann et al. 2008, Michalski and Niles 2010; Morris et al. 2010), and solution-related carbonate minerals are indicated by solution-chemistry measurements at the *Phoenix* landing site (Boynton et al. 2009, Hecht et al. 2009, Smith et al. 2009). Bridges and Grady (2000) modeled the sequence of evaporite-mineral precipitation from hypothetical brines in the three longest-known and therefore best-studied nakhlites. They concluded that the Ca-sulfate-and-halite–absent product-mineral assemblage of Lafayette formed from the least-evaporated solutions and that the Ca-sulfate + halite assemblage of Nakhla formed from the most-evaporated brines, with Governador Valadares intermediate.

After MER mission data for the sulfate-rich sedimentary section at Meridiani Planum became available (Grotzinger et al. 2005, McLennan et al. 2005), Tosca and McLennan (2006) built upon thermodynamic reaction-progress models and developed a Mars-relevant framework of chemical divides conceptually similar to those used to explain the causes of mineralogic variations in terrestrial evaporites but designed for cation contributions in basalt-dominated waters. At Meridiani Planum, product minerals were precipitated as primary chemical sediments, which were then reworked by eolian processes to form detrital accumulations of evaporite-mineral grains, as at White Sands, New Mexico (Grotzinger et al. 2005, McLennan et al. 2005). Similar evaporite minerals were also precipitated as cements in Mars' sedimentary rocks (McLennan et al. 2005), on exposed surfaces of Mars' sedimentary rocks (Knoll et al. 2008), and in weathered rinds on igneous boulders at the surface of Mars (McSween et al. 2004). Tosca and McLennan (2006) proposed that both Martian meteorite solutions and Mars' surface solutions pass through broadly similar Ca-poor, SO_4-rich states at intermediate degrees of evaporation and arrive at similar HCO_3-poor, Cl-rich brines at advanced stages. However, the relative abundances of solutes required to produce the evaporite mineral assemblages of Martian meteorites and Mars' surface materials differ (Tosca and McLennan 2006). During evaporation of solutions with $HCO_3^- > SO_4^{2-}$, $HCO_3^- > Fe^{2+}$, $SO_4^{2-} > Ca^{2+}$, $HCO_3^- < Ca^{2+}+Mg^{2+}$, and $Mg^{2+} > SO_4^{2-}$, evaporitic minerals would form in the sequence siderite–gypsum–epsomite–halite (Tosca and McLennan 2006). This assemblage is the same as observed in Nakhla (Gooding et al. 1991, Bridges and Grady 2000, Tosca and McLennan 2006). Different chemical divides—$SO_4^{2-} > HCO_3^-$, $HCO_3^- << Ca^{2+} < SO_4^{2-}$, $SO_4^{2-} > Fe^{2+}$, and $Mg^{2+} > SO_4^{2-}$—are required to produce the gypsum–melanterite–epsomite–halite sequence that conforms to the generally carbonate-absent, sulfate-dominated evaporites observed at the surface of Mars, which include Fe-dominated salts (e.g., ferricopiapite, jarosite; see also Morris et al. 2006, Lane et al. 2008) (Tosca and McLennan 2006). The aqueous alteration products in the most complete Martian meteorite evaporite-mineral assemblages formed from solutions with higher initial HCO_3^-/SO_4^{2-} than the solutions that affected most Mars' surface materials (Tosca and McLennan 2006). This difference in initial solution compositions resulted in Fe being incorporated into sideritic–ankeritic carbonates in nakhlites prior to precipitation of sulfates and Fe-sulfates within the overall sequence of sulfate-mineral precipitation in most Mars' surface materials.

The aqueous alteration products of Martian meteorites record a slightly different set of geochemical conditions (carbonic-acid dominated) than is recorded in the alteration minerals of surface materials on Mars (sulfuric-acid dominated). This may be because Mars' surface materials have been directly accessible to volcanogenic sulfuric acid, whereas Martian meteorites were exhumed (during impact and ejection) from subsurface depths (tens of meters up to ~200 m; Fritz et al. 2005) possibly sufficient to isolate those rocks from atmospheric sulfuric-acid acidity (Newsom 2005, Tosca and McLennan 2006, McLennan and Grotzinger 2008). This would allow acid-consuming reactions (e.g., reaction Eq. 2) with basaltic minerals (with reactions confined to igneous grain boundaries, and low water–rock ratios) to raise local pH sufficiently to allow shifts in carbonate equilibria and increases in bicarbonate abundances (Tosca and McLennan 2006).

Chassignites have the same (igneous) crystallization age as the nakhlites. Some previous evaporation models for nakhlite alteration minerals (Bridges and Grady 2000) and some evaporation sequences on the carbonate-dominant side of the chemical divide diagram of Tosca and McLennan (2006) are consistent with the crystallization sequence (sulfate first, then carbonate) implied by the sulfate–carbonate disks in Chassigny reported by Wentworth et al. (2005). However, observed carbonate minerals are more diverse and complicated than current models can address. Calcite occurs along with magnesite and hydromagnesite in the chassignite fall Chassigny (Wentworth and Gooding 1994), and along with aragonite (identified by Raman spectroscopy) in the recently recovered chassignite NWA 2737 (Beck et al. 2006, Treiman et al. 2007). Calcite occurs as apparently the only carbonate in the shergottite find EETA 79001 (identified as calcite by XRD; Gooding et al. 1988). Calcite occurs along with Mg-, Fe-, and Mn-carbonates in the unique orthopyroxenite find ALH 84001 (e.g., Harvey and McSween 1996, Corrigan and Harvey 2004). These occurrences of calcite, especially those in falls, are consistent with the overall bicarbonate-dominated branch of Tosca and McLennan (2006). If the carbonates in ALH 84001 and EETA 79001 are preterrestrial (the data, while not unequivocal, are strong for both meteorites), Martian meteorites of a broad range of crystallization and alteration ages (not just nakhlites) all formed their carbonate and alkaline-earth sulfate minerals by evaporation from an initially bicarbonate-dominated solution, in contrast to sulfate-dominated Mars' surface conditions (Tosca and McLennan 2006). However, the initial solute conditions and chemical divides required to produce calcite differ from those required to produce the sideritic–ankeritic carbonates of the nakhlites (Tosca and McLennan 2006). This may be important for nakhlites, in which siderite, magnesite, and a wide range of carbonates with partial Fe- and Mn-substitution all occur at various locations in different meteorites and samples, and it is particularly significant for ALH 84001, in which calcite, magnesite, and a wide range of carbonates with partial Fe- and Mn-substitution all coexist in rosettes. The causes of the solution-composition variations that produced carbonates from different chemical-divide sequences of Tosca and McLennan (2006) in different Martian meteorites from different groups (e.g., calcite in EETA 79001 as opposed to sideritic–ankeritic carbonates in nakhlites), different meteorites from the same group (nakhlites), and especially different parts of the same rosettes in the same meteorite (ALH 84001) require further investigation.

SUMMARY

Meteorites of several igneous rock types have a variety of chemical and isotopic indicators of formation on Mars. Martian meteorites complement data from missions to Mars. Microscopic (petrographic) analysis of Martian meteorites in terrestrial laboratories reveals features much smaller in scale than almost all the smallest-scale imagery from Mars' surface missions. Textural evidence from terrestrial microscopy also establishes paragenetic reaction and sequence relationships that cannot be distinguished in bulk mineral assemblages as detected by spectroscopic instruments from orbital or surface missions. Variations in hydration states among otherwise similar minerals can be better distinguished from samples in meteorites than from orbiter, lander, or rover spectroscopy, but even in Martian meteorites, hydration states of such minerals may have been modified from their Martian state during sample arrival at Earth, recovery, and curatorial processing. In addition, geochronologic analyses of Martian meteorite samples in terrestrial laboratories place absolute age-dating constraints on igneous crystallization and aqueous alteration processes on Mars that cannot yet be replicated by remote spacecraft instrumental methods.

The paucity of abundant and mobile solutions undersaturated with respect to primary minerals manifests itself as widespread persistence of olivine and pyroxene both in Martian meteorites and at the surface of Mars. Unlike Earth, where a vigorous hydrologic cycle results in widespread leaching of soluble products from weathering sites and eventual precipitation elsewhere as chemical sediments, the dissolved products of weathering reactions in Martian meteorites were not removed far from the reaction sites. Solutes released by primary mineral weathering in Martian meteorites precipitated locally in transmineral and grain-boundary fractures as cation-rich clays, carbonates, and evaporite minerals, rather than being leached away.

Primary mineral corrosion and secondary minerals occur in Martian meteorites of a range of ages, from ca. 3.9 Ga (approximately mid-Noachian; opx corrosion and associated carbonate formation in ALH 84001), through one or more episodes after ca. 1.3Ga (approximately mid–late Amazonian; pyroxene corrosion and diverse alteration minerals in nakhlites), and possibly until quite recently (late Amazonian; preterrestrial carbonate and sulfate in the ca. 170 Ma shergottite EET 79001). These occurrences record aqueous alteration processes over a broad range of times in the paleoenvironmental history of Mars' surface.

Both Martian meteorites and Mars' surface materials exhibit corrosion features on the same primary minerals (pyroxene and olivine), and many of the same alteration-product minerals occur, including clay minerals and mineraloids, carbonates, sulfates, and halite. However, aqueous alteration in Martian meteorites is not identical to the aqueous alteration inferred to have affected Mars' surface materials. The main secondary host minerals for Fe differ; in Martian meteorites, sideritic–ankeritic carbonate is the main secondary host mineral for Fe, whereas in Mars' surface materials, ferric oxides and ferric sulfates are the main secondary host minerals for Fe. This is most likely a consequence of differences in the initial compositions of the altering solutions, with carbonate/bicarbonate dominating in the solutions that altered Martian meteorites and sulfate dominating the solutions that altered most Mars' surface materials. Martian meteorites may have been exhumed from depths on the order of tens of meters, depths that may have isolated Martian meteorites from large quantities of Mars' surface solutions and created more basalt-mineral–buffered high pHs in the fracture microenvironments within which Martian meteorite aqueous alteration occurred. If so, Martian meteorites provide insight into the basic nature of low-temperature mineral–water interactions on Mars, and they direct attention to potentially promising subsurface targets for future exploration of potentially habitable environments on Mars (Grotzinger 2009).

ACKNOWLEDGMENTS

Jim Gooding stimulated and informed my research on meteorite alteration. The late Jean Delvigne provided considerable insight into natural alteration textures of pyroxenes and amphiboles. Helpful discussions with colleagues Susan J. Wentworth, Kathie Thomas-Keprta, David S. McKay, Tom Pike, Urs Staufer, Walter Goetz, Morton Bo Madsen, Mike Hecht, Danita S. Brandt, Nilton O. Renno, Lina C. Patino, Susan Stipp, Kathy Nagy, Siggy Gislason, Allan Treiman, Hort Newsom, John C. Dixon, Art White, Aric M. Velbel, Theresa Longazo, and Lauren M. Spencer are appreciated, as are discussions with and assistance in the laboratory from former Michigan State University graduate students Jason R. Price, Jennifer A. Wade, Angela R. Donatelle, Daniel R. Snyder, Anna I. Losiak, Kathleen A. Jeffery, and Allison R. Pluda. I thank Kathie Thomas-Keprta for sharing microscopic imagery from ALH 84001. Ewa Danielewicz (Michigan State University Center for Advanced Microscopy) and Angela R. Donatelle ably operated the scanning electron microscopes. Detailed, thorough reviews by Joel Hurowitz and Allan Treiman are greatly appreciated. This research was supported by NASA Mars Fundamental Research Program grant NNG05GL77G (M.A. Velbel, P.I.).

REFERENCES

Alkattan M, Oelkers EH, Dandurand J-L, Schott J. 1997. Experimental studies of halite dissolution kinetics: 1. The effect of saturation state and the presence of trace metals. *Chemical Geology* 137:201–219.

Ash RD, Knott SF, Turner G. 1996. A 4-Gyr shock age for a Martian meteorite and implications for the cratering history of Mars. *Nature* 380:57–59.

Banfield JF, Veblen DR, Jones BF. 1990. Transmission electron microscopy of subsolidus oxidation and weathering of olivine. *Contributions to Mineralogy and Petrology* 106:110–123.

Banin A, Clark BC, Wanke H. 1992. Surface chemistry and mineralogy. *In* Kieffer HH et al. (Editors). *Mars*: The University of Arizona Press, Tucson. p. 594–625.

Barber DJ, Scott ERD. 2003. Transmission electron microscopy of minerals in the Martian meteorite Allan Hills 84001. *Meteoritics and Planetary Science* 38:831–848.

Barber DJ, Scott ERD. 2006. Shock and thermal history of Martian meteorite Allan Hills 84001 from transmission electron microscopy. *Meteoritics and Planetary Science* 41:643–662.

Barrat JA, Gillet P, Lecuyer C, Sheppard SMF, Lesourd M. 1998. Formation of carbonates in the Tatahouine meteorite. *Science* 280:412–414.

Barrat JA, Gillet P, Lesourd M, Blichert-Toft J, Poupeau GR. 1999. The

Tatahouine diogenite: Mineralogical and chemical effects of sixty-three years of terrestrial residence. *Meteoritics and Planetary Science* 34:91–97.

Beck P, Barrat JA, Gillet P, Wadhwa M, Franchi IA, Greenwood RC, Bohn M, Cotton J, van der Moortèle B, Reynard B. 2006. Petrography and geochemistry of the chassignite Northwest Africa 2737 (NWA 2737). *Geochimica et Cosmochimica Acta* 70:2127–2139.

Becker RH, Pepin RO. 1984. The case for a Martian origin of the shergottites: Nitrogen and noble gases in EETA 79001. *Earth and Planetary Science Letters* 69:225–242.

Bell MS. 2007. Experimental shock decomposition of siderite and the origin of magnetite in Martian meteorite ALH84001. *Meteoritics and Planetary Science* 42:935–949.

Benzerara K, Chapon V, Moreira D, López-García P, Guyot F, Heulin T. 2006. Microbial diversity on the Tatahouine meteorite. *Meteoritics and Planetary Science* 41:1249–1265.

Benzerara K, Menguy N, Guyot F, Dominici C, Gillet P. 2003. Nanobacteria like calcite single crystals at the surface of the Tatahouine meteorite. *Proceedings of the National Academy of Sciences of the USA* 100:7438–7442.

Benzerara K, Menguy N, Guyot F, Vanni C, Gillet P. 2005a. High-resolution study of silicate–carbonate-microorganism interface prepared by focused ion beam (FIB). *Geochimica et Cosmochimica Acta* 69:1413–1422.

Benzerara K, Yoon TH, Menguy N, Tyliszczak T, Brown G Jr. 2005b. Nanoscale environments associated with bioweathering of a meteoritic Mg-Fe-pyroxene. *Proceedings of the National Academy of Sciences of the USA* 102:979–982.

Berner EK, Berner RA. 1996. *Global Environment: Water, Air, and Geochemical Cycles*: Prentice-Hall, Englewood Cliffs, New Jersey. 376 p.

Berner RA, Schott J. 1982. Mechanism of pyroxene and amphibole weathering—II. Observations of soil grains. *American Journal of Science* 282:1214–1231.

Berner RA, Sjöberg EL, Velbel MA, Krom MD. 1980. Dissolution of pyroxenes and amphiboles during weathering. *Science* 207:1205–1206.

Bevan A, de Laeter JR. 2002. *Meteorites: A Journey Through Space and Time*: Smithsonian Institution Press, Washington, DC. 215 p.

Bibring J-P, Langevin Y. 2008. Mineralogy of the Martian surface from Mars Express OMEGA observations. *In* Bell JF III (Editor). *The Martian Surface: Composition, Mineralogy, and Physical Processes*: Cambridge University Press, Cambridge, UK. p. 153–168.

Bibring J-P, Langevin Y, Mustard JF, Poulet F, Arvidson R, Gendrin A, Gondet B, Mangold N, Pinet P, Forget F, the OMEGA Team. 2006. Global mineralogical and aqueous Mars history derived from OMEGA/Mars Express data. *Science* 312:400–404.

Bogard DD, Johnson P. 1983. Martian gases in an Antarctic meteorite? *Science* 221:651–654.

Borg LE, Connelly JN, Nyquist LE, Shih C-Y, Wisemann H, Reese Y. 1999. The age of carbonates in Martian meteorite ALH84001. *Science* 286:90–94.

Boynton WV, Ming DW, Kounaves SP, Young SMM, Arvidson RE, Hecht MH, Hoffman J, Niles PB, Hamara DK, Quinn RC, Smith PH, Sutter B, Catling DC, Morris RV. 2009. Evidence for calcium carbonate at the Mars Phoenix landing site. *Science* 325:61–64.

Bradley JP. 2005. Interplanetary dust particles. *In* Davis AM (Editor). *Meteorites, Comets, and Planets, Treatise on Geochemistry, Vol. 1*: Elsevier, Amsterdam. p. 689–711.

Brearley AJ. 2003. Magnetite in ALH 84001: An origin by shock induced thermal decomposition of iron carbonate. *Meteoritics and Planetary Science* 38:849–870.

Bridges JC, Catling DC, Saxton JM, Swindle TD, Lyon IC, Grady MM. 2001. Alteration assemblages in Martian meteorites: Implications for near-surface processes. *Space Science Reviews* 96:365–392.

Bridges JC, Grady MM. 1999. A halite–siderite–anhydrite–chlorapatite assemblage in Nakhla: Mineralogical evidence for evaporites on Mars. *Meteoritics and Planetary Science* 34:407–415.

Bridges JC, Grady MM. 2000. Evaporite mineral assemblages in the nakhlite (Martian) meteorites. *Earth and Planetary Science Letters* 176:267–279.

Bridges JC, Warren PH, Lee MR. 2004. Olivine decomposition features in the Y-000593 and NWA 998 nakhlites. *In* Abstracts of the 67th Annual Meeting of the Meteoritical Society: *Meteoritics and Planetary Science* 39:A18.

Brownlee D, Tsou P, Aléon J, Alexander CMOD, Araki T, Bajt S, Baratta GA, Bastien R, Bland P, Bleuet P, Borg J, Bradley JP, Brearley A, Brenker F, Brennan S, Bridges JC, Browning ND, Brucato JR, Bullock E, Burchell MJ, Busemann H, Butterworth A, Chaussidon M, Cheuvront A, Chi M, Cintala MJ, Clark BC, Clemett SJ, Cody G, Colangeli L, Cooper G, Cordier P, Daghlian C, Dai Z, D'Hendecourt L, Djouadi Z, Dominguez G, Duxbury T, Dworkin JP, Ebel DS, Economou TE, Fakra S, Fairey SAJ, Fallon S, Ferinni G, Ferroir T, Fleckenstein H, Floss C, Flynn G, Franchi IA, Fries M, Gainsforth Z, Gallien J-P, Genge M, Gilles MK, Gillet P, Gilmour J, Glavin DP, Gounelle M, Grady MM, Graham GA, Grant PG, Green SF, Grossemy F, Grossman L, Grossman JN, Guan Y, Hagiya K, Harvey R, Heck P, Herzog GF, Hoppe P, Hörz F, Huth J, Hutcheon ID, Ignatyev K, Ishii H, Ito M, Jacob D, Jacobsen C, Jacobson S, Jones S, Joswiak D, Jurewicz A, Kearsley AT, Keller LP, Khodja H, Kilcoyne ALD, Kissel J, Krot A, Langenhorst F, Lanzirotti A, Le L, Leshin LA, Leitner J, Lemelle L, Leroux H, Liu M-C, Luening K, Lyon I, MacPherson G, Marcus MA, Marhas K, Marty B, Matrajt G, McKeegan K, Meibom A, Mennella V, Messenger K, Messenger S, Mikouchi T, Mostefai S, Nakamura T, Nakano T, Newville M, Nittler LR, Ohnishi I, Ohsumi K, Okudaira K, Papanastassiou DA, Palma R, Palumbo ME, Pepin RO, Perkins D, Perronnet M, Pianetta P, Rao W, Rietmeijer FJM, Robert F, Rost D, Rotundi A, Ryan R, Sandford SA, Schwandt CS, See TH, Schlutter D, Sheffield-Parker J, Simionovici A, Simon S, Sitnitsky I, Snead CJ, Spencer MK, Stademann FJ, Steele A, Stephan T, Stroud R, Susini J, Sutton SR, Suzuki Y, Taheri M, Taylor S, Teslich N, Tomeoka K, Tomioka N, Toppani A, Trigo-Rodríguez JM, Troadec D, Tsuchiyama A, Tuzzolino AJ, Tyliszczak T, Uesugi K, Velbel M, Vellenga J, Vicenzi E, Vincze L, Warren J, Weber I, Weisberg M, Westphal AJ, Wirick S, Wooden D, Wopenka B, Wozniakiewicz P, Wright I, Yabuta H, Yano H, Young ED, Zare RN, Zega T, Ziegler K, Zimmerman L, Zinner E, Zolensky M. 2006. Comet 81P/Wild 2 Under a Microscope: *Science* 314(5806):1711–1716.

Carr M. 2006. *The Surface of Mars*: Cambridge University Press, Cambridge, UK. 307 p.

Carr MH, Greeley R, Blasius KR, Guest JE, Murray JB. 1977. Some Martian volcanic features as viewed from the Viking orbiters. *Journal of Geophysical Research* 82:3985–4015.

Carr RH, Grady MM, Wright IP, Pillenger CT. 1985. Martian atmospheric carbon dioxide and weathering products in Mars meteorites. *Nature* 314:248–250.

Changela HG, Bridges JC. 2010. Alteration assemblages in the nakhlites: Variation with depth on Mars. *Meteoritics and Planetary Science* 45:1847–1867.

Christensen PR, Wyatt MB, Glotch TD, Rogers AD, Anwar S, Arvidson RE, Bandfield JL, Blaney DL, Budney C, Calvin WM, Fallacaro A, Fergason RL, Gorelick N, Graff TG, Hamilton VE, Hayes A, Johnson JR, Knudson AT, McSween HY Jr, Mehall GL, Mehall LK, Moersch JE, Morris RV, Smith MD, Squyres SW, Ruff SW, Wolff MJ. 2004. Mineralogy at Meridiani Planum from the Mini-TES experiment on the Opportunity rover. *Science* 306:1733–1739.

Clark BC. 1993. Geochemical components in Martian soil. *Geochimica et Cosmochimica Acta* 57:4575–4581.

Clark BC III, Baird AK, Rose HJ Jr, Toulmin P III, Christian RP, Kelliher WC, Castro AJ, Rowe CD, Keil K, Huss GR. 1977. The Viking X-ray fluorescence experiment: Analytical methods and early results. *Journal of Geophysical Research* 82:4577–4594.

Clark BC, Baird AK, Weldon RJ, Tsusaki DM, Schnabel L, Candelaria MP. 1982. Chemical composition of Martian fines. *Journal of Geophysical Research* 87:10059–10067.

Clark BC, Morris RV, McLennan SM, Gellert R, Jolliff B, Knoll AH, Squyres SW, Lowenstein TK, Ming DW, Tosca NJ, Yen A, Christensen PR, Gorevan S, Brückner J, Calvin W, Dreibus G, Farrand W, Klingelhoefer G, Waenke H, Zipfel J, Bell JF III, Grotzinger J, McSween HY, Rieder R. 2005. Chemistry and mineralogy of outcrops at Meridiani Planum. *Earth and Planetary Science Letters* 240:73–94.

Clark BC, van Hart D. 1981. The salts of Mars. *Icarus* 45:370–378.

Clayton RN. 1993. Oxygen isotope analysis. *Antarctic Meteorite Newsletter* 16(3):4.

Clayton RN, Mayeda TK. 1983. Oxygen isotopes in eucrites, shergottites, nakhlites, and chassignites. *Earth and Planetary Science Letters* 62:1–6.

Clayton RN, Onuma N, Mayeda TK. 1976. A classification of meteorites based on oxygen isotopes. *Earth and Planetary Science Letters* 30:10–18.

Corrigan CM, Harvey RP. 2004. Multi-generational carbonate assemblages in Martian meteorite Allan Hills 84001: Implications for nucleation, growth, and alteration. *Meteoritics and Planetary Science* 39:17–30.

Delvigne J. 1998. *Atlas of Micromorphology of Mineral Alteration and Weathering*: The Canadian Mineralogist, Special Publication 3, Ottawa, Canada. 495 p.

Delvigne J, Bisdom EBA, Sleeman J, Stoops G. 1979. Olivines: Their pseudomorphs and secondary products. *Pedologie* 39:247–309.

Dodd RT. 1981. *Meteorites: A Petrologic-Chemical Synthesis*: Cambridge University Press, Cambridge, UK. 368 p.

Eggleton RA. 1984. Formation of iddingsite rims on olivine: A transmission electron microscope study. *Clays and Clay Minerals* 32:1–11.

Eggleton RA. 1986. The relation between crystal structure and silicate weathering rates. *In* Colman SM, Dethier DP (Editors). *Rates of Chemical Weathering of Rocks and Minerals*: Academic Press, Orlando, Florida. p. 21–40.

Ehlmann BL, Mustard JF, Murchie SL, Poulet F, Bishop JL, Brown AJ, Calvin WM, Clark RN, Des Marais DJ, Milliken RE, Roach LH, Roush TL, Swayze GA, Wray JJ. 2008. Orbital identification of carbonate-bearing rocks on Mars. *Science* 322:1828–1832.

Eugster O, Herzog GF, Marti K, Caffee MW. 2006. Irradiation records, cosmic-ray exposure ages, and transfer times of meteorites. *In* Lauretta DS, McSween HY Jr (Editors). *Meteorites and the Early Solar System II*: The University of Arizona Press, Tucson. p. 829–851.

Farquhar J, Thiemans MH, Jackson T. 1998. Atmospheric-surface interactions on Mars: $\Delta^{17}O$ measurements of carbonate from ALH84001. *Science* 280:1580–1582.

Farrand WH, Glotch TD, Rice JW Jr, Hurowitz JA, Swayze GA. 2009. Discovery of jarosite within the Mawrth Vallis region of Mars: Implications for the geologic history of the region. *Icarus* 204:478–488.

Fisk MR, Popa R, Mason OU, Storrie-Lombardi MC, Vicenzi EP. 2006. Iron-magnesium silicate bioweathering on Earth (and Mars?). *Astrobiology* 6:48–68.

Frank TD, Gui Z, the ANDRILL SMS Science Team. 2010. Cryogenic origin for brine in the subsurface of southern McMurdo Sound, Antarctica. *Geology* 38:587–590.

Friedman Lentz RC, Taylor GJ, Treiman AH. 1999. Formation of a Martian pyroxenite: A comparative study of the nakhlite meteorites and Theo's Flow. *Meteoritics and Planetary Science* 34:919–932.

Fritz J, Artemieva N, Greshake A. 2005. Ejection of Martian meteorites. *Meteoritics and Planetary Science* 40:1393–1411.

Garrels RM, Mackenzie FT. 1971. *Evolution of Sedimenary Rocks*: W.W. Norton & Company, Inc., New York. 397 p.

Garrison DH, Bogard DD. 1998. Isotopic composition of trapped and cosmogenic noble gases in several Martian meteorites. *Meteoritics and Planetary Science* 33:721–736.

Gendrin A, Mangold N, Bibring J-P, Langevin Y, Gondet B, Poulet F, Bonello G, Quantin C, Mustard J, Arvidson R, LeMouélic S. 2005. Sulfates in Martian layered terrains: The OMEGA/Mars Express view. *Science* 307:1587–1591.

Gillet P, Barrat JA, Deloule E, Wadhwa M, Jambon A, Sautter V, Devouard B, Neuville D, Benzerara K, Lesourd M. 2002. Aqueous alteration in the Northwest Africa 817 (NWA 817) Martian meteorite. *Earth and Planetary Science Letters* 203:431–444.

Gillet P, Barrat JA, Heulin T, Achouak W, Lesourd M, Guyot F, Benzerara K. 2000. Bacteria in the Tatahouine meteorite: Nanometric-scale life in rocks. *Earth and Planetary Science Letters* 175:161–167.

Golden DC, Ming DW, Schwandt CS, Morris RV, Yang SV, Lofgren GE. 2000. An experimental study on kinetically-driven precipitation of calcium–magnesium–iron carbonates from solution: Implications for the low-temperature formation of carbonates in Martian meteorite Allan Hills 84001. *Meteoritics and Planetary Science* 35:457–466.

Gooding JL. 1986a. Clay-mineraloid weathering products in Antarctic meteorites. *Geochimica et Cosmochimica Acta* 50:2215–2223.

Gooding JL. 1986b. Weathering of stony meteorites in Antarctica. *In* Annexstad JO, Schultz L, Wänke H (Editors). *International Workshop on Antarctic Meteorites*: Lunar and Planetary Institute, Houston, Texas, Technical Report 86–01, p. 48–54

Gooding JL. 1992. Soil mineralogy and chemistry on Mars: Possible clues from salts and clays in SNC meteorites. *Icarus* 99:28–41.

Gooding JL, Arvidson RE, Zolotov M Yu. 1992. Physical and chemical weathering. *In* Kieffer HH et al. (Editors). *Mars*: The University of Arizona Press, Tucson. p. 626–651.

Gooding JL, Muenow DW. 1986. Martian volatiles in shergottite EETA 79001: New evidence from oxidized sulfur and sulfur-rich aluminosilicates. *Geochimica et Cosmochimica Acta* 50:1049–1059.

Gooding JL, Wentworth SJ, Zolensky ME. 1988. Calcium carbonate and sulfate of possible extraterrestrial origin in the EETA 79001 meteorite. *Geochimica et Cosmochimica Acta* 52:909–915.

Gooding JL, Wentworth SJ, Zolensky ME. 1991. Aqueous alteration of the Nakhla meteorite. *Meteoritics* 26:135–143.

Gounelle M, Zolensky ME. 2001. A terrestrial origin for sulfate veins in CI1 chondrites. *Meteoritics and Planetary Science* 36:1321–1329.

Grady MM. 2000. *Catalog of Meteorites,* 5th ed.: Cambridge University Press, Cambridge, UK. 689 p.

Grady MM, Wright IP, Pillenger CT. 1998. A nitrogen and argon stable isotope study of Allan Hills 84001: Implications for the evolution of the Martian atmosphere. *Meteoritics and Planetary Science* 33:795–802.

Grantham JH, Velbel MA. 1988. Influence of climate and topography on rock fragment abundance in modern fluvial sands of the southern Blue Ridge Mountains, North Carolina. *Journal of Sedimentary Petrology* 58:219–227.

Grotzinger J. 2009. Beyond water on Mars. *Nature Geoscience* 2:231–233.

Grotzinger JP, Arvidson RE, Bell JF III, Calvin W, Clark BC, Fike DA, Golombek M, Greeley R, Haldemann A, Herkenhoff KE, Jolliff BL, Knoll AH, Malin M, McLennan SM, Parker T, Soderblom L, Sohl-Dickstein JN, Squyres SW, Tosca NJ, Watters WA. 2005. Stratigraphy and sedimentology of a dry to wet eolian depositional system, Burns formation, Meridiani Planum, Mars. *Earth and Planetary Science Letters* 240:11–72.

Grotzinger J, Jordan TH. 2010. *Understanding Earth*, 6th ed.: W.H. Freeman and Company, New York. 654 p.

Hartmann WK. 2005. Martian cratering 8: Isochron refinement and the chronology of Mars. *Icarus* 174:294–320.

Harvey RP, McSween HY, Jr. 1996. A possible high-temperature origin for the carbonates in the martian meteorite ALH84001. *Nature* 382:49–51.

Hecht MH, Kounaves SP, Quinn RC, West SJ, Young SMM, Ming DW, Catling DC, Clark BC, Boynton WV, Hoffman J, DeFlores LP, Gospodinova K, Kapit J, Smith PH. 2009. Detection of perchlorate and the soluble chemistry of Martian soil at the Phoenix Lander site. *Science* 325:64–67.

Herkenhoff KE, Squyres SW, Anderson R, Archinal BA, Arvidson RE, Barrett JM, Becker KJ, Bell JF III, Budney C, Cabrol NA, Chapman MG, Cook D, Ehlmann BL, Farmer J, Franklin B, Gaddis LR, Galuska DM, Garcia PA, Hare TM, Howington-Kraus E, Johnson JR, Johnson S, Kinch K, Kirk RL, Lee EM, Leff C, Lemmon M, Madsen MB, Maki JN, Mullins KF, Redding BL, Richter L, Rosiek MR, Sims MH, Soderblom LA, Spanovich N, Springer R, Sucherski RM, Sucharski T, Sullivan R, Torson JM, Yen A. 2006. Overview of the Microscopic Imager Investigation during Spirit's first 450 sols in Gusev crater. *Journal of Geophysical Research* 111(E02S04). DOI:10.1029/2005JE002574

Herkenhoff KE, Squyres SW, Arvidson R, Bass DS, Bell JF III, Bertelsen P, Ehlmann BL, Farrand W, Gaddis L, Greeley R, Grotzinger J, Hayes AG, Hviid SF, Johnson JR, Jolliff B, Kinch KM, Knoll AH, Madsen MB, Maki JN, McLennan SM, McSween HY, Ming DW, Rice JW Jr, Richter L, Sims M, Smith PH, Soderblom LA, Spanovich N, Sullivan R, Thompson S, Wdowiak T, Weitz C, Whelley P. 2004. Evidence from Opportunity's Microscopic Imager for water on Meridiani Planum. *Science* 306:1727–1730.

Herut B, Starinsky A, Katz A, Bein A. 1990. The role of seawater freezing in subsurface brines. *Geochimica et Cosmochimica Acta* 54:13–21.

Herzog GF. 2005. Cosmic-ray exposure ages of meteorites. *In* Davis AM (Editor). *Meteorites, Comets, and Planets, Treatise on Geochemistry, Vol. 1*: p. 347–380.

Hurowitz JA, McLennan SM. 2007. A ~ 3.5 Ga record of water-limited, acid weathering conditions on Mars. *Earth and Planetary Science Letters* 260:432–443.

Hutchins KS, Jakosky BM. 1997. Carbonates in Martian meteorite ALH84001: A planetary perspective on formation temperature. *Geophysical Research Letters* 24:819–822.

Hutchison R. 2004. *Meteorites: A Petrologic, Chemical and Isotopic Synthesis*: Cambridge University Press, Cambridge, UK. 506 p.

Imae N, Ikeda Y. 2007. Petrology of the Miller Range 03346 nakhlite in comparison with the Yamato-000593 nakhlite. *Meteoritics and Planetary Science* 42:171–184.

Imae N, Ikeda Y, Shinoda K, Kojima H, Iwata N. 2003. Yamato nakhlites: Petrography and mineralogy. *Antarctic Meteorite Research* 16:13–33.

Johnsson MJ. 1990. Tectonic versus chemical-weathering controls on the composition of fluvial sands in tropical environments. *Sedimentology* 37:713–726.

Johnsson MJ. 1992. Chemical weathering controls on sand composition. *In* Nierenberg, WA (Editor). *Encyclopedia of Earth System Science, Vol. 1*: Academic Press, San Diego, California. p. 455–466.

Johnsson MJ, Stallard RF. 1989. Physiographic controls on the composition of sediments derived from volcanic and sedimentary terrains on Berro Colorado Island, Panama. *Journal of Sedimentary Petrology* 59:768–781.

Johnsson MJ, Stallard RF, Meade RH. 1988. First-cycle quartz arenites in the Orinoco River basin, Venezuela and Colombia. *Journal of Geology* 96:263–277.

Jull AJT, Cheng S, Gooding JL, Velbel MA. 1988. Rapid growth of magnesium–carbonate weathering products in a stony meteorite from Antarctica. *Science* 242:417–419.

Kerr RC. 1995. Convective crystal dissolution. *Contributions to Mineralogy and Petrology* 121:237–246.

Kirschvink JL, Maine AT, Vali H. 1997. Paleomagnetic evidence of a low-temperature origin of carbonate in the Martian meteorite ALH84001. *Science* 275:1629–1633.

Klingelhöfer G, Morris RV, Bernhardt B, Schröder C, Rosionov DS, de Souza PA Jr, Yen A, Gellert R, Evlanov EN, Zubkov B, Foh J, Bonnes U, Kankeleit E, Gütlich P, Ming DW, Renz F, Wdowiak T, Squyres SW, Arvidson RE. 2004. Jarosite and hematite at Meridiani Planum from Opportunity's Mössbauer spectrometer. *Science* 306:1740–1745.

Knoll AH, Jolliff BL, Farrand WH, Bell JF III, Clark BC, Gellert R, Golombek MP, Grotzinger JP, Herkenhoff KE, Johnson JR, McLennan SM, Morris R, Squyres SW, Sullivan R, Tosca NJ, Yen A, Learner Z. 2008. Veneers, rinds, and fracture fills: Relatively late alteration of sedimentary rocks at Meridiani Planum, Mars. *Journal of Geophysical Research* 113(E06S16). DOI:10.1029/2007JE002949

Knott SF, Ash RD, Turner G. 1995. ^{40}Ar-^{39}Ar dating of ALH84001: Evidence for the early bombardment of Mars. *In Twenty-sixth Lunar and Planetary Science Conference*: Lunar and Planetary Institute, Houston. p. 765–766.

Lane MD, Bishop JL, Dyar MD, King PL, Parente M, Hyde BC. 2008. Mineralogy of Paso Robles soils on Mars. *American Mineralogist* 93:728–739.

Lapen TJ, Righter M, Brandon AD, Debaille V, Beard BL, Shafer JT, Pestier AH. 2010. A younger age for ALH84001 and its geochemical link to shergottite sources in Mars. *Science* 328:347–351.

Lasaga AC. 1998. *Kinetic Theory in the Earth Sciences*: Princeton University Press, Princeton, New Jersey. 811 p.

Leshin LA, Vicenzi E. 2006. Aqueous processes recorded by Martian meteorites: Analyzing Martian water on Earth. *Elements* 2:157–162.

Lofgren GE. 1989. Dynamic crystallization of chondrule melts of porphyritic olivine composition: Textures experimental and natural. *Geochimica et Cosmochimica Acta* 53:461–470.

Lofgren GE. 1996. A dynamic crystallization model for chondrule melts. *In* Hewins RH, Jones RH, Scott ERD (Editors). *Chondrules and the Protoplanetary Disk*: Cambridge University Press, Cambridge, UK. p. 187–196.

Lofgren GE, Lanier AB. 1990. Dynamic crystallization study of barred olivine chondrules. *Geochimica et Cosmochimica Acta* 54:3537–3551.

Lofgren GE, Russell WJ. 1986. Dynamic crystallization of chondrule melts of porphyritic and radial pyroxene composition. *Geochimica et Cosmochimica Acta* 50:1715–1726.

Losiak AI, Velbel MA. 2011. Evaporite formation during the weathering of Antarctic meteorites—A weathering census analysis based on the ANSMET database. *Meteoritics and Planetary Science* 46:443–458. DOI: j.1945-5100.2010.01166.x

Marian GM, Farren RE, Komrowski AJ. 1999. Alternative pathways for seawater freezing. *Cold Regions Science and Technology* 29:259–266.

Marti K, Kim JS, Thakur AN, McCoy TJ, Keil K. 1995. Signatures of the Martian atmosphere in glass of the Zagami meteorite. *Science* 267:1981–1984.

Mason B, Melson WG. 1970. *The Lunar Rocks*: John Wiley & Sons, New York. 179 p.

Mathew KJ, Kim JS, Marti K. 1998. Martian atmosphere and indigenous components of xenon and nitrogen in Shergotty, Nakhla, and Chassigny group meteorites. *Meteoritics and Planetary Science* 33:655–664.

McCubbin FM, Tosca NJ, Smirnov A, Nekvasil H, Steele A, Fries M, Lindsley DH. 2009. Hydrothermal jarosite and hematite in a pyroxene-hosted melt inclusion in Martian meteorite Miller Range (MIL) 03346: Implications for magmatic-hydrothermal fluids on Mars. *Geochimica et Cosmochimica Acta* 73:4907–4917.

McKay DS, Gibson EK, Jr, Thomas-Keprta KL, Vali H, Romanek CS, Clemett SJ, Chillier XDF, Maechling CR, Zare RN. 1996. Search for past life on Mars: Possible relic biogenic activity in Martian meteorite ALH84001. *Science* 273:924–930.

McKay G, Schwandt C, Le L, Mikouchi T. 2007. Minor elements in nakhlite pyroxenes: Does Cr record changes in REDOX conditions during crystallization? *In Lunar and Planetary Science Conference XXXVIII*: Lunar and Planetary Institute, Houston. LPI Contribution No. 1109 (CD-ROM), abstract #1721.

McLennan SM, Bell JF III, Calvin WM, Christensen PR, Clark BC, de Souza PA, Farmer J, Farrand WH, Fike DA, Gellert R, Ghosh A, Glotch TD, Grotzinger JP, Hahn B, Herkenhoff KE, Hurowitz JA, Johnson JR, Johnson SS, Jolliff B, Klingelhöfer G, Knoll AH, Learner Z, Malin MC, McSween HY Jr, Pocock J, Ruff SW, Soderblom LA, Squyres SW, Tosca NJ, Watters WA, Wyatt MB, Yen A. 2005. Provenance and diagenesis of the evaporite-bearing Burns formation, Meridiani Planum, Mars. *Earth and Planetary Science Letters* 240:95–121.

McLennan SM, Grotzinger JP. 2008. The sedimentary rock cycle of Mars. *In* Bell JF III (Editor). *The Martian Surface: Composition, Mineralogy, and Physical Processes*: Cambridge University Press, Cambridge, UK. p. 541–577.

McSween HY Jr. 1984. SNC meteorites: Are they Martian rocks? *Geology* 12:3–6.

McSween HY Jr. 1985. SNC Meteorites: Clues to Martian petrologic evolution? *Reviews in Geophysics* 23:391–416.

McSween HY Jr. 1994. What we have learned about Mars from SNC meteorites. *Meteoritics* 29:757–779.

McSween HY Jr. 1999. *Meteorites and Their Parent Planets*, 2nd ed.: Cambridge University Press, Cambridge, UK. 310 p.

McSween HY Jr. 2002. The rocks of Mars, from far and near. *Meteoritics and Planetary Science* 37:7–25.

McSween HY Jr. 2008. Martian meteorites as crustal rocks. *In* Bell JF III (Editor). *The Martian Surface: Composition, Mineralogy, and Physical Processes*: Cambridge University Press, Cambridge, UK. p. 383–395.

McSween HY Jr, Arvidson RE, Bell JF III, Blaney D, Cabrol NA, Christensen PR, Clark BC, Crisp JA, Crumpler LS, Des Marais DJ, Farmer JD, Gellert R, Ghosh A, Gorevan S, Graff T, Grant J, Haskin LA, Herkenhoff KE, Johnson JR, Jolliff BL, Klingelhoefer G, Knudson AT, McLennan S, Milam KA, Moersch JE, Morris RV, Rieder R, Ruff SW, de Souza PA Jr, Squyres SW, Wänke H, Wang A, Wyatt MB, Zipfel J. 2004. Basaltic rocks analyzed by the Spirit rover in Gusev Crater. *Science* 305:842–845.

McSween HY Jr, Harvey RP. 1998. An evaporation model for formation of carbonates in the ALH 84001 Martian meteorite. *International Geology Reviews* 9:840–853.

McSween HY Jr, Stolper EM, Taylor LA, Muntean RA, O'Kelley GD, Eldridge JS, Biswas S, Ngo HT, Lipschutz ME. 1979. Petrogenetic relationship between Allan Hills 77005 and other achondrites. *Earth and Planetary Science Letters* 45:275–284.

McSween HY Jr, Treiman AH. 1998. Martian meteorites. *In* Papike JJ (Editor). *Planetary Materials*: Mineralogical Society of America Reviews in Mineralogy 36:6-1-6–53.

Meyer C. 2005. Mars Meteorite Compendium, 2nd ed. http://www-curator.jsc.nasa.gov/curator/antmet/mmc/mmc.htm. Accessed.

Michalski JR, Niles PB. 2010. Deep crustal carbonate rocks exposed by meteor impact on Mars. *Nature Geoscience* 3:751–755.

Mikouchi T, Koizumi E, Monkawa A, Ueda Y, Miyamoto M. 2003. Mineralogy and petrology of Yamato 000593: Comparison with other Martian nakhlite meteorites. *Antarctic Meteorite Research* 16:34–57.

Mikouchi T, Miyamoto M, Koizumi E, Makishima J, McKay G. 2006. Relative burial depths of nakhlites: An update. *In Lunar and Planetary Science Conference XXXVII*: Lunar and Planetary Institute, Houston (CD-ROM). LPI Contribution No. 1303, abstract #1865.

Mikouchi T, Monkawa A, Koizumi E, Chokai J, Miyamoto M. 2005. MIL03346 nakhlite and NWA2737 ("Diderot") chassignite: Two new Martian cumulate rocks from hot and cold deserts. *In Lunar and Planetary Science Conference XXXVI*: Lunar and Planetary Institute, Houston (CD-ROM). LPI Contribution No. 1234, abstract #1944.

Milliken RE, Fischer WW, Hurowitz JA. 2009. Missing salts on early Mars. *Geophysical Research Letters* 36(L11202). DOI:10.1029/2009GL038558

Ming DW, Morris RV, Clark BC. 2008. Aqueous alteration on Mars. *In* Bell JF III (Editor). *The Martian Surface: Composition, Mineralogy, and Physical Processes*: Cambridge University Press, Cambridge, UK. p. 519–540.

Mittlefehldt DW. 1994. ALH84001, a cumulate orthopyroxenite member of the Martian meteorite clan. *Meteoritics* 29:214–221.

Morin S, Marion GM, von Glasow R, Voisin D, Bouchez J, Savarino J. 2008. Precipitation of salts in freezing seawater and ozone depletion events: A status report. *Atmospheric Chemistry and Physics* 8:7317–7324.

Morris RV, Klingelhöfer G, Bernhardt B, Schröder C, Rosionov DS, de Souza PA Jr, Yen A, Gellert R, Evlanov EN, Foh J, Kankeleit E, Gütlich P, Ming DW, Renz F, Wdowiak T, Squyres SW, Arvidson RE. 2004. Mineralogy at Gusev Crater from the Mössbauer spectrometer on the Spirit Rover. *Science* 305:833–836.

Morris RV, Klingelhöfer G, Schröder C, Rodionov DS, Yen A, Ming DW, de Souza PA Jr, Fleischer I, Wdowiak T, Gellert R, Bernhardt B, Evlanov EN, Zubkov B, Foh J, Bonnes U, Kankeleit E, Gütlich P, Renz F, Squyres SW, Arvidson RE. 2006. Mössbauer mineralogy of rock, soil, and dust at Gusev crater, Mars: Spirit's journey through weakly altered olivine basalt on the plains and pervasively altered basalt in the Columbia Hills. *Journal of Geophysical Research* 111(E02S13). DOI: 10.1029/2005JE002584.

Morris RV, Ruff SW, Gellert R, Ming DW, Arvidson RE, Clark BC, Golden DC, Siebach K, Klingelhöfer G, Schröder C, Fleischer I, Yen AS, Squyres SW. 2010. Identification of carbonate-rich outcrops on Mars by the Spirit rover. *Science* 329:421–424.

Mutch TA, Arvidson RE, Head JW III, Jones KL, Saunders RS. 1976. *The Geology of Mars*: Princeton University Press, Princeton, New Jersey. 400 p.

Nahon DB. 1991. *Introduction to the Petrology of Soils and Chemical Weathering*: John Wiley & Sons, Inc., New York. 313 p.

Navarro-González R, Vargas E, de la Rosa J, Raga AC, McKay CP. 2010. Reanalysis of the Viking results suggests perchlorate and organics at mid-latitudes on Mars. *Journal of Geophysical Research* 115(E12010). DOI:10.1029/2010JE003599

Newsom H. 2005. Clays in the history of Mars. *Nature* 438:570–571.

Niles PB, Michalski J. 2009. Meridiani Planum sediments on Mars formed through weathering in massive ice deposits. *Nature Geoscience* 2:215–220.

Niles PB, Zolotov MYu, Leshin LA. 2009. Insights into the formation of Fe- and Mg-rich aqueous solutions on early Mars provided by the ALH 84001 carbonate. *Earth and Planetary Science Letters* 286:122–130.

Noguchi T, Nakamura T, Misawa K, Imae N, Aoki T, Toh S. 2009. Laihunite and jarosite in the Yamato 00 nakhlites: Alteration products on Mars? *Journal of Geophysical Research* 114(E10004). DOI:10.1029/2009JE003364

Nyquist LE, Bogard DD, Shih C-Y, Park J, Reese YD, Irving AJ. 2009. Concordant Rb-Sr, Sm-Nd, and Ar-Ar ages for Northwest Africa 1460: A 346 Ma old basaltic shergottite related to "lherzolitic" shergottites. *Geochimica et Cosmochimica Acta* 73:4288–4309.

Osborn W, Matty D, Velbel M, Brown P, Wacker J. 1997. Fall and recovery of the Coleman chondrite and its associated fireball. *Meteoritics and Planetary Science* 32:781–790.

Papike JJ, Ryder G, Shearer CK. 1998. Lunar samples. *In* Papike JJ (Editor). *Planetary Materials*: Mineralogical Society of America Reviews in Mineralogy 36:5-1-5–234.

Peterson RC, Nelson W, Madu B, Shurvell HF. 2007. Meridianiite: A new mineral species observed on Earth and predicted to exist on Mars. *American Mineralogist* 92:1756–1759.

Railsback LB. 1993. A geochemical view of weathering and the origin of sedimentary rocks and natural waters. *Journal of Geoscience Education* 41:404–411.

Reid AM, Bunch TE. 1975. The nakhlites—II: Where, when, and how. *Meteoritics* 10:317–324.

Rieder R, Gellert R, Anderson RC, Brückner J, Clark BC, Dreibus G, Economou T, Klingelhöfer G, Lugmair GW, Ming DW, Squyres SW, d'Uston C, Wänke H, Yen A, Zipfel J. 2004. Chemistry of rocks and soils at Meridiani Planum from the Alpha Particle X-ray Spectrometer. *Science* 306:1746–1749.

Rietmeijer FJM. 1998. Interplanetary dust particles. *In* Papike JJ (Editor). *Planetary Materials*: Mineralogical Society of America Reviews in Mineralogy 36:2-1-2–95.

Romanek CS, Grady MM, Wright IP, Mittlefehldt DW, Socki RA, Pillenger CT, Gibson EK Jr. 1994. Records of fluid–rock interactions on Mars from the meteorite ALH84001. *Nature* 372:655–657.

Schwandt CS. 2005. Evolution of meteorite chip samples during typical storage methods: A seven and a half year ALH 84001 case study. *In Lunar and Planetary Science Conference XXXVI*: Lunar and Planetary Institute, Houston (CD-ROM). LPI Contribution No. 1234, abstract #1910.

Scott ERD. 1999. Origin of carbonate–magnetite–sulfide assemblages in Martian meteorite ALH84001. *Journal of Geophysical Research* 104(E2):3803–3813.

Scott ERD, Krot AN, Yamaguchi A. 1998. Carbonates in fractures of Martian meteorite Allan hills 84001: Petrologic evidence for impact origin. *Meteoritics and Planetary Science* 33:709–719.

Scott ERD, Yamaguchi A, Krot AN. 1997. Petrological evidence for shock melting of carbonate in the Martian meteorite ALH84001. *Nature* 387:377–379.

Simon B. 1981. Dissolution rates of NaCl and KCl in aqueous solutions. *Journal of Crystal Growth* 52:789–794.

Smith KL, Milnes AR, Eggleton RA. 1987. Weathering of basalt: Formation of iddingsite. *Clays and Clay Minerals* 35:418–428.

Smith PH, Tamppari LK, Arvidson RE, Bass D, Blaney D, Boynton WV, Carswell A, Catling DC, Clark BC, Duck T, DeJong E, Fisher D, Goetz W, Gunnalugsson HP, Hecht MH, Hipkin V, Hoffman J, Hviid SF, Keller HU, Kounaves SP, Lange CF, Lemmon MT, Ming DW, Morris RV, Pike WT, Renno N, Staufer U, Stoker C, Taylor P, Whiteway JA, Zent AP. 2009. H_2O at the Phoenix landing site. *Science* 325:58–61.

Soderblom LA, Bell JF III. 2008. Exploration of the Martian surface. *In* Bell JF III (Editor). *The Martian Surface: Composition, Mineralogy, and Physical Processes*: Cambridge University Press, Cambridge, UK. p. 3–19.

Squyres SW, Arvidson RE, Bell JF III, Brückner J, Cabrol NA, Calvin W, Carr MH, Christensen PR, Clark BC, Crumpler L, Des Marais DJ, d'Uston C, Economou T, Farmer J, Farrand W, Folkner W, Golombek M, Gorevan S, Grant JA, Greeley R, Grotzinger J, Haskin L, Herkenhoff KE, Hviid S, Johnson J, Klingelhöfer G, Knoll AH, Landis G, Lemmon M, Li R, Madsen MB, Malin MC, McLennan SM, McSween HY, Ming DW, Moersch J, Morris RV, Parker T, Ricce JW Jr, Richter L, Rieder R, Sims M, Smith M, Smith P, Soderblom LA, Sullivan R, Wänke H, Wdowiak T, Wolff M, Yen A. 2004. The Opportunity Rover's Athena science investigation at Meridiani Planum, Mars. *Science* 306:1698–1703.

Squyres SW, Knoll AH, Arvidson RE, Clark BC, Grotzinger JP, Jolliff BL, McLennan SN, Tosca N, Bell JF III, Calvin WM, Farrand WH, Glotch TD, Golombek MP, Herkenhoff KE, Johnson JR, Klingelhöfer G, McSween HY, Yen A. 2006. Two years at Meridiani Planum: Results from the Opportunity Rover. *Science* 313:1403–1407.

Starinsky A, Katz A. 2003. The formation of natural cryogenic brines. *Geochimica et Cosmochimica Acta* 67:1475–1484.

Steele A, Fries MD, Amundsen HEF, Mysen BO, Fogel ML, Schweizer M, Boctor NZ. 2007. Comprehensive imaging and Raman spectroscopy of carbonate globules from Martian meteorite ALH 84001 and a terrestrial analog from Svalbard. *Meteoritics and Planetary Science* 42:1549–1566.

Swindle TD, Olson EK. 2004. ^{40}Ar-^{39}Ar studies of whole rock nakhlites: Evidence for the timing of formation and aqueous alteration on Mars. *Meteoritics and Planetary Science* 39:755–766.

Swindle TD, Treiman AH, Lindstrom DJ, Burkland MK, Cohen BA, Grier J, Li

B, Olson EK. 2000. Noble gases in iddingsite from the Lafayette meteorite: Evidence of liquid water on Mars in the last few hundred million years. *Meteoritics and Planetary Science* 35:107–115.

Taylor G, Eggleton RA. 2001. *Regolith Geology and Geomorphology*: John Wiley & Sons, New York. 375 p.

Taylor SR. 1975. *Lunar Science: A Post-Apollo View*: Pergamon Press, Inc., New York. 372 p.

Terribilini D, Eugster O, Burger M, Jakob A, Krähenbühl U. 1998. Noble gases and chemical composition of Shergotty mineral fractions, Chassigny, and Yamato 793605: The trapped argon-40/argon-36 ration and ejection times of Martian meteorites. *Meteoritics and Planetary Science* 33:677–684.

Thomas-Keprta KL, Clemett SJ, McKay DS, Gibson EK, Wentworth SJ. 2009. Origins of magnetite nanocrystals in Martian meteorite ALH84001. *Geochimica et Cosmochimica Acta* 73:6631–6677.

Tosca NJ, McLennan SM. 2006. Chemical divides and evaporite assemblages on Mars. *Earth and Planetary Science Letters* 241:21–31.

Tosca NJ, McLennan SM, Clark BC, Grotzinger JP, Hurowitz JA, Knoll AH, Schröder C, Squyres SW. 2005. Geochemical modeling of evaporation processes on Mars: Insight from the sedimentary record and Meridiani Planum. *Earth and Planetary Science Letters* 240:122–148.

Treiman AH. 1996. To see a world in 80 kilograms of rock. *Science* 272:1447–1448.

Treiman AH. 1998. The history of Allan Hills 84001 revised; multiple shock events. *Meteoritics and Planetary Science* 33:753–764.

Treiman AH. 2003. Submicron magnetite grains and carbon compounds in Martian meteorite ALH84001: Inorganic, abiotic formation by shock and thermal metamorphism. *Astrobiology* 3:369–392.

Treiman AH. 2005. The nakhlite meteorites: Augite-rich igneous rocks from Mars. *Chemie der Erde* 65:203–270.

Treiman AH, Barrett RA, Gooding JL. 1993. Preterrestrial aqueous alteration of the Lafayette (SNC) meteorite. *Meteoritics* 28:86–97.

Treiman AH, Dyar MD,McCanta M, Noble SK, Pieters CM. 2007. Martian dunite NWA 2737: Petrographic constraints on geological history, shock events, and olivine color. *Journal of Geophysical Research* 112(E04002). DOI:10.1029/2006JE002777

Treiman AH, Gleason JD, Bogard DD. 2000. The SNC meteorites are from Mars. *Planetary and Space Science* 48:1213–1230.

Treiman AH, Lindstrom DJ. 1997. Trace element geochemistry of Martian iddingsite in the Lafayette meteorite. *Journal of Geophysical Research* 102:9153–9163.

Treiman AH, Romanek CS. 1998. Bulk and stable isotopic compositions of carbonate minerals in Martian meteorite Allan Hills 84001: No proof of high formation temperature. *Meteoritics and Planetary Science* 33:737–742.

Valley JW, Eiler JM, Graham CM, Gibson EK, Romanek CS, Stolper EM. 1997. Low-temperature carbonate concretions in the Martian meteorite ALH84001: Evidence from stable isotopes and mineralogy. *Science* 275:1633–1638.

Vaniman DT, Bish DL, Chipera SJ, Fialips CI, Carey JW, Feldman WC. 2004. Magnesium sulphate salts and the history of water on Mars. *Nature* 431:663–665.

Vaniman DT, Chipera SJ. 2006. Transformations of Mg- and Ca-sulfate hydrates in Mars regolith. *American Mineralogist* 91:1628–1642.

Velbel MA. 1984. Natural weathering mechanisms of almandine garnet. *Geology* 12:631–634.

Velbel MA. 1988. The distribution and significance of evaporitic weathering products on Antarctic meteorites. *Meteoritics* 23:151–159.

Velbel MA. 1989a. Effect of chemical affinity on feldspar hydrolysis rates in two natural weathering systems. *Chemical Geology* 78:245–253.

Velbel MA. 1989b. Weathering of hornblende to ferruginous products by a dissolution-reprecipitation mechanism: Petrography and stoichiometry. *Clays and Clay Minerals* 37:515–524.

Velbel MA. 1993. Formation of protective surface layers during silicate-mineral weathering under well-leached, oxidizing conditions. *American Mineralogist* 78:408–417.

Velbel MA. 1999. Bond strength and the relative weathering rates of simple orthosilicates. *American Journal of Science* 299:679–696.

Velbel MA. 2004. Laboratory and homework exercises in the geochemical kinetics of mineral–water reaction: Rate law, Arrhenius activation energy, and the rate-determining step in the dissolution of halite. *Journal of Geoscience Education* 52:52–59.

Velbel MA. 2007. Surface textures and dissolution processes of heavy minerals in the sedimentary cycle: Examples from pyroxenes and amphiboles. *In* Mange M, Wright D (Editors). *Heavy Minerals in Use*: *Developments in Sedimentology* 58:113–150.

Velbel MA. 2009, Dissolution of olivine during natural weathering. *Geochimica et Cosmochimica Acta* 73:6098–6113.

Velbel MA. 2011. Microdenticles on naturally weathered hornblende. *Applied Geochemistry* 26:1594–1596. DOI:10.1016/j.apgeochem.2011.05.008

Velbel MA, Barker WW. 2008. Pyroxene weathering to smectite: Conventional and low-voltage cryo-field emission scanning electron microscopy, Koua Bocca ultramafic complex, Ivory Coast. *Clays and Clay Minerals* 56:111–126.

Velbel MA, Donatelle AR, Formolo MJ. 2009. Reactant-product textures, volume relations, and implications for major-element mobility during natural weathering of hornblende, Tallulah Falls Formation, Georgia Blue Ridge, U.S.A. *American Journal of Science* 309(10):661–688.

Velbel MA, Gooding JL. 1990. Terrestrial weathering of Antarctic stony meteorites—Developments 1985–1989. *In* Koeberl C, Cassidy WA (Editors). *Workshop on Differences Between Antarctic and Non-Antarctic Meteorites, Vienna, Austria, July, 1989*: Lunar and Planetary Institute, Houston, Texas, Technical Report 90–01, p. 94–98.

Velbel MA, Long DT, Gooding JL. 1991. Terrestrial weathering of Antarctic stone meteorites: Formation of Mg-carbonates on ordinary chondrites. *Geochimica et Cosmochimica Acta* 55:67–76.

Velbel MA, Losiak AI. 2010. Denticles on chain silicate grain surfaces and their utility as indicators of weathering conditions on Earth and Mars. *Journal of Sedimentary Research* 80:771–780. DOI: 10.2110/jsr.2010.074

Velbel MA, Matty DJ, Wacker JF, Linke M. 2002. The Worden meteorite: A new ordinary chondrite fall from Michigan, U.S.A. *Meteoritics and Planetary Science* 37(Supplement):B25–B29.

Velbel MA, McGuire JT, Madden AS. 2007. Scanning electron microscopy of garnet from southern Michigan soils: Etching rates and inheritance of pre-glacial and pre-pedogenic grain–surface textures. *In* Mange M, Wright D (Editors). *Heavy Minerals in Use*: *Developments in Sedimentology* 58:413–432.

Walker D, Kiefer WS. 1985. Xenolith digestion in large magma bodies. *In* Proceedings of the Fifteenth Lunar and Planetary Science Conference, Part 2. *Journal of Geophysical Research* 90(Supplement):C585–C590.

Warren PH. 1998. Petrologic evidence for low-temperature, possibly flood-evaporitic origin of carbonates in the ALH 84001 meteorite. *Journal of Geophysical Research* 103:16759–16773.

Warren PH. 2005. The Moon. *In* Davis AM (Editor). *Meteorites, Comets, and Planets*, *Treatise on Geochemistry, Vol. 1*: Elsevier, Amsterdam. p. 559–599.

Wasson JT. 1985. *Meteorites: Their Record of Early Solar System History*: W.H. Freeman and Company, New York. 267 p.

Wasson JT, Wetherill GW. 1979. Dynamical, chemical, and isotopic evidence regarding the formation locations of asteroids and meteorites. *In* Gehrels T (Editor). *Asteroids*: The University of Arizona Press, Tucson, Arizona. p. 926–974.

Wentworth SJ, Gibson EK, Velbel MA, McKay DS. 2005. Antarctic Dry Valleys and indigenous weathering in Mars meteorites: Implications for water and life on Mars. *Icarus* 174:382–395.

Wentworth SJ, Gooding JL. 1993. Weathering features and secondary minerals in Antarctic shergottites ALHA77005 and LEW88516. *In Lunar and Planetary Science Conference XIV*: Lunar and Planetary Institute, Houston, Texas. Abstracts, p. 1507–1508.

Wentworth SJ, Gooding JL. 1994. Carbonates and sulfates in the Chassigny meteorite: Further evidence for aqueous chemistry on the SNC parent planet. *Meteoritics* 29:860–863.

White AF, Brantley SL. 2003. The effect of time on the weathering of silicate minerals: Why do weathering rates differ in the laboratory and field? *Chemical Geology* 202:479–506.

Wiens RC, Becker RH, Pepin RO. 1986. The case for a Martian origin of the shergottites: II. Trapped and indigenous gas components in EETA 79001 glass. *Earth and Planetary Science Letters* 77:149–158.

Wood CA, Ashwal LD. 1981. SNC meteorites: Igneous rocks from Mars? *In*

Proceedings of the Twelfth Lunar and Planetary Science Conference, Vol. 12B: Lunar and Planetary Institute, Houston, Texas. p. 1359–1375.

Yano H, Kubota T, Miyamoto H, Okada T, Scheeres D, Takagi Y, Yoshida K, Abe M, Abe S, Barnouin-Jha O, Fujiwara A, Hasegawa S, Hashimoto T, Ishiguro M, Kato M, Kawaguchi J, Mukai T, Saito J, Sasaki S, Yoshikawa M. 2006. Touchdown of the Hayabusa spacecraft at the Muses Sea on Itokawa. *Science* 312:1350–1353.

Zolensky ME, Zega TJ, Yano H, Wirick S, Westphal AJ, Weisberg MK, Weber I, Warren JL, Velbel MA, Tsuchiyama A, Tsou P, Toppani A, Tomioka N, Tomeoka K, Teslich N, Taheri M, Susini J, Stroud R, Stephan T, Stadermann FJ, Snead CJ, Simon SB, Simionovici A, See TH, Robert F, Rietmeijer FJM, Rao W, Perronnet MC, Papanastassiou DA, Okudaira K, Ohsumi K, Ohnishi I, Nakamura-Messenger K, Nakamura T, Mostefaui S, Mikouchi T, Meibom A, Matrajt G, Marcus MA, Leroux H, Lemelle L, Le L, Lanzirotti A, Langenhorst F, Krot A, Keller LP, Kearsley A, Joswiak D, Jacob D, Ishii H, Harvey R, Hagiya K, Grossman L, Grossman JN, Graham GA, Gounelle M, Gillet P, Genge MJ, Flynn GJ, Ferroir T, Fallon S, Ebel DS, Dai ZR, Cordier P, Chi M, Butterworth AL, Brownlee DE, Browning N, Bridges JC, Brennan S, Brearley A, Bradley JP, Bland P, Bastien R. 2006. Mineralogy and petrology of comet 81P/Wild 2 nucleus samples. *Science* 314(5806):1735–1739.

GEOCHEMISTRY OF SEDIMENTARY PROCESSES ON MARS

SCOTT M. MCLENNAN

Department of Geosciences, State University of New York at Stony Brook, Stony Brook, New York 11794-2100 USA

e-mail: Scott.McLennan@sunysb.edu

Abstract: Mars has an extensive, long-lived sedimentary record that is complimentary to the terrestrial record, bearing both first-order similarities and first-order differences. The igneous record is composed of basaltic rocks, in fundamental contrast to the granodioritic upper continental crust of the Earth, which in turn dominates the provenance of clastic and chemical sedimentary rocks. The crust and sedimentary mass of Mars on average are older than the terrestrial records, and Mars provides exceptional potential for understanding processes that were active during the earliest history (>3.5 Gyr) of the solar system. Numerous sedimentary minerals have been identified both from orbit and by rovers/landers and include a variety of clays, sulfates, amorphous silica, minor carbonates, and possibly chlorides. The Martian sedimentary mineralogical record is Fe- and Mg-enriched and Na- and K-depleted compared to the terrestrial record, reflecting differing crustal compositions and differing aqueous surficial environments. There is evidence for three distinct sedimentary mineralogical epochs: an early clay-rich era, intermediate sulfate-rich era, and a younger era dominated by secondary iron oxides. This mineralogical evolution likely records desiccation, acidification, and oxidation of the surface over geological time. There is also evidence that surficial processes were controlled by a sulfur cycle, rather than the carbon cycle, over much of Martian geological time, leading to low-pH aqueous conditions. The nature of this S cycle changed over time as volcanic sulfur sources and amounts of near-surface water diminished. There is a linkage between the S cycle and iron/oxygen cycles related to diagenetic oxidation of iron sulfates to form iron oxides. Where studied in detail, weathering is dominated by low pH, with mobility of ferric iron being common. Lack of evidence for expected aluminum mobility indicates that low water–rock ratio conditions prevailed. In Noachian terrains, where clay minerals are common, it is more likely that aqueous conditions were closer to circum-neutral, but detailed study awaits future landed missions. Numerous depositional environments are recognized, including fluvial, deltaic, lacustrine, eolian, and glacial settings. Evaporitic rocks appear common and are characterized by distinctive suites of Mg-, Ca-, and Fe-sulfates and possibly chlorides. A system of chemical divides can be constructed and indicates that the range of observed evaporite minerals can be explained by typical water compositions derived from acidic weathering of Martian crust, and with variable initial pH controlled by HCO_3^-/SO_4^{2-} ratios. Several diagenetic processes have also been identified, including complex groundwater diagenetic histories. One process, consistent with experimental studies, that explains the correlation between sulfate and iron oxide minerals seen from orbit, as well as formation of hematitic concretions in the Burns Formation on Meridiani Planum, is oxidation of iron sulfates to form iron oxides. In general, the diagenetic record that has been identified, including incomplete iron sulfate oxidation, limited clay mineral transformations, and absence of amorphous silica recrystallization, indicates highly water-limited postdepositional conditions. Among the most important outstanding questions for sedimentary geochemistry are those related to the quantification of the size and lithological distribution of the sedimentary record, the detailed history of near-surface water, and the origin and history of acidity in the aqueous environment.

Key words: Mars, geochemistry, sedimentary rocks, weathering

INTRODUCTION

The first decade of the 21st century has witnessed what must be described as a revolution in our understanding of the geology of Mars. During this time, there have been no fewer than four spacecraft orbiting the planet (Mars Surveyor, Mars Odyssey, Mars Express, Mars Reconnaissance Orbiter) and three more exploring its surface (*Spirit, Opportunity, Phoenix*), each returning large volumes of mostly complementary geological, geophysical, geochemical, and mineralogical data. With committed international engagement and launch opportunities approximately every 26 months, continued exploration of the Martian stratigraphic record including, in due course, the return to Earth of Martian sedimentary samples, appears reasonably assured (e.g., Pratt et al. 2010). Recent reviews of the general geology and geochemistry of Mars can be found in Nimmo and Tanaka (2005), Carr (2006), Soderblom and Bell (2008), Taylor and McLennan (2009), Carr and Head (2010), Grotzinger et al. (2011), Fassett and Head (2011), and McSween and McLennan (2012).

Among the many discoveries that have been made is the growing recognition of a very long-lived, highly dynamic sedimentary rock cycle that has both first-order similarities and first-order differences with regard to the terrestrial record (McLennan and Grotzinger 2008). Indeed, a well-preserved sedimentary record on Mars can be traced back into the Early Noachian era, over 4.0 billion years ago. Eolian and glacial processes are ongoing at the Martian surface today, and, accordingly, the preserved stratigraphic record of Mars very likely represents a longer interval of geological time than that recorded in the stratigraphic record of the Earth (Fig. 1). Most of the crust of Mars—and thus a large fraction of the preserved Martian sedimentary record—is very ancient (>3 Gyr), and for those interested in comparative planetology, this record is highly complementary to the stratigraphic record of Earth, which is mostly young, and indeed holds promise for providing a higher-resolution record of the early history of the solar system than may ever be available in the terrestrial record.

Although sedimentary geologists have found much that is very familiar, as described elsewhere in this volume, there are also important first-order differences between the geological and stratigraphic records of the two planets (e.g., McLennan and Grotzinger 2008, Taylor and McLennan 2009, Grotzinger et al. 2011). For example, there is little evidence that plate tectonics were ever active on Mars, and, accordingly, the development and evolution of sedimentary basins proceeded within a very different tectonic regime. The great antiquity of much of the stratigraphic record means that impact processes must have played a dominant role in the formation and distribution of sedimentary particles (e.g., Barnhart and Nimmo 2011) and in the development of sedimentary basins. Mars is a "basaltic planet" with no evidence for significant exposures of evolved granitic crust, so characteristic of the terrestrial upper continental crust, and thus the chemical and mineralogical composition of the primary sources of both clastic and chemical sediment differ fundamentally. Because of the low gravity and atmospheric pressure on Mars, volcanic processes also differ (Wilson and Head 1994, Carr 2006), one effect being that despite

Sedimentary Geology of Mars
SEPM Special Publication No. 102,
SEPM (Society for Sedimentary Geology), Print ISBN 978-1-56576-312-8, CD/DVD ISBN 978-1-56576-313-5, p. 119–138.

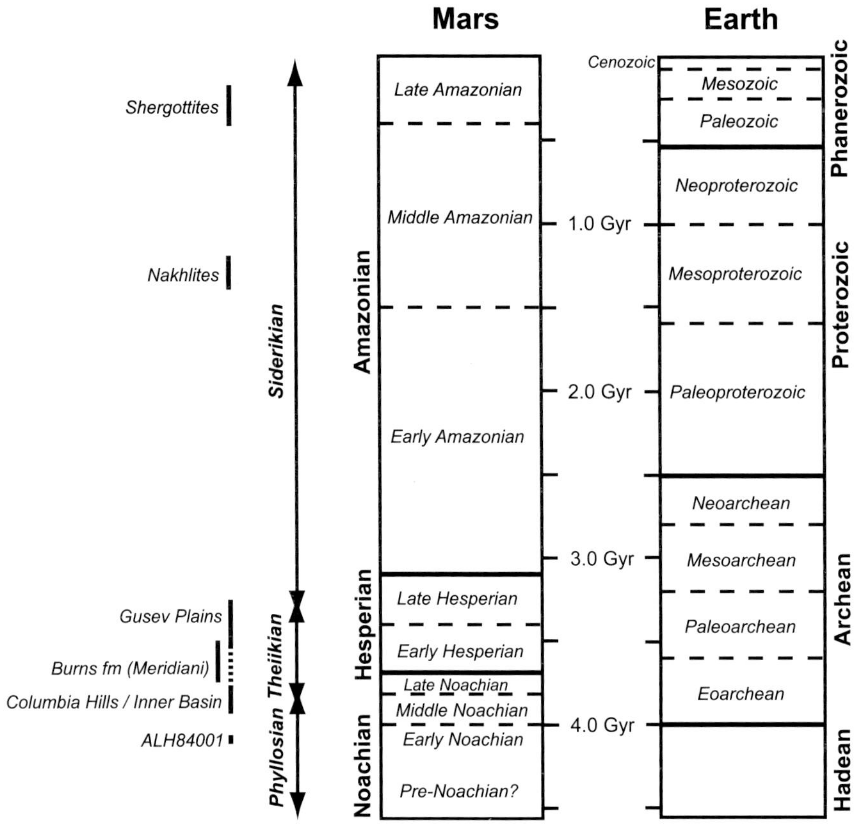

FIG. 1.—Comparison of the geological timescales for Mars and Earth. Also shown are the approximate ages of the mineralogical epochs defined by Bibring et al. (2006), sedimentary rocks studied by the rovers *Spirit* (Gusev Plains, Columbia Hills, Inner Basin) and *Opportunity* (Meridiani), and the radiometric ages of the different Martian meteorites. The overall preserved stratigraphic record for Mars dates from present-day sedimentological activity into the Early Noachian, equivalent to the terrestrial Hadean Eon, which has no preserved sedimentary record. Note that the absolute ages of Martian chronostratigraphic boundaries, taken from Head et al. (2001) and Hartmann (2005), are highly uncertain and are especially so for younger periods.

basaltic igneous compositions, pyroclastic deposits should be more common on Mars than on Earth. Finally, there is evidence that some version of a sulfur cycle, rather than a carbon cycle, may have dominated surficial processes on Mars during significant parts of its geological history, resulting in very different aqueous conditions that influenced weathering, sedimentation, and diagenesis.

McLennan and Grotzinger (2008) reviewed many of the constraints on our current understanding of the physical and chemical processes involved with the Martian sedimentary rock cycle. However, even since then new insights and some growing consensus have been forthcoming (e.g., Grotzinger et al. 2011). For example, we have a better understanding of the diversity and bulk composition of the upper crust that acts as the ultimate source of sediment (McSween et al. 2009, Taylor and McLennan 2009). Recognition of the importance of clay mineralogy, especially in the Early to Middle Noachian, and identification of long–sought-after carbonate minerals are very recent. Finally, there is a growing appreciation that some version of a sulfur cycle may at times have dominated Martian surficial processes, analogous to the way the carbon cycle dominates terrestrial processes. Accordingly, this article attempts to review a geochemical perspective of the Martian sedimentary rock cycle, focusing on recent advances in our understanding of sedimentary geochemistry/mineralogy, including the possible role that the sulfur cycle may play in governing surficial processes and thus the Martian stratigraphic record.

COMPOSITION AND EVOLUTION OF THE MARTIAN CRUST

On the Earth, the relatively evolved magmatic rocks of the upper continental crust and their metamorphic and sedimentary descendants dominate the ultimate provenance of both terrigenous and chemical components of sedimentary rocks. The composition and evolution of the upper crust is thus a critical factor in determining the chemical and mineralogical nature of the sedimentary record (Taylor and McLennan 1985), and so this is an appropriate place to begin.

The terrestrial upper continental crust is dominated by granitic rocks (sensu lato) and their extrusive and metamorphic equivalents and on average approximates to the igneous rock type granodiorite (Taylor and McLennan 1985). This crust is thus silica-rich, with elevated levels of incompatible elements (e.g., K, Th, U, and REE) and relatively low levels of ferro-magnesian elements (e.g., Fe, Mg, Cr, and Ni). This chemistry results in an average igneous mineralogy dominated by quartz, plagioclase, K-feldspar, and micas (Nesbitt and Young 1984). In contrast, basaltic rocks of mainly subalkaline to mildly alkaline affinity and their intrusive equivalents appear to dominate the Martian crust (Taylor et al. 2008, McSween et al. 2009). Estimates of bulk composition, based on orbital gamma ray chemical mapping of the Martian surface (e.g., Boynton et al. 2007) and surficial soil and dust compositions, indicate that the average Martian crust has a mildly incompatible element-enriched basalt composition (Taylor and McLennan 2009). Such a composition is likely to result in an average mineralogy dominated by plagioclase, olivine, pyroxene, and Fe-Ti oxides. Table 1 compares selected chemical and mineralogical estimates for the Martian upper crust and terrestrial upper continental crust, and Figure 2 illustrates these differences in terms of total alkali elements vs. silica.

Planetary crusts in general can be divided into the following three fundamental types based on their ultimate origin (Taylor 1989; Taylor and McLennan 2009):

1. "Primary" crusts that form as a result of initial planetary differentiation, typically through magma ocean processes, the lunar highland crust being the type example. Any primary crust that may have formed on the Earth has been lost as a result of various recycling processes. On Mars, primary crust likely exists but is profoundly disrupted by later impact processes and substantially covered by younger magmatism and sediments.

2. "Secondary" crusts that form over longer periods of time by partial melting of the mantle, the Earth's basaltic oceanic crust and lunar maria being the best-understood examples. Although secondary crusts are invariably basaltic in composition, in detail the nature of secondary crusts is highly variable throughout the solar system. Mars has a long history of magmatism, and a significant fraction of the surface is likely covered by secondary crust.

3. "Tertiary" crusts that form because of dehydration or partial melting of secondary crusts. The terrestrial continental crust is the only known example in the solar system, formed as a result of plate tectonic cycling through the mantle of the oceanic crust. The continental crust is further differentiated, predominantly by intra-crustal melting, into a more granitic upper crust (the ultimate source of most sediment) and a more mafic lower crust. Growth of continental crust has proceeded in an episodic fashion since the earliest Archean and has a mean age of about 2.5 Gyr (Taylor and McLennan 1985).

TABLE 1.—Chemical and mineralogical comparison of the upper Martian igneous crust and terrestrial upper continental crust (adapted from Nesbitt and Young [1984], McLennan and Grotzinger [2008], and Taylor and McLennan [2009]).[1]

	Mars crust	Earth upper crust
SiO_2	49.3	65.9
TiO_2	0.98	0.65
Al_2O_3	10.5	15.2
FeO_T	18.2	4.52
MnO	0.36	0.08
MgO	9.06	2.21
CaO	6.92	4.20
Na_2O	2.97	3.90
K_2O	0.45	3.36
P_2O_5	0.90	0.16
Sum	99.6	100.2
Cr	2600	83
Ni	337	44
Zn	320	71
Ba	55	550
Rb	12.5	112
La	5.5	30
Ce	13.9	64
Y	18	22
Th	0.70	10.7
U	0.18	2.8
Olivine	17	0
Pyroxene	22	1
Fe-Ti oxides	10	1
Glass	21	13
Plagioclase	29	35
K-feldspar	0	11
Sheet silicates	0	14
Quartz	0	20
Other	0	4

[1] Major elements and minerals in weight percent; trace elements in parts per million. Total iron as FeO_T.

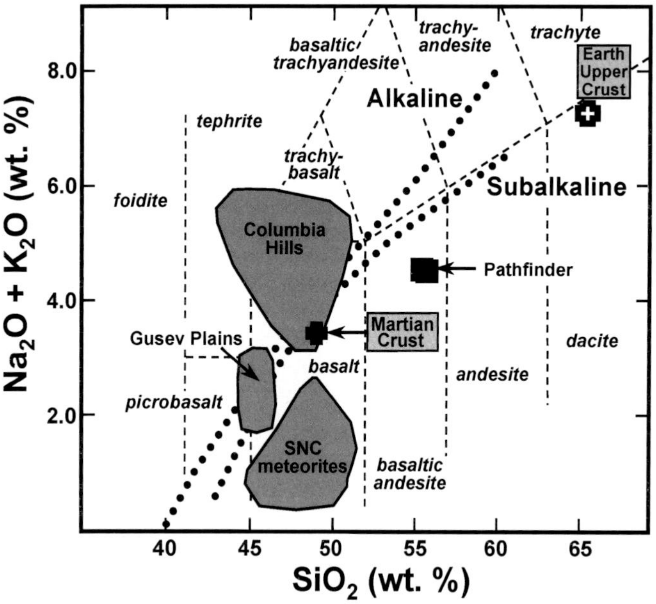

FIG. 2.—Plot of (Na_2O+K_2O) vs. SiO_2 for igneous rocks from Mars, including fields for Martian meteorites and rock samples analyzed at the Pathfinder and Spirit landing sites superimposed onto a classification scheme for volcanic rocks. Heavy dotted line shows two estimates of the boundary between alkaline and subalkaline compositions. The two large crosses are estimates of the bulk composition of the Martian upper crust and composition for the Earth's upper continental crust. The Martian crust is basaltic and has a chemical and mineralogical character that fundamentally differs from the "granodioritic" terrestrial upper crust. Adapted from Taylor and McLennan (2009) and McSween et al. (2009).

A variety of both long-lived ($^{238,235}U$-$^{206,207}Pb$, ^{187}Re-^{187}Os, ^{176}Lu-^{176}Hf, ^{147}Sm-^{143}Nd, ^{87}Rb-^{87}Sr) and short-lived (^{182}Hf-^{182}W, ^{146}Sm-^{142}Nd, ^{129}I-^{129}Xe) radiogenic isotopes indicates that Mars differentiated very early (~4.5 Gyr) into core, mantle, and primary crust, most likely as a result of magma ocean processes (e.g., Halliday et al. 2001, Elkins-Tanton et al. 2005). The oldest known example of Martian crust is the meteorite ALH84001, recently dated by Lu-Hf methods at 4.09 ± 0.03 Gyr (Lapen et al. 2010), significantly younger than previously thought (earlier estimates indicated an age of ~4.5 Gyr). Isotopic systematics are consistent with derivation from a mantle source, with characteristics similar to much younger Martian meteorites (shergottites), and so this rock is likely an example of ancient secondary crust. Crater counting techniques provide evidence that magmatic activity has continued throughout Martian history, essentially through to the present, but at drastically reduced rates over geological time (Werner 2009, Carr and Head 2010). Accordingly, the crust of Mars is mostly very ancient and dominated by a combination of old mafic primary crust and later (but still mostly old) secondary basaltic crust. Taylor and McLennan (2009) estimated that most of the Martian crust was primary, although likely highly disrupted by intense early impact bombardment (e.g., Frey 2006), with about 20 ± 10% being secondary crust. The mean age of the Martian crust then is >4.0 Gyr and much older on average than terrestrial crust. A comparison of models for the growth history of the Martian crust and terrestrial continental and oceanic crusts is provided in Figure 3.

The ages of the crusts of Earth and Mars in turn influence the age distribution of their respective sedimentary masses. The cumulative stratigraphic age distribution of the terrestrial sedimentary mass is plotted in Figure 3 (Veizer and Jansen 1985). The much younger age of the sedimentary mass compared to the continental crust has long been recognized as being related mainly to various intracrustal and sedimentary recycling processes (e.g., Garrels and Mackenzie 1971; Veizer and Jansen 1979, 1985; McLennan 1988; Veizer and Mackenzie 2003). The age distribution of the Martian sedimentary mass is very poorly constrained, not in the least as a result of the dearth of measured stratigraphic sections. Tanaka et al. (1988) quantified the surface ages (based on crater counting statistics) of different geological terrain types using Viking-era maps, and the cumulative area curve for sedimentary terrains is also shown in Figure 3. These data pre-date the recognition of abundant sedimentary rocks on Mars, which are mostly Early Hesperian and older (Malin and Edgett 2000). In addition, the regolith is likely significantly thicker in ancient terrains than in younger ones (e.g., Hartmann and Barlow 2006). Accordingly, the cumulative curve shown in Figure 3 likely represents an overall minimum age limit for the Martian sedimentary mass as a whole. Nevertheless, it is clear that the Martian sedimentary record is very ancient and on average probably less dominated by cannibalistic sedimentary recycling processes, compared to the terrestrial record, especially after about the Noachian–Hesperian boundary.

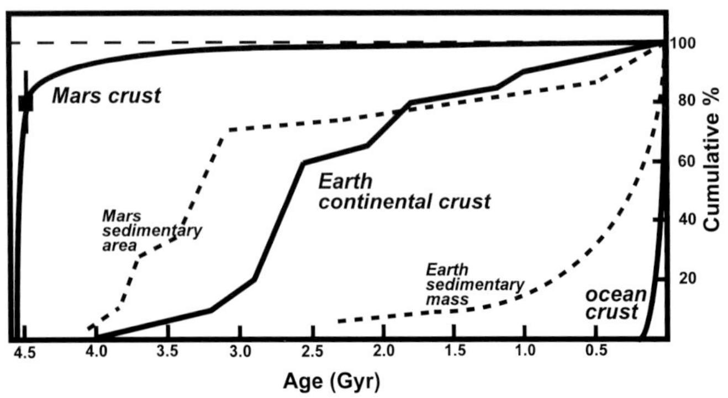

FIG. 3.—Comparison of crustal growth histories for Mars and Earth (adapted from Taylor and McLennan 2009). Also shown are cumulative growth histories for the terrestrial sedimentary mass (Veizer and Jansen 1985) and areas dominated by sedimentary deposits on Mars (Tanaka et al. 1988). The square symbol and vertical bar reflect the conclusion that 20 ± 10% of the Martian crust is secondary and formed after the crystallization of the primary crust, during solidification of an early magma ocean. The Martian crust and sedimentary mass are substantially older than Earth's continental crust and sediment. See text for further discussion.

THE SEDIMENTARY RECORD: GEOCHEMISTRY AND MINERALOGY

The present-day surface of Mars is characterized by a vanishingly thin atmosphere averaging about 6 millibars and composed mainly of CO_2, global surface temperatures averaging about −60° C and only rarely rising above 0° C, and gravitational acceleration (*g*) that is 0.38 times that of Earth. Although seasonally dynamic H_2O–CO_2 ice caps exist at both poles, conditions permitting liquid water at or near the surface only rarely exist. Nevertheless, the surface of Mars has long been recognized as having been sculpted by sedimentary processes that, in addition to the expected pyroclastic, volcaniclastic, and impact deposits, include eolian, glacial, and subaqueous activity, the latter in the form of erosional channels, branching valley networks, gullies, and lakes (see recent reviews by Carr [2006] and Carr and Head [2010]). This apparent contradiction is generally explained by proposals that Mars once had a more substantial atmosphere and more clement surface conditions, resulting from an early greenhouse effect. This early atmosphere was lost either by meteorite impact and solar wind erosion (Mars has not had a protective magnetic field since about the Early–Mid-Noachian) and/or by sequestration of CO_2 into a subsurface carbonate mineral reservoir of either sedimentary and/or hydrothermal origins. Another factor may be that in the absence of a large moon (Mars' two moons, Diemos and Phobos, measure ~12 km and ~22 km, respectively), obliquity changes are far more dramatic than on Earth, with Mars currently being in a period of exceptionally cold conditions.

Although most subaqueous activity is thought to be ancient (Noachian to Early Hesperian), there is evidence that some large channels may have formed as recently as ~100 Myr, and there is time-lapse photographic evidence for rare gully formation today (e.g., Malin and Edgett 2006). Suggestions that Mars once had an extensive ocean-sized body of water in the northern hemisphere lowlands, based primarily on geomorphological evidence, continue to appear (e.g., Di Achille and Hynek 2010) but have yet to achieve widespread consensus.

Recognition that ancient layered deposits include true lithified sedimentary rocks is recent (Malin and Edgett 2000). High-resolution spectral imaging from the Mars Express and Mars Reconnaissance Orbiter spacecrafts has revealed extensive successions of indurated layered deposits, often with complex stratal geometries, within the walls of the large canyons (e.g., Valles Marineris) and within craters in the ancient southern highlands. Layered deposits may have formed by impact, pyroclastic, volcaniclastic, and eolian processes, but recognition of a variety of secondary sedimentary minerals, including clays, sulfates, Fe-oxides, and minor carbonates, also points to a role for aqueous processes (e.g., weathering, chemical sedimentation). In some craters (e.g., Eberswalde, Holden) there is strong geomorphological and stratigraphic evidence for fans, subaqueous deltas, terraces, and outlet channels. The physical attributes of such deposits are described elsewhere in this volume, and the reader is directed there for this discussion.

In addition to orbital observations, the rovers *Spirit* and *Opportunity* evaluated ancient sedimentary rocks and soils[1] on the surface of Mars. The chemical compositions of some selected samples are listed in Table 2. A key feature of all these analyses, in addition to the basically basaltic nature of the compositions, is the elevated levels of S and Cl. The most thoroughly studied is the Late Noachian to Early Hesperian Burns Formation exposed at Meridiani Planum. The Burns Formation consists of weakly indurated, well-sorted eolian sandstones composed of sulfate-bearing grains derived from a weathered basaltic source and

[1]The term "soils" (or "regolith") is used here in the planetary sense of being an unconsolidated surficial deposit of predominantly impact and eolian origins with no implied connotation, or indeed likelihood, of biological processes or substantial vertical in situ alteration processes necessarily being involved in their formation.

TABLE 2.—Chemical composition of selected sedimentary rocks and unconsolidated surficial sediment (soils) on Mars.[1]

	Sedimentary rocks				Unconsolidated sediments (soils)				
	Meridiani Planum average Burns Formation	Columbia Hills (Peace)	Columbia Hills (Alligator)	Home Plate (Crawford)	Pathfinder average soil	Meridiani Planum basaltic soil	Gusev Plains basaltic soil	Columbia Hills basaltic soil	Average dust-rich soil
SiO_2	37.0	37.3	41.8	46.6	42.1	46.4	46.2	46.3	45.7
TiO_2	0.78	0.45	0.53	1.11	0.87	1.02	0.86	0.87	0.97
Al_2O_3	6.16	2.24	5.49	9.98	9.5	9.46	10.1	10.3	9.53
FeO_T	15.9	20.4	18.3	15.4	21.6	18.3	16.3	15.5	17.0
Cr_2O_3	0.20	0.75	0.63	0.34	0.29	0.40	0.38	0.30	0.35
MnO	0.32	0.47	0.33	0.29	0.31	0.37	0.33	0.31	0.34
MgO	7.61	21.53	16.27	10.3	7.78	7.29	8.64	8.67	7.98
CaO	5.00	4.90	4.72	6.74	6.37	7.07	6.45	6.20	6.48
Na_2O	1.69	n.d.	1.6	3.36	2.84	2.22	2.89	3.21	2.61
K_2O	0.58	n.d	0.19	0.32	0.60	0.49	0.43	0.44	0.47
P_2O_5	1.05	0.49	0.29	1.27	0.74	0.83	0.75	0.92	0.86
SO_3	22.48	10.6	8.48	2.91	6.27	5.45	5.66	6.18	6.83
Cl	1.05	0.72	1.26	1.35	0.76	0.63	0.69	0.73	0.78
Σ	**99.8**	**99.9**	**99.9**	**100.0**	**100.0**	**99.9**	**99.7**	**99.9**	**99.9**
Cr	1350	5130	4310	2330	1980	2740	2600	2050	2395
Ni	589	774	506	297	–	445	424	466	565
Zn	448	64	205	314	–	273	264	274	384
Br	102	71	217	91	–	65	52	68	64

[1] Details: Burns Formation average of 34 analyses; Pathfinder average of 7; Meridiani basaltic soil average of 8; Gusev basaltic soil average of 15; Columbia Hills average of 9; Dust-rich soil average of 8 from Gusev Plains (4) and Meridiani Planum (4). Notes: Samples in Burns Formation average, Peace, and Alligator analyses all from abraded surfaces; Crawford from brushed surface. Crawford is from the upper part of an eolian reworked pyroclastic deposit found at Home Plate. Data sources: Burns Formation from Brückner et al. (2008) and Planetary Data System; Peace and Alligator from Gellert et al. (2006); Home Plate from Squyres et al. (2007). Unconsolidated sediments from compilations reported in Taylor and McLennan (2009).

FeO_T = total iron as FeO; – = no analysis available; n.d. = not detected.

cemented by later sulfate-dominated chemical components (Squyres et al. 2004; Grotzinger et al. 2005, 2006; McLennan et al. 2005; Metz et al. 2009). To date, more than 25 m of stratigraphic section have been measured over about 10 km of lateral distance. The sands, possibly derived from a desiccating playa lake, were deposited in a dry to wet dune–interdune setting. Syn- to early postdepositional recharge of very high–ionic strength acidic groundwater, which periodically breached the surface to form local subaqueous environments, resulted in a pervasive diagenetic overprint with formation of secondary porosity, multiple generations of cements, and hematitic concretions. The Burns Formation comprises, on an anhydrous basis, approximately 40% siliciclastic components (in decreasing abundance, poorly characterized altered igneous material, plagioclase, pyroxene, and olivine) and 60% chemical constituents (amorphous silica, Mg-sulfate, Ca-sulfate, jarosite, hematite, and possibly chlorides). The presence of jarosite $[(Na,K,H_3O)Fe_3^{3+}(SO_4)_2(OH)_6]$, which typically is stable at a pH of <4, has been cited as prime evidence for low pH conditions.

The Spirit Rover has also identified older (Noachian) sedimentary rocks in Gusev Crater on the opposite side of the planet. These deposits are less well characterized because the geology and stratigraphy are more complex, exposures are less extensive, and far less detailed work has been carried out on each exposure. Included are layered deposits that have detrital textures and have been interpreted to mostly represent pyroclastic, volcaniclastic, impact, and eolian deposits (Squyres et al. 2006). Notable examples include the rock Peace, exposed in the Columbia Hills, that appears to be a Mg-sulfate–cemented mafic–ultramafic sandstone and a cross-bedded well-sorted sandstone unit preserved above a coarse pyroclastic unit (e.g., containing a bomb sag) at a large outcrop called Home Plate. The Home Plate sandstone has essentially the same chemical composition as the underlying pyroclastics and, accordingly, is interpreted to have been deposited by eolian reworking of the underlying unit (Squyres et al. 2007, Lewis et al. 2008).

An important result of orbital remote sensing and in situ Mössbauer and visible-infrared spectroscopy coupled with chemical data is that a wide diversity of secondary sedimentary minerals exists on the surface of Mars (see recent review in Murchie et al. [2009]). Common secondary detrital constituents include a range of smectites, from Fe/Mg-rich to relatively aluminous and other phyllosilicates, including the more aluminous kaolin-group minerals, as well as mixed-layer clay minerals, illite, chlorite, and possibly prehnite. Chemical constituents that have been identified include hydrated Mg-, Ca-, and Fe-sulfates; amorphous silica; secondary iron oxides; chlorides; and Mg-carbonates. Compared to terrestrial settings, this sedimentary mineral suite tends to be Fe- and Mg-rich and Na- and K-poor, likely reflecting both the differing composition of the crustal sources and differing aqueous conditions of alteration and deposition (Table 3; see below).

Orbital spectral mapping of Martian sedimentary deposits has led to the suggestion that there may be distinctive sedimentary mineralogical epochs through Martian geological history (Bibring et al. 2006). Such an evolution perhaps should not be surprising and, although differing greatly in detail, could be analogous to the terrestrial mineralogical evolution, for which as many as 10 distinct mineralogical stages over ~4 Gyr of Earth history have been proposed (Hazen et al. 2008). For Mars, mineralogical changes are interpreted to reflect an evolution in the amount and composition of near-surface water available for surficial processes. In this model, the Martian surface has effectively

TABLE 3.—Comparison of selected common sedimentary mineralogy observed on Earth with known sedimentary mineralogy on Mars with idealized chemical formulas.[1]

Earth		Mars	
Sand-sized detrital			
Quartz	SiO_2	Plagioclase	$(Ca,Na)(Si,Al)_4O_8$
K-feldspar	$KAlSi_3O_8$	Olivine	$(Fe,Mg)_2SiO_4$
Plagioclase	$(Ca,Na)(Si,Al)_4O_8$	Pyroxene	$XY(Si,Al)_2O_6$
(Intermediate-Felsic lithics)		(Mafic lithics)	
Clays			
Kaolinite	$Al_2Si_2O_5(OH)_4$	Kaolinite	$Al_2Si_2O_5(OH)_4$
Illite	$K_{0.8}Al_{2.8}Si_{3.2}O_{10}(OH)_2$	–	
Smectite[2]		Fe-Mg-smectite[2]	
e.g., $Ca_{0.17}(Al,Mg,Fe)_2$ $(Si,Al)_4O_{10}(OH_2)2H_2O$		e.g., $(Ca_{0.5},Na)_{0.3}Fe_2(Si,Al)_4O_{10}$ $(OH)_2{\cdot}nH_2O$	
Sulfates			
Gypsum	$CaSO_4{\cdot}2H_2O$	Gypsum	$CaSO_4{\cdot}2H_2O$
Anhydrite	$CaSO_4$	Kieserite	$MgSO_4{\cdot}H_2O$
		Jarosite[3]	$(Na,K,H_3O)Fe^{3+}_3(SO_4)_2(OH)_6$
		Ferricopiapite[3]	$Fe^{3+}_{0.66}Fe^{3+}_4(SO_4)_6(OH)_2{\cdot}20(H_2O)$
		"Polyhydrated sulfates"[4]	
Carbonates			
Aragonite	$CaCO_3$	Mg-carbonate	$?MgCO_3$
Calcite	$CaCO_3$	Calcite	$CaCO_3$
Dolomite	$(Mg,Ca)CO_3$		
Secondary iron oxide phases			
Hematite	Fe_2O_3	Hematite	Fe_2O_3
Goethite	$FeO{\cdot}OH$	Goethite	$FeO{\cdot}OH$
		Nanophase iron oxides	
Secondary silica phases			
Opal-A	$SiO_2{\cdot}nH_2O$	Amorphous silica	$SiO_2{\cdot}nH_2O$
Opal-CT	$SiO_2{\cdot}nH_2O$		
Micro-, mega-quartz	SiO_2		

[1] Idealized formulas for Martian minerals are highly uncertain.
[2] Smectite compositions in terrestrial sediments are highly variable, and this is also likely the case for Mars. Formulas given here are examples. Spectral evidence does point to most Martian smectites being Fe- and Mg-rich (e.g., nontronite, saponite).
[3] Jarosite and ferricopiapite are two examples of what is likely a significantly larger number of complex ferric sulfates.
[4] Nonspecific term commonly used by spectroscopists to include higher hydrates of Mg-sulfates ($MgSO_4{\cdot}nH_2O$) and other sulfate salts with multiple waters of hydration.

desiccated, acidified, and oxidized over geological history. Thus, the Martian sedimentary record has been divided into three epochs, including the following (Fig. 1):

1. Phyllosian, during the Early and Mid-Noachian, during which clay minerals are common, sulfates rare, and conditions thought to be water-rich, with near-surface waters at circum-neutral pH.

2. Theiikian, during the Late Noachian through Early Hesperian, during which hydrated sulfate minerals and low pH conditions dominate. The amount of water available for surficial processes is more controversial, with estimates ranging from water-rich (Bibring et al. 2006) to highly water-limited (e.g., Hurowitz and McLennan 2007, Tosca and Knoll 2009).

3. Siderikian, during the Late Hesperian through Amazonian, during which water-limited, low-pH, and oxidizing conditions gave rise to secondary ferric oxides dominating secondary mineralogy.

Although this model has been influential in the interpretation of Martian geological history, it has not yet achieved full consensus, and a number of aspects of the model are under vigorous investigation (Grotzinger et al. 2011). As mentioned above, detailed studies relating mineralogy to physical stratigraphy are not abundant, and the amount of water involved in surficial processes, especially during the Noachian and Early Hesperian, is controversial. The origin of acidity during the Late Noachian is also unclear, with suggestions including processes such as increased volcanic activity (e.g., Bibring et al. 2006), for which there seems to be little geological evidence (e.g., Werner 2009), and hydrolysis associated with iron oxidation (e.g., Hurowitz et al. 2010). In addition, it has been noted that an early clay-rich epoch, without a complementary reservoir of chemical constituents (e.g., carbonate, sulfate, and/or chloride salts), presents global geochemical mass balance problems (e.g., Milliken et al. 2009).

THE MARTIAN SULFUR CYCLE

Terrestrial surficial processes are dominated by the long-term and short-term carbon cycles, which give rise to the acidity required for rock weathering. These acids, including carbonic acid (H_2CO_3) and a variety of organic acids, are weak (i.e., low solubility and dissociation constants), leading to the vast majority of terrestrial aqueous environments being circum-neutral, with pH mostly in the range of 5.5 to 8.5. Carbonate minerals dominate the salts that result from the weathering and breakdown of primary igneous minerals, and a large amount of carbon is stored as organic material in sedimentary rocks (e.g., Garrels and Mackenzie 1971).

The situation on Mars appears to be different. It is widely assumed that the early secondary atmosphere of Mars was dominated by CO_2, as it was for Earth and Venus. Unlike Earth, where carbon is stored in the rock record, or Venus, where CO_2 remains in the atmosphere, the fate of the Martian CO_2 reservoir is not understood. Carbonate minerals, although identified in the geological record (e.g., Morris et al. 2010) and the polar regolith (Boynton et al. 2009), do not appear to play a dominant role in observed sedimentary mineralogy. A previously unrecognized reservoir of CO_2 ice was recently identified in the south polar cap but is only the equivalent of a few millibars of atmospheric pressure (Phillips et al. 2011). Possible explanations include loss of CO_2 to space or preservation of large amounts of unobserved carbonates at depth within the crust. The former suggestion is consistent with the loss of magnetic shielding during the Noachian, when meteorite bombardment was still intense, and the latter with the presence of secondary carbonate minerals in fractures within Martian meteorites (e.g., Bridges and Grady 1999). Of course, both processes may have been operating.

The sulfur-rich character of surface soil and dust deposits has been known since the first Viking chemical analyses (Clark et al. 1976) and has been confirmed by all subsequent landed missions (see Table 2). Orbital mapping by the Mars Odyssey spacecraft has confirmed that elevated sulfur in surficial deposits is a global feature (King and McLennan 2010, McLennan et al. 2010). There is also geochemical and cosmochemical evidence to indicate that the silicate portion of Mars is significantly enriched in S compared to the Earth (e.g., Clark and Baird 1979, Wänke and Dreibus 1994, Gaillard and Scaillet 2009, Righter et al. 2009). Coupled with evidence for extensive ancient sulfate-rich sedimentary rocks, paucity of carbonates, and evidence for widespread mobility of Fe^{3+} under oxidizing conditions, these observations have led to suggestions that a sulfur cycle, characterized by low pH conditions (~pH2–pH5), rather than the carbon cycle has dominated surficial processes on Mars throughout a significant portion of its geological history (e.g., Clark and Baird 1979, Settle 1979, Fairén et al. 2004, King et al. 2004, McLennan et al. 2006, Halevy et al. 2007, Johnson et al. 2008, McLennan and Grotzinger 2008, King and McLennan 2010).

An important point to keep in mind is that a sedimentary regime dominated by the sulfur cycle does not preclude the presence of significant amounts of CO_2 in the atmosphere. Dissociation constants of both sulfurous and sulfuric acid are many orders of magnitude greater than the dissociation constants of carbonic acid (e.g., Halevy et al. 2007). Accordingly, even very low concentrations of SO_2 in the atmosphere ($<10^{-5}$ bar) can lead to acidic-pH (<4) aqueous environments within a CO_2-dominated atmosphere.

Sulfur Reservoirs on Mars

Estimates of the Martian primitive mantle sulfur concentration typically are ≥400 ppm, or about twice that of Earth (e.g., Gaillard and Scaillet 2009, Righter et al. 2009). The composition of the Martian basaltic crust is difficult to estimate, but meteorite data and experimental results are consistent with a value of ~2000 ppm, or about twice that of terrestrial oceanic crust and four times that of the continental crust. Can we estimate the size of the sedimentary sulfur reservoir? Measurements from orbital gamma ray mapping indicate that the S content of the upper half-meter of the planet averages about 2% (King and McLennan 2010; W. Boynton, personal communication, 2009), a value broadly consistent with analyses of soil deposits by landed spacecraft (Table 2). However, there is little understanding of the mass and distribution of these deposits in space (i.e., vertically) and time (e.g., Hartmann and Barlow 2006). Sulfate minerals have been recognized from orbit in sedimentary rocks but the concentrations are poorly constrained and the overall distributions are basically unknown. The sedimentary rocks analyzed by the Mars rovers indicate S contents that are generally high but varying by over an order of magnitude in just a few locations. Accordingly, at this stage, there appears to be no way to directly estimate the average sulfur content of the sedimentary mass.

A model dependent approach to estimating the mass of the sedimentary sulfur reservoir is to examine constraints from planetary degassing. Righter et al. (2009) and Gaillard and Scaillet (2009) each estimated the amount of S degassed from the Martian mantle over geological time, and estimates ranged from 4.5×10^{19} g to 5.4×10^{21} g, respectively, with both being considered conservative estimates. The lower value of Righter et al. (2009) adopted volcanic rates from Greeley and Schneid (1991), but if more recent estimates of Noachian volcanism are considered (e.g., McEwen et al. 1999), post-magma ocean Martian volcanism could be as much as an order of magnitude greater, and, thus, the S degassing value would also increase by about an order of magnitude, bringing these S degassing estimates into closer agreement, at about 10^{21} g.

Another approach is to assume that Mars has outgassed sulfur proportionately the same as Earth. This also seems like a conservative assumption since sulfur is an incompatible element and on average Mars has differentiated more of its other incompatible elements into the crust than has the Earth (Taylor and McLennan 2009). Canfield (2004) estimated that approximately 1.1×10^{23} g, or 11% of the S in the Earth's primitive mantle, has outgassed over geological time, although much of this has also been involved with plate tectonic recycling processes, and so the amount of S at the surface at any given time is much less. If we assume that 11% of the Martian primitive mantle S has also been outgassed and if we correct for differing primitive mantle concentrations, a value of 2.3×10^{22} g of S results. In the absence of plate tectonics, any S degassed is expected to remain in the near-surface sedimentary environment.

For scale, a Martian sulfur reservoir on the order of 10^{21} to 10^{22} g is approximately equivalent to a planetary-wide sedimentary layer, with S content equal to average soil (~6% SO_3), that is between ~200 m and 2 km thick. Another way to think of it is that this mass of sulfur would equate to a planetary-wide thickness of between ~50 and 500 m of typical hydrated sulfate minerals (e.g., mixture of gypsum, kieserite, epsomite, melanterite, jarosite, etc.). Thus, the known and reasonably inferred sedimentary record appears to readily provide a reasonable sink for this amount of sulfur.

Early Sulfur Cycle

The sulfur cycle on Mars can be divided into distinctive early and late versions (Fig. 4), with the separation between them likely corresponding to the point in time during which Martian crustal growth rates subsided, some time around the end of the Hesperian (Figs. 1, 3). The early phase of the sulfur cycle would be when sulfur was actively outgassing into the surficial environment and thus providing potential acidity for aqueous weathering processes. Most magmatic activity on Mars, the ultimate source of sulfur at the surface, took place early in Martian history (>3 Gyr), during the Noachian and Hesperian (Taylor and McLennan 2009). This was also the time when

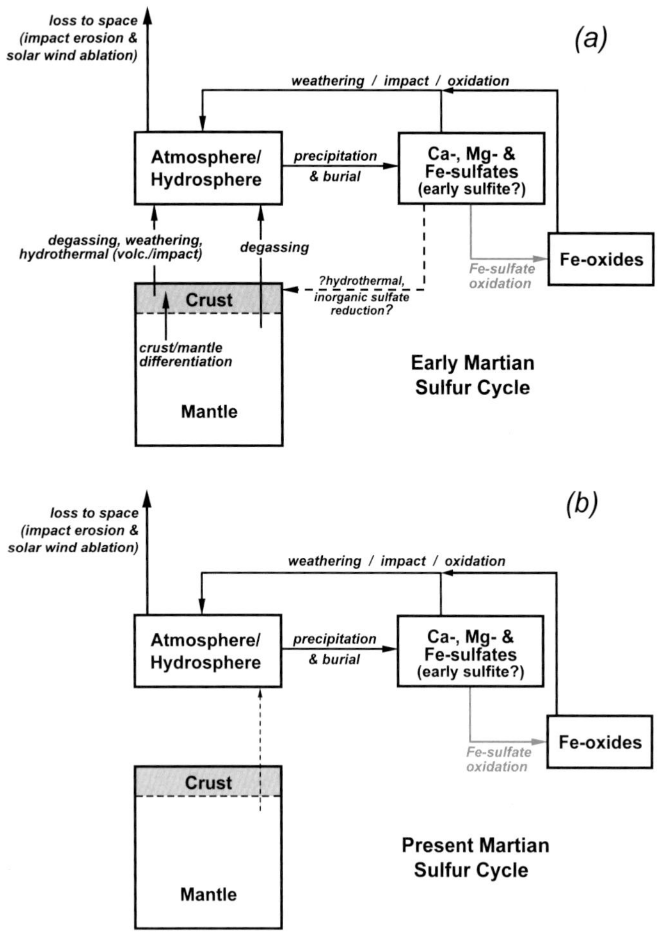

FIG. 4.—Schematic model of the sulfur cycle on Mars. (a) Possible sulfur cycle during the early history of Mars, when volcanic activity was relatively high and sulfur was delivered to the surface from mantle sources. Evidence for young volcanism, such as that represented by nakhlite and shergottite meteorites (Fig. 1), indicates that this cycle may have continued to contribute sulfur to the surface, but in a highly diminished form (Fig. 3), through later Martian history. (b) Possible sulfur cycle dominant during the youngest part of Martian geological history, when volcanic activity had substantially subsided and sulfur species taking part in surficial processes were derived mainly by recycling processes, such as impact recycling, Fe-sulfate oxidation, and sulfate weathering. Note that the sulfur cycle is linked to the iron cycle through oxidation of iron sulfates to iron oxides, shown in gray (also see Fig. 5). See text for further discussion.

relatively widespread subaqueous conditions prevailed, probably facilitated by an early greenhouse effect. Experimental and modeling studies and in situ measurements of mineralogical/chemical compositions of Martian rocks and soils can be combined to identify a variety of processes involved in the surficial S cycle (see below).

Sulfuric acid alteration of basaltic rocks and minerals is reasonably well established on the Martian surface. Processes likely included both low temperature alteration, such as that required to produce the high–ionic strength fluids responsible for evaporitic minerals at Meridiani Planum (e.g., McLennan et al. 2005, Tosca et al. 2005), and higher temperature epithermal to hydrothermal fluids, such as those responsible for a variety of mixed (including ferric) sulfates in the vicinity of Home Plate (e.g., Squyres et al. 2007, 2008). The widespread occurrence of sulfate and chloride minerals on the Martian surface observed from orbit (Murchie et al. 2009) and within rocks and soils at the landing sites points to widespread production of a variety of evaporite minerals, dominated by Ca-, Mg-, and Fe-sulfates, likely of mixed hydration state. Finally, a variety of recycling processes has been identified that have the potential for recycling sulfur through the surficial environment, including impact processes, weathering of sulfate minerals, and oxidation of ferrous and ferric sulfates to form iron oxides (see below).

One concern resulting from the low pH conditions predicted from a dominant S cycle early in Mars history is the widespread occurrence of clay and apparent dearth of sulfate minerals during the Early Noachian (e.g., Milliken et al. 2009). In order to explain this, Halevy et al. (2007) have proposed an even earlier version of a sulfur cycle. In their model, reducing conditions inhibited sulfur oxidation and, accordingly, sulfurous acid was formed from degassed SO_2 rather than being rapidly oxidized to form sulfuric acid. This scenario results in acidic but far more modest pH conditions (pH ~4.5–5.5) that led to suppression of carbonate mineral precipitation but permitted formation of clay minerals. On the other hand, such a model predicts widespread formation of a variety of sulfite (i.e., $X^{2+}S^{4+}O_3$) minerals during the Early Noachian, for which there is no evidence. Indeed, it is unlikely that such minerals could survive the near-surface environment of Mars over geological time (Halevy et al. 2007).

Present-Day Sulfur Cycle

Although rates of volcanism and aqueous activity were greatly reduced after about 3 Gyr, there is evidence that surficial processes influenced by a sulfur cycle continued beyond this time. Rock surfaces analyzed by Spirit in the Gusev Crater show clear evidence of acid alteration on the scale of up to several millimeters depth in spite of the fact that they were continually, though very slowly, physically abraded by eolian processes (e.g., Hurowitz et al. 2006). Thus, later in Martian geological history, it appears that the S cycle continued, but on a greatly diminished scale and at a greatly diminished rate.

There are a number of possibilities to explain the origin of the acidity required for younger alteration. Evidence that some level of volcanism continued essentially through to the recent period includes Late Amazonian magmatic crystallization ages for most Martian meteorites and crater counts indicating volcanic surfaces perhaps as young as a few million years old. Some degassing would likely take place during such activity, although formation of acids from these reduced gases might be limited by availability of atmospheric oxidants (Zolotov 2007). Another possible source is through impact recycling of sulfate-bearing soils and sedimentary rocks (McLennan et al. 2006, Zolotov 2007). On Earth, the Late Cretaceous Chicxulub impact, which includes sulfate minerals in the target rocks, is thought to have resulted in formation of sulfuric acid vapor that was widely distributed in the atmosphere (e.g., Pierazzo et al. 1998), and a comparable event on Mars could produce as much as a 2-mm layer of H_2SO_4 over the entire Martian surface. The S-rich Martian regolith is one potential target, although the regolith is likely everywhere less than a few hundred meters in depth (e.g., Hartmann and Barlow 2006), and S recycling from impact into regolith would accordingly be less efficient, on average. Another potential target could be ancient S-rich sedimentary rocks, such as the Burns Formation. A final potential source of acidity could be oxidation of ferrous sulfates to ferric sulfates and iron oxides, as discussed in the next section.

Fe–S–O Cycles

Using orbital spectroscopy a number of workers have noted a spatial correlation between iron oxide and sulfate minerals on the Martian surface, indicating a possible genetic link (e.g., Bibring et al. 2007, Murchie et al. 2009, Roach et al. 2010). This relationship is also consistent with experimental and modeling studies that indicate a potential diagenetic transformation of ferrous sulfates to iron oxides under Martian surface conditions (McLennan et al. 2005; Tosca et al. 2005; 2008a; Sefton-Nash and Catling 2008). Tosca et al. (2008a) experimentally examined ferrous sulfate (melanterite; $FeSO_4{\cdot}7H_2O$) oxidation in high–ionic strength fluids and demonstrated a number of mineralogical pathways involving dissolution/precipitation, oxidation, hydrolysis, and dehydration, leading ultimately to ferric oxide (hematite) formation. Depending on specific conditions, intermediate steps included a diverse array of mixed valence sulfates (e.g., copiapite), ferric sulfates (e.g., ferricopiapite), and complex ochres (e.g., jarosite, schwertmannite, ferrihydrite). A simplified version of the reaction pathways is shown in Figure 5. These observations and experiments thus indicate a fundamental linkage between the Fe, O, and S cycles during surficial processes on Mars.

One example of a possible reaction pathway involving oxidation of aqueous ferrous iron ⇒ ferric iron ⇒ schwertmannite [$FeO(OH)_{0.75}(SO_4)_{0.125}$] ⇒ goethite [FeO(OH)] ⇒ hematite (Fe_2O_3) is illustrated in reactions 1 through 5 below (Tosca et al. 2008a, Hurowitz et al. 2010):

$$0.5H_2O + h\nu \Leftrightarrow 0.25O_2 + \mathbf{H^+} + e^- \quad (1)$$

$$Fe^{2+}_{aq} + 0.25O_2 + \mathbf{H^+} \Leftrightarrow Fe^{3+}_{aq} + 0.5H_2O \quad (2)$$

$$Fe^{3+}_{aq} + \mathbf{0.125SO_4^{2-}} + 1.75H_2O \Leftrightarrow FeO(OH)_{0.75}(SO_4)_{0.125} + \mathbf{2.75H^+} \quad (3)$$

$$FeO(OH)_{0.75}(SO_4)_{0.125} + 0.25H_2O \Leftrightarrow FeO(OH) + \mathbf{0.25H^+} + \mathbf{0.125SO_4^{2-}} \quad (4)$$

$$FeO(OH) \Leftrightarrow 0.5Fe_2O_3 + 0.5H_2O \quad (5)$$

These reactions provide mechanisms both for producing acidity and for recycling sulfur in surficial environments. For example, in the above reactions, for every mole of iron oxidized, converted to schwertmannite, and then oxidized/dehydrated to form hematite, 3.0 moles of H^+ are produced and 0.125 moles of S are recycled through the system. Accordingly, such reaction pathways provide potential sources of acidity both for early groundwater diagenesis, such as that observed at Meridiani Planum (e.g., Hurowitz et al. 2010), and both acidity and sulfur for formation of younger alteration surfaces on exposed rocks (e.g., Hurowitz et al. 2006).

SEDIMENTARY PROCESSES

Constraints on sedimentary geochemical processes that have operated on the surface of Mars over geological time come from four basic approaches. The first approach employs mission data that provide direct evidence for mineralogy and geochemistry in addition to stratigraphic relationships and, for the MER rovers, ~100-μm–resolution microtextural relationships from a microscopic imager. However, such data have several significant drawbacks. The *Spirit* and *Opportunity* rovers provide abundances for major and a few trace elements, quantitative constraints on iron mineralogy from Mössbauer spectroscopy, and additional but less well-constrained mineralogical information from thermal emission and visible through near-infrared spectroscopy. However, these methods "interrogate" rock surfaces on different lateral and depth scales, and, of fundamental significance, none of the methods allows for correlating geochemical or mineralogical data to textural information. Orbital spectroscopy and imaging allow for evaluation of correlations between mineralogy and stratigraphic relationships at ever-increasing resolution, but in many cases, quantitative evaluation of mineralogical abundances is difficult. Geochemical mapping by gamma ray and neutron spectroscopy operates at orders of magnitude lower spatial resolution but interrogates up to several tens of centimeters in depth, whereas infrared spectroscopy can have <10 m of spatial resolution but only interrogates the uppermost few microns of surfaces.

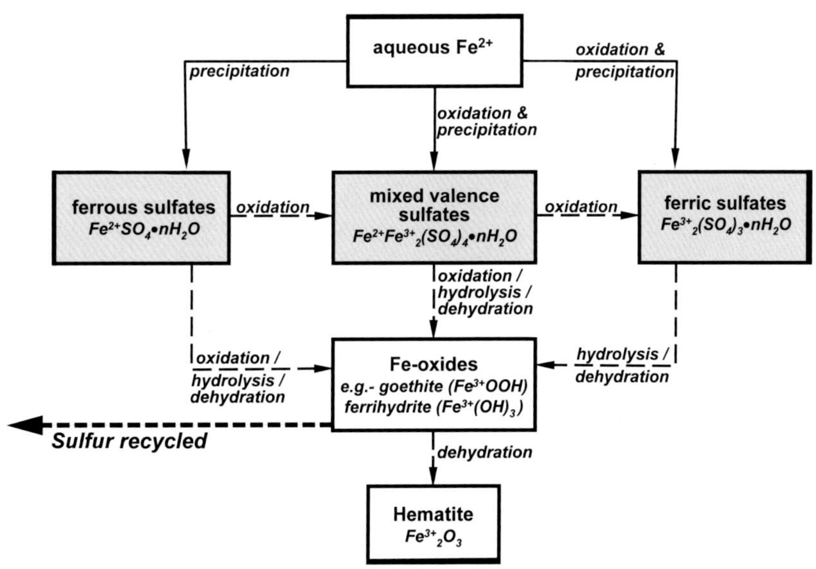

FIG. 5.—Major reaction pathways that may have been operating on Mars involving iron oxidation and precipitation in low-pH and sulfate-rich environments (adapted from Tosca et al. 2008a). Depositional processes are shown with solid arrows, and diagenetic reactions are shown with dashed arrows. Note that the transformation of iron sulfates to oxides involves liberation of sulfur that could be recycled back into the sedimentary system.

The second approach relies on a variety of laboratory experiments that have examined aqueous alteration, evaporation, and diagenetic processes. In these studies, experiments attempt to simulate rock, mineral, and fluid compositions and physico-chemical conditions thought to be relevant to Mars. The experimental results can then be used to test models by comparing results to chemical and mineralogical data obtained from the Martian surface. To date, there have been a number of studies that have explored acid alteration under Mars conditions (e.g., Tosca et al. 2004, Golden et al. 2005, Hurowitz et al. 2005, Tosca and McLennan 2009) but none that have yet explored the more modest conditions implied by recent discovery of clay minerals.

A third approach is to use basic thermodynamic and kinetic aqueous modeling tools to help constrain sedimentary processes under conditions relevant to Mars. Many of the modeling tools that are widely available often require significant adjustments to thermodynamic databases to accommodate differing conditions on Mars. For example, evaporation modeling of terrestrial systems does not typically consider iron because of the low concentrations in most aqueous fluids. However, on Mars, where low pH conditions prevailed, it is important to incorporate thermodynamic data for a variety of potential ferrous and ferric sulfate, carbonate, and chloride minerals (Tosca et al. 2005). Specialized aqueous modeling tools, such as FREZCHEM, have also been developed to deal with very low temperature conditions that prevail on Mars today (e.g., Marion et al. 2010).

A final approach is to examine geological and geochemical relationships in terrestrial settings that are thought to represent good analogs for comparable settings on Mars. The physico-chemical conditions that operate on Mars are likely sufficiently different (e.g., Tosca et al. 2011) that terrestrial settings constitute only partial analogs of some aspects of the chemistry and/or mineralogy being considered. Good examples are acid mine drainage systems and acid lakes that have been used as analogs for low pH conditions on Mars (e.g., Fernándes-Remolar et al. 2005, Baldridge et al. 2009). Important similarities in chemistry and mineralogy are observed, but in these cases the geological settings are not good analogs for what is observed on Mars.

Weathering

The red coloration of the Martian surface has long provided an obvious indication that some form of secondary alteration has influenced the surface of the planet (e.g., Burns 1993). The chemical compositions of soils at the Viking landing sites, including high S and Cl contents, were best modeled as representing mixtures of silicate minerals and sulfate/chloride salts (Clark 1993), further indicating that some form of aqueous alteration had been at work. More recent work (McSween et al. 2010) used a combination of chemical mass balance and spectroscopic observations to model Meridiani and Gusev soils as mixtures of approximately 70 to 85% igneous components (mainly plagioclase, pyroxene, olivine, oxides, and phosphates) and approximately 15 to 30% of an alteration assemblage (variable mixtures of sulfates, silica, clays, secondary oxides, and chlorides).

The exact processes involved in aqueous alteration have long been a matter of contention, with a variety of low-temperature (e.g., weathering, evaporites, iron formation, crater lakes) through intermediate- and high-temperature (e.g., acid fogs, palagonitization, hydrothermal alteration) models being proposed. More recent identification of Mg-, Ca-, and Fe-sulfate–cemented sandstones containing amorphous silica and altered basalt components at Meridiani Planum (Squyres et al. 2004, Clark et al. 2005, McLennan et al. 2005, Glotch et al. 2006), and with a mineralogy similar to that predicted by experimental and theoretical modeling studies of low-pH basalt alteration, coupled with orbital remote sensing observations for a wide variety of secondary minerals of likely sedimentary origin (sulfates, amorphous silica, clays, and possibly chlorides) pointed to a strong influence from a chemically dynamic sedimentary rock cycle (McLennan and Grotzinger 2008).

The chemical composition of Martian sedimentary materials strongly indicates a number of fundamental differences from terrestrial sedimentary environments. The issue of an overwhelmingly basaltic provenance was discussed above. This has profound implications both for the chemical and mineralogical composition of clastic sediments and for the composition of Martian natural waters. Figure 6 shows a ternary diagram plotting mole fractions Al_2O_3–(CaO+Na_2O+K_2O)–(FeO_T+MgO), or A-CNK-FM, which is commonly used to evaluate weathering and mixing processes in basaltic systems (Nesbitt and Wilson 1992). Under the circum-neutral pH conditions typical of terrestrial weathering, aluminum and ferric iron are insoluble, leading to Al- and Fe-depleted natural waters and weathering residues that are concentrated in these elements (Fig. 6a, c). Siliciclastic sediments in turn are also enriched in weathering residues (i.e., clays and iron oxides), with chemical sediments forming from the dissolved constituents (for basalts, dominated by Ca, Mg, and Na). Thus, terrestrial weathering profiles and the siliciclastic sediments derived from them typically have compositions that scatter above the feldspar–(FeO_T+MgO) join on the A-CNK-FM diagram.

In contrast, and with only a very small number of exceptions, Martian sedimentary materials that have been analyzed to date, including soils, altered rock surfaces, and sedimentary rocks, scatter parallel to but below the feldspar–(FeO_T+MgO) join. This trend has been used to suggest that acidic alteration dominated, such that mafic mineral dissolution (e.g., dissolution of olivine, Fe-Ti oxides, pyroxene, glass), rather than mineral alteration to clays, was the major process (Fig. 6d). The lack of evidence for significant Al mobility, for example, as Al-sulfates, that would be expected in low-pH environments led Hurowitz and McLennan (2007) to suggest that low water–rock ratios were involved, such that only the most soluble minerals (i.e., olivine but not plagioclase) were mainly affected. Such an interpretation is also consistent with low-pH aqueous alteration experiments on Martian basalt compositions (e.g., Tosca et al. 2004; Golden et al. 2005; Hurowitz et al. 2005, 2006).

One potential complication to these interpretations is that most of the materials shown in Figure 6b, unlike those shown for the terrestrial examples (Fig. 6a), include both siliciclastic and chemical (mainly sulfate) constituents that are mixed together in various proportions, thus potentially obscuring any terrestrial-like geochemical weathering trends that might exist (McLennan 2010). Mineralogical constraints are simply too uncertain and nonquantitative to unambiguously distinguish small amounts of an aluminous residual component. Another factor, described above, is that more recent evidence indicates that sulfate-rich sedimentary rocks appear to be characteristic of younger Late Noachian–Hesperian terrains, whereas clay-rich exposures dominate in the earlier Noachian. If correct, this might also indicate that weathering regimes have evolved over geological time.

Regardless of exact aqueous conditions, the discovery of extensive exposures of clay-bearing strata in Noachian terrains indicates that weathering processes have been operating in some form essentially throughout recorded Martian geological time. There is considerable diversity in the apparent composition of Noachian clays, but they appear to be dominated by Fe-Mg smectites (e.g., nontronite) with lesser aluminous smectites and highly aluminous kaolinite. The exact origin of the Noachian clay mineral suites is not well constrained and could be due to some form of pedogenic weathering, in situ alteration (e.g., analogous to terrestrial bentonites), diagenesis, and/or hydrothermal alteration associated with volcanic centers and large impact craters. Accordingly, the implications for the nature of Noachian weathering processes are also not yet well constrained. Assuming that these clays are at least in part the result of surficial weathering processes, the predominance of Fe-Mg smectites might indicate

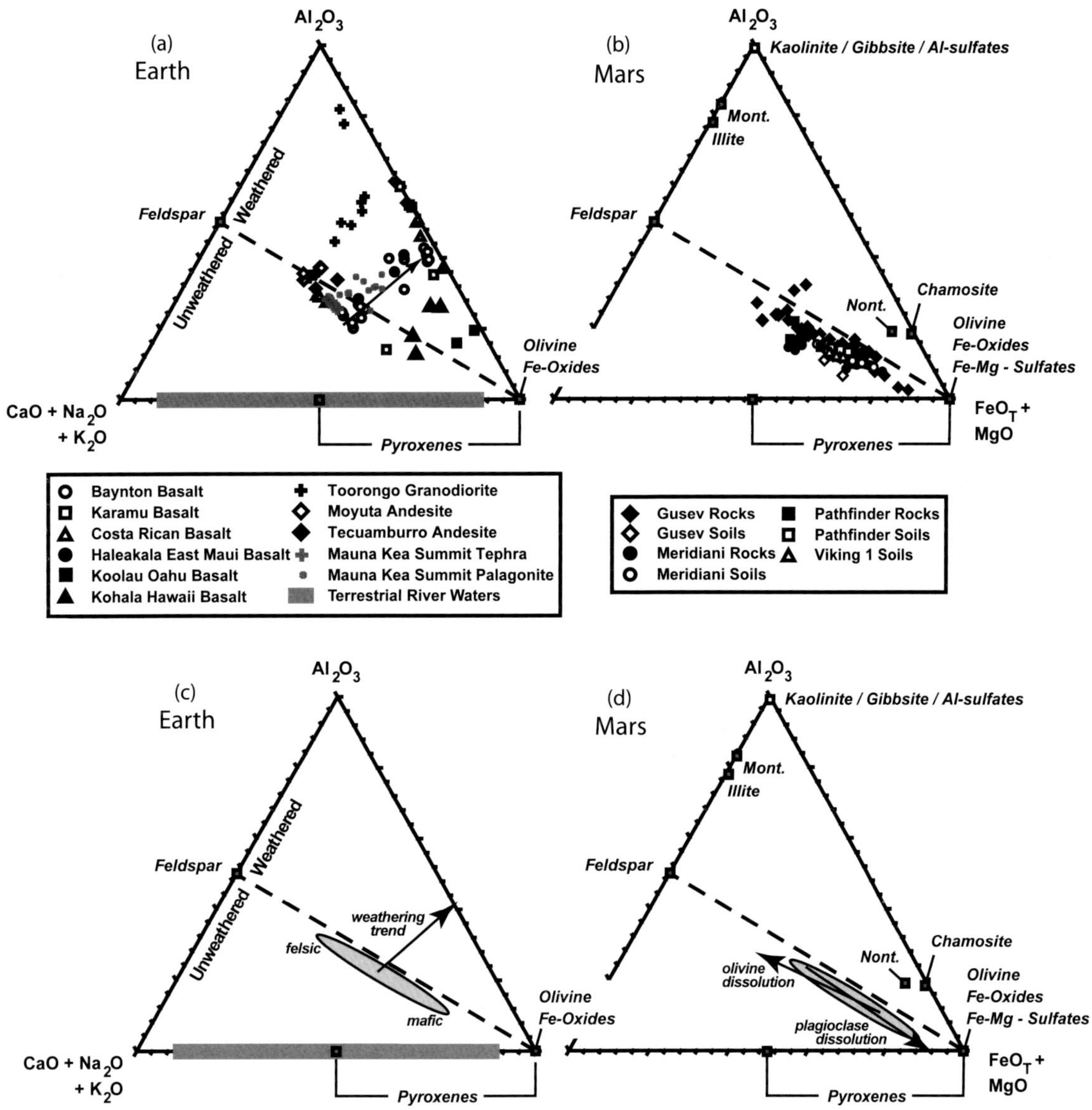

FIG. 6.—A-CNK-FM ternary diagrams plotting molar proportions of Al_2O_3–(CaO+Na_2O+K_2O)–(FeO_T+MgO). Typical compositions for some major igneous and sedimentary minerals are plotted; note that the dashed diagonal line in each diagram separates regions of the diagram dominated by primary igneous silicate minerals (unweathered) from the area dominated by clay minerals (weathered). (a) Selected samples from weathering and alteration profiles in terrestrial settings, emphasizing basaltic–andesitic terrains. Note that in all cases increased alteration results in enrichment in Al and Fe as a result of their insoluble character and concentration in residual clays and oxides. (b) Available data for Martian soils, rocks, altered rock surfaces, and sedimentary rocks (n ~ 200). Note that almost all Martian data fall below and parallel to the diagonal dashed line in spite of evidence for alteration in many of the samples. (c) Field illustrating compositional variation in primary terrestrial igneous rocks, with an arrow showing a typical terrestrial weathering trend. (d) Field illustrating compositional variation in primary Martian mafic to ultramafic igneous rocks. Arrows show alteration trends for olivine dissolution and plagioclase dissolution from typical basalt composition. The differing trends observed for terrestrial and Martian global alteration patterns have been explained by low-pH, low water–rock ratio conditions on Mars, such that mineral dissolution dominates over alteration to clays. Adapted from Hurowitz and McLennan (2007).

relatively low water–rock ratios since in well-drained basaltic weathering environments, aluminous clays, notably kaolinite, would be expected (e.g., White 1995, van der Weijden and Pacheco 2003). Yet another possibility is that such smectites, although ultimately formed in low water–rock environments, could be later transported and deposited in water-dominated fluvial or lacustrine environments.

Depositional Processes

Many contributions, including a number of those to this volume, illustrate the wide diversity of both subaerial and subaqueous depositional environments that have been recognized on Mars, including eolian, fluvial, deltaic, glacial, lacustrine, and possibly fan settings. However, in situ major and trace element data are only available for clastic sedimentary rocks at two landing sites (Gusev and Meridiani), with only Meridiani being well characterized for sedimentology and stratigraphy, and, accordingly, the constraints that geochemistry provide are limited. Some insight regarding the influence of various sedimentary processes (e.g., weathering, sorting) can be obtained from the study of unconsolidated Martian soils (e.g., McLennan 2000, Yen et al. 2005, Karunatillake et al. 2010, McSween et al. 2010), but soils are of broadly uniform composition (although with some influence from local geology), and the limited variation makes it difficult to unravel any competing influences of provenance from processes. It is also worth pointing out that the number of studies of terrestrial systems that have examined source-to-sink geochemistry of sedimentary basins that are dominated by basaltic provenance are remarkably few (e.g., van de Kamp and Leake 1985, 1995; Saha et al. 2010), and from the perspective of understanding Mars, this represents a fruitful area for further research.

One area that has received considerable attention is the nature of evaporitic depositional processes on Mars.[2] This is because fluid evaporation, unlike clastic sedimentation, involves to a much larger degree equilibrium processes and accordingly is amenable to both experimental and thermodynamic modeling approaches. One of the major findings of the past decade is the recognition that evaporative processes have been a major depositional (and diagenetic) process for much of Martian geological time. Ancient sedimentary sequences, studied by both *Opportunity* and *Spirit*, possess abundant sulfate minerals, and from orbit, widespread sulfates have been identified in layered sequences; in addition, even exposures of possible chloride minerals have been observed.

Tosca and colleagues (Tosca et al. 2005, 2007, 2008b; Tosca and McLennan 2006) used Pitzer ion interaction thermodynamic models to constrain evaporation processes on Mars. Although such modeling techniques are well established in terrestrial settings, in order to apply them to Martian conditions it is necessary to incorporate an internally consistent thermodynamic database for a wide variety of Fe^{2+} and Fe^{3+} species. Figure 7 illustrates one example of such modeling and shows the mineralogical pathways (so-called chemical divides; see Hardie and Eugster 1970, Eugster and Jones 1979) for the evaporation of a single basaltic brine having a composition derived from acid weathering of olivine-bearing basalt and with varying initial pH governed by variable HCO_3^-/SO_4^{2-} ratios. Because Martian brines are likely sulfate-rich; enriched in Mg, Ca, and Fe; and depleted in Na and K—unlike terrestrial seawater, which is chloride-rich; elevated in Ca, Mg, Na, and K; and devoid of Fe—the evaporation pathways for typical Martian brines and terrestrial seawater differ greatly (Table 4).

It is notable that the numerous possible evaporation pathways in Figure 7 lead only to a limited number of late-stage brine compositions, which is similar to what is seen in terrestrial systems. Also of note is the fact that the two distinct evaporite mineral associations observed on Mars, within the Burns Formation (Mg-, Ca-, Fe^{3+}-sulfate; possibly halite) and within fractures in Martian nakhlite meteorites (siderite, gypsum, sulfates, chlorides) (Bridges and Grady 1999), can be explained by the same basaltic brine but with differing initial anion chemistry (HCO_3^-/SO_4^{2-}) and thus differing initial pH. Some caution is warranted since neither of these examples represent equilibrium assemblages. On the other hand, they probably do provide some constraints on the nature of the fluids involved. Tosca and McLennan (2006) suggested that the evaporite minerals in Burns Formation were formed by low-pH acid–sulfate surface waters, whereas the assemblages within fractures in Martian basaltic meteorites likely formed from fluids out of contact with the surface environment and thus buffered by basalt–fluid interaction.

McLennan (2003) pointed out that aqueous alteration of basaltic rocks efficiently produces excess silica in solution, and so sedimentary silica should be common on Mars. Simple mass balance of typical mafic mineral weathering reactions indicates that there should be very approximately 1 mole of SiO_2 for each mole of sulfate formed from reaction of sulfate with various cations released by the alteration process (Fig. 8). Although spectral libraries did not at the time adequately represent amorphous silica, quartz, the typical diagenetic product of opaline silica maturation (e.g., Knauth 1994), had not been detected at any significant level. Spectral libraries have greatly improved and secondary amorphous silica has now been identified, using in situ geochemistry and infrared spectroscopy, in the sedimentary rocks of the Burns Formation (Glotch et al. 2006) and as likely hydrothermal deposits in Gusev Crater (Squyres et al. 2008, Ruff et al. 2011). Orbital remote sensing has also identified amorphous silica in a variety of geological settings on the Martian surface (Milliken et al. 2008, Bishop et al. 2009, Murchie et al. 2009).

The scale, nature, and depositional processes of any carbonate sedimentary reservoir remains largely unknown. Fe-Mg-Ca carbonates have been observed in Martian meteorites, notably the 1.3 Gyr nakhlites and the 4.1 Gyr ALH84001, as secondary mineral assemblages in veins and interstitial areas (McSween and Harvey 1998, Bridges and Grady 1999, Bridges et al. 2001). Polar soils at the Phoenix landing site show evidence for about 3 to 5% $CaCO_3$ (Boynton et al. 2009), and a small carbonate component (2–5%) of uncertain mineralogy, but possibly $MgCO_3$, has been identified in Martian dust from orbital spectroscopy (see review in Christensen et al. [2008]). Other identifications of carbonate minerals both from orbit (Ehlmann et al. 2008) and from in situ observations by the Spirit Rover (Morris et al. 2010) indicate the presence of carbonates, but most likely in hydrothermal settings.

Diagenetic Processes and Recycling

It is now reasonably well established that water-mediated diagenesis has taken place on Mars, perhaps over much of its geological history, although at highly variable rates. One of the major findings from the Burns Formation at Meridinai Planum was unambiguous textural, mineralogical, and geochemical evidence for syn- to postdepositional diagenetic overprints involving a fluctuating groundwater table (Squyres et al. 2004, Grotzinger et al. 2005, McLennan et al. 2005). Among the major processes recorded (from oldest to youngest) were syndepositional crystallization, perhaps at the capillary fringe, of a cross-cutting highly soluble evaporite mineral; formation of pore-filling cement; formation of millimeter-scale uniformly distributed hematitic concretions; dissolution of the previously precipitated cross-cutting evaporite mineral and early-formed cements to form secondary crystal mold and sheet-like vug secondary porosity, respectively; formation of late cements overgrowing concretions; and nodular features related to recrystallization. When diagenetic stratigraphy is coupled with sedimentological constraints (Grotzinger et al. 2005, 2006; Metz et al. 2009), at least four distinct recharge events can be recognized during deposition and diagenesis. Tosca et al. (2004; also see Sefton-Nash and Catling [2008]) carried out geochemical modeling that further indicated that the groundwater was of very high ionic strength, low pH, and derived from the weathering of olivine-bearing basalt.

As pointed out above, a number of workers have noted a correlation between the distribution of iron oxides and sulfate minerals on the Martian surface using orbital spectroscopy (e.g., Bibring et al. 2007, Murchie et al. 2009, Roach et al. 2010). This relationship is consistent

[2]In the context of Mars, evaporites could form when natural waters either evaporate or freeze. Exact mineral and geochemical evolution pathways may differ slightly for these processes, but for this work only evaporation is considered, both because early Mars is often thought to have been "warmer and wetter" and because most experimental and theoretical modeling research has been carried out on the evaporation process.

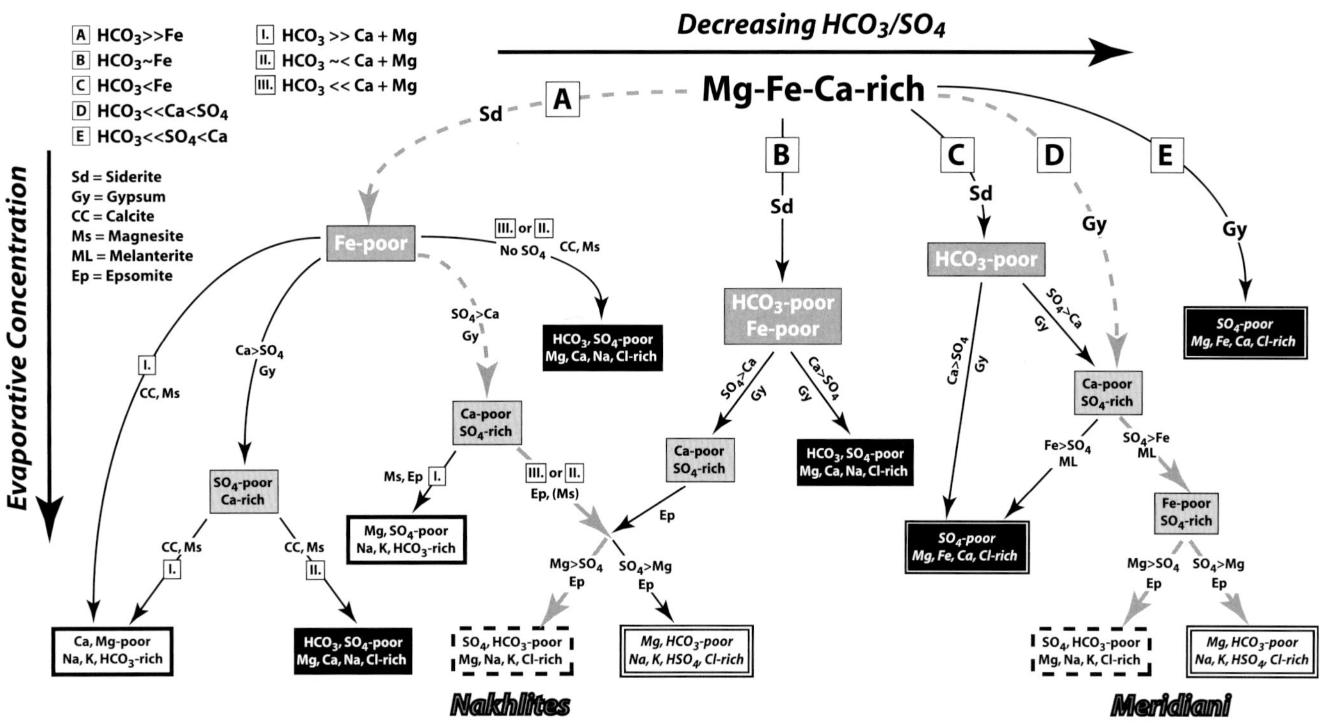

FIG. 7.—Flowchart illustrating chemical and mineralogical evolution (chemical divides) of an evaporating fluid derived by low-pH weathering of an olivine-bearing basalt. A single brine cation composition is modeled, with evaporative concentration increasing downward and the ratio of HCO_3^-/SO_4^{2-} (i.e., initial pH) decreasing to the right. The chemical characteristics of the latest-stage brines are described in the boxes that terminate each evaporation pathway. Although nearly 20 evaporation pathways are possible, only five distinctive end-stage brines result, which is similar to that seen for the evaporation of terrestrial brines. Possible evaporation/mineral precipitation pathways for Meridiani Planum evaporite mineralogy (heavy dashed gray lines on the right-hand side) and evaporite mineral assemblages in the Nakhla meteorites (heavy dashed gray lines on the left-hand side of the diagram) are shown. Figure adapted from Tosca and McLennan (2006).

TABLE 4.—Comparison of the order of precipitation of evaporite mineral sequences for terrestrial seawater and a potential Martian brine.[1]

Seawater			Mars brine	
		First crystallized		
Aragonite	$(CaCO_3)$		[Siderite][2]	$[(FeCO_3)]$
Gypsum	$(CaSO_4{\cdot}2H_2O)$		Gypsum	$(CaSO_4{\cdot}2H_2O)$
Halite	(NaCl)		Melanterite[3]	$(FeSO_4{\cdot}7H_2O)$
Epsomite	$(MgSO_4{\cdot}7H_2O)$		[Magnesite][2]	$[(MgCO_3)]$
Kainite	$(KClMgSO_4{\cdot}3H_2O)$		Epsomite	$(MgSO_4{\cdot}7H_2O)$
Carnallite	$(KMgCl_3{\cdot}3H_2O)$		Halite	(NaCl)
Bischofite	$(MgCl_2{\cdot}6H_2O)$		Other late-stage bittern salts[4]	
Other late-stage bittern salts				
		Last crystallized		

[1] Various dehydration effects (e.g., gypsum ⇒ anhydrite; epsomite ⇒ hexahydrite ⇒ kieserite, etc.) not included.
[2] Siderite and magnesite are predicted to precipitate only for fluids with relatively high HCO_3^-/SO_4^{2-} ratios and thus higher initial pH.
[3] Melanterite forms when brine is nonoxidizing. Depending on how oxidized the brine is, various ferric sulfates (e.g., copiapite, bilinite, jarosite) are also predicted to precipitate both before and/or after epsomite precipitation (Tosca et al. 2008).
[4] Evaporation modeling of Martian brines has not been taken beyond halite precipitation.
References: Warren (2006), Tosca and McLennan (2006), and Tosca et al. (2008b).

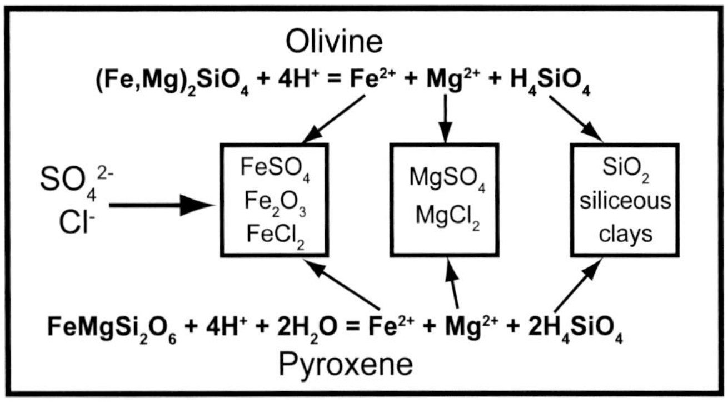

FIG. 8.—Schematic diagram illustrating the mass-balance arguments to indicate that secondary silica should be common on Mars if basalt is altered and sulfate/chloride minerals are precipitated from resulting fluids. Alteration of olivine and pyroxene results in the release of approximately 1 to 2 moles of silica (typically as silicic acid or related species) for each mole of mineral dissolved. Adapted from McLennan (2003).

with experimental and modeling studies that indicate a diagenetic transformation of ferrous sulfates to ferric sulfates and ultimately to iron oxides under Martian surface conditions (McLennan et al. 2005; Tosca et al. 2005, 2008a; Sefton-Nash and Catling 2008; also see Navrotsky et al. [2005], Golden et al. [2008]). Tosca et al. (2008a) showed one likely mineralogical pathway with schwertmannite and jarosite as intermediate steps (Fig. 5). As discussed further below, the observation that ferric sulfates have not more completely transformed to iron oxides appears to indicate a water-limited process. Nevertheless, in addition to providing a recycling mechanism for sulfur (see above), such a process could represent a pathway for very slow, incremental oxidation of the Martian near-surface environment over geological time (i.e., during the Siderikian era of Bibring et al. [2006]).

Although it is clear that aqueous diagenesis has occurred, an equally interesting characteristic of the Martian sedimentary record is the apparent dearth of diagenetic changes under conditions during which such reactions might be expected. Tosca and Knoll (2009) described sedimentary assemblages as being "juvenile and diagenetically under-developed," based on modeling of temperature–time integrals (Siever 1983) for Martian sedimentary basins under a wide range of possible burial conditions. Three observations are especially notable:

1. Persistence of iron sulfates, notably of jarosite within the Late Noachian–Early Hesperian Burns Formation, in spite of it being metastable with respect to goethite and hematite over a broad pH range (Elwood Madden et al. 2004; also see Berger et al. [2009]). Comparisons with laboratory and field data, such as at Rio Tinto, indicate that groundwater could only have been present for $<10^4$ years, and perhaps as little as 10^1 years, after jarosite formation.

2. Predominance of Fe- and Mg-rich smectitic clays (e.g., nontronite, saponite) over more mature aluminous clays (e.g., kaolinite, montmorillonite) in Noachian terranes. In terrestrial settings, smectites diagenetically transform in the presence of water to either illite or chlorite, depending on composition, with timescales depending on thermal conditions of burial, but typically <400 Myr. The persistence of these diagenetically immature clay mineral assemblages for about an order of magnitude more time than this is thus consistent with limited postdepositional interaction with water.

3. Perhaps the most remarkable observation related to sediment diagenesis is the apparent absence of any recrystallization of amorphous silica to form quartz, despite being in places inferred to be of Noachian–Hesperian age (e.g., Glotch et al. 2006, Squyres et al. 2008, Ruff et al. 2011). Quartz has distinctive infrared spectral properties such that it can be identified from remote sensing with confidence at the few percent level (e.g., Bandfield et al. 2004). In terrestrial settings, amorphous silica rarely survives more than a few million years, with only very few occurrences being $>10^8$ years, before undergoing the Opal-A ⇒ Opal-CT ⇒ quartz transition. Modeling temperature–time integrals for basin evolution on Mars indicates that even under the most favorable circumstances, amorphous silica should survive on Mars for no more than about 400 Myr, or about an order of magnitude less than the oldest known occurrences. The most likely mechanism allowing for long-term persistence of amorphous silica is the absence of aqueous conditions (Tosca and Knoll 2009).

An area that has not received much attention is the strong potential for diagenetic recycling of evaporites on Mars. On Earth, evaporite deposits are dominated by gypsum and anhydrite, which are relatively insoluble. Late-stage evaporite minerals, including halite and the various bittern salts, are far less abundant and highly prone to dissolution and recycling back into the aqueous environment. Analogous processes have not been considered for Mars. Figure 9 shows the effects of interacting a typical Martian evaporite assemblage with dilute waters, and, not surprisingly, very small amounts of water–mineral interaction result in complete dissolution of Mg- and Fe-sulfates and halite, with virtually no effect on gypsum.

One implication of this is that the relatively high abundance of highly soluble sulfates and possibly chlorides exposed at the surface of Mars limits the amount of interaction with relatively fresh waters since the time of deposition. Any later fluids that interacted with earlier-

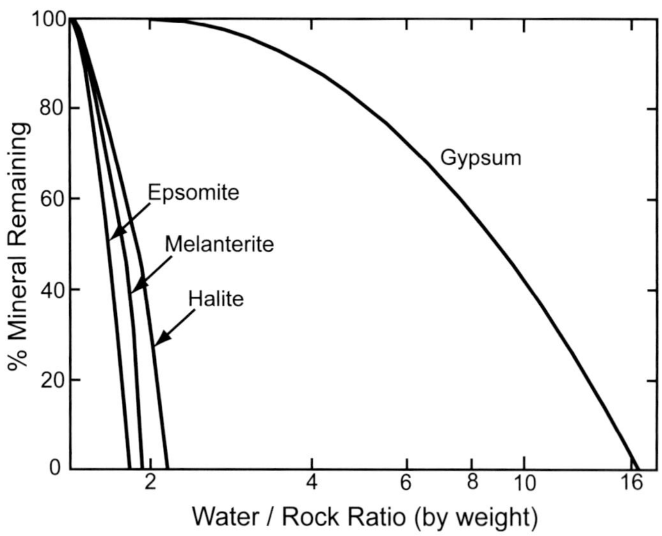

FIG. 9.—Plot of amount of mineral remaining vs. water–rock ratio for the interaction of a potential Martian evaporite mineral assemblage with pure water, using published mineral dissolution rates (see Hurowitz and McLennan [2007] for general approach). Apart from gypsum, Martian evaporite minerals are highly soluble, and at the point of complete dissolution of epsomite, melanterite and halite, <1% of the gypsum has dissolved. Accordingly, diagenetic interaction of Martian evaporites with any relatively fresh water after deposition has the potential to result in strong mineralogical fractionation of gypsum from the other evaporite minerals.

formed Martian evaporite assemblages would also have mostly evolved to late-stage brines. This in turn could also have astrobiological implications. Thus, Tosca et al. (2008b) suggested that the implied low water activity (a_{H2O}) of late-stage Martian brines inhibits habitability.

A second implication is that wherever diagenetic water–mineral interaction did take place, it could provide an efficient mechanism for separating relatively insoluble gypsum from other evaporite minerals. Effectively, gypsum becomes a residual mineral, which then could be preferentially recycled and concentrated by physical processes. One example where this might be relevant is in the enormous gypsum-bearing north polar ergs (Langevin et al. 2005, Fishbaugh et al. 2007, Horgan et al. 2009, Szynkiewicz et al. 2010). These deposits have been mostly interpreted as forming by depositional processes during which gypsum is initially produced in playas or near-surface groundwater and then transported by eolian processes, such as at White Sands, New Mexico (e.g., Szynkiewicz et al. 2010), or gypsum is produced by direct precipitation (related to either alteration or evaporation) from fluids percolating through the dunes themselves (e.g., Fishbaugh et al. 2007). However, another possible scenario could be related to diagenetic processes, whereby gypsum is separated from other evaporitic minerals in the near-surface environment through interaction with very small quantities of water and is then physically recycled by eolian processes.

DISCUSSION

Although remarkable progress has been made in understanding the nature of the Martian sedimentary record, especially over the past decade, there is still much that is unknown and even more that is uncertain. For example, on Earth there are good quantitative constraints on the lithological and chemical composition of the sedimentary rock record through geological time, and this provides crucial evidence for understanding how the terrestrial sedimentary mass forms and evolves (e.g., Garrels and Mackenzie 1971; Veizer and Jansen 1979, 1985; Ronov 1983; Taylor and McLennan 1985, 2009; Veizer and Mackenzie 2003). For Mars, we are just beginning to take an inventory of the sedimentary materials, and anything approaching a quantitative understanding is likely many years away. Nevertheless, considerable insight into conditions of Martian surficial processes is possible. Accordingly, in the following sections, several of the major issues related to the Martian sedimentary rock cycle that can be constrained from the geochemical data are discussed. Each of these issues provides rich opportunities for further research.

Size and Constitution of the Martian Sedimentary Mass

Is it possible to place some geochemical constraints on the overall size of the sedimentary mass and its various lithological reservoirs? As discussed above, the amount of sulfur in the surficial environment is likely on the order of at least 10^{21} to 10^{22} g. For the following discussion, the lower sulfur value of 10^{21} g is adopted to simplify calculations and to provide what is considered a reasonable lower limit. However, the likely range allows for values an order of magnitude or more higher. Assuming that all of this surficial sulfur is tied up in sulfate minerals with an average global hydrated sulfate mineral formula of $Fe_{0.4}Mg_{0.4}Ca_{0.2}SO_4 \cdot 2H_2O$, with cation proportions based on their relative amounts in Martian crust (Taylor and McLennan 2009), this translates into a sulfate mineral reservoir of approximately 5.5×10^{21}g, or about 0.2% of the terrestrial sedimentary mass (for reference, the terrestrial sedimentary mass is $\sim 2.7 \times 10^{24}$ g; Veizer and Mackenzie 2003). Adopting the average Martian soil S/Cl mass ratio of 3.6 (Taylor and McLennan 2009), a chloride reservoir (assuming a 50:50 mix of NaCl and $Mg_{0.67}Ca_{0.33}Cl_2$) is calculated to be 2×10^{20} g. The formation of secondary silica is strongly tied to the breakdown of mafic minerals, such as that shown in Figure 8 (McLennan 2003). Assuming that approximately 1 mole of silica is formed for each mole of sulfate or chloride that is formed, a secondary silica (assuming $SiO_2 \cdot 1.5H_2O$) reservoir of approximately 2×10^{21} g is also estimated. Although it is known that carbonate minerals exist on Mars, their overall abundances thus far appear to be trivial, and, accordingly, a carbonate sedimentary reservoir is assumed to be negligible. Thus, the total amount of sulfate, chloride, and silica is calculated as being between about 8×10^{21} and 8×10^{22} g, and this represents a likely minimum estimate of the chemical constituents in the Martian sedimentary mass.

In terrestrial sedimentary systems, there are several lines of evidence that can be used to constrain the relative proportions of the chemical constituents to the overall size of the sedimentary mass. For example, Millot et al. (2002) evaluated the empirical relationship between chemical (R_{Chem}) and physical (R_{Phys}) weathering rates in a wide variety of geological settings such that

$$R_{Chem} = 0.39R_{Phys}^{0.66} \tag{6}$$

which, if applied to the overall Martian sedimentary record (an assumption, since no such relationship has yet been documented for Mars), indicates a ratio of detrital to chemical constituents of about 3:1. This is comparable to the relative proportions of detrital to chemical constituents in modern terrestrial rivers and in the overall terrestrial sedimentary rock record, for which the ratios are estimated to be about 4:1 and 6.5:1, respectively (Garrels and Mackenzie 1971, Henderson and Henderson 2009). In addition, Stewart (1993) used chemical mass balance calculations to estimate the ratio for 13 rivers draining sedimentary rock terrains and derived a value of approximately 6:1. Impact processes, which would be dominated by particulate material, likely played a much more important role in forming the preserved sediment on Mars compared to Earth, where the >4.0 Gyr sedimentary record is missing. Finally, it is likely that pyroclastic activity, also generating large volumes of particulate debris, has been much more active on Mars than on Earth (Wilson and Head 1994). Accordingly, if we adopt a fairly conservative clastic to chemical ratio of 5:1, this indicates an overall minimum Martian sedimentary rock record of between about 5×10^{22} and 5×10^{23} g, or between about 2 and 20% of the size of the terrestrial sedimentary mass. These calculations, though highly preliminary, indicate that Mars likely has a very substantial sedimentary record, especially given that the surface area of Mars is only about one-third that of Earth and that sediment preservation associated with plate tectonic processes (e.g., Veizer and Jansen 1985) has not occurred.

A Fundamental Paradox: How Much Water?

The surface of Mars provides compelling geomorphological evidence, in the form of channels and subaqueous deltas, for large amounts of liquid water flowing on the surface early in its history. Secondary minerals (e.g., sulfates, clays, silica, chlorides) observed both within sedimentary rocks at the *Spirit* and *Opportunity* landing sites and from orbit also indicate that water-mediated surficial alteration processes were active in the formation and deposition of Martian sedimentary rocks. Thus, Bibring et al. (2006) proposed that the Martian surface was water-rich through to about the Mid-Hesperian. The Burns Formation preserved at Meridiani Planum shows clear evidence for a groundwater influence during both deposition and diagenesis, and a likely hydrothermal system has been recognized in the Inner Basin of the Columbia Hills. Thus, at first glance, early in Mars history it would appear that near-surface liquid water may have been rather abundant.

However, when the chemistry and mineralogy of the Martian sedimentary record are examined more closely, a very different picture emerges, leading to an apparent paradox. Chemical and mineralogical evidence indicates that most sedimentary materials that have been examined to date are best explained by low water–rock ratios. Some of this evidence can be summarized as follows:

1. Although the Burns Formation was influenced by groundwater, including evidence for phreatic (i.e., saturated) conditions (McLennan et al. 2005), other observations point to low water–rock ratios. Inferred mineralogy includes Mg-sulfates and possibly chlorides (e.g., Clark et al. 2005). The evaporation modeling of Tosca et al. (2008b) indicates that these minerals precipitate at very low water activities (i.e., $a_{H2O} < 0.78$ for epsomite; <0.5 for halite) and accordingly derive from concentrated brines. The mineralogy of these sedimentary rocks is thus highly labile, and preservation of submillimeter-scale laminations and textures (Grotzinger et al. 2005) indicates that they have never interacted with waters less dilute than those with a_{H2O} greater than about 0.8 since deposition (also see Fig. 9). The presence of jarosite also indicates that the groundwater system must have been short-lived (Elwood Madden et al. 2004, Berger et al. 2009).
2. The chemical data shown in Figure 6 indicate that, in comparison with Earth, the nature of surficial processes is fundamentally different on Mars (Hurowitz et al. 2006, Hurowitz and McLennan 2007), at least for certain parts of Martian geological time. The widespread apparent mobility of ferric iron implies low pH conditions. However, at these conditions aluminum is far more soluble than ferric iron. The Martian sedimentary record is replete with evidence for secondary iron phases that imply mobility (iron sulfates, secondary iron oxides) but thus far little evidence for secondary aluminum phases (e.g., aluminum sulfates) at the times represented by the rocks and soils plotted in Figure 6. Hurowitz and McLennan (2007) interpreted this to indicate that low-pH aqueous alteration processes operated at very low water–rock ratios, such that only the most soluble phases (e.g., olivine, phosphates, Fe-oxides, but not plagioclase) were dissolved before altering solutions were effectively buffered and became nonreactive.
3. Occurrences of ancient diagenetically immature mineral assemblages (iron sulfates, smectitic clays, amorphous silica) described above all point to very low water–rock ratios since the time of deposition (Tosca and Knoll 2009).

Paradoxes rarely reflect the true operations of nature but instead reflect insufficient understanding of the geological record. What are some possible explanations? One possibility is that liquid water flow on the early Martian surface was intermittent rather than continuous. This might indicate that the record of clay minerals preserved in the Noachian does not reflect a ubiquitous water-rich environment, as generally thought. Another possibility is that the transition from relatively water-rich (i.e., Phyllosian) to water-restricted (i.e., Theiikian or Siderikian) conditions was extremely abrupt, such that aqueous-mediated diagenetic processes were effectively inoperative or at least extremely sluggish once sediment had been deposited. This does not appear to be consistent with the detailed stratigraphic relationships that are observed. Thus, the Burns Formation, which records a dynamic groundwater table, likely was deposited during the Late Noachian to Early Hesperian, after the so-called Phyllosian era. Regardless of the exact explanation, these apparently contradictory observations demonstrate the need for further work both in the laboratory and on the surface of Mars.

Mars: A Planet on Acid?

It is now widely agreed that for much of Martian geological time, surficial processes operated under low-pH, acid-sulfate conditions. Some of the basic evidence (in no particular order) includes highly elevated sulfur concentrations in the near-surface environment; a dearth of carbonate minerals, especially in the ancient sedimentary record; widespread evidence for ferric iron mobility, despite the Martian surface appearing relatively oxidizing; the presence of evaporite mineral associations consistent with evaporation of low-pH, sulfate-rich waters; and igneous rock chemical/mineralogical alteration trends consistent with acid alteration experiments. However, the presence of widespread Noachian clays indicates more modest pH conditions at that time, and this provides the fundamental evidence for the concept of mineralogical eras proposed by Bibring et al. (2006). Trying to understand the global chemical regimes that gave rise to these differing conditions is a daunting task.

If indeed the sulfur cycle dominated the surficial processes on Mars it is perplexing to consider how that process could have been so different in the earlier Noachian. For example, there is no strong evidence for sedimentary carbonates associated with the clay-bearing sedimentary rocks (Milliken et al. 2009). Halevy et al. (2007) proposed that conditions were also more reducing, resulting in sulfite, rather than sulfate, being the dominant aqueous sulfur species. Under reducing conditions the sulfur cycle would give rise to acidic but more modest pH (i.e., sulfurous rather than sulfuric acids), thus balancing on the tightrope by being basic enough for clay formation but acidic enough to inhibit carbonate formation. Although somewhat ad hoc, the model in principle is testable since it predicts the formation of minerals that have not yet been detected (i.e., sulfites), but perhaps with better spectral libraries could be. On the other hand, this family of minerals may not survive through geological time (Halevy et al. 2007).

Another possibility is that the geological context, and thus the implications, of Noachian clays may not be fully understood. For example, it is not obvious that significant amounts of water are necessarily required to form the most abundant Noachian clay mineral suites, dominated by Fe/Mg-smectites. During typical basalt weathering, smectites tend to form during the earliest stages of weathering, with kaolinite dominating more advanced stages (e.g., Nesbitt and Wilson 1992). Accordingly, perhaps the Noachian clay assemblages represent incipient alteration involving small amounts of water. Such conditions might also be more consistent with the apparent absence of the complementary chemical constituents (Milliken et al. 2009).

A long-standing question related to the idea of low-pH aqueous conditions is that chemical interaction of such fluids with basaltic rocks should rapidly lead to a buffered circum-neutral pH, and, consequently, it is not possible to sustain acidic conditions. Tosca et al. (2005; also see Tosca and McLennan [2006]) pointed out that the very low pH expected for acid-sulfate waters requires orders of magnitude more rock interaction to buffer them than do the weakly acidic waters associated with the carbon cycle. In addition, evaporation of such fluids further lowers pH, and the overall evolution of the pH of the system is a competition between acid generation from sulfur addition (e.g., from volcanic gases or recycling) and evaporation and the buffering actions of weathering reactions.

Recent work by Hurowitz et al. (2010) on the Burns Formation indicates that low pH conditions may in fact be more localized than previously thought. Water–rock interaction of basalts, in addition to driving the system toward circum-neutral pH, leads to reducing conditions. Relatively reducing conditions would allow for the transport of ferrous iron in solution, and subsequent oxidation of iron produces significant amounts of acidity in the near-surface environment (also see Hurowitz et al. [2009]).

Perhaps the most important conclusion that can be drawn from the above discussion is that Martian sedimentary processes involve a wide variety of complex geochemistry and mineralogy that have varied considerably over both space and time. Such an emerging picture perhaps should not be surprising, given that the terrestrial sedimentary cycle, although differing greatly in detail, also exhibits considerable variability in both space and time.

ACKNOWLEDGMENTS

I am grateful to students, post-docs, and faculty collaborators at Stony Brook, past and present, who worked with me on the sedimentary geochemistry of Mars: Shannon Arlaukas, Lauren Beavon, Brian Hahn, Joel Hurowitz, Suniti Karunatillake, Don Lindsley, Scott Perl, Martin Schoonen, Nick Tosca, Janette Wilson, and Sara Zhao. I am also grateful to colleagues on both the Mars Exploration Rover (MER) and the Mars Odyssey gamma ray spectrometer science teams. My ongoing collaboration with Ross Taylor, on the composition and evolution of the crusts of planets, also provided many of the insights described in this article. I also thank Dave Fike and Dave Vaniman for very thorough and helpful comments on the manuscript. This work has been supported by grants from the NASA Mars Fundamental Research and Mars Data Analysis programs and NASA/JPL subcontracts for participation in MER and Mars Odyssey science teams.

REFERENCES

Baldridge AM, Hook SJ, Crowley JK, Marion GM, Kargel JS, Michalski JL, Thomson BJ, de Souza CR, Bridges NT, Brown AJ. 2009. Contemporaneous deposition of phyllosilicates and sulfates: Using Australian acidic saline lake deposits to describe geochemical variability on Mars. *Geophysical Research Letters* 36:L19291, doi:10.1029/2009GL040069

Bandfield JL, Hamilton VE, Christensen PR, McSween HY. 2004. Identification of quartzofeldspathic materials on Mars. *Journal of Geophysical Research*, 109:E10009, doi:10.1029/2004JE042290

Barnhart CJ, Nimmo F. 2011. Role of impact excavation in distributing clays over Noachian surfaces. *Journal of Geophysical Research* 116:E01009, doi:10.1029/2010JE003629

Berger G, Toplis MJ, Treguier E, d'Uston C, Pinet P. 2009. Evidence in favor of small amounts of ephemeral and transient water during alteration at Meridiani Planum, Mars. *American Mineralogist* 94:1279–1282.

Bibring J-P, Arvidson RE, Gendrin A, Gondet B, Langevin Y, Le Mouelic S, Mangold N, Morris RV, Mustard JF, Poulet F, Quantin C, Sotin C. 2007. Coupled ferric oxides and sulfates on the Martian surface. *Science* 317:1206–1210.

Bibring J-P, Langevin Y, Mustard JF, Poulet F, Arvidson R, Gendrin A, Gondet B, Mangond N, Pinet P, Forget F, and the OMEGA Team. 2006. Global mineralogical and aqueous Mars history derived from OMEGA/Mars Express data. *Science* 312:400–404.

Bishop JL, Parente M, Weitz CM, Dobrea EZN, Roach LH, Murchie SL, McGuire PC, McKeown NK, Rossi CM, Brown AJ, Calvin WM, Milliken R, Mustard JF. 2009. Mineralogy of Juventae Chasma: Sulfates in the light-toned mounds, mafic minerals in the bedrock, and hydrated silica and hydroxylated ferric sulfate on the plateau. *Journal of Geophysical Research* 114:E00D09, doi:10.1029/2009JE003352

Boynton WV, Ming DW, Kounaves SP, Young SMM, Arvidson RE, Hecht MH, Hoffman J, Niles PB, Hamara DK, Quinn RC, Smith PH, Sutter B, Catling DC, Morris RV. 2009. Evidence for calcium carbonate at the Mars Phoenix landing site. *Science* 325:61–64.

Boynton WV, Taylor GJ, Evans LG, Reedy RC, Starr R, Janes DM, Kerry KE, Drake DM, Kim KJ, Williams RMS, Crombie MK, Dohm JM, Baker V, Metzger AE, Karunatillake S, Keller JM, Newsom HE, Arnold JR, Brückner J, Englert PAJ, Gasnault O, Sprague AL, Mitrofanov I, Squyres SW, Trombka JI, d'Uston L, Wänke H, Hamara DK. 2007. Concentration of H, Si, Cl, K, Fe, and Th in low- and mid-latitude regions of Mars. *Journal of Geophysical Research* 112:E12S99, doi:10.1029/2007JE002887

Bridges JC, Catling DC, Saxton JM, Swindle TD, Lyon IC, Grady MM. 2001. Alteration assemblages in Martian meteorites: Implications for near-surface processes. *Space Science Reviews* 96:365–392.

Bridges JC, Grady M. 1999. A halite-siderite-anhydrite-chlorapatite assemblage in Nakhla: Mineralogical evidence for evaporates on Mars. *Meteoritics and Planetary Science* 34:407–415.

Brückner J, Dreibus G, Gellert R, Squyres SW, Wänke H, Yen A, Zipfel J. 2008. Mars Exploration Rovers: Chemical compositions by the APXS. *In* Bell JF III (Editor). *The Martian Surface: Composition, Mineralogy, and Physical Properties*: Cambridge University Press, Cambridge, UK. p. 58–101.

Burns RG. 1993. Rates and mechanisms of chemical weathering of ferromagnesian silicate minerals on Mars. *Geochimica et Cosmochimica Acta* 47:4555–4574.

Canfield DE. 2004. The evolution of the Earth surface sulfur reservoir. *American Journal of Science* 304:839–861.

Carr MH. 2006. *The Surface of Mars*. Cambridge University Press, Cambridge, UK. 307 p.

Carr MH, Head JW. 2010. Geologic history of Mars. *Earth and Planetary Science Letters* 294:185–203.

Christensen PR, Bandfield JL, Rogers AD, Glotch TD, Hamilton VE, Ruff SW, Wyatt MB. 2008. Global mineralogy mapped from the Mars Global Surveyor Thermal Emission Spectrometer. *In* Bell JF III (Editor). *The Martian Surface: Composition, Mineralogy, and Physical Properties*: Cambridge University Press, Cambridge, UK. p. 195–220.

Clark BC. 1993. Geochemical components in Martian soil. *Geochimica et Cosmochimica Acta* 57:4575–4581.

Clark BC, Baird AK. 1979. Is the Martian lithosphere sulfur rich? *Journal of Geophysical Research* 84:8395–8403.

Clark BC, Baird AK, Rose HJ, Toulmin P, Keil K, Castro AJ, Kelliher WC, Rowe CD, Evans PH. 1976. Inorganic analyses of Martian surface samples at Viking landing sites. *Science* 194:1283–1288.

Clark BC, Morris RV, McLennan SM, Gellert R, Jolliff B, Knoll AH, Squyres SW, Lowenstein TK, Ming DW, Tosca NJ, Yen A, Christensen PR, Gorevan S, Brückner J, Calvin W, Dreibus G, Farrand W, Klingelhoefer G, Waenke H, Zipfel J, Bell JF III, Grotzinger J, McSween HY, Rieder R. 2005. Chemistry and mineralogy of outcrops at Meridiani Planum. *Earth and Planetary Science Letters* 240:73–94.

Di Achille G, Hynek BM. 2010. Ancient ocean on Mars supported by global distribution of deltas and valleys. *Nature Geoscience* 3:459–463.

Ehlmann BL, Mustard JF, Murchie SL, Poulet F, Bishop JL, Brown AJ, Calvin WM, Clark RN, Des Marais DJ, Milliken RE, Roach LH, Roush TL, Swayze GA, Wray JJ. 2008. Orbital identification of carbonate-bearing rocks on Mars *Science* 322:1828–1832.

Elkins-Tanton LT, Hess PC, Parmentier EM. 2005. Possible formation of ancient crust on Mars through magma ocean processes. *Journal of Geophysical Research* 110:E12S01, doi:10.1029/2005JE002480

Elwood Madden ME, Bodnar RJ, Rimstidt JD. 2004. Jarosite as an indicator of water-limited chemical weathering on Mars. *Nature* 431:821–823.

Eugster HP, Jones BF. 1979. Behavior of major solutes during closed-basin brine evolution. *American Journal of Science* 279:609–631.

Farién AG, Fernández-Remolar D, Dohm JM, Baker VR, Amils R. 2004. Inhibition of carbonate synthesis in acidic oceans on early Mars. *Nature* 431:423–426.

Fassett CI, Head JW. 2011. Sequence and timing of conditions on early Mars. *Icarus* 211:1204–1214.

Fernándes-Remolar DC, Morris RV, Gruener JE, Amils R, Knoll AH. 2005. The Rio Tinto basin, Spain: Mineralogy, sedimentary geobiology, and implications for interpretation of outcrop rocks at Meridiani Planum, Mars. *Earth and Planetary Science Letters* 240:149–167.

Fishbaugh KE, Poulet F, Chevrier V, Langevin Y, Bibring J-P. 2007. On the origin of gypsum in the Mars polar region. *Journal of Geophysical Research* 112:E07002, doi:10.1029/2006JE002862

Frey HV. 2006. Impact constraints on, and a chronology for, major events in early Mars history. *Journal of Geophysical Research* 111:E08ZS91, doi:10.1029/2005JE002449

Gaillard F, Scaillet B. 2009. The sulfur content of volcanic gases on Mars. *Earth and Planetary Science Letters* 279:34–43.

Garrels RM, Mackenzie FT. 1971. *Evolution of Sedimentary Rocks*: Norton, New York. 397 p.

Gellert R, Rieder R, Brückner J, Clark B, Dreibus G, Klingelhöfer G, Lugmair G, Ming D, Wänke H, Yen A, Zipfel J, Squyres S. 2006. The Alpha Particle X-Ray Spectrometer (APXS): Results from Gusev crater and calibration report. *Journal of Geophysical Research* 111:E02S05, doi:10.1029/2005JE002555

Glotch TD, Bandfield JL, Christensen PR, Calvin WM, McLennan SM, Clark BC, Rogers AD, Squyres SW. 2006. Mineralogy of the light-toned outcrop rock at Meridiani Planum as seen by the Miniature Thermal Emission Spectrometer and implications for its formation. *Journal of Geophysical Research* 111:E12S03, doi:10.1029/2005JE002762

Golden DC, Ming DW, Morris RV, Graff TG. 2008. Hydrothermal synthesis of hematite spherules and jarosite: Implications for diagenesis and hematite spherule formation in sulfate outcrops at Meridiani Planum, Mars. *American Mineralogist* 93:1201–1214.

Golden DC, Ming DW, Morris RV, Mertzmann SM. 2005. Laboratory-simulated acid-sulfate weathering of basaltic materials: Implications for formation of sulfates at Meridiani Planum and Gusev crater, Mars. *Journal of Geophysical Research* 110:E12S07, doi:10.1029/2005JE002451

Greeley R, Schneid BD. 1991. Magma generation on Mars: Amounts, rates, and comparisons with Earth, Moon, and Venus. *Science* 254:996–998.

Grotzinger J, Beaty D, Dromart G, Gupta S, Harris M, Hurowitz J, Kocurek G, McLennan S, Milliken R, Ori GG, Sumner D. 2011. Mars sedimentary geology: Key concepts and outstanding questions. *Astrobiology*, v. 11:77–87.

Grotzinger J, Bell J III, Herkenhoff K, Johnson J, Knoll A, McCartney E, McLennan S, Metz J, Moore J, Squyres S, Sullivan R, Ahronson O, Arvidson R, Jolliff B, Golombek M, Lewis K, Parker T, Soderblom J. 2006. Sedimentary textures formed by aqueous processes, Erebus crater, Meridiani Planum, Mars. *Geology* 34:1085–1088.

Grotzinger JP, Arvidson RE, Bell JF III, Calvin W, Clark BC, Fike DA, Golombek M, Greeley R, Haldemann A, Herkenhoff KE, Jolliff BL, Knoll AH, Malin M, McLennan SM, Parker T, Soderblom L, Sohl-Dickstein JN, Squyres SW, Tosca NJ, Watters WA. 2005. Stratigraphy and sedimentology of a dry to wet eolian depositional system, Burns Formation, Meridiani Planum, Mars. *Earth and Planetary Science Letters* 240:11–72.

Halevy I, Zuber MT, Schrag DP. 2007. A sulfur dioxide climate feedback on early Mars. *Science* 318:1903–1907.

Halliday AN, Wänke H, Birck J-L, Clayton RN. 2001. The accretion, composition and early differentiation of Mars. *Space Science Reviews* 96:197–230.

Hardie LA, Eugster HP. 1970. The evolution of closed-basin brines. *Mineralogical Society of America Special Publication* 3:273–290.

Hartmann WK. 2005. Martian cratering 8: Isochron refinement and the chronology of Mars. *Icarus* 174:294–320.

Hartmann WK, Barlow NG. 2006. Nature of the Martian uplands: Effect on Martian meteorite age distribution and secondary cratering. *Meteoritics and Planetary Science* 41:1453–1467.

Hazen RM, Papineau D, Bleeker W, Downs RT, Ferry JM, McCoy TJ, Sverjensky DA, Yang H. 2008. Mineral evolution. *American Mineralogist* 93:1693–1720.

Head JW, Greely R, Golombek MP, Hartmann WK, Hauber E, Jaumann R, Masson P, Neukum G, Nyquist LE, Carr MH. 2001. Geological processes and evolution. *Space Sciences Reviews* 96:263–292.

Henderson P, Henderson GM. 2009. *The Cambridge Handbook of Earth Science Data*: Cambridge University Press, Cambridge, UK. 277 p.

Horgan BH, Bell JF, Noe Bobrea EZ, Cloutis EA, Bailey DT, Craig MA, Roach LH, Mustard JF. 2009. Distribution of hydrated minerals in the north polar region of Mars. *Journal of Geophysical Research* 114:E01005, doi:10.1029/2008JE003187

Hurowitz JA, Fischer WW, Tosca NJ, Milliken RE. 2010. Origin of acidic surface waters and the evolution of atmospheric chemistry on early Mars. *Nature Geoscience* 3:323–326.

Hurowitz JA, McLennan SM. 2007. A ~3.5 Ga record of water-limited, acidic conditions on Mars. *Earth and Planetary Science Letters* 260:432–443.

Hurowitz JA, McLennan SM, Lindsley DH, Schoonen MAA. 2005. Experimental epithermal alteration of synthetic Los Angeles meteorite: Implications for the origin of Martian soils and the identification of hydrothermal sites on Mars. *Journal of Geophysical Research* 110:E07002, doi:10.1029/2004JE002391

Hurowitz JA, McLennan SM, Tosca NJ, Arvidson RE, Michalski JR, Ming DW, Schöder C, Squyres SW. 2006. In-situ and experimental evidence for acidic weathering on Mars. *Journal of Geophysical Research* 111:E02S19, doi:10.1029/2005JE002515

Hurowitz JA, Tosca NJ, Dyar MD. 2009. Acid production by $FeSO_4•nH_2O$ dissolution and implications for terrestrial and Martian aquatic systems. *American Mineralogist* 94:409–414.

Johnson SS, Mischna MA, Grove TL, Zuber MT. 2008. Sulfur-induced greenhouse warming on early Mars. *Journal of Geophysical Research* 113:E08005, doi:10.1029/2007JE002962

Karunatillake S, McLennan SM, Herkenhoff K. 2010. Grain mixing influences on the geochemistry of soil at Gusev Crater, Mars. *Journal of Geophysical Research* 115:E00F04, doi:10.1029/2010JE003637

King PL, Lescinsky DT, Nesbitt HW. 2004. The composition and evolution of primordial solutions on Mars, with application to other planetary bodies. *Geochimica et Cosmochimica Acta* 68:4993–5008.

King PL, McLennan SM. 2010. Sulfur on Mars. *Elements* 6:107–112.

Knauth LP. 1994 Petrogenesis of chert. *Reviews in Mineralogy* 29:233–258.

Langevin Y, Poulet F, Bibring J-P, Bondet B. 2005. Sulfates in the north polar region of Mars detected by MOEGA/Mars Express. *Science* 307:1584–1586.

Lapen TJ, Righter M, Brandon AD, Debaille V, Beard BL, Shafer JT, Peslier AH. 2010. A younger age for ALH84001 and its geochemical link to shergottite sources in Mars. *Science* 328:347–351.

Lewis KW, Aharonson O, Grotzinger JP, Squyres SW, Bell JF, Crumpler LS, Schmidt ME. 2008. Structure and stratigraphy of Home Plate from the Spirit Mars Exploration Rover. *Journal of Geophysical Research* 113:E12S36, doi:10.1029/2007JE003025

Malin MC, Edgett KS. 2000. Sedimentary rocks of early Mars. *Science* 290:1927–1937.

Malin MC, Edgett KS. 2006. Present-day impact cratering rate and contemporary gully activity on Mars. *Science* 314:1573–1577.

Marion GM, Mironenko MV, Roberts MW. 2010. FREZCHEM: A geochemical model for cold aqueous solutions. *Computers and Geosciences* 36:10–15.

McEwen AS, Malin MC, Carr MH, Hartmann WK. 1999. Voluminous volcanism on early Mars revealed in Valles Marineris. *Nature* 397:584–586.

McLennan SM. 1988. Recycling of the continental crust. *Pure and Applied Geophysics* 128:683–724.

McLennan SM. 2000. Chemical composition of Martian soil and rocks: Evidence of complex mixing and sedimentary transport. *Geophysical Research Letters* 27:1335–1338.

McLennan SM. 2003. Sedimentary silica on Mars. *Geology* 31:315–318.

McLennan SM. 2010. Sedimentary provenance studies on Mars with an example from the Burns Formation, Meridiani Planum. *In First International Conference on Mars Sedimentology and Stratigraphy,* LPI Contribution No. 1547, The Lunar & Planetary Institute, Houston. p. 42.

McLennan SM, Bell JF III, Calvin W, Christensen PR, Clark BC, de Souza PA, Farmer J, Farrand WH, Fike DA, Gellert R, Ghosh A, Glotch TD, Grotzinger JP, Hahn B, Herkenhoff KE, Hurowitz JA, Johnson JR, Johnson SS, Jolliff B, Klingelhöfer G, Knoll AH, Learner Z, Malin MC, McSween HY Jr, Pocock J, Ruff SW, Soderblom LA, Squyres SW, Tosca NJ, Watters WA, Wyatt MB, Yen A. 2005. Provenance and diagenesis of the evaporite-bearing Burns Formation, Meridiani Planum, Mars. *Earth and Planetary Science Letters* 240:95–121.

McLennan SM, Boynton WV, Karunatillake S, Hahn BC, Taylor GJ, the Mars Odyssey GRS Team. 2010. Distribution of sulfur on the surface of Mars determined by the 2001 Mars Odyssey Gamma-ray spectrometer. *Lunar and Planetary Science* XLI, Abstract 2174, The Lunar and Planetary Institute, Houston (CD-ROM).

McLennan SM, Grotzinger JP. 2008. The sedimentary rock cycle of Mars. *In* Bell JF III (Editor). *The Martian Surface: Composition, Mineralogy, and Physical Properties*. Cambridge University Press, Cambridge, UK. p. 541–577.

McLennan SM, Grotzinger JP, Hurowitz JA, Tosca NJ. 2006. Sulfate geochemistry and the sedimentary rock record of Mars. *In Workshop on Martian Sulfates as Records of Atmospheric-fluid-Rock Interactions*, LPI Contribution No. 1331, The Lunar & Planetary Institute, Houston. p. 54.

McSween HY, Harvey RP. 1998. An evaporation model for formation of carbonates in ALH84001 Martian meteorite. *International Geological Review* 40:774–783.

McSween HY, McGlynn IO, Rogers AD. 2010. Determining the modal mineralogy of Martian soils. *Journal of Geophysical Research* 115:E00F121, doi:10.1029/2010JE003582

McSween HY, McLennan SM. 2012. Mars. *In* Holland HD, Turekian KK (Editors). Treatise on Geochemistry, Volume 1, Davis AM ed. *Meteorites, Comets and Planets*. Elsevier, Amsterdam. **In press.**

McSween HY, Taylor GJ, Wyatt MB. 2009. Elemental composition of the Martian crust. *Science* 324:736–739.

Metz JM, Grotzinger JP, Rubin DM, Lewis KW, Squyres SW, Bell JF. 2009. Sulfate-rich eolian and wet interdune deposits, Erebus crater, Meridiani Planum, Mars. *Journal of Sedimentary Research* 79:247–264.

Milliken RE, Fischer WW, Hurowitz JA. 2009. Missing salts on early Mars. *Geophysical Research Letters* 36:L11202, doi:10.1029/2009GL038558

Milliken RE, Swayze GA, Arvidson RE, Bishop JL, Clark RN, Ehlmann BL, Green RO, Grotzinger JP, Morris RV, Murchie SL, Mustard JF, Weitz C. 2008. Opaline silica in young deposits on Mars. *Geology* 36:847–850.

Millot R, Gaillardet J, Dupré B, Allègre JC. 2002. The global control of silicate weathering rates and the coupling with physical erosion: New insights from rivers on the Canadian Shield. *Earth and Planetary Science Letters* 196:83–98.

Morris RV, Ruff SW, Gellert R, Ming DW, Arvidson RE, Clark BC, Golden DC, Siebach K, Klingelhöfer G, Schröder C, Fleischer I, Yen AS, Squyres SW. 2010. Identification of carbonate-rich outcrops on Mars by the Spirit Rover. *Science* 329:421–424.

Murchie SL, Mustard JF, Ehlmann BL, Milliken RE, Bishop JL, McKeown NK, Noe Dobrea EZ, Seelos FP, Bucxkowski DL, Wiseman SM, Arvidson RE, Wray JJ, Swayze G, Clark RN, Des Marais DJ, McEwen AS, Bibring J-P. 2009. A synthesis of Martian aqueous mineralogy after 1 Mars year of observations from the Mars Reconnaissance Orbiter. *Journal of Geophysical Research* 114:E00D06, doi:10.1029/2009JE003342

Navrotsky A, Forray FL, Drouet C. 2005. Jarosite stability on Mars. *Icarus* 176:250–253.

Nesbitt HW, Wilson HW. 1992. Recent chemical weathering of basalts. *American Journal of Science* 292:740–777.

Nesbitt HW, Young GM. 1984. Prediction of some weathering trends of plutonic and volcanic rocks based on thermodynamic and kinetic considerations. *Geochimica et Cosmochimica Acta* 48:1523–1534.

Nimmo F, Tanaka KL. 2005. Early crustal evolution of Mars. *Annual Reviews of Earth and Planetary Science* 33:133–161.

Phillips RJ, Davis BJ, Tanaka KL, Byrne S, Mellon MT, Putzig NE, Haberle RM, Kahre MA, Campbell BA, Carter LM, Smith IB, Holt JW, Smrekar SE, Nunes DC, Plaut JJ, Egan AF, Titus TN, Seu R. 2011. Massive CO_2 ice deposits sequestered in the south polar layered deposits of Mars. *Science* 332:838–841, doi:10.1126/science.1203091

Pierazzo E, Kring DA, Melosh HJ. 1998. Hydrocode simulation of the Chicxulub impact event and the production of climatically active gases. *Journal of Geophysical Research* 103:28607–28625.

Pratt LM, Allen C, Allwood A, Anbar A, Atreya S, Carr M, Des Marais D, Grant J, Glavin D, Hamilton V, Herkenhoff K, Hipkin V, Lollar BS, McCollom T, McEwen A, McLennan S, Milliken R, Ming D, Ori GG, Parnell J, Poulet F, Westall F. 2010. The Mars Astrobiology Explorer-Cacher (MAX-C): A potential rover mission for 2018. *Astrobiology* 10:127–163.

Righter K, Pando K, Danielson LR. 2009. Experimental evidence for sulfur-rich Martian magmas: Implications for volcanism and surficial sulfur sources. *Earth and Planetary Science Letters* 288:235–243.

Roach LH, Mustard JF, Lane MD, Bishop JL, Murchie SL. 2010. Diagenetic hematite and sulfate assemblages in Valles Marineris. *Icarus* 207:659–674.

Ronov AB. 1983. *The Earth's Sedimentary Shell: Quantitative Patterns in its Structure, Compositions and Evolution*. AGI Reprint Series V, American Geological Institute, Falls Church, Virginia. 80 p.

Ruff SW, Farmer JD, Calvin WM, Herkenhoff KE, Johnson JR, Morris RV, Rice MS, Arvidson RE, Bell JF III, Christensen PR, Squyres SW. 2011. Characteristics, distribution, origin, and significance of opaline silica observed by the Spirit Rover in Gusev Crater, Mars. *Journal of Geophysical Research* 116:E00F23, doi:10.1029/2010JE003767

Saha S, Banerjee S, Burley SD, Ghosh A, Saraswati PK. 2010. The influence of flood basalt source terrains on the efficiency of tectonic setting discrimination diagrams: An example from the Gulf of Khambhat, western India. *Sedimentary Geology* 228:1–13.

Sefton-Nash E, Catling DC. 2008. Hematitic concretions at Meridiani Planum, Mars: Their growth timescale and possible relationship with iron sulfates. *Earth and Planetary Science Letters* 269:365–375.

Settle M. 1979. Formation and deposition of volcanic sulfate aerosols on Mars. *Journal of Geophysical Research* 84:8343–8354.

Siever R. 1983. Burial history and diagenetic reaction kinetics. *American Association of Petroleum Geologists Bulletin* 67:684–691.

Soderblom LA, Bell JF. 2008. Exploration of the Martian surface: 1992–2007. *In* Bell JF III (Editor). *The Martian Surface: Composition, Mineralogy, and Physical Properties*. Cambridge University Press, Cambridge, UK. p. 3–19.

Squyres SW, Aharonson O, Clark BC, Cohen BA, Crumpler LA, de Souza PA, Farrand WH, Gellert R, Grant J, Grotzinger JP, Haldemann AFC, Johnson JR, Klingelhöfer G, Lewis KW, Li R, McCoy T, McEwen AS, McSween HY, Ming DW, Moore JM, Morris RV, Parker TJ, Rice JW, Ruff S, Schmidt M, Schröder C, Soderblom LA, Yen A. 2007. Pyroclastic activity at Home Plate in Gusev Crater, Mars. *Science* 316:738–742.

Squyres SW, Arvidson RE, Blaney DL, Clark BC, Crumpler L, Farrand WH, Gorevan S, Herkenhoff KE, Hurowitz J, Kusack A, McSween HY, Ming DW, Morris RV, Ruff S, Wang A, Yen A. 2006. Rocks of the Columbia Hills, *Journal of Geophysical Research* 111:E02S11, doi:10.1029/2005JE002562

Squyres SW, Arvidson RE, Ruff S, Gellert R, Morris RV, Ming DW, Crumpler L, Farmer J, Des Marais DJ, Yen A, McLennan SM, Calvin W, Bell JF III, Clark BC, Wang A, McCoy TJ, Schmidt ME, de Souza PA Jr. 2008. Detection of silica-rich deposits on Mars. *Science* 320:1063–1067.

Squyres SW, Grotzinger JP, Arvidson RE, Bell JF III, Calvin W, Christensen PR, Clark BC, Crisp JA, Farrand WH, Herkenhoff KE, Johnson JR, Klingelhöfer G, Knoll AH, McLennan SM, McSween HY Jr, Morris RV, Rice JW Jr, Rieder R, Soderblom LA. 2004. In-situ evidence for an ancient aqueous environment on Mars. *Science* 306:1709–1714.

Stewart AD. 1993. The ratio of mechanical to chemical denudation in alluvial systems, derived from chemical mass balance. *Transactions of the Royal Society of Edinburgh: Earth Sciences* 84:73–78.

Szynkiewicz A, Moore CH, Glamoclija M, Bustos D, Pratt LM. 2010. Origin of terrestrial gypsum dunes: Implications for Martian gypsum-rich dunes of Olympia Undae. *Geomorphology* 121:69–83.

Tanaka KL, Isbell NK, Scott DH, Greeley R, Guest JE. 1988. The resurfacing history of Mars: A synthesis of digitized, Viking-based geology. *In* Proceedings of the 18th Lunar and Planetary Science Conference; Cambridge University Press, New York; Lunar and Planetary Institute, Houston. p. 665–678.

Taylor GJ, McLennan SM, McSween HY, Wyatt MB, Lentz RCF. 2008. Implications of observed primary lithologies. *In* Bell JF III (Editor). *The Martian Surface: Composition, Mineralogy, and Physical Properties*. Cambridge University Press, Cambridge, UK. p. 501–518.

Taylor SR. 1989. Growth of planetary crusts. *Tectonophysics* 161:147–156.

Taylor SR, McLennan SM. 1985. *The Continental Crust: Its Composition and Evolution*: Blackwell Scientific, Cambridge, UK. 312 p.

Taylor SR, McLennan SM. 2009. *Planetary Crusts: Their Composition, Origin, and Evolution*: Cambridge University Press, Cambridge, UK. 378 p.

Tosca NJ, Knoll AH. 2009. Juvenile chemical sediments and the long term persistence of water at the surface of Mars. *Earth and Planetary Science Letters* 286:379–386.

Tosca NJ, Knoll AH, McLennan SM. 2008b. Water activity and the challenge for life on early Mars. *Science* 320:1204–1207.

Tosca NJ, McLennan SM. 2006. Chemical divides and evaporite assemblages on Mars. *Earth and Planetary Science Letters* 241:21–31.

Tosca NJ, McLennan SM. 2009. Experimental constraints on the evaporation of partially oxidized acid-sulfate waters at the Martian surface. *Geochimica et Cosmochimica Acta* 73:1205–1222.

Tosca NJ, McLennan SM, Clark BC, Grotzinger JP, Hurowitz JA, Knoll AH, Schröder C, Squyres SW. 2005. Geochemical modeling of evaporation processes on Mars: Insight from the sedimentary record at Meridiani Planum. *Earth and Planetary Science Letter* 240:122–148.

Tosca NJ, McLennan SM, Dyar MD, Sklute EC, Michel FM. 2008a. Fe oxidation processes at Meridiani Planum and implications for secondary Fe mineralogy on Mars. *Journal of Geophysical Research* 113:E05005, doi:10.1029/2007JE003019

Tosca NJ, McLennan SM, Lamb MP, Grotzinger JP. 2011. Physico-chemical properties of concentrated Martian surface waters. *Journal of Geophysical Research* 116:E05004, doi:10.1029/2010JE003700

Tosca NJ, McLennan SM, Lindsley DH, Schoonen MAA. 2004. Acid-sulfate weathering of synthetic Martian basalt: The acid fog model revisited. *Journal of Geophysical Research* 109:E05003, doi:10.1029/2003JE002218

Tosca NJ, Smirnov A, McLennan SM. 2007. Application of the Pitzer ion interaction model to isopiestic data for the $Fe_2(SO_4)_3$-H_2SO_4-H_2O system at 298.15K and 323.15K. *Geochimica et Cosmochimica Acta* 71:2680–2698.
van de Kamp PC, Leake BE. 1985. Petrography and geochemistry of feldspathic and mafic sediments of the northeastern Pacific margin. *Transactions of the Royal Society of Edinburgh: Earth Sciences* 76:411–449.
van de Kamp PC, Leake BE. 1995. Petrology and geochemistry of siliciclastic rocks of mixed feldspathic and ophiolitic provenance in the Northern Apennines, Italy. *Chemical Geology* 122:1–20.
van der Weijden CH, Pacheco FAL. 2003. Hydrochemistry, weathering and weathering rates on Madeira Island. *Journal of Hydrology* 283:122–145.
Veizer J, Jansen SL. 1979. Basement and sedimentary recycling and continental evolution. *Journal of Geology* 87:341–370.
Veizer J, Jansen SL. 1985. Basement and sedimentary recycling–2: Time dimension to global tectonics. *Journal of Geology* 93:625–643.
Veizer J, Mackenzie FT. 2003. Evolution of Sedimentary Rocks. *In* Holland HD, Turekian KK (Editors). Treatise on Geochemistry, Volume 7, Mackenzie FT, ed., *Sediments, Diagenesis, and Sedimentary Rocks.* Elsevier, Amsterdam. p. 369–407.
Wänke H, Dreibus G. 1994. Chemistry and accretion history of Mars. *Philosophical Transaction of the Royal Society of London, Series A* 349:285–293.
Warren JK. 2006. *Evaporites: Sediments, Resources and Hydrocarbons*: Springer, Berlin. 1035 p.
Werner SC. 2009. The global Martian volcanic evolutionary history. *Icarus* 201:44–68.
White AF. 1995. Chemical weathering rates of silicate minerals in soils. *Reviews in Mineralogy* 31:407–461.
Wilson L, Head JW III. 1994. Mars: Review and analysis of volcanic eruption theory and relationships to observed landforms. *Reviews of Geophysics* 32:221–263.
Yen AS, Gellert R, Schröder C, Morris RV, Bell JF III, Knudson AT, Clark BC, Ming DW, Crisp JA, Arvidson RE, Blaney D, Brückner J, Christensen PR, Des Marais DJ, de Souza PA Jr, Economou TE, Ghosh A, Hahn BC, Herkenhoff KE, Haskin LA, Hurowitz JA, Jolliff BL, Johnson JR, Klingelhöfer G, Madsen MB, McLennan SM, McSween HY, Richter L, Rieder R, Rodionov D, Soderblom L, Squyres SW, Tosca NJ, Wang A, Wyatt M, Zipfel J. 2005. An integrated view of the chemistry and mineralogy of Martian soils. *Nature* 436:49–54.
Zolotov MY. 2007. Origin of acid fluids on Mars: Impacts vs. volcanism. Lunar and Planetary Science, Vol. 38, Abstract 1343. The Lunar and Planetary Institute, Houston (CD-ROM).

WERE AQUEOUS RIPPLES ON MARS FORMED BY FLOWING BRINES?

MICHAEL P. LAMB, JOHN P. GROTZINGER
Department of Geological and Planetary Sciences, California Institute of Technology, Pasadena, California 91125 USA
e-mail: mpl@gps.caltech.edu

JOHN B. SOUTHARD
Department of Earth, Atmospheric, and Planetary Sciences, Massachusetts Institute of Technology, Cambridge, Massachusetts 02139 USA

AND

NICHOLAS J. TOSCA
Department of Earth Sciences, University of Cambridge, Cambridge CB2 3EQ UK

ABSTRACT: The discovery in 2004 by Mars exploration rover *Opportunity* of sedimentary rocks with centimeter-scale trough cross-bedding is one of the compelling lines of evidence for flowing water on the Martian surface. The rocks contain a significant evaporite component mixed with weathered mafic silicates, suggesting that the aqueous fluid in contact with the sediments must have been of very high ionic strength because dissolution features are not observed. Recent thermodynamic modeling indicates that these brines could have had higher densities (by up to a factor of 1.3) and significantly higher viscosities (by up to a factor of 40) than pure water. Because fluid density and viscosity can significantly affect sediment transport mechanics, herein we analyze whether ripples could have been stable bed forms under flowing Martian brines. To this end, we compiled bed form stability diagrams with an emphasis on those studies that have considered high-viscosity fluids. For the case of viscous Martian brines, we find that ripples are stable under modest Shields numbers and low particle Reynolds numbers. These conditions translate into sediment sizes ranging from sand to gravel, and they are substantially coarser than sediment sizes for equivalent ripple-forming flows in freshwater. It is likely that ripples might also form in silt sizes under viscous brines, but these conditions (i.e., particle Reynolds numbers < 0.1) have not yet been explored in flume experiments, motivating future work. Using flow-resistance equations and assuming steady uniform flow, we calculate that Marian brines must have had flow depths ranging from 0.01 to 1 m and flow velocities of 0.01 to 1 m/s, and been driven by gravity on slopes of 10^{-4} to 10^{-2} in order to generate the bed stresses necessary to produce ripples. These conditions seem reasonable given the interdune environment that has been proposed for the Burns formation. In addition to the potential for ripples in much coarser sediments, ripples formed by viscous brines also might be larger in height and wavelength than their freshwater counterparts by as much as a factor of 12. Thus, large ($>$10 cm heights) and fine-grained ($<$1 mm particle diameter) cross strata would be compelling physical evidence for flowing brines in the Martian past, provided that independent evidence could be provided for a subaqueous (i.e., not eolian) origin of the cross-stratification. Smaller centimeter-scale ripples can also be formed by brines due to flow-depth limitations or lower-viscosity fluids, and therefore the physical sedimentological evidence in support of brines versus freshwater flows may be ambiguous in these cases.

KEY WORDS: Mars, brines, ripples, high viscosity, bed forms

INTRODUCTION

The 2004 discovery of outcropping sedimentary bedrock formed and altered in the presence of water by the Mars exploration rover *Opportunity* opened a new chapter in the field of sedimentary geology (Squyres et al. 2004). The investigation of these deposits, informally known as the Burns formation, demonstrated that basic stratigraphic and sedimentologic tools that have been used to evaluate the geological history of Earth may be applied to Mars with surprising rigor. For the first time on Mars, it was possible to make observations of grain-scale textures and bed-scale sedimentary structures, and to place these within a stratigraphic framework defined by distinct facies and significant surfaces (Grotzinger et al. 2005). Furthermore, it was possible to evaluate diagenetic overprints that included evidence for recrystallization, precipitation, and later dissolution of displacive evaporite minerals, and the precipitation of concretions (McLennan et al. 2005). Initial observations and interpretations were built on analysis of Eagle and Endurance craters; subsequent analysis of Erebus (Grotzinger et al. 2006, Metz et al. 2009) and Victoria craters (Squyres et al. 2009, Hayes et al. 2011, Edgar et al. this volume) supported the general applicability of the model and its relevance in the broader interpretation of Martian environmental history (McLennan and Grotzinger 2008).

A major discovery was that the Burns formation is composed of sandstone formed of composite grains that have a significant evaporite component mixed with weathered mafic silicates. The evaporitic component in these grains makes up at least 50% of the rock by weight (Clark et al. 2005) and is considered to have been in equilibrium with coeval groundwater, which necessarily must have been of very high ionic strength (McLennan et al. 2005). This would have been required to avoid pervasive leaching of the bedrock by undersaturated fluids, which might have formed large, irregular-shaped vugs, caverns, or collapse features, all of which are unobserved in the Burns formation (McLennann et al. 2005). Given the extremely high calculated solubility of the Mg, Fe, Al, and Ca sulfate salts that are inferred to constitute the Burns formation (Tosca et al. 2005), it is highly unlikely that freshwater interacted with these rocks.

The required brines have such high ionic strength that their physical properties are calculated to be substantially different from freshwater (Tosca et al. 2011), especially their densities and viscosities. Because these properties play a fundamental role in aqueous sediment transport

Sedimentary Geology of Mars
SEPM Special Publication No. 102, Copyright © 2012
SEPM (Society for Sedimentary Geology), Print ISBN 978-1-56576-312-8, CD/DVD ISBN 978-1-56576-313-5, p. 139–150.

(e.g., Vanoni 1974, Southard 1991), it makes sense to ask if such brines would have been capable of generating the ripple cross-stratification preserved in the Burns formation. Previous work has considered bed form stability only under the influence of lowered Martian gravity (e.g., Komar 1979, Southard and Boguchwal 1990b, Grotzinger et al. 2005) and has not considered changes in the viscosity and density of the flow, particularly as driven by changes in brine concentration for flows of high ionic strength. Recent experiments have formed sediment fans by viscous sulfuric acid solution (Benison et al. 2008), but these were not of sufficient scale to produce ripples or cross-bedding.

The focus of this paper is to address this knowledge gap. First, we briefly review observations of the Burns formation and the centimeter-scale cross-stratification within it. Second, we discuss the mechanism by which this cross-stratification was likely produced; i.e., by flowing brines of high ionic strength. Third, we compile existing bed form stability diagrams and confirm that they hold true for high-viscosity and high-density fluids. Fourth, the bed form stability diagrams for conditions of brines on Mars are translated into dimensional space to elucidate the flow depths, velocities, and grain sizes in which ripples are expected to be stable. Finally, we discuss the potential effects of high-viscosity brines on ripple size and broader implications for aqueous flows on Mars.

THE BURNS FORMATION AND CENTIMETER-SCALE CROSS-STRATIFICATION

The Burns formation is the informal name (Grotzinger et al. 2005) given to the sequence of well-sorted, moderately indurated sandstones preserved immediately beneath the surface of the Meridiani plains and examined in detail at exposures in the vicinity and within Eagle, Endurance, Erebus, and Victoria craters. Regional geology suggests that the deposit is of Late Noachian age and thus likely in excess of 3.5 Ga in age. A composite stratigraphy for Eagle and Endurance craters indicates that eolian and locally subaqueous processes deposited these sedimentary rocks as part of a "wetting-upward" dune–sand sheet–interdune depression sequence (Grotzinger et al. 2005). A composite stratigraphy for Erebus and Victoria craters demonstrates that a "drying-upward" trend followed the "wetting-upward" sequence, resulting in widespread sand dune accumulation (Metz et al. 2009, Hayes et al. 2011, Edgar et al. this volume).

The salient attributes of the depositional model for these rocks can be summarized as follows. Eolian dune facies are characterized by the occurrence of large-scale cross-bedded well-sorted sandstones (Grotzinger et al. 2005, Metz et al. 2009, Hayes et al. 2011). These sandstones have bedset thicknesses that suggest moderately large dunes in excess of several meters; eolian sand-sheet facies characterized by planar-laminated to low-angle cross-stratified, well-sorted sandstones were formed by migrating impact ripples. These facies are interpreted to represent a migrating dune system of unknown extent that was deposited under dry conditions such that the sediment was noncohesive and thus transported in migrating dune fields. Wet to evaporitic interdune facies characterized by centimeter-scale trough (or "festoon") ripple cross-lamination are diagnostic of subaqueous current transport in the lower-flow regime (Grotzinger et al. 2005). At Endurance and Eagle craters, this subaqueous facies is found in the upper part of the upper unit of the Burns formation and is interpreted to mark the location where the groundwater table breached the surface and gravity-driven aqueous currents transported the sediment. Centimeter-scale trough cross-lamination is particularly well developed at Erebus crater, where additional features suggestive of desiccation also are present (Grotzinger et al. 2006). Prism cracks are interpreted to have formed during multiple wetting and drying events, and soft-sediment deformation features are consistent with sediment liquefaction (Metz et al. 2009).

Centimeter-scale trough (or festoon) cross-stratification is observed at Eagle crater (Squyres et al. 2004, Grotzinger et al. 2005), Endurance crater (Grotzinger et al. 2005), and Erebus crater (Grotzinger et al. 2006). The best examples occur at the Overgaard and Cornville outcrops in Erebus crater (Grotzinger et al. 2006). The bedding sequence in the upper part of Overgaard contains three units: a lower cross-stratified unit, a middle planar-laminated unit, and an upper unit with centimeter-scale festoon cross-lamination (Fig. 1A). Laminae that define the middle planar-stratified unit onlap a scour surface from right to left. Note that five laminae define the middle unit on the right part of the rock (Fig. 1A, yellow bar); however, due to onlap, these decrease in number to one or two laminae at the left part of the rock. Centimeter-scale cross-laminae of the upper unit scour down into the middle unit, from left to right. Note additional, stratigraphically higher, planar laminae in the middle unit at the far left (Fig. 1A, green arrows), where the scour surface rises up through the section.

Centimeter-scale trough cross-stratification in the upper unit forms a bedset with at least 3 to 4 cm of apparent thickness. Blue arrows point to three distinct troughs, as indicated by basal truncation and concave-upward geometry. Several superimposed sets show basal scouring and backfilling by concave-upward to occasionally convex-upward cross-laminae (Fig. 1A, red arrows). Truncation surfaces and backfilling of subjacent sets are well expressed in the left-center part of the upper unit. Individual troughs are 3 to 4 cm wide. A mosaic of Microscopic Imager (MI) frames was constructed to coincide with the middle and upper units of upper Overgaard; this provides additional supporting detail of the grain-scale variability in primary lamination (Fig. 1B). Furthermore, the MI mosaic was used to construct a digital elevation model (DEM) of the outcrop topography (Fig. 1C); the rock surface is flat at the relevant scale across the left-center region where the upper unit is exposed, and otherwise bedding shows no systematic relationship with respect to the minor local slopes. This demonstrates that the cross-lamination is a primary attribute of the rock rather than an artifact of erosion.

Centimeter-scale trough cross-lamination is pervasive at the Cornville outcrop, covering its entire exposure (Fig. 2). This represents a true stratigraphic thickness of close to 20 to 30 cm (Grotzinger et al. 2006). Centimeter-scale trough cross-lamination is well developed near the center of the rock. The geometry and scale of these cross-laminae sets are nearly indistinguishable from terrestrial analogs formed in aqueous flows (Fig. 3). Cornville festoon cross-laminae sets are up to 1 to 2 cm thick and up to 2 to 3 cm wide in sections that are most likely transverse to flow. The rock surface exposes a cut that is largely transverse to flow, with a small component of climb from right to left.

RATIONALE FOR RIPPLE FORMATION BY AQUEOUS FLOWS

When subjected to shear stresses created by shallow, freshwater flows with moderate current velocities, fine- to medium-grained sand spontaneously forms highly sinuous, crested ripples with amplitudes up to a few centimeters (Southard and Boguchwal 1973, Southard and Boguchwal 1990a). The diagnostic attribute of such ripples is exposed in cuts transverse to flow, where laminae have a trough-shaped geometry (Rubin 1987). Cross-lamination of this type *and* scale is not known to develop in eolian flows (see discussion in Grotzinger et al. 2005) and, to our knowledge, has not been observed in deposits resulting from other types of flows such as volcanic- or impact-generated base surges (e.g., Fralick et al. this volume). Very likely, small-scale ripples are not stable under the conditions of high sedimentation rates and rapidly decelerating currents that are characteristic of base-surge flows. For these reasons, the presence of centimeter-scale cross-lamination in the Burns formation has been regarded as strong evidence for overland water flow across the ancient surface of Mars (Squyres et al. 2004, Grotzinger et al. 2005).

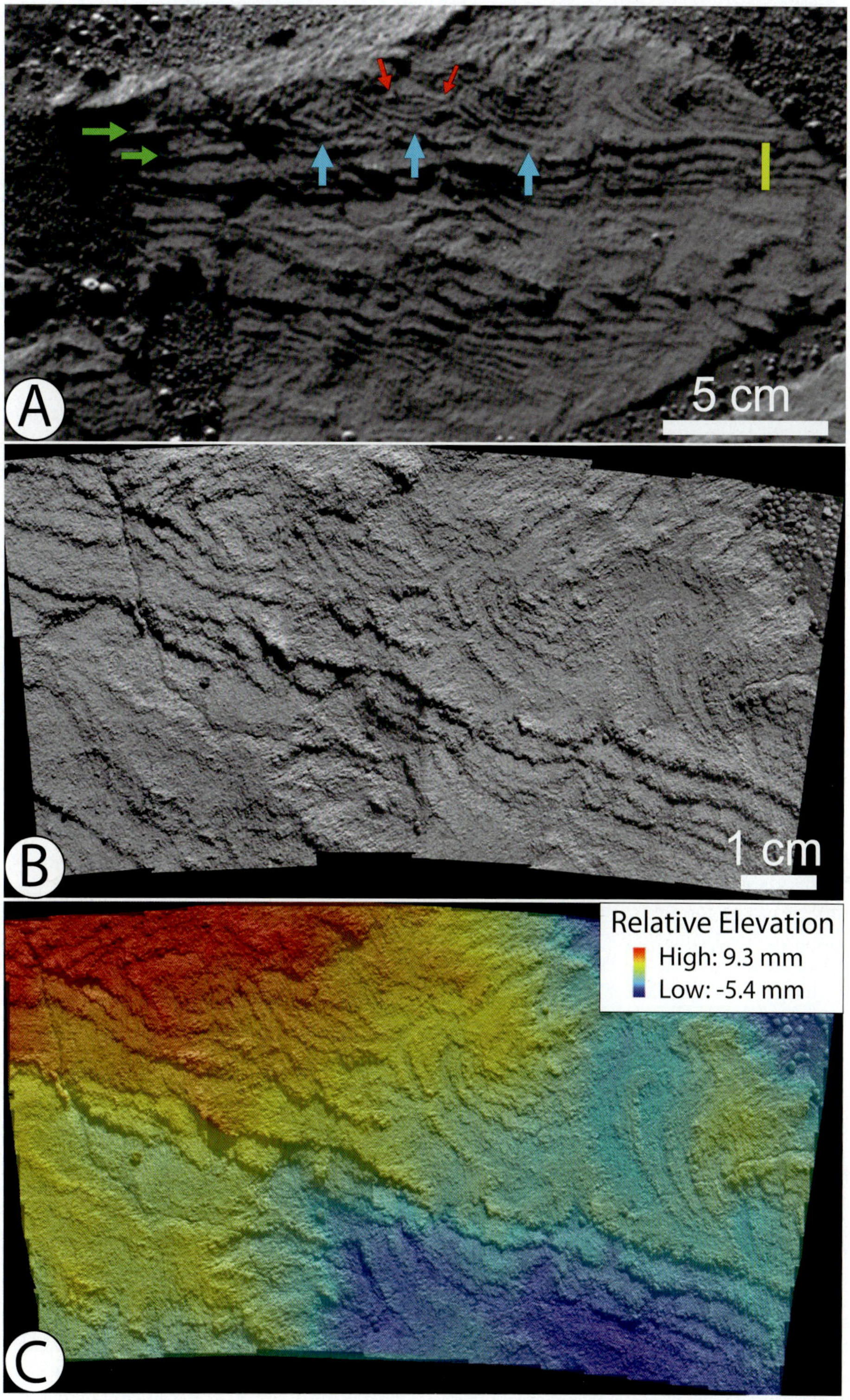

FIG. 1.—Centimeter-scale trough cross-lamination at Overgaard outcrop, Burns formation, Erebus crater, Mars. **A)** Cross-stratified lower unit, overlain by a planar-laminated middle unit (indicated by yellow bar) and a trough cross-laminated upper unit. See text for discussion. This low-angle light image was acquired on sol 716 at 16:15:46 LTST, sequence id p2593, 432 nanometer filter. **B)** Microscopic Imager mosaic of the middle and upper units. Note generally granular texture and excellent sorting, right-to-left pinch-out of the middle unit, and stratal truncation and downlap of the cross-laminae in upper unit (compare with 2a, red arrows). The mosaic was constructed from images obtained on sols 721 and 723. **C)** Digital elevation model (DEM) of MI mosaic. Note the lack of correlation between topography and bedding in left-central part of the upper unit, indicating a primary origin of stratal (lamina) geometries. See Grotzinger et al. (2006) for further details on the image attributes and technical details on DEM construction. After Grotzinger et al. (2006).

FIG. 2.—Centimeter-scale trough cross-lamination is abundant at the Cornville outcrop. The center part is highlighted and approximates the area of Figure 3A. Exposed surface is largely transverse to flow, with a small component of climb from right to left. Note abundant scour surfaces, which truncate the underlying concave-up laminae, overlain by downlapping cross-laminae of the next set. Note the moderately steep dips of laminae exposed along the left side of the crack at the center part of the lower boundary of the rock. These indicate that the true thickness of this cross-laminated unit must be less than the apparent thickness, which equals the long dimension of the rock (~50 cm); correcting for dip (~30 degrees), the true thickness of centimeter-scale cross-lamination is likely still to be on the order of several tens of centimeters, the thickest observed at Meridiani. Superres image of Cornville acquired sol 705 at 14:15:22 LTST, using Pancam's 482 nanometer filter (sequence id p2576).

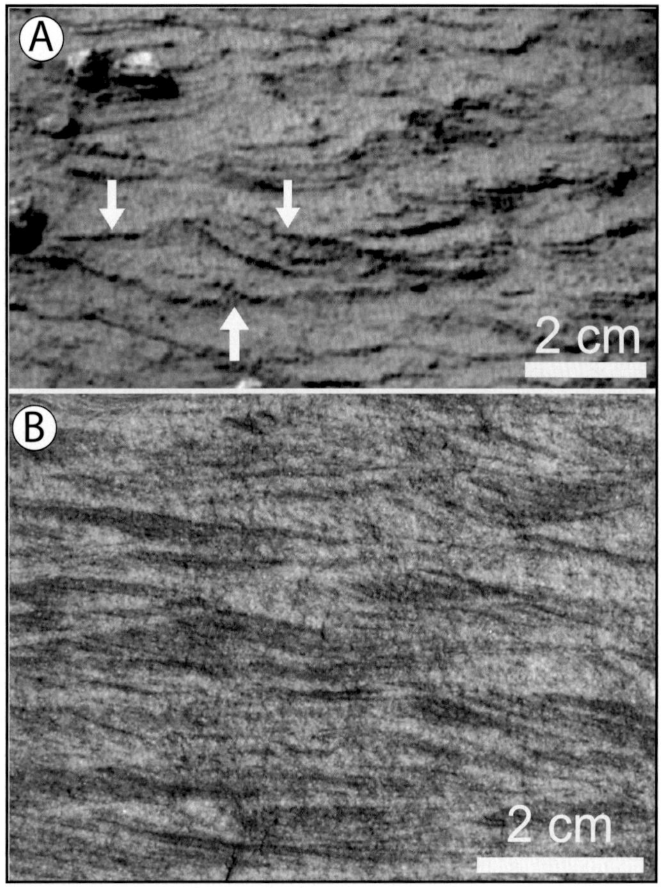

FIG. 3.—**A)** Enlargement of the center part of Figure 2. Well-developed centimeter-scale trough cross-lamination. In the left-center of the image, the set has a classic concave-up lower bounding surface (up arrow), and the upper boundaries are marked by two intersecting concave-up surfaces (down arrows). Internal cross-laminae are truncated by the upper bounding surface on the right, and that bounding surface is then downlapped by cross-laminae of the superjacent set. **B)** Terrestrial centimeter-scale trough cross-lamination (Cretaceous Mesa Verde Group) shows strong geometric similarity with that preserved in Cornville. Compare with part A.

RATIONALE TO INVOKE BRINES

In addition to the diagenetic textures that place constraints on the composition of groundwater that interacted with the Burns formation (general absence of rock dissolution features), the primary textures also strongly indicate brines. Recall that the each grain is interpreted to consist of a mixture of >50% sulfate plus altered silicate minerals (Clark et al. 2005). These minerals are so fine grained that they cannot be distinguished with the *Opportunity* rover's microscopic imager (MI) (McLennan et al. 2005). The grains themselves are typically sand size, 0.3 to 0.8 mm in diameter, yet they are composed of particles likely to be of silt size and finer. Furthermore, the minerals that compose these grains consist largely of Mg-sulfates (e.g., originally epsomite), which in concentrated Martian surface waters have very high solubility, even higher than halite (Tosca et al. 2005, 2011).

Therefore, it is conservative to assume the solubility of halite to provide a rough estimate of dissolution rates in freshwater for the sand grains of the Burns formation. In natural systems, the dissolution rates of highly soluble minerals (e.g., Mg-sulfates and halite) are frequently limited by the transport of dissolved components from the dissolving mineral surface (Lasaga 1998). For example, experimental work on the dissolution of sand-size halite spheres in water indicates that such grains cannot survive in fresh stagnant water for more than about 2 minutes at 20° C (Simon 1981, Kerr 1995, Alkattan 1997, Velbel 2004) (Fig. 4). Decreasing temperature slows the dissolution rate (Velbel 2004), but the timescale required to completely dissolve sand-size grains still would be only a few minutes at 0° C (Fig. 4). Temperatures much below 0 to −3° C would have produced a significant amount of water ice, which is not supported as a sedimentary component (but cannot be entirely ruled out) based on outcrop observations to date. Allowing the grains to fall through freshwater greatly increases the dissolution rate, so that survival times are measured in tens of seconds rather than minutes (Velbel 2004); in the presence of fully turbulent flows, dissolution rates are even further increased (Kerr 1995). Finally, it is worth pointing out—estimations of dissolution rates aside—that it

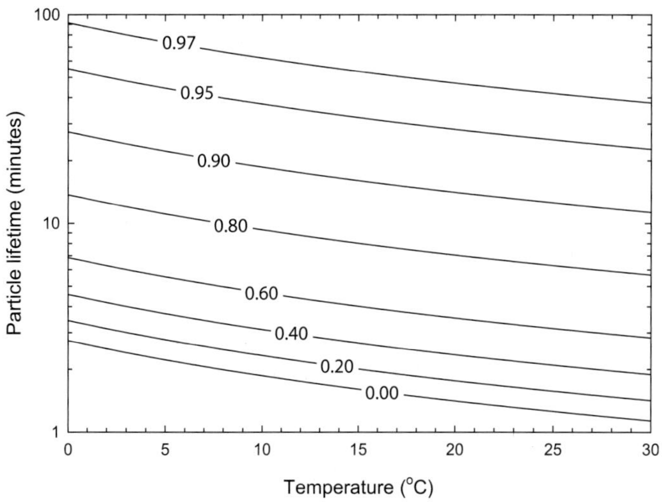

FIG. 4.—The timescale for a 1-mm-diameter particle of halite to completely dissolve as a function of temperature based on experimental data from Alkattan et al. (1997). The contours are the saturation state (i.e., concentration normalized by the equilibrium concentration), where zero represents freshwater, and unity represents a fully saturated brine. Although the dissolution timescales for magnesium sulfate, which is a better analog for Martian sediments than halite, are not as well known, the relative effects of temperature and concentration should be equivalent to halite (Lasaga 1998).

is logical to assume that groundwater migrating through a regionally extensive sulfate-rich unit should be saturated with respect to those mineral phases. Regional stratigraphy and mineralogy favor this.

The Burns formation shows multiple stacked sets of ripple cross-lamination with an apparently small angle of climb over a thickness of 20 to 30 cm, indicating ripple migration under conditions of sustained flow (Grotzinger et al. 2006). Because the timescale of ripple formation can be expected to be at least tens of minutes, if not days or longer (Southard 1991, Baas 1999, Betat et al. 2002), we need to seek other effects to explain the anomalous survival of such soluble grains.

One possible influence is given by the observation that the minor mineralogical component of the sand grains of the Burns formation is likely to be altered silicate minerals. These would have orders of magnitude lower solubility than the sulfates, and this might increase their survival in dilute flows. However, the dissolution of the sulfate component would still dramatically weaken the grain, making it much more susceptible to impact-driven disintegration.

Another possible explanation, which is the focus of the work herein, is that the aqueous ripple-forming flows were concentrated brines of high ionic strength. Dissolution modeling suggests that the brines must have been nearly saturated in order to extend particle lifetimes to tens of minutes or longer (Fig. 4). Recent thermodynamic modeling, which is based in part on thousands of laboratory measurements, suggests that brines could have existed on the Martian surface in chemical equilibrium with the host rock (Tosca et al. 2011). The modeled brines are of such high ionic strength, however, that their physical properties are substantially different than pure water (Tosca et al. 2011). In some cases, the fluid viscosity of the modeled brines is 40-fold that of pure water. Given that, among other variables, fluid density and viscosity influence the stability of bed forms (Vanoni 1974, Boguchwal and Southard 1990, Southard and Boguchwal 1990b), it is important to verify whether ripples could indeed form as a result of flowing brines on Mars.

DIMENSIONLESS REPRESENTATION OF BED FORM STABILITY FIELDS

There have been several attempts to formulate a graphical framework representing the conditions of flow, sediment, and fluid for the stable existence of current ripples. In steady, uniform, open-channel flow of a Newtonian fluid, dimensional analysis indicates that at least three independent dimensionless variables are needed to describe stable bed form states (Vanoni 1974, Van Rijn 1984, Southard 1991, van den Berg and van Gelder 1993, Garcia 2007). These dimensionless variables provide a set of scale-modeling parameters that can take into account a wide range of sedimentologically interesting behaviors, such as the effects of changing grain size, current velocity, flow depth, fluid viscosity, grain density, and the acceleration of gravity. The dimensionless numbers are nonunique, however; dimensional analysis indicates that there are many equally valid sets of three nondimensional variables. For example, Boguchwal and Southard (1990) prefer to express their experimental bed form stability results in terms of a dimensionless flow depth, dimensionless flow velocity, and a dimensionless particle size. Herein, we choose another common and equally valid approach to describe bed states by representing the flow strength (or intensity of sediment transport) through the dimensionless Shields parameter,

$$\tau_* = \frac{\tau_b}{(\rho_s - \rho)gD}, \tag{1}$$

where τ_b is the shear stress at the bed, ρ_s is the density of sediment, ρ is the fluid density, g is the acceleration due to gravity, and D is the particle diameter. The effective sediment size is accounted for using a particle Reynolds number,

$$\mathrm{Re}_p = \frac{(RgD)^{1/2}D}{\nu}, \tag{2}$$

where $R = (\rho_s - \rho)/\rho$ is the submerged specific density of sediment, and ν is the kinematic viscosity of the fluid (e.g., Garcia 2007). The final dimensionless number needed to describe bed form stability in this framework is a dimensionless flow depth, which, by assuming that R and a bed friction coefficient are constant (e.g., Garcia 2007), is often written as a Froude number,

$$\mathrm{Fr} = U/\sqrt{gh}, \tag{3}$$

where U is the mean flow velocity, and h is the flow depth. Numerous flume experiments have shown this to be a tractable strategy. For Fr greater than about unity, flows are considered supercritical, and antidunes form rather than ripples or dunes (Vanoni 1974, Van Rijn 1984, Southard and Boguchwal 1990a). For $\mathrm{Fr} < 1$, bed form stability is independent of Fr, and therefore it can be neglected as a variable. Thus, Eq. 1 and Eq. 2 alone are sufficient to elucidate the stability of ripple bed forms for the subcritical flows of concern here (e.g., van den Berg and van Gelder 1993).

We compiled results from five sediment-transport and bed form studies to build a comprehensive bed form stability diagram that spans a larger range in Re_p than any of the individual studies alone (Fig. 5). Each study contains many data points (in some cases, hundreds) that each represent a single flume experiment or field observation, and these are not repeated here. Instead, we show the dividing lines that were empirically found to bound the bed form stability fields from each study. The compilation includes a lower boundary separating no motion from the fields of sediment transport; i.e., the Shields curve (Shields 1936). For coarse particle sizes, we represent the Shields curve

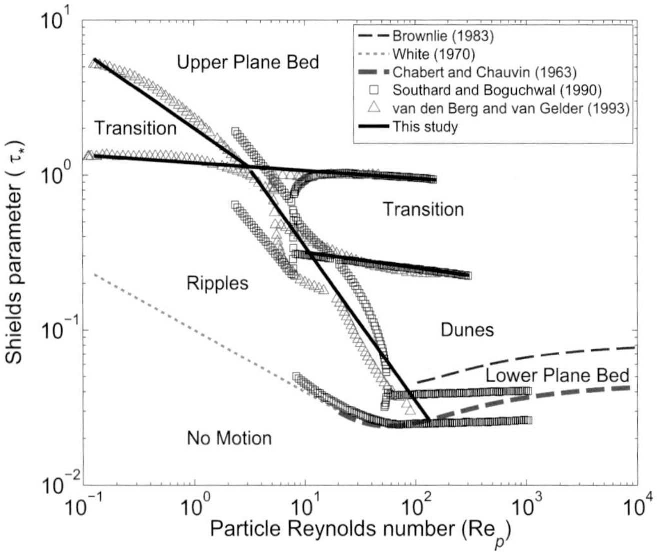

FIG. 5.—Phases of bed form stability as a function of Shields parameter and particle Reynolds number. Summary boundaries from five different studies are shown, separating no motion from lower plane bed, ripples, dunes, and upper plane bed. Results from previous studies were given in terms of τ_* and Re_p, except the study of Southard and Boguchwal (1990a), from which we translated their Figure 11 into the present coordinate system. The solid black line represents our power-law fits (by eye).

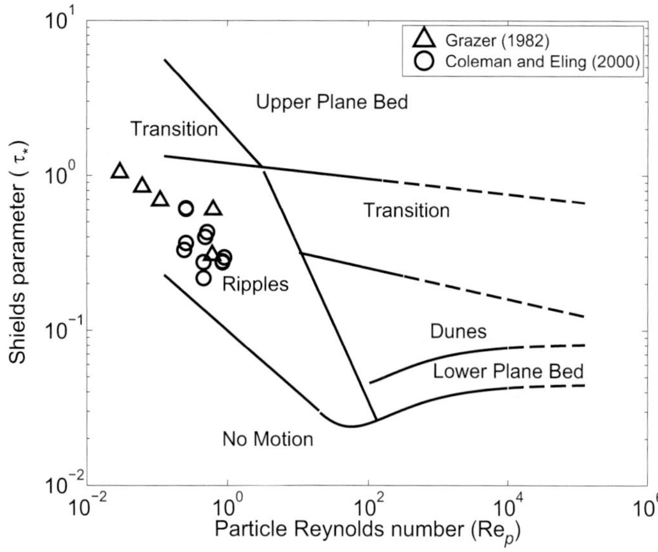

FIG. 6.—Summary bed form stability diagram. The solid lines are fits to the bed form transition boundaries as shown in Figure 5. The dashed lines are extrapolations of these boundaries to higher particle Reynolds numbers. Also shown are data points from two studies that produced ripples in a series of experiments with high-viscosity fluids.

following the semi-empirical fit to incipient motion data of Parker (1991) and Brownlie (1983). For small Re_p, we use our own fit to the experimental data of White (1970) because he used a high-viscosity fluid ($\nu = 8 \times 10^{-6}$ m^2/s, approximately eightfold that of freshwater on Earth's surface) to achieve small Re_p, rather than the more common approach of reducing D, which avoids the complicating cohesive effects common to clay-size sediment. In the regimes of mobile sediment, ripples form at low Re_p and moderate values of τ_*, and ripples wash out to the upper plane bed at τ_* greater than approximately unity. For large Re_p, there exists a stable regime of lower plane bed, followed by dunes and upper plane bed with increasing Shields parameter. Ripples are similar to dunes but are usually smaller and form in the presence of a viscous sublayer (i.e., small Re_p) (e.g., Garcia 2007). For these bed form stability transitions, we show the results of Chabert and Chauvin (1963), as well as the comprehensive studies of Southard and Boguchwal (1990a) and van den Berg and van Gelder (1993). For the case of Southard and Boguchwal, we recast their dimensional diagram (their Fig. 11) into the dimensionless variables given by Eq. 1 and Eq. 2.

It is encouraging to note that the stability field boundaries are relatively consistent despite the fact that the databases contain both field and flume data, measurements were made by different workers, and data span a wide range of flow velocities, flow depths, particle sizes, particle densities, fluid densities, and fluid viscosities (Fig. 5). This illustrates the robustness of the dimensionless framework.

In order to explore bed form stability for Martian brines, we need a summary bed form stability diagram that can be translated into dimensional space. To accomplish this, and given the ambiguity of some of the transitions from the different studies in Figure 5, we fit power-law boundaries (by eye) to the bed form stability transitions (Fig. 5). The resulting compilation shows solid lines where the stability fields have been explored either experimentally or with field data, and dashed lines represent power-law extrapolations (Fig. 6). Thus, the parameter space for $\mathrm{Re}_p < 10^{-1}$, for example, has yet to be explored experimentally. The reason for this is that, for freshwater flows typical of Earth-surface conditions, $\mathrm{Re}_p < 10^{-1}$ corresponds to clay-size particles, for which cohesive effects become important and bed form stability can no longer be described completely by Eq. 1 and Eq. 2 (e.g., Dzulynski and Sanders 1962, Grazer 1982, Mehta 1991, Best and Leeder 1993). For Martian brines, however, small values of Re_p are expected for coarser, noncohesive particles owing to the inverse dependence of Re_p on viscosity (Eq. 2).

For most experiments and field observations, Re_p has been varied through varying the particle size (Eq. 2), which is a practical and relevant strategy for most natural flows on Earth. However, we are concerned instead with variations in Re_p through the kinematic viscosity, so it is important to verify that the dimensionless framework holds true under varying viscosity. This was precisely the objective of the experimental study of Boguchwal and Southard (1990) and Southard and Boguchwal (1990b), which explored bed form stability under water flows with temperatures ranging from 10 to 75° C, corresponding to viscosities that span 3.7×10^{-7} to 1.4×10^{-6} m^2/s. They found the dimensionless framework to hold true, and their data were used in part to construct the boundaries of Southard and Boguchwal (1990a) (Fig. 5). Less work exists on sediment transport under high-viscosity fluids, but two studies are of note. Grazer (1982) produced ripples on a silt bed at very low Re_p using water–sucrose solutions with viscosities as high as 1.05×10^{-5} m^2/s, which approach the highest viscosity condition modeled by Tosca et al. (2011) of $\nu = 4 \times 10^{-5}$ m^2/s for Martian brines. More recently, Coleman and Eling (2000) produced ripples in eleven different experiments on a sand bed using hydraulic oil with a viscosity as high as $\nu = 9 \times 10^{-5}$ m^2/s; i.e., twice that inferred for the most viscous case on Mars. The viscosity was so great in the experiments of Coleman and Eling that the flows were laminar. We recast the data presented in these studies to calculate τ_* and Re_p following Eq. 1 and Eq. 2 (Fig. 6). The resulting data points

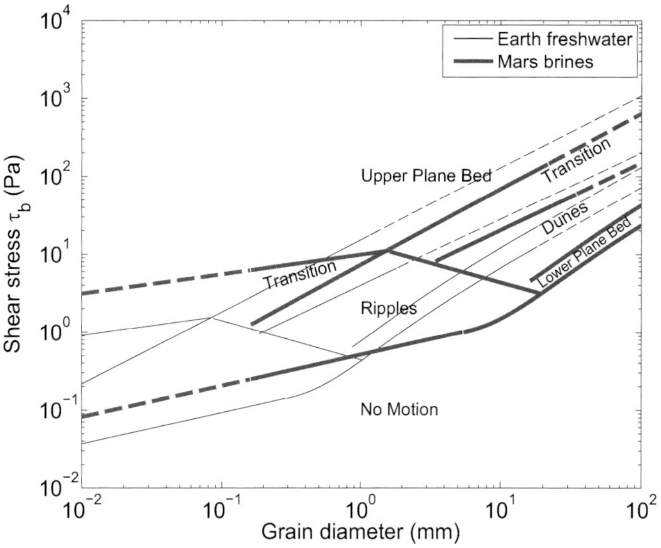

FIG. 7.—Translation of Figure 6 into dimensional space for the case of freshwater flows on Earth and a dense, viscous brine on Mars. See text for specific fluid properties. Note the dramatic shift in bed form space to coarser sediment for the brine case. Solid lines represent boundaries that have been explored experimentally following Figures 5 and 6, and dashed lines are extrapolations. Labels correspond to the Martian case.

plot within the ripple stability field, confirming that the dimensionless framework is capturing the appropriate physics, regardless of the particle size or the fluid viscosity.

FLOW CONDITIONS FOR RIPPLE FORMATION BY MARTIAN BRINES

Equations 1 and 2 reveal that ripple formation depends on fluid density, viscosity, and gravity, all of which are different for brines on Mars compared to freshwater flows on Earth. To illustrate this effect, we can replot the stability fields of Figure 6 in terms of the flow stress on the bed (τ_b) versus particle diameter (D). To accomplish this, estimates are needed for the physical properties of the brines as well as the density of sediment and acceleration due to gravity. Work by Tosca et al. (2011) indicates that Martian brines might have had viscosities as high as $\nu = 4 \times 10^{-5}$ m²/s and densities as high as $\rho = 1300$ kg/m³. Herein, these upper estimates are used to analyze the possible existence of ripples. We assume particle densities of $\rho_s = 2650$ kg/m³, which seems reasonable given that evaporite minerals range from 1900 to 2400 kg/m³ and there is potential for a mixture of basaltic grains (3000 kg/m³), and set $g = 3.73$ m/s². A reference case for freshwater flows transporting siliceous material on Earth is also shown where $\rho = 1000$ kg/m³, $\rho_s = 2650$ kg/m³, $\nu = 10^{-6}$ m²/s, and $g = 9.81$ m/s².

There is a significant shift in bed form stability for the Martian brines as compared to freshwater flows on Earth (Fig. 7). This is primarily due to the 40-fold increase in fluid viscosity. For example, in comparison to freshwater on Earth, finer-grained sediment ($D < 2$ mm) moves at higher stresses by flowing Martian brines. In contrast, gravel sizes and larger ($D > 2$ mm) move at lower stresses for Martian brines than freshwater on Earth. This is because of the strong particle Reynolds number dependence of incipient motion for fine sediment. Heightened viscosity increases the size of the viscous sublayer near the bed and decreases the local velocities about grains for a given bed shear stress. Coarse particles, on the other hand, protrude above the viscous sublayer (i.e., hydraulically transitional or rough flow) and are less sensitive to the particle Reynolds number and viscosity (Yalin 1977).

Ripples on Earth are stable in freshwater for particles in the silt and sand size ranges, with an upper maximum size of about $D = 1$ mm (Fig. 7). In contrast, ripples formed by Martian brines are predicted to form in much coarser particle sizes, ranging from sand to coarse gravel, up to $D \approx 20$ mm. Unlike freshwater flows on Earth, where dunes are common in sand, under conditions of viscous Martian brines, dunes are predicted to form only in gravel sizes and coarser ($D > 2$ mm). It is likely that ripples form under Martian brines for silt-size grains as well, but the particle Reynolds number regime corresponding to $D < 0.2$ mm for Martian brines has yet to be explored experimentally (Fig. 7). Thus, ripples have the potential to be a stable bed form over a wide range of particle sizes when formed under viscous brines, including sand, gravel, and potentially silt. However, in order for ripples to form, the flows must generate the necessary bed stresses of ~0.1 to 10 Pa.

To investigate the flow dynamics that could result in the necessary bed stresses to generate ripples, we assume steady and uniform flow and calculate the depth-averaged flow velocity, U, from

$$U = \frac{u_*^2 h}{3\nu}, \quad \text{for Re} < 5 \times 10^2, \tag{4A}$$

$$U = \frac{1}{h}\int_{z_0}^{h} \frac{u_*}{\kappa}\ln\left(\frac{z}{z_0}\right)dz, \quad \text{for Re} > 10^3, \tag{4B}$$

where $u_* \equiv \sqrt{\tau_b/\rho}$ is the bed shear velocity, κ is von Kármán's constant (0.4), z is the height above the bed, z_0 is the bed roughness length scale, and $\text{Re} = Uh/\nu$ is the flow Reynolds number, which determines whether the flow is laminar (Eq. 4A) or turbulent (Eq. 4B) (e.g., Chow 1959). For turbulent flow, we set the roughness length scale equal to

$$z_0 = \nu/(9u_*), \quad \text{for Re}_p < 5, \tag{5A}$$

$$z_0 = D/10, \quad \text{for Re}_p > 100, \tag{5B}$$

corresponding to hydraulically smooth (Eq. 5A) and hydraulically rough (Eq. 5B) basal boundary conditions (e.g., Yalin 1977).

The resulting calculations are shown in Figure 8 for $D = 0.5$ mm size sediment and for the most viscous brine modeled by Tosca et al. (2011) (i.e., $\rho = 1300$ kg/m³, $\nu = 4 \times 10^{-5}$ m²/s). Note, for transitionally turbulent ($5 \times 10^2 < \text{Re} < 10^3$) or transitionally rough ($5 < \text{Re}_p < 10^2$) flow, we matched the end-member solutions by eye for simplicity. Also shown is the boundary for supercritical flow (Fr = 1), which is approximately where ripples should give way to antidunes. As predicted from Figure 7, ripples are stable for basal shear stresses ranging from 0.1 to 10 Pa. This corresponds to flow velocities of 0.01 to 1 m/s and flow depths in the range of 0.01 to 1 m. For freshwater flows on Earth, ripples in sand are almost always formed under turbulent flows. In contrast, the predicted properties of Martian brines are so viscous that ripples are predicted to form under a wide range of laminar and transitionally turbulent flows, consistent with recent experimental and theoretical findings (e.g., Coleman and Eling 2000, Devauchelle et al. 2010, Lajeunesse et al. 2010).

For a given flow depth, there is a unique value of bed slope ($S = \sin\theta$, where θ is the bed-slope angle from horizontal) that is needed to generate the appropriate bed stresses to form ripples, assuming steady and uniform flow (i.e., $\tau_b = \rho g h S$) (Fig. 8). The resulting values of slope of $10^{-4} < S < 10^{-2}$ seem reasonable for the interdune environment interpreted by Grotzinger et al. (2005) for the Burns formation. Nonuniform flow is also possible, for example, if brines flowed from one interdune into an adjacent depression.

By the line of reasoning presented above, we conclude that ripples

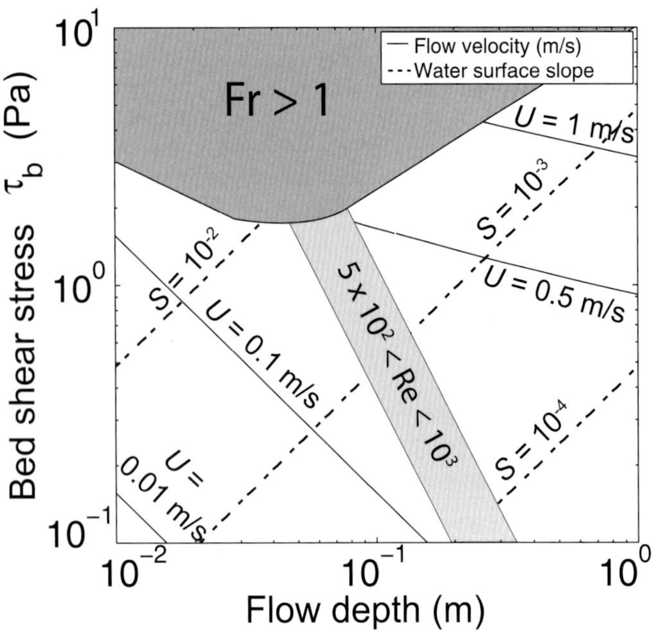

FIG. 8.—Combinations of depth-averaged flow velocity and depth necessary to generate ripples by flowing brines on Mars. The bed shear stress needed to generate ripples ranges from approximately 10^{-1} to 10 Pa (Fig. 7). Assuming steady and uniform flow, the flow velocity (U) and depth combinations have a unique value of bed slope gradient (S), which is also shown. Ripples are not expected to form where Fr $>$ 1 due to standing waves that generate antidunes. The zone of transitionally turbulent flow (i.e., $5 \times 10^2 < \mathrm{Re} < 10^3$) is also shown, with laminar flow occurring for smaller flow depths and velocities and turbulent flow occurring for larger flow depths and velocities.

should be stable over a wide range of particle sizes and bed shear stresses, even for the most viscous brine considered by Tosca et al. (2011). In fact, ripples are expected to be stable over a much larger range in conditions under Martian brines, as compared to freshwater flows on Earth, including most of the gravel size range. Moreover, viscous brines are predicted to be capable of generating the bed stresses necessary to generate flows under reasonable flow conditions, including centimeter- to decimeter-scale flow depths, velocities of centimeters per second, and modest bed slopes. However, the flow conditions for ripple formation in sand sizes by viscous, dense brines have not yet been explored experimentally. This motivates the need for experiments on bed form stability for $\mathrm{Re}_p < 10^{-1}$.

EFFECT OF BRINE VISCOSITY ON RIPPLE SIZE

It is distinctive of current ripples that for flows of a given fluid, like water at room temperature, the size of current ripples (average spacing or average height) varies by less than a factor of three or so over the entire range of sediment sizes for which ripples are the stable bed configuration. For example, in freshwater flows ripple heights (η_f) and wavelengths (λ_f) are typically about $\eta_f = 1.5$ cm and $\lambda_f = 9$ cm (e.g., Southard 1991). The same is not true, however, for the kinematic viscosity of the fluid. Dimensional analysis indicates that the kinematic viscosity ν enters any given dimensionless length variable to the inverse two-thirds power (Southard and Boguchwal 1990a). Thus, dynamically similar-sized ripples in a viscous brine would have heights of $\eta_b = (\nu_b/\nu_f)^{2/3}\eta_f$ and wavelengths of $\lambda_b = (\nu_b/\nu_f)^{2/3}\lambda_f$, where ν_b is the viscosity of the brine, and ν_f is the viscosity of freshwater (Grazer 1982, Southard and Boguchwal 1990a). For a brine with viscosity 40 times that of pure water, the ripples should be larger than equivalent ripples formed in freshwater by a factor of about 12, with predicted ripple heights of 18 cm and wavelengths of 1.08 m! When comparing ripple sizes for freshwater flows on Earth to viscous brines on Mars, there is also a dependence on the relative difference in gravitational acceleration to the one-third power (or a length-scale factor of $\sim$ 1.3) (Southard and Boguchwal 1990b), which is minor given the potential viscosity scale effects considered here. Note that despite their large size, these bed forms would still be considered ripples and not dunes because they formed in the presence of a viscous sublayer (e.g., Garcia 2007).

Finding large current ripples in the Martian sedimentary record would be conclusive evidence for flows with viscosity much greater than that of pure water (the small ripples recorded in Burns formation strata are consistent with, but do not prove, this point, as discussed later herein). First, however, it would need to be established that the bed forms in question were not eolian dunes (e.g., due to an absence of translatent cross strata, pin-stripe lamination, massive grainfall deposits, or other contextual evidence; see Grotzinger et al. [2005] for discussion). If the bed forms can be shown to be subaqueous in origin, there remains the problem of subaqueous ripples versus subaqueous dunes. It is clear from observations both in laboratory flumes and in natural flows on Earth that with increasing sediment size as well as increasing flow velocity, ripples give way abruptly to dunes, which are in large measure similar to ripples in their geometry but are typically an order of magnitude larger in deep freshwater flows (e.g., Southard 1991, Garcia 2007). If we found dune-scale bedforms from liquid flows on Mars, how could we be sure that we were dealing with current ripples (larger by virtue of the greater kinematic viscosities) rather than dunes generated by aqueous flows with viscosities not much greater than that of pure water?

There is a definite minimum dimensionless sediment size for the existence of dunes, and likewise there is a definite maximum sediment size for the existence of ripples (Fig. 7). These scale breaks follow theoretical considerations for the existence of a laminar sublayer, which separates ripples from dunes (Garcia 2007). Three cases can be recognized: (1) the sediment size in the Martian bedforms is sufficiently fine as to be smaller than the minimum sediment size for the existence of subaqueous dunes (i.e., $D < 1$ mm in Fig. 7); (2) the sediment size in the Martian bedforms is sufficiently large as to be larger than the maximum sediment size for the existence of subaqueous ripples (i.e., $D > 20$ mm in Fig. 7); and (3) the sediment size of the Martian bedforms falls between the sizes noted in cases 1 and 2. It is only in case 3 (i.e., $2 < D < 20$ mm) that we cannot be certain whether we are observing either ripples or dunes.

Owing to the relatively fine particle sizes found in cross-stratification in the Burns formation (0.3–0.8 mm; Grotzinger et al. 2005), it is likely that these bed forms are indeed ripples and not dunes. If in fact they were formed by the highly viscous end-member brine modeled by Tosca et al. (2011), we might expect ripple heights of tens of centimeters, whilst observations indicate centimeter-scale cross-stratification. Although preservation of full ripple heights in cross-stratification requires a high angle of climb (Rubin and Hunter 1982), it seems highly unlikely that the ripples were decimeters high, given the observed centimeter-scale widths of trough cross-beds.

An important caveat to the ripple-size scaling discussed here is that the flows must have had sufficient depth and duration for the ripples to reach their equilibrium form. In the experiments of Coleman and Eling (2000), for example, ripple heights under viscous laminar flows were only $\sim$0.5 to 1 cm, whereas we might have predicted that their heights should have been tens of centimeters. However, these experiments were only run for 30 to 90 minutes, a duration that was probably not long enough for the ripples to reach their equilibrium sizes (Coleman and Eling 2000). Moreover, ripple heights are significantly limited when they reach one-third to one-half the flow depth (e.g., Garcia 2007). In the experiments of Coleman and Eling (2000), flow depths ranged from 3 to 6 cm, which would preclude ripple heights larger than ~1 to 3 cm.

A reanalysis of the experiments of Grazer (1982) shows similar results. Using only the viscosity scale factor discussed previously, and assuming $\eta_f = 1.5$ cm and $\lambda_f = 9$ cm, we calculated predicted ripple heights and wavelengths for the higher-viscosity fluids used in the experiments and compared these to the measurements of Grazer. The results show that measured ripple heights were much smaller than the predicted values (Fig. 9A). These experiments were run for 25 to 30 hours to ensure that the timescale needed to generate equilibrium ripples was not the limiting factor (Grazer 1982). Indeed, the predictions for larger ripple wavelength with increasing viscosity match the measurements well (Fig. 9B). Thus, it appears that the anomalously small ripple heights are a result of a flow-depth limitation. The predictions match the observations much better when the flow depth is at least twice the predicted ripple height (Fig. 9C). Because the predicted flow depths necessary to generate ripples on Mars by flowing brines might be as low as 1 cm (Fig. 8), ripple heights could necessarily be small regardless of flow viscosity. Thus, like dunes in freshwater flows, ripples formed by Martian brines may in cases be depth limited; these bed forms are still ripples, however, due to the presence of a viscous sublayer (i.e., small Re_p).

IMPLICATIONS FOR BRINES ON MARS

The analysis here indicates that ripples could be formed from even the most viscous brine modeled by Tosca et al. (2011) under a wide range of sediment sizes, and that the ripples themselves could have heights ranging from a few to tens of centimeters, depending on the flow depth. This notwithstanding, it is very possible that brines might have existed on the Martian surface with substantially lower viscosities than the end-member case considered here. Although the end-member sulfate-dominated case, "brine 6b" of Tosca et al. (2011), is predicted to have a viscosity nearly 40 times that of freshwater at 0° C, its viscosity is significantly less at warmer temperatures (e.g., 1.6×10^{-5} m^2/s at 25° C or 16 times that of freshwater). Composition also influences viscosity. For example, "brine 5b" of Tosca et al. (2011), which is far less acidic than brine 6b, has a viscosity at equilibrium of 1.1×10^{-5} m^2/s at 0° C and 5.3×10^{-6} m^2/s at 25° C (i.e., 11-fold and 6-fold that of freshwater). For these lower-viscosity brines, the ripple stability field in Figure 7 would not extend as far into the coarse sediment sizes, and the predicted equilibrium ripple heights would be substantially less than those discussed previously herein.

The Burns formation at Meridiani Planum presents a case study in ancient subaqueous and eolian sediment transport on the Late Noachian surface of Mars. The evidence for subaqueous sediment transport is very limited, consisting only of thin, decimeter-thick beds of centimeter-scale trough cross-stratification that signify likely brief, shallow, low-discharge overland flows. Evidence for longer-term discharge or more vigorous flows has not been observed, nor has any facies indicative of standing bodies of water such as playas or other lacustrine environments. More recent sedimentological work (Hayes et al. 2011, Edgar et al. this volume) reinforces initial hypotheses formulated at early stages in the mission (Grotzinger et al. 2005, 2006; McLennan et al. 2005) that the Burns formation accumulated in an environment that was generally dry, and only rarely inundated by gentle flows of what must have been very high-concentration brines. This is consistent with our calculations that ripples could be generated by highly viscous brines under only modest flow conditions. For example, with a bed slope of 10^{-3}, a centimeter-scale-deep laminar flow with velocities of ~0.1 m/s would be capable of generating ripples in 1 mm sand (Fig. 8). Such a flow environment could have

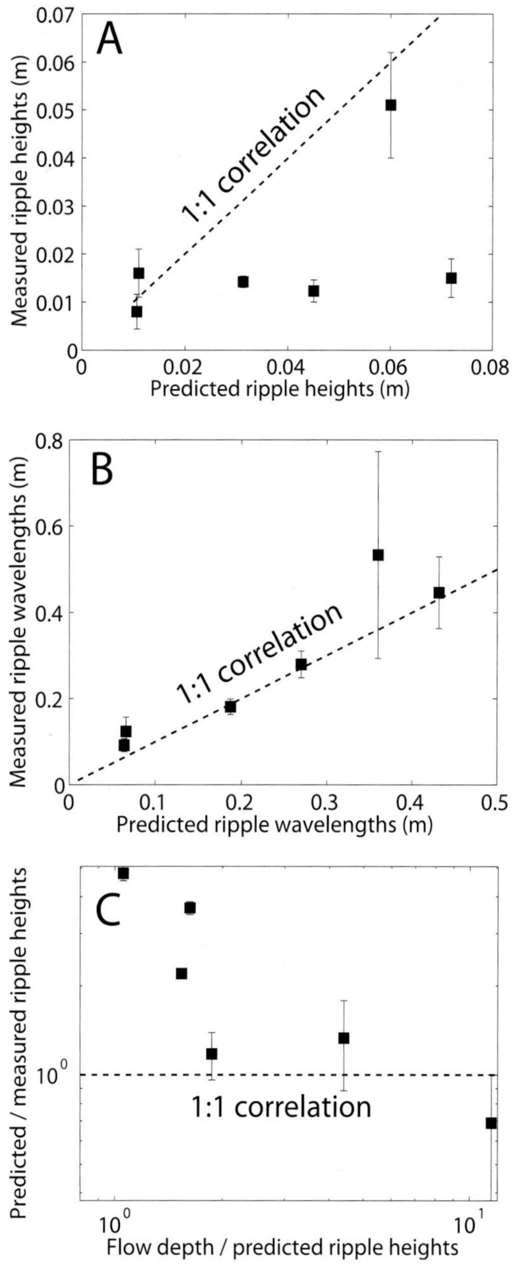

FIG. 9.—Predicted versus measured **A)** ripple heights and **B)** wavelengths. Data points are the average ripple dimensions reported by Grazer (1982), and the error bars represent ±one standard deviation. Ripple dimensions were predicted by assuming freshwater ripple heights of 1.5 cm and freshwater ripple wavelengths of 9 cm, and adjusting these values only by multiplying by the fluid viscosity scale factor (i.e., relative fluid viscosity raised to the two-thirds power). **C)** Predicted versus measured ripple heights from part A versus the measured flow depth normalized by the predicted ripple height. All dashed lines represent a perfect match between prediction and measurement.

FIG. 10.—Centimeter-scale ripples developed in a shallow flow that emerges from a large pile of sulfide-rich mine tailings at Rio Tinto, Spain. The groundwater is very acidic (pH 1–2) and sulfate rich, and it emerges in small channels that collect to form larger channels such as the one shown here. Ripples are formed of sand-size particles that are aggregates of weathered silicate minerals, plus Ca-Mg-Fe-Al sulfate minerals. This is almost identical to what is inferred for the composition of sand particles that compose the Burns formation (Clark et al. 2005, McLennan et al. 2005). The brines are close chemical analogs as well, with high concentrations of Fe^{2+}, Mg, and Al, but they are toward the dilute end of those calculated in Tosca et al. (2011) for Mars. The ripples have three-dimensional crest lines and well-developed scour pits that will create centimeter-scale cross-lamination during migration. Flow is from right to left, and the fully submerged channel in the foreground is about 0.75 m in width.

been created by one interdune depression filling with brine relative to an adjacent, lower depression, allowing the brine to move from the former to the latter. If the filling events were brief enough, sediment could be transported without the creation of a lacustrine facies per se.

Farther afield, substantial sulfate deposits are observed elsewhere on Mars, such as the Vallis Marineris and adjacent chasmata (Bibring et al. 2005, Murchie et al. 2009). Sulfates are also being discovered in other types of topographic depressions, such as large craters (Milliken et al. 2010). Most recently, sulfates are being seen as an important component of the oldest terrains on Mars (e.g. Nili Fossae, NE Syrtis [Ehlmann et al. 2009]; Mawrth Vallis [Wray et al. 2010]). Given that these sulfate minerals are hydrated, it is reasonable to assume that sediment transport at these sites may have—at least in part—been driven by subaqueous sediment transport. It is also possible that in some cases, these sequences were subjected to even greater volumes of flowing water than what is inferred for the ancient record at Meridiani. If so, then this also would have required the fluid to be a brine of high concentration since epsomite is inferred to be one of the most common minerals (Bibring et al. 2005, Murchie et al. 2009).

This leads to at least two possible scenarios for the creation of such a brine, and the way in which it could have led to deposition of sulfate-rich sediments. In the first case, dissolution of basalt under acidic conditions may have led to the creation of sulfate-rich pore fluids (Tosca et al. 2005) that discharged as briny groundwater in topographic depressions, perhaps along the lines of what has been suggested for Meridiani (Andrews-Hanna et al. 2007). Evaporation of the brine might have produced a series of evaporitic playas; flow between basins might have induced ripples and other sedimentary bed forms to form. In the second scenario, the sulfate-rich deposits were generated via some other mechanism, such as hydrothermal alteration. The deposits then interacted with freshwater, so that the deposit was leached and/or eroded in the topographically updip portions of the aquifer as the freshwater reached saturation with respect to the sulfate mineral phases. Farther downdip, the sulfate deposit would exist in equilibrium with the fluid, experiencing no net gain or loss of sulfate from solution. In the most downdip positions, the fluid may have emerged, creating surface flows of highly concentrated brines. A similar process occurs today at Rio Tinto, Spain, where emerging sulfate-rich groundwater precipitates to create sediment particles that are then transported by the flow to form centimeter-scale ripples (Fig. 10). These ripples are possible analogs for those inferred to have formed in the Burns formation.

CONCLUSIONS

Cross-stratification preserved in the Burns formation suggests formation by flowing aqueous fluids. Given the extremely high solubility of the minerals observed there, it seems likely that these flows must have been brines of high ionic strength, which necessarily have densities and viscosities that exceed that of freshwater. A compilation of previous studies on bed form stability and ripple formation under high-viscosity fluids indicates that ripples generated by flowing viscous brines on Mars would be stable over a wide range of bed shear stresses and particle sizes, including sand and gravel; i.e., sizes much larger than the stability limit for ripples in freshwater on Earth. It is likely that ripples might also form in silt sizes under viscous brines, but these conditions (i.e., particle Reynolds numbers less than 0.1) have not yet been explored in flume experiments, motivating future work. To generate the necessary bed stresses, the brines must have had flow depths ranging from 0.01 to 1 m and flow velocities of 0.01 to 1 m/s, and been driven by gravity on slopes of 10^{-4} to 10^{-2}. These conditions seem reasonable given the interdune environment proposed for the Burns formation. Ripples formed by viscous brines might be larger than their freshwater counterparts, potentially by as much as a factor of twelve for the highest viscosity case considered. Thus, large (>10 cm), fine-grained (<1 mm) cross strata of subaqueous origin would be compelling physical evidence for flowing brines in the Martian past. This notwithstanding, smaller ripples might also be produced by flowing brines if ripple growth was limited by the flow depth, or if equilibrium brine viscosity was not as high as the end-member brine considered. For these cases, the physical sedimentological evidence in support of brines versus freshwater flows may be ambiguous. Reconstruction of small ripples in the Burns formation supports these inferences.

ACKNOWLEDGMENTS

We thank two reviewers whose comments added clarity to the final manuscript and Ralph Milliken for serving as editor. J.P.G. acknowledges support from the NASA Astrobiology Institute and the Mars Exploration Rover Project. M.P.L. acknowledges the Donors of the American Chemical Society Petroleum Research Fund for partial support of this research.

REFERENCES

Alkattan M. 1997. Experimental studies of halite dissolution kinetics: 1. The effect of saturation state and the presence of trace metals. *Chemical Geology* 137(3–4):201–219.

Andrews-Hanna JC, Philips RJ, Zuber MT. 2007. Meridiani Planum and the global hydrology of Mars. *Nature* 443:163–166.

Baas JH. 1999. An empirical model for the development and equilibrium morphology of current ripples in fine sand. *Sedimentology* 46:123–138.

Benison KC, LaClair D, Walker J. 2008. Physical sedimentology experiments with sulfuric acid solutions: Implications for Mars?: *Earth and Planetary Science Letters* 270:330–337.

Best JL, Leeder M. 1993. Drag reduction in turbulent muddy seawater flows and some sedimentary consequences. *Sedimentology* 40:1129–1137.

Betat A, Kruelle CA, Frette V, Rehberg I. 2002. Long-time behavior of sand ripples induced by water shear flow. *European Physical Journal E* 8:465–476.

Bibring J-P, Langevin Y, Gendrin A, Gondet B, Poulet F, Berthe M, Soufflot A, Arvidson R, Mangold N, Mustard J, Drossart P, Team ATO. 2005. Mars surface diversity as revealed by the *OMEGA/Mars Express* observations. *Science* 307:1576–1581.

Boguchwal LA, Southard JB. 1990. Bed configurations in steady unidirectional water flows: 1. Scale-model study using fine sands. *Journal of Sedimentary Petrology* 60:649–657.

Brownlie WR. 1983. Flow depth in sand-bed channels. *Journal of Hydraulic Engineering* 109:959–990.

Chabert J, Chauvin JL. 1963. Formation de dunes et de rides dans les modeles fluviaux. *Bulletin du Centre de Recherche et d'Essais de Chatou* 4:31–51.

Chow VT. 1959. *Open Channel Hydraulics*: McGraw Hill, New York. 680 p.

Clark BC, Morris RV, McLennan SM, Gellert R, Jolliff B, Knoll AH, Squyres SW, Lowenstein TK, Ming DW, Tosca NJ, Yen A, Christensen PR, Gorevan S, Bruckner J, Calvin W, Dreibus G, Farrand W, Klingelhoefer G, Waenke H, Zipfel J, Bell J.F, Grotzinger J, McSween HY, Rieder R. 2005. Chemistry and mineralogy of outcrops at Meridiani Planum. *Earth and Planetary Science Letters* 240:73–94.

Coleman SE, Eling B. 2000. Sand wavelets in laminar open-channel flows. *Journal of Hydraulic Research* 38:331–338.

Devauchelle O, Malverti L, Lajeunesse E, Lagree PY, Josserand C, Thu-Lam KDN. 2010. Stability of bedforms in laminar flows with free surface: From bars to ripples. *Journal of Fluid Mechanics* 642:329–348.

Dzulynski S, Sanders JE. 1962. Current marks on frim mud bottoms: *Transactions of the Connecticut Academy of Arts and Sciences* 42:57–96.

Ehlmann BL, Mustard JF, Swayze GA, Clark RN, Bishop JL, Poulet F, Des Marais D, Roach LH, Milliken RE, Wray J, Barnouin-Jha O, Murchie SL. 2009. Identification of hydrated silicate minerals on Mars using MRO-CRISM: Geologic context near Nili Fossae and implications for aqueous alteration. *Journal of Geophsical Research* 114, doi:10.1029/2009JE003339.

Garcia MH. 2007. *Sedimentation Engineering: Process, Measurement, Modeling, and Practice*: American Society of Civil Engineers, Reston, Virginia, 1132 p.

Grazer RA. 1982. Experimental study of current ripples using medium silt [unpublished MS thesis]: Massechusetts Institute of Technology, Cambridge, Massachusetts, 131 p.

Grotzinger JP, Arvidson RE, Bell JF, Calvin W, Clark BC, Fike DA, Golombek M, Greeley R, Haldemann A, Herkenhoff KE, Jolliff BL, Knoll AH, Malin M, McLennan SM, Parker T, Soderblom L, Sohl-Dickstein JN, Squyres SW, Tosca NJ, Watters WA. 2005. Stratigraphy and sedimentology of a dry to wet aeolian depositional system, Burns formation, Meridiani Planum, Mars. *Earth and Planetary Science Letters* 240:11–72.

Grotzinger JP, Bell J, Herkenhoff K, Johnson J, Knoll A, McCartney E, McLennan S, Metz J, Moore J, Squyres S, Sullivan R, Aharonson A, Arvidson R, Joliff B, Golombek M, Lewis K, Parker T, Soderblom J. 2006. Sedimentary textures formed by aqueous processes, Erebus crater, Meridiani Planum, Mars. *Geology* 34:1085–1088.

Hayes AG, Grotzinger JP, Edgar LA, Squyres SW, Watters WA, Sohl-Dickstein J, Athena Science Team. 2011. Reconstruction of eolian beforms and paleocurrents from cross-bedded strata at Victoria crater, Meridiani Planum, Mars. *Journal of Geophysical Research* 116, doi:10.1029/2010JE003688.

Kerr RC. 1995. Convective crystal dissolution. *Contributions to Mineralogy and Petrology* 121:237–246.

Komar PD. 1979. Comparisons of the hydraulics of water flows in Martian outflow chanels with flows of similar scale on Earth. *Icarus* 37:156–181.

Lajeunesse E, Malverti L, Lancien P, Armstrong L, Metivier F, Coleman S, Smith CE, Davies T, Cantelli A, Parker G. 2010. Fluvial and submarine morphodynamics of laminar and near-laminar flows: A synthesis. *Sedimentology* 57:1–26.

Lasaga AC. 1998. *Kinetic Theory in the Earth Sciences*: Princeton University Press, Princeton, New Jersey, 811 p.

McLennan SM, Bell JF, Calvin WM, Christensen PR, Clark BC, de Souza PA, Farmer J, Farrand WH, Fike DA, Gellert R, Ghosh A, Glotch TD, Grotzinger JP, Hahn B, Herkenhoff KE, Hurowitz JA, Johnson JR, Johnson SS, Jolliff B, Klingelhofer G, Knoll AH, Learner Z, Malin MC, McSween HY, Pocock J, Ruff SW, Soderblom LA, Squyres SW, Tosca NJ, Watters WA, Wyatt MB, Yen A. 2005. Provenance and diagenesis of the evaporite-bearing Burns formation, Meridiani Planum, Mars. *Earth and Planetary Science Letters* 240:95–121.

McLennan SM, Grotzinger JP. 2008. The sedimentary rock cycle of Mars. *In* Bell, J (Editor), *The Martian Surface: Composition, Mineralogy, and Physical Properties*. Cambridge University Press, Cambridge, UK. p. 541–577.

Mehta AJ. 1991. Understanding fluid mud in a dynamic environment. *Geo-Marine Letters* 11:113–118.

Metz JM, Grotzinger JP, Rubin DM, Lewis KW, Squyres SW, Bell JF. 2009. Sulfate-rich eolian and wet interdune deposits, Erebus crater, Meridiani Planum, Mars. *Journal of Sedimentary Research* 79:247–264.

Milliken RE, Grotzinger JP, Thomson BJ. 2010. Paleoclimate of Mars as captured by the stratigraphic record of Gale crater. *Geophysical Research Letters* 37, doi:10.1029/2009GL041870.

Murchie SL, Mustard JF, Ehlmann BL, Milliken RE, Bishop JL, McKeown NK, Dobrea EZN, Seelos FP, Buczkowski DL, Wiseman SM, Arvidson RE, Wray JJ, Swayze G, Clark RN, Marais DJD, McEwen AS, Bibring J.-P. 2009. A synthesis of Martian aqueous mineralogy after 1 Mars year of observations from the Mars Reconnaissance Orbiter. *Journal of Geophysical Research* 114, doi: 10.1029/2009JE003342

Parker G. 1991. Selective sorting and abrasion of river gravel. II: Applications. *Journal of Hydraulic Engineering* 117:150–171.

Rubin DM. 1987. Cross-Bedding, Bedforms, and Paleocurrents: Society for Sedimentary Geology, Tulsa, Oklahoma. SEPM, Concepts in Sedimentology and Paleontology 1, 187 p.

Rubin DM, Hunter RE. 1982. Bedform climbing in theory and nature. *Sedimentology* 29:121–138.

Shields A. 1936. Application of Similarity Principles and Turbulence Research to Bed-Load Movement: California Institute of Technology, U.S. Dept of Agriculture Soil Conservation Service Cooperative Laboratory, Pasadena, California. Hydrodynamics Laboratory Publication 167, 47 p.

Simon B. 1981. Dissolution rates of NaCl and KCl in aqueous solutions. *Journal of Crystal Growth* 52:789–794.

Southard JB. 1991. Experimental determination of bed-form stability. *Annual Review of Earth and Planetary Sciences* 19:423–455.

Southard JB, Boguchwal LA. 1973. Flume experiments on the transition from ripples to lower flat bed with increasing sand size. *Journal of Sedimentary Research* 43:1114–1121.

Southard JB, Boguchwal LA. 1990a. Bed configurations in steady unidirectional water flows: 2. Synthesis of flume data. *Journal of Sedimentary Petrology* 60:658–679.

Southard JB, Boguchwal LA. 1990b. Bed configurations in steady unidirectional water flows: 3. Effects of temperature and gravity. *Journal of Sedimentary Petrology* 60:680–686.

Squyres SW, Grotzinger JP, Arvidson RE, Bell JF, Calvin W, Christensen PR, Clark BC, Crisp JA, Farrand WH, Herkenhoff KE, Johnson JR, Klingelhofer G, Knoll AH, McLennan SM, McSween HY, Morris RV, Rice JW, Rieder R, Soderblom LA. 2004. In situ evidence for an ancient aqueous environment at Meridiani Planum, Mars. *Science* 306:1709–1714.

Squyres SW, Knoll AH, Arvidson RE, Ashley JW, Bell JF III, Calvin WM, Christensen PR, Clark BC, Cohen BA, de Souza PA Jr, Edgar L, Farrand WH, Fleischer I, Gellert R, Golombek MP, Grant J, Grotzinger J, Hayes A, Herkenhoff KE, Johnson JR, Jolliff B, Klingelhöfer G, Knudson A, Li R, McCoy TJ, McLennan SM, Ming DW, Mittlefehldt DW, Morris RV, Rice JW Jr, Schröder C, Sullivan RJ, Yen A, Yingst RA. 2009. Exploration of Victoria Crater by the rover *Opportunity. Science* 324, doi: 10.1126/science.1170355.

Tosca NJ, McLennan SM, Clark BC, Grotzinger JP, Hurowitz JA, Knoll AH, Schroder C, Squyres SW. 2005. Geochemical modeling of evaporation processes on Mars: Insight from the sedimentary record at Meridiani Planum. *Earth and Planetary Science Letters* 240:122–148.

Tosca NJ, McLennan SM, Lamb MP, Grotzinger JP. 2011. Physico-chemical properties of concentrated Martian surface waters. *Journal of Geophysical Research—Planets* 116, doi:10.1029/2010JE003700.

van den Berg JH, van Gelder A. 1993. A new bedform stability diagram, with emphasis on the transition of ripples to plane bed in flows over fine sand and silt. *In* Marzo M, Puigdefábregas C (Editors), *Alluvial Sedimentation*: International Association of Sedimentologists, Special Publication 17, Blackwell Scientific Publications, Oxford, UK. p. 11–21.

Vanoni VA. 1974. Factors determining bed form of alluvial streams. *Journal of the Hydraulics Division* 100:363–378.

Van Rijn LC. 1984. Sediment transport: Part III. Bed forms and alluvial roughness. *Journal of Hydraulic Engineering* 110:1613–1641.

Velbel MA. 2004. Laboratory and homework exercises in the geochemical kinetics of mineral-water reaction: Rate law, Arrhenius activation energy, and the rate-determining step in the dissolution of halite. *Journal of Geological Education* 52:52–59.

White SJ. 1970. Plane bed thresholds of fine grained sediments. *Nature* 228:152–153.

Wray JJ, Squyres SW, Roach LH, Bishop JL, Mustard JF, Noe Dobrea EZ. 2010. Identification of the Ca-sulfate bassanite in Mawrth Vallis, Mars. *Icarus* 209:416–421, doi:10.1016/j.icarus.2010.06.001.

Yalin MS. 1977. *Mechanics of Sediment Transport*: Pergamon Press, Oxford, UK. 298 p.

SOURCE-TO-SINK: AN EARTH/MARS COMPARISON OF BOUNDARY CONDITIONS FOR EOLIAN DUNE SYSTEMS

GARY KOCUREK

Department of Geological Sciences, University of Texas, Austin, Texas 78712 USA
e-mail: garyk@mail.utexas.edu

AND

RYAN C. EWING

Division of Geological & Planetary Sciences, California Institute of Technology, MC 170–25, Pasadena, California 91125 USA
**Present address: University of Alabama, Department of Geological Sciences, Tuscaloosa, Alabama 35487 USA*

ABSTRACT: Eolian dune fields on Earth and Mars evolve as complex systems within a set of boundary conditions. A source-to-sink comparison indicates that although differences exist in sediment production and transport, the systems largely converge at the dune-flow and pattern-development levels, but again differ in modes of accumulation and preservation. On Earth, where winds frequently exceed threshold speeds, dune fields are sourced primarily through deflation of subaqueous deposits as these sediments become available for transport. Limited weathering, widespread permafrost, and the low-density atmosphere on Mars imply that sediment production, sediment availability, and sand-transporting winds are all episodic. Possible sediment sources include relict sediments from the wetter Noachian; slow physical weathering in a cold, water-limited environment; and episodic sediment production associated with climatic cycles, outflow events, and impacts. Similarities in dune morphology, secondary airflow patterns over the dunes, and pattern evolution through dune interactions imply that dune stratification and bounding surfaces on Mars are comparable to those on Earth, an observation supported by outcrops of the Burns formation. The accumulation of eolian deposits occurs on Earth through the dynamics of dry, wet, and stabilizing eolian systems. Dry-system accumulation by flow deceleration into topographic basins has occurred throughout Martian history, whereas wet-system accumulation with a rising capillary fringe is restricted to Noachian times. The greatest difference in accumulation occurs with stabilizing systems, as manifested by the north polar Planum Boreum cavi unit, where accumulation has occurred through stabilization by permafrost development. Preservation of eolian accumulations on Earth typically occurs by sediment burial within subsiding basins or a relative rise of the water table or sea level. Preservation on Mars, measured as the generation of a stratigraphic record and not time, has an Earth analog with infill of impact-created and other basins, but differs with the cavi unit, where preservation is by burial beneath layered ice with a climatic driver.

KEY WORDS: Eolian, Earth, Mars, complex systems, boundary conditions

INTRODUCTION

The growing body of high-resolution images from satellites and surface missions shows that the surface of Mars is a cold, hyperarid desert and that eolian processes have played a dominant role in development of its geomorphic surface and stratigraphic record. Because eolian and cryogenic processes on Mars are commonly interrelated, the closest Earth analog is the McMurdo Dry Valleys of Antarctica, and significant insights have been gained from studies of eolian (Bourke et al. 2009; Bristow et al. 2010a, 2010b), permafrost (e.g., Marchant et al. 2002, Levy et al. 2006), weathering (e.g., Gibson et al. 1983, Marchant and Head 2007), and other dynamic systems in Antarctica. There are, however, important differences between Antarctica and the environmentally far more extreme Mars, and limits exist for analogs per se, because no two environments, separated by time or space, are ever exactly the same on Earth, and this is even less so for planetary comparisons.

A more conceptual approach to understanding eolian systems on Mars, one that does not use Earth analogs, is to consider these as source-to-sink complex systems operating within a set of boundary conditions. Complex systems are typically dissipative systems involving a large number of elements governed by nonlinear dynamics, which operate within a set of external environmental variables or boundary conditions. The boundary conditions within which the system evolves affect the geomorphic expression and stratigraphic record of the system over a wide range of spatial and temporal scales (Werner 2003). Because no two systems ever share exactly the same boundary conditions, no two systems are ever exactly the same. Based on this thinking, each unique solution is a case study, but broad comparisons can be made based upon planetary-scale boundary conditions. Conversely, evolving environmental conditions on Mars can be reconstructed from the geomorphic and stratigraphic record of eolian systems. Because of the dominance of eolian processes on Mars, the eolian record may be the primary vehicle for interpreting planetary history, and, conceivably, planetary resurfacing may well be driven by eolian cycles.

This article explores the geomorphic and stratigraphic nature of eolian dune systems on Mars by way of a source-to-sink comparison of Earth/Mars boundary conditions (Fig. 1). The suspended (i.e., dust) component of the Martian eolian transport system is not considered. The overall conclusion is that although sediment sourcing and grain transport on Mars and Earth are significantly different, these planetary eolian systems largely converge at the dune and dune-field pattern levels, but again differ in modes of accumulation and preservation. The primary product of this article is a body of questions that arise from the Earth/Mars comparison, questions that may help to focus future work. In addition, the comparison of eolian systems evolving within planetary environments as different as Earth and Mars elucidates the dynamics of eolian systems themselves.

SOURCING EOLIAN SYSTEMS

Earth

The boundary conditions on Earth that yield sediment for eolian dunes include abundant water, dynamic atmospheric transfer of water

Sedimentary Geology of Mars
SEPM Special Publication No. 102, Copyright © 2012
SEPM (Society for Sedimentary Geology), Print ISBN 978-1-56576-312-8, CD/DVD ISBN 978-1-56576-313-5, p. 151–168.

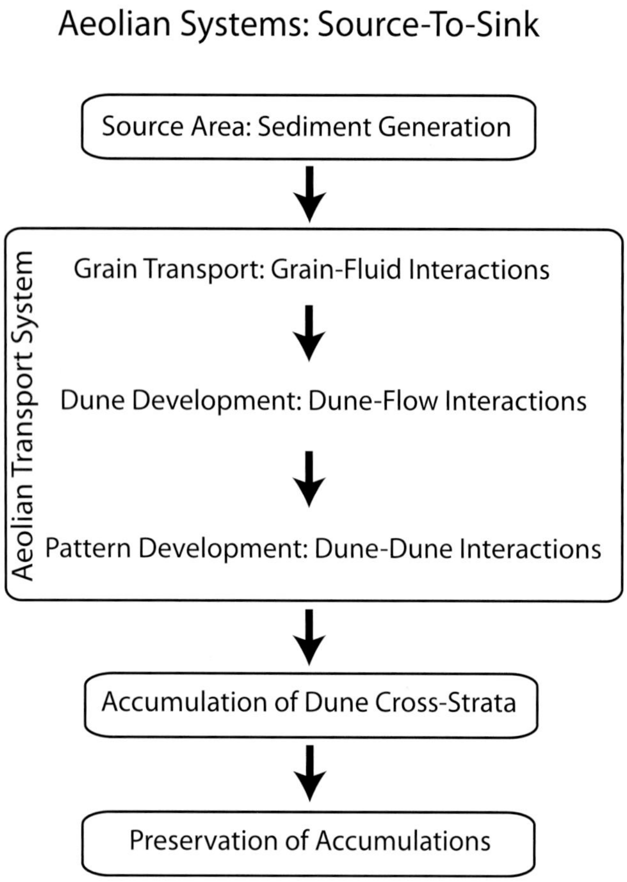

FIG. 1.—Flow diagram for source-to-sink eolian dune systems. At each stage of the source-to-sink route, differing sets of boundary conditions affect the system dynamics.

from oceans to continents, and warm temperatures, such that C-cycle chemical weathering, subaqueous erosion, and transport by fluvial systems dominate the surface. Because of the dominant role of water, most eolian systems are sourced directly or indirectly (via coastal processes) by fluvial systems (see discussion in Lancaster [1995; p. 88–90]). Eolian deflation primarily becomes significant once sources of sand have been emplaced by subaqueous processes. A minority of dune fields have been sourced directly by wind erosion of poorly consolidated sedimentary strata, but few (if any) dune fields have arisen solely through wind deflation of bedrock.

Weathering, erosion, and fluvial transport operate within a larger, slower-paced set of boundary conditions of tectonism, sea level, and climate (Fig. 2A). Changes in these boundary conditions cause temporal and spatial changes in source areas and dune fields. Because of these changes and the internal dynamics of dune fields, the antiquity of deserts on Earth is measured in 10^2 to 10^7 years, specific dune fields exist for 10^1 to 10^5 years, and individual dunes persist for $<10^0$ to 10^5 years. The Namib (Ward et al. 1983) and the Atacama (Alpers and Brimhall 1988) are probably the oldest deserts on Earth, dating to at least the Miocene. Many major dune fields date to only the Last Glacial Maximum, which was a time of major resurfacing in the eolian realm (e.g., Kocurek 1999). The oldest dated dunes are linear dunes in Australia (~200 ka), although numerous cycles of activation and stabilization are evident internally (Nanson et al. 1992).

The temporal relationship between sediment source and dune-field construction can be conceptualized by the "sediment state" of the system (Kocurek and Lancaster 1999), and this relationship is described as one of the three following types: (1) contemporaneous, (2) lagged in time, or (3) a mix of contemporaneous and lagged sediment influx (Fig. 3A–C). For example, throughout its long history, the Namib has been sourced contemporaneously by the Orange River via coastal processes (Lancaster and Ollier 1983, Bluck et al. 2007). White Sands in New Mexico has been sourced by a mix of lagged sediment influx that occurred with deflation of lake-bed accumulations during regressions and contemporaneous influx from shorelines and exposed playa flats (Langford 2003). The Sahara has arguably received a lagged fluvial source at two scales. Quaternary Milankovitch cycles cause humid periods during which fluvial sediments are stored; during subsequent arid periods these stored sediments are reworked and source dune fields as lagged influx. The original major sourcing of the Sahara, however, probably lay with Pliocene reworking of extensive Miocene fluvial deposits, which were emplaced during the coincidence of active tectonism and the last major humid period of the region (Kocurek 1998).

Two other boundary conditions complete the sediment state of any eolian system: (1) the transport capacity of the wind and (2) the volume of sediment available on the surface for eolian transport (Fig. 3A–C). On Earth, the frequency of sand-transporting winds is high, and it is only the magnitude of the wind that varies spatially over the globe and temporally through climatic cycles (e.g., Milankovitch). Sediment availability is most commonly limited by vegetation, but a near-surface water table, surface cementation, and surface armoring by coarse grains are also common factors. Because of the high frequency of sand-transporting winds, once sediment is made available for transport by, for example, the onset of aridity or marine regression, if the sediment source is not regularly replenished it is generally exhausted rapidly, and dune-field constructional events are relatively short-lived (Kocurek and Lancaster 1999).

Mars

Significant volumes of sediment have been generated on Mars, as evident by (1) eolian, fluvial, and other depositional features on the surface, (2) the presence of a sedimentary stratigraphic record, and (3) the fact that there is little exposed bedrock on Mars (Edwards et al. 2009). The emerging image of sediment generation on Mars, however, indicates very different boundary conditions than prevailed on Earth during the Phanerozoic. The common occurrence of very ancient (i.e., Archean equivalent) rock on the surface of Mars in itself attests to very slow weathering rates. The potential for and nature of sediment generation on Mars, however, has evolved. There is significant evidence that during the Noachian (~>3.5 b.y.) Mars was wetter, and possibly warmer, with a denser atmosphere. Global change occurred during the late Noachian toward an increasingly cold, desiccated planet during the Hesperian and Amazonian. Whereas spot estimations of Noachian weathering rates are comparable to slow denudation rates on Earth, estimated Hesperian/Amazonian rates are orders of magnitude lower (Golombek and Bridges 2000, Golombek et al. 2006).

Satellite (e.g., Edgett and Malin 2000, Malin and Edgett 2006, Bridges et al. 2007, Bandfield and Feldman 2008) and surface (e.g., Greeley et al. 2000, 2004; Bell et al. 2004; Levy et al. 2009) images clearly show evidence for physical weathering driven by eolian abrasion, impact shattering, thermal stress, permafrost processes, gravity-driven mass wasting, and salt weathering. The relative roles of the major processes of physical weathering through time can be speculated upon with planetary evolution (Fig. 2B).

Chemical weathering has occurred overall within a cold, water-limited, low-pH S-cycle (Hurowitz and McLennan 2007, McLennan and Grotzinger 2008), during which chemical reactions were incomplete (Madden et al. 2004, Tosca and Knoll 2009) and perhaps

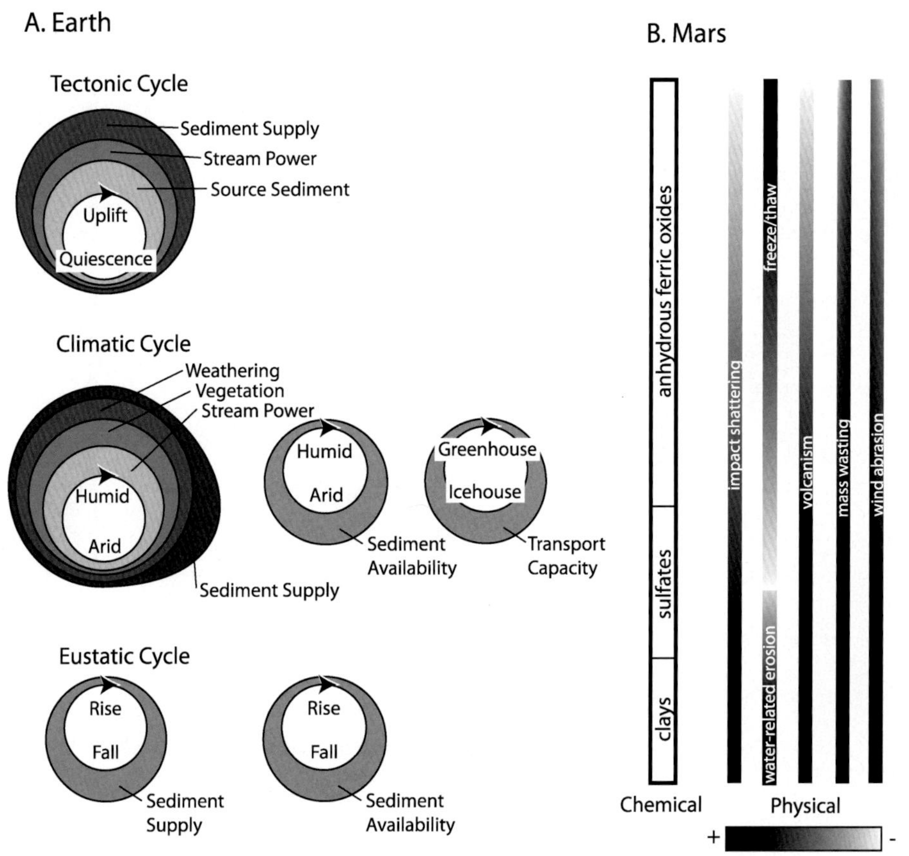

FIG. 2.—Sources of eolian sediment. (A) Earth, where rates of sediment generation vary with tectonic, climatic, and eustatic cycles. Tectonic uplift creates enhanced source-strata exposure and stream power for erosion to yield greater sediment supply. During a climatic cycle, greatest sediment supply typically occurs with the transition from humid to arid (Langbein and Schumm 1958), whereas sediment availability is greatest with aridity, and wind transport capacity is greatest during Icehouse stages. Both sediment supply and availability increase with lowered relative sea level because of enhanced erosion with lowered base level and increased shelf exposure, respectively. Modified from Kocurek (1999). (B) Mars showing eras of chemical weathering from Bibring et al. (2006) and postulated variations in types of physical weathering. Impact shattering, water processes, and volcanism may have dominated Noachian physical sediment production, whereas permafrost, Eolian, and mass-wasting processes probably dominate during the Amazonian.

short-lived episodes terminated by freezing and/or evaporation (Zolotov and Mironenko 2007). The most comprehensive interpretation of the general evolution of chemical weathering on Mars comes from global mineralogical mapping in which (1) a Noachian "phyllosian era" characterized by aqueous alteration and clay-mineral generation, yields to (2) a Hesperian "theiikian era" characterized by sulfate generation and (3) a late Hesperian–Amazonian "siderikian era" characterized by the formation of anhydrous ferric oxides in slow, superficial weathering without liquid water (Fig. 2B; Bibring et al. 2006).

Combining the potential for sediment generation from both physical and chemical weathering over time on Mars, three distinct eolian sediment states can be postulated (Fig. 3D–F). First, following an Earth analog, the primary episode of dune-field construction would have been contemporaneously or somewhat lagged in time with Noachian enhanced sediment generation (Fig. 3D). In this scenario, dune fields would have been sourced by deflation of fluvial, deltaic, and shoreline deposits, as has occurred during Earth history. Reworking of these original dune fields would have continued, but few additional dune fields would have been emplaced once Noachian sediment sources were exhausted. Given the abundance of dune fields on Amazonian terrain, however, this sediment-state scenario alone cannot account for the Martian eolian record.

In a sediment state that departs significantly from an Earth analog, eolian dune-field construction has occurred throughout Martian history, including those dune fields associated with Noachian events, but continuing through the Amazonian via sediment generated by very slow weathering processes (Fig. 3E). Unlike on Earth, where sediment is removed from the terrestrial surface through tectonism, burial, or transport to ocean basins, this "slow-cumulative" model allows for the aggregate of billions of years of slow weathering to remain on the surface. In this scenario, new dune fields would have arisen where sufficient sediment collected on the surface, albeit over protracted time intervals. This sediment state also falls short in explaining the occurrence of eolian dune fields because the post-Noachian Martian surface appears to have been far more dynamic, especially in response to specific events.

The sediment state that best incorporates enhanced Noachian sediment production and subsequent slow "background" weathering, but also envisions significant periods of dune-field generation on post-Noachian Mars, is an "episodic" model in which sediment generation, wind transport, and sediment availability are all episodic on a variety of

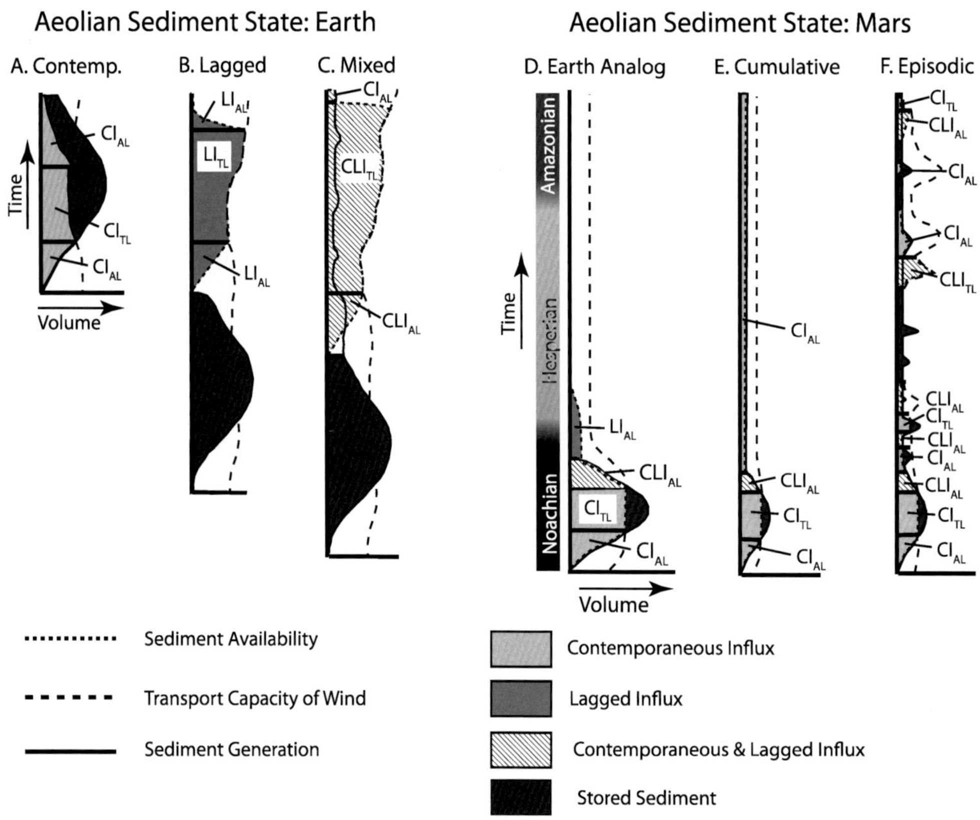

FIG. 3.—Sediment state diagrams for Earth (A–C) and Mars (D–F), in which sediment generation, transport capacity of the wind, and sediment availability (all in volumes) are simultaneously plotted against time to yield all possible dune-field configurations. (A) Dune field sourced by contemporaneous sediment influx (CI), which is limited by sediment availability (CI_{AL}) or the transport capacity of the wind (CI_{TL}). (B) Dune field sourced from previously stored sediment (lagged influx, LI). Dune-field construction may be availability-limited (LI_{AL}) or transport-limited (LI_{TL}). (C) Dune field sourced from both previously stored sediment and contemporaneous influx (CLI) and may be availability-limited (CLI_{AL}) or transport-limited (CLI_{TL}). (D) Earth analog for Mars, in which dune fields are contemporaneous or somewhat lagged in time with enhanced Noachian sediment generation. (E) Slow cumulative model, in which Noachian dune-field construction is followed by continued, slow dune-field construction, which is strongly availability-limited owing to the slow rate of sediment production. (F) Episodic model for dune-field construction incorporating Noachian construction and subsequent slow sediment production, but this model postulates episodic enhanced sediment production and periodic sediment availability and sand-transporting wind events. The results are episodes of dune-field construction of various types. See Kocurek and Lancaster (1999) for additional details on diagram construction and term definitions.

timescales that are not necessarily correlated (Fig. 3F). For example, sediment generation for intercrater dune fields is initiated by short-lived, local impact events (i.e., Fenton 2005). On a longer timescale, the eolian Planum Boreum cavi unit may have arisen through reworking of older Planum Boreum units (Rupes Tenuis and Scandia unit), which were emplaced as distal deposits of outflow channels in which large volumes of sediment were deposited on the northern plains during the Hesperian (Tanaka et al. 2010). Unlike Earth, where sand-transporting winds are taken as a "given" (Fig. 3A–C), in the Martian "episodic" model sand-transporting winds represent episodic events, as gauged by the infrequency of sand-transporting winds. Similarly, sediment availability may be episodic because during much of Martian history frost (both H_2O and CO_2) and permafrost may have limited the ability of the wind to entrain sediment (c.f., Schatz et al. 2006, Schorghofer and Edgett 2006, Bourke et al. 2008, Feldman et al. 2008, Fenton and Hayward 2010). Where ice is a control on sediment availability, the potential for grain entrainment may vary by seasons or Milankovitch cycles. Dune-field construction is episodic because it relies upon the favorable coincidence of all three components of sediment state.

In another significant departure from Earth, planetary resurfacing on Mars probably includes direct eolian deflation of indurated dunes and sedimentary strata. Earth-style cementation (i.e., lithification), which typically occurs in the subsurface with the passage of large volumes of pore waters, must be very restricted on Mars because of the water-limited state. Rather, Martian strata are probably poorly lithified, and ice/frost cementation is widespread (e.g., Schatz et al. 2006, Bandfield and Feldman 2008, Bourke et al. 2008, Byrne et al. 2009). Given ice cementation, sediment generation (recycling) would occur with sublimation driven by climatic cycles at a variety of temporal scales (i.e., daily to Milankovitch). Probable examples include the deflation of indurated dunes (e.g., Edgett and Malin 2000) and, in particular, the wholesale reworking of the Planum Boreum cavi unit in the North Polar Region to source the Olympia Undae Dune Field (Byrne and Murray 2002, Tanaka et al. 2008).

Questions

Weathering, erosion, and sediment production are very limited on the current Martian surface, yet large volumes of sediment have been generated on Mars. The fundamental questions are (1) by what processes were these sediments generated, and (2) how have these processes varied over time? What is the relative balance between physical and chemical weathering in sediment production? What proportion of eolian sediments can be accredited to Noachian vs. later Hesperian and Amazonian processes? Is sediment abundant on the Martian surface simply because it is rarely removed? At what rates and by what processes are indurated dunes and sedimentary strata recycled on the planet?

GRAIN TRANSPORT (GRAIN–FLUID INTERACTIONS)

Earth

Eolian sediment-transport systems on Earth develop wherever there is a sediment source, this sediment is available for entrainment by the wind, and wind speeds occur that are capable of sand transport. The friction speed (u_*) necessary to initiate movement (i.e., static or fluid threshold) is only ~0.15 to 0.25 m/s for quartz grains 0.1 to 0.3 mm in diameter, which form the bulk of sand dunes on Earth (Bagnold 1941; p. 88). Using grain roughness only, wind speeds required to initiate grain movement are only 4.7 to 7.2 m/s at 1 m above the surface (i.e., a 10–16-mph breeze). For continued saltation (i.e., dynamic or impact threshold), u_* falls to 0.12 to 0.20 m/s, or a wind speed of 2.8 to 4.0 m/s at 1 m above the surface. Most dune sand moves by saltation, with the coarser fraction traveling in surface creep via impacts from saltating grains.

Sand transport occurs as blowing sand, or as ripples, and then as dunes with growing surface availability of sand. Dunes, therefore, fundamentally represent packages of sand in transport, and a dune-field constructional event requires only sufficient available sand and transporting winds. Measured dune migration rates and interpreted rates from the rock record based upon annual cycles (e.g., Hunter and Rubin 1983) range from less than a meter to a few meters per year for large dunes to tens of meters per year for small dunes. Because the volume of sand transported by the wind increases as a cubic function of friction speed, significant dune activity occurs during less-common high-energy events.

Wilson (1973) was first to recognize that the eolian sediment-transport system on Earth could be desert-wide, with major dune fields typically forming within areas of wind deceleration, such as in topographic basins, where the flow expands vertically. Because of the potential long-distance transport of sand, it is commonly difficult to identify the source of eolian dune fields in even modern systems.

Mars

Compared to Earth, the basic boundary conditions on Mars that govern eolian transport are a lower surface gravity (3.7 m/s^2 on Mars vs. 9.8 m/s^2 on Earth) and a much lower atmospheric density (0.02 kg/m^3 on Mars vs. 1.23 kg/m^3 on Earth). Combined, these boundary conditions indicate that the initiation of sand transport by the wind is much more difficult than on Earth, but a continuation of transport occurs more readily. The most common grain mineralogy on Mars is basalt, which is somewhat denser than quartz, the most common eolian grain on Earth (3.0 g/cm^3 for basalt vs. 2.65 g/cm^3 for quartz). Thermal inertia data from space-borne imagery of Mars indicates that Martian dunes have an average grain size that is coarser (medium- to coarse-grained sand) than on Earth (Edgett and Christensen 1991, Presley and Christensen 1997). Limited surface observations (e.g., Gusev Crater by Greeley et al. [2006] and Sullivan et al. [2008]; Meridiani Planum by Jerolmack et al. [2006]) indicate eolian sands that are similar in size or slightly finer than those on Earth.

Wind tunnel experiments (e.g., Greeley et al. 1980, Iverson and White 1982) indicate that the threshold (i.e., fluid or static) friction speed to initiate sand movement on Mars is about an order of magnitude greater than that on Earth (i.e., 2.2 m/s on Mars vs. 0.22 m/s on Earth for 0.2-mm grains in the Greeley et al. experiments). This is a direct result of the low atmospheric density, which reduces the fluid drag on grains. Using relationships derived from Iverson and White (1982), Jerolmack et al. (2006) determined a u_* of 2.5 to 3.5 m/s for formation of wind ripples traversed by *Opportunity* on Meridiani Planum. These ripples consist of very coarse (1.3-mm), hematitic (ρ_s = 4.1 g/cm^3) sand crests and fine-grained (~0.05–0.2-mm) basaltic sand interiors. Accounting for surface roughness, formative winds at 1 m above this surface ranged from 49 to 70 m/s, or Earth hurricane category 3 to 4 on the Saffir–Simpson scale.

The lower gravity on Mars and a lower vertical fluid drag (caused by the lesser atmospheric density), however, cause a much larger hysteresis on Mars vs. Earth between the threshold and dynamic friction speeds (Claudin and Andreotti 2006, Kok 2010). Once initiated, saltation may continue at wind speeds an order of magnitude less than required to initiate it (Kok 2010), and transport itself has been suggested to be more energetic than on Earth (Almeida et al. 2008).

Given the potential for eolian transport, the frequency of transport and the conditions required for transport to occur are not clear. The Martian atmosphere is clearly dynamic, dust storms are common (e.g., Cantor et al. 2006), small dunes appear to have been deflated (Bourke et al. 2008), sand deposits have been modified at Gusev Crater (Sullivan et al. 2008), complex wind regimes are evident (e.g., Tsoar et al. 1979, Greeley et al. 2000, Ewing et al. 2010, Silvestro et al. 2010a), and sand has accumulated on the decks of the rover Spirit (Greeley et al. 2006). Most notably, Silvestro et al. (2010b) report ~1.7 m of migration within 4 months for ripples on the stoss slope of a barchan dune in Nili Patera. In contrast, Golombek et al. (2010) show that no migration has occurred in hundreds of thousands of years for a field of ripples at Meridiani Planum, and calculated dune migration rates under present conditions are orders of magnitude slower than on Earth (Claudin and Andreotti 2006, Parteli and Herrmann 2007). Currently, an understanding of the frequency of sand transport on Mars is hindered by both the limited time-series of dune images needed to determine morphological change and the unknown ages of eolian features globally.

In addition, the nature of the sand-transporting winds is not known. Although the occurrence of major dune fields appears broadly consistent with global wind patterns (Anderson et al. 1999), and although Hayward et al. (2007) found a general agreement between crestline orientations and predicted GCM wind patterns, more localized winds may well be more significant in sand transport. Extensive documentation of wind-formed surface features in Gusev Crater (as seen in Spirit images) shows the dominance of topography-modified surface winds (Greeley et al. 2006). The short radiative timescales and lower inertia of the thin CO_2 Martian atmosphere make nighttime katabatic winds stronger than on Earth (Spiga 2011), and modeling argues for these denser, shallow flows emanating from uplands (Maria et al. 2006). Katabatic winds are paramount in the formation of eolian features on the polar ice caps (e.g., Howard 2000) and are thought to be a significant sand-transporting wind in the Olympia Undae Dune Field (Ewing et al. 2010). As with a determination of wind frequency, the lack of long-term data hinders a determination of what components of the wind system transport sand as well as determination of the conditions required for these transport events to occur.

Sediment availability on the surfaces poses an additional issue that must be considered in evaluating eolian transport on Mars.

Calculations of sand transport based upon shear stress assume that sediment is available on the surface for transport. As noted above, surface stabilization by frost and permafrost is probably widespread and a primary cause for current dune stability, especially in the higher latitudes (Schatz et al. 2006, Bourke et al. 2008, Feldman et al. 2008, Ewing et al. 2010, Fenton and Hayward 2010). Surface stabilization would not only restrict actual sediment flux but also cause potential sand-transporting winds to be undersaturated and erosional, manifested by common surface features such as grooves and yardangs.

Given the current infrequency of sand transport on Mars, it is remarkable that, as first recognized by Breed et al. (1979), most dunes appear in terminal constructional settings such as basins, typically with no obvious sand-transport paths. This includes crater basins (e.g., Fenton et al. 2003, Fenton 2005, Hayward et al. 2007), valleys and troughs (Bourke et al. 2004), and the large north polar topographic basin (Tanaka et al. 2008). One possibility is that most dune fields are derived from in-basin sources and that transport paths are short (Fenton et al. 2003, Fenton 2005, Fenton and Hayward 2010). The second possibility is that most dune fields were emplaced during distant times and their sources and transport corridors have been long erased. Silvestro et al. (2010a) demonstrated sand-transport corridors connecting dune fields over a scale of tens of kilometers, but no long-distance transport comparable to that of Earth has yet been documented.

Questions

There is a dichotomy on Mars between the current apparent infrequency of sand transport and the commonness of eolian geomorphic and stratigraphic features, especially the occurrence of dune fields within terminal basins. When does eolian transport occur? Are eolian features on Mars largely the product of rare high-energy events? Are transport events associated with specific seasons, global conditions, or portions of Milankovitch cycles? Has the frequency of sand transport changed over Martian history? Are dune fields relicts of a distant, denser atmosphere? What components of the wind system drive sand transport? Are these winds part of the global circulation, local orographic winds, or shallow katabatic winds from ice caps and highlands? Do factors that increase flow density (i.e., very cold katabatic winds) significantly increase transport capacity? Is current dune stability primarily a function of the infrequency of high winds or a result of surface stabilization? Are dune fields on Mars primarily locally derived from in-basin sources, or has evidence for past long-distance transport corridors been lost?

DUNE DEVELOPMENT (DUNE–FLOW INTERACTIONS)

Earth

As is the case for all bedforms, eolian dunes develop through interactions between the flow field and the sediment bed (Fig. 4). Within these complex-system interactions, the "primary" flow (i.e., the flow before the bedforms) becomes slaved to the emergent bedform topography (and not the other way around) to produce a characteristic "secondary" flow over the bedforms. This characteristic secondary flow differs in speed, direction, shear stress, and level of turbulence from the primary flow as a function of position on the bedform (for eolian dunes, see Frank and Kocurek [1996a, 1996b] and Walker and Nickling [2003]). The most basic flow field over dunes consists of flow acceleration and erosion on the upstream or stoss slope and flow deceleration and deposition on the downstream or lee slope. The resultant dune migration causes the formation of a set of cross-strata, which is a record of lee-face deposition over time with dune migration.

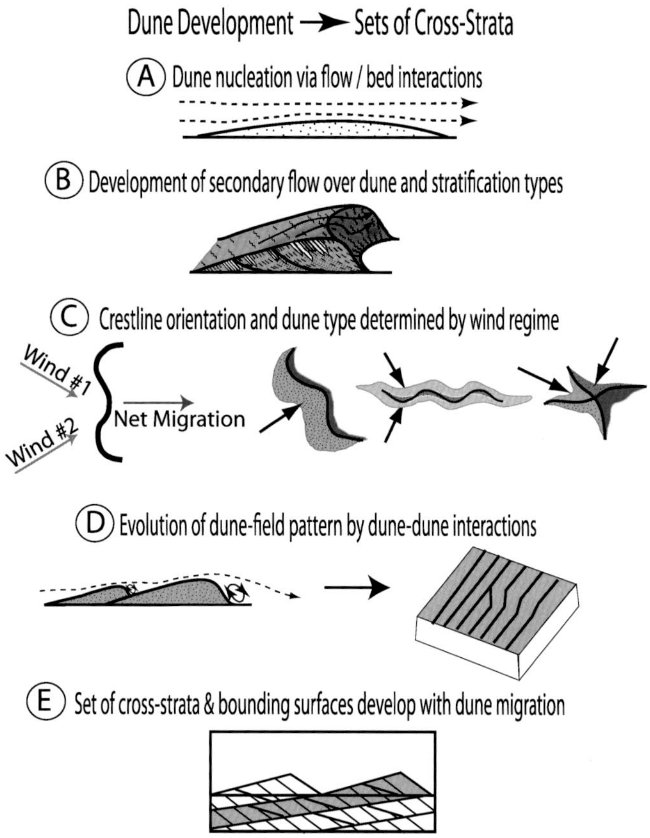

FIG. 4.—Schematic representation of stages from dune development through generation of sets of cross-strata. (A) Dune nucleation through flow and sediment bed interactions. (B) Development of secondary flow over emergent dune topography with resultant surface processes and stratification types. (C) Crestline orientation as a function of the wind regime. (D) Field-scale pattern emerges and continues to evolve through dune–dune interactions. (E) Generation of sets of cross-strata with bounding surfaces as dunes migrate and accumulations form. Stages A through D occur within the eolian transport system of Fig. 1.

The three-dimensional geometry of a set of cross-strata is a function of dune shape and behavior over time (e.g., Rubin 1987).

On Earth, dune development commonly occurs within a wind regime that varies in both direction and speed, typically as seasonal cycles (Fig. 4). Because of this boundary condition, as eolian dunes grow and become too large to reorient with changing wind directions, the crestline assumes an orientation that is as perpendicular as possible to all sand-transporting wind directions (i.e., gross bedform-normal transport of Rubin and Hunter [1987]). It is this variability in wind directions that yields the diversity of "simple" dunes on Earth (e.g., crescentic, linear, star).

Because dune crestlines are not straight and because the overall crest orientation is the gross bedform-normal orientation, a given primary wind direction may be transverse, oblique, or longitudinal to specific segments of the crestline, as defined by the incidence angle, which is the angle between a wind direction and the orientation of a segment of crestline (Fig. 5). Characteristic secondary flow conditions develop for each of these categories of incidence angles (Sweet and Kocurek 1990, Walker and Nickling 2002). Each characteristic secondary flow

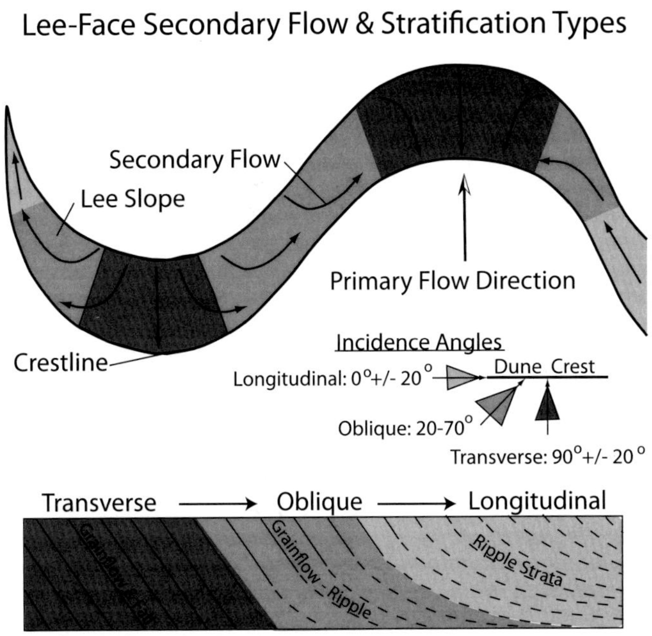

FIG. 5.—Idealized secondary flow over the lee face of a curved dune crestline, with characteristic dune stratification. Local incidence angles between the primary flow direction and the crest segment determine the lee secondary flow. In turn, the secondary flow gives rise to surface processes and their stratification. Other stratification types occur (see Kocurek 1991), primarily reflecting temporal change (e.g., annual cycles that yield transverse to oblique repetitive couplets).

condition, in turn, gives rise to surface processes that result in a specific configuration of stratification types with deposition (Kocurek 1991). The basic stratification types from Hunter (1977) are (1) wind-ripple laminae, which occur with traction transport; (2) grainfall strata formed by grains blown past the dune brink to settle under the influence of gravity and lee turbulence (Nickling et al. 2002); and (3) grainflow strata that form by the gravity-driven avalanching of grainfall deposits once these reach the angle of initial yield. The spatial (i.e., along the crestline) and temporal (in the migration direction) configuration of stratification types within a set of cross-strata is the primary tool in reconstructing paleo-flow conditions (e.g., Hunter and Rubin 1983, Kocurek et al. 1991b).

In addition to the gross geometry of sets of cross-strata and distribution of stratification types within a set, bounding surfaces are the third component in reconstructing dunes and flow conditions (Fig. 6). Bounding surfaces occur with the migration of bedforms (Brookfield 1977) and consist of (1) reactivation surfaces within a set of cross-strata that form with any erosional reconfiguration of the lee face during migration, (2) superposition surfaces that form with the migration of superimposed dunes over the larger host bedform, and (3) interdune surfaces that separate sets or co-sets of cross-strata (see discussion in Kocurek [1996]). At the regional or basin scale, yet more extensive bounding surfaces cap entire eolian sequences ("super surfaces" of Kocurek [1988]) and may arise for a variety of reasons, including the onset of a more humid climate, in which dunes are stabilized by vegetation, development of an armoring lag surface on the trailing margin of a migrating dune field, and exhaustion of the sediment supply with deflation to the capillary fringe of the water table.

The large body of works on eolian units from the stratigraphic record on Earth shows that overall boundary conditions for dune development on Earth have been similar during the entire Phanerozoic. Rephrased, the range of grain sizes, stratification types, architecture created by dune migration, and creation of regional bounding surfaces can be interpreted by observable processes, ranges, and rates. The potential to detect differences in earlier Earth history is limited by the rarity of

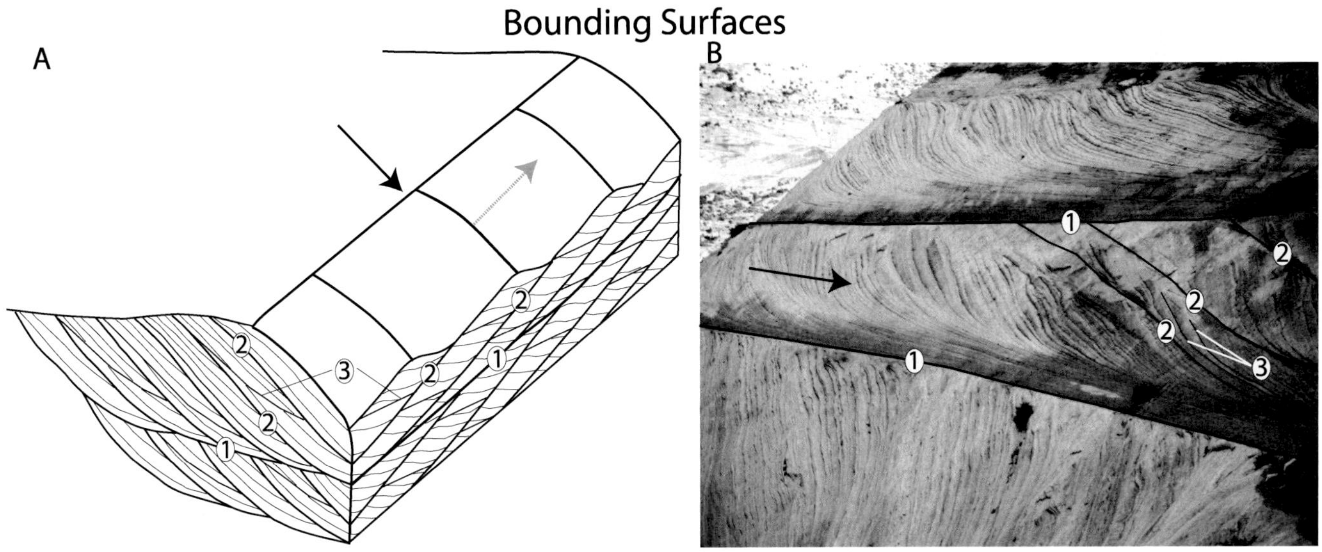

FIG. 6.—Generation of bounding surfaces with dune migration. (A) Diagrammatic compound dune with lee superimposed dunes migrating alongslope the larger host bedform; migration directions shown by arrows. This configuration is the most common compound dune type. Drawn and modified from computer-generated cross-strata in Rubin (1987). (B) Compound cross-strata in Jurassic Entrada Sandstone near Vernal, Utah, interpreted as representing a compound bedform. Labeled set has a maximum thickness of 2 m. Arrow shows migration direction of the main bedform; superimposed bedforms migrated into the page. Some bounding surfaces labeled in A and B: (1) interdune surface, (2) superposition surfaces, (3) reactivation surfaces, as described in the text.

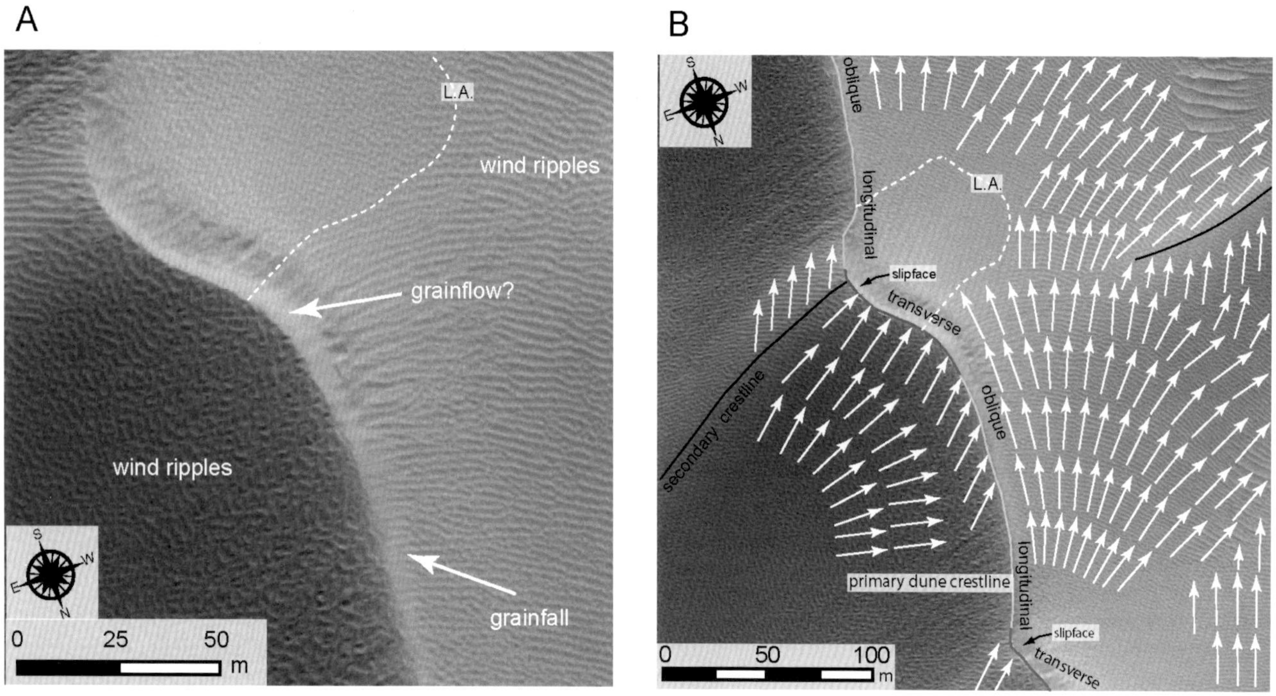

FIG. 7.—Secondary flow and surface processes on a dune segment in Olympia Undae. (A) Portion of crestline in (B) showing surface features arising from the dune-modified secondary airflow and line of attachment (L.A.) that defines the lee separation cell. Note poor ripple development within the separation cell. (B) Secondary flow-field based upon wind ripple orientations; white arrows indicate transport direction. Modified primary flow up the stoss slope is from the northeast. Range of incidence angles (transverse, oblique, longitudinal) defined by local crest orientation to the primary wind. Lee secondary flow field show deflected along-slope transport, separation cell, flow recovery leeward, and deflection along a yet-leeward dune spur. Note that the transverse portions of the crestline are coincident with grainflow, with a widening wedge of grainfall extending from the oblique segment into the transverse segment. From Ewing et al. (2010).

surviving eolian strata. Eolian strata on Earth older than 1.8 Ga (Hesperian–Amazonian boundary on Mars) are rare (Eriksson and Simpson 1998), meaning that many of the events and processes that affected eolian strata on Mars pre-date the earliest eolian records on Earth.

Mars

Based upon the Mars Global Digital Dune Database (Hayward et al. 2007) and earlier views from Mariner and Viking images (Greeley et al. 1992), most dunes on Mars are some variety of crescentic dune (i.e., crescentic ridges and barchans, the latter of which represent a lower sand availability). Other dune types have been recognized, including linear and star dunes (e.g., Edgett and Blumberg 1994, Lee and Thomas 1995), a variety of dune features associated with topography (Bourke et al. 2004), and some dune forms that are more enigmatic in classification (e.g., transverse eolian ridges; Zimbelman 2010). The overwhelming dominance of crescentic dunes on Mars, however, contrasts with the much greater variety of dunes on Earth and the commonality of linear dunes. Because crescentic dunes form in a more unidirectional wind regime than linear dunes, which require an obtuse bidirectional wind, and star dunes, which can form in multidirectional winds or a bimodal regime where the winds are roughly perpendicular to each other (e.g., Rubin and Hunter 1987, Rubin and Ikeda 1990), the simplest explanation is that the constructive wind regimes on Mars are less complex than on Earth.

Where secondary airflow has been mapped over dunes by the orientation of wind ripples in HiRISE images (Ewing et al. 2010), the spatial distribution of surface processes and the predicted airflow behavior is consistent with Earth examples (Fig. 7). In this Olympia Undea example, lee-slope deflection of the flow occurs with an oblique incidence angle, and a slipface and separation cell occur in the lee of areas with a transverse incidence angle.

The most detailed outcrop analysis of eolian cross-strata on Mars is for the Burns formation, as viewed by *Opportunity* in the Endurance Crater (Grotzinger et al. 2005) and in the Erebus Crater (Metz et al. 2009). Stratification types, especially the distinctive wind-ripple laminae, are clearly visible, as are reactivation surfaces and an interpreted super bounding surface (Wellington). The presence of reactivation surfaces and the dominance of wind-ripple laminae indicate that the winds were not unidirectional.

Questions

In complex-system hierarchy theory (e.g., Werner 2003), lower-level variables become "slaved" to emergent higher-level emergent behavior as the system evolves (Fig. 1). For the Martian eolian transport system, although there are significant differences from Earth at the fluid–grain level, these may become subordinate to emergent dune topography at the higher flow–bedform level. Martian dunes and dune dynamics would, therefore, be very similar to those of Earth, although differences may exist because of local boundary conditions (i.e., the shape of grainflow may differ as a result of unequal sublimation over the surface). The promise, therefore, is that dune geomorphic and stratigraphic features can be interpreted by principles evident on Earth. What, then, do current dune fields of different ages and latitudes

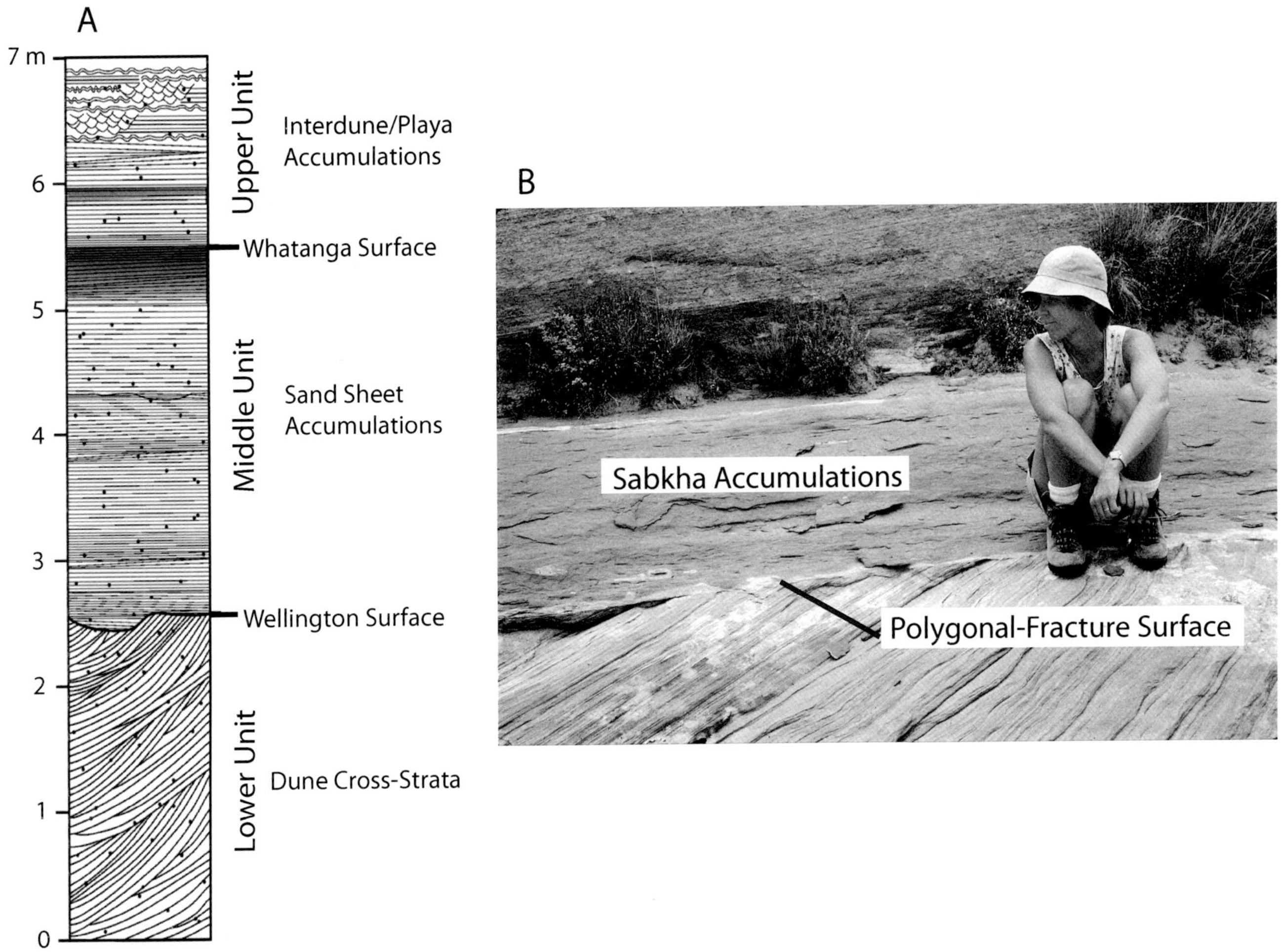

FIG. 10.—Comparison of accumulations in sequences in the Burns formation as seen in Endurance Crater (A) and the Jurassic Page Sandstone, Utah (B). In both cases, interpreted dry-system dune cross-strata underlie water-table–controlled deflation surfaces (Wellington, Mars; polygonally fractured surface, Page). Overlying accumulations are interpreted to have formed with a relative rising water table. Burns section modified from Grotzinger et al. (2005).

Accumulation requires a positive sand budget, in which more sand enters the system than exits, so that the surface over which the dunes are migrating, the accumulation surface, rises over time (Fig. 9). In contrast, bypass along the accumulation surface results in a neutral sand budget, and a negative sand budget characterizes erosional systems in which the accumulation surface falls over time (Kocurek and Havholm 1994). With migrating bedforms, rise of the accumulation surface occurs as bedforms leave behind their lower portions as they migrate (i.e., a set of cross-strata). The relation between bedform migration and behavior of the accumulation surface over time can be characterized by the angle of bedform climb (θ), defined as $\tan \theta = V_y/V_x$, where V_y is the vertical accumulation rate and V_x is the bedform migration rate (Brookfield 1977, Rubin and Hunter 1982). Accumulation occurs with a positive angle of climb, whereas the angle of climb is zero for bypass systems and negative for erosional systems. In natural bedform systems, spatial variation in the depth of scour as bedforms migrate complicates the simple image of migrating and climbing bedforms (Paola and Borgman 1991), but the general application of climb theory is robust.

Except in uncommon cases (e.g., dunes buried beneath lava flows), the eolian rock record on Earth is necessarily constructed by dunes that were leaving an accumulation. The causes of accumulation, as well as bypass or erosion, lay with the dynamics of the three basic types of eolian systems: (1) dry, (2) wet, and (3) stabilizing (Kocurek and Havholm 1994, Kocurek 1999) (Fig. 9). Dry eolian systems are those in which aerodynamics alone determine the behavior of the accumulation surface over time. Accumulation most typically occurs in these systems because of a regional deceleration of the wind along pressure gradients or as a result of flow into topographic basins where deceleration occurs with vertical flow expansion (Wilson 1973, Rubin and Hunter 1982). Wet eolian systems are those in which the water table is at or near the surface; accumulation in these systems occurs because of a rise of the capillary fringe over time. These systems are not restricted to areas of decelerating flow. Stabilizing eolian systems are those in which some form of surface stabilization fosters accumulation, such as vegetation that causes wind deceleration.

Modern and ancient eolian systems on Earth have most commonly been dry systems, including the Sahara (c.f., Wilson 1973, Kocurek 1998), the Jurassic Navajo (Kocurek 2003), Jurassic Page (Havholm et al. 1993), and Permian Cedar Mesa (Mountney and Jagger 2004). The Jurassic Entrada may be the best example of a regional wet eolian system (Crabaugh and Kocurek 1993, Crabaugh and Kocurek 1998), but wet eolian systems are common in coastal areas (e.g., Padre Island, Texas [Kocurek et al. 1992]; Guerrero Negro, Baja [Fryberger et al. 1990]). The vegetated Thar sand sea may be an example of a regional stabilizing eolian system (Wasson et al. 1983, Singhvi and Kar 2004). Hybrid systems are also common, such as White Sands, New Mexico, where the water table and surface stabilization by gypsum cements are both important controls on the behavior of the accumulation surface over time (Kocurek et al. 2007). In nearly all cases the angle of climb is very low, typically tenths of a degree (e.g., Crabaugh and Kocurek 1993, Mountney and Jagger 2004, Kocurek et al. 2007), because nearly always the rate of dune migration (V_x) is much greater than the vertical accumulation rate (V_y).

Mars

Given the water-limited boundary conditions on Mars, dry eolian systems are expected to dominate, and, indeed, most dune fields on the surface occur in craters, troughs, and topographic basins where dry eolian systems are expected to accumulate. Stratigraphic analysis of the Burns formation in Endurance Crater (Grotzinger et al. 2005) shows that the Lower Unit consists of sharp (i.e., dry) bounding surfaces between and within sets of cross-strata and that wet/damp-surface interdune deposits are absent, thus indicating that these dunes accumulated in a dry system (c.f., fig. 19 in Grotzinger et al. [2005]). However, interpreted accumulation of eolian sand sheets (Middle Unit) and interdune/playa deposits (Upper Unit) in an upward-wetting sequence by a rise of groundwater characterizes accumulation in a wet eolian system. The overall unit (fig. 3 in Grotzinger et al. [2005]), in which dry-system dune sets of cross-strata are truncated by a water-table–controlled super surface (Wellington Contact) with overlying sand-sheet and sabkha strata, is identical to many eolian sequences on the Colorado Plateau, such as the Jurassic Page Sandstone (e.g., Havholm et al. 1993) (Fig. 10). For these Earth examples, dry-system dunes accumulated during low-stands, but sand availability was apparently terminated with the onset of marine transgression. Subsequent deflation occurred to the rising water table, which was the inland response to the marine transgression (see Kocurek et al. 2001). Continued rise of the water table allowed for the accumulation of sabkha and related deposits overlying the surface.

Contrasting with the dry/wet system accumulations in the Burns formation are accumulations of the much younger (Amazonian) Planum Boreum cavi unit in the north polar region of Mars. The overall unit architecture consists of preserved dune topography in which (1) stacked dune stoss-slope deposits are bounded by erosional surfaces or surfaces showing minimal permafrost development, as manifested by polygonal fractures, (2) down-lapping lee-face deposits are bounded by surfaces of pronounced permafrost polygonal fractures, and (3) interdune areas show the amalgamation of pronounced permafrost surfaces and ice deposits (Fig. 11). This architecture argues for cycles of dune reactivation and stabilization by freezing. During reactivation stoss slopes were deflated, and lee deposition occurred as downlapping, progradational wedges that tapered onto the interdune floors. During stabilization, the entire dune likely developed a permafrost surface, and wind-blown ice accumulated in interdune hollows. Only the stoss slopes show evidence of deflation; permafrost surfaces appear largely undisturbed on the lee slopes and within interdune hollows. Because dune topography was preserved during stabilization, it served as the antecedent boundary condition during subsequent reactivation such that renewed deposition conformed to the existing dune topography. Remnant dune topography is carried upward through the cavi unit, resulting in exceptionally high angles of bedform climb. Because accumulation of dune strata occurred through freezing, the Cavi dune field is interpreted as a stabilizing system, analogous to Earth systems, within which the most common stabilizing agent is vegetation.

Questions

The causes of accumulation of eolian strata can be understood by the dynamics that characterize the three basic eolian system types (dry, wet, stabilizing). Planetary boundary conditions dictate the form these dynamics take within these systems. The dynamics of dry eolian systems appear to be readily transferable from Earth to Mars. The dynamics of stabilizing systems may be best manifested on Mars through intergranular freezing and permafrost development. Documentation of wet eolian systems is strong evidence for sustained near-surface groundwater on Mars. Can the causes of eolian accumulation be recognized on Mars, and have these changed through time? Do these change by latitude? Is accumulation of eolian strata currently occurring in craters and other basins? Is accumulation itself cyclic on Mars?

PRESERVATION OF EOLIAN ACCUMULATIONS

Earth

The boundary conditions that cause accumulation of eolian strata on Earth are not entirely the same as those that cause preservation of the accumulations (i.e., incorporation into the rock record). For example, some dune fields of the western Sahara have built significant accumulations, but these rest upon Precambrian basement and have little preservation potential (e.g., Kocurek et al. 1991a). The accumulations within Chott Rharsa in Tunisia, although occupying a foreland topographic basin with a surface elevation below sea level, are in a current state of deflation (Blum et al. 1998). Most eolian accumulations preserved in the rock record on Earth have occurred (1) in subsiding basins where the continued addition of sediment has buried the eolian accumulations, (2) in inland basins that experienced a sustained absolute or relative rise in the water table, or (3) as dune fields adjacent to marine bodies in which transgressions resulted in a rise of the continental water table and/or the eolian accumulations themselves were marine transgressed (Fig. 12; Kocurek and Havholm 1994, Kocurek 1999). For example, the Jurassic sand seas of the Colorado Plateau (Navajo, Temple Cap, Page, Entrada) were all flanked to the west by marine environments, and the eolian record is an eastward back-stepping of vertically stacked eolian units overlain by transgressive marine units within a subsiding foreland basin (see fig. 11.7 in Kocurek [1999]). At the formation level, the water-table–controlled deflation surfaces in the Page Sandstone (described above; Fig. 10) all correspond to flood surfaces within the equivalent marine/sabkha Carmel Formation to the west (Havholm et al. 1993). Whereas the eolian accumulations in the western Sahara have little preservation potential, these same accumulations are preserved on the continental shelf, where they were transgressed by sea-level rise following the last glacial maximum (Sarnthein 1978). The dune field at White Sands, New Mexico, has a several-meter-thick subsurface record that apparently formed with the relative rise of the water table within a subsiding basin (Kocurek et al. 2007). On Earth, the dynamics of tectonism, eustasy, and climate drive the temporal and spatial nature of preservation. With planetary resurfacing, essentially no preservation is permanent (e.g., the once preserved Jurassic record is largely uplifted and erosional now).

Mars

Preservation of eolian accumulations on Mars has been infrequently addressed, and the geometries of bodies of strata and their "containers" are poorly known. In the absence of tectonism and eustasy, preservation on Mars can be envisioned by burial by lava flows, impact debris, and continued eolian and other accumulations in topographic basins, such as craters. It is plausible that initial preservation of the Burns formation occurred through an areal rise in

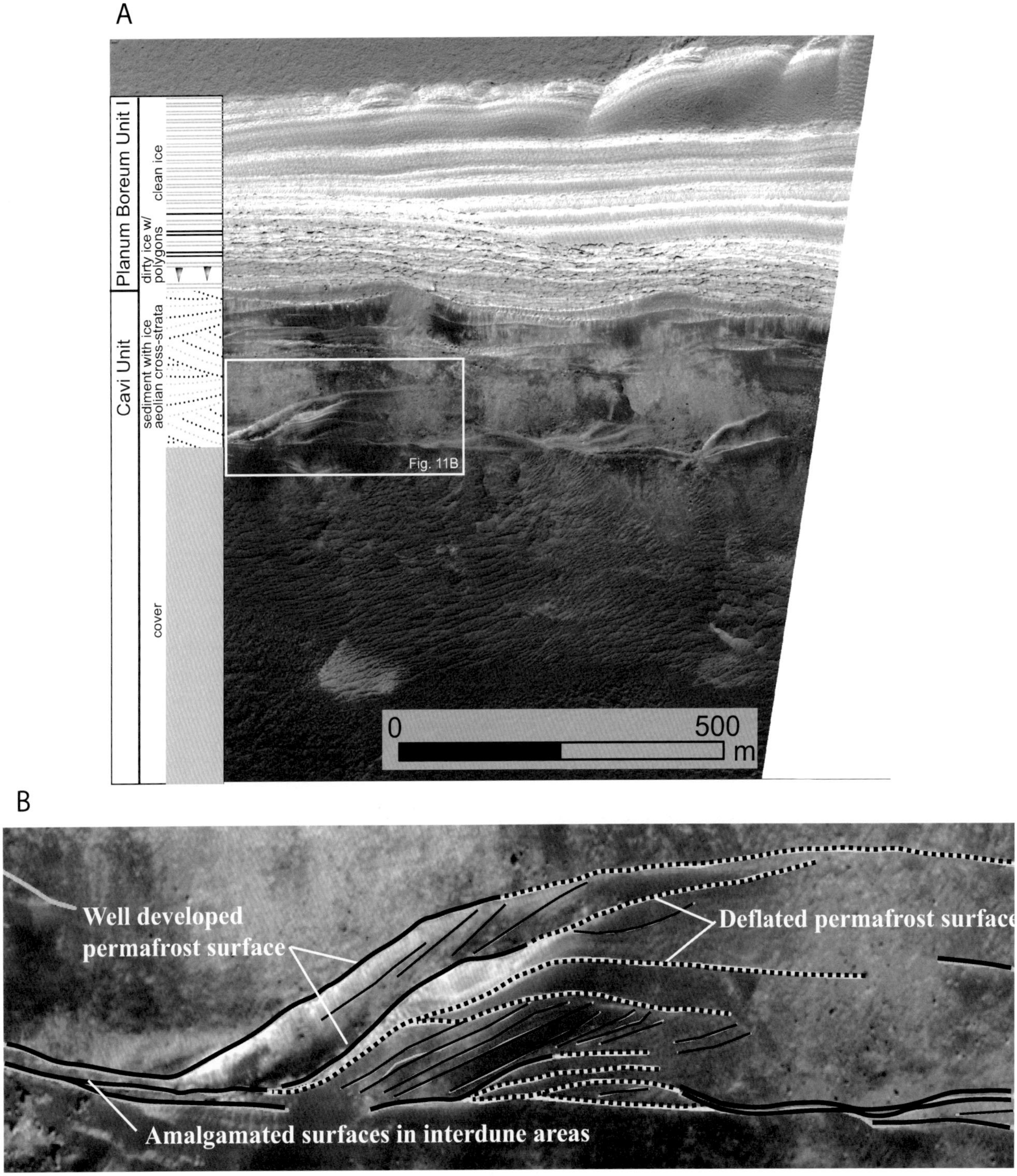

FIG. 11.—Accumulation and preservation of eolian cross-strata in the Planum Boreum cavi unit, north polar region of Mars. (A) Section shows upward gradation of sediment dune topography into layers of dirty ice that conform to underlying dune topography, then to layered ice. (B) Dune topography showing the stacking of dune stoss and lee deposits separated by permafrost surfaces, interpreted to represent cyclic dune activation and stabilization. Accumulation occurs because of the stabilizing effects of freezing. Preservation of the accumulations occurred with continued ice–sediment layering. Preservation space is currently being lost as the ice cap retreats and the underlying cavi unit erodes. HiRISE Image TRA_000863_2640.

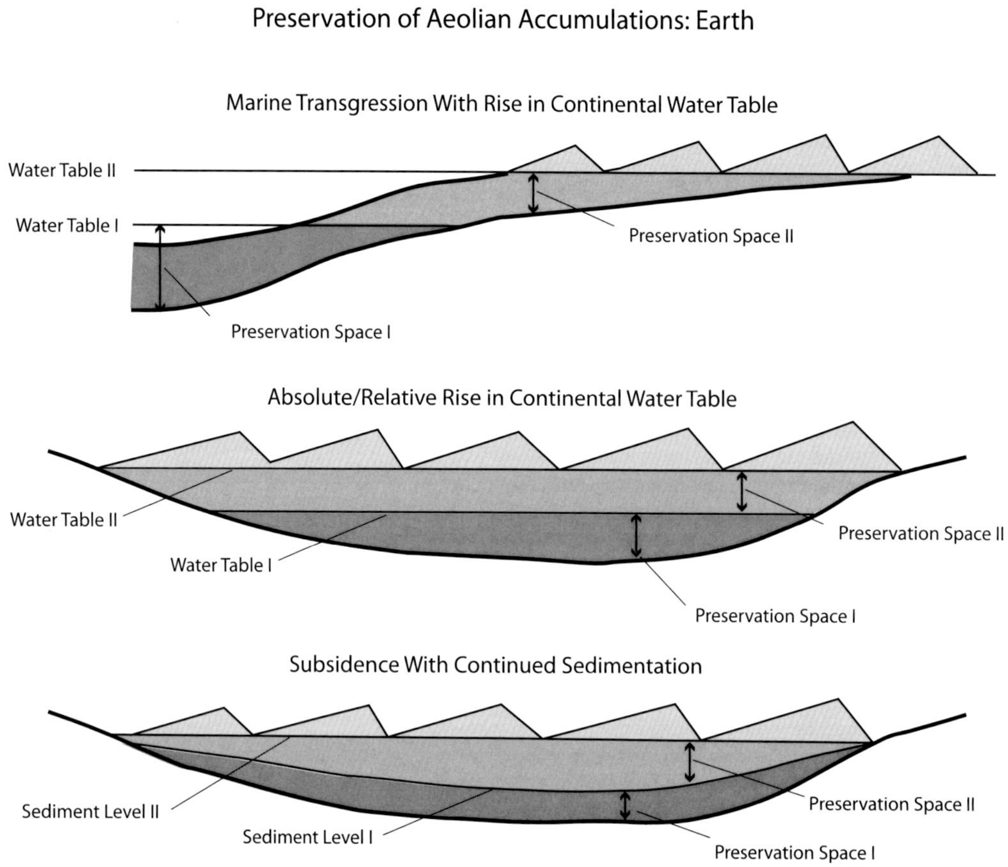

FIG. 12.—Modes of preservation of eolian accumulation on Earth. See text for discussion.

the water table, but the regional context of the unit is unknown. Grotzinger et al. (2005) suggested that the surface bounding the region around the *Opportunity* landing site is a prominent deflated stratigraphic contact. Preservation and loss of "preservation space" of the cavi unit in the north polar region of Mars has been explained as a function of climatic change (Byrne and Murray 2002, Fishbaugh and Head 2005). As described above, accumulation of the cavi unit is interpreted to have occurred through the cyclic stabilization of dunes through freezing and permafrost development. These dune accumulations are clearly transitional upward into the overlying layered ice cap accumulation (Planum Boreum 1 Unit; Fishbaugh and Head 2005, Tanaka et al. 2008). Preservation of the cavi accumulations was, therefore, by burial beneath the growing ice cap. Retreat of the ice cap has resulted in loss of preservation space and extensive erosion of the unit (Byrne and Murray 2002, Fishbaugh and Head 2005).

Questions

Given the absence of plate tectonism and eustasy on Mars, preservation must be envisioned differently on Mars than on Earth. Crater tectonism and preservation of eolian accumulations within crater basins is one fairly ready substitution. Climatic change, however, emerges as a significant driver of preservation dynamics, as evident by the cavi unit and probably elsewhere on Mars. At the planetary scale, however, preservation of eolian and other accumulations on Mars takes on a different relevance than on Earth. Where even Noachian depositional systems are still evident on the surface and planetary resurfacing occurs at Earth "deep time" scales, what defines preservation? Are surface dune fields stabilized by freezing or permafrost preserved because they remain undisturbed for millions of years? The Olympia Undae Dune Field may well be contemporary to the Jurassic Navajo Sandstone. Should preservation of Mars not be considered in time units, but rather be evident only by the creation of a stratigraphic record? Are there planetary rates and processes of resurfacing that define preservation on Mars?

CONCLUSIONS

Mars is the most eolian-dominated planet in the solar system, and eolian processes, along with cryogenic processes, may well be the primary drivers in planetary resurfacing. A simplified, source-to-sink comparison of eolian systems on Earth and Mars serves to underscore how little is understood about Martian eolian systems, even as Earth systems are far from totally understood. More importantly, however, this comparison demonstrates how the governing principles of eolian systems are manifested differently on Earth and Mars because of different planetary boundary conditions.

The planetary boundary conditions that yield sediment for eolian systems on Earth are absent, at least currently, on Mars, thus raising the fundamental question of how sediment is generated on Mars. Martian boundary conditions do not include loss of sediment from the surface through tectonism, burial in deep subsiding basins, or transport to oceanic basins. Sediment on Mars probably reflects early planetary history, when Mars was more similar to Earth, subsequent to very slow weathering processes and episodic generation during climatic cycles, impacts, and outflow events. Differences in planetary rock cycles may be manifested on Mars by direct deflation of grains from indurated strata where ice is the primary cementing agent and weathering occurs through sublimation during climatic cycles.

Regardless of how sediment is generated on Mars, eolian transport events are probably episodic because they require the coincidence of a wind capable of transporting sand and the availability of this sediment. The primary impact of the low-density atmosphere is that winds capable of sand transport are relatively rare. The cause of these winds is poorly known, but factors that increase air density may be more important than on Earth. Local katabatic winds that form over the highlands and polar ice caps, for example, may be more important for sediment transport than are planetary wind belts. Because of widespread permafrost, sediment availability is probably episodic, only occurring with periods of sublimation.

In spite of planetary differences in sediment production and grain transport, the eolian systems of Earth and Mars appear to largely converge with the emergence of dune topography. Within the hierarchy of the eolian transport system, dune–flow and dune–dune interactions are very similar on Earth and Mars, as manifested by dune morphologies, secondary flow over the bedforms, and pattern ordering through dune interactions. The implication is that dunes and their rock record can be interpreted by principles used on Earth, allowing for spatial comparisons and the development of a chronology of planetary change.

It is likely that most accumulation of eolian strata on Mars has occurred as dry systems, in which flow decelerates into topographic basins, which on Mars are commonly craters. Wet-system accumulation, as shown by the Burns formation at Endurance Crater, should be restricted to early planetary history. The most striking manifestation of a stabilizing system on Mars is accumulation within the cavi unit. The dynamics of stabilization by permafrost development in the cavi are analogous to the niche occupied by vegetation in stabilizing systems on Earth.

Although Earth-style preservation of eolian accumulations occurs on Mars with burial within topographic basins—albeit many of the basins arise through impact and not subsidence—the concept of preservation is challenged on Mars, where "deep-time" events can remain pristine on the surface. In restricting preservation to generation of a stratigraphic record, the preservation of cavi-style eolian accumulation by burial beneath the polar ice cap is significant because climatic cycles, and not tectonic nor eustatic cycles, have driven the preservation dynamics.

ACKNOWLEDGMENTS

This material is partly based upon work supported by the National Science Foundation under Award No. 0846233 to R.C. Ewing. We are grateful to Ken Tanaka and Ed Simpson for their constructive comments in reviewing the manuscript.

REFERENCES

Almeida MP, Parteli EJR, Andrade JS, Herrmann HJ. 2008. Giant saltation on Mars. *Proceedings of the National Academy of Science* 105:6222–6226.

Alpers CN, Brimhall GH. 1988. Middle Miocene climatic change in the Atacama Desert, northern Chile: Evidence from supergene mineralization at La Escondida. *Bulletin Geological Society of America* 100:1640–1656.

Anderson FS, Greeley R, Xu P, Lo E. 1999. Assessing the Martian surface distribution of aeolian sand using a Mars general circulation model. *Journal of Geophysical Research* 104:18991–19002.

Bagnold RA. 1941. *The Physics of Blown Sand and Desert Dunes*: Chapman and Hall, London.

Bandfield JL, Feldman WC. 2008. Martian high latitude permafrost depth and surface cover thermal inertia distributions. *Journal of Geophysical Research* 113, E08001, doi:10.1029/2007JE003007

Bell JF, Squyres SW, Arvidson RE, Arneson HM, Bass D, Blaney D, Cabrol N, Calvin W, Farmer J, Farrand WH, Goetz W, Golombek M, Grant JA, Greeley R, Guinness E, Hayes AG, Hubbard MYH, Herkenhoff KE, Johnson MJ, Johnson JR, Joseph J, Kinch KM, Lemmon MT, Li R, Madsen MB, Maki JN, Malin M, McCartney E, McLennan S, McSween HY, Ming DW, Moersch JE, Morris RV, Noe Dobrea EZ, Parker TJ, Proton J, Rice JW, Seelos F, Soderblom J, Soderblom LA, Sohl-Dickstein JN, Sullivan RJ, Wolff MJ, Wang A. 2004. Pancam multispectral imaging results from the Spirit rover at Gusev crater. *Science* 305:800–806.

Beveridge C, Kocurek G, Ewing RC, Lancaster N, Morthekai P, Singhvi AK, Mahan SA. 2006. Development of spatially diverse and complex dune-field patterns: Gran Desierto Dune Field, Sonora, Mexico. *Sedimentology* 53:1391–1409.

Bibring JP, Langevin Y, Mustard J, et al. 2006. Global mineralogical and aqueous Mars history derived from OMEGA/Mars Express data. *Science* 312:400–404.

Bishop MA. 2007. Point pattern analysis of north polar crescentic dunes, Mars: A geography of dune self-organization. *Icarus* 191:151–157.

Bluck BJ, Ward JD, Cartwright J, Swart R. 2007. The Orange River, southern Africa: An extreme example of a wave-dominated sediment dispersal system in the south Atlantic Ocean. *Journal of the Geological Society* 164:341–351.

Blum M, Kocurek G, Deynoux M, Lancaster N, Price D, Pion JC. 1998. Quaternary wadi, lacustrine, aeolian depositional cycles and sequences, Chott Rharsa Basin, southern Tunisia. *In* Alsharhan A, Glennie K, Whittle G, Kendall C (Editors). *Quaternary Deserts and Climatic Change:* A.A. Balkema, Rotterdam, The Netherlands. p. 539–552.

Bourke MC, Bullard JE, Barbouin-Jha OS. 2004. Aeolian sediment transport pathways and aerodynamics at troughs on Mars. *Journal of Geophysical Research* 109, E07005, doi:10.1029/2003JE002155

Bourke MC, Edgett KS, Cantor BA. 2008. Recent aeolian dune changes on Mars. *Geomorphology* 94:247–255.

Bourke MC, Ewing RC, Finnegan D, McGowan HA. 2009. Sand dune movement in the Victoria Valley, Antarctica. *Geomorphology* 109:148–160.

Breed CS, Grolier MJ, McCauley JF. 1979. Morphology and distribution of common 'sand' dunes on Mars: Comparison with the Earth. *Journal of Geophysical Research* 84:8183–8204.

Bridges NT, Geissler PE, McEwen AS, Thomson BJ, Chuang FC, Herkenhoff KE, Keszthelyi LP, Alonso-Martinez S. 2007. Windy Mars: A dynamic planet as seen by the HiRise camera. *Geophysical Research Letters* 34, L23205, doi:10.1029/2007GL031445

Bristow CS, Augustinus PC, Wallis IC, Jol HM, Rhodes EJ. 2010a. Investigation of the age and migration of reversing dunes in Antarctica using GPR and OSL, with implications for GPR on Mars. *Earth and Planetary Science Letters* 289:30–42.

Bristow CS, Jol HM, Augustinus P, Wallis I. 2010b. Slipfaceless 'whaleback' dunes in a polar desert, Victoria Valley, Antarctica: Insights from ground penetrating radar. *Geomorphology* 114:361–372.

Brookfield ME. 1977. The origin of bounding surfaces in ancient aeolian sandstones. *Sedimentology* 24:303–332.

Byrne S, Dundas CM, Kennedy MR, et al. 2009. Distribution of mid-latitude ground ice on Mars from new impact craters. *Science* 325:1674–1676.

Byrne S, Murray BC. 2002. North polar stratigraphy and the paleo-erg of Mars. *Journal of Geophysical Research* 107, E6, 5044, doi:10.1029/2001JE001615

Cantor B, Kanak AKM, Edgett KS. 2006. Mars Orbiter Camera observations of Martian dust devils and their tracks (September 1997 to January 2006) and evaluation of theoretical vortex models. *Journal of Geophysical Research* 111: E12002, doi:10.1029/2006JE002700

Claudin P, Andreotti B. 2006. A scaling law for aeolian dunes on Mars, Venus, Earth, and for subaqueous ripples. *Earth and Planetary Science Letters* 252:30–44.

Crabaugh M, Kocurek G. 1993. Entrada Sandstone: An example of a wet aeolian system. *In* Pye K (Editor). *The Dynamics and Environmental Context of Aeolian Sedimentary Systems:* Geological Society, London. Special Publication 72. p. 103–126.

Crabaugh M, Kocurek G. 1998. Continental sequence stratigraphy of a wet eolian system: A key to relative sea-level change. *In* Shanley KW, McCabe PJ (Editors). *Relative Role of Eustasy, Climate, and Tectonism in Continental Rocks:* SEPM Society for Sedimentary Geology, Tulsa, OK, Special Publication 59. p. 213–228.

Derickson D, Kocurek G, Ewing RC, Bristow C. 2008. Origin of a complex and spatially diverse dune-field pattern, Algodones, southeastern California. *Geomorphology* 99:186–204.

Edgett KS, Blumberg DG. 1994. Star and linear dunes on Mars. *Icarus* 112:448–464.

Edgett KS, Christensen PR. 1991. The particle size of Martian aeolian dunes. *Journal of Geophysical Research* 96:22765–22776, doi:10.1029/91JE02412

Edgett KS, Malin MC. 2000. New views of Mars eolian activity, materials, and surface properties: Three vignettes from the Mars Global Surveyor Mars orbiter camera. *Journal of Geophysical Research* 105(E1):1623–1650.

Edwards CS, Bandfield JL, Christensen PR, Fergason RL. 2009. Global distribution of bedrock exposures on Mars using THEMIS high-resolution thermal inertia. *Journal of Geophysical Research* 114: E11001, doi:10.1029/2009JE003363

Elbelrhiti H, Andreotti B, Claudin P. 2008. Barchan dune corridors: Field characterization and investigation of control parameters. *Journal of Geophysical Research* 113: F02S15, doi:10.1029/2007JF000767

Eriksson KA, Simpson EL. 1998. Controls on spatial and temporal distribution of Precambrian eolianites. *Sedimentary Geology* 120:275–294.

Ewing RC, Kocurek G. 2010a. Dune interactions and dune-field pattern development: White Sands Dune Field, New Mexico, USA. *Sedimentology* 57:1199–1219.

Ewing RC, Kocurek G. 2010b. Aeolian dune-field pattern boundary conditions. *Geomorphology* 114:175–187.

Ewing RC, Kocurek G, Lake L. 2006. Pattern analysis of dune-field parameters. *Earth Surface Processes and Landforms* 31:1176–1191.

Ewing RC, Peyret A, Kocurek G, Bourke M. 2010. Recent transporting winds and dune-field pattern formation in the Olympia Undea Dune Field, Mars. *Journal of Geophysical Research* 115: E08005, doi:10.1029/2009JE003526

Feldman WC, Bourke MC, Elphic RC, Maurice S, Bandfield J, Prettyman TH, Diez B, Lawrence DJ. 2008. Hydrogen content of sand dunes within Plympia Undae. *Icarus* 196:422–432, doi:10.1016/j.icarus.2007.08.044

Fenton LK. 2005. Potential sand sources for dune fields in Noachis Terra, Mars. *Journal of Geophysical Research* 110: E11004, doi:10.1029/2005JE002436

Fenton LK, Bandfield JL, Ward JL. 2003. Aeolian processes in Proctor crater on Mars: Sedimentary history as analyzed from multiple data sets. *Journal of Geophysical Research* 108: E12, 5129, doi:10.1029/2002JE002015

Fenton LK, Hayward RK. 2010. Southern high latitude dune fields on Mars: Morphology, aeolian inactivity, and climate change. *Geomorphology* 121:98–121.

Fishbaugh KE, Head JW. 2005. Origin and characteristics of the Mars north polar basal unit and implications for polar geologic history. *Icarus* 174:444–474.

Forrest SP, Haff PK. 1992. Mechanics of wind ripple stratigraphy. *Science* 255:1240–1243.

Frank A, Kocurek G. 1996a. Airflow up the stoss slope of sand dunes: Limitations of current understanding. *Geomorphology* 17:47–54.

Frank A, Kocurek G. 1996b. Toward a model for airflow on the lee side of aeolian dunes. *Sedimentology* 43:451–458.

Fryberger SG, Krystinik LF, Schenk CJ. 1990. Tidally flooded back barrier dunefield, Guerrero Negro area, Baja California, Mexico. *Sedimentology* 37:23–43.

Gibson EK, Wentworth SJ, McKay DS. 1983. Chemical weathering and diagenesis of a cold desert soil from Wright Valley, Antarctica: An analog of Martian weathering processes. *Proceedings of the Lunar Science Conference* 13:A912–A928.

Golombek MP, Bridges NT. 2000. Erosion rates on Mars and implications for climate change: Constraints from the Pathfinder landing site. *Journal Geophysical Research* 105:1841–1853.

Golombek MP, Grant JA, Crumpler LS, Greeley R, Arvidson RE, Bell JF, Weitz CM, Sullivan R, Christensen PR, Soderblom LA, Squyres SW. 2006. Erosion rates at the Mars Exploration Rover landing sites and long-term climate change on Mars. *Journal of Geophysical Research* 111: E12S10, doi:10.1029/2006JE002754

Golombek MP, Robinson K, McEwen A, Bridges N, Ivanov B, Tornabene L, Sullivan R. 2010. Constraints on ripple migration at Meridiani Planum from Opportunity and HiRISE observations of fresh craters. *Journal of Geophysical Research* 115: E00F08, doi:10.1029/2010JE003628.

Greeley R, Arvidson RE, Barlett PW, et al. 2006. Gusev crater: Wind-related features and processes observed by the Mars Exploration Rover Spirit. *Journal of Geophysical Research* 111: E02S09, doi:10.1029/2005JE002491

Greeley R, Kraft MD, Kuzmin RO, Bridges NT. 2000. Mars Pathfinder landing site: Evidence for a change in wind regime from lander and orbitor data. *Journal of Geophysical Research* 105(E1):1829–1840.

Greeley R, Lancaster N, Lee S, Thomas P. 1992. Martian aeolian processes, sediments and features. *In* Kieffer HH, Jakowsky B, Snyder C, Matthews M (Editors). *Mars:* The University of Arizona Press, Tucson. p. 730–766.

Greeley R, Leach R, White B, Iverson J, Pollack J. 1980. Threshold wind speeds for sand on Mars: Wind tunnel experiments. *Geophysical Research Letters* 7:121–124.

Greeley R, Squyres SW, Arvidson RE, et al. 2004. Wind-related processes detected by the Spirit rover at Gusev crater, Mars. *Science* 305:810–813.

Grotzinger JP, Arvidson RE, Bell JF, et al. 2005. Stratigraphy and sedimentology of a dry to wet eolian depositional system, Burns Formation, Meridiani Planum, Mars. *Earth and Planetary Science Letters* 240:11–72.

Havholm KG, Blakey RC, Capps M, Jones LS, King DD, Kocurek G. 1993. Aeolian genetic stratigraphy: An example from the Middle Jurassic Page Sandstone, Colorado Plateau. *In* Pye K, Lancaster N (Editors). *Aeolian Sediments: Ancient and Modern* Blackwell, Oxford, UK. p. 87–107.

Hayward RK, Mullins KF, Fenton LK, Hare TM, Titus TN, Bourke MC, Colaprete A, Christensen PR. 2007. Mars global digital dune database and initial science results. *Journal of Geophysical Research* 112: E11007, doi:10.1029/2007JE002943

Hersen P, Douady S. 2005. Collisions of barchans dunes as a mechanism of size regulation. *Geophysical Research Letters* 32: L21404, doi:10.1029/2005GL024179

Howard AD. 2000. The role of eolian processes in forming surface features of the Martian Polar Layered Deposits. *Icarus* 144:267–288.

Hunter RE. 1977. Basic types of stratification in small eolian dunes. *Sedimentology* 24:361–387.

Hunter RE, Rubin DM. 1983. Interpreting cyclic cross-bedding, with an example from the Navajo Sandstone. *In* Brookfield ME, Ahlbrandt TS (Editors). *Eolian Sediments and Processes, Developments in Sedimentology* 38:429–454.

Hurowitz JA, McLennan SM. 2007. A ~3.5 Ga record of water-limited, acidic weathering conditions on Mars. *Earth and Planetary Science Letters* 260:432–443.

Iverson JD, White BR. 1982. Saltation threshold on Earth, Mars and Venus. *Sedimentology* 29:111–119.

Jerolmack DJ, Mohrig D, Grotzinger JP, Fike DA, Watters WA. 2006. Spatial grain size sorting in eolian ripples and estimation of wind conditions on planetary surfaces: Application to Meridiani Planum, Mars. *Journal of Geophysical Research* 111: E12S02, doi:10/1029/2005JE002544

Kocurek G. 1988. First-order and super bounding surfaces in eolian sequences—bounding surfaces revisited. *Sedimentary Geology* 56:193–206.

Kocurek G. 1991. Interpretation of ancient eolian sand dunes. *Annual Review of Earth and Planetary Science* 19:43–75.

Kocurek G. 1996. Desert Aeolian systems. *In* Reading HG (Editor). *Sedimentary Environments: Processes, Facies and Stratigraphy*, 3rd ed.: Blackwell, Oxford, UK. p. 125–153.

Kocurek G. 1998. Aeolian system response to external forcing factors—A sequence stratigraphic view of the Sahara region. *In* Alsharhan AS, Glennie KW, Whittle GL, Kendall CG StC (Editors). *Quaternary Deserts and Climatic Change:* A.A. Balkema, Rotterdam, The Netherlands. p. 327–337.

Kocurek G. 1999. The aeolian rock record (Yes, Virginia, it exists, but it really is rather special to create one). *In* Goudie A, Livingston I (Editors). *Aeolian Environments, Sediments and Landforms:* John Wiley, London. p. 239–259.

Kocurek G. 2003. Limits on extreme Eolian systems: Sahara of Mauritania and Jurassic Navajo Sandstone examples. *In* Chan M, Archer A (Editors). *Extreme Depositional Environments: Mega End Members in Geologic Time:* Geological Society of America Special Paper 370. p. 43–52.

Kocurek G, Carr M, Ewing R, Havholm KG, Nagar YC, Singhvi AK. 2007. White Sands Dune Field, New Mexico: Age, dune dynamics and recent accumulations. *Sedimentary Geology* 197:313–331.

Kocurek G, Deynoux M, Blakey RC, Havholm KG. 1991a. Amalgamated accumulations resulting from climatic and eustatic changes, Akchar Erg, Mauritania. *Sedimentology* 38:751–772.

Kocurek G, Ewing RC. 2005. Aeolian dune field self-organization—implications for the formation of simple versus complex dune-field patterns. *Geomorphology* 72:94–105.

Kocurek G, Ewing RC, Mohrig D. 2010. How do bedform patterns arise? New views on the role of bedform interactions within a set of boundary conditions. *Earth Surface Processes and Landforms* 35:51–63.

Kocurek G, Havholm K. 1994. Eolian sequence stratigraphy—a conceptual framework. *In* Weimer P, Posamentier H (Editors). *Siliciclastic Sequence Stratigraphy:* American Association of Petroleum Geologists Memoir 58. p. 393–409.

Kocurek G, Knight J, Havholm K. 1991b. Outcrop and semi-regional three-dimensional architecture and reconstruction of a portion of the eolian Page Sandstone (Jurassic). *In* Miall A, Tyler N (Editors). *Three-Dimensional Facies Architecture of Terrigenous Clastic Sediments and its Implications for Hydrocarbon Discovery and Recovery*: SEPM Society for Sedimentary Geology Concepts in Sedimentology and Paleontology 3, Tulsa, OK. p. 25–43.

Kocurek G, Lancaster N. 1999. Aeolian system sediment state: Theory and Mojave Desert Kelso dune field example. *Sedimentology* 46:505–515.

Kocurek G, Robinson NI, Sharp JM. 2001. The response of the water table in coastal aeolian systems to changes in sea level. *Sedimentary Geology* 139:1–13.

Kocurek G, Townsley M, Yeh E, Havholm K, Sweet ML. 1992. Dune and dune-field development on Padre Island, Texas, with implications for interdune deposition and water-table–controlled accumulation. *Journal of Sedimentary Petrology* 62:622–635.

Kok JF. 2010. Differences in the wind speed required for initiation versus continuation of sand transport on Mars: Implications for dunes and dust storms. *Physical Review Letters* 104: doi:10.1103/PRL104.074502

Lancaster N. 1995. *Geomorphology of Desert Dunes:* Routledge, London. 290 p.

Lancaster N, Ollier CD. 1983. Sources of sand for the Namib Sand Sea. *Zeitschrift fur Geomorphologie Supplementband* 45:71–83.

Landry W, Werner BT. 1994. Computer simulations of self-organized wind ripple patterns. *Physica D* 77:238–260.

Langbein W, Schumm S. 1958. Yield of sediment in relation to mean annual precipitation. *Transactions of the American Geophysical Union* 39:1076–1084.

Langford R. 2003. The Holocene history of the White Sands Dune Field and influence on eolian deflation and playa lakes. *Quaternary International* 104:31–39.

Lee P, Thomas P. 1995. Longitudinal dunes on Mars: Relation to current wind regimes. *Journal of Geophysical Research* 36(E3):5381–5395.

Levy JS, Head JW, Marchant DR. 2009. Cold and dry processes in the Martian Arctic: Geomorphic observations at the Phoenix landing site and comparisons with terrestrial cold desert landforms. *Geophysical Research Letters* 36: L21203, doi:10.1029/2009GL04634

Levy JS, Marchant DR, Head JW. 2006. Distribution and origin of patterned ground on Mullins Valley debris-covered glacier, Antarctica: The roles of ice flow and sublimation. *Antarctica Science* 18:385–397.

Madden MEE, Bodnar RJ, Rimstidt JD. 2004. Jarosite as an indicator of water-limited chemical weathering on Mars. *Nature* 431:821–823.

Malin MC, Edgett KS. 2006. Present-day impact cratering rates and contemporary gully activity on Mars. *Science* 314:1573–1577.

Marchant DR, Head JW. 2007. Antarctic dry valleys: Microclimate zonation, variable geomorphic processes, and implications for assessing climate change on Mars. *Icarus* 192:187–222.

Marchant DR, Lewis GH, Phillips WC, Moore EJ, Souchez R, Landis GP. 2002. Formation of patterned-ground and sublimation till over Miocene glacier ice in Beacon Valley, Antarctica. *Geological Society America Bulletin* 114:718–730.

Maria MRV Sta, Rafkin SCR, Michaels TI. 2006. Numerical simulation of atmospheric bore waves on Mars. *Icarus* 185:383–394.

McLennan SM, Grotzinger JP. 2008. The sedimentary rock cycle of Mars. *In* Bell JF (Editor). *The Martian Surface: Composition, Mineralogy, and Physical Properties:* Cambridge University Press, Cambridge, UK. p. 541–577.

Metz JM, Grotzinger JP, Rubin DM, Lewis KW, Squyres SW, Bell JF. 2009. Sulfate-rich eolian and wet interdune deposits, Erebus Crater, Meridiani Planum, Mars. *Journal Sedimentary Research* 79:247–264.

Mountney NP, Jagger A. 2004. Stratigraphic evolution of an aeolian erg margin system: The Permian Cedar Mesa Sandstone, SE Utah, USA. *Sedimentology* 51:713–743.

Nanson GC, Price DM, Short SA. 1992. Wetting and drying of Australia over the past 300 ka. *Geology* 20:791–794.

Nickling WG, McKenna Neuman C, Lancaster N. 2002. Grainfall processes in the lee of transverse dunes, Silver Peak, Nevada. *Sedimentology* 49:191–209.

Paola C, Borgman L. 1991. Reconstructing random topography from preserved stratification. *Sedimentology* 38:553–565.

Parteli EJR, Herrmann HJ. 2007. Dune formation on present Mars. *Physical Review E—Statistical, Nonlinear, and Soft Matter Physics* 76(4) 041307–041315.

Presley MA, Christensen PR. 1997. Thermal conductivity measurements of particle material 2. Results. *Journal of Geophysical Research* 102(E2):6551–6566.

Rubin DM. 1987. Cross-bedding, bedforms, and paleocurrents: SEPM Society for Sedimentary Geology, Concepts in Sedimentology and Paleontology 1. Tulsa, OK. 187 pp.

Rubin DM, Hunter RE. 1982. Bedform climbing in theory and nature. *Sedimentology* 29:121–138.

Rubin DM, Hunter RE. 1987. Bedform alignment in directionally varying flows. *Science* 237:276–278.

Rubin DM, Ikeda H. 1990. Flume experiments on the alignment of transverse, oblique, and longitudinal dunes in directionally varying flows. *Sedimentology* 37:673–684.

Sarnthein M. 1978. Sand deserts during glacial maximum and climatic optimum. *Nature* 272:43–46.

Schatz V, Tsoar H, Edgett KS, Parteli EJR, Herrmann H. 2006. Evidence for indurated sand dunes in the Martian north polar region. *Journal of Geophysical Research* 111: E04006, doi:10.1029/2005JE002514

Schorghofer N, Edgett KS. 2006. Seasonal surface frost at low latitudes on Mars. *Icarus* 180:321–334.

Silvestro S, Di Achille G, Ori GG. 2010a. Dune morphology, sand transport pathways and possible source areas in east Thaumasia region (Mars). *Geomorphology* 121:84–97.

Silvestro S, Fenton LK, Vaz DA, Bridges NT, Ori GG. 2010b. Ripple migration and dune activity on Mars: Evidence for dynamic wind processes. *Geophysical Research Letters* 37: L20203, doi:10.1029/2010GL044743

Singhvi AK, Kar A. 2004. The aeolian sedimentation record of the That Desert. *Proceedings of the Indian Academy of Sciences* 11:371–401.

Spiga A. 2011. Elements of comparison between Martian and terrestrial mesoscale meteorological phenomena: Katabatic winds and boundary layer convection. *Planetary and Space Science,* 59:915–922.

Sullivan R, Arvidson R, Bell JE, Gellert R, Golombek M, Greeley R, Herkenhoff K, Johnson J, Thompson S, Wheeley P, Wray J. 2008. North polar region of Mars: Wind-driven particle mobility on Mars: Insights from Mars exploration rover observations at "El Dorado" and surroundings at Gusev crater. *Journal of Geophysical Research* 113: E06S07, doi:10.1029/2008JE003101

Sweet ML, Kocurek G. 1990. An empirical model of eolian dune lee-face dynamics. *Sedimentology* 37:1023–1038.

Tanaka KL, Fortezzo CM, Hayward RK, Rodriguez JA, Skinner JA. 2010. History of plains resurfacing in the Scandia region of Mars. *Planetary and Space Science,* doi:10.1016/J.PSS.2010.11.004

Tanaka KL, Rodriguez JA, Skinner JA, Bourke MC, Fortezzo CM, Herkenhoff KE, Kolb EJ, Okubo CH. 2008. North polar region of Mars: Advances in stratigraphy, structure, and erosional modification. *Icarus* 196:318–358.

Tosca NJ, Knoll AH. 2009. Juvenile chemical sediments and the long term persistence of water at the surface of Mars. *Earth and Planetary Science Letters* 286:379–386.

Tsoar H, Greeley R, Peterfreund AR. 1979. Mars: The north polar sand sea and related wind patterns. *Journal of Geophysical Research* 84(B14):8167–8180.

Walker IJ, Nickling WG. 2002. Dynamics of secondary airflow and sediment transport over and in the lee of transverse dunes. *Progress in Physical Geography* 26:47–75.

Walker IJ, Nickling WG. 2003. Simulation and measurement of surface shear stress over isolated and closely spaced transverse dunes in a wind tunnel. *Earth Surface Processes and Landforms* 28:1111–1124.

Ward JD, Seely MK, Lancaster N. 1983. On the antiquity of the Namib. *South African Journal of Science* 79:175–183.

Wasson RJ, Rajaguru SN, Misra VN, Agrawal DP, Dhir RP, Singhvi AK, Rao KK. 1983. Geomorphology, Late Quaternary stratigraphy, and paleoclimatology of the Thar dunefield. *Journal of Geomorphology* 45:117–151.

Werner BT. 1995. Eolian dunes: Computer simulations and attractor interpretation. *Geology* 23:1107–1110.

Werner BT. 1999. Complexity in natural landform patterns. *Science* 284:102–104.

Werner BT. 2003. Modeling landforms as self-organized, hierarchical dynamical systems. *In* Wilcock PR, Iverson RM (Editors). *Predictions in Geomorphology, Geophysical Monograph 135:* American Geophysical Union: Washington. p. 133–150.

Wilson I. 1973. Ergs. *Sedimentary Geology* 10:77–106.

Zimbelman JR. 2010. Transverse aeolian ridges on Mars: First results from HiRISE images. *Geomorphology* 121:22–29.

Zolotov MY, Mironenko MV. 2007. Timing of acid weathering on Mars: A kinetic-thermodynamic assessment. *Journal of Geophysical Research* 112: E07006, doi:10:1029/2006JE002882

DUSTSTONES ON MARS: SOURCE, TRANSPORT, DEPOSITION, AND EROSION

NATHAN T. BRIDGES
Applied Physics Laboratory, MP3-E171, 11101 Johns Hopkins Road, Laurel, Maryland 20723 USA
e-mail: nathan.bridges@jhuapl.edu

AND

DANIEL R. MUHS
US Geological Survey, MS 980, Box 25046, Federal Center, Denver, Colorado 80225 USA

Abstract: Dust is an abundant material on Mars, and there is strong evidence that it is a contributor to the rock record as "duststone," analogous in many ways to loess on Earth. Although a common suite of dust formation mechanisms has operated on the two planets, fundamental differences in environments and geologic histories have resulted in vastly different weighting functions, causing distinct depositional styles and erosional mechanisms. On Earth, dust is derived predominantly from glacial grinding and, in nonglacial environments, by other processes, such as volcanism, eolian abrasion, and fluvial comminution. Hydrological and biological processes convert dust accumulations to loess deposits. Active hydrology also acts to clean dust from the atmosphere and convert loess into soil or erode it entirely. On Mars, glacial production of dust has been minor, with most fine particles probably produced from ancient volcanic, impact, and fluvial processes. Dust is deposited under arid conditions in which aggregate growth and cementation are the stabilizing agents. Thick accumulations result in duststone.

Key words: dust, loess, Mars, eolian processes, analogs

INTRODUCTION

The geologic record of a planetary body is reflective of its dominant rock-forming processes, preservation mechanisms, and degrees of erosion and exposure. In the sense that Earth and Mars are both "terrestrial" planets and the laws of physics are universal, analogs are useful because they are amenable to detailed study that can be applied to other environments. However, in many cases, only "partial" analogs are appropriate given the vastly different physical, chemical, and geological conditions between planets. Just as the suite of metamorphic facies on Earth resulting from plate tectonics is rare or absent within Mars' contiguous lithosphere, unique processes in the Martian environment have produced rocks with no direct terrestrial analog. As will be shown in this paper, the vast quantities of dust on Mars have likely formed "duststone." Although analogous to terrestrial loess in its mode of deposition and primary particle size and to yardangs in its erosion style, it stands on its own as a probable newly recognized rock type in the Solar System. Its geologic extent may be as significant as sandstone is on Earth. Major questions concern its genesis, evolution, and longevity. The intent of this paper is to show that duststones exist on Mars and that they are a component of that planet's global dust cycle. By integrating observational- and theoretical-based arguments, we discuss duststone formation and compare it to analogous processes on Earth. This paper is partly a review of the duststone hypothesis proposed by Bridges et al. (2010), but with the addition of a detailed comparison to the dust cycle and deposits on Earth.

To properly frame our discussion and arguments, definitions must be established. For the purposes of this paper, dust is defined as any relatively fine-grained particle capable of being suspended in a planet's atmosphere for a considerable length of time. Theoretically, the maximum size for suspension can be derived by balancing the threshold friction speed (u^*_t) against the terminal fall velocity, with larger-size material predicted to saltate (Greeley and Iversen 1985). In practice, this predicts suspended material grain sizes that are too large. For example, the saltation/suspension boundary for basalt on Mars should be 175 μm (Bridges et al. 2010), despite the fact that eolian ripples, features formed by saltation and creep processes, contain particles less than 100 μm in diameter, which theory predicts should be suspended (Sullivan et al. 2008). This discrepancy between theory and observation may exist because of the lower atmospheric density on Mars compared to Earth, which causes particles to respond more slowly to turbulent eddies that drive suspension (Sullivan et al. 2005, 2011). The size range of suspended dust on Mars is estimated at 2 to 5 μm (Kahn et al. 1992, Lemmon et al. 2004), which is on the lesser side of the distribution of sizes found for loess, but similar to dust aerosols on Earth (Fig. 1). As will be discussed more comprehensively later, dust on the Martian surface commonly aggregates into larger particles and reaches sizes such that it can only move by saltation, or, upon further growth, it becomes immobile. Therefore, we define "dust" as particles that are ones to tens of micrometers in diameter that are suspended or deposited and "dust aggregates" as particles ~100 μm and larger that are on the surface and derived from dust. "Sand" is defined as grains ~100 μm and larger that undergo saltation and are the direct by-product of disintegrated rock.

TERRESTRIAL DUST AND LOESS: ORIGIN AND DEPOSITS

Definitions and Geomorphic Expressions

Although there has been considerable research comparing Martian and terrestrial dunes, the ways in which Martian dust and its depositional character contrast to that on Earth have received less attention. The general lack of geomorphic expression in terrestrial dust and loess deposits is at least part of the reason such analogs have not been extensively recognized or discussed. With increased sophistication in imaging of the Martian surface and atmosphere, it is now possible to make at least some comparisons between fine-grained eolian deposits on the two planets. In this section, we review dust and loess deposits on Earth, including their genesis, geologic record, and relation to climate forcing through time.

Fine-grained eolian deposits on Earth are classified as "loess" and

Sedimentary Geology of Mars
SEPM Special Publication No. 102, Copyright © 2012
SEPM (Society for Sedimentary Geology), Print ISBN 978-1-56576-312-8, CD/DVD ISBN 978-1-56576-313-5, p. 169–182.

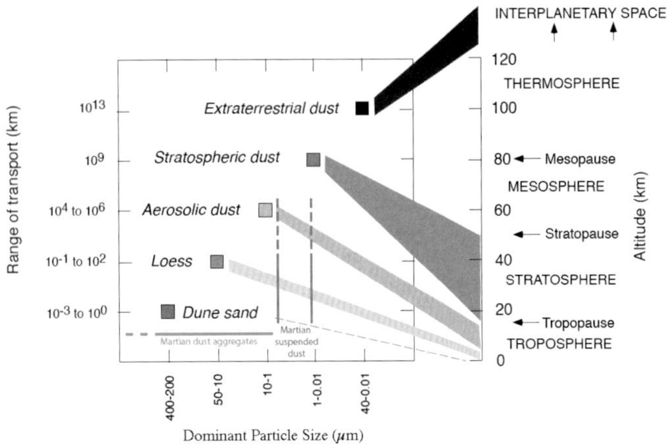

FIG. 1.—Diagram showing the continuum of dust on Earth (solid boxes), with range of transport in the initial phase (in contrast to multiple cycles of movement over long periods of time) as a function of dominant particle size and generalized zone of transport in the atmosphere. The vertical bars show the approximate particle size range of Martian suspended dust, and the horizontal line shows the size of dust aggregates on the planet. The Earth portion of this diagram was generated using concepts in Syers et al. (1969).

"aerosolic dust," unlike the simpler classification of all fine-grained material on Mars as "dust" (see previous). On Earth, these fine-grained materials can be viewed as intermediate members of a continuum of eolian deposits that range from sand to very-fine-grained (<1-μm-diameter) particles that are transported in the stratosphere (Fig. 1). Loess is dominated by silt-sized particles (2–50 μm particle diameters) and is coarser than the material referred to as aerosolic dust (particles <10 μm diameter), which is capable of long-range transport. Typically, loess contains 60–90% silt-sized particles and smaller amounts of sand and clay (< 2μm); it may cover as much as 5–10% of Earth's land surface (Pye 1987), including small areas of Africa and the Middle East, attesting to non-glacial origins for some deposits.

Loess deposits are expressed as distinctive sedimentary bodies. Thickness is highly variable and can range from a few centimeters to several hundred meters, with thickness and mean particle size decreasing downwind away from a source (Pye 1987; Muhs and Bettis 2000, 2003; Muhs et al. 2004, 2008). The deposits commonly have little or no geomorphic expression (Pye 1987), being draped over preexisting landforms as a mantle, with thickest accumulations in protected, low-lying areas or on flat, stable upland divides and thinnest accumulations occurring on narrow, rounded hillcrests (Fig. 2). One exception to the general rule of little geomorphic expression of loess may be "pahas," which are streamlined loess landforms aligned with paleowinds that may be yardang-like features (Flemal et al. 1972). However, similar morphologies also have been tied to primary deposition (Lewis 1960), so, at best, geomorphic expressions of eolian erosion of loess are limited.

Unlike eolian sand or fluvial–marine sediments, primary structures in loess are subtle or absent altogether. Faint, horizontal laminations and cross-bedding are only rarely apparent (Fig. 3). Rather, most loess deposits are characterized by a massive (as opposed to structured) condition. Interparticle binding by phyllosilicate clay minerals or secondary calcite accumulations commonly result in a significant amount of material strength, allowing for the formation of vertical faces along riverbanks or stream banks and road cuts. Secondary structures in loess are more common than primary structures, and these consist of fractures, burrows, rhizoliths (root casts composed of Fe-oxides or carbonate), carbonate nodules or concretions, oxidation or reduction streaks or bands, and paleosols. Lithification or cementation of loess particles to form siltstone is unknown in Quaternary (past 2.6 Ma) loess deposits on Earth, but it has been hypothesized to explain the origin of some Paleozoic and Mesozoic siltstones (Johnson 1989; Soreghan et al. 1997, 2002; Chan 1999).

Loess Composition

Loess has a mineralogy that usually includes quartz, plagioclase, K-feldspar, mica, calcite (and sometimes dolomite), and phyllosilicate clay minerals (smectite, chlorite, mica, and kaolinite) (Pye 1987). Heavy minerals are usually present, but in small amounts. Bulk geochemical studies show that the dominant constituent in loess is SiO_2, ranging from ~45% to 75%, but more typically in the range 55% to 65% (Pye 1987, Muhs and Bettis 2003, Muhs et al. 2008). The high SiO_2 content of loess deposits reflects a dominance of quartz, but smaller amounts of feldspars and clay minerals also contribute to this value. Clay mineralogy in loess shows considerable variability. In North America, for example, clay minerals of loess are a direct reflection of source sediment. In Alaska, loess deposits are ultimately derived from metamorphic rocks in major mountain ranges and are rich in mica and chlorite (Muhs and Budahn 2006). In the midcontinent, loess in the Mississippi River drainage basin is derived to a great extent from Paleozoic carbonate rocks (explaining the high calcite and dolomite) and shales, which explain the high mica content in the clay fraction (Grimley 2000). Loess in the Great Plains region is rich in smectite, derived mainly from Tertiary volcaniclastic siltstone that is also rich in smectite (Aleinikoff et al. 1999, 2008; Muhs et al. 1999, 2008).

Origin of Loess

The "glacial" model of loess formation is a traditional view that silt-sized particles are produced dominantly by glacial grinding of rocks, deposition in till, reworking by fluvial processes as outwash, and finally entrainment, transportation, and deposition by wind (Fig. 4a; see also discussion in Muhs and Bettis 2003). This model has led to the view that loess deposits are primarily markers of global glacial periods. Support for the classical glacial model of silt production and later eolian entrainment as loess comes from modern observations in regions such as Alaska, Canada, and Iceland. In these regions and elsewhere, glaciers exist now, and it is possible to observe loess entrainment from rivers that drain glaciated terrain (e.g., Muhs et al. 2004).

There is also strong evidence for nonglacial loess (Wright et al. 1998; Wright 2001a, 2001b; Smith et al. 2002; Crouvi et al. 2010). "Desert" loess is a term used loosely to describe eolian silt generated in and derived from arid or semiarid regions that were not glaciated. The debate on desert vs. glacial loess centers on whether silt-sized particles can be produced by mechanisms other than glacial grinding, particularly in deserts. This has particular applicability for Mars, where glacial activity has been at most a minor contributor to dust-size particle production over the history of the planet (see following). A variety of mechanisms can, in principle, produce silt-sized particles in arid regions, and these are summarized in a highly simplified model (Fig. 4b). These processes include frost shattering at high altitudes, comminution (particle size reduction by crushing or grinding) by fluvial and mass-movement transport, chemical weathering, salt weathering, eolian abrasion, and ballistic impacts (Wright et al. 1998; Wright 2001a, 2001b; Smith et al. 2002; Crouvi et al. 2010). As with the glacial environment, it is possible to observe silt entrainment in modern desert environments. A good example is the yearly transport of fine silts and clays out of Africa by wind, derived from the Sahara

FIG. 2.—Gallery of loess deposits on Earth. (A) Upper part of last-glacial-age (Peoria) loess exposed in a quarry near Morrison, Illinois; thickness of section exposed is ~8 m. (B) Holocene (upper ~1 m) and last-glacial-age (lower ~8 m) loess exposed in a road cut south of the Yukon River, central Alaska. (C) Multiple loess units (gray bands) with intercalated calcic paleosols (white bands) spanning ~170 ka, exposed in the Ruhama Badlands, northern Negev Desert, Israel (see Wieder et al. 2008); thickness of section is ~15 m. (D) Loess-like, African-dust–derived eolian silt deposits (whitish, upper part of exposure) overlying basaltic tephra of Quaternary age (gray band, in foreground and lower part of exposure), near Famara, Lanzarote, Canary Islands, Spain; thickness of exposure is ~4 m. All photographs are by D.R. Muhs.

and Sahel regions, where glaciation has played no role in silt particle formation. Indeed, although fine particle transport by wind is highly seasonal in Africa, this continent is, overall, the largest source of dust on Earth today.

In the glacial loess-versus-desert loess debate, there is generally little consideration of silt particle inheritance from sedimentary rocks or volcanic tephra. Notable exceptions include sites in Argentina, where an Andean volcaniclastic source has been recognized (Zárate and Blasi 1993), and Australia, where siltstones may be the ultimate source of much of the silt-sized dust in arid basins (McTainsh 1989). Silt is abundant in the sedimentary rock record, and estimates show that fully half of the detrital quartz in the world's sedimentary rocks is composed of silt-sized particles (Blatt 1987). In the Great Plains of North America, sedimentary rock (volcaniclastic siltstone) is the most important source of silt-sized particles in Peoria Loess of last-glacial age (Aleinikoff et al. 1999, 2008; Muhs et al. 1999, 2008). Tephra is commonly composed of silt-sized particles, and where there are active volcanoes, silt-sized tephra is one of the components of loess, such as is the case in Holocene loess in southern Alaska (Muhs et al., 2004). Thus, in both the "glacial" and "desert" models of silt particle formation, inheritance from tephra or siltstone can be added as processes (Fig. 4).

It is therefore likely that loess in many regions has origins from both glacial and nonglacial processes. There is no question that glaciers are efficient producers of silt. Drainage basins in Alaska that have glaciers have significantly higher sediment yield than nearby basins that lack glaciers (Hallet et al. 1996). In Iceland, several meters of loess, mostly from glaciogenic sources, have been produced entirely within Holocene time. On the other hand, laboratory experiments demonstrate that eolian abrasion and fluvial comminution are very efficient silt producers (Wright 2001b), and it has been shown that at least some component of loess in deserts is derived from abrasion of sand-sized

FIG. 3.—Bignell Hill, in the Loess Hills area of western Nebraska, where a road cut (A) exposes the thickest known exposure (~48 m) of last-glacial (ca. 35 ka to ca. 12 ka) loess in the world, overlain by ~2 m of Holocene loess (B). Middle part of last-glacial loess displays rare occurrence of primary structures (laminations) in loess (C). See Muhs et al. (2008) for details of Bignell Hill locality; all photographs are by D.R. Muhs.

particles by wind to silt-sized particles (Crouvi et al. 2010). This issue is relevant for studies of Martian dust and dust deposits because some processes of silt production on Earth do not occur now on Mars and perhaps only operated billions of years ago.

Geologic Records of Loess and Dust

The geologic record of loess on Earth consists of sequences of relatively unaltered loess with intercalated buried soils, or paleosols. Soil formation takes place when vegetation stabilizes the surface, usually because of a decrease in sedimentation rate. Such conditions may arise because of a decrease in wind strength, elimination of the loess–sediment source, or both. In Europe and much of North America, where loess has a glaciogenic origin, it is no surprise that loess in the stratigraphic record represents glacial periods, and paleosols represent interglacial periods (Muhs and Bettis 2003, Muhs 2007). In China, although the immediate sources of loess are desert basins upwind from the Loess Plateau, the loess is actually of glaciogenic origin, derived from mountain glaciers that surround the desert basins (Sun 2002). Thus, in China as well as in regions of Europe and North America near glaciated terrain, loess represents glacial periods, and paleosols represent interglacial periods that can be correlated with the deep-sea oxygen isotope record of glacial–interglacial cycles. Elsewhere in North America (e.g., the Great Plains), where loess is nonglaciogenic, ages of loess also correspond to glacial periods (Muhs et al. 1999, 2008; Roberts et al. 2003). Therefore, although loess in this region is not dependent on glacial sources, climatic conditions during glacial periods are apparently favorable for loess entrainment, transportation, and deposition.

Geologic records of aerosolic dust flux occur in a number of depositional settings for sediments, including deep-ocean basins, lakes, large ice sheets, and soils. Deep-sea sediments are one of the great archives of glacial–interglacial cycles of the Quaternary. Foraminifera in deep-sea cores faithfully record the ocean temperature changes and fluxes in the oxygen isotope composition of seawater that accompany the shifts between glacial and interglacial climates. The same deep-sea cores may also retain long-range–transported aerosolic dust, if fluvial transport and other means of sediment origin can be ruled out or accounted for. Several decades of study have shown that both the Pacific and Atlantic Oceans have long records of eolian sediment transport from Asia and Africa, respectively (Rea 2007). The coincidence of dust flux maxima with glacial periods indicates that glacial-climate conditions favored greater dust production and transport.

Thus, loess records from both glacial and nonglacial sources, the eolian record in deep-sea sediments, and the eolian record in polar ice caps all indicate that glacial periods on Earth correspond to times of greater dust flux. It is likely that stronger winds, lower vegetation density, drier climates in dust source areas, and a decreased intensity of the hydrological cycle all combined to enhance dust flux repeatedly in glacial periods of the Quaternary (Mahowald et al. 1999). During such times, which approach conditions on Mars, Earth was a far dustier planet than it is today. Quaternary glacial–interglacial cycles on Earth are controlled, to a great extent, by astronomical cycles of precession and tilt of Earth's axis and, to a lesser extent, orbital eccentricity (Berger 1992). These cycles are particularly critical for the distribution of solar insolation in summer and therefore affect ice-sheet growth and decay in the Northern Hemisphere of Earth. As discussed later herein, astronomical cycles also exert controls on Martian dust cycles.

THE ORIGIN OF MARTIAN DUST

The first geologic process discovered on Mars is also its most pervasive: Global dust storms were first observed telescopically in the nineteenth century and tend to occur annually (Martin et al. 1992, Newman 2001). Spacecraft observations show that smaller-scale dust storms and vortices are also frequent. It is estimated that 2.9×10^{12} kg yr^{-1} of dust is exchanged between the surface and atmosphere (Pollack et al. 1977). If evenly coated over the surface, this would form a layer just a few micrometers thick. However, as on Earth, the distribution and thickness of dust are nonuniform. Based on radar measurements, accumulations of dust in high-albedo, low-thermal inertia regions are estimated at up to 1 to 2 m from radar measurements (Christensen 1986), and in Arabia, thicknesses of 20 m and greater are based on the lack of small impact craters (Mangold et al. 2009) (a map showing Arabia and other Mars locations presented in the text is shown in Fig. 5). The polar layered deposits at both poles, with a thickness of a few kilometers, have a significant dust component (Kahn et al. 1992, Thomas et al. 1992, Smith et al. 1999, Phillips et al. 2008). Fundamental questions include: How did so much dust form, and how is it preserved on the surface today?

Based on remote sensing and in situ measurements, the composition of Martian dust is dominated by framework silicates, mostly feldspar, with lesser amounts of sulfate, pyroxene, olivine, amorphous phases, hematite, and magnetite—a mineralogy consistent with the mechanical

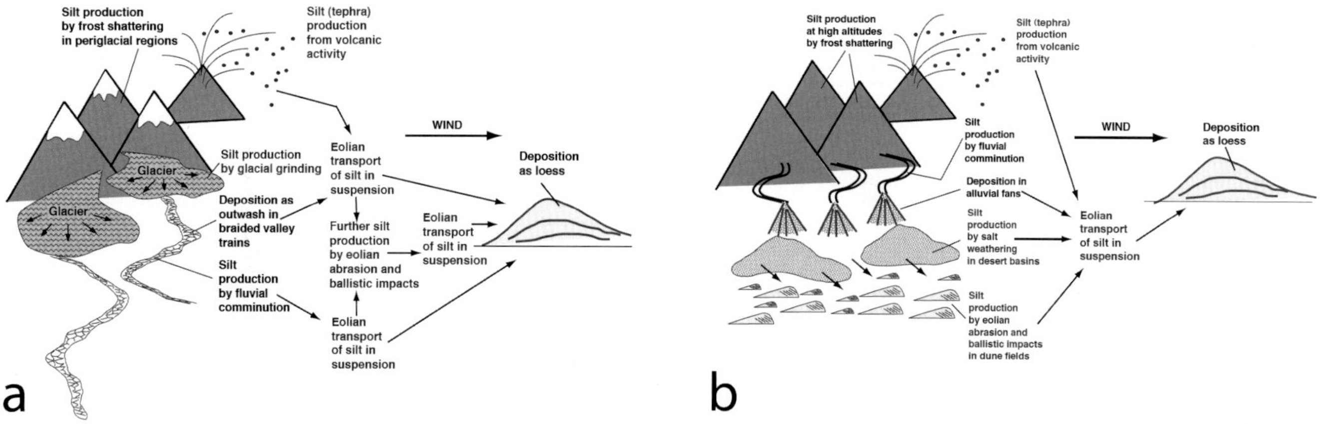

FIG. 4.—(a) Classical model of "glacial" loess formation on Earth wherein silt-sized particles are produced primarily by glacial grinding, delivered to outwash streams, and finally entrained by wind. (b) Model of "desert" loess formation on Earth wherein silt-sized particles are produced by a variety of nonglaciogenic processes before eventual entrainment by wind. Processes hypothesized to be active on both Earth and Mars are shown in gray text. Redrawn from Muhs and Bettis (2003).

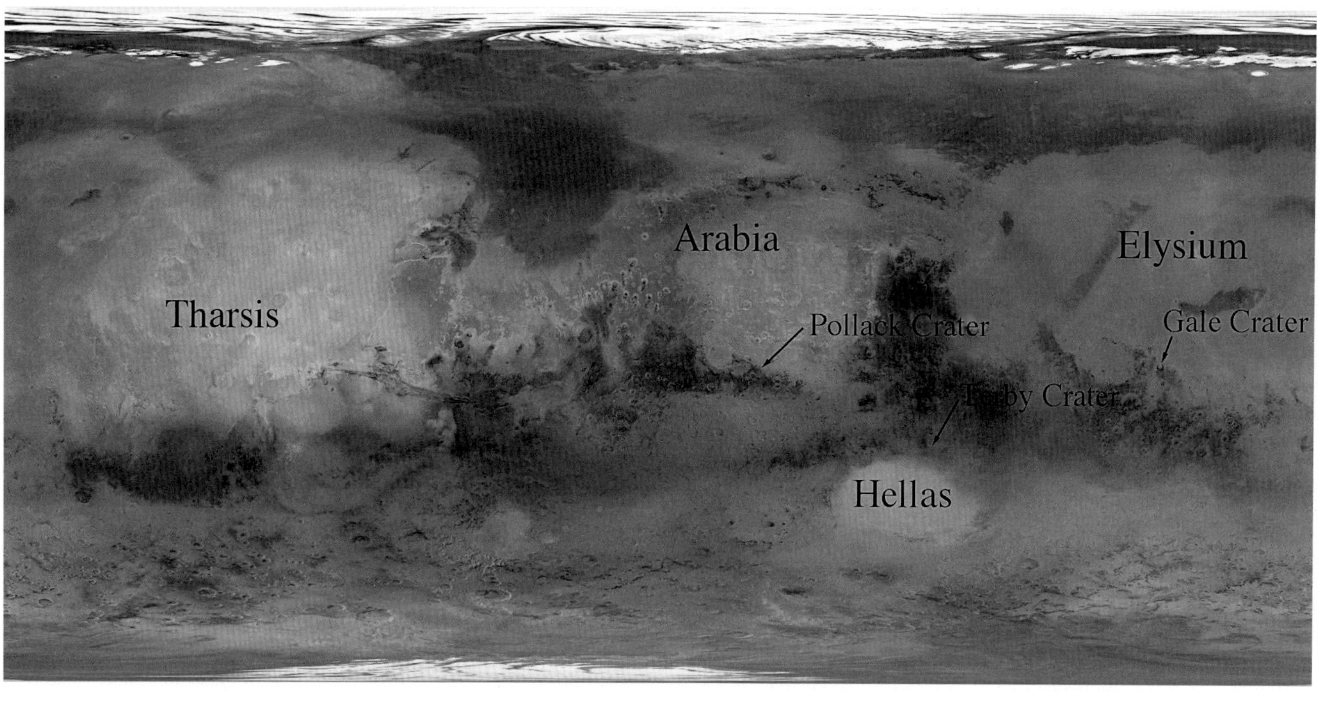

FIG. 5.—Mars map showing geographic features mentioned in the text.

weathering of basalt (Hamilton et al. 2005). This is distinct from the dominantly granitic (or rocks derived from granites) source for the majority of terrestrial loess. The presence of olivine in Martian dust in particular indicates that water did not play a dominant role in dust formation, with genesis occurring during dry periods of Martian history (Goetz et al. 2005).

There are several processes that can potentially produce Martian dust (Fig. 6). Mars has abundant volcanoes and lava flows, some of which have a morphology consistent with pyroclastic eruptions, such that ancient tephra was probably a significant contributor. Impacts have pulverized the surface, releasing large quantities of ejecta, including dust-size material. Meteoritic infall provides a steady source of particulate materials. Based on the impact cratering record, ~8.3 cm of material has been added to Mars, if distributed evenly over the surface, since the early–mid Noachian (>4 Ga) (Yen et al. 2006). The flux of nonmelted meteoritic material that reaches the Martian surface is estimated at 8.6×10^6 kg yr^{-1} (Flynn 1996), yielding ~8 cm of material (assuming basalt density), consistent with the cratering-derived estimate. Erosion associated with outflow channels, valley networks, other fluvial features, and possibly glacial grinding has disaggregated rocks and coarse soil components.

All of these processes were largely confined to the Noachian and

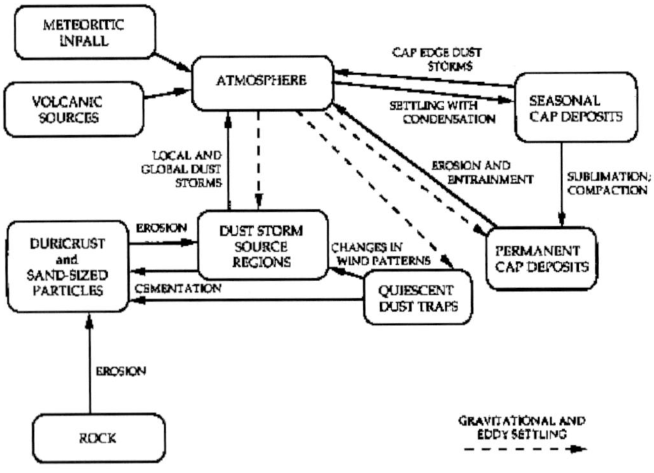

FIG. 6.—Schematic diagram of the Martian dust cycle. Reprinted with permission from Kahn et al. (1992).

Hesperian Periods before ca. 3.5 to 1.8 Ga (Tanaka et al. 1992). These high-energy processes (except for minor meteoritic infall) do not occur today, such that current rates of dust production are considerably lower. Probably the most significant dust-forming mechanism in the modern era is the abrasion and deflation of the relatively soft rock that constitutes yardangs and eolian mantles over much of Mars. Much of this material is likely duststone, the erosion of which recycles dust back into the Martian eolian system. More minor processes include eolian abrasion of rocks and soil by sand, rounding and disintegration of this sand in saltation and abrasion, thermal cycling of rocks, and possibly salt weathering (Wells and Zimbelman 1997). On Earth, precipitation cleanses dust from the atmosphere, and oceans and other water bodies serve as permanent sinks. On land, vegetation acts to stabilize dust and form soils. With Mars lacking these characteristics, dust is not easily removed from the eolian system.

MARTIAN DUST TRANSPORT AND DEPOSITION

Numerous wind tunnel and theoretical studies have demonstrated that the threshold for lofting dust that is ones to tens of micrometers in size by wind, regardless of the planetary surface, exceeds that needed for fine sand (Greeley and Iversen 1985). This is because dust is more easily sheltered by roughness elements, it resides partially within a laminar sublayer, and its low weight makes it more susceptible to electrostatic and other cohesive forces (Iversen et al. 1976, Shao and Lu 2000). Friction speeds needed to raise dust on Mars are on the order of 10 m s^{-1}, which is about an order of magnitude greater than on Earth (Greeley and Iversen 1985). These speeds correspond to free-stream winds over 100 m s^{-1} up to several kilometers per second—conditions rarely, if ever, achieved. So, under standard conditions, linear winds on their own cannot raise dust, and other mechanisms must be considered to account for the significant load of suspended dust in the atmosphere.

Some of the highest wind speeds on Mars are attained by katabatic flow off the retreating seasonal polar cap, combined with diurnal winds flowing from high elevations (Lee et al. 1982, Magalhaes and Gierasch 1982, Kahn et al. 1992). The magnitude of these winds is enhanced in the Southern Hemisphere, where the proximity of perihelion in late spring to the summer solstice produces strong gradients (Wells and Zimbelman 1997). These winds can exceed the threshold friction speeds of ~2 m s^{-1} necessary to saltate sand. The flux of dust lofted by saltating sand is proportional to $(u^*/u^*_t)^3$ (White 1979, Newman et al. 2002) or $(u^*/u^*_t)^4$ (Westphal et al. 1987, Haberle et al. 2003), such that even winds slightly above sand threshold can cause large dust fluxes. Vortices induced by surface–atmosphere thermal instabilities can also saltate sand and provide enhanced lift to raise dust, producing dust devils (Greeley et al. 2004). More minor contributors to dust suspension include removal of dust on pebbles and rocks, which, being higher in the wind speed profile and not sheltered by nearby roughness elements, are subjected to greater shear stress (White et al. 1997), and small-scale surface disruptions such as dust-rich landslides (Sullivan et al. 2001). On the other hand, dust aggregates with sizes equivalent to sand are easily moved (Fig. 7). This likely accounts for the presence of ripples in dust-rich regions (Bridges et al. 2010) and is a component of the duststone model that will be presented herein.

In local events, such as dust devils, dust is transported only a few kilometers. However, global storms can raise dust as high as 30 to 40 km or more, where it is subjected to transport by Hadley cells from the southern latitudes in the summer to the middle and temperate northern latitudes (Toon et al. 1980, Pollack et al. 1981, Magalhaes 1987, Haberle et al. 1993). In the latter stages of global storms, the high-altitude dust acts to dampen the lapse rate (temperature change with altitude), decreasing convective turbulence and causing dust to fall out (Haberle 1986, Wells and Zimbelman 1997). Grains can also serve as condensation nuclei for CO_2 and H_2O, enhancing dust deposition in polar regions (Kahn et al. 1992).

On Mars today, the coincidence of perihelion with southern spring results in global storms originating in the Southern Hemisphere and net dust deposition in the equatorial and northern latitudes via the cross-equatorial Hadley cell. Dust is spread nonuniformly, with enhanced deposition in preexisting high-albedo, low-thermal-inertia, low-rock-abundance regions, specifically, Tharsis, Elysium, and Arabia (Fig. 5). Why these regions serve as dust sinks is not completely understood. One hypothesis is that they may have initially formed in areas of low wind velocity, allowing dust to bury sand and rocks, thereby removing sand saltation triggers and roughness elements that could enhance turbulence to otherwise drive dust into suspension, thereby increasing the net accumulation rate (Christensen 1988).

The argument of perihelion has a period of ~51,000 years, such that the hemisphere where global storms originate changes every 25,000 years or so, with alternate epochs perhaps having a net north–south transport (Christensen 1986, Kahn et al. 1992, Kieffer and Zent 1992). Under this model, net dust sources and sinks alternate in hemispheric polarity in response to orbital precession. If true, then remnant, thick dust/duststone deposits may exist over much of Mars. These and other hypotheses of Martian dust removal and deposition are simplifications, as terrestrial experience and detailed models show that erosion and deposition are fundamentally driven by the mass balance per unit volume of surface (Exner 1925, Kubatko and Westerink 2007). In detail, it may be that some areas of "dust sinks" on Mars undergo periods of erosion and abrasion, and some "sources" may locally accumulate sediment. In our study of Mars, we are presently limited to observations that reflect broad-scale processes integrated over long time periods.

EVIDENCE FOR DUSTSTONE

Three classes of observations provide compelling evidence for the existence of duststone on Mars. Those are outlined next, along with interpretations based on these observations.

(1) Thermophysical and Geochemical Properties in Some Regions of Mars Are Consistent with Duststone

Observations: Cohesion and geochemical properties are correlated in Martian soil, with cohesive cloddy materials and surface crusts enriched in sulfur and chlorine (Clark et al. 1982; Rieder et al. 1997, 2004; Moore et al. 1999). The formation of this "duricrust" is believed

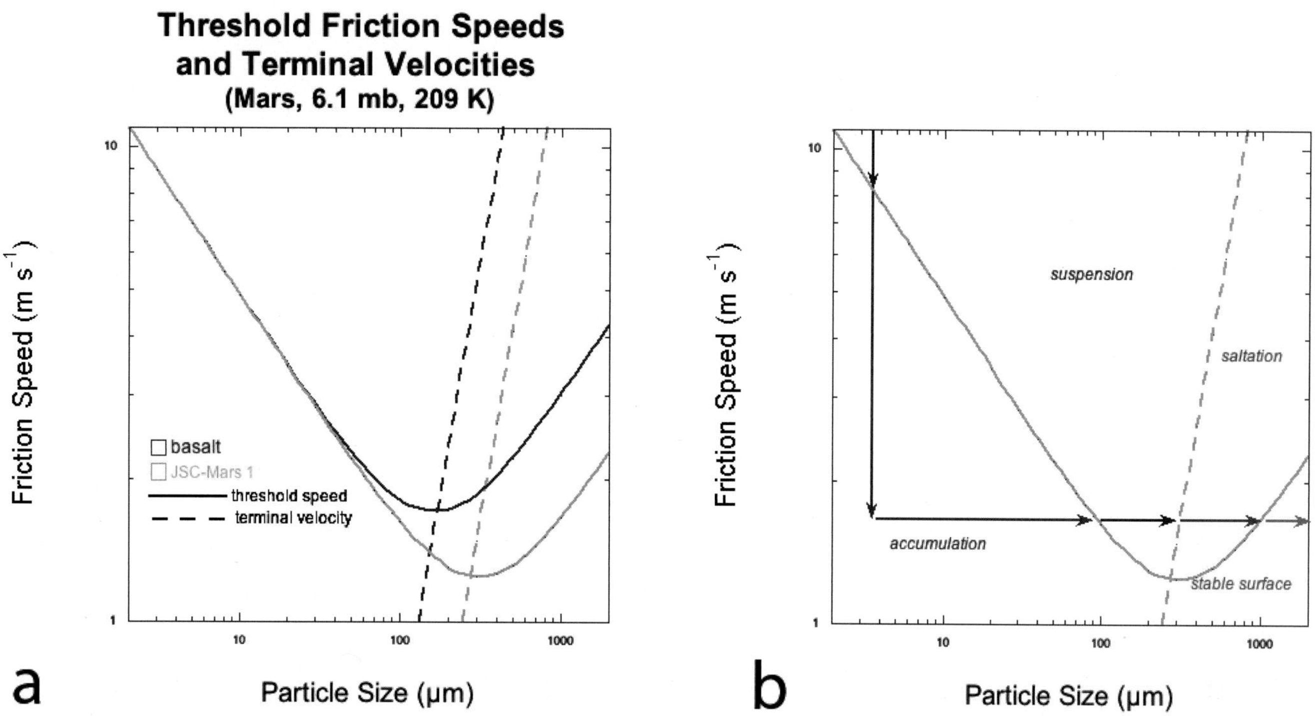

FIG. 7.— (a) Friction speed versus particle size for basalt and JSC-Mars-1, a Martian soil stimulant (Allen et al. 1998) with properties similar to that of Martian dust (Bridges et al. 2010). The solid curves are calculated threshold friction speeds using the parameterization of Shao and Lu (2000). The dotted lines represent the terminal fall velocity that theoretically corresponds to the suspension–saltation boundary. Details of these calculations are in Bridges et al. (2010). (b) Curves for JSC-Mars 1, with labels designating the dominant dynamic behavior of the particles within zones bounded by the threshold curve and terminal fall line. The bulk density of the particles is assumed to be constant. When winds fall below threshold, dust comes out of suspension and accumulates. Similarly, any saltating sand will stop moving, resulting in a stable surface. Arrows represent the evolution of dust particles: (1) Dust stays suspended in the atmosphere, falling to the surface in quiescent periods when friction speeds drop below threshold (blue and red vertical arrows, with blue showing zone at which the dust can be resuspended and red representing lower wind speeds for which dust accumulates). (2) Dust on the surface clumps into aggregates (red horizontal arrow). (3) The dust grows in size until it is above the threshold curve and can be resuspended (blue horizontal arrow). (4) Eventually the aggregates reach sufficient size such that they are too heavy to get suspended (so below the terminal velocity line, yet above the threshold curve). These particles, shown by the green arrow, will undergo saltation. (5) Further aggregation will again put the particles below the threshold curve, such that they remain stable (purple arrow). Cementation and burial harden the aggregates into "duststone."

to occur via the mobilization of S and Cl by water-vapor diffusion through the soil (Jakosky and Christensen 1986a) and may account for variations in thermal inertia outside the lowest inertia areas (Jakosky and Christensen 1986a, 1986b; Christensen and Moore 1992). Mapped at coarse scale by the Thermal Emission Spectrometer (TES), duricrust is not exposed everywhere, but rather it makes up one of four distinct thermophysical units, with the corresponding "Unit C" having moderate to high thermal inertia and intermediate albedo (Mellon et al. 2000). This unit consistently borders "Unit A," in which low inertia and high albedo indicate a dust cover. Tharsis is relatively rich in Cl and depleted in Si and Fe in its upper ~50 cm, as indicated by Gamma Ray Spectrometer (GRS) data (Karunatillake et al. 2009).

Interpretations: The thermophysical properties of Unit C, combined with its association with Unit A, are consistent with thick cemented dust (Mellon et al. 2000). In the dusty Unit A regions, a dust cover thicker than a couple of centimeters (the diurnal skin depth) can mask any underlying indurated material. Therefore, cemented dust could exist in these regions as well, or be exposed as outcrops at spatial scales finer than the TES data (~3 km per pixel). In support of this hypothesis, Tharsis, within Mellon et al.'s Unit A, exhibits large seasonal variability and diurnal differences in the apparent thermal inertia that can be attributed to a layered surface, such as loose dust covering an indurated layer, although horizontal mixing and subpixel slope effects are other possible causes (Putzig and Mellon 2007a, 2007b). Finally, in Tharsis, the chlorine-rich composition detected by GRS is consistent with an indurated soil composition.

(2) Dust Aggregates, the Precursors to Duststone, Are Common on Mars

Observations: Images from the Microscopic Imager on the Mars Exploration Rovers (MERs) show dust aggregates that reach sizes of several hundred micrometers (Herkenhoff et al. 2004, 2006, 2008; Sullivan et al. 2008), equivalent to that of sand. The much thinner and drier atmosphere on Mars makes dust much more prone to electrostatic forces than in the Earth environment (Greeley 1979, Krinsley and Leach 1981), such that aggregates and "cloddy material" (as classified at the *Viking* landing sites) can form (Shorthill et al. 1976, Greeley 1979). In addition to those at the landing sites, aggregates may make up low-thermal-inertia dunes and materials of the north polar erg (Herkenhoff and Vasavada 1999), a region also proposed to contain

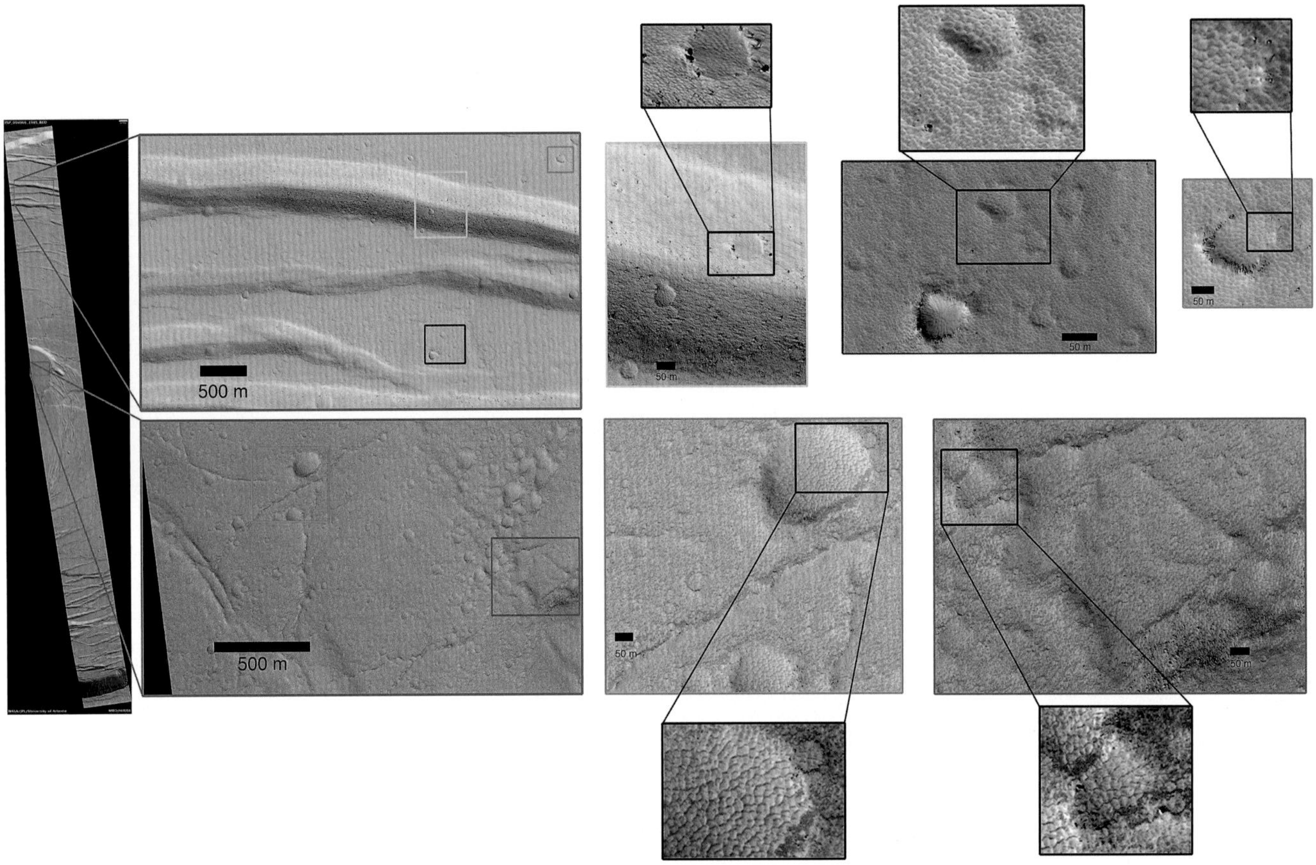

FIG. 8.—Reticulate bed forms within the Olympus Mons caldera. Note that the bedforms superpose underlying structural, volcanic, and crater topography. HiRISE footprint is at left. Boxes at left show locations of enlargements to the right. Black boxes within those show outline of stretched enlargements above and below (HiRISE image PSP_004966_1985).

"loess" deposits (Tanaka et al. 2008). Layered deposits in the Hellas Basin and nearby Terby Crater have been proposed to form from fine-grained material such as dust or loess (Wilson et al. 2007). It has also been conjectured that "parna sheets," analogous to clay aggregate lunette duneforms on Earth, may exist on Mars, although none have been definitively identified (Greeley and Williams 1994).

Interpretation: Dust aggregates are common on Mars, especially in areas for which remote-sensing observations indicate dust-rich compositions. Aggregates will continue to grow until they are disrupted by wind forces. In other words, dust that is ones to tens of micrometers in size that is found in the atmosphere is not likely to remain in this state on the surface, but rather to clump together with other dust particles and aggregates.

(3) There Is Geomorphological and Remote-Sensing Evidence for Thick Dust Deposits

Observations: High-resolution orbital images from the High Resolution Imaging Science Experiment (HiRISE) camera show that regions on Mars classified as dusty–Unit A (e.g., Tharsis, Arabia, and Elysium) contain suites of fine-scale reticulate bedforms intermixed with wind-sculpted, bright bedrock (Bridges et al. 2010) (Figs. 8, 9). The bedrock is commonly composed of blunt-to-tapered outcrops for which plan-view shape and texture are analogous to terrestrial yardangs. Earth-based radar measurements restrict the depth of a cover of fines to 1 to 2 m on average (Christensen 1986), consistent with a dusty surface above a more indurated substrate. In Tharsis, images of stratigraphic cross sections showing contacts with underlying basalt demonstrate that the light-toned, presumably dust-rich, bedform surfaces sit atop a mantle at least several meters thick (Keszthelyi et al. 2008) (Fig. 10). In Arabia, thickness of dusty material is estimated at 20 m or more (Mangold et al. 2009).

Interpretations: Because the thermophysical properties of these areas indicate a dust-rich composition, and particles with diameter of ones to tens of micrometers are suspended by winds above threshold, Bridges et al. (2010) proposed that the reticulate bedforms were formed from the saltation of dust aggregates hundreds of micrometers in size (Fig. 7b). Furthermore, because aggregates approaching a millimeter in size are below threshold, it was also suggested that continued growth would also act to stabilize the particles. This, combined with the formation of duricrust and subsequent burial by younger dust deposits, would form a semidurable "duststone" that, when exposed, would be subjected to wind abrasion processes.

The putative duststone rock may exist in other areas of Mars (Fig. 11). Buried and exhumed terrain is common on the planet (Malin and Edgett 2001, Frey et al. 2002, Edgett 2005), and much of the middle latitudes is covered by a mantle composed of ice-permeated dust that is many meters thick (Mustard et al. 2001). In some craters that were probably wholly or partially exhumed, light-toned rock masses are apparent. This includes Pollack Crater (containing "White Rock"; Fig.

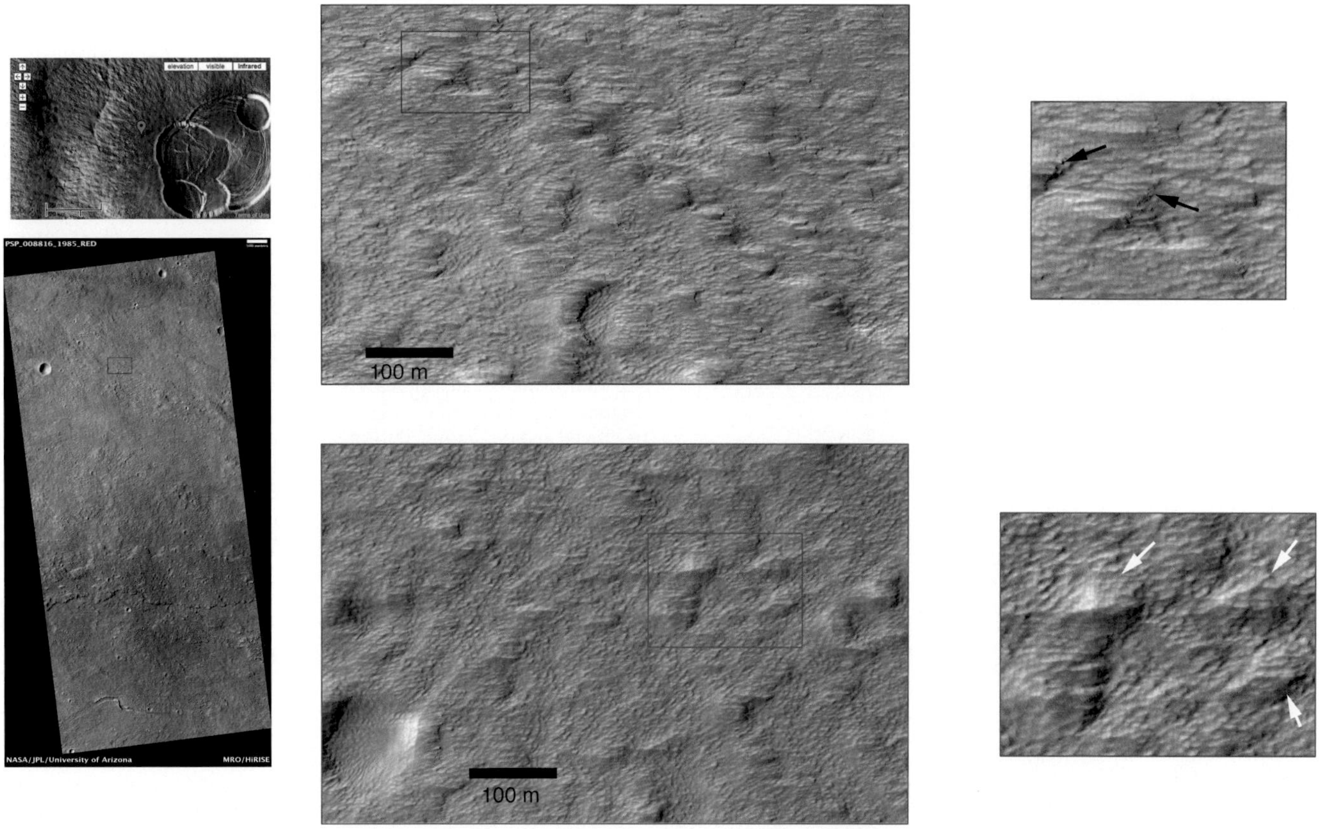

FIG. 9.—Details of yardangs on the western flank of Olympus Mons in HiRISE image PSP_008816_1985. Note the blunt to tapered edge of most yardangs and the wind tails in the lee of rocks, going from right to left, consistent with downslope winds. Black arrows in the upper right subframe show rocks and linear ridges to the west interpreted as wind tails. The white arrows in the lower right subframe show yardangs. Upper-left frame shows the location of the HiRISE footprint, at lower left.

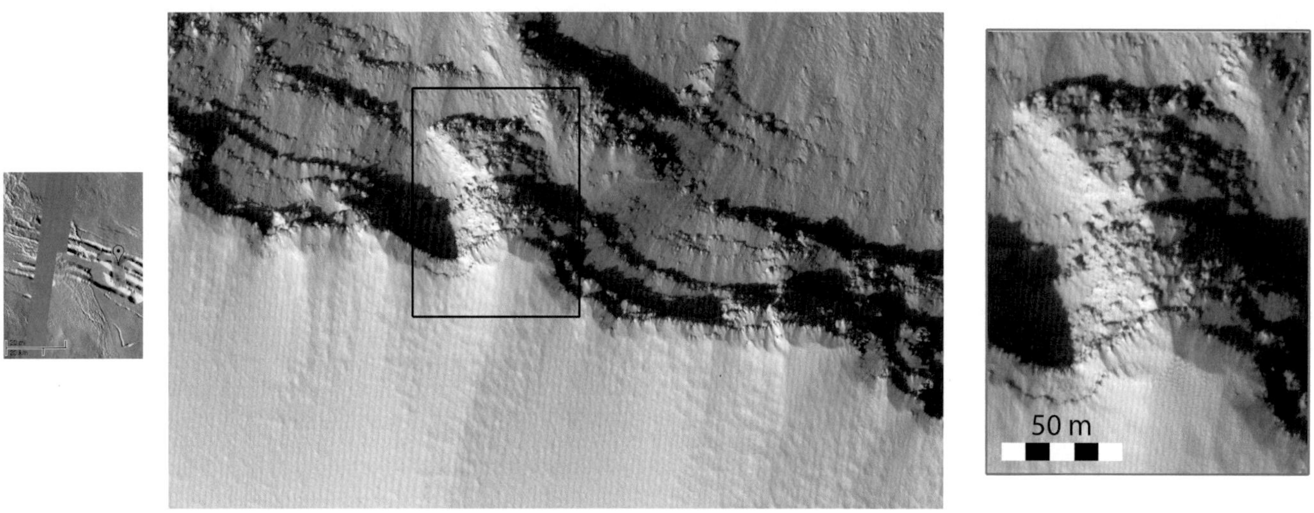

FIG. 10.—HiRISE image PSP_005651_1840 at the edge of the northern flank of Pavonis Mons. Left frame shows the location of the HiRISE footprint. Middle image shows details of bedrock outcrops, with enlargement of the red boxed area at right showing a thick dust cover.

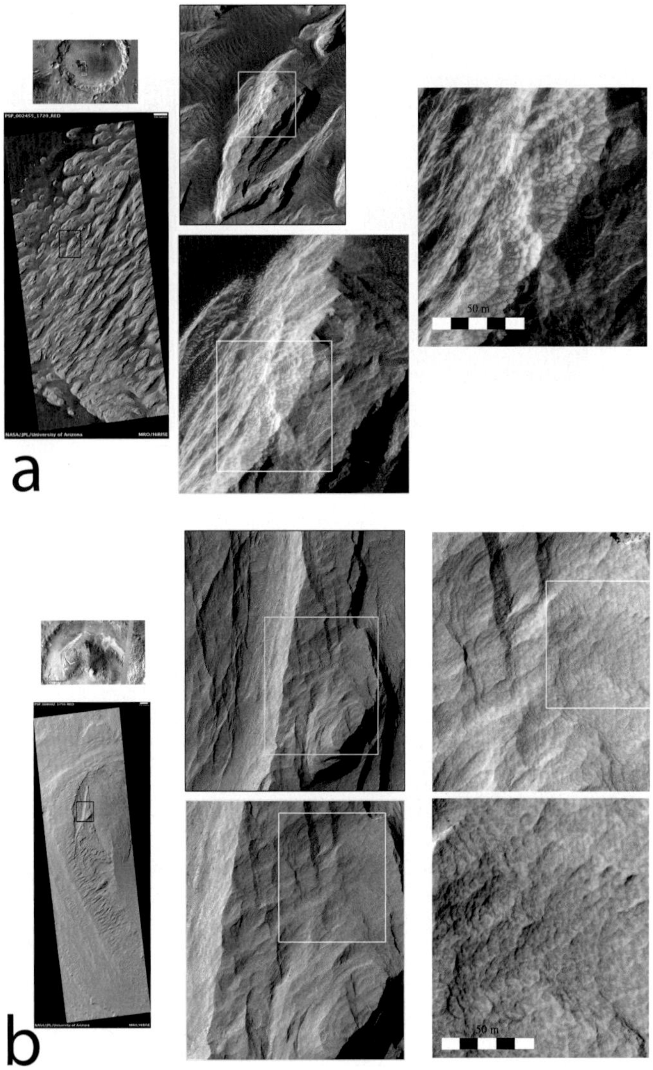

FIG. 11.—Two examples of possible duststone outside of low-thermal-inertia regions of Tharis, Arabia, and Elysium: (a) "White Rock" in Pollack Crater. Upper-left frame shows the location of the HiRISE footprint at lower left. A succession of boxes shows increasing detail of the rock texture (HiRISE image PSP_002455_1720). (b) The upper mound of Gale Crater. Upper-left frame shows the location of the HiRISE footprint at lower left. A succession of boxes shows increasing detail of the rock texture (HiRISE image PSP_008002_1750). Note that both White Rock and the Gale upper mound have a scalloped surface, with patterns similar to the reticulate bed forms and scoured surfaces seen in Figures 8 and 9, respectively.

11a) and the upper rock masses of the Gale Crater mound (Fig. 11b), both of which have remote-sensing properties consistent with indurated dust (Ruff et al. 2001, Milliken et al. 2010). A light-toned outcrop in Becquerel Crater and the upper Gale mound also contain nonconformable, rhythmically bedded rock masses with stratigraphic bundles consistent with astronomically-forced mechanisms (Lewis et al. 2008, Milliken et al. 2010) such as is expected for cyclical air-fall dust.

THE MARTIAN DUSTSTONE MODEL

The large volume of dust on Mars and the unique environment of its deposition compared to Earth account for the formation of duststone on that planet as opposed to our own. We present a hypothesis, modified after Bridges et al. (2010), for the formation of duststone that we believe is consistent with current data and suggest several measurements that can be performed by future missions to test it. The major stages of formation are:

(1) Dust is deposited on the Martian surface from atmospheric fallout. In "sinks," accumulation exceeds removal, and thick deposits form. Dust clumps into aggregates via electrostatic processes. It eventually reaches a large enough size that it can no longer be suspended when subjected to high-speed winds and instead saltates and forms small-scale ripples, such as those currently seen in the dusty Tharsis, Elysium, and Arabia regions. Many aggregates will be destroyed and comminuted to smaller particles. Depending on the pore space of the growing aggregates, others may remain above the threshold curve and be removed by the winds. However, a fraction, those with great enough density and those not destroyed, will grow to sufficient size such that they are below the threshold curve, resulting in a stable surface. This stable population increases over time (Fig. 7b).

(2) Once the surface is dominated by particles having a size below threshold, it becomes further stabilized through the formation of duricrust through multiple cycles of water-vapor diffusion (Jakosky and Christensen 1986a).

(3) Some stabilized surfaces will be buried by new dust deposits, which are, in turn, stabilized as well. Compaction of the underlying dust converts the cemented material into harder duststone.

(4) If subsequently exposed, the duststone will be subjected to wind abrasion, depending on the thickness and extent of local particulate materials. In many settings, this will be variable, as seen in Tharsis today, with some exposed duststone undergoing erosion and other units being sheltered. A similar scenario is envisioned for putative duststones in craters such as Pollack and Gale, although in these cases the nature of the former burying material is less certain as these areas are not currently dust sinks, although they may have been in past epochs. Sand serves as the main abrasive agent where it is present, given the much greater effectiveness of its ability to erode rock (Laity and Bridges 2009). However, in areas like Tharsis, where sand appears lacking, duststone should only be abraded if it is weakly cemented, because the main abrading material is probably dust aggregates.

(5) Over time, lithified duststone is eroded, forming scalloped yardangs and remnant rock masses. Many of the eroded particulates probably rejoin the eolian system as suspended grains or as saltating sand-size particles that are incorporated into the soil or break apart and go into suspension.

This model does not consider the role of ice. It has been proposed that the desiccated mantle covering much of the Martian midlatitudes is air-fall dust cemented by ice derived from diffused atmospheric water vapor under favorable orbital configurations (Mustard et al. 2001). This is analogous to the polar layered deposits being composed of dust–ice mixtures (Thomas et al. 1992). Such an ice-rich mantle is more like permafrost than bedrock, although desiccation is predicted to produce loose loess-like materials (Mustard et al. 2001), so that remnant masses could be considered duststone.

Finally, not all light-toned rocks on Mars are duststone. For example, the morphology of the wind-eroded Medusae Fossae Formation and its location near the Tharsis and Elysium volcanic constructs are consistent with a pyroclastic origin (Ward 1979, Scott and Tanaka 1982, Bradley et al. 2002, Mandt et al. 2008, de Silva et al. 2010). Fluvial processes have probably contributed to the formation of the lower parts of the light-toned layer mound in Gale Crater (Milliken et al. 2010, Thomson et al. 2011). It is likely that other light-toned rocks on Mars have similarly diverse origins. Our hypothesis is that many, but not all, light-toned rocks, especially those associated with thick dust deposits, are duststone.

CONTRIBUTION TO THE MARTIAN ROCK RECORD

The rates of erosion of duststone, and the ways in which these balance against rates of formation, are not known. Given that the dust cycle on Mars seems to have been occurring since the Noachian, it is likely that cycles of deposition, stability, and erosion have been ongoing ever since. Duststones will be protected from erosion for the duration over which they are buried, such that there could be very old formations beneath the surface today or recently exposed formations. Given the low rate of primary dust production on Mars in the current epoch, the rates of duststone formation versus erosion are directly tied into the question of whether dust abundance overall is increasing or decreasing. Answers to this and related questions cannot be found until more data on rock formation and exposure ages are provided by samples or in situ measurements.

With greater confidence we can say that the range of rock types currently forming on Mars is far more limited than it was in the Noachian and Hesperian. Volcanic and fluvial processes that produced suites of igneous and water-based sedimentary rocks are infrequent today. Thus, any preserved Noachian and Hesperian duststones are probably intermixed with strata formed by other geologic processes. Given their air-fall nature, duststones could serve as temporal marker beds analogous to ash and ash-derived clays on Earth. In the modern era, Martian rock production is limited to the movement of sand and the dust cycle, combined with processes to stabilize and lithify these materials. We therefore consider duststone, perhaps with limited sandstone production, as the dominant rock-forming processes in the current era. This idea can be tested by future in situ missions that visit areas believed to contain duststone.

COMPARISONS BETWEEN THE DUST CYCLES ON EARTH AND MARS, AND CONSIDERATIONS OF OTHER PLANETARY BODIES

When we define the dust cycle as describing the production of particles with diameters of ones to tens of micrometers, followed by their deposition and removal or incorporation into the geologic record, there are important commonalities and differences between Mars and Earth. Building upon the previous sections, we compare and contrast the dust cycles on the two planets and conclude with some general statements about the role of dust on other planetary bodies. Some of the ideas here are broad generalizations and are presented as working models that, we hope, will spur further research into the role of dust in the planetary context.

Some geologic processes that Earth and Mars have in common are those that act to comminute rock into finer material that can subsequently be suspended and deposited into accumulations of many meters or more. Although hydrological and particularly glacial processes have been more prevalent on Earth, with the latter considered one of the main producers of loess, the mechanism by which dust is formed has little impact on its subsequent evolution. In other words, it is sufficient to state that dust is formed in abundance on Earth and Mars, and that the modes of deposition, stabilization, accumulation, and removal are the steps in the process where environmental effects exert the strongest biases.

The deposition of dust on Mars is much more uniform than on Earth, where thicknesses are correlated to distance from source. Because Martian dust can be transported vast distances without being flushed out by precipitation or being affected by ocean-induced global circulation patterns, deposition can occur on global scales. Even the net south–north hemispheric bias in the current epoch is counterbalanced, over the long term, by the Martian precessional cycle.

Processes that remove dust compete against those acting to stabilize it, and these mechanisms are significantly different on Mars and Earth. Wind can obviously remove dust on both planets, but the much higher threshold speeds needed in Mars' low-density atmosphere make areas of low wind intensity or surface roughness much more prone to serve as dust sinks than is the case on Earth, where dust in virtually all locations can be resuspended. On Earth, fluvial erosion is an additional and significant removal agent and one that is completely absent on Mars. Therefore, unconsolidated dust is much more likely to remain stable on Mars.

On the other hand, dust stabilization on Mars seems to require some combination of aggregation, cementation, and burial. No liquid water is involved. In contrast, the active terrestrial hydrosphere and biosphere acts to alter primary dust and lesser amounts of sand and clay into a weakly cemented sediment, commonly bound by carbonate cement and clays. Such a "loessification" process is not expected on Mars today. Pre-Quaternary loessite, though probably requiring water for its diagenesis, may be a closer analog to Martian dust deposits in that it is better cemented and buried.

The erosion of dust deposits on the two planets is likewise quite different. Although loess can be derived from arid settings on Earth, it rarely accumulates in significant amounts in deserts, tending to get transported to areas where it can be stabilized by vegetation. Subsequently, pedogenesis begins, and soil formation dominates until further loess deposition takes place. Even pre-Quaternary loessite has identifiable paleosols (Soreghan et al. 1997, 2002). As such, the erosional expression of loess on Earth is poorly defined geomorphically, although there is no question that there are stratigraphic records of loess erosion seen as unconformities (see Muhs et al. [2003] for examples in Alaska). This is in marked contrast to Mars, where all dust accumulates in dry settings, and subsequent erosion is confined to eolian abrasion and deflation, forming yardangs. In Earth's deserts, yardangs can be composed of sandstone (quartz, carbonate, or gypsum-dominated) or fine-grained sediments, such as silts and clays.

The role of dust on Mars is unique in the Solar System. Unlike on Earth, it is almost entirely unaffected by hydrological and biological alteration processes that can accumulate into thick sequences. Once deposited, it becomes part of the sedimentary strata, as dust or duststone, or reenters the eolian system through deflation or abrasion. On Earth, about 23% of the surface is sandstone (Ronov 1983), or about twice the amount of loess coverage. Given the dustiness on Mars, the volume of duststone could be as proportionally significant as sandstone is on Earth. This can be verified or refuted by future missions and data analysis.

The differences between Mars and the other worlds in our Solar System with solid surfaces and significant atmospheres (e.g., Venus and Titan) are even more distinct. Venus lacks a hydrosphere, and dust production from volcanic sources is strongly inhibited by the great atmospheric pressure (93 bars), which reduces volatile exsolution (Head and Wilson 1986). Although *Magellan* radar shows evidence for wind features associated with dust (Greeley et al. 1992), production of fine particles is probably limited to impact processes. Vast deposits of dust like those on Mars or in loess on Earth are not seen. Titan has dunes that are probably composed of frozen organics (Barnes et al. 2008). Dust-size organics or other materials may be present on the surface, but the methane "hydrological" cycle (Tokano et al. 2006) may act to flush it out. Expanding our horizons to extra-solar planets, Mars-like bodies with thin atmospheres having limited or no hydrologic cycles could also have dust that is continually produced and accumulates over time. These may yet be discovered in future investigations.

CONCLUSIONS

Dust formation, transport, deposition, stabilization, and erosion are active geologic processes on Earth and Mars. Dust accumulates where it can be stabilized and will not be subjected to resuspension. On Earth, stabilization is driven by hydrological and biological processes. On Mars, accumulation occurs in dust sinks that may switch hemispheric polarity in response to the precessional cycle. The comparison to loess is therefore informative in understanding the ways in which dust on two terrestrial planets can accumulate and then have divergent evolutionary paths. On Mars, which lacks liquid water and biology, dust is stabilized by growth of micrometer-sized dust into aggregates and cementation. Continual accumulation will compact these materials, forming duststone. The role of dust on the two planets is similar in that thick accumulations can form, but the different environments result in divergent evolutionary paths.

ACKNOWLEDGMENTS

Comments from D. Rubin, J. Pigati, and an anonymous reviewer significantly improved this paper.

REFERENCES

Aleinikoff JN, Muhs DR, Bettis EA III, Johnson WC, Fanning CM, Benton R. 2008. Isotopic evidence for the diversity of late Quaternary loess in Nebraska: glaciogenic and non-glaciogenic Sources. *Geological Society of America Bulletin* 120:1362–1377.

Aleinikoff JN, Muhs DR, Sauer R, Fanning, CM. 1999. Late Quaternary loess in northeastern Colorado, Part II—Pb isotopic evidence for the variability of loess sources. *Geological Society of America Bulletin* 111: 1876–1883.

Allen CC, Morris RV, Jager KM, Golden DC, Lindstrom DJ, Lindstrom MM, Lockwood JP. 1998. Martian regolith simulant JSC Mars-1. *Lunar and Planetary Science* XXIX:1690.

Barnes JW, Brown RH, Soderblom L, Sotin C, Le Mouèic S, Rodriguez S, Jaumann R, Beyer RA, Buratti BJ, Pitman K, Baines KH, Clark R, Nicholson P. 2008. Spectroscopy, morphometry, and photoclinometry of Titan's dunefields from Cassini/VIMS. *Icarus* 195:400–414.

Berger AL. 1992. Astronomical theory of paleoclimates and the last glacial–interglacial cycle. *Quaternary Science Reviews* 11:571–581.

Blatt H. 1987. Oxygen isotopes and the origin of quartz. *Journal of Sedimentary Petrology* 57:373–377.

Bradley BA, Sakimoto SEH, Frey H, Zimbelman JR. 2002. Medusae Fossae Formation: new perspectives from Mars Global Surveyor. *Journal of Geophysical Research–Planets* 107. DOI:10.1029/2001JE001537

Bridges NT, Banks ME, Beyer RA, Chuang FC, Noe Dobrea EZ, Herkenhoff KE, Keszthelyi LP, Fishbaugh KE, McEwen AS, Michaels TI, Thomson BJ, Wray JJ. 2010. Eolian bedforms, yardangs, and indurated surfaces in the Tharsis Montes as seen by the HiRISE camera: evidence for dust aggregates. *Icarus* 205:165–182.

Chan MA. 1999. Triassic loessite of north-central Utah: stratigraphy, petrophysical character and paleoclimate implications. *Journal of Sedimentary Research* 69:477–485.

Christensen PR. 1986. Regional dust deposits on Mars: physical properties, age, and history. *Journal of Geophysical Research* 91:3533–3545.

Christensen PR. 1988. Global albedo variations on Mars: implications for active eolian transport, deposition, and erosion. *Journal of Geophysical Research* 93:7611–7624.

Christensen PR, Moore HJ. 1992. The Martian surface layer. *In* Kieffer HH, Jakosky BM, Snyder CW, Matthews MS (Editors). *Mars*: University of Arizona Press, Tucson, Arizona. p. 686–729.

Clark BC, Baird AK, Weldon RJ, Tsusaki DM, Schnabel L, Candelaria MP. 1982. Chemical composition of Martian fines. *Journal of Geophysical Research* 87:10059–10067.

Crouvi O, Amit R, Enzel Y, Gillespie AR. 2010. Active sand seas and the formation of desert loess. *Quaternary Science Reviews* 29:2087–2098.

de Silva SL, Bailey JE, Mandt KE, Viramonte JM. 2010. Yardangs in terrestrial ignimbrites: synergistic remote and field observations on Earth with applications to Mars. *Planetary and Space Science* 58:459–471.

Edgett KS. 2005. The sedimentary rocks of Sinus Meridiani: five key observations from data acquired by the Mars Global Surveyor and Mars Odyssey orbiters. *Mars: The International Journal of Mars Science and Exploration* 1:5–58.

Exner FM. 1925. Uber die Wechselwirkung zwischen Wasser und Geschiebe in Flussen. *Sitzunsberichte Adademie der Wessenschaften in Wien, Mathematische–Naturwissenschaftliche Klasse Abt. 2A*, 134:183–228.

Flemal RC, Odom IE, Vail RG. 1972. Stratigraphy and origin of paha topography of northwestern Illinois. *Quaternary Research* 2:232–243.

Flynn GJ. 1996. The delivery of organic matter from asteroids and comets to the early surface of Mars. *Earth, Moon, and Planets* 72:469–474.

Frey HV, Roark JH, Shockey KM, Frey EL, Sakimoto SEH. 2002. Ancient lowlands of Mars. *Geophysical Research Letters* 29. DOI:10.1029/2001GL013832

Goetz W, Bertelsen P, Binau CS, Gunnlaugsson HP, Hviid SF, Kinch KM, Madsen DE, Madsen MB, Olsen M, Gellert R, Klingelhöfer G, Ming DW, Morris RV, Rieder R, Rodionov DS, de Souza PA Jr, Schröder C, Squyres SW, Wdowiak T, Yen A. 2005. Indication of drier periods on Mars from the chemistry and mineralogy of atmospheric dust. *Nature* 436:62–65.

Greeley R. 1979. Silt–clay aggregates on Mars. *Journal of Geophysical Research* 84:6248–6284.

Greeley R, Arvidson RE, Elachi C, Geringer MA, Plaut JJ, Saunders RS, Schubert G, Stofan ER, Thouvenot EJP, Wall SD, Weitz CM. 1992. Aeolian features on Venus: preliminary Magellan results. *Journal of Geophysical Research* 97:13319–13345.

Greeley R, Iversen JD. 1985. *Wind as a Geological Process*. Cambridge University Press, Cambridge, UK. 333 p.

Greeley R, Whelley PL, Neakrase LDV. 2004. Martian dust devils: directions of movement inferred from their tracks. *Geophysical Research Letters* 31. DOI:10.1029/2004GL021599

Greeley R, Williams SH. 1994. Dust deposits on Mars: the "Parna" Analog. *Icarus* 110:165–177.

Grimley DA. 2000. Glacial and nonglacial sediment contributions to Wisconsin Episode loess in the central United States. *Geological Society of America Bulletin* 112:1475–1495.

Haberle RM. 1986. Interannual variability of global dust storms on Mars. *Science* 234:459–461.

Haberle RM, Murphy JR, Schaeffer J. 2003. Orbital change experiments with a Mars general circulation model. *Icarus* 161:66–89.

Haberle RM, Pollack JB, Barnes JR, Zurek RW, Leovy CB, Murphy JR, Lee H, Schaeffer J. 1993. Mars atmospheric dynamics as simulated by the NASA Ames general circulation model, 1. The zonal–mean circulation. *Journal of Geophysical Research* 98:3093–3123.

Hallet B, Hunter L, Bogen J. 1996. Rates of erosion and sediment evacuation by glaciers: A review of field data and their implications. *Global and Planetary Change* 12:213–235.

Hamilton VE, McSween HY, Hapke B. 2005. Mineralogy of Martian atmospheric dust inferred from thermal infrared spectra of aerosols. *Journal of Geophysical Research* 110. DOI:10.1029/2005JE002501

Head JW, Wilson L. 1986. Volcanic processes and landforms on Venus—theory, predictions, and observations. *Journal of Geophysical Research* 91:9407–9446.

Herkenhoff KE, Golombek MP, Guinness EA, Johnson JB, Kussack A, Richter L, Sullivan R, Gorevan S. 2008. In-situ observations of the physical properties of the Martian surface. *In* Bell JF III (Editor), *The Martian Surface: Composition, Mineralogy, and Physical Properties*: Cambridge University Press, Cambridge, UK. p. 451–467.

Herkenhoff KE, Squyres SW, Arvidson R, Bass DS, Bell JF III, Bertelsen P, Cabrol NA, Gaddis L, Hayes AG, Hviid SF, Johnson JR, Kinch KM, Madsen MB, Maki JN, McLennan SM, McSween HY, Rice JW Jr, Sims M, Smith PH, Soderblom LA, Spanovich N, Sullivan R, Wang A. 2004. Textures of the soils and rocks at Gusev Crater from Spirit's Microscopic Imager. *Science* 305:824–826.

Herkenhoff KE, Vasavada AR. 1999. Dark material in the polar layered deposits and dunes on Mars. *Journal of Geophysical Research* 104:16487–16500.

Herkenhoff KE, Squyres SW, Anderson R, Archinal BA, Arvidson RE, Barrett

JM, Becker KJ, Bell JF III, Budney C, Cabrol NA, Chapman MG, Cook D, Ehlmann BL, Farmer J, Franklin B, Gaddis LR, Galuska DM, Garcia PA, Hare TM, Howington-Kraus E, Johnson JR, Johnson S, Kinch K, Kirk RL, Lee EM, Leff C, Lemmon M, Madsen MB, Maki JN, Mullins KF, Redding BL, Richter L, Rosiek MR, Sims MH, Soderblom LA, Spanovich N, Springer R, Sucherski RM, Sucharski T, Sullivan R, Torson JM, Yen A. 2006. Overview of Microscopic Imager investigation during Spirit's first 450 sols in Gusev Crater. *Journal of Geophysical Research* 111. DOI:10.1029/JE002574.

Iversen JD, Pollack JB, Greeley R, White BR. 1976. Saltation threshold on Mars: the effect of interparticle force, surface roughness, and low atmospheric density. *Icarus* 29:381–393.

Jakosky BM, Christensen PR. 1986a. Are the Viking Lander sites representative of the surface of Mars? *Icarus* 66:125–133.

Jakosky BM, Christensen PR. 1986b. Global duricrust of Mars - Analysis of remote-sensing data. *Journal of Geophysical Research* 91:3547–3559.

Johnson SY. 1989. Significance of loessite in the Maroon Formation (Middle Pennsylvanian to Lower Permian), Eagle Basin, northwest Colorado. *Journal of Sedimentary Petrology* 59:782–791.

Kahn RA, Martin TZ, Zurek RW, Lee SW. 1992. The Martian dust cycle. *In* Kieffer HH, Jakosky BM, Snyder CW, Matthews MS (Editors). *Mars*: University of Arizona Press, Tucson, Arizona. p. 1017–1053.

Karunatillake S, Wray JJ, Squyres SW, Taylor GJ, Gasnault O, McLennon SM, Boynton W, El Maary MR, Dohm JM. 2009. Chemically striking regions on Mars and stealth revisited. *Journal of Geophysical Research* 114. DOI:10.1029/2008JE003303

Keszthelyi L, Jaeger W, McEwen A, Tornabene L, Beyer RA, Dundas C, Milazzo M. 2008. High Resolution Imaging Science Experiment (HiRISE) images of volcanic terrains from the first 6 months of the Mars Reconnaissance Orbiter Primary Science Phase. *Journal of Geophysical Research* 113. DOI:10.1029/2007JE002968

Kieffer HH, Zent AP. 1992. Quasi-periodic climate change on Mars. *In* Kieffer HH, Jakosky BM, Snyder CW, Matthews MS (Editors). *Mars*: University of Arizona Press, Tucson, Arizona. p. 1180–1218.

Krinsley D, Leach R. 1981. Properties of electrostatic aggregates and their possible presence on Mars. *Precambrian Research* 14:167–178.

Kubatko EJ, Westerink JJ. 2007. Exact discontinuous solutions to Exner's bed evolution model: simple theory for sediment bores. *Journal of Hydraulic Engineering* 133. DOI:10.1061/(ASCE)0733-9429(2007)133:3(305)

Laity JE, Bridges NT. 2009. Ventifacts on Earth and Mars: analytical, field, and laboratory studies supporting sand abrasion and windward feature development. *Geomorphology* 105:202–217.

Lee SW, Thomas PC, Veverka J. 1982. Wind streaks in Tharsis and Elysium: implications for sediment transport by slope winds. *Journal of Geophysical Research* 87:10025–10041.

Lemmon MT, Wolff MJ, Smith MD, Clancy RT, Banfield D, Landis GA, Ghosh A, Smith PH, Spanovich N, Whitney B, Whelley P, Greeley R, Thompson S, Bell JF III, Squyres SW. 2004. Atmospheric imaging results from the Mars Exploration Rovers: Spirit and Opportunity. *Science* 204:1753–1756.

Lewis KW, Aharonson O, Grotzinger JP, Kirk RL, McEwen AS, Suer TA. 2008. Quasi-periodic bedding in the sedimentary rock record of Mars. *Science* 322:1532–1535.

Lewis PF. 1960. Linear topography in the southwestern Palouse, Washington–Oregon. *Annals of the Association of American Geographers* 50:98–111.

Magalhaes JA. 1987. The Martian Hadley circulation: comparison of 'viscous' model predictions to observations. *Icarus* 70:442–468.

Magalhaes J, Gierasch P. 1982. A model of Martian slope winds: implications for eolian transport. *Journal of Geophysical Research* 87:9975–9984.

Mahowald N, Kohfeld K, Hansson M, Balkanski Y, Harrison SP, Prentice IC, Schulz M, Rodhe H. 1999. Dust sources and deposition during the last glacial maximum and current climate: A comparison of model results with paleodata from ice cores and marine sediments. *Journal of Geophysical Research* 104:15895–15916.

Malin MC, Edgett KS. 2001. Mars Global Surveyor Mars Orbiter Camera: interplanetary cruise through primary mission. *Journal of Geophysical Research* 106:23429–23570.

Mandt K, de Silva SL, Zimbelman JR, Crown DA. 2008. The origin of the Medusae Fossae Formation, Mars: insights from a synoptic approach. *Journal of Geophysical Research* 113. DOI:10.1029/2008JE003076

Mangold N, Ansan V, Masson P, Vincendon C. 2009. Estimate of the eolian dust thickness in Arabia Terra, Mars: implications of a thick mantle ($\geq$20 m) for hydrogen detection. *Géomorphologie: Relief, Processus, Environnement*:23–32.

Martin LJ, James PB, Dollfus A, Iwasaki K, Beish JD. 1992. Telescopic observations: visual, photographic, polarimetric. *In* Kieffer HH, Jakosky BM, Snyder CW, Matthews MS (Editors). *Mars*: University of Arizona Press, Tucson, Arizona, p. 34–70.

McTainsh GH. 1989. Quaternary aeolian dust processes and sediments in the Australian region. *Quaternary Science Reviews* 8:235–253.

Mellon MT, Jakosky BM, Kieffer HH, Christensen PR. 2000. High-resolution thermal inertia mapping from the Mars Global Surveyor Thermal Emission Spectrometer. *Icarus* 148:437–455.

Milliken RE, Grotzinger JP, Thomson BJ. 2010. Paleoclimate of Mars as captured by the stratigraphic record of Gale Crater. *Geophysical Research Letters* 37. DOI:10.1029/2009GL041870

Moore HJ, Bickler DB, Crisp JA, Eisen HJ, Gensler JA, Haldemann AFC, Matijevic JR, Reid LK, Pavlics F. 1999. Soil-like deposits observed by Sojourner, the Pathfinder rover. *Journal of Geophysical Research* 104:8729–8746.

Muhs DR. 2007. Loess deposits, origins, and properties. *In* Elias S (Editor). *The Encyclopedia of Quaternary Sciences*: Elsevier, Amsterdam. p. 1405–1418.

Muhs DR, Ager TA, Bettis EA III, McGeehin J, Been JM, Begét JE, Pavich MJ, Stafford TW Jr, Stevens DSP. 2003. Stratigraphy and paleoclimatic significance of late Quaternary loess–paleosol sequences of the last interglacial–glacial cycle in central Alaska. *Quaternary Science Reviews* 22:1947–1986.

Muhs DR, Aleinikoff JN, Stafford TW Jr, Kihl R, Been J, Mahan SA, Cowherd S. 1999. Late Quaternary loess in northeastern Colorado, part I—age and paleoclimatic significance. *Geological Society of America Bulletin* 111:1861–1875.

Muhs DR, Bettis EA III. 2000. Geochemical variations in Peoria Loess of western Iowa indicate paleowinds of midcontinental North America during last glaciations. *Quaternary Research* 53:49–61.

Muhs DR, Bettis EA III. 2003. Quaternary loess–paleosol sequences as examples of climate-driven sedimentary extremes. *In* Chan MA, Archer AW (Editors). *Extreme Depositional Environments: Mega End Members in Geologic Time:* Geological Society of America, Boulder, Colorado, Special Paper 370, p. 53–74.

Muhs DR, Bettis EA III, Aleinikoff J, McGeehin JP, Beann J, Skipp G, Marshall BD, Roberts HM, Johnson WC, Benton R. 2008. Origin and paleoclimatic significance of late Quaternary loess in Nebraska: evidence from stratigraphy, chronology, sedimentology, and geochemistry. *Geological Society of America Bulletin* 120:1378–1407.

Muhs DR, Budahn R. 2006. Geochemical evidence for the origin of late Quaternary loess in central Alaska. *Canadian Journal of Earth Sciences* 43:323–337.

Muhs DR, McGeehin JP, Beann J, Fisher E. 2004. Holocene loess deposition and soil formation as competing processes, Matanuska Valley, southern Alaska. *Quaternary Research* 61:265–276.

Mustard JF, Cooper CD, Rifkin MK. 2001. Evidence for recent climate change on Mars from the identification of youthful near-surface ground ice. *Nature* 412:411–414.

Newman CE. 2001. Modeling the dust cycle in the Martian atmosphere [unpublished Ph.D. thesis]: St. Anne's College, University of Oxford, UK, 494 p.

Newman CE, Lewis SR, Read PL, Forget F. 2002. Modeling the Martian dust cycle 1. Representations of dust transport processes. *Journal of Geophysical Research* 107. DOI:10.1029/2002JE001910

Phillips RJ, Zuber MT, Smrekar SE, Mellon MT, Head JW, Tanaka KL, Putzig NE, Milkovich SM, Campbell BA, Plaut JJ, Safaeinili A, Seu R, Biccari D, Carter LM, Picardi G, Orosei R, Surdas Mohit P, Heggy E, Zurek RW, Egan AF, Giacomoni E, Russo F, Cutigni M, Pettinelli E, Holt JW, Leuschen CJ, Mariangeli L. 2008. Mars north polar deposits: stratigraphy, age, and geodynamical response. *Science* 320:1182–1185.

Pollack JB, Colburn D, Kahn R, Hunter J, Van Camp W, Carlston CE, Wolf MR.

1977. Properties of aerosols in the Martian atmosphere, as inferred from Viking Lander imaging data. *Journal of Geophysical Research* 82:4479–4496.

Pollack JB, Leovy CB, Greiman PW, Mintz Y. 1981. A Martian general circulation experiment with large scale topography. *Journal of Atmospheric Science* 38:3–29.

Putzig NE, Mellon MT. 2007a. Apparent thermal inertia and the surface heterogeneity of Mars. *Icarus* 191:68–94. DOI:10.1016/j.icarus.2007.05.013

Putzig NE, Mellon MT. 2007b. Thermal behavior of horizontally mixed surfaces on Mars. *Icarus* 191:52–67. DOI:10.1016/j.icarus.2007.03.022

Pye K. 1987. *Aeolian Dust and Dust Deposits*: Academic Press, San Diego, California. 334 p.

Rea DK. 2007. Eolian records, deep-sea sediments. *In* Elias S (Editor). *The Encyclopedia of Quaternary Sciences*: Elsevier, Amsterdam. p. 643–649.

Rieder R, Economou T, Wanke H, Turkevich A, Crisp J, Brückner J, Dreibus G, McSween HY. 1997. The chemical composition of Martian soil and rocks returned by the Mobile Alpha Proton X-ray Spectrometer: preliminary results from the X-ray mode. *Science* 278:1771–1774.

Rieder R, Gellert R, Anderson RC, Brückner J, Clark BC, Dreibus G, Economou T, Klingelhöfer G, Lugmair GW, Ming DW, Squyres SW, d'Uston C, Wänke H, Yen A, Zipfel, J. 2004. Chemistry of rocks and soils at Meridiani Planum from the Alpha Particle X-ray Spectrometer. *Science* 306:1746–1749.

Roberts HM, Muhs DR, Wintle AG, Duller GAT, Bettis EA. 2003. Unprecedented last glacial mass accumulation rates determined by luminescence dating of loess from western Nebraska. *Quaternary Research* 59:411–419.

Ronov AB. 1983. *The Earth's Sedimentary Shell: Quantitative Patterns of its Structure, Composition and Evolution*: American Geological Institute, Alexandria, Virginia, Reprint Series 5, 80 p.

Ruff SW, Christensen PR, Clark RN, Kieffer HH, Malin MC, Bandfield JL, Jakosky BM, Lane MD, Mellon MT, Presley MA. 2001. Mars "White Rock" feature lacks evidence of an aqueous origin: results from Mars Global Surveyor. *Journal of Geophysical Research* 106:23921–23927.

Scott DH, Tanaka KL. 1982. Ignimbrites of Amazonis Planitia region of Mars. *Journal of Geophysical Research* 87:1170–1190.

Shao Y, Lu H. 2000. A simple expression for wind erosion threshold friction velocity. *Journal of Geophysical Research* 105:22437–22443.

Shorthill RW, Hutton RE, Moore HJ, Scott RF, Spitzer CR. 1976. Physical properties of Martian surface from Viking 1 Lander: preliminary results. *Science* 193:805–809.

Smith BJ, Wright JS, Whalley WB. 2002. Sources of non-glacial, loess-size quartz silt and the origins of "desert loess." *Earth-Science Reviews* 59:1–26.

Smith DE, Zuber MT, Solomon SC, Phillips RJ, Head JW, Garvin JB, Banerdt WB, Muhleman DO, Pettengill GH, Neumann GA, Lemoine FG, Abshire JB, Aharonson, O, Brown CD, Hauck SA, Ivanov AB, McGovern PJ, Zwally HJ, Duxbury TC. 1999. The global topography of Mars and implications for surface evolution. *Science* 284:1495–1503.

Soreghan GS, Elmore RD, Katz B, Cogoini M, Banerjee S. 1997. Pedogenically enhanced magnetic susceptibility variations preserved in Paleozoic loessite. *Geology* 25:1003–1006.

Soreghan GS, Elmore RD, Lewchuk MT. 2002. Sedimentologic–magnetic record of western Pangean climate in upper Paleozoic loessite (lower Cutler beds, Utah). *Geological Society of America Bulletin* 114:1019–1035.

Sullivan R, Arvidson R, Bell JF III, Gellert R, Golombek M, Greeley R, Herkenhoff K, Johnson J, Thompson S, Whelley P, Wray J. 2008. Wind-driven particle mobility on Mars: insight from Mars Exploration Rover observations at "El Dorado" and surroundings at Gusev Crater. *Journal of Geophysical Research* 113. DOI:10.1029/2008JE003101

Sullivan R, Banfield D, Bell JF, Calvin W, Fike D, Golombek M, Greeley R, Grotzinger J, Herkenhoff K, Jerolmack D, Malin M, Ming D, Soderblom LA, Squyres SW, Thompson S, Watters WA, Weitz CM, Yen A. 2005. Aeolian processes at the Mars Exploration Rover Merididiani Planum landing site. *Nature* 436. DOI:10.1038/nature03641.

Sullivan R, Bandfield D, Collins LR, Heineck JT, Korda DT. 2011. Determining the minimum saltation grain size on Mars. *Lunar and Planetary Science* XLXII:2651.

Sullivan R, Thomas P, Veverka J, Malin M, Edgett KS. 2001. Mass movement slope streaks imaged by the Mars Orbiter Camera. *Journal of Geophysical Research* 106:23607–23633.

Sun J. 2002. Provenance of loess material and formation of loess deposits on the Chinese Loess Plateau. *Earth and Planetary Science Letters* 203:845–859.

Syers JK, Jackson ML, Berkheiser VE, Clayton RN, Rex RW. 1969. Eolian sediment influence on pedogenesis during the Quaternary. *Soil Science* 107:421–427.

Tanaka KL, Rodriguez JAP, Skinner JA, Bourke MC, Fortezzo CM, Herkenhoff KE, Kolb EJ, Okubo CH. 2008. North polar region of Mars: advances in stratigraphy, structure, and erosional modification. *Icarus* 196:318–358.

Tanaka KL, Scott DH, Greeley R. 1992. Global stratigraphy. *In* Kieffer HH, Jakosky BN, Snyder CW, Matthews MS (Editors). *Mars*: University of Arizona Press, Tucson, Arizona. p. 345–382.

Thomas P, Squyres S, Herkenhoff K, Howard A, Murray B. 1992. Polar deposits of Mars. *In* Kieffer HH, Jakosky BM, Snyder CW, Matthews MS (Editors), *Mars*, University of Arizona Press, Tucson, 767–795.

Thomson BJ, Bridges NT, Milliken R, Baldridge A, Hook SJ, Crowley JR, Marion GM, de Souza Filho CR, Brown AJ, Weitz CM. 2011. Constraints on the origin and evolution of the layered mound in Gale Crater, Mars using Mars Reconnaissance Orbiter data. *Icarus* 214:413–432. DOI:10.1016/j.icarus.2011.05.002

Tokano T, McKay CP, Neubauer FM, Atreya SK, Ferri F, Fulchignoni M, Niemann HB. 2006. Methane drizzle on Titan. *Nature* 442:432–435.

Toon OB, Pollack JB, Ward W, Burns JA, Bilski K. 1980. The astronomical theory of climatic change on Mars. *Icarus* 44:552–607.

Ward AW. 1979. Yardangs on Mars: evidence of wind erosion. *Journal of Geophysical Research* 84:8147–8166.

Wells GL, Zimbelman JR. 1997. Extraterrestrial arid surface processes. *In* Thomas DSG (Editor). *Arid Zone Geomorphology: Process, Form, and Change in Drylands*, 2nd ed.: John Wiley & Sons Ltd, Chichester, West Sussex, UK. p. 559–690.

Westphal DL, Toon OB, Carlson TN. 1987. A two-dimensional numerical investigation of the dynamics and microphysics of Saharan dust storms. *Journal of Geophysical Research* 92:3027–3049.

White BR. 1979. Soil transport by winds on Mars. *Journal of Geophysical Research* 84:4643–4651.

White BR, Lacchia BM, Greeley R, Leach RN. 1997. Aeolian behavior of dust in a simulated Martian environment. *Journal of Geophysical Research* 102:25629–25640.

Wieder M, Gvirtzman G, Porat N, Dassa M. 2008. Paleosols of the southern coastal plain of Israel. *Journal of Plant Nutrition and Soil Science* 171:533–541.

Wilson SA, Howard AD, Moore JM, Grant JA. 2007. Geomorphic and stratigraphic analysis of Crater Terby and layered deposits north of Hellas basin, Mars. *Journal of Geophysical Research* 112. DOI:10.1029/2006JE002830

Wright J. 2001a. "Desert" loess versus "glacial" loess: quartz silt formation, source areas and sediment pathways in the formation of loess deposits. *Geomorphology* 36:231–256.

Wright J. 2001b. Making loess-sized quartz silt: data from laboratory simulations and implications for sediment transport pathways and the formation of 'desert' loess deposits associated with the Sahara. *Quaternary International* 76/77:7–19.

Wright J, Smith BJ, Whalley WB. 1998. Mechanisms of loess-sized quartz silt production and their relative effectiveness: laboratory simulations. *Geomorphology* 23:15–34.

Yen AS, Mittlefehldt DW, McLennan SM, Gellert R, Bell JF, McSween HY, Ming DW, McCoy TJ, Morris RV, Golombek M, Enonomou T, Madsen MB, Wdowiak T, Clark BC, Jolliff BL, Schröder C, Brückner J, Zipfel J, Squyres SW. 2006. Nickel on Mars: constraints on meteoritic material at the surface. *Journal of Geophysical Research* 111. DOI:10.1029/2006JE002797

Zárate M, Blasi A. 1993. Late Pleistocene–Holocene eolian deposits of the southern Buenos Aires Province, Argentina: a preliminary model. *Quaternary International* 17:15–20.

FOCUSING THE SEARCH FOR BIOSIGNATURES ON MARS: FACIES PREDICTION WITH AN EXAMPLE FROM ACIDALIA PLANITIA

DOROTHY Z. OEHLER AND CARLTON C. ALLEN

Astromaterials Research and Exploration Science Directorate, National Aeronautics and Space Administration, Johnson Space Center, 2101 NASA Parkway, Houston, Texas 77058 USA

e-mail: dorothy.z.oehler@nasa.gov

Abstract: The search for martian biosignatures can be enhanced by focusing exploration on locations most likely to contain organic-rich shales. Such shales both concentrate and preserve organic matter and are major repositories of organic geochemical biomarkers in sediments of all ages on Earth. Moreover, it has been suggested that for Mars, accumulations of organic matter may be the most easily detected and least ambiguous of possible biosignatures (Summons et al. 2010). Since current surface conditions on Mars are unfavorable for preservation of organic matter, focusing exploration on locations predicted to contain ancient organic-rich shales would offer one of the best chances of detecting evidence of life—if it ever evolved on the planet.

Orbital data can be used to evaluate regional sediment sources and sinks on Mars, and, based on that, facies can be predicted and locations identified that are most likely to contain organic-rich sediments. An example is presented from Acidalia Planitia, in the martian lowlands, where this approach led to the conclusion that facies in southern Acidalia were likely to be dominated by fine-grained, muddy sediments. That conclusion added weight to the hypothesis that mounds in Acidalia are martian versions of mud volcanoes as well as the suggestion that organic materials, if present, would have been deposited in the same area as the mounds. This allowed speculation that potential mud volcano clasts in Acidalia could include preserved, organic biosignatures and, thus, that the mounds in Acidalia constitute an untested class of exploration target for Mars.

Facies prediction using orbital data is particularly applicable to planetary exploration where ground truth is most often lacking but orbital data sets are increasingly available. This approach is well suited to the search for potential geochemical biomarkers in organic-rich shales. The approach additionally could be applied to exploration for other categories of biosignatures (such as stromatolites or morphologically preserved microfossils) and to more general planetary objectives, such as the search for hydrothermal sediments, carbonates, or any particular type of geologic deposit.

Key words: Mars, Acidalia, mud volcano, facies, biosignature

INTRODUCTION

One of the primary goals of the Mars exploration program is to determine if living systems ever arose on that planet by seeking evidence of past or present life. Water is a critical requirement for life, but today Mars is extremely dry. Nevertheless, martian river valleys attest to a wetter past, with surface accumulations of liquid water likely in the early part of martian history (e.g., Squyres and Kasting 1994, Craddock and Howard 2002). Accordingly, the first phase of martian exploration was focused on a "follow the water" strategy (Briggs 2000).

An adjunct to the "follow the water" strategy has been a focus on areas with clay minerals (phyllosilicates), since clays form in the presence of water and therefore are an indicator of past water. Phyllosilicates on Mars have been located with near-infrared spectra acquired from the Compact Reconnaissance Imaging Spectrometer for Mars (CRISM) on the National Aeronautics and Space Administration's (NASA's) Mars Reconnaissance Orbiter (MRO) (Bishop et al. 2008, Ehlmann et al. 2008, Mustard et al. 2008) and the Observatoire pour la Mineralogie, l'Eau, les Glaces et l'Activite' (OMEGA) spectrometer on the European Space Agency's Mars Express Orbiter (Bibring et al. 2006). Using OMEGA data, Bibring et al. (2006) reported that global distributions of martian phyllosilicates correspond to the oldest terrains, indicating that earliest Mars was most likely to have had the most habitable (i.e., the wettest) conditions. That, plus the fact that organic compounds can be adsorbed within smectite clays (Kennedy et al. 2002), has stimulated interest in searching for clays as potential repositories of organic biosignatures from the early—most habitable—time on Mars (Ehlmann et al. 2008).

A complexity with this approach is that the presence of phyllosilicates does not necessarily imply associated organics. Clays can form in settings lacking organic matter. They can also accumulate and persist in environments where organics—though initially present—may have been destroyed by postdepositional processes such as oxidation, radiolysis, or photocatalytic decomposition. This is a concern on Mars, where surface oxidation appears to be significant over much of the planet's surface, and oxidants such as perchlorates and possibly peroxides have been postulated (e.g., Encrenaz et al. 2004, 2008; Ten Kate et al. 2006; Hurowitz et al. 2007; Hecht et al. 2009). While adsorption of organic matter within smectite interlayers plays a role in organic preservation on Earth (Kennedy et al. 2002), it is not certain that comparable preservation would occur on Mars, since peroxides are stronger oxidants than terrestrial oxygen. In addition, Shkrob et al. (2010) have suggested that potential organics on Mars may be destroyed by photocatalytic decomposition as a result of the reaction with particulate iron (III) oxides that are abundant in the martian surface. They conclude that there may be "no safe haven" for organics from this process since it involves small, mobile, and highly reactive radicals. Finally, preservation of organic materials on Earth is enhanced by burial, which isolates sediments from surface oxidants and reduces permeability and porosity, thereby minimizing interaction with surficial fluids and gases. So burial on Mars might be part of the equation for preservation of organics.

Others have focused on different attributes of living systems, and there have been suggestions that Mars exploration should "follow the nitrogen" (Capone et al. 2006), "follow the energy" (Hoehler et al. 2007), "follow the carbon" (National Research Council 2007), or "follow the chemistry" (Hecht 2009). Each of these strategies captures important aspects of habitability or life, and each has different advantages and challenges. However, all of the above can be incorporated within a strategy focused on predicting settings in which

Sedimentary Geology of Mars
SEPM Special Publication No. 102, Copyright © 2012
SEPM (Society for Sedimentary Geology), Print ISBN 978-1-56576-312-8, CD/DVD ISBN 978-1-56576-313-5, p. 183–194.

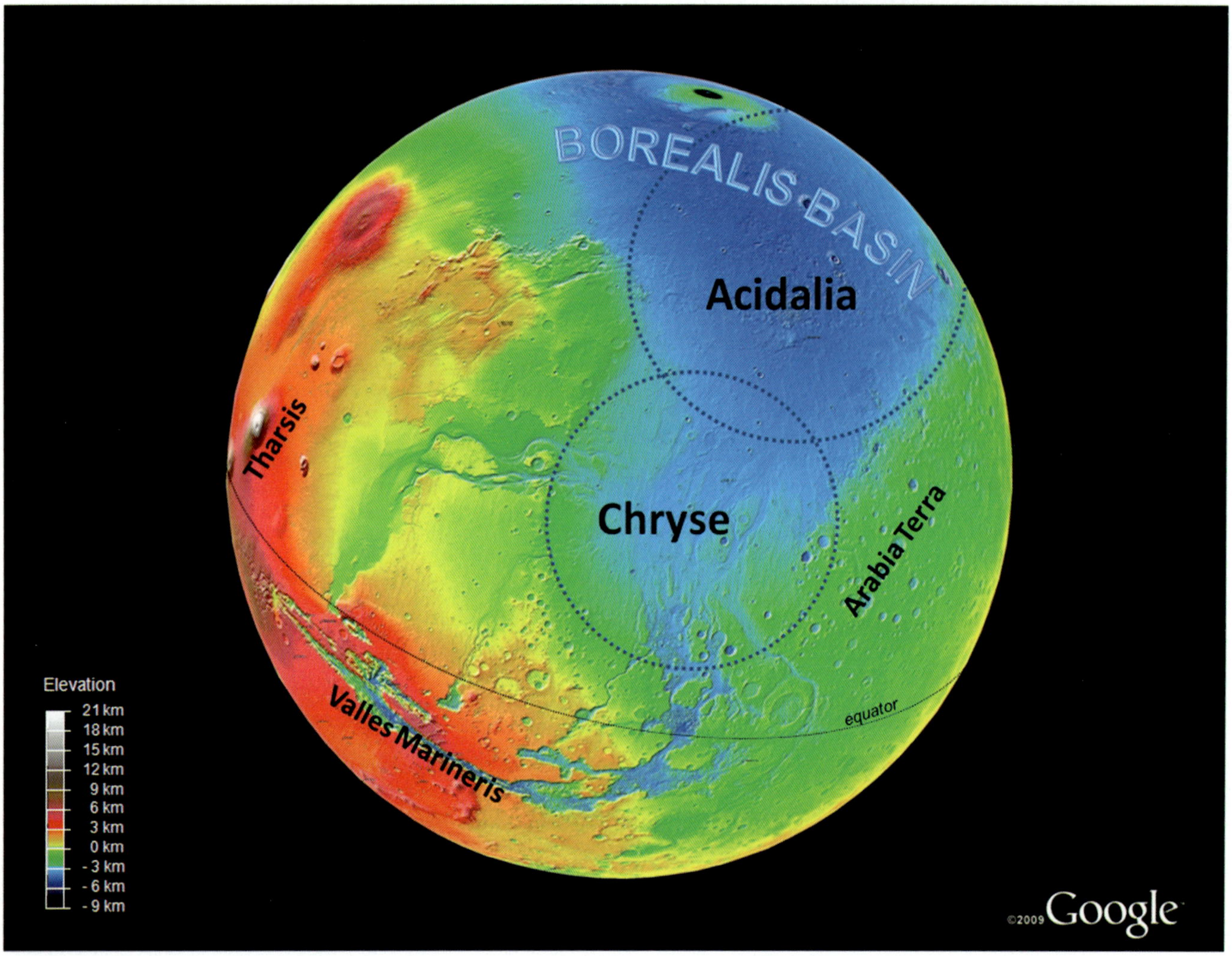

FIG. 1.—Global view of Acidalia and Chryse Planitiae on a basemap of topography from the Mars Orbiter Laser Altimeter (MOLA). Dashed, circular features show approximate locations of Quasi-Circular-Depressions (QCDs), which have been proposed to be the ancient Chryse and Acidalia impact basins (Frey 2006, 2008). Together, Chryse and Acidalia form an embayment in the northern lowlands. Major channels, many of which originate in Valles Marineris, can be seen flowing into the Chryse–Acidalia Embayment, mainly from the southern and western perimeters of Chryse. These are the channels associated with the Hesperian outflow floods. Basemap from Google Mars.

martian life not only may have thrived but also where it is likely to have been both concentrated and preserved within sediments.

POTENTIAL FOR BIOSIGNATURES ON MARS

Present-day Mars has a surface that is relatively hostile to life—as it is very cold, very dry, and lacks an atmosphere that could shield organisms from ultraviolet and cosmic rays. There are no indications that the planet is teeming with life, though it is conceivable that microbial life may exist in the subsurface. In contrast, early Mars may have had surface conditions more conducive to life, with a wetter, possibly warmer climate and a denser CO_2-rich atmosphere that could have provided protection from ultraviolet and cosmic radiation and contributed to a warming greenhouse effect (Molina-Cuberos et al. 2001, Lammer et al. 2002, Bibring et al. 2006, Fairén et al. 2009, Tian et al. 2009, Morris et al. 2010, Warner et al. 2010).

Of the three Martian eras (Noachian 4.65 to 3.7 Ga, Hesperian 3.7 to 3.0 Ga, and Amazonian ~3.0 Ga to present) (Hartmann and Neukum 2001, Neukum et al. 2001, Neukum 2008), the most habitable time is considered to be in the Noachian and earlier parts of the Hesperian. It has been suggested further that the Late Noachian–Early Hesperian (~3.5 Ga) may have been the most favorable time for development of life, as by that time, major impact bombardment would have ceased but Mars may have still possessed an atmosphere that contributed to warmer, wetter conditions and afforded protection from radiation (Grady and Wright 2006).

On Earth, the paleontological and geochemical records indicate that life was thriving by at least ~3 Ga and perhaps by ~3.5 Ga. Evidence derives from (1) stromatolites (Walter et al. 1980; Hofmann et al. 1999; Allwood et al. 2006, 2009), (2) organically preserved microfossils (Rasmussen 2000; Altermann and Kazmierczak 2003; Schopf 2006; Schopf and Walter 2007; Schopf et al. 2007; Sugitani et al. 2007; Oehler et al. 2009, 2010; Javaux et al. 2010), and (3) organic chemical biomarkers in cherts and shales (Duck et al. 2007, Marshall et al. 2007, Van Zuilen et al. 2007, Derenne et al. 2008, De Gregorio et al. 2009, Oehler et al. 2009). This indicates that life developed and diversified

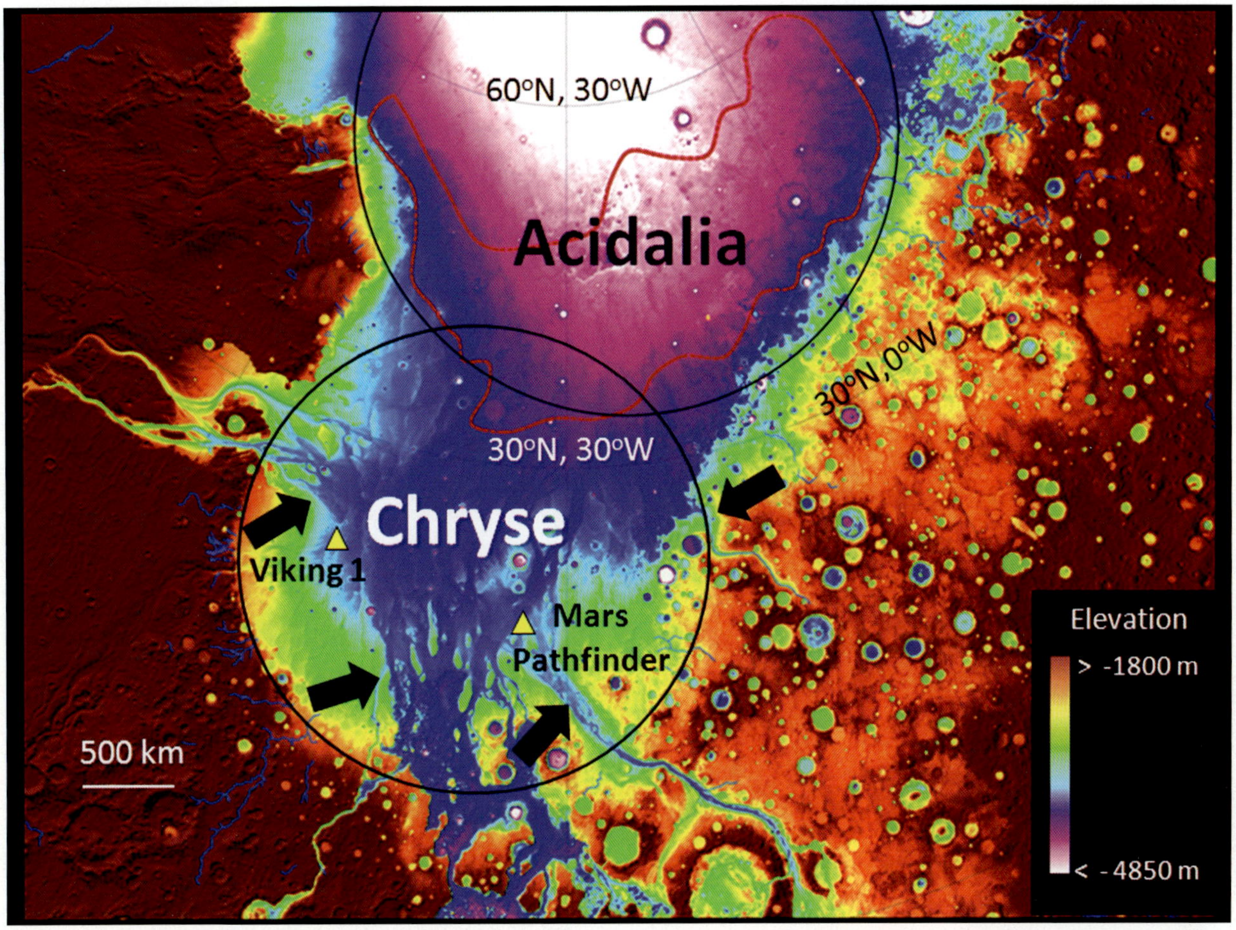

FIG. 2.—Chryse–Acidalia Embayment on a basemap of stretched MOLA data (polar projection). Black arrows point to several of the major outflow channels from the Hesperian outflow floods. Approximate locations of the proposed Chryse and Acidalia impact basins (Frey 2006, 2008) are shown by the black circles. The red outline in southern Acidalia indicates the area within which 18,000+ mud-volcano–like mounds have been mapped (Oehler and Allen 2010). Yellow triangles indicate the Mars Pathfinder and Viking 1 landing sites.

relatively early in the history of the Earth (Oehler et al. 2010) and may have done so on a similar timescale on Mars—during its most habitable period.

Biosignatures include both inorganic and organic remains of living systems. Inorganic biosignatures consist of biologically influenced minerals as well as inorganic morphological fossils or bio-sedimentary structures like stromatolites. Organic biosignatures include morphologically preserved remains of microbial life and geochemical remnants (organic biomarkers) of biological materials. For Mars, it has been suggested (Summons et al. 2010) that accumulations of sedimentary organic matter (including potential organic geochemical biomarkers) may be the most easily detected and the least ambiguous of potential biosignatures. On Earth, organic-rich shales and cherts go back in the geologic record to nearly 3.5 Ga and are commonly present even when morphologically recognizable microfossils are not preserved.

In fact, organic-rich shales are major repositories for terrestrial organic matter of all ages. These organic-rich sedimentary rocks represent environments in which organic matter has been both concentrated and preserved. These are the sorts of environments that should be sought in martian exploration, as they would enhance the chances of finding organic biosignatures—if life were ever present on Mars.

PREDICTIVE APPROACH

Organic-rich shales are common source rocks for oil and gas, and petroleum companies have developed methods for predicting their occurrence in areas with limited ground truth. Using these methods, regional structure is typically assessed first from satellite-derived topography and bathymetry using radar, gravity, and magnetics coupled with paleogeographic reconstructions. From that assessment, basin configurations are identified along with likely sediment sources and sinks; sedimentary facies are predicted from analysis of sediment input and basin geometry. Since much is known about the settings in which organic-rich sediments occur, source rock accumulations can be predicted from the mapped facies (see Allen and Allen 1990, Klemme and Ulmishek 1991, Gautier 2005, Tommeras and Mann 2008). Examples of these types of predictive studies can be found in the works of Christ et al. (2003) and Dickson et al. (2003, 2005, in press).

In general, organic-rich shales are found in both marine and lacustrine, basinal settings within distal, quiet-water facies (e.g., Macquaker and Curtis 1989, Tyson 1996, Gautier 2005, Jenkyns 2007, Davies-Vollum and Smith 2008). These are environments in which in situ living systems may have proliferated, in which organic material is concentrated by transport processes, and in which burial with associated muds and sometimes evaporites enhances preservation of

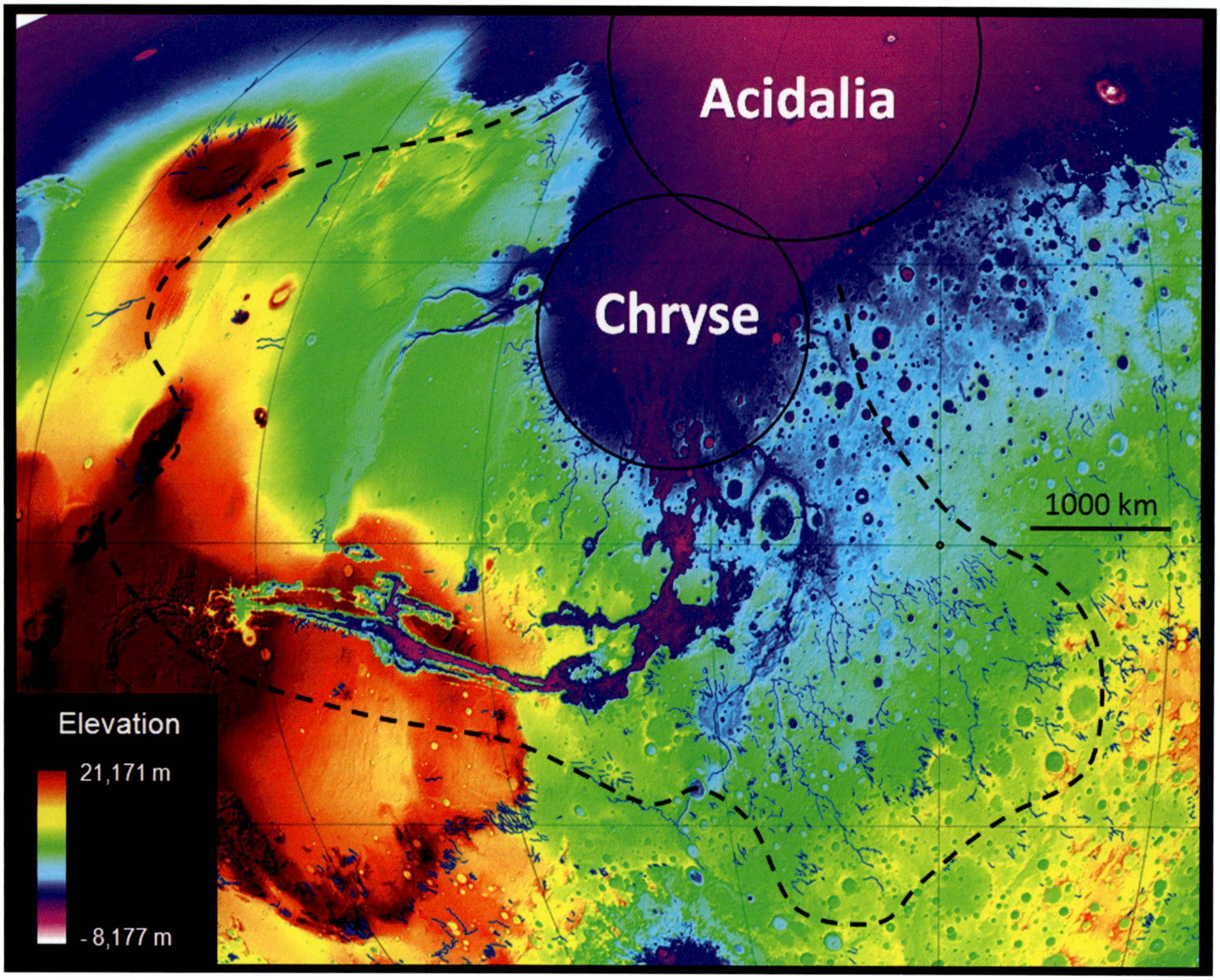

FIG. 3.—Estimate of catchment area (black dashed line) for the Chryse–Acidalia Embayment on a basemap of MOLA data (Mollweide projection). Approximate locations of the proposed Chryse and Acidalia impact basins (Frey 2006, 2008) are shown by the black circles. Thin blue lines show channels in the highlands mapped by Carr and provided by the US Geological Survey in their MarsGIS DVD v. 1.4.

organic matter by isolating it from destructive surface processes. The sedimentary processes resulting in accumulation and preservation of organic-rich shales are applicable to sediments of any age and can be applied beyond oil exploration to the general search for organics as biosignatures of early life.

Similar predictive approaches can be applied to the search for organic biosignatures on Mars. Orbital topographic data can be used for first-pass designations of basin architecture, sediment sources, and sediment sinks, since Mars has had a comparatively stable structural framework since ~3.7 Ga (Golombek 2005, Golombek and Phillips 2009). Based on that assessment, facies can be predicted where organic-rich shales would be expected. The regional evaluation can be fine-tuned with geomorphic and hyperspectral data from the higher resolution imaging spectrometers now orbiting Mars. This approach would allow prediction of facies and expected sediment composition for many of the areas on Mars that are lacking ground truth. It is rationally a first step that should be taken in identifying exploration targets.

EXAMPLE FROM ACIDALIA PLANITIA

An example from Acidalia Planitia, in the martian lowlands (Fig. 1), is provided for which facies have been predicted from orbital data (Oehler and Allen 2010). Results were applied to assessment of the mounds in Acidalia, supporting an interpretation that the mounds represent a form of mud volcanism. Multiple alternatives were considered, but the geologic context added considerable weight to the interpretation of mud volcanism. The analysis further allowed prediction of locations for accumulation of potential organics as well as the suggestion that the mounds in Acidalia represent an untested class of exploration target for Mars.

The Acidalia example is presented here in abbreviated form, as it is not the intention of this article to make the argument for mud volcanism but rather to illustrate the methods used for determining geologic context and the value of this approach in interpreting planetary features. The Acidalia example provides a case study for which facies have been predicted that could contain organic-rich shales, including potential biosignatures. A comprehensive study of the

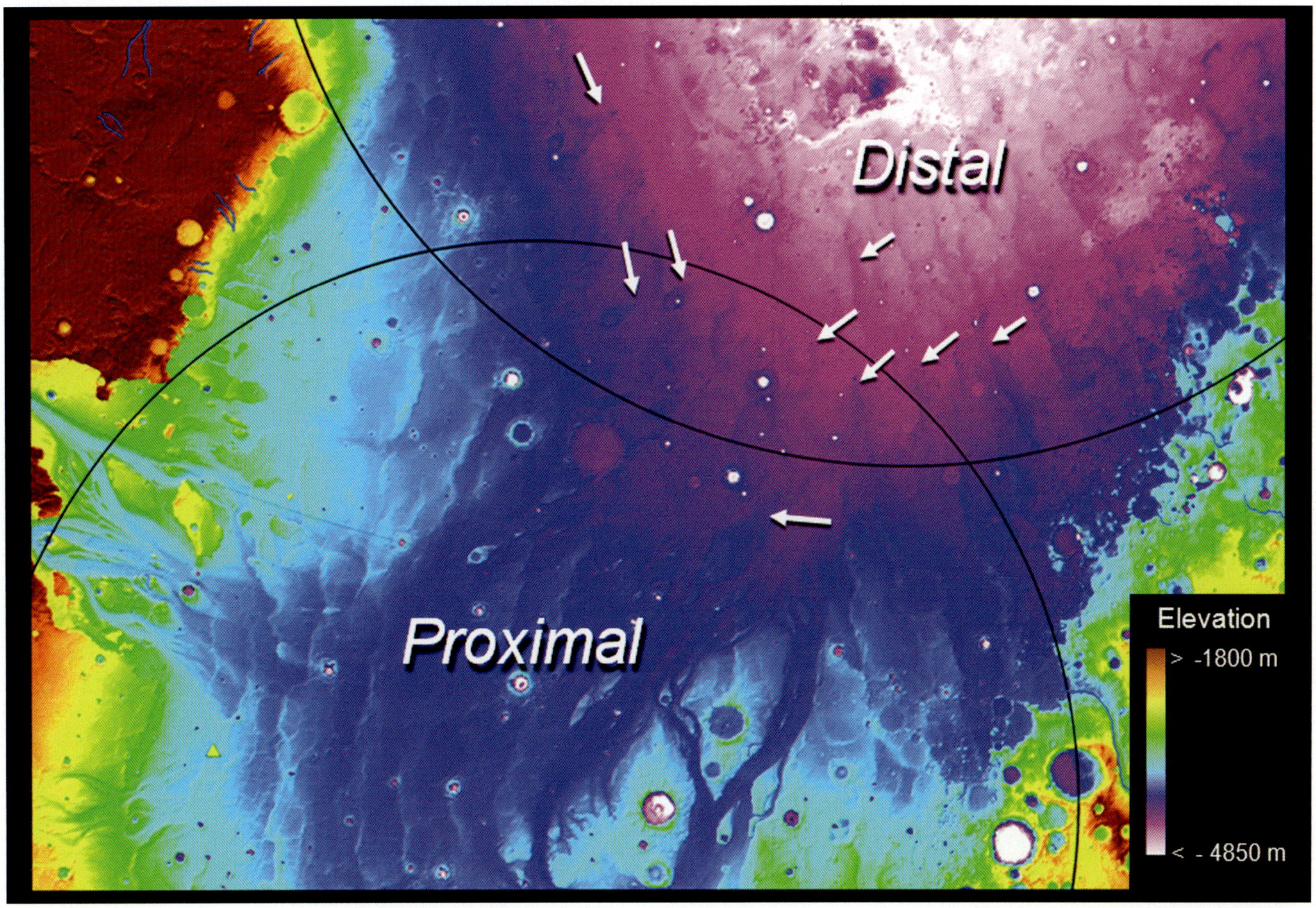

FIG. 4.—Detail of the region of overlap between the proposed Chryse and Acidalia impact basins (Frey 2006, 2008) on a basemap of stretched MOLA data (polar projection). Locations of predicted distal and proximal facies with respect to outflow sedimentation are indicated. Approximate locations of the proposed Chryse and Acidalia impact basins in this area of overlap are shown by the black arcs. White arrows point to some of the streamlined islands that indicate that outflow waters continued flowing northward from Chryse into Acidalia.

mounds in Acidalia is reported in Oehler and Allen (2010), and the reader is referred to that article for detailed treatment of the mapping methods, geomorphology and spectral signatures of the mounds, facies prediction, alternative hypotheses, timing of events, potential relationships to methane, triggers for sediment mobilization, differences between the Acidalia mounds and those elsewhere in the lowlands, and expected differences between terrestrial and martian mud volcanism.

Regional Setting

Acidalia Planitia is part of the northern lowlands of Mars (Fig. 1); it extends nearly 3000 km in an east–west direction and lies between Chryse Planitia to the south, the Tharsis volcanic province to the southwest, Arabia Terra to the southeast, and the North Polar province to the north (Fig. 1). Recent work (Andrews-Hanna et al. 2008, Marinova et al. 2008) indicates that the martian lowlands were formed very early in the history of the planet, perhaps >4 Ga, by a giant impact that formed the huge northern Borealis Basin. Frey (2006, 2008, 2010a, 2010b) proposes that the lowlands were subsequently affected by several very large impacts, perhaps during the martian equivalent of a late heavy bombardment ~3.9 Ga, and that remnants of those partially buried impact basins can be mapped as quasi-circular depressions (QCDs). Chryse and Acidalia are both proposed to be large impact basins of this type.

Analysis of the tectonic history on Mars (Golombek 2005, Golombek and Phillips 2009) indicates that development of the enormous Tharsis volcanic edifice prior to ~3.7 Ga may have contributed to a warm and wet climate of early Mars. This early emplacement of Tharsis would have altered the elliptical shape of the Borealis Basin (Andrews-Hanna et al. 2008), leaving the Chryse–Acidalia region as a relatively restricted embayment (Figs. 1, 2; Oehler and Allen 2010). The tectonic analysis further indicates that the lithospheric structure of Mars has been relatively unchanged since ~3.7 Ga, and this, plus the observation that ancient river valleys flow down the present topographic gradient, combine to argue that major, regional topographic trends on Mars have not varied substantially since ~3.7 Ga.

There is considerable discussion as to whether the northern lowlands were the site of a past ocean on Mars (proposed by Parker et al. [1987, 1989, 1993] and summarized by Dohm et al. [2009]). The evidence that the northern plains have remained a planet-wide topographic low since the late Noachian indicates that, at a minimum, the lowlands were a focal point for fluid accumulation during much of martian history. Moreover, there is evidence that massive floods occurred in the Hesperian and that the floodwaters debouched into the Chryse Basin

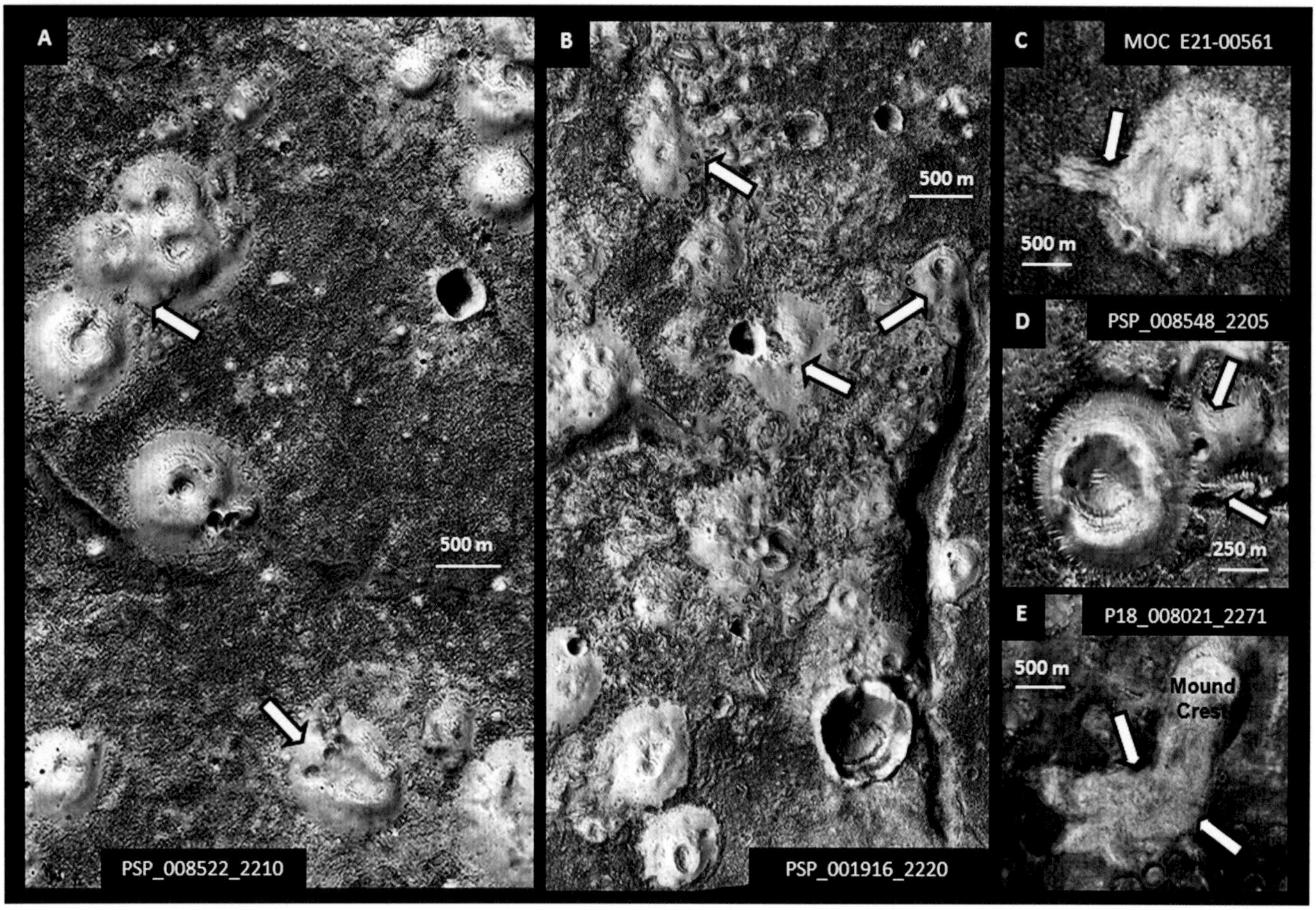

FIG. 5.—Mounds in Acidalia. (A, B) examples of morphological variability and irregular clustering. Arrows point to flow-like features. (C–E) mounds with significant flow-like features (arrows); (D) illustrates a well-developed moat. A, B, and D are HiRISE images; C is a MOC image; E is a CTX image.

(Fig. 2) from its southern and western perimeters (Golombek et al. 1995a, 1995b; Rice and Edgett 1997). These floods appear to have been unique on the planet and would have introduced significant sediment and water into Chryse and southern Acidalia. Some even consider that as a result of the Hesperian floods, Chryse, Acidalia, and perhaps the entire northern lowlands were under water (Parker et al. 1989, 1993; Baker et al. 1991; Scott et al. 1991, 1995; Rice and Edgett 1997; Head et al. 1999; Boyce et al. 2005; Dohm et al. 2009).

Orbital Data

There are multiple data sets available for Mars from orbiting satellites. Those used in analysis of the mounds in Acidalia (Oehler and Allen 2010) included topography from the Mars Orbiter Laser Altimeter (MOLA) on NASA's Mars Global Surveyor orbiter (MGS), nighttime and daytime infrared data from the Thermal Emission Imaging Spectrometer on NASA's Mars Odyssey orbiter, and mosaics for these data sets provided by the US Geological Survey in their MarsGIS DVD v. 1.4. For detailed geomorphology, image data were used from the Mars Orbiter Camera (MOC) on MGS and from the Context (CTX) and High Resolution Imaging Science Experiment (HiRISE) cameras on MRO. For reference geology, the geologic map of the Northern Plains (Tanaka et al. 2005) was used. CRISM data from MRO were incorporated to provide information on mineralogy.

Facies

In this article, the term facies is used to describe distinctive rock types that broadly correspond to certain environments or modes of origin. This concept is sometimes referred to as "sedimentary facies" or "lithofacies."

For the Chryse–Acidalia Embayment, analysis of topographic slopes and mapped river valleys predict a catchment area that covers a large portion of the southern highlands (Fig. 3). A depositional sink in Chryse is clear from the river and outflow channels. Moreover, streamlined islands in northeastern Chryse (Fig. 4) indicate that water and sediments from the outflow floods spilled from Chryse into the deeper Acidalia Basin (Scott et al. 1991, Rotto and Tanaka 1995, Rice and Edgett 1997). Within this framework, proximal channel facies (with boulders, pebbles, gravels, and sands) would be expected in Chryse, and distal, fine-grained facies (with mainly muds) would be expected in Acidalia (Oehler and Allen 2010).

Mounds

Tens of thousands of high-albedo mounds (Fig. 5) occur in Acidalia. These were first noted in Viking imagery (Allen 1979, Frey et al. 1979, Frey and Jarosewich 1982) and have been interpreted as pseudocraters,

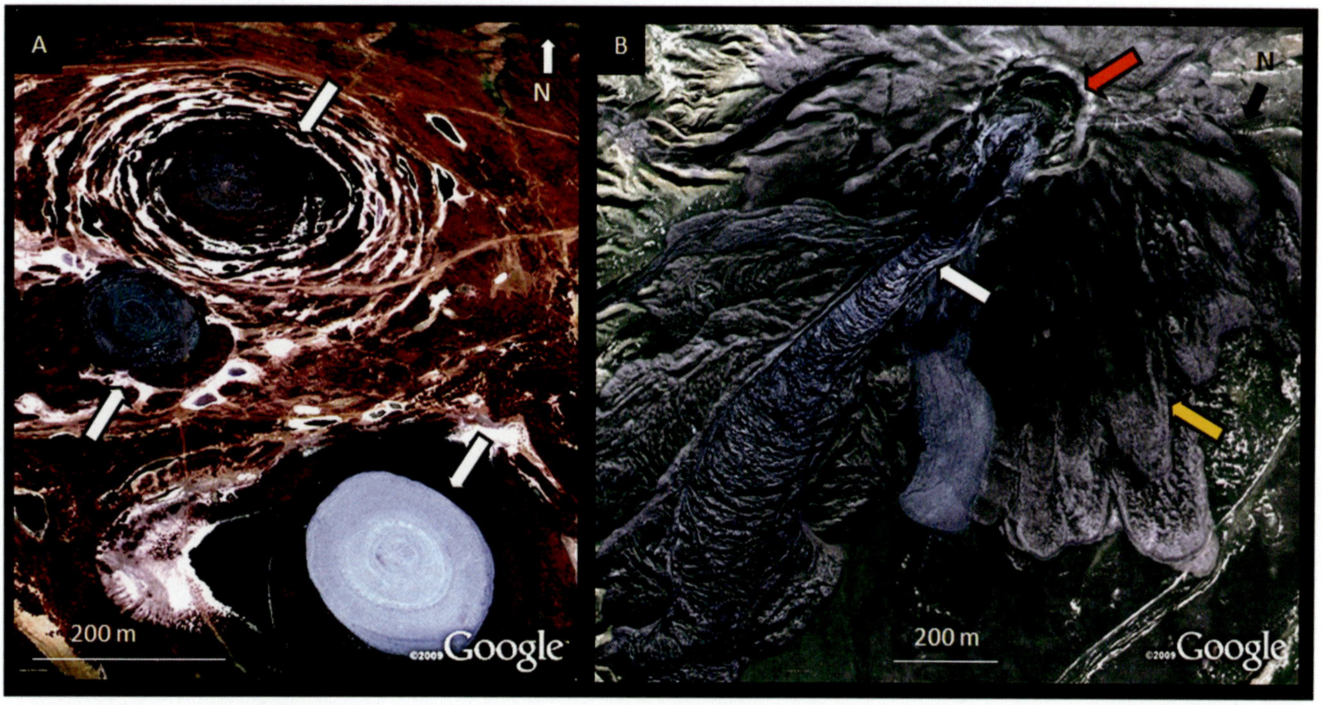

FIG. 6.—Morphological variability in terrestrial mud volcanoes from Azerbaijan. (A) Mud volcanoes with concentric crestal structures and few obvious, lengthy mud flows onto the plains. Arrows point to crests and high albedo portions. Center point, 40.36°N, 49.07°E. (B) Mud volcano showing extended mud flow onto the plains (white arrow), crestal vent (red arrow), and older flows (orange arrow). Center point, 39.98°N; 49.36°E.

cinder cones, tuff cones, pingos, ice disintegration features, mud volcanoes, or a combination of mud volcanoes and evaporites (Tanaka 1997; Tanaka and Banerdt 2000; Tanaka et al. 2003, 2005; Farrand et al. 2005; Rodríguez et al. 2007; McGowan 2009; Skinner and Mazzini 2009).

Oehler and Allen (2010) mapped more than 18,000 of the mounds and estimated that more than 40,000 occur in southern Acidalia (Fig. 2). The mounds generally overlie early Amazonian-aged units mapped by Tanaka et al. (2005); mound diameters average ~800 m and analysis of one HiRISE stereo pair yielded a height of tens of meters. Many alternative hypotheses were considered to explain their origin (see Oehler and Allen [2010] for discussion of all hypotheses), but the data were most consistent with an analog of mud volcanism (e.g., Fig. 6). This interpretation was strongly supported by the geologic context developed from regional analysis.

Requirements for mud volcanism are (1) subaqueous deposition of thick sequences of fine-grained sediments, (2) development of overpressure (Kopf 2002, Deville et al. 2003, Van Rensbergen et al. 2003, Deville 2009), and (3) subsequent triggering events to initiate sediment mobilization. These requirements are likely to have been met in southern Acidalia. The embayment-like geometry of the Chryse–Acidalia region coupled with its large catchment would have fostered accumulation of sediments in that region from as early as the late Noachian, with the finer-grained muds accumulating in the distal facies in Acidalia. Overpressure is likely to have developed as a result of rapid and massive Hesperian outflow sedimentation, and triggering events could have included such things as hydrothermal or compressive pulses from Tharsis, sublimation of ice and resulting loss of overburden, or destabilization of clathrates, with consequent release of gas (Oehler and Allen 2010).

Exploration Targets

The geologic setting of Acidalia provides advantages for potential biosignature concentration, preservation, and access to rovers. For example, the proposed catchment (Fig. 3) could have drained a substantial portion of the highlands (Fig. 3), carrying organic microfossils or organic geochemical biomarkers derived from any life in that large area. Since organic materials are of relatively low density, they tend to stay in suspension during fluvial transport and accumulate with the muddy fraction of sediments. Accordingly, organic remnants of possible life in the Chryse–Acidalia watershed are most likely to have been deposited with muds in Acidalia.

There is the additional possibility that the subsurface in Acidalia could have supported in situ life forms—in microhabitats within fluid-filled porosity (Fig. 7C). On Earth, microbial assemblages occur in the subsurface of mud volcano systems of the Gulf of Mexico, supported by upwelling fluids containing dissolved organic matter (Joye et al. 2009). Similarly, mud volcanism in Acidalia could have promoted upward migration of fluids from depth. Such fluids could have contained dissolved organics from Hesperian or Noachian strata or from any subsurface geological or biological sources of methane. These fluids could have supplied a continuing source of nutrients for potential microbial life (e.g., Niederberger et al. 2010, Pohlman et al. 2011) and also may have provided a source of warmth on an increasingly cold planet. Andrews-Hanna et al. (2007) have suggested that Acidalia may have been a region of enhanced upwelling from regional hydrological considerations. If this were the case, then these upwelling waters could have been channeled to the surface through conduits created by mud volcanism (Fig. 8) (e.g., Van Rensbergen et al. 2005a, 2005b; Van Rooij et al. 2005), as analogous conduits on Earth

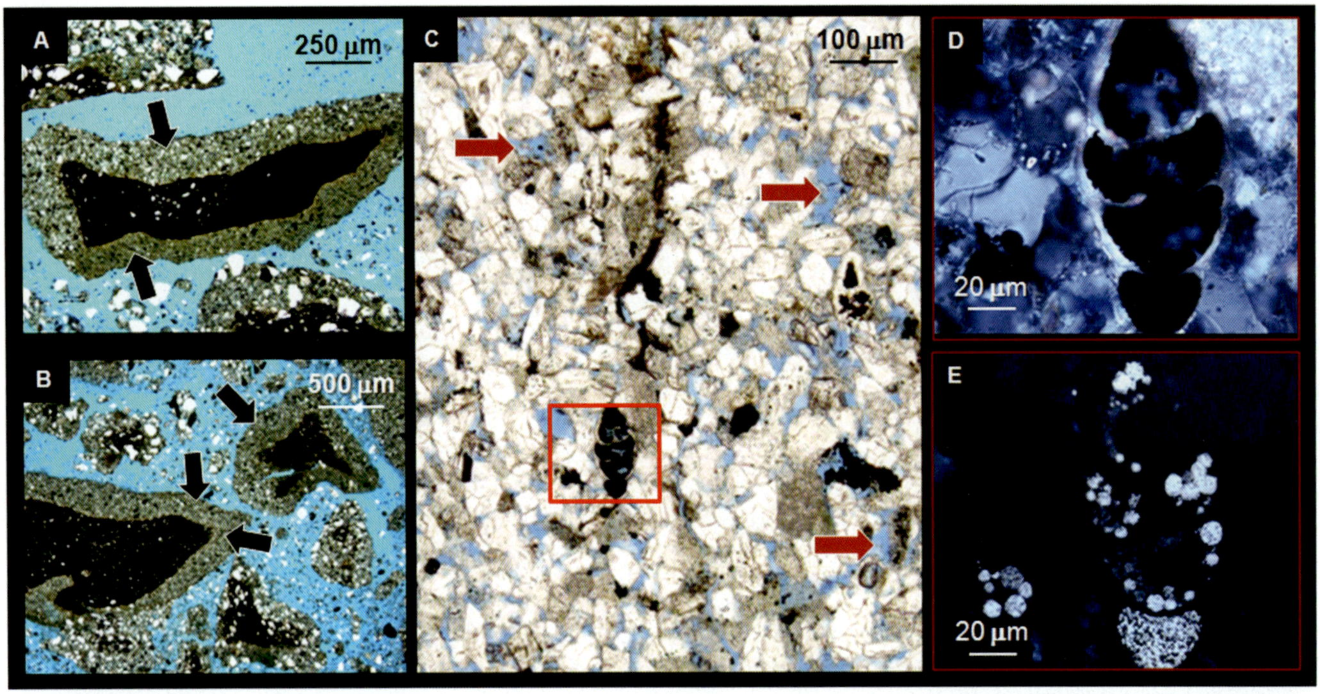

FIG. 7.—Clasts brought to the surface by terrestrial mud volcanoes. (A, B) Breccia clasts from Wushanting Mud Volcano, Kaohsiung Prefecture, Taiwan (sample provided by Dr. K. Tanaka, Yamaguchi University, Japan). Optical photomicrographs in transmitted light. Arrows indicate alteration rinds around clasts of fine-grained shales. Blue is tinted epoxy. (C–E) Breccia clast from Bozdag-Gyuzdek Mud Volcano in Azerbaijan (sample provided by Dr. A. Feyzullayev, Azerbaijan National Academy of Sciences). (C) Optical photomicrograph in transmitted light. Red rectangle encompasses a calcareous microfossil with internal chambers filled with pyrite (black opaque material). Red arrows point to examples of porosity (indicated by blue-tinted epoxy). (D, E) High-magnification images of area of red rectangle of C, showing detail of the calcareous microfossil and pyrite inclusions; (D) Optical photomicrograph in transmitted light with crossed nicols; (E) Optical photomicrograph in reflected light, illustrating bright pyrite accumulations.

provide long-lived pathways (up to kilometers deep) for fluid migration (Etiope et al. 2010).

Finally, burial of transported or in situ biosignatures with outflow muds could have protected entrained organics from destructive processes on the surface of Mars, and such preserved materials could be contained in rock clasts brought to the surface by mud volcanism. Terrestrial mud volcanoes extrude clasts of lithified rock that are ripped from the walls of subsurface strata during mud eruption. Many of these clasts contain well-preserved organic biomarkers and nannofossils (Fig. 7) (Robertson and Kopf 1998, Kopf 2002, Stewart and Davies 2006). If burial in Acidalia were adequate for lithification of Noachian and early Hesperian strata, rock clasts containing potential biosignatures might be present on the surface in this region. Since no Mars mission has analyzed structures interpreted as mud volcanoes, the mounds of Acidalia can be considered to represent a new class of exploration target.

SUMMARY AND CONCLUSIONS

Mars exploration has been centered on the goal of determining if life was ever present on that planet, with the primary emphasis being the search for organic geochemical biosignatures. However, the surface of Mars currently appears to be unfavorable for preservation of organic materials as a result of potential oxidation from peroxides and perchlorates and decomposition from photocatalytic processes. One way to improve the chances of finding organic biosignatures (if they are present) would be to seek facies in which ancient, organic-rich sediments would be expected. Additionally, and importantly for Mars, if sites could be located where predicted organic-rich sediments are likely to have been compacted and lithified into shales, then the preservation potential of any organic constituents would be enhanced.

Orbital data can be used to evaluate regional sediment sources and sinks, and from this, predictions can be made of locations likely to have basinal, quiet-water, distal facies. These represent settings in which organic-rich sediments typically occur. An example is presented in which this approach was used in an evaluation of mounds in Acidalia Planitia, in the martian lowlands. In that study (Oehler and Allen 2010), regional assessment and facies prediction provided geologic context that supported interpretation of the mounds as martian equivalents of mud volcanoes. The facies model further predicted that organic materials would have been deposited coincidently with the muds. This allowed speculation that the mud volcano clasts erupted to the surface from depth in Acidalia might include preserved biosignatures.

This approach could also be used to search for other types of potential martian biosignatures, such as morphologically preserved, organic microfossils or stromatolites—the macroscopic, mineralized remnants of microbial mats. Each of these provides significant information about the Archean biosphere on Earth and could represent signatures of early life on Mars. A search for microfossils could be focused on cherts, shales, or evaporite deposits, all of which preserve organic microfossils on Earth (Schopf 2006; Schopf and Walter 2007; Schopf et al. 2007, 2010; Javaux et al. 2010). Similarly, exploration for martian stromatolites could be directed to locations predicted to have facies analogous to the carbonate-, silica-, or evaporite-precipitating settings in which terrestrial stromatolites occur (e.g., Walter et al. 1980, Jones et al. 2005, Allwood et al. 2006, Allwood and Kanik 2010).

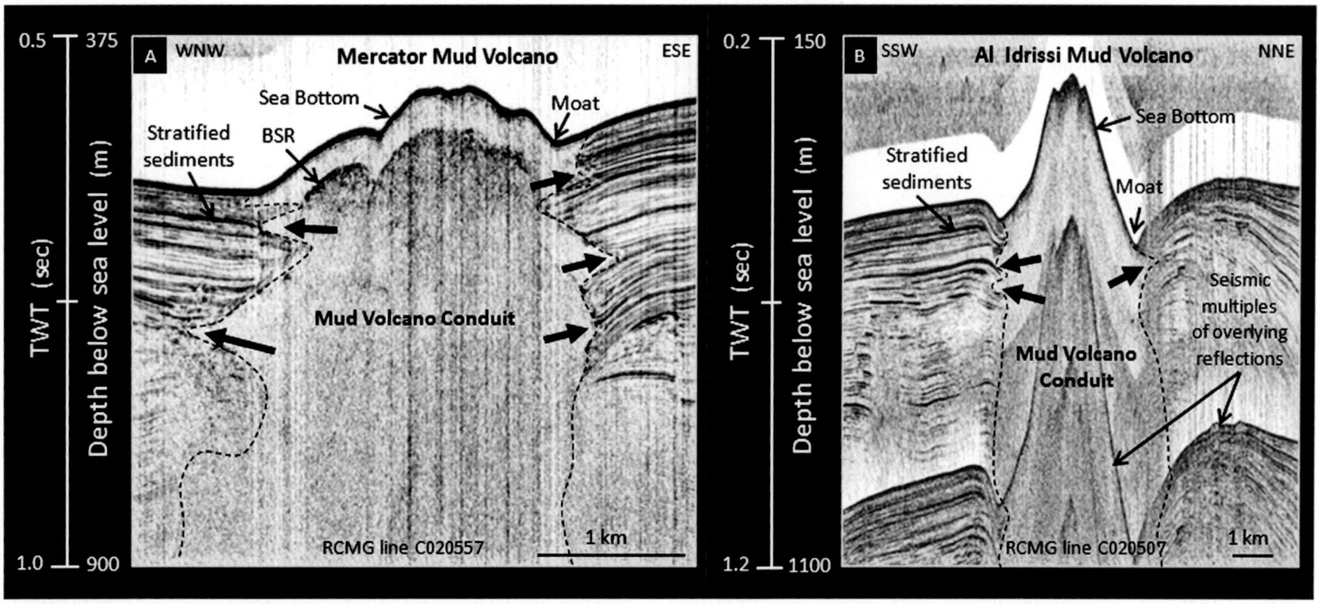

FIG. 8.—High-resolution, single-channel reflection seismic profiles across Mercator (A) and Al Idrissi (B) mud volcanoes in the Gulf of Cadiz, offshore Morocco. These profiles illustrate the mud conduits, which appear as columnar zones lacking seismic reflections. Thick arrows point to levels where mud flows interfinger with the stratified sediments. BSR = Bottom Simulating Reflector. TWT is the two-way travel time of seismic acoustic wave from its source to actual depth and back. For more details, see Van Rensbergen et al. (2005a, 2005b) and Van Rooij et al. (2005). Seismic images courtesy of Dr. D. Van Rooij, Renard Centre of Marine Geology (RCMG), Ghent University, Belgium.

Geologic assessment from orbital data is a rational first step in the exploration process. It is commonly used in pursuit of petroleum and mineral deposits on Earth in remote areas, incorporating satellite-derived topography/bathymetry, gravity, and magnetics for assessment of tectonic structure, basin architecture, and facies. The same approach should be particularly useful for planetary exploration where ground truth is commonly lacking but where orbital data sets are increasingly available. The approach is well suited to the search for extraterrestrial biosignatures and could be applied to more general planetary objectives, such as exploration for hydrothermal sediments, carbonates, or any particular type of geologic deposit.

ACKNOWLEDGMENTS

This article grew out of a presentation made at the 1st International Conference on Mars Sedimentology and Stratigraphy in April 2010. We are grateful to Drs. D. Beaty (Jet Propulsion Laboratory) and J. Grotzinger (California Institute of Technology) for encouraging development of these ideas and to Drs. G. Etiope (Istituto Nazionale di Geofisica e Vulcanologia), A. Feyzullayev (Azerbaijan National Academy of Sciences), and K. Tanaka (Yamaguchi University) for providing mud volcano samples and insights into the process of mud volcanism. We are also appreciative of valuable discussions with Dr. P. Van Rensbergen (Shell Oil) regarding subsurface sediment mobilization. Dr. D. Van Rooij (Ghent University) provided the seismic images for Fig. 8. We thank the reviewers for many helpful suggestions and the Astromaterials Research and Exploration Science Directorate at Johnson Space Center for providing facilities and support.

REFERENCES

Allen CC. 1979. Volcano/ice interactions on Mars. *Journal of Geophysical Research* 84:8048–8059.

Allen PA, Allen JR. 1990. *Basin Analysis; Principles and Applications*: Blackwell Publishing Company, Oxford UK, 451 p.

Allwood AC, Grotzinger JP, Knoll AH, Burch IW, Anderson MS, Coleman ML, Kanik I. 2009. Controls on development and diversity of Early Archean stromatolites. *Proceedings of the National Academy of Sciences* 106:9548–9555.

Allwood AC, Kanik I. 2010. Formation of stromatolites and other potentially microbially influenced structures and textures in terrestrial (and martian?) chemical sediments. *In 1st International Conference on Mars Sedimentology and Stratigraphy*; April 19-21, 2010; El Paso, Texas. Abstract No. 6039.

Allwood AC, Walter MR, Kamber BS, Marshall CP, Burch IW. 2006. Stromatolite reef from the Early Archaean era of Australia. *Nature* 441:714–718.

Altermann W, Kazmierczak J. 2003. Archean microfossils: a reappraisal of early life on Earth. *Research in Microbiology* 154:611–617.

Andrews-Hanna JC, Phillips RJ, Zuber MT. 2007. Meridiani Planum and the global hydrology of Mars. *Nature* 446:163–166.

Andrews-Hanna JC, Zuber MT, Banerdt WB. 2008. The Borealis basin and the origin of the martian crustal dichotomy. *Nature* 453:1212–1216.

Baker VR, Strom RG, Gulick VC, Kargel JS, Komatsu G, Kale VS. 1991. Ancient oceans, ice sheets, and the hydrological cycle of Mars. *Nature* 352:589–594.

Bibring J-P, Langevin Y, Mustard JF, Poulet F, Arvidson R, Gendrin A, Gondet B, Mangold N, Pinet P, Forget F, and the OMEGA team. 2006. Global mineralogical and aqueous Mars history derived from OMEGA/Mars Express Data. *Science* 312:400–404.

Bishop JL, Noe Dobrea E, McKeown N, Parente M, Ehlmann B, Michalski JR, Milliken RE, Poulet F, Swayze G, Mustard J, Murchie SL, Bibring J-P. 2008. Phyllosilicate diversity and past aqueous activity revealed at Mawrth Vallis, Mars. *Science* 321:830–833.

Boyce JM, Mouginis-Mark P, Garbeil H. 2005. Ancient oceans in the northern lowlands of Mars: evidence from impact crater depth/diameter relationships. *Journal of Geophysical Research* 110:E03008, doi:10.1029/2004JE002328

Briggs G. 2000. "Follow the water." *Meteoritics & Planetary Science* 35(5):892–893.

Capone DG, Popa R, Flood B, Nealson KH. 2006. Follow the Nitrogen. *Science* 312:708–709.

Christ JM, Dickson WG, Granath JW. 2003. *Finding the Pearl in the Pearl River Mouth Basin, Offshore China: A deepwater turbidite oil play?* American Association of Petroleum Geologists, Search and Discovery Article No. 10044, 6 p.

Craddock RA, Howard AD. 2002. The case for rainfall on a warm, wet early Mars. *Journal of Geophysical Research* 107, doi:10.1029/2001JE001505

Davies-Vollum KS, Smith ND. 2008. Factors affecting the accumulation of organic-rich deposits in a modern avulsive floodplain: examples from the Cumberland marshes, Saskatchewan, Canada. *Journal of Sedimentary Research* 78:683–692.

De Gregorio BT, Sharp TG, Flynn GJ, Wirick S, Hervig RL. 2009. Biogenic origin for Earth's oldest putative microfossils. *Geology* 37:631–634.

Derenne S, Robert F, Skrzypczak-Bonduelle A, Gourier D, Binet L, Rouzaud J-N. 2008. Molecular evidence for life in the 3.5 billion year old Warrawoona chert. *Earth Planetary Science Letters* 272:476–480.

Deville E. 2009. Mud volcano systems. *In* Lewis N, Moretti A (Editors), *Volcanoes: Formation, Eruptions and Modelling*: Nova Science Publishers, New York, p. 95–125.

Deville E, Battani A, Griboulard R, Guerlais S, Herbin JP, Houzay JP, Muller C, Prinzhofer A. 2003. The origin and process of mud volcanism: new insights from Trinidad. *In* Van Rensbergen P, Hillis RR, Maltman AJ, Morley CK (Editors), *Subsurface Sediment Mobilization*: Geological Society of London, London, Special Publication 216, p. 475–490.

Dickson W, Schiefelbein C, Odegard M. Defining a supergiant petroleum system in Brazil's Santos Basin with multi-disciplinary methods and (mostly) non-seismic data. Society of Exploration Geophysicists, Special Volume, "Non-seismic detection of hydrocarbons." **In press.**

Dickson W, Schiefelbein C, Zumberge J, Odegard M. 2005. Basin analysis in Brazilian and West African conjugates: combining disciplines to deconstruct petroleum systems. *In* 25th Annual Gulf Coast SEPM Foundation Bob F. Perkins Research Conference, 8 p.

Dickson WG, Danforth A, Odegard M. 2003. Gravity signatures of sediment systems: prediction reservoir distribution in Angolan and Brazilian basins. *In* Arthur TJ, MacGregor DS, Cameron N (Editors), *Petroleum Geology of Africa: New Themes and Developing Technologies*: Geological Society, London. Special Publication 207, p. 241–256.

Dohm JM, 20 co-authors. 2009. GRS evidence and the possibility of paleooceans on Mars. *Planetary and Space Science* 57:664–684.

Duck LJ, Glikson M, Golding SD, Webb RE. 2007. Microbial remains and other carbonaceous forms from the 3.24 Ga Sulfur Springs black smoker deposit, Western Australia. *Precambrian Research* 154:205–220.

Ehlmann BL, Mustard JF, Fassett CI, Schon SC, Head JW III, Des Marais DJ, Grant JA, Murchie SL. 2008. Clay minerals in delta deposits and organic preservation potential on Mars. *Nature Geoscience* 1:355–358.

Encrenaz T, Bézard B, Greathouse TK, Richter MJ, Lacy JH, Atreya SK, Wong AS, Lebonnois S, Lefévre F, Forget F. 2004. Hydrogen peroxide on Mars: evidence for spatial and seasonal variations. *Icarus* 170:424–429.

Encrenaz T, Greathouse TK, Bitner M, Kruger A, Richter MJ, Lacy JH, Bézard B, Fouchet T, Lefévre F, Forget F, Atreya SK. 2008. Ground-based infrared observations of water vapor and hydrogen peroxide in the atmosphere of Mars, Mars Atmosphere. Modeling and Observations, Abstract No. 9018.

Etiope G, Oehler DZ, Allen CC. 2010. Methane emissions from Earth's degassing: implications for Mars. *Planetary and Space Science* doi:10.1016/j.pss.2010.06.003

Fairén AG, Davila AF, Gago-Duport L, Amils R, McKay CP. 2009. Stability against freezing of aqueous solutions on early Mars. *Nature* 459:401–404.

Farrand WH, Gaddis LR, Keszthlyi L. 2005. Pitted cones and domes on Mars: observations in Acidalia Planitia and Cydonia Mensae using MOC, THEMIS, and TES data. *Journal of Geophysical Research* 110, doi:10.1029/2004JE002297

Frey H. 2008. Ages of very large impact basins on Mars: implications for the late heavy bombardment in the inner solar system. *Geophysical Research Letters* 35:L13203, doi:10.1029/2008GL033515

Frey H, Jarosewich M. 1982. Subkilometer martian volcanoes—properties and possible terrestrial analogs. *Journal of Geophysical Research* 87:9867–9879.

Frey H, Lowry BL, Chase SA. 1979. Pseudocraters on Mars. *Journal of Geophysical Research* 84:8075–8086.

Frey HV. 2006. Impact constraints on, and a chronology for, major events in early Mars history. *Journal of Geophysical Research* 111:E08S91, doi:10.1029/2005JE002449

Frey HV. 2010a. A minimum crater retention age for the proposed "Borealis Basin" on Mars. Lunar and Planetary Science Conference XLI, Abstract No. 1136.

Frey HV. 2010b. Ages of large impact basins on Mars: implications for inner solar system chronology, late heavy bombardment, and the possibility of early life. Astrobiology Science Conference 2010, Abstract No. 5679.

Gautier DL. 2005. *Kimmeridgian Shales Total Petroleum System of the North Sea Graben Province*: USGS Bulletin 2204-C, 24 p.

Golombek MP. 2005. Tectonics of Mar: Geological Society of America, Abstracts Program, v. 37, p. 312, No. 137–3.

Golombek MP, Edgett KS, Rice JW Jr (Editors). 1995a. *Mars Pathfinder Landing Site Workshop II: Characteristics of the Ares Vallis Region and Field Trips in the Channeled Scabland, Washington*: LPI Technical Report 95–01, Part 1, 63 p.

Golombek MP, Edgett KS, Rice JW Jr (Editors). 1995b. *Mars Pathfinder Landing Site Workshop II: Characteristics of the Ares Vallis Region and Field Trips in the Channeled Scabland, Washington*: LPI Technical Report 95–01, Part 2, 47 p.

Golombek MP, Phillips RJ. 2009. Mars tectonics. *In* Watters TR, Schultz RA (Editors), *Planetary Tectonics*: Cambridge University Press, Cambridge, UK. p. 183–232.

Grady MM, Wright I. 2006. The carbon cycle on early Earth—and on Mars? *Philosophical Transactions of the Royal Society B* 361:1703–1713.

Hartmann WK, Neukum G. 2001. Cratering chronology and the evolution of Mars. *Space Science Reviews* 96:165–194.

Head JW, Hiesinger H, Ivanov MA, Kreslavsky MA, Pratt S, Thomson BJ. 1999. Possible ancient oceans on Mars: evidence from Mars Orbiter Laser Altimeter data. *Science* 286:2137–2143.

Hecht MH. 2009. Introduction to New Martian Chemistry Workshop: "Follow the Chemistry" The New Martian Chemistry Workshop, Abstract No. 8030.

Hecht MH, Kounaves SP, Quinn RC, West SJ, Young SMM, Ming DW, Catling DC, Clark BC, Boynton WV, Hoffman J, DeFlores LP, Gospodinova K, Kapit J, Smith PH. 2009. Detection of perchlorate and the soluble chemistry of martian soil at the Phoenix Lander Site. *Science* 325:64–67.

Hoehler TM, Amend JP, Shock EL. 2007. A "Follow the Energy" approach for astrobiology. *Astrobiology* 7(6):819–823.

Hofmann HJ, Grey K, Hickman AH, Thorpe RI. 1999. Origin of 3.45 Ga coniform stromatolites in Warrawoona Group, Western Australia. *Geological Society of America Bulletin* 111:1256–1262.

Hurowitz JA, Tosca NJ, McLennan SM, Schoonen MAA. 2007. Production of hydrogen peroxide in martian and lunar soil. *Earth and Planetary Science Letters* 255:41–52.

Javaux EJ, Marshall CP, Bekker A. 2010. Organic-walled microfossils in 3.2-billion-year-old shallow-marine siliciclastic deposits. *Nature* 463:934–938.

Jenkyns HC. 2007. The genesis of condensed sequences in the Tethyan Jurassic. *Lethaia* 4:327–352.

Jones B, Renaut RW, Konhauser KO. 2005. Genesis of large siliceous stromatolites at Frying Pan Lake, Waimangu geothermal field, North Island, New Zealand. *Sedimentology* 52:1229–1252.

Joye SB, Samarkin VA, Orcutt BN, MacDonald IR, Hinrichs K-U, Elvert M, Teske AP, Lloyd KG, Lever MA, Montoya JP, Meile CD. 2009. Metabolic variability in seafloor brines revealed by carbon and sulphur dynamics. *Nature Geoscience* doi:10.1038/NGEO475

Kennedy JJ, Pevear DR, Hill RJ. 2002. Mineral surface control of organic carbon in black shale. *Science* 295:657–660.

Klemme HD, Ulmishek GF. 1991. Effective petroleum source rocks of the world: stratigraphic distribution and controlling depositional factors. *American Association of Petroleum Geologists Bulletin* 75:1809–1851.

Kopf AJ. 2002. Significance of mud volcanism. *Reviews of Geophysics* 40:2–52.

Lammer H, Selsis F, Molina-Cuberos GJ, Stumpner W, Berces A, Kerekgyarto T, Ronto G. 2002. Was the ancient martian surface sterilized by radiation? *In* Montesinos B, Gimenez A, Guinan EF (Editors), *The Evolving Sun and its*

Influence on Planetary Environments: San Francisco, Astronomical Society of the Pacific, ASP Conference Proceedings, v. 269, 151 p.

Macquaker JHS, Curtis CD. 1989. The sedimentology and diagenesis of organic-rich sediments. *Journal of the Geological Society* 146:271–272.

Marinova MM, Aharonson O, Asphaug E. 2008. Mega-impact formation of the Mars hemispheric dichotomy. *Nature* 453:1216–1219.

Marshall CP, Love GD, Snape CE, Hill AC, Allwood AC, Walter MR, Van Kranendonk MJ, Bowden SA, Sylvia SP, Summons RE. 2007. Structural characterization of kerogen in 3.4 Ga Archaean cherts from the Pilbara Craton, Western Australia. *Precambrian Research* 155:1–23.

McGowan E. 2009. Spatial distribution of putative water related features in Southern Acidalia/Cydonia Mensae, Mars. *Icarus* 202:78–89.

Molina-Cuberos GJ, Stumptner W, Lammer H, Kömle NI, O'Brien K. 2001. Cosmic ray and UV models on the ancient martian surface. *Icarus* 154:216–222.

Morris RV, Ruff SW, Gellert R, Ming DW, Arvidson RE, Clark BC, Golden DC, Siebach K, Klongelhofer G, Schroder C, Fleischer I, Yen AS, Squyres SW. 2010. Identification of carbonate-rich outcrops on Mars by the Spirit Rover. *Sciencexpress* doi:10.1126/science.1189667

Mustard JF, Murchie SL, Pelkey SM, Ehlmann BL, Milliken RE, Grant JA, Bibring J-P, Poulet F, Bishop J, Roach L, Seelos F, Humm D, and the CRISM Science Team. 2008. Hydrated silicate minerals on Mars observed by the CRISM instrument on MRO. *Nature* 454:305–309.

Neukum G. 2008. The lunar and martian cratering records and chronologies. Lunar and Planetary Science Conference XXXIX, Abstract No. 2509.

Neukum G, Ivanov BA, Hartmann WK. 2001. Cratering records in the inner solar system in relation to the lunar reference system. *Space Science Reviews* 96:55–86.

Niederberger TD, Perreault NN, Tille S, Sherwood Lollar B, Lacrampe-Couloume G, Andersen D, Greer CW, Pollarld W, Whyte LG. 2010. Microbial characterization of a subzero, hypersaline methane seep in the Canadian High Arctic. *The ISME Journal* doi:10:1038/ismej.2010.57

National Research Council, Committee on an Astrobiology Strategy for the Exploration of Mars. 2007. *An Astrobiology Strategy for the Exploration of Mars*: The National Academies Press, Washington, DC. 130 p.

Oehler DZ, Allen CC. 2010. Evidence for pervasive mud volcanism in Acidalia Planitia, Mars. *Icarus* 208:636–657.

Oehler DZ, Robert F, Walter MR, Sugitani K, Allwood A, Meibom A, Mostefaoui S, Selo MR, Thomen A, Gibson EK. 2009. NANOSIMS: insights to biogenicity and syngeneity of Archaean carbonaceous structures. *Precambrian Research* 173:70–78.

Oehler DZ, Robert F, Walter MR, Sugitani K, Meibom A, Mostefaoui S, Gibson EK. 2010. Diversity in the Archean Biosphere: new insights from NanoSIMS. *Astrobiology* 10:413–424.

Parker TJ, Gorsline DS, Saunders RS, Pieri DC, Schneeberger DM. 1993. Coastal geomorphology of the Martian northern plains. *Journal of Geophysical Research* 98:11061–11078.

Parker TJ, Saunders RS, Schneegerger DM. 1989. Transitional morphology in west Deuteronilus Mensae, Mars: implications for modification of the lowland/upland boundary. *Icarus* 82:111–135.

Parker TJ, Schneeberger DM, Pieri DC, Saunders RS. 1987. Geomorphic evidence for ancient seas on Mars. *In MECA Symposium on Mars: Evolution of its Climate and Atmosphere*: Lunar and Planetary Institute, Houston. LPI Technical Report 87–01, p. 96–98.

Pohlman JW, Bauer JE, Waite WF, Osburn CL, Chapman NR. 2011. Methane hydrate-bearing seeps as a source of aged dissolved organic carbon to the oceans. *Nature Geoscience* 4:37–41.

Rasmussen B. 2000. Filamentous microfossils in a 3235-milliion-year-old volcanogenic massive sulphide deposit. *Nature* 405:676–679.

Rice JW Jr, Edgett KS. 1997. Catastrophic flood sediments in Chryse Basin, Mars, and Quincy Basin, Washington: application of sandar facies model. *Journal of Geophysical Research* 102:4185–4200.

Robertson AHF, Kopf A. 1998. Tectonic setting and processes of mud volcanism on the Mediterranean Ridge Accretionary complex: evidence from Leg 160. *In* Robertson AHF, Emeis K–C, Richter C, Camerlenghi A (Editors), *Proceedings of the Ocean Drilling Program, Scientific Results,* v. 160, p. 665–680.

Rodríguez JAP, Tanaka KL, Kargel JS, Dohm JM, Kuzmin R, Fairén AG, Sasaki S, Komatsu G, Schulze-Makuch D, Jianguo Y. 2007. Formation and disruption of aquifers in southwestern Chryse Planitia, Mars. *Icarus* 191:545–567.

Rotto SL, Tanaka KL. 1995. *Geologic/geomorphologic Map of the Chryse Planitia Region of Mars*: US Geological Survey Miscellaneous Investigations Series, Map I-2441, scale 1:5,000,000.

Schopf JW. 2006. Fossil evidence of Archaean life. *Philosophical Transactions of the Royal Society B* 361:869–885.

Schopf JW, Kudryavtsev AB, Czaja D, Tripathi AB. 2007. Evidence of Archean life: stromatolites and microfossils. *Precambrian Research* 158:141–155.

Schopf JW, Kudryavtsev AB, Foster IS, Farmer JD, Butterfield NJ. 2010. Neoproterozoic Lagerstätten (and the search for past life on Mars). Geological Society of America, Abstracts with Programs, Paper 167–13, v. 42, p. 410.

Schopf JW, Walter MR. 2007. Earliest evidence of life on Earth. *Precambrian Research* 158:139–140.

Scott DH, Dohm JM, Rice JW Jr. 1995. *Map of Mars Showing Channels and Possible Paleolake Basins*: US Geological Survey Miscellaneous Investigations Series, Map I–2461, scale 1:30,000,000.

Scott DH, Rice JW Jr, Dohm JM. 1991. Martian paleolakes and waterways: exobiologic implication. *Origins of Life and Evolution of the Biosphere* 21:189–198.

Shkrob IA, Chemerisov SD, Marin TW. 2010. Photocatalytic decomposition of carboxylated molecules on light-exposed martian regolith and its relation to methane production on Mars. *Astrobiology* 10:425–436.

Skinner JA Jr, Mazzini A. 2009. Martian mud volcanism: terrestrial analogs and implications for formational scenarios. *Marine and Petroleum Geology* doi:10.1016/j.marpetgeo.2009.02.006

Squyres SW, Kasting JF. 1994. Early Mars: how warm and how wet? *Science* 265:744–749.

Stewart SA, Davies RJ. 2006. Structure and emplacement of mud volcano systems in the South Caspian Basin. *American Association of Petroleum Geologists Bulletin* 90:771–786.

Sugitani K, Grey K, Allwood A, Nagaoka T, Mimura K, Minami M, Marshall CP, Van Kranendonk MJ, Walter MR. 2007. Diverse microstructures from Archaean chert from the Mount Goldsworthy-Mount Grant area, Pilbara Craton, Western Australia: microfossils, dubiofossils, or pseudofossils? *Precambrian Research* 158:228–262.

Summons RE, Amend JP, Buick R, Cody GD, Dromart G, Eigenbrode JL, Knoll AH. 2010. Identification of biosignatures preservation windows to guide MSL landing site selection. Astrobiology Science Conference 2010, Abstract No. 5256.

Tanaka KL. 1997. Sedimentary history and mass flow structures of Chryse and Acidalia Planitiae, Mars. *Journal of Geophysical Research* 102:4131–4150.

Tanaka KL, Banerdt WB. 2000. The interior lowland plains unit of Mars: evidence for a possible mud ocean and induced tectonic deformation. Lunar and Planetary Science XXXI, Abstract No. 2041.

Tanaka KL, Skinner JA Jr, Hare TM. 2005. *Geologic Map of the Northern Plains of Mars*: US Geological Survey Scientific Investigations, Map 2888.

Tanaka KL, Skinner JA Jr, Hare TM, Joyal T, Wenker A. 2003. Resurfacing history of the northern plains of Mars based on geologic mapping of Mars Global Surveyor data. *Journal of Geophysical Research* 108, 8043, doi:10.1029/2002JE001908

Ten Kate IL, Garry JRC, Peeters Z, Foing B, Ehrenfruend P. 2006. The effects of martian near surface conditions on the photochemistry of amino acids. *Planetary and Space Science* 54:296–302.

Tian F, Kasting JE, Solomon SC. 2009. Thermal escape of carbon from the early martian atmosphere. *Geophysical Research Letters* 36:L02205, doi:10.1029/2008GL036513

Tommeras A, Mann U. 2008. Improved hydrocarbon charge prediction by source-rock modeling. *Petroleum Geoscience* 14:291–299.

Tyson RV. 1996. Sequence stratigraphical interpretation of organic facies variations in marine siliclastic systems. *In* Hesselbo SP, Parkinson DN (Editors), *Sequence Stratigraphy in British Geology*: Geological Society, London. Special Publication 103, p. 75–96.

Van Rensbergen P, Deprieter D, Pannemans B, Henriet J-P. 2005a. Seafloor expression of sediment extrusion and intrusion at the El Arraiche mud volcano field, Gulf of Cadiz. *Journal of Geophysical Research* 110:F02010, doi:10.1029/2004JF000165

Van Rensbergen P, Deprieter D, Pannemans B, Moerkerke G, Van Rooij D, Marsset B, Akhmanov G, Blinova V, Ivanov M, Rachidi M, Magalhaes V, Pinheiro L, Cunha M, Henriet J-P. 2005b. The El Arraiche mud volcano field at the Moroccan Atlantic slope, Gulf of Cadiz. *Marine Geology* 219:1–17.

Van Rensbergen P, Hillis RR, Maltman AJ, Morley CK. 2003. Subsurface sediment mobilization: introduction. *In* Van Rensbergen P, Hillis RR, Maltman AJ, Morley CK (Editors), *Subsurface Sediment Mobilization*: Geological Society, London. Special Publication 216, p. 1–8.

Van Rooij D, Depreiter D, Bouimetarhan I, De Boever E, De Rycker K, Foubert A, Huvenne V, Réveillaud J, Staelens P, Vercruysse J, Versteeg W, Henriet J-P. 2005. First sighting of active fluid venting in the Gulf of Cadiz. *EOS Transactions* 86:509–511.

Van Zuilen MA, Chaussidon M, Rollion-Bard C, Marty G. 2007. Carbonaceous cherts of the Barberton Greenstone Belt, South Africa: isotopic, chemical and structural characteristics of individual microstructures. *Geochimica et Cosmochimica Acta* 71:655–669.

Walter MR, Buick R, Dunlop JSR. 1980. Stromatolites 3400–3500 Myr old from the North Pole area, Western Australia. *Nature* 284:443–445.

Warner N, Gupta S, Kim J-R, Lin S-Y, Muller JP. 2010. Hesperian equatorial thermokarst lakes in Ares Vallis as evidence for transient warm conditions on Mars. *Geology* 38:71–74.

STRATIGRAPHIC ARCHITECTURE OF BEDROCK REFERENCE SECTION, VICTORIA CRATER, MERIDIANI PLANUM, MARS

LAUREN A. EDGAR, JOHN P. GROTZINGER, AND ALEX G. HAYES

Division of Geological and Planetary Sciences, California Institute of Technology, Pasadena, California 91125 USA

e-mail: ledgar@caltech.edu

DAVID M. RUBIN

US Geological Survey Pacific Science Center, Santa Cruz, California 95060 USA

STEVE W. SQUYRES AND JAMES F. BELL

Astronomy Department, Cornell University, Ithaca, New York 14853 USA

AND

KEN E. HERKENHOFF

US Geological Survey Astrogeology Science Center, Flagstaff, Arizona 86001 USA

ABSTRACT: The Mars Exploration Rover *Opportunity* has investigated bedrock outcrops exposed in several craters at Meridiani Planum, Mars, in an effort to better understand the role of surface processes in its geologic history. *Opportunity* has recently completed its observations of Victoria crater, which is 750 m in diameter and exposes cliffs up to ~15 m high. The plains surrounding Victoria crater are ~10 m higher in elevation than those surrounding the previously explored Endurance crater, indicating that the Victoria crater exposes a stratigraphically higher section than does the Endurance crater; however, Victoria strata overlap in elevation with the rocks exposed at the Erebus crater. Victoria crater has a well-developed geomorphic pattern of promontories and embayments that define the crater wall and that reveal thick bedsets (3–7 m) of large-scale cross-bedding, interpreted as fossil eolian dunes. *Opportunity* was able to drive into the crater at Duck Bay, located on the western margin of Victoria crater. Data from the Microscopic Imager and Panoramic Camera reveal details about the structures, textures, and depositional and diagenetic events that influenced the Victoria bedrock. A lithostratigraphic subdivision of bedrock units was enabled by the presence of a light-toned band that lines much of the upper rim of the crater. In ascending order, three stratigraphic units are named Lyell, Smith, and Steno; Smith is the light-toned band. In the Reference Section exposed along the ingress path at Duck Bay, Smith is interpreted to represent a zone of diagenetic recrystallization; however, its upper contact also coincides with a primary erosional surface. Elsewhere in the crater the diagenetic band crosscuts the physical stratigraphy. Correlation with strata present at nearby promontory Cape Verde indicates that there is an erosional surface at the base of the cliff face that corresponds to the erosional contact below Steno. The erosional contact at the base of Cape Verde lies at a lower elevation, but within the same plane as the contact below Steno, which indicates that the material above the erosional contact was built on significant depositional paleotopography. The eolian dune forms exposed in Duck Bay and Cape Verde, combined with the geometry of the erosional surface, indicate that these outcrops may be part of a larger-scale draa architecture. This insight is possible only as a result of the larger-scale exposures at Victoria crater, which significantly exceed the more limited exposures at the Erebus, Endurance, and Eagle craters.

KEY WORDS: Mars, eolian, cross-bedding, stratigraphy

INTRODUCTION

Sedimentary rocks on Mars provide insight into past aqueous and atmospheric processes, climate regimes, and potential habitability. The Mars Exploration Rover *Opportunity* has investigated several impact craters in Meridiani Planum, Mars (Fig. 1), studying the exposed sedimentary rocks of the Burns formation[1] in an effort to better understand the role of aqueous activity in its geologic history. In doing so, *Opportunity* discovered in situ evidence for an aqueous depositional environment at the Eagle, Endurance, and Erebus craters (Squyres et al. 2004b; Grotzinger et al. 2005, 2006; Squyres et al. 2006; Metz et al. 2009). Furthermore, these results also indicate that the stratigraphic architecture of sedimentary rocks on Mars is similar (though not identical) to that of Earth, indicating that the processes that govern facies deposition and alteration on Mars can be reasonably inferred through reference to analogous terrestrial depositional systems (Grotzinger et al. 2005).

As the mission progressed *Opportunity* explored increasingly larger outcrops over a greater spatial scope, and it became clear that both depositional and diagenetic processes acted regionally in extent (Squyres et al. 2009). Here, diagenesis refers to any low-temperature, low-pressure alteration after deposition, which may include weathering. One important interpretation is that most primary sedimentary bedforms—a key element in the sedimentology of the Burns formation—are not only regional in extent (found in many bedrock outcrops explored by the rover), they are also of very large magnitude. This is important because it facilitates a greater understanding of the processes controlling deposition as well as the scale of the depositional environment. The outcrop exposed at *Opportunity*'s landing site in Eagle crater was only 35 cm thick (Squyres et al. 2004a). At Erebus crater, exposed outcrops were up to 5 m thick (Grotzinger et al. 2006, Metz et al. 2009), and at Endurance crater, the rover analyzed more than 7 m of stratigraphy (Grotzinger et al. 2005). These latter localities

[1]Note that the location and formation names used in this study are informal and have not been approved by the International Astronomical Union.

Sedimentary Geology of Mars
SEPM Special Publication No. 102, Copyright © 2012
SEPM (Society for Sedimentary Geology), Print ISBN 978-1-56576-312-8, CD/DVD ISBN 978-1-56576-313-5, p. 195–209.

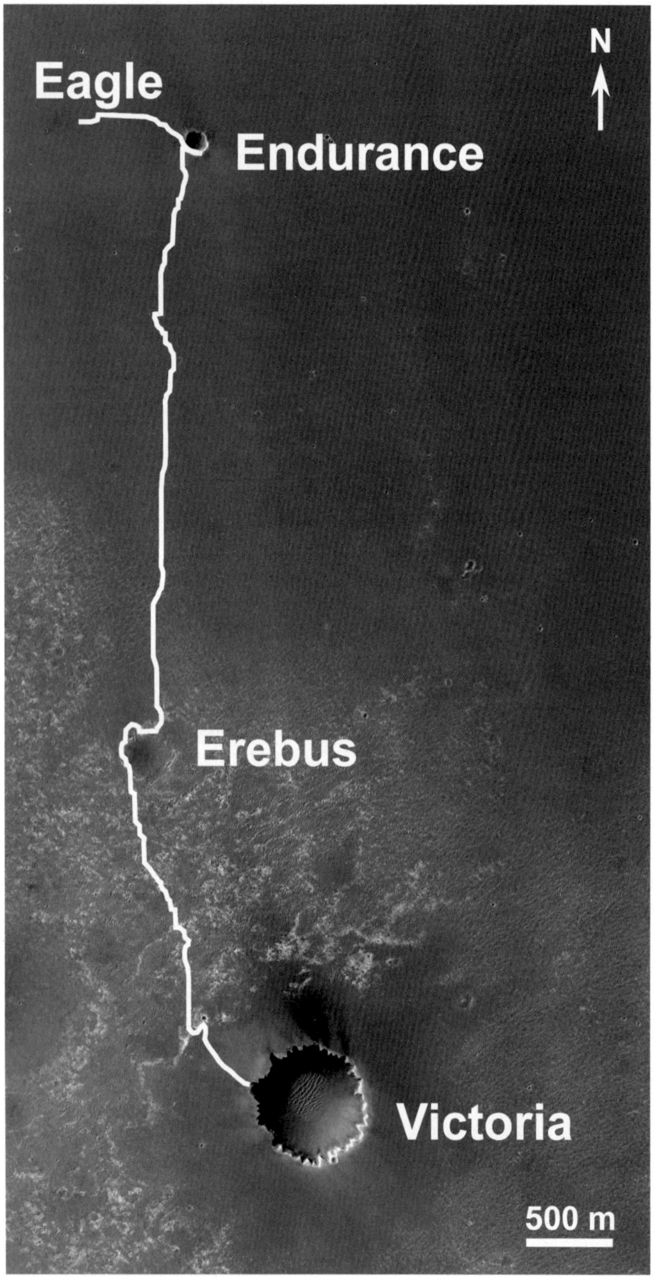

FIG. 1.—*Opportunity* traverse map as of Sol 952 plotted on HiRISE image (PSP_001414_1780). Victoria crater is approximately 6 km southeast of the original *Opportunity* landing site in Eagle crater. Sunlight illuminates scene from upper left. Image credit: NASA/JPL/University of Arizona.

provided evidence for meter-scale cross-stratification. In contrast, outcrops at Victoria crater provide exposures of on the order of 15 m of true stratigraphic thickness, which permits analysis of larger scales of cross-stratification, and, thus, larger-scale primary bedforms can now be reconstructed.

Victoria crater lies 6 km southeast from the original *Opportunity* landing site in Eagle crater. At ~750 m in diameter (Grant et al. 2008), it is the largest crater yet explored by *Opportunity*. Victoria crater offers the opportunity to test the existing models for bedrock formation and provides new insight into larger-scale sedimentary bedforms. This study examines the stratigraphy and eolian architecture along the ingress path at Duck Bay using data from the Microscopic Imager and Panoramic Camera to elucidate the structures, textures, and depositional and diagenetic events that formed and modified the Victoria bedrock. Several depositional and diagenetic hypotheses are explored, and a Reference Section exposed along the ingress path at Duck Bay is interpreted in the context of the regional geologic history.

GEOLOGIC SETTING

Rocks exposed in the region of the *Opportunity* landing site reveal a complex sedimentary history, involving eolian sediment transport and deposition, followed by episodic inundation by shallow surface water, evaporation, exposure, and desiccation (Squyres et al. 2004b; Grotzinger et al. 2005, 2006; Metz et al. 2009). The bedrock exposed at Eagle crater, combined with that in Endurance crater, reveals a stratigraphic thickness of approximately 7 m, known as the Burns formation. The Burns formation refers to a sequence of well-sorted, moderately indurated sandstones, interpreted to be deposited by eolian and locally subaqueous processes (Grotzinger et al. 2005, McLennan and Grotzinger 2008). Represented by dune, sand sheet, and interdune facies, this formation records a progressive increase in the influence of groundwater and surface water during deposition (Grotzinger et al. 2005). This wetting-upward trend contrasts with the overall drying-upward trend as seen in the Erebus crater, at a slightly higher stratigraphic level (Metz et al. 2009). The series of outcrops explored by *Opportunity* at Meridiani Planum may comprise a full climatic cycle, from dry to wet to dry conditions, as one moves stratigraphically upward from the strata at Eagle crater through the strata at Erebus crater (Metz et al. 2009). After completing its exploration of the outcrops at Erebus crater, *Opportunity* set out to explore Victoria crater.

Victoria crater is a degraded simple crater at Meridiani Planum (1.9483°S, 354.4742°E). Orbital images show that stratified rocks beneath the Meridiani plains lie disconformably on Middle to Late Noachian cratered terrains, indicating that the Meridiani plains might be as much several billion years old (Arvidson et al. 2003). Topographic measurements made by the Mars Orbiter Laser Altimeter indicate that the plains surrounding Victoria crater are ~10 m higher in elevation than those surrounding the previously explored Endurance crater, indicating that Victoria crater exposes a stratigraphically higher section (assuming flat dip of strata). The outcrop exposed at Victoria crater may lie at the same elevation as the uppermost unit in Erebus crater, allowing for possible stratigraphic correlation between these two locations. Victoria crater has a scalloped rim produced by erosion and downhill movement of crater wall material. Most of the erosion is attributed to eolian processes, which are actively widening and filling the crater and elongating the crater across a WNW–ESE axis (Grant et al. 2008). The scalloped rim consists of a series of alcoves and promontories around the crater, exposing more than 10 m of well-bedded sedimentary rocks.

Prior to ingress at Duck Bay, *Opportunity* spent several months traversing the rim of the crater. Observations of outcrops at several promontories revealed thickly bedded units containing large-scale cross-stratification with bedsets of at least several meters in thickness (Squyres et al. 2009). Analysis of cross-bedding geometry indicates a paleo-wind direction oriented in a north–south direction (Hayes et al. 2011). A distinct light-toned band lines much of the upper rim of the crater. This band served as the basis for defining stratigraphic units (Squyres et al. 2009) named after prominent geologists and stratigraphers: Lyell, Smith, and Steno, in ascending order. Smith is the light-toned band, which is overlain by the Steno unit and underlain by the Lyell unit. Steno's upper contact is the base of the breccia

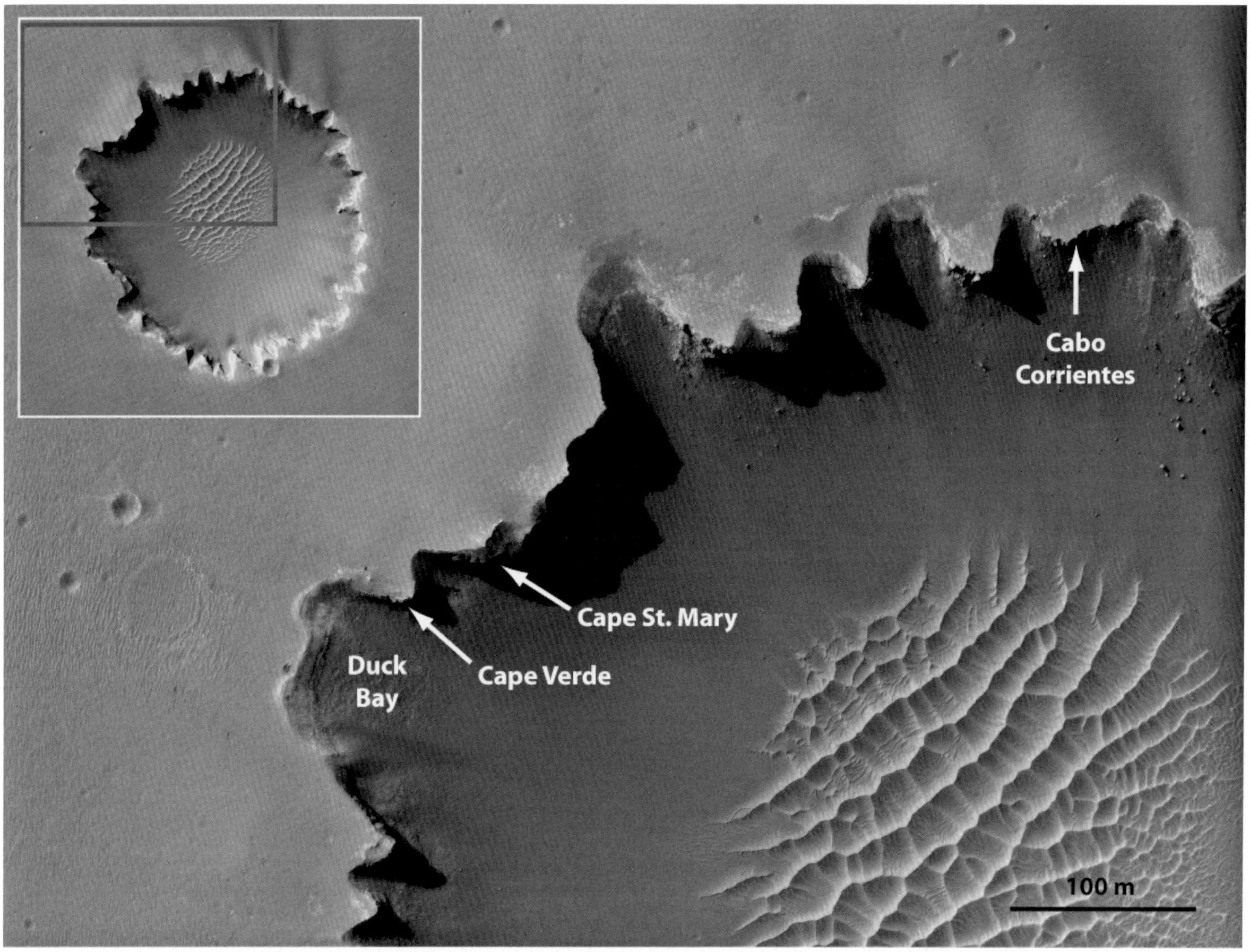

FIG. 2.—Duck Bay and nearby promontories Cape Verde and Cape St. Mary, seen in HiRISE image (TRA_000873_1780). Inset shows the crater's scalloped rim and the location of Duck Bay on the western margin of Victoria crater. Image credit: NASA/JPL/University of Arizona.

deposit generated during Victoria crater formation; Lyell's lower contact is not exposed, buried beneath the modern sands that fill the floor of the crater.

Duck Bay is located on the western margin of Victoria crater (between the promontories Cape Verde and Cabo Frio) with a slope shallow enough (~15–25°) for *Opportunity* to ingress (Fig. 2). Detailed measurements of the three stratigraphic units were taken as *Opportunity* descended into the crater.

METHODOLOGY

Instruments

The stratigraphy and bedrock sedimentary structures at Duck Bay were observed by the Panoramic Camera (Pancam) and Microscopic Imager (MI). Pancam is a multispectral imaging system, which consists of two digital cameras mounted on the rover's mast (1.5 m above the ground). Pancam is capable of both stereoscopic measurements (using parallax between the two camera "eyes," which are spaced 30 cm apart) and panoramic measurements (with the ability to image 360° in azimuth and ±90° in elevation). Pancam has a focal length of 43 mm and a field of view of 1024 by 1024 pixels (16° by 16°). Each of the two camera "eyes" includes a small eight-position filter wheel, allowing for multispectral mineralogical studies in the 400 to 1100 nm wavelength range (Bell et al. 2003).

The MI is a high-resolution imaging system mounted on the rover's "arm," or Instrument Deployment Device (IDD). The MI acts as both a microscope and a camera. With a camera body identical to those of Pancam, it also has a field of view of 1024 by 1024 pixels (Herkenhoff et al. 2003). The MI has a focal length of 20 mm, a ±3-mm depth of field, and a field of view of 31 mm by 31 mm at a working distance of 63 mm (measured from the front of the lens to the object). MI produces images with a resolution of 30 μm/pixel (Herkenhoff et al. 2003). As a result of highly variable topography, MI images are usually taken as a stack of images, with the IDD moving along a path normal to the surface, pausing every few millimeters to acquire an image (Herkenhoff et al. 2003).

The rover also has the capability to grind away dust and surface alteration to expose fresh rock surfaces for chemical and textural analyses. This is accomplished by the Rock Abrasion Tool (RAT). The RAT is positioned against a rock by the rover's arm and uses a diamond-tipped grinding wheel to remove a cylindrical area 4.5 cm in diameter and 0.5 cm deep (Gorevan et al. 2003). The RAT was designed to enhance petrologic textures of the fresh rock surfaces so

TABLE 1.—Summary of MI observations.

Unit	Target name	Sols acquired	Stack	RATed
Steno	Steno	1311	2×2×5	none
	Steno	1313	1×1×5	brushed
	Steno	1316	2×2×5	RATed
	Steno	1320	1×1×10	RATed
Smith	Smith_Hall_1	1322	1×1×18	none
	Smith	1332	2×2×5	none
	Smith	1348	2×2×5	brushed
	Smith2*	1359	2×2×5	none
	Smith2 take2	1366	3×1×5	none
	Smith2	1373	2×2×3	RATed
	Smith_Lyell_Smith_side	1409	4×1×5	none
	Smith_Lyell_Smith_side	1414	1×1×5	none
Lyell	Lyell_1	1384	2×2×14	none
	Lyell_1	1395	2×2×14	RATed
	Smith_Lyell_Lyell_side	1411	4×1×5	none
	Lyell_B	1421	2×2×5	none
	Lyell_Buckland	1430	2×2×5	RATed
	Lyell_Buckland	1435	1×1×1	RATed
	Lyell_Buckland	1436	1×1×4	RATed
	Lyell_Exeter	1443	2×2×14	none

* Joint 1 stall, observation was not completed.

that they could be viewed effectively using the MI (Gorevan et al. 2003). However, overuse issues[2] (which resulted in approximately 60% of the RAT holes covered in RAT grindings) significantly degraded the quality of the MI images taken along the ingress path Reference Section. The presence of RAT grindings in the holes obscures some features, so caution must be used when evaluating MI images of RATed targets.

Data Collection

Opportunity began its ingress into Victoria crater on Sol 1293. After driving 11 m downslope into the crater at Duck Bay, making detailed stratigraphic measurements of the exposed outcrop, the rover drove laterally ~20 m to investigate several scientific targets at the nearby promontory Cape Verde. At its closest approach to Cape Verde, *Opportunity* maintained an ~10-m stand-off distance. While detailed chemical and textural analyses could not be acquired at this distance, super-resolution imaging allowed substantial observations of the stratification geometry and textures exposed at the cliff face. On Sol 1607, *Opportunity* began its egress, following nearly the same route as it did during ingress and eventually exiting the crater on Sol 1634.

Opportunity acquired many Pancam and MI observations over the course of nearly 150 sols on its 11-m traverse downslope. Three stratigraphic units were distinguished based on sedimentary structures, diagenetic textures, and color, as revealed in Pancam images. Pancam observations were made almost every sol. The MI provided additional information on small-scale textures and diagenetic features. The RAT was used to expose fresh surfaces in some of the IDD targets by grinding away the outermost several millimeters of rock. However, after Sol 1348, the motor encoder failure resulted in partial covering of RAT holes, which obscured some of the features of interest, so our quantitative survey of diagenetic textures used only un-RATed images. Tables 1 and 2 summarize the key MI and Pancam observations of the rock units and the sols on which they were acquired.

Data Processing

Structural attitudes were obtained using Pancam stereo data. Range and topographic data can be derived from Pancam stereo image pairs, producing range maps out to 100 m or more from the rover (Squyres et al. 2003, Lewis et al. 2008). Occasionally, stereo images from the Navigation Camera (Navcam) also were used to produce range maps. Following the methods described by Lewis et al. (2008), linear segments along bedding planes were traced manually, and the corresponding topographic data were extracted. The natural curvature of the crater in the horizontal direction provided constraints on the three-dimensional geometry. A best-fit plane was calculated for each segment, and mathematical criteria ensured that the layers were well fit by a plane. Following the method of Lewis et al. (2008), we determined principle component eigenvalues to describe the variance in three orthogonal directions (the first and second components describe the plane, and the third component lies out of the plane). The variance in the first principal component was limited to less than 99%, which excludes measurements for which the topography is too linear to determine the second dimension. Additionally, we excluded data for

[2]Encoders are attached to each of the three RAT motors and provide precise position information with which to monitor the penetration process. Another wheel holds a set of spinning brushes to wipe away dust and grit so that the abraded material is moved outward to form a dust skirt around the RAT hole. However, as the mission progressed many times beyond its planned 90-sol lifetime, problems associated with overuse have developed. On Sol 1045, the RAT grind motor encoder failed. The MER science and engineering team developed a way to run the RAT without this encoder feedback, but on Sol 1334, the RAT revolve motor encoder failed. This meant that the RAT tool would run without precise monitoring (without knowing its exact position). On Sol 1348, the team first noticed an issue with the brush spot. Rather than clearing the grindings out of the RAT hole, only a small portion of the hole was cleared, and much of the hole contained RAT grindings. This was attributed to running the brush in the opposite direction, which resulted in a reconfiguration of the brush bristles—rather than being in line with the grinding bit, the bristles were twisted and some aligned with the z-axis, serving to erode material more than clear material out of the hole.

TABLE 2.—Pancam observations.

Unit	Location	Sols on which most useful images were acquired	Sols on which additional images were acquired
Steno	Duck Bay	1319, 1335, 1338, 1353, 1363, 1385	1302, 1305, 1315, 1316, 1318, 1320, 1321, 1333, 1334, 1339, 1377
Smith	Duck Bay	1332, 1335, 1337, 1349, 1350, 1351, 1353, 1366, 1374, 1385, 1426	1327, 1333, 1338, 1344, 1348, 1371, 1377, 1390
Lyell	Duck Bay	1385, 1423, 1426, 1432, 1433, 1438, 1465	1398, 1419, 1425, 1428, 1511, 1512
	Cape Verde	952, 970–991, 1006, 1329, 1342, 1356, 1487, 1574–1576, 1579–1580, 1581	1329, 1342, 1346, 1349, 1358, 1457, 1472, 1473, 1487, 1570, 1572, 1599, 1611

which the ratio of the variance described by the second principal component to that of the third was greater than 15 to ensure that the planar fit was significantly larger than the out-of-plane error. Strike and dip values were then calculated from the best-fit plane. The stratigraphic thicknesses reported below are true thickness, corrected for a few degrees of westward dip, obtained using the methods described above.

Super-resolution images exposed fine details in targets that the rover was unable to reach. Super resolution is a method of combining information from a series of slightly offset images to produce a single image of the scene at higher resolution (Bell et al. 2006). The camera acquires 15 to 20 images in rapid succession at slightly offset pointings, and coregistration is performed. The images are first forward-mapped into high-resolution scratch space, and then a back-propagation algorithm is repeatedly applied to the scratch space, which iteratively improves the high-resolution image. Additional details are described in Bell et al. (2006).

Simulations

Several bedform simulations were run and cross-bedding geometries were compared to those observed in outcrop to test potential bedform morphologies and behaviors. We used a code that mathematically simulates bedforms and cross-bedding, which was developed by Rubin and Carter (2005). The code approximates bedform surfaces as sine curves. Bedform migration is created by displacement of the sine curves, height is produced by changing the amplitude, and superimposed bedforms can be modeled as combined sets of sine curves. The resulting block diagrams can be compared to the cross-bedding observed in rock outcrops in a variety of geometries.

DUCK BAY SEDIMENTOLOGY AND STRATIGRAPHY

Opportunity's ingress path in Duck Bay intersects three stratigraphic units named Lyell, Smith, and Steno, in ascending stratigraphic order (Fig. 3). These units, as observed along the rover ingress path, define the "Reference Section" for Victoria crater. The definitions and descriptions presented here form a set of observations that can be used for the purpose of comparison to other locations in the crater and for interregional correlations between craters. Changes in stratigraphic attributes relative to a defined Reference Section form the traditional basis for understanding lateral and temporal variations in depositional environment.

Lyell Unit

The lowermost unit, designated "Lyell," consists of ~1.8 m of sulfate-rich sandstone (Squyres et al. 2009). Lyell is characterized by its darker tone, with well-defined pinstripe laminations. The average bed thickness is ~2 mm. It is a well-sorted, fine-grained sandstone, with meter-scale cross-bedding (Fig. 4E). Strike and dip measurements, calculated from Pancam stereo data, indicate that Lyell dips approximately 2° to the west—away from the center of the crater. Measurements of cross-strata indicate dip directions to the southwest (Fig. 5).

The MI provided additional information on small-scale textures and diagenetic features, including abundant hematite-rich spherules and vugs indicative of crystal molds (Fig. 4F). Volumetric densities of spherules were estimated following the method of McLennan et al. (2005), using the relation $V = (\pi m l^2/6A)$, where m is the number of spherules on a planar rock surface with area A and l is the mean spherule diameter. Spherules in Lyell range from 0.5 mm in diameter up to 1.5 mm, with an average diameter of 0.90 mm (standard deviation [SD] = 0.27, 378 spherules). Lyell has more embedded spherules than does Smith or Steno, with a volumetric density of 1.04% (SD = 0.46%, n = 16 frames). Lyell also contains abundant vugs, with tabular-prismatic shape, and random orientations. The vugs in Lyell have a typical width of ~0.25 mm and length of 1 to 2 mm. While the vugs show similar geometry and similar proportions to those seen in the Eagle crater (Squyres et al. 2004b), they are an order of magnitude smaller. The vugs may contribute to the darker tone of this unit, helping to trap dark, likely basaltic sand grains (Squyres et al. 2009).

Smith Unit

The light-toned Smith unit directly overlies the Lyell unit. The contact between Smith and Lyell is gradational, and the darker tone and well-defined stratification of Lyell gradually fade upward. Smith is characterized by its smoother texture and lighter tone, which contrasts with its bounding units (Fig. 4C). The Smith unit at first appears massive, but images taken at low solar incidence angles reveal fine lamination, with lamina thicknesses similar to those in Lyell, on the order of ~1 to 2 mm. Smith dips approximately 2° to the west and has a thickness of ~0.8 m (consistent with other exposures around the crater [Hayes et al. 2011]).

Smith has fewer spherules than does Lyell (Fig. 4D), with a volumetric density of 0.35% (SD = 0.15%, n = 14 frames). Spherules in Smith have an average diameter of 0.73 mm (SD = 0.50 mm; 162 spherules). The lighter tone and poor expression of lamination of Smith indicates that Smith is recrystallized.

Steno Unit

Along the ingress path, a clear erosional contact distinguishes Smith from the overlying Steno unit. Inclined stratification in Smith is truncated by the lower bounding surface of Steno. Strike and dip measurements indicate that this truncation surface between Steno and

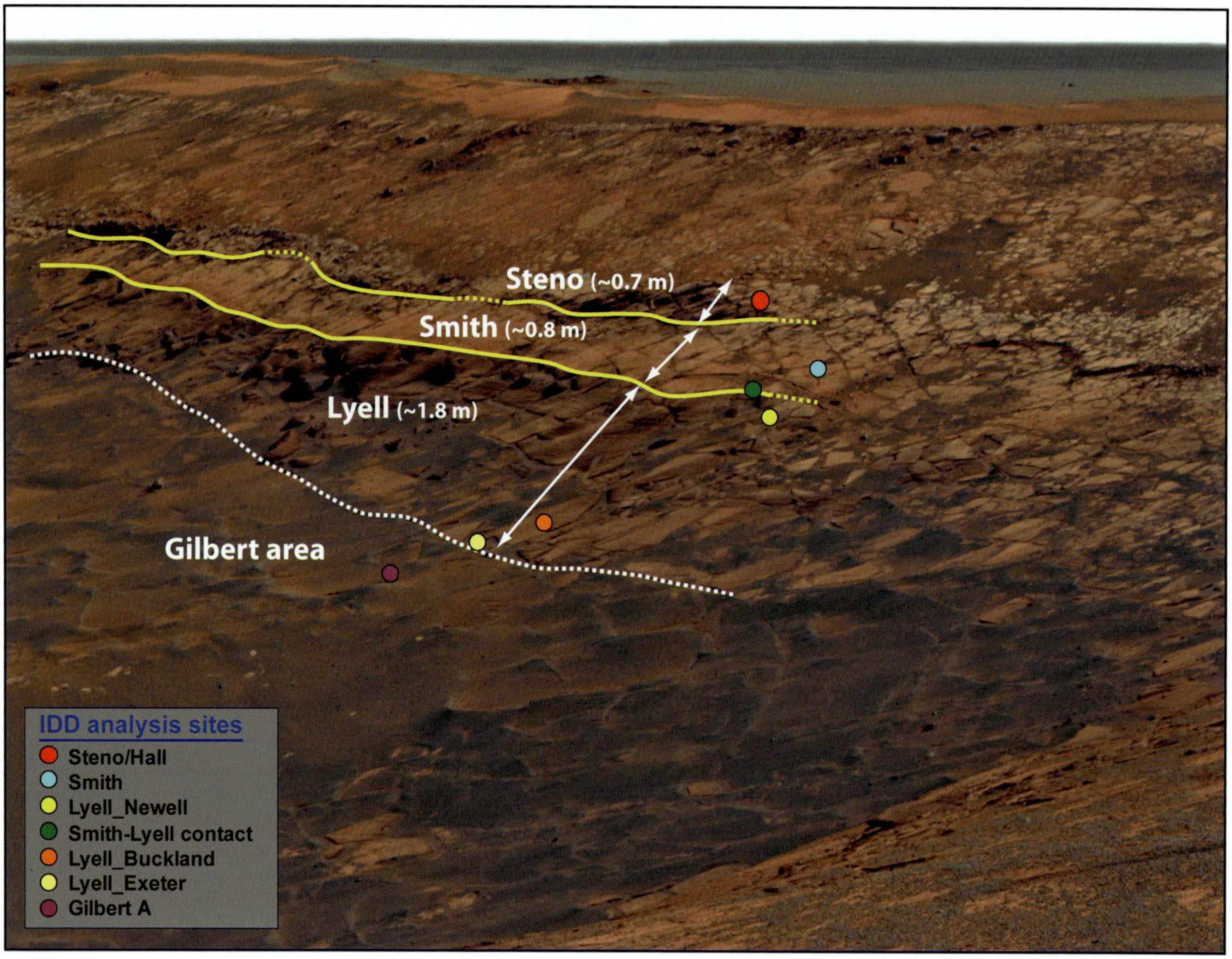

FIG. 3.—Ingress path as viewed from Cape Verde promontory (Cape Verde panorama, L257 false color mosaic, acquired during Sols 970–991). Three distinct units are visible in Duck Bay, named Lyell, Smith, and Steno, in ascending order. The Gilbert area corresponds with brecciated bedrock below the intact Lyell unit. Stratigraphic thicknesses (reported in parentheses) were derived from Pancam stereo images. Colored circles correspond to locations of detailed analyses by the MI as well as other tools on the IDD.

Smith has a dip of ~10° to the southeast. The beds above the erosional contact conform to the surface and dip predominantly to the southeast (Fig. 5). Measured as a true stratigraphic thickness, Steno consists of ~0.7 m of sandstone. This measurement represents a minimum thickness due to erosion at the top of the unit, which coincides with the Amazonian weathering surface of Meridiani; the unit is overlain by impact breccia.

The Steno unit is composed of fine- to medium-grained sandstone, with well-defined laminae, on the order of ~2 mm in thickness. Centimeter-to-meter scale cross-bedding is visible in places. In the ingress path, Steno shows a varied appearance when weathered—appearing as a more resistant step in topography (Fig. 4A). Often it has a rougher surface texture and has a darker tone, but sometimes it displays a smoother texture and lighter tone.

The spherules in Steno are relatively uniform in size and shape, with a mean diameter of 0.72 mm (SD = 0.19 mm; 80 spherules). Steno has more spherules than Smith but fewer than Lyell, with a volumetric density of 0.46% (SD = 0.1%; n = 5 frames). Steno also contains crystal-form vugs similar to those in Lyell, but less abundant.

Chemostratigraphy

In addition to observing the physical stratigraphy and diagenetic textures of the different rock units in Duck Bay, *Opportunity* made measurements of chemical and mineralogic compositions, which we will briefly summarize here. Locations of the chemical measurements are represented by colored circles in Figure 3, corresponding to labeled points in the stratigraphic column (Fig. 6). Overall, the rock compositions measured in Duck Bay fall within the range of rocks observed at previous craters studied by *Opportunity* and indicate that the rock units are composed of sulfates and silicates in slightly varying abundances (Mittlefehldt et al. 2008). Notable compositional differences include more S in Steno than was previously observed at Meridiani (Squyres et al. 2009) and more Zn in Smith (24% higher) than in any other rock observed to date (Mittlefehldt et al. 2008). Smith

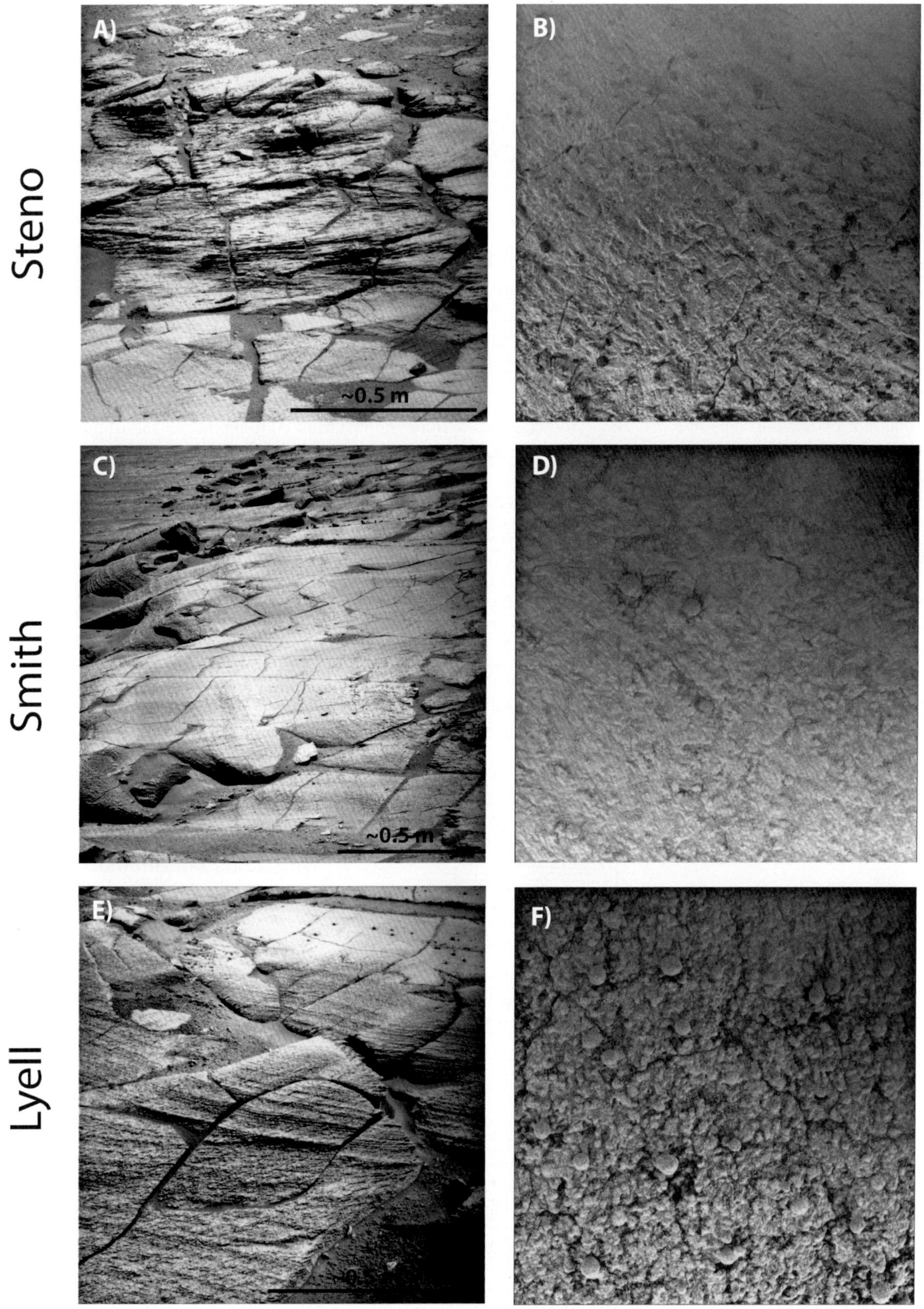

FIG. 4.—Pancam (left) and MI (right) images of Lyell, Smith, and Steno, documenting the large- and small-scale characteristics of the three different units in the ingress path. For scale, MI images are 31 mm on a side. MI images were taken with target fully shadowed. Note that the dark upper right corner of each MI image is the result of dust contamination of the MI optics. (A) In the ingress path Steno appears as a more resistant step in topography. (B) Steno consists of fine- to medium-grained sandstone with abundant hematite spherules. (C, D) Smith is characterized by its lighter tone and smoother texture at both scales. (E) Lyell is characterized by its darker tone (Smith is visible in the upper right portion of the image for comparison) and well-defined pinstripe laminations. (F) MI images of Lyell reveal numerous hematite spherules and vugs. Spherules in Lyell are larger and more abundant than those in Steno or Smith.

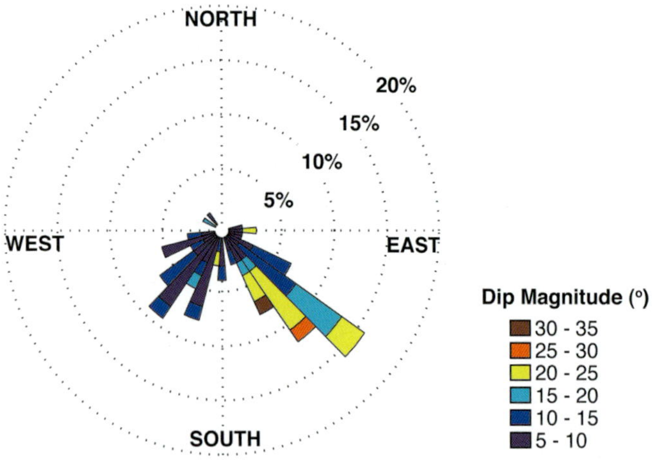

FIG. 5.—Rose diagram of bedding dips at Duck Bay, following the methods of Hayes et al. (2011). Colors correspond to dip magnitudes. Dip directions to the southwest correspond to measurements of cross-bedding in Lyell, while dip directions to the southeast correspond to measurements of cross-bedding in Steno. Bedding was not measured in Smith as a result of the poor expression of bedding.

and Lyell both show an increase in Cl content compared to Steno—Lyell reveals more than a factor of two increase in Cl compared to Steno (Mittlefehldt et al. 2008). The chemostratigraphic trends in Duck Bay are similar to those measured at Endurance crater—both sections reveal a decrease in S, Fe, and Mg down section and corresponding Al and Si enrichment (Squyres et al. 2009). The chemostratigraphic similarities between Endurance and Victoria may reflect the recurrence of depositional and diagenetic processes through time (Squyres et al. 2009), or they may reflect downward-penetrating diagenesis related to surface exposure (Amundson et al. 2008).

CORRELATION TO CAPE VERDE

The three stratigraphic units are visible around much of Duck Bay, but they cannot be directly correlated with the nearest promontory, Cape Verde (Fig. 7). A large area of brecciated bedrock separates the ingress path stratigraphy from that at the promontory, prohibiting the direct tracing of beds. Although the rover was unable to reach the cliff face at Cape Verde, detailed observations of several science targets were made through super-resolution imaging at the rover's closest approach (~10-m stand-off distance). The strata exposed at Cape Verde contain a light-toned band similar in thickness to that of Smith. Here it overprints well-laminated sandstone with low-angle cross-bedding. Strata near the base of the cliff face have planar lamination to low angle cross-stratification, with dips that steepen upward in the upper half of the cliff face. Super-resolution imaging also reveals small climbing ripples superimposed on the larger dune cross-stratification (Fig. 7C).

The base of the cliff face contains a truncation surface (Fig. 7). *Opportunity* was unable to do a detailed analysis on the bedrock underlying this truncation surface, although a few exposures of bedrock indicate that the strata are inclined relative to those that overlie the surface (Fig. 7B). The truncation surface has a dip of ~10° to the southeast (Hayes et al. 2011); given that the erosional contact at the base of Steno also has a ~10° dip, it seems likely that these surfaces are related and likely represent scour by the same migrating bedform. This

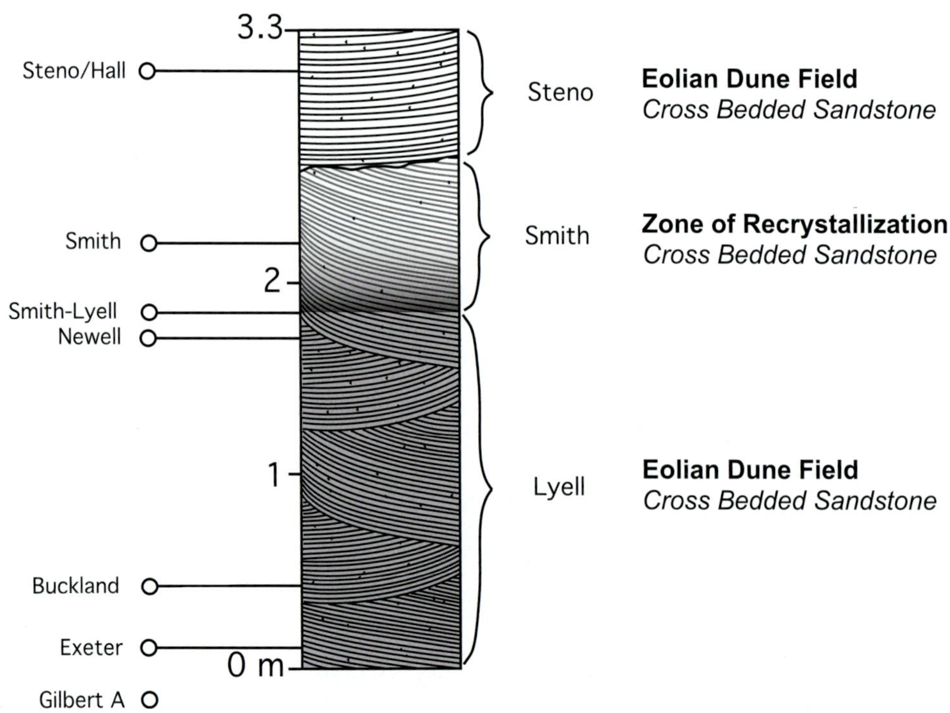

FIG. 6.—Duck Bay stratigraphic column. Thicknesses are corrected for a 2° dip to the west. Left-hand side shows locations of IDD targets (Gilbert A was sampled from a brecciated block below the intact unit Lyell). Smith is interpreted as a diagenetic band superimposed on cross-bedded sandstone. In the ingress path Smith is also bounded by an erosional contact with Steno.

FIG. 7.—Cape Verde erosional surface. (A) White dashed line shows location of erosional surface at the base of Cape Verde, dipping approximately 10° to the southeast (to the right in the above image) (Cape Verde super-resolution mosaic acquired during Sols 1342–1356). White boxes show locations of B and C. (B) Super-resolution image of erosional surface. Below the erosional surface, bedrock is exposed in the slope, with angled beds truncated by low-angle cross-stratification in the cliff face (Cape Verde super-resolution mosaic acquired during Sols 1574–1576). (C) Cross-strata represent ripples climbing toward the upper right portion of the image (arrow) (Cape Verde mosaic acquired during Sols 1579–1580).

is further supported by the observation that the two exposures of the truncation surface lie in the same plane.

PALEO-DEPOSITIONAL ENVIRONMENT

The primary sedimentary structures and diagenetic textures preserved in the ingress Reference Section at Victoria Crater define two facies. These include cross-bedded sandstone and planar-laminated to low angle cross-stratified sandstone. These facies require transport of sand under turbulent flows to generate bedforms ranging in scale from centimeters to meters. The dunes that deposited the cross-strata observed at Duck Bay and Cape Verde were probably an order of magnitude greater than 1 m, likely on the order of tens of meters, as suggested by Rubin and Hunter (1982), as calculated from the preservation ratio [(mean thickness of preserved sets)/(mean topographical height)] for dunes (0.12) proposed by Paola and Borgman (1991) or inferred from the ratio of preserved set thickness to complete dune height that Mountney et al. (1999) observed for eolian dunes and cross-strata buried by lava. The abundance of planar, low-angle cross-stratification, meter-scale cross-stratification, and pinstripe laminae (diagnostic of deposition by wind ripples (Hunter 1977) are most consistent with an eolian depositional environment. These observations and their inferred depositional environment are similar to what has been proposed previously for other locations studied by *Opportunity*.

However, in the interpretation of cross-stratified facies in Meridiani Planum, it is necessary to consider two alternative depositional models. Knauth et al. (2005) propose that the Meridiani strata represent an impact-induced base surge, and McCollom and Hynek (2005) propose that the strata were deposited during a volcanic base surge. These depositional models can be discounted on several grounds: (1) lack of evidence for contemporaneous sources of volcanism or impacts of sufficient size to produce base surge deposits (Squyres et al. 2006, McLennan and Grotzinger 2008); (2) spherules are dispersed relative to bedding (McLennan et al. 2005) rather than concentrated along bedding planes, as would be expected if they were accretionary lapilli

FIG. 8.—Cabo Corrientes false color mosaic acquired on Sol 1108. The false color stretch brings out subtle color differences in the outcrop, highlighting small blocks of the diagenetic band that are reworked into the impact breccia (white arrows point to several blocks). The diagenetic band ("Smith" as described in the ingress section) is interpreted to have formed after deposition but before the crater impact.

or iron condensation spherules; and (3) the scale of the cross-bedding exposed in Victoria crater (reconstructing to bedforms on the order of meters to tens of meters) has not been observed in any known base surge deposits. Therefore, the most probable explanation for the origin of cross-stratified facies at Victoria crater is that they were produced in an eolian depositional environment.

Deposition and Diagenesis at Duck Bay

Lyell and Smith, as defined in the Reference Section at Duck Bay, were likely deposited as part of the same eolian dune package, as evidenced by the gradational contact between them. Steno is a distinct stratigraphic unit marked by an erosional contact, which separates it from the light-toned Smith unit. While Smith is bounded by the erosional contact below Steno, its presence above the erosional contact at Cape Verde indicates that Smith crosscuts the primary bedding. This relationship is consistent with observations made at other exposures around the crater (Hayes et al. 2011). Small-scale diagenetic features indicate that Smith is recrystallized. Lyell shows the highest density of vugs and spherules, followed by Steno, and Smith shows the lowest density. The lighter tone, smooth appearance, and lack of vugs are the result of recrystallization. It is clear that Smith owes its unique identity to diagenesis, but the key question involves when that diagenesis occurred.

The observation that the diagenetic band has a low dip and is found near the rim of the crater raises hypotheses that it may be due to processes related to the Meridiani plains (intersection with the ground water table?), or it may be due to processes related to the crater impact (impact-induced melting of ice?). The fact that the diagenetic band is brecciated and in some outcrops reworked upward into the impact breccia (Fig. 8) indicates that the diagenetic band was not formed during or after the crater impact. The diagenetic band must postdate deposition of the eolian strata but predate the crater impact.

We also observe a low density of spherules in the diagenetic band. This may be explained by recrystallization prior to the precipitation of spherules. Early recrystallization of Smith may have eliminated pore space, so subsequent groundwater infiltration would not precipitate as many spherules in Smith as there are in Steno and Lyell.

INTERPRETATION OF EOLIAN STRATIGRAPHIC ARCHITECTURE

The recognition of cross-bedding geometry and bounding surfaces in ancient eolian deposits can be used to reconstruct wind direction, the nature of dune migration, and, ultimately, the sequence of events that produced the deposit. In turn, this may permit some understanding of changes in climate, sand supply, and possible subsidence. However, we are working with much more limited data than are available in most terrestrial studies of this nature, constrained by the few bedrock outcrops observed by the rover and by limited stereo coverage. Given the available data, we will consider a list of possibilities for the origin of the stratification exposed at Duck Bay and Cape Verde.

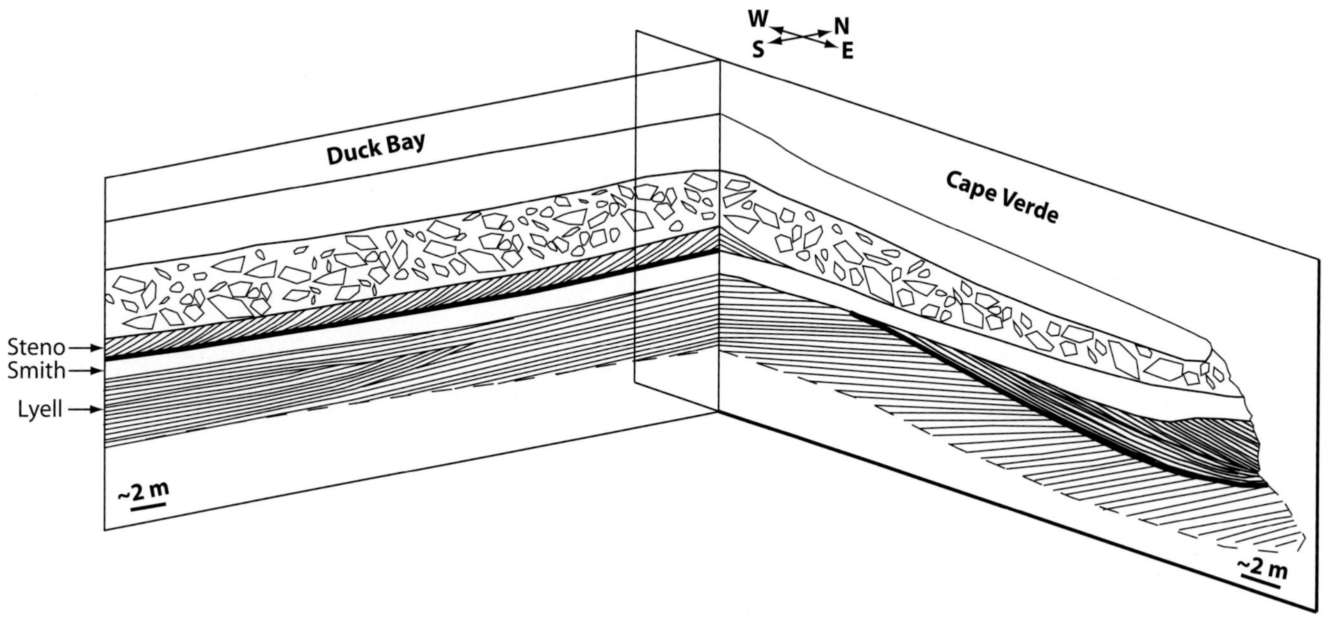

FIG. 9.—Schematic fence diagram depicting the three-dimensional geometry of the stratigraphy exposed at Duck Bay and Cape Verde. The diagenetic band Smith (white) crosscuts the erosional surface (heavy black lines). The erosional contact between Steno and Smith in the ingress path lies in the same plane as the erosional surface at the base of the Cape Verde cliff face. Note: the region at the intersection of these planes is inferred as a result of a lack of intact bedrock connecting Duck Bay and Cape Verde.

We begin by considering the truncation surface between Steno and Smith in the ingress path. As discussed above, we infer that the erosional contact at the base of Steno correlates with the erosional surface at the base of the Cape Verde cliff face. Although these surfaces lie at different elevations, they have a similar dip of approximately 10° into the crater, and projection of this dip shows the potential continuity of the surface between elevations. The surface spans an area of least 300 sq m (limited by the few outcrops observed by the rover) and may represent an architectural element larger than the scale of the dune cross-bedding. In this case the erosional surface may be interpreted as a bounding surface produced by migration of dunes on a larger bedform (cf., Fryberger 1993). The compound bedform (smaller dunes migrating over a larger dune) may be termed a "draa." Figure 9 provides a reconstruction of the stratigraphy between the reference section along the ingress path and at Cape Verde. Note the crosscutting relationship of the diagenetic unit Smith and the position of the truncation surface.

Further consideration of cross-bedding geometry produces additional insight. Strata below the erosional surface, both at Duck Bay and at Cape Verde, dip to the southwest, while the strata above the surface dip to the southeast. This leads to two possible interpretations. The different dip directions may be interpreted as remnants of three-dimensional sinuous crested dunes migrating southward. In this scenario, the strata dipping to the southwest would represent deposition on the west-dipping surface of the east side of a southward-migrating topographic depression. The strata dipping to the southeast would represent deposition on the east-dipping surface of the west side of a southward-migrating topographic depression (Fig. 10). Trough cross-bedding produced by three-dimensional sinuous crested dunes is observed at nearby promontory Cape St. Mary, although the scale of the bedsets is much smaller (Hayes et al. 2011).

Alternatively, the different dip directions could be indicative of different dune-migration directions, from which paleo-wind directions maybe inferred. This interpretation is consistent with observations of terrestrial draas, which may contain reactivation surfaces representing the migration of dunes across a draa in different directions (McKee 1966). Work by Hayes et al. (2011) indicates that a pattern of reversing transport direction is observed at other locations around Victoria crater, and also with juxtaposition of cross-bedsets across larger-scale surfaces.

Figure 11 illustrates this shift in wind directions and the resulting cross-stratification. Using the model of Rubin and Carter (2005), we produce a large transverse bedform migrating from north to south, with superimposed bedforms that reverse migration direction. To model a reversal in wind direction, we allow the superimposed bedforms to reverse migration direction and asymmetry. While the superimposed bedforms reverse migration direction, their heights are decreased and then increased. The superimposed bedforms initially migrate from east to west, then reverse direction and migrate from west to east. The resulting stratification shows a subset of crossbeds at the base that dip to the southwest and an overlying subset of crossbeds that dip to the southeast (this is illustrated most clearly at the front corner of the block). The bounding surface that separates the two subsets dips to the south (note: in actuality, the bounding surface at Duck Bay and Cape Verde dips more to the southeast). A snapshot of the preserved cross-stratification may reveal a bounding surface with cross-strata above and below the surface dipping in different directions, indicating a potential reversal in wind direction. This is analogous to the stratal geometry observed at the base of Cape Verde.

Given the available data, we are unable to distinguish between these two hypotheses. However, in both scenarios, the inferred net transport direction is roughly north to south, which is consistent with the inferred paleowind directions measured at other locations around the crater (Hayes et al. 2011). In the case of a large bedform with superimposed bedforms that reverse migration direction, the orientation of the erosional surface (rather than the cross-strata above the surface) may

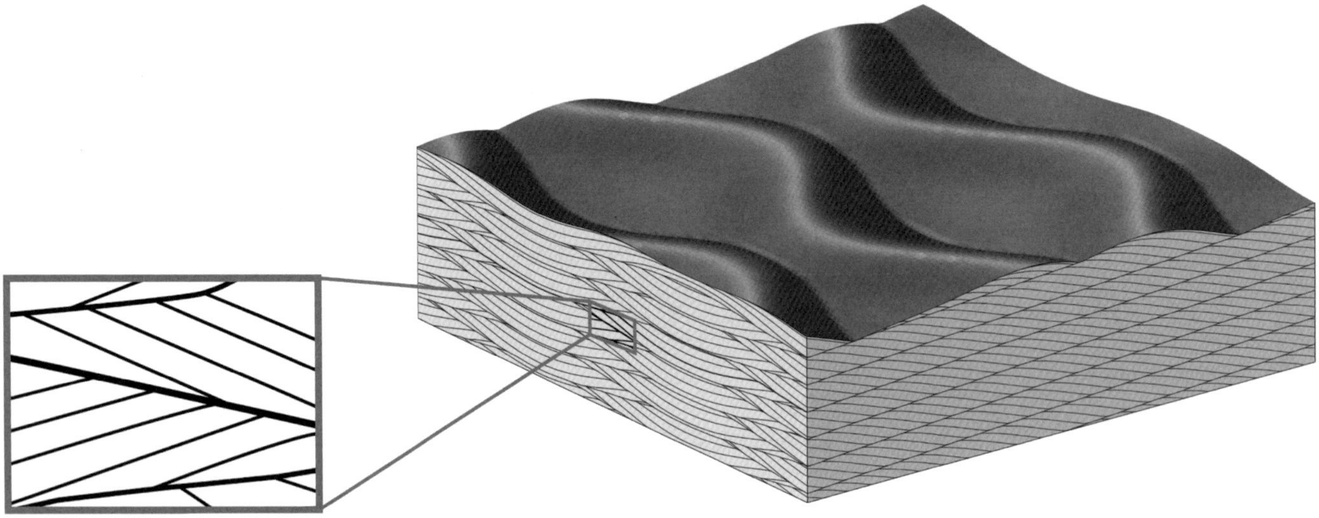

FIG. 10.—Three-dimensional sinuous crested bedforms, modified from Rubin and Carter (2005). The magnified box highlights a section perpendicular to the flow direction, showing a bounding surface (bold line) with strata above and below the bounding surface dipping in opposite directions. In the case of Cape Verde, the strata below the bounding surface may have been deposited on the east side of a southward migrating depression (strata below the bold line), whereas the strata above the erosional surface may have been deposited on the west side of a southward migrating depression (strata above the bold line).

serve as a better indicator of the local orientation of the bedform surface when it was formed. Rubin and Hunter (1983) have shown that in compound cross-stratification, the orientation of a bounding surface scoured by superimposed bedforms serves as a more accurate indicator of draa orientation than does the dip orientation of cross-strata (note that this result holds for situations in which the surface slope of the main bedform is relatively steep; 10° is likely steep enough for this to be a better approximation than the crossbeds deposited by the superimposed bedforms). Cross-stratification—especially trough cross-stratification—can display a wide range of dip directions (up to 180°) even in a unidirectional wind; if only a small sample of these dip directions are measured—as is the case for Victoria crater—then the results might not be representative of the actual range of dips. In contrast, the dip direction of bounding surfaces scoured by superimposed bedforms is oriented roughly perpendicular to the draa brink and therefore serves as a more accurate indicator of the primary dune migration direction.

The Algodones dune field in southern California provides an excellent terrestrial example of a draa oriented roughly perpendicular—but not perfectly perpendicular—to the long-term resultant transport direction, with superimposed features driven by the secondary airflow (Havholm and Kocurek 1988). Havholm and Kocurek (1988) found that on the lee slope of a draa, surface airflow and the migration of superimposed bedforms is typically oblique or parallel to the draa brinkline, highlighting the importance of these bounding surfaces in reconstructing paleowind directions in the rock record. At Victoria crater, the erosional surface dips 10° to the southeast. If this surface is a bounding surface scoured by a superimposed bedform, then the bounding surface may reflect the true orientation of the draa brinkline, migrating from northwest to southeast.

The erosional surface at Duck Bay and Cape Verde is also important in that it exposes a larger-scale bedform than had previously been seen in Meridiani Planum. If it represents the migration of dunes across a draa, then three orders of bedforms are observed at Duck Bay and Cape Verde: ripples, dunes, and draas. Draa deposits appear to be a common feature in the terrestrial rock record. Draas are typical of modern ergs (sand seas), often occurring in the centers of well-developed ergs, where sand cover is thickest (Wilson 1971, Havholm and Kocurek 1988). These conditions favor preservation, so draa deposits would be expected to be widespread in the rock record of eolian deposits on Earth (Havholm and Kocurek 1988) and in similarly arid environments on Mars. The three orders of bedforms observed at Cape Verde indicate that the ancient dune field glimpsed by *Opportunity* may have been a large sand sea (erg) (although the size of the reconstructed bedforms is somewhat smaller than that of terrestrial bedforms thought to have been deposited in eolian environments, such as the Navajo Sandstone [Rubin and Hunter 1982]).

The dunes in the center of Victoria crater provide a modern example of a dune field with multiple orders of bedforms, analogous to the ancient dune field preserved in outcrop. Pancam images reveal intersecting dunes in the center of the crater, with smaller superimposed dunes riding on the backs of the larger dunes (Fig. 12). The crater provides a confined basin for deposition, and complex wind patterns shape the sediment into star dunes. Modified secondary airflow builds smaller dunes on the backs of the larger dunes. Using a Digital Elevation Model (DEM) produced from stereo images from the orbiting High Resolution Imaging Science Experiment (HiRISE), we obtained estimates of dune heights ranging from 2 to 5 m. While these are somewhat smaller than the estimated height of the ancient dunes preserved in the Victoria bedrock, this modern example indicates that the processes that governed ancient deposition are still acting on the modern Martian surface.

One feature that is notably absent from the bedrock outcrops at Victoria crater is evidence for brief periods of aqueous deposition, as observed at Eagle, Endurance, and Erebus craters (Grotzinger et al. 2005, Metz et al. 2009). Metz et al. (2009) suggest that as the rover has moved stratigraphically higher through Eagle, Endurance, and Erebus craters that it has observed a full climatic cycle, from dry to wet to dry. The bedrock outcrops at Victoria crater represent a stratigraphically higher section than that observed at Eagle and Endurance craters and may represent even drier climatic conditions and increased sediment

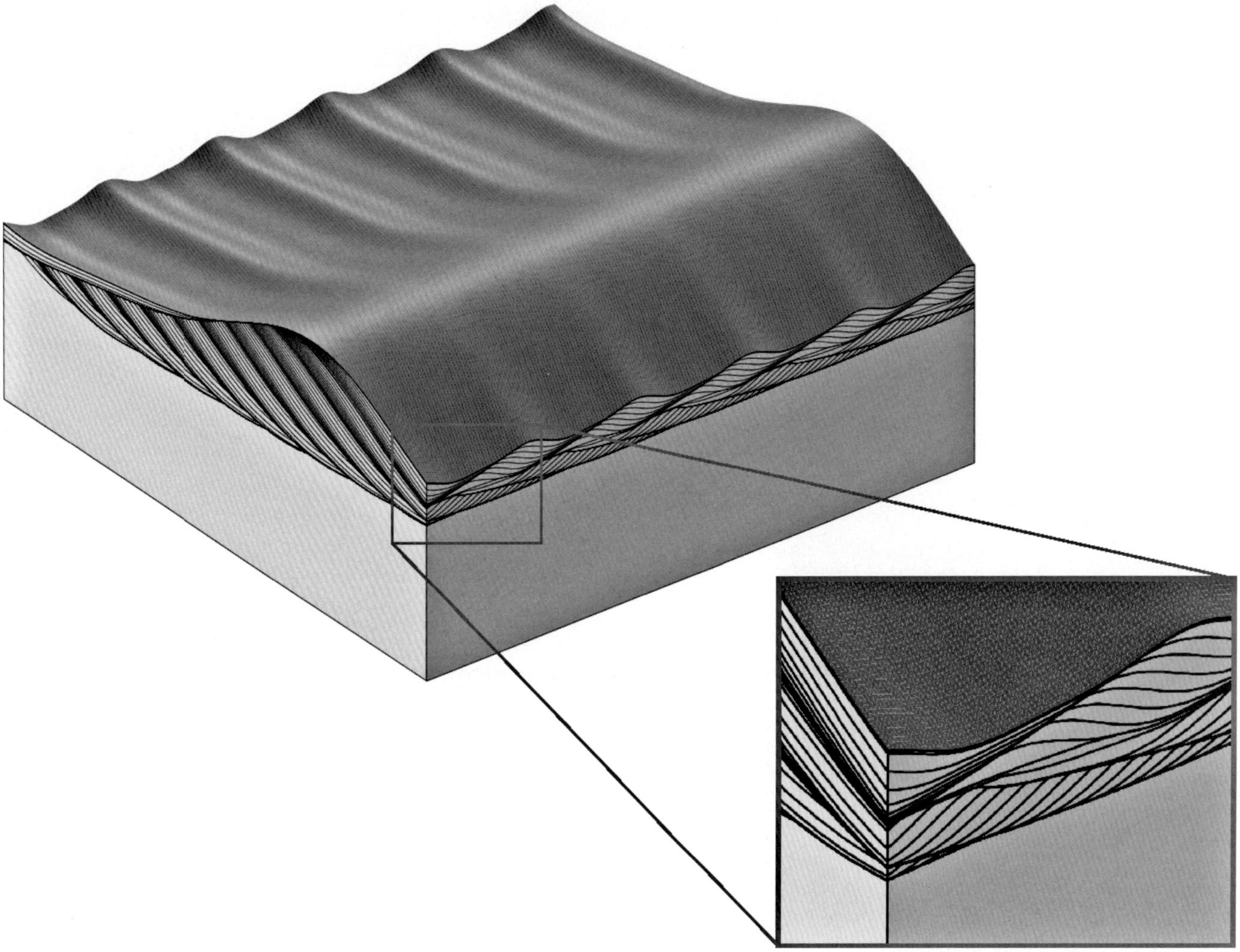

FIG. 11.—Depositional model for the stratal geometry at the base of Cape Verde, using code modified from Rubin and Carter (2005). Model illustrates the stratification produced by a large transverse bedform migrating from N to S, with superimposed bedforms that initially migrate from E to W, then reverse direction and migrate from W to E. The resulting stratification shows a subset of crossbeds at the base that dip to the SW, an overlying subset of crossbeds that dip to the SE, and a bounding surface separating the two subsets that dips to the S. A snapshot of the preserved cross-stratification may reveal a bounding surface with cross-strata above and below the surface dipping in different directions, indicating a potential change in wind direction, analogous to the stratal geometry observed at the base of Cape Verde.

flux. This allogenetic hypothesis indicates that the conditions at a given point changed through time as a result of a changing climate. Alternatively, the lack of evidence for aqueous deposition may indicate lateral variations in time-equivalent facies. Victoria crater may expose strata from the center of a well-developed sand sea, whereas strata in Eagle and Endurance craters may represent a depression (or depressions) within the sand sea and may be more prone to the emergence of groundwater. In this autogenetic scenario, the degree of wetness of the depositional environment would depend more on local elevation relative to the water table, with changes in wetness controlled by the thickness of dry sand. The outcrop exposed in the Reference Section at Duck Bay appears to lie at a similar elevation to the uppermost unit exposed in Erebus crater. The upper unit of the Yavapai formation is interpreted to represent sandsheet/dune field facies (Metz et al. 2009). It is possible that the upper unit at Erebus represents the same period of deposition observed in the ingress path at Duck Bay, but the larger scale of the outcrops at Victoria crater exposes the stratal geometries consistent with construction of larger-scale eolian bedforms.

CONCLUSIONS

The strata exposed at Duck Bay and Cape Verde indicate deposition in an eolian dune environment, with further modifications though diagenesis. Images from the Pancam and MI reveal three distinct stratigraphic units in the Reference Section traced by the rover ingress path: Lyell, Smith, and Steno, in ascending order. In the Reference Section, Smith is interpreted as a secondary, diagenetic unit, which also is bounded by a primary, erosional contact, but elsewhere in the crater the diagenetic band crosscuts the primary stratigraphic surfaces. Evidence of the diagenetic unit reworked in the impact breccia indicates that Smith formed prior to the crater impact. Correlation with nearby promontory Cape Verde reveals that there is an erosional surface at the base of the cliff face that likely corresponds to the

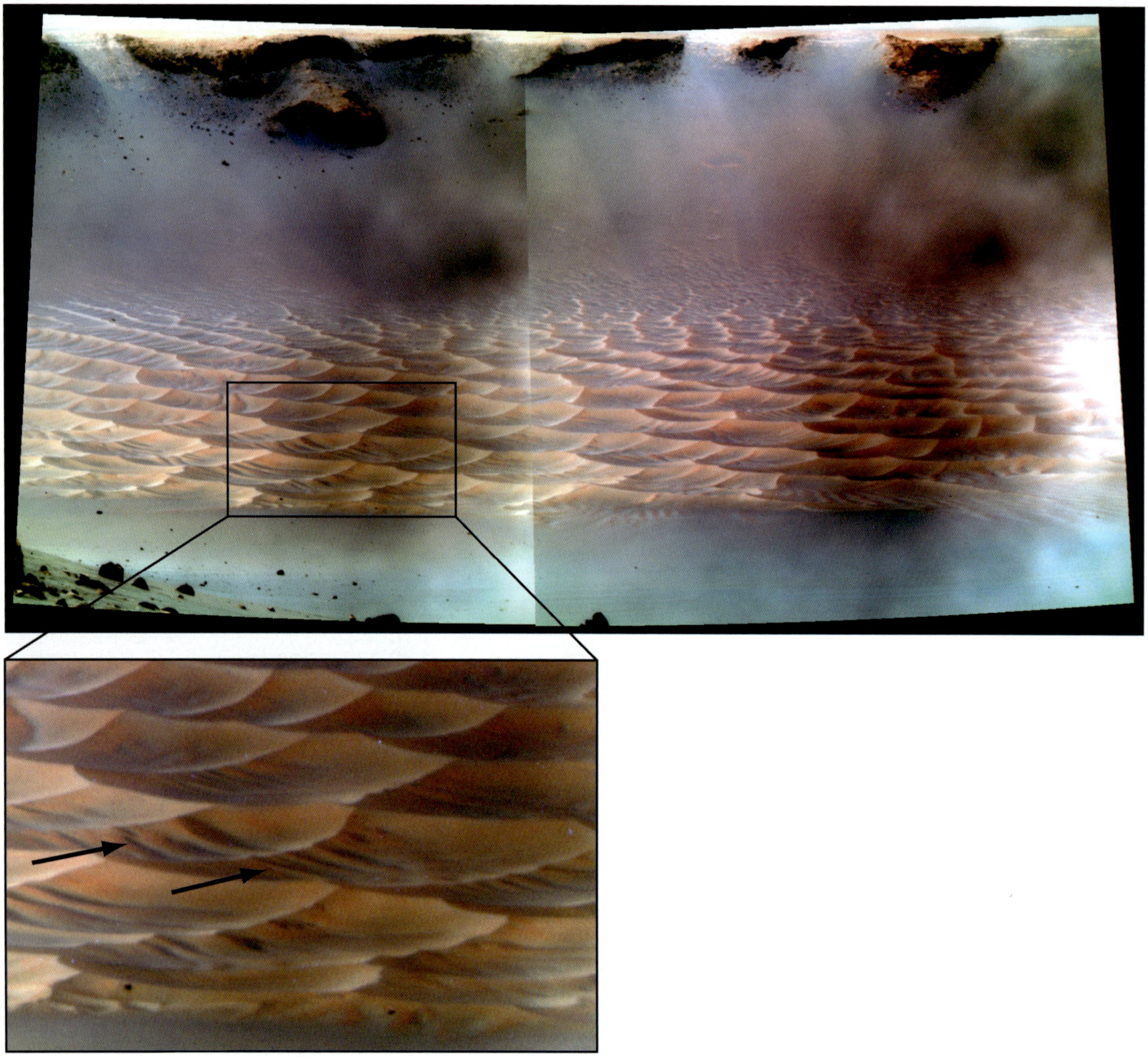

FIG. 12.—Modern dunes in the center of Victoria crater (Pancam false color image acquired on Sol 1437). Maximum dune heights range from 2 to 5 m, as determined from a HiRISE DEM. Close-up image shows small superimposed dunes (black arrows) migrating across larger star dunes. While these superimposed dunes may be analogous to the ancient bedforms inferred at Duck Bay and Cape Verde, they are an order of magnitude smaller.

erosional contact below Steno. This surface is interpreted to represent the migration of dunes across a draa, and its orientation indicates that the draa was migrating from northwest to southeast at the time that the surface was formed. The stratal geometry above and below the erosional surface indicates dune migration in opposing directions or by southward migrating three-dimensional bedforms. Additionally, the presence of three orders of bedforms and a complex wind regime indicates that the strata may have been part of a large sand sea, with no evidence for aqueous deposition, as observed at Eagle and Endurance craters. Victoria crater not only reveals the regional extent of processes seen elsewhere in Meridiani Planum, but the greater size of its outcrop exposures reveals the building of ever-larger eolian bedforms.

REFERENCES

Amundson R, Ewing S, Dietrich W, Sutter B, Owen J, Chadwick O, Nishiizumi K, Walvoord M, McKay C. 2008. On the in situ aqueous alteration of soils on Mars. *Geochimica Et Cosmochimica Acta* 72:3845–3864.

Arvidson ER, Anderson RC, Haldemann AFC, Landis FA, Li R, Lindemann RA, Matijevic JR, Morris RV, Richter L, Squyres SW, Sullivan RJ, Snider NO. 2003. Physical properties and localization investigations associated with the 2003 Mars Exploration rovers. *Journal of Geophysical Research–Planets* 108, doi:10.1029/2002JE002041

Bell JF III, Squyres SW, Herkenhoff KE, Maki JN, Arneson HM, Brown D, Collins SA, Dingizian A, Elliot ST, Hagerott EC, Hayes AG, Johnson MJ,

Johnson JR, Joseph J, Kinch K, Lemmon MT, Morris RV, Scherr L, Schwochert M, Shepard MK, Smith GH, Sohl-Dickstein JN, Sullivan RJ, Sullivan WT, Wadsworth M. 2003. Mars Exploration Rover Athena Panoramic Camera (Pancam) investigation. *Journal of Geophysical Research–Planets* 108, doi: 10.1029/2003JE002070

Bell JF III, Joseph J, Sohl-Dickstein JN, Arneson HM, Johnson MJ, Lemmon MT, Savransky D. 2006. In-flight calibration and performance of the Mars Exploration Rover Panoramic Camera (Pancam) instruments. *Journal of Geophysical Research–Planets* 111, doi:10.1029/2005JE002444

Fryberger SG. 1993. A review of aeolian bounding surfaces, with examples from the Permian Minnelusa Formation, USA. *In* North CP, Prosser DJ (Editors), *Characterization of Fluvial and Aeolian Reservoirs*: Geological Society of London. Special Publications 73, p. 167–197.

Gorevan SP, Myrick T, Davis K, Chau JJ, Bartlett P, Mukherjee S, Anderson R, Squyres SW, Arvidson RE, Madsen MB, Bertelsen P, Goetz W, Binau CS, Richter L. 2003. The Rock Abrasion Tool: Mars Exploration Rover Mission. *Journal of Geophysical Research* 108, doi:10.1029/2003JE002061

Grant JA, Wilson SA, Cohen BA, Golombek MP, Geissler PE, Sullivan RJ, Kirk RL, Parker TJ. 2008. Degradation of Victoria crater, Mars. *Journal of Geophysical Research–Planets* 113, doi:10.1029/2008JE003155

Grotzinger JP, Arvidson RE, Bell JF, Calvin W, Clark BC, Fike DA, Golombek M, Greeley R, Haldemann A, Herkenhoff KE, Jolliff BL, Knoll AH, Malin M, McLennan SM, Parker T, Soderblom L, Sohl-Dickstein JN, Squyres SW, Tosca NJ, Watters WA. 2005. Stratigraphy and sedimentology of a dry to wet eolian depositional system, Burns formation, Meridiani Planum, Mars. *Earth and Planetary Science Letters* 240:11–72.

Grotzinger J, Bell J, Herkenhoff K, Johnson J, Knoll A, McCartney E, McLennan S, Metz J, Moore J, Squyres SW, Sullivan R, Ahronson O, Arvidson R, Jolliff B, Golombek M, Lewis K, Parker T, Soderblom J. 2006. Sedimentary textures formed by aqueous processes, Erebus crater, Meridiani Planum, Mars. *Geology* 34:1085–1088.

Havholm KG, Kocurek G. 1988. A preliminary-study of the dynamics of a modern draa, Algodones, Southeastern California, USA. *Sedimentology* 35:649–669.

Hayes AG, Grotzinger JP, Edgar LA, Squyres SW, Watters WA, Sohl-Dickstein J. 2011. Reconstruction of eolian bedforms and paleocurrents from cross-bedded strata at Victoria Crater, Meridiani Planum, Mars. *Journal of Geophysical Research* 116, doi:10.1029/2010JE003688

Herkenhoff KE, Squyres SW, Bell JF III, Maki JN, Arneson HM, Bertelsen P, Brown DI, Collins SA, Dingizian A, Elliott ST, Goetz W, Hagerott EC, Hayes AG, Johnson JM, Kirk RL, McLennan S, Morris RV, Scherr LM, Schwochert MA, Shiraishi LR, Smith GH, Soderblom LA, Sohl-Dickstein JN, Wadsworth MV. 2003. Athena Microscopic Imager investigation. *Journal of Geophysical Research–Planets* 108, doi:10.1029/2003JE002076

Hunter RE. 1977. Basic types of stratification in small eolian dunes. *Sedimentology* 24:361–387.

Knauth LP, Burt DM, Wohletz KH. 2005. Impact origin of sediments at the opportunity landing site on Mars. *Nature* 438:1123–1128.

Lewis KW, Aharonson O, Grotzinger JP, Squyres SW, Bell JF III, Crumpler LS, Schmidt M. 2008. Structure and stratigraphy of Home Plate from the Spirit Mars Exploration Rover. *Journal of Geophysical Research–Planets* 113, doi:10.1029/2007JE003025

McCollom TM, Hynek BM. 2005. A volcanic environment for bedrock diagenesis at Meridiani Planum on Mars. *Nature* 438:1129–1131.

McKee ED. 1966. Structure of dunes at White Sands National Monument, New Mexico. *Sedimentology* 7:1–61.

McLennan SM, Bell JF, Calvin WM, Christensen PR, Clark BC, de Souza PA, Farmer J, Farrand WH, Fike DA, Gellert R, Ghosh A, Glotch TD, Grotzinger JP, Hahn B, Herkenhoff KE, Hurowitz JA, Johnson JR, Johnson SS, Jolliff B, Klingelhofer G, Knoll AH, Learner Z, Malin MC, McSween HY, Pocock J, Ruff SW, Soderblom LA, Squyres SW, Tosca NJ, Watters WA, Wyatt MB, Yen A. 2005. Provenance and diagenesis of the evaporite-bearing Burns formation, Meridiani Planum, Mars. *Earth and Planetary Science Letters* 240:95–121.

McLennan SM, Grotzinger JP. 2008. The sedimentary rock cycle of Mars. *In* Bell JF III (Editor), *The Martian Surface: Composition, Mineralogy, and Physical Properties*: Cambridge University Press, UK, p. 541–577.

Metz JM, Grotzinger JP, Rubin DM, Lewis KW, Squyres SW, Bell JF III. 2009. Sulfate-rich eolian and wet interdune deposits, Erebus Crater, Meridiani Planum, Mars. *Journal of Sedimentary Research* 79:247–264.

Mittlefehldt DW, Schröder C, Gellert R, Klingelhöfer G, Jolliff BL, Morris RV, and the Athena Science Team. 2008. Chemostratigraphy and Fe Mineralogy of the Victoria Crater Duck Bay Section: *Opportunity* APXS and Mössbauer Results. Lunar and Planetary Science, Abstract 2404.

Mountney N, Howell J, Flint S, Jerram D. 1999. Relating eolian bounding-surface geometries to the bed forms that generated them: Etjo Formation, Cretaceous, Namibia. *Geology* 27:159–162.

Paola C, Borgman L. 1991. Reconstructing random topography from preserved stratification. *Sedimentology* 38:553–565.

Rubin DM, Carter CL. 2005. *Bedforms 4.0: MATLAB Code for Simulating Bedforms and Cross-Bedding:* US Geological Survey Open-File Report 2005–1272.

Rubin DM, Hunter RE. 1982. Bedform climbing in theory and nature. *Sedimentology* 29:121–138.

Rubin DM, Hunter RE. 1983. Reconstructing bedform assemblages from compound cross-bedding. *In* Brookfield ME, Ahlbrandt TS (Editors). *Eolian Sediments and Processes*: Elsevier Science Publishing Company, New York, p. 407–428.

Squyres SW, Arvidson RE, Baumgartner ET, Bell JF, Christensen PR, Gorevan S, Herkenhoff KE, Klingelhofer G, Madsen MB, Morris RV, Rieder R, Romero RA. 2003. Athena Mars rover science investigation. *Journal of Geophysical Research–Planets* 108, doi:10.1029/2003JE002121

Squyres SW, Arvidson RE, Bell JF, Bruckner J, Cabrol NA, Calvin W, Carr MH, Christensen PR, Clark BC, Crumpler L, Des Marais DJ, d'Uston C, Economou T, Farmer J, Farrand W, Folkner W, Golombek M, Gorevan S, Grant JA, Greeley R, Grotzinger J, Haskin L, Herkenhoff KE, Hviid S, Johnson J, Klingelhofer G, Knoll AH, Landis G, Lemmon M, Li R, Madsen MB, Malin MC, McLennan SM, McSween HY, Ming DW, Moersch J, Morris RV, Parker T, Rice JW, Richter L, Rieder R, Sims M, Smith M, Smith P, Soderblom LA, Sullivan R, Wanke H, Wdowiak T, Wolff M, Yen A. 2004a. The *Opportunity* Rover's Athena science investigation at Meridiani Planum, Mars. *Science* 306:1698–1703.

Squyres SW, Grotzinger JP, Arvidson RE, Bell JF, Calvin W, Christensen PR, Clark BC, Crisp JA, Farrand WH, Herkenhoff KE, Johnson JR, Klingelhofer G, Knoll AH, McLennan SM, McSween HY, Morris RV, Rice JW, Rieder R, Soderblom LA. 2004b. In situ evidence for an ancient aqueous environment at Meridiani Planum, Mars. *Science* 306:1709–1714.

Squyres SW, Aharonson O, Arvidson RE, Bell JF, Christensen PR, Clark BC, Crisp JA, Farrand W, Glotch T, Golombek MP, Grant J, Grotzinger J, Herkenhoff KE, Johnson JR, Jolliff BL, Knoll AH, McLennan SM, McSween HY, Moore JM, Rice JW, Tosca N. 2006. Bedrock formation at Meridiani Planum. *Nature* 443:E1–E2.

Squyres SW, Knoll AH, Arvidson RE, Ashley JW, Bell JF, Calvin WM, Christensen PR, Clark BC, Cohen BA, de Souza PA, Edgar L, Farrand WH, Fleischer I, Gellert R, Golombek MP, Grant J, Grotzinger J, Hayes A, Herkenhoff KE, Johnson JR, Jolliff B, Klingelhofer G, Knudson A, Li R, McCoy TJ, McLennan SM, Ming DW, Mittlefehldt DW, Morris RV, Rice JW, Schroder C, Sullivan RJ, Yen A, Yingst RA. 2009. Exploration of Victoria Crater by the Mars Rover Opportunity. *Science* 324:1058–1061.

Wilson IG. 1971. Desert sandflow basins and a model for development of ergs. *Geographical Journal* 137:180–199.

POTENTIAL RECOGNITION OF ACCRETIONARY LAPILLI IN DISTAL IMPACT DEPOSITS ON MARS: A FACIES ANALOG PROVIDED BY THE 1.85 GA SUDBURY IMPACT DEPOSIT

PHILIP FRALICK
Department of Geology, Lakehead University, Thunder Bay, Ontario, P7B 5E1 Canada
e-mail: philip.fralick@lakeheadu.ca

AND

JOHN GROTZINGER AND LAUREN EDGAR
Division of Geological and Planetary Sciences, California Institute of Technology, Pasadena, California 91125 USA

ABSTRACT: Our understanding of the significance and abundance of sedimentary strata on Mars has increased considerably during the last decade. The highly cratered surface of Mars leads to the prediction that impact ejecta deposits, possibly containing accretionary lapilli, should be part of the sediment record. While no impact-induced base surge deposits have been confirmed on Mars, it is likely that they will one day be discovered, and it is important to establish criteria for their recognition in the rock record. The recognition of ejecta deposits containing accretionary lapilli on Mars requires reliable facies models developed from known impact-generated strata on Earth. Sections through ejecta layers formed by the 1850 Ma Sudbury impact event provide data to begin development of such models. These deposits are laterally variable but generally show a vertical decrease in lithic clast size and, where present, an upward fining in accretionary lapilli. In thicker deposits, the accretionary lapilli–bearing portion of sections generally progresses upward from decimeter-scale beds of clast-supported lapilli interlayered with centimeter-scale sandstone beds, to parallel and undulatory laminated lapilli, and sandstones. These are overlain by lapilli stringers and isolated lapilli in parallel-laminated to cross-stratified sandstone. Both grain size and sedimentary structures indicate a succession deposited by an impact-generated base surge during decelerating flow. Thinner deposits of ejecta, possibly laid down on topographic highs, are commonly massive with reverse and normal grading. We compare the accretionary lapilli–bearing strata in the Sudbury ejecta deposits to proposed impact-generated base surge deposits in the Burns formation at Meridian Planum, Mars. Units comprising the Burns formation do not have the internal organization of spherule-bearing layers exhibited by the Sudbury ejecta deposits. Comparison with Sudbury ejecta layers and theoretical considerations indicate that the spherules developed in the Burns formation do not represent grains deposited by a base surge and are most likely diagenetic in origin. However, impact ejecta layers should be present in the sedimentary successions on Mars, and comparison with similar strata on Earth may lead to their eventual identification.

INTRODUCTION

Sedimentary rocks were first recognized on Mars only a decade ago based on observations by the *Mars Global Surveyor* spacecraft (Malin and Edgett 2000). Recent interpretations of data collected by the *Mars Express* and *Mars Reconnaissance Orbiter* (MRO) spacecraft have confirmed the abundance of these sedimentary rocks, and the past role of water on the Martian surface. Various depositional mechanisms, including fluvial, deltaic, lacustrine, eolian, volcaniclastic, and impact processes, have been suggested for these strata (see summary in McLennan and Grotzinger 2008). Continuing evaluation of the current stream of data being returned from active missions, supplemented by future landed missions, will further refine our understanding of each one of these different surface processes in creating a stratigraphic record.

In observing the highly cratered surface of Mars, it is logical to ask if sediment formation, transportation, and deposition processes caused by impacts may have been an important factor in the development of its sedimentary deposits. The discovery of impact megabreccia (Grant et al. 2008) in the crater walls of Holden Crater plus quantitative studies of ejecta blankets (see Barlow [2005] and references therein; Mouginis-Mark and Garbeil 2007) provide evidence that impact-related deposits do exist on Mars. Furthermore, eroding sedimentary deposits expose ghost craters, formed at the time of deposition of these deposits (Malin and Edgett 2001), and, therefore, their associated ejecta must be part of the sedimentary record. Knauth et al. (2005) and Burt et al. (2008) stress the likely existence of impact-related deposits and specifically point to the hematite-rich spherules in the bedrock at Meridiani Planum as examples of condensation melt droplets or accretionary lapilli formed from impact-generated base surges. The formation of these spherules also has been ascribed to: precipitation from groundwater to form concretions (Squyres et al. 2004b, Grotzinger et al. 2005, McLennan et al. 2005, Squyres et al. 2006a), precipitation from hydrothermal fluids (Golden et al. 2008), reaction of sulfur-rich fluids with pyroclastic rocks (McCollom and Hynek 2005), and resedimentation of spherules formed by the interaction of volcanic deposits and acidic, hydrothermal fluids (Fan et al. 2010).

Clearly, criteria are necessary for evaluating the genesis of specific deposits in order to determine if occurrences of spherules similar to those found at Meridiani were formed by impact-related processes, or if they are of another origin. Regardless of their origin at Meridiani, which seems very likely to have been diagenetic, it is probably inevitable that impact deposits—possibly bearing spherules—will someday be discovered on the surface of Mars. If so, then the important question that must be answered in the course of in situ lander observations becomes: What are the characteristic features of spherule-bearing units produced by major impact-generated base surges that allow their identification? Unfortunately, descriptions of spherule-bearing deposits produced by impact with projectiles in the 10-km-diameter and larger size range, which are thus capable of producing aerially extensive ejecta layers, are rare in the literature (Graup 1981; Melosh 1989; Newsom et al. 1990; Pope et al. 1999, 2005; Warme et al. 2002; Masaitis 2003; Koeberl et al. 2007; Branney and Brown 2011). However, reinterpreted rock units in the western Lake Superior region now furnish examples of these types of deposits. In this case, the spherules formed as ash pellets, coated ash pellets, and accretionary lapilli.

The recent discovery of strata formed as a result of the Sudbury impact event (Addison et al. 2005) provides field-based evidence that can more accurately define the attributes of spherule-bearing deposits

Sedimentary Geology of Mars
SEPM Special Publication No. 102, Copyright © 2012
SEPM (Society for Sedimentary Geology), Print ISBN 978-1-56576-312-8, CD/DVD ISBN 978-1-56576-313-5, p. 211–227.

related to large collision events. At 1850 Ma (Krogh et al. 1984), an object, probably in excess of 10 km in diameter, struck the edge of the Superior Province in the area that is now Sudbury, Canada. The impact produced the second largest known crater on the surface of the Earth (Earth Impact Database 2004, Spray et al. 2004), propelling an immense amount of material into the atmosphere. The region 400 to 900 km west of the impact site consisted of a peneplained terrain that was subaerial over most of its extent, with lithified carbonate and chert bedrock (Fralick and Burton 2008). It was possibly flooded with very shallow marine conditions in its western portions. The ejecta cloud swept across this setting, entraining locally derived material (Addison et al. 2010). In places, the impact layer deposited by this base surge was altered by possible tsunami activity (Pufahl et al. 2007), subaerial reworking, and diagenesis (Fralick and Burton 2008), and, 15 million years later, it was buried by marine sediments (Addison et al. 2005).

We have developed a facies model for the deposits associated with very large impacts based on the Sudbury ejecta blanket. This model is suitable for comparison to the spherule-rich deposits observed by the *Opportunity* rover at Meridiani Planum, Mars, and at other possible future planetary occurrences. The spherule-rich deposits at Sudbury indicate strong hydrodynamic sorting of grain size, and development of specific relationships between grain types and associated sedimentary structures such as cross-bedding.

SUDBURY IMPACT DEPOSIT FIELD DESCRIPTIONS

Overview

The Sudbury impact deposit was first discovered and documented in Ontario and Minnesota by Addison et al. (2005) (Fig. 1). Subsequently, other areas containing the layer have been identified (Pufahl et al. 2007, Jirsa 2008, Addison et al. 2010, Cannon et al. 2010). Over 20 outcrops and cored intervals through the impact-generated deposits, over an area of approximately 50,000 km^2, were examined for this study. The layer is extremely laterally variable, ranging from totally absent in some locations to tens of meters thick in others. Locations separated by only a few hundred meters can be composed of different materials with contrasting textures and sedimentary structures. Even with these limitations, the outcrops and cored holes through the deposit provide abundant information on sedimentary structures, textures, and architectural organization of subaerially deposited spherules (accretionary lapilli) formed by the large impact event. For the sake of brevity, the term accretionary lapilli, as used here, includes ash pellets, coated ash pellets, and accretionary lapilli (see classification of Brown et al. 2010, Branney and Brown, 2011). All sites described contain a mixture of these three types of ash aggregates.

The Sudbury impact deposit includes all strata containing material that was moved laterally due to the impact event. Not all successions through the Sudbury impact deposit contain accretionary lapilli. Areas further than approximately 800 km from the impact site commonly do not contain accretionary lapilli. Sites closer to the crater generally do not contain accretionary lapilli in areas of the deposit where boulders of underlying lithologies or lapilli-sized devitrified glass are common. Sections containing abundant accretionary lapilli overlie disrupted bedrock, commonly either lithified carbonate layers or chert that was fractured by the impact-generated earthquake (Addison et al. 2010). At many locations, a zone of displaced rubble, which may contain boulder-sized clasts, immediately overlies the fractured basement, separating it from the lapilli-bearing units.

The impact-produced deposit may contain: (1) pebble- to sand-sized devitrified glass, where the larger pieces are commonly vesicular; (2) accretionary lapilli with diameters of 3 to 20 mm; (3) pebbles to boulders of predominantly locally derived chert and carbonate; (4) unshocked quartz and feldspar grains; and (5) quartz with planar deformation features (PDFs) (Addison et al. 2010). The deposit also is associated with highly elevated amounts of iridium (Pufahl et al. 2007).

Three outcrop areas containing well-developed ejecta-bearing successions with accretionary lapilli are described in detail here. These were chosen because they are typical of the range of accretionary lapilli–bearing units that exist throughout the region.

Site #1

The Connors Creek site (site #1; Fig. 1) is located in the western portion of the Dead River Basin, near Marquette, Michigan (Cannon et al. 2010). A curving outcrop here allowed a series of sections to be compiled from the underlying basement to the top of the accretionary lapilli–rich layer (Fig. 2). In this outcrop, the Sudbury impact assemblage rests on chert, which is highly fractured in places. The lowermost impact-generated bed consists of matrix-supported, rounded to angular, granule- to pebble-sized chert fragments in coarse to very coarse sand. The sand is composed of: (1) devitrified glass, some of which is vesicular (Cannon et al. 2010); (2) chert grains; and (3) quartz grains. The pebbles, and more rarely up to boulder-sized clasts, are randomly orientated in this matrix. An irregular contact with protruding pebbles is draped by the overlying coarse- to very coarse-grained, parallel-laminated, pebble-bearing sandstone, which is interstratified with well-sorted, medium-grained sandstone. The coarser sandstone beds are, in places, laterally discontinuous, lensing out into the surrounding medium-grained sandstone. The bed boundaries in the upper portion of this sandy succession become undulatory in places, and erosive truncations show an upward increase in abundance. Some downcutting bed boundaries have their shapes mimicked by overlying beds, creating a succession of curving in-phase bed forms. This assemblage of bed forms is somewhat similar to the progression of upper-flow-regime, parallel-lamination to type V and type III dunes described by Schmincke et al. (1973, see their Fig. 6) from base surge deposits in the Laacher See area of Germany.

The first appearance of accretionary lapilli generally occurs as one-clast-thick stringers or small lenses, of 1.2-cm-diameter lapilli, in the sandstones. This is overlain by beds of clast-supported accretionary lapilli alternating with beds of coarse-grained sandstone with some isolated, matrix-supported accretionary lapilli (Fig. 2A, C, and E). Both textures are parallel-laminated to undulatory parallel-laminated (Fig. 2E). The tops of lapilli protrude into overlying sandstones, whereas the tops of the sandstones are smooth and flat. The matrix in the accretionary lapilli layers is commonly coarser grained than the adjacent sandstone, being dominated by very coarse-grained sand to granules (Fig. 2F). Accretionary lapilli beds are laterally transitional into very coarse-grained sandstone layers, and sand stringers and lenses exist in some lapilli beds (Fig. 2E). The accretionary lapilli themselves decrease in size upward through the unit from 0.7 cm in the middle of the accretionary lapilli–bearing assemblage to 0.4 cm at the top. The upper portion of this unit contains bundles of laminae separated by low- to moderate-angle erosive scour surfaces backfilled with accretionary lapilli and sand-dominated beds. Backfilling strata fill scours, building in an upstream direction, and extend laterally outside the scour on the downstream side to drape the underlying scoured beds (Fig. 2C, D). This layering is similar to what Schmincke et al. (1973) termed "dune types III and IV."

The uppermost portion of the accretionary lapilli–bearing interval is denoted by the increased abundance of coarse-grained, upper-flow-regime, parallel-laminated sandstone beds in which lapilli are restricted to thin, discontinuous beds, lenses, and single-clast-thick stringers in the sandstones. In places, 10- to 40-cm-thick, festoon (trough) cross-stratified sandstones fill erosive cuts into the lapilli-bearing units (Fig. 2A, B). This is overlain by a massive, coarse-grained sandstone bed with chert pebbles and cobbles and scattered lapilli. The bed is at least 30 cm thick, limited by the upper extent of the outcrop. X-ray

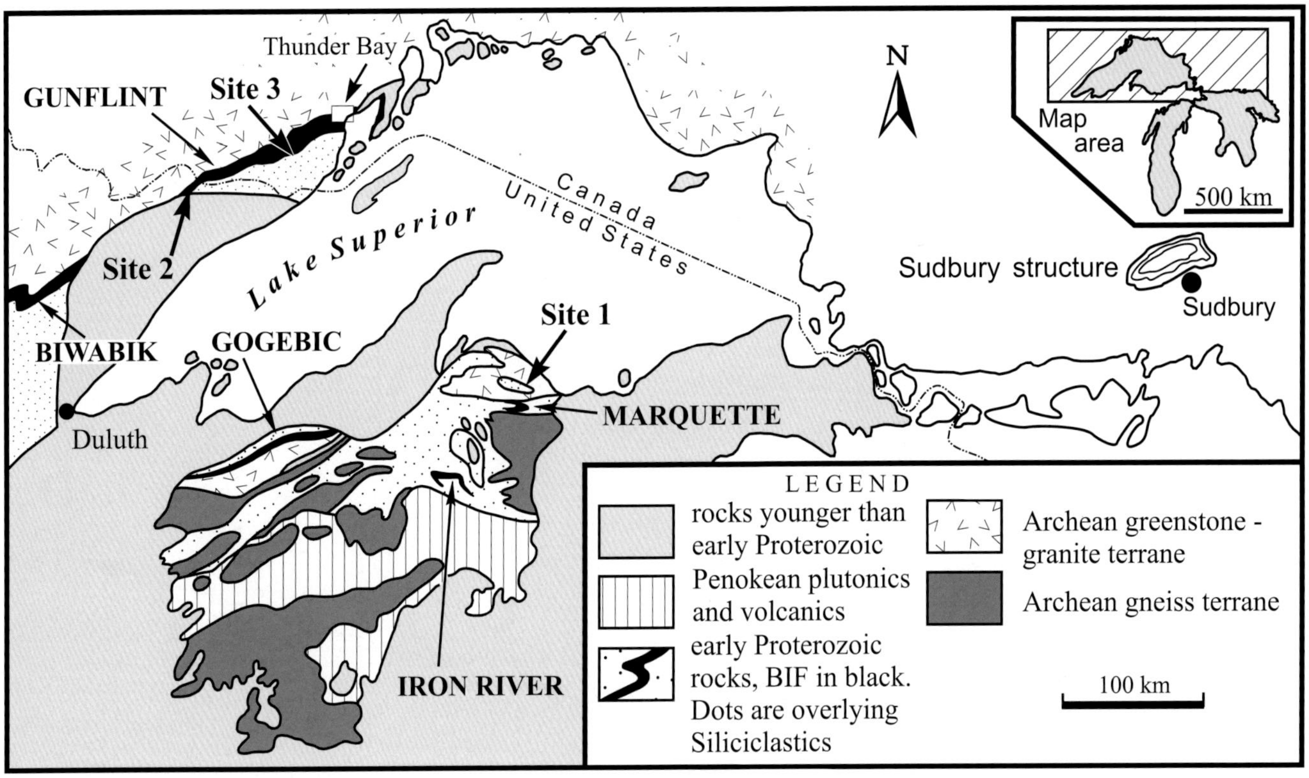

FIG. 1.—Location of the Sudbury structure, the remnant of the crater formed by the impact, and the three outcrop sites discussed in detail in the text. The oval outline of the Sudbury structure was caused by deformation during the Grenville Orogeny. The layer formed by the impact occurs at the contact between the iron formation, in black, and a thick succession of overlying shales and turbiditic sandstones (small dots).

diffraction (XRD) studies have shown that the lapilli are composed of 45–60% quartz, 15–30% dolomite (possibly secondary), and lower concentrations of microcline, plagioclase, muscovite, and chlorite (Cannon et al. 2010).

Another outcrop of the Sudbury impact deposit exists approximately 200 m from the one described previously (Cannon et al. 2010). Here, the lower portion of the impact deposit is not exposed, and the succession begins with 24 cm of upper-flow-regime, parallel-laminated, medium- to coarse-grained sandstone. This is followed by 26 cm of massive, coarse-grained sandstone with scattered, angular to well-rounded, pebble-sized chert fragments. A couple of round pebbles may be accretionary lapilli, approximately 0.7 cm in diameter. The overlying bed is 50 cm thick, and it is composed of coarse-grained sand to granules organized into what is probably one set of moderate-angle, sigmoidal-shaped cross-stratification. Scattered pebbles and rare accretionary lapilli, in matrix-support, occur at the base of this bed. The uppermost bed in the succession consists of parallel-laminated, medium- to coarse-grained sandstone, approximately 80 cm thick. Bedded chert overlies the ejecta deposit and represents commencement of non-ejecta-related sedimentation.

Site #2

Site #2 occurs southwest of Gunflint Lake, Minnesota, approximately 750 km from the impact crater (Jirsa 2008) (Fig. 1). Deposits here appear to be consistently thicker than in other areas, even though these sites are more distal than those in Michigan and Ontario. The impact deposits at sites further away from the crater than Gunflint Lake are much thinner, and lapilli are only rarely present. This led Addison et al. (2010) to hypothesize that the Gunflint Lake deposits may represent thick ramparts, as described for end-of-flow Martian base surge deposits (Kenkmann and Schonian 2006, Osinski 2006, Mouginis-Mark and Garbeil 2007). Most of the deposit at Gunflint Lake is made up of disorganized-bedded boulder breccia, with clasts composed of lithologies present in rocks of the underlying Gunflint Formation. The boulder breccia is commonly overlain by a 10 to 20-cm-thick, matrix-supported pebble breccia to massive, pebbly sandstone. Rare, scattered accretionary lapilli occur in this unit. The pebble breccia is overlain in some sections by accretionary lapilli beds with clast-supported textures. These fill shallow scours in the top of the pebble breccia, or deeper scours that remove strata all the way down to the boulder breccia. The base of the scours is commonly overlain by a 1-cm-thick layer of coarse-grained sandstone, followed vertically by the accretionary lapilli. In other locations, in an area covering a few square kilometers, layers of upper-flow-regime, parallel-laminated, medium-grained sandstone separate the breccia from the accretionary lapilli–rich beds (lapillistone) (Fig. 3). In places, cross-stratified accretionary lapilli fill scours into the sandstone at the base of the lapillistone, giving a paleocurrent direction of 260°. The down-current bearing from Sudbury to Gunflint Lake is 280°. At other locations, where individual smaller scours at the base of the lapillistone are not present, the basal, clast-supported lapillistone bed drapes a shallow erosive scour (Fig. 3). The lowermost accretionary lapilli–rich bed is massive-textured, as are overlying accretionary lapilli–rich beds, except where rare, small-scale, low-angle cross-stratification dipping toward 060° is visible. The basal bed is generally less than 15 cm thick and is succeeded upward by 1- to 5-cm-thick accretionary lapilli beds alternating with coarse- and medium-grained sandstone beds (Fig. 3). The sandstones are parallel-laminated and composed of a mixture of locally derived chert and iron formation grains and more far-traveled

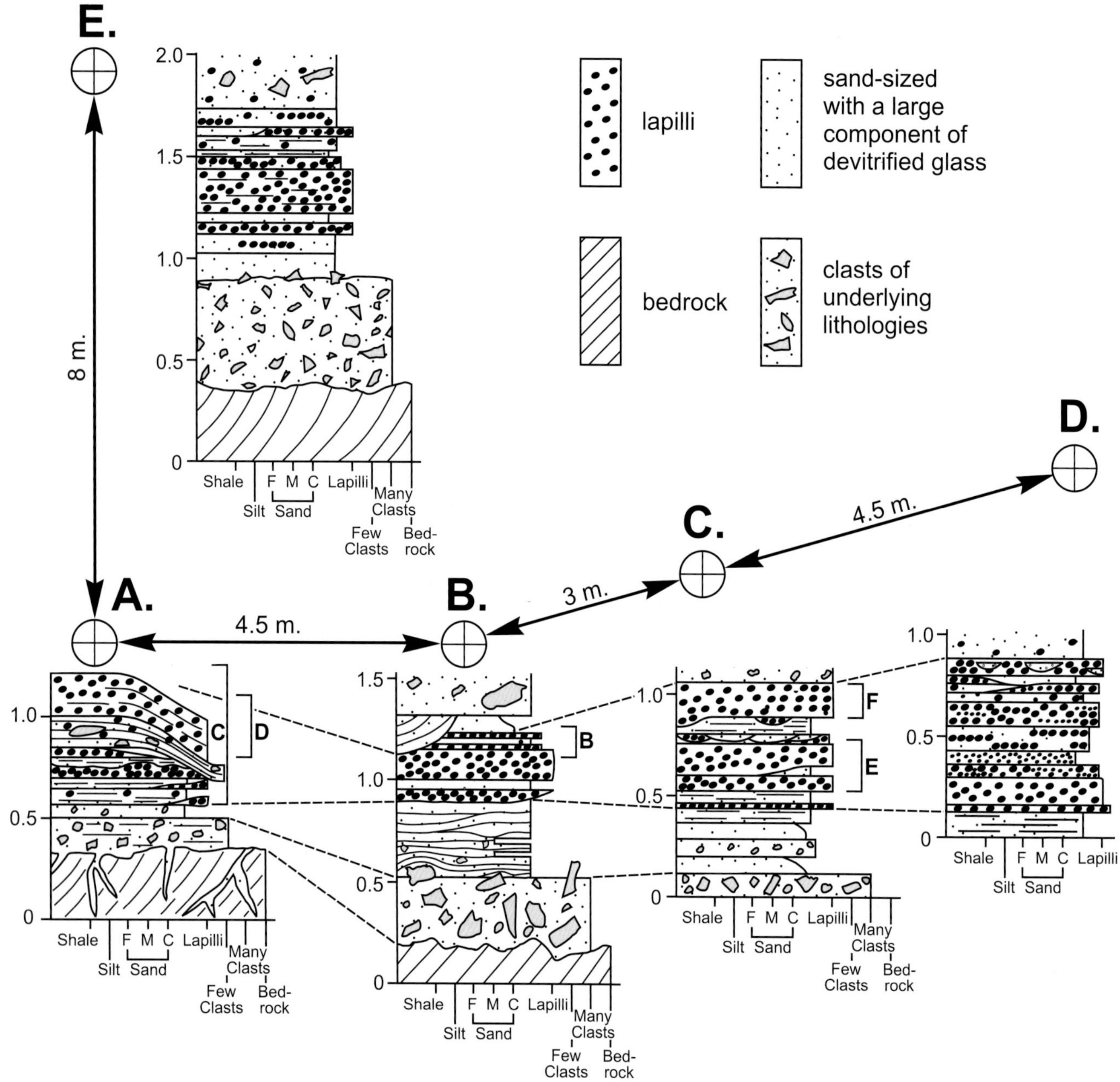

FIG. 2.—**Left)** Stratigraphic architecture of the units present at site #1. The upper sand-rich portion of the impact layer is not present at this outcrop. Locations of photographs are denoted. **Right)** Photographs of units present at site #1. **A)** Photograph of stratigraphic section B. **B)** Enlargement of the upper left of outcrop in A showing a trough-shaped erosive scour cutting interlayered parallel- to wavy-laminated accretionary lapilli and sand-rich beds, and filled by trough cross-stratified sandstone. Flow is from right to left. The oval shape of the lapilli is due to deformation in this area. Coin is 26 mm. **C)** Photograph of the upper half of section A. **D)** Enlargement of the upper central portion of C showing parallel- to wavy-laminated accretionary lapilli and sand-rich layers that are truncated and overlain by an antidune-like structure similar to Schmincke et al.'s (1973) dune type III. The uppermost laminations visible in the upper right of C may represent an upward transition to chute and pool deposition, though the extent of the outcrop is not large enough to discern this reliably. This would denote either increasing velocity and/or decreasing flow depth with time, resulting in upper-flow-regime bed forms progressing from steady flow to shooting flow with hydraulic jumps. Flow is from right to left. Coin is 19 mm. **E)** Lapilli-dominated area of the outcrop. The wavy lamination, especially visible in the upper central portion of the photograph, probably reflects deposition in the transitional zone from plane bed to antidunes. Coin is 26 mm. **F)** Polished slab showing accretionary lapilli– and coarse sand-bearing beds. Note that the matrix of the lapilli beds is coarser than the sandstone beds, reflecting higher velocity and shear stress conditions during deposition of the former.

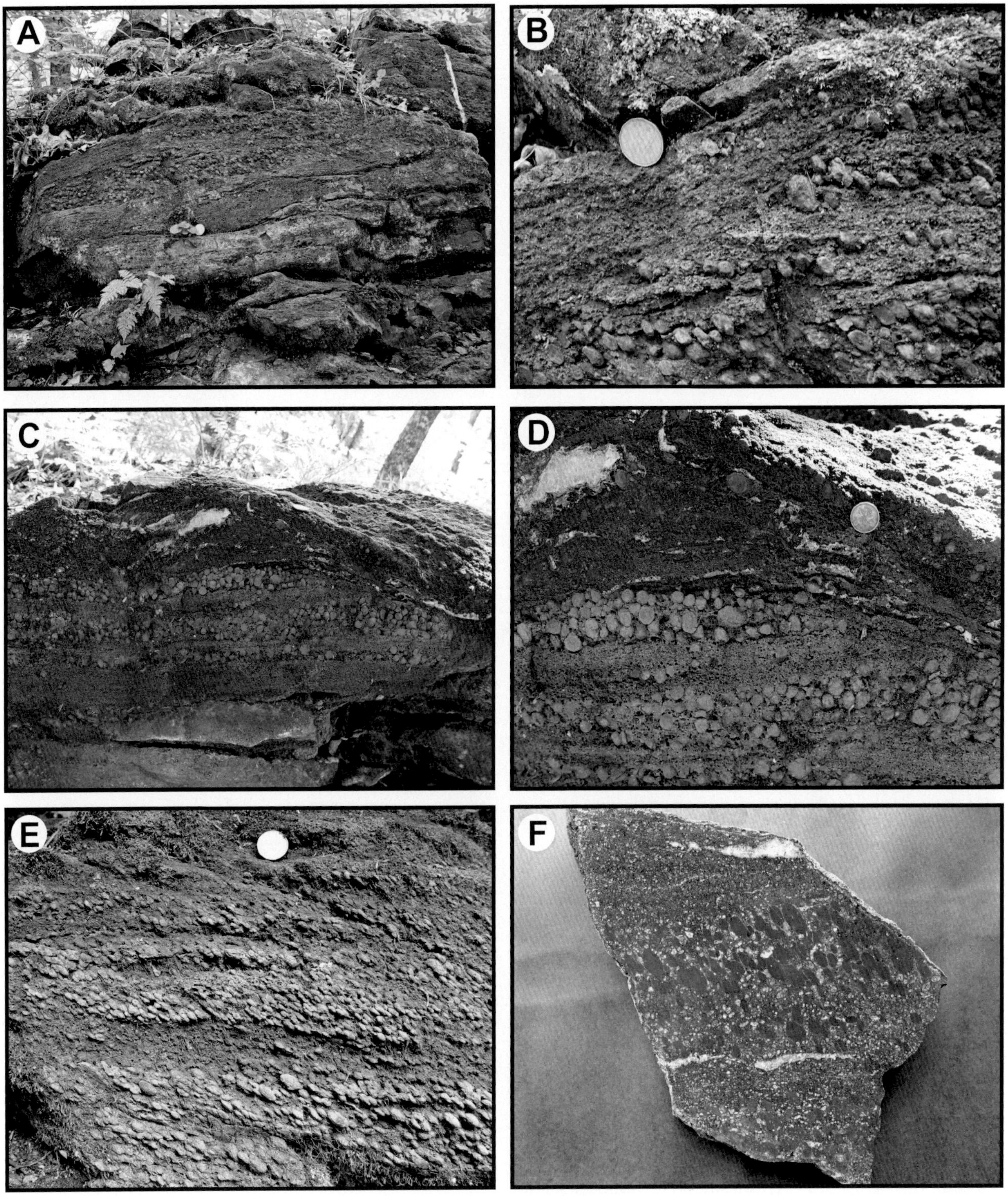

FIG. 2.—*Continued.*

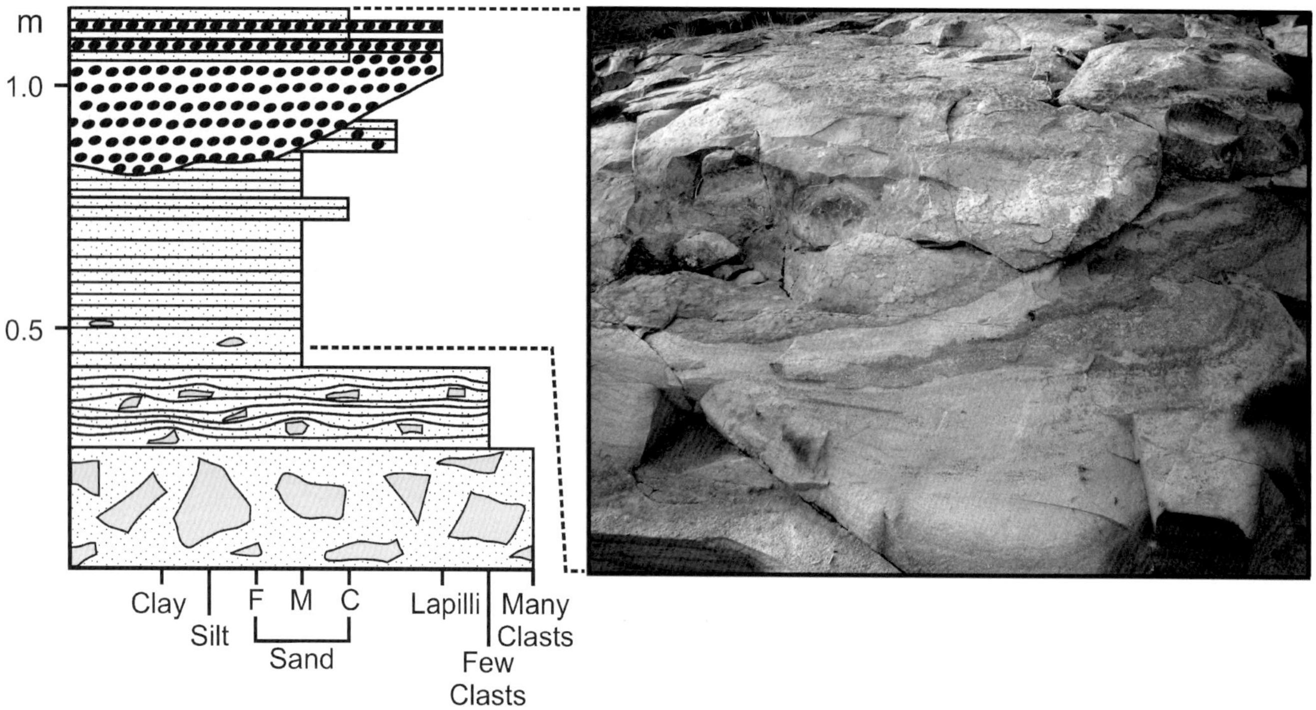

FIG. 3.—Section through one of the outcrops in the area demarked site #2 on Figure 1. Note that one fining-upward succession is erosively truncated by an accretionary lapilli-rich second fining-upwards succession.

grains of devitrified glass. Accretionary lapilli in the bed at the base of the lapilli-rich interval average 0.7 to 0.8 cm, and those higher in the section and interbedded with sandstone are 0.2 to 0.4 cm. The sandstone beds become more dominant in the next few decimeters. Here, they are medium- to fine-grained, with stringers and patches of small accretionary lapilli. Some beds are massive with abundant, isolated lapilli. Parallel-lamination to undulating parallel-lamination is common in the non-massive beds. An approximately 10 cm interval of thinly laminated siltstone caps the impact deposit.

Site #3

Site #3 is located west of Thunder Bay and is described by Addison et al. (2010) as the Highway 588 site (Fig. 1). The outcrop exposure here was blasted during road improvements, which led to an abundance of large blocks of the impact deposit, providing good three-dimensional views (this is termed site #3A). To add to the database, two cored holes were drilled through the unit approximately 200 m from the road exposure (site #3B).

The main exposure consists of lithified carbonate grainstones and stromatolites that were erosively planed off and are overlain by impact-derived material (Fig. 4A–C). Scattered pebbles to boulders of the underlying lithologies occur in the impact deposit (Fig. 4E). The lowest lapilli-bearing unit lies directly on the stromatolites and grainstones in places (Fig. 4A, D), but in other locations, it is separated from the substrate by a thin bed of very coarse-grained sandstone with granules (Fig. 4C). The sandstone is also present in depressions in the bedrock surface and was scoured prior to deposition of the overlying lapilli-bearing unit (Fig. 4A, B). The lower 30-cm-thick bed of lapillistone appears massive in most samples, but a crude cross-stratification is present in some portions of the outcrop that are highlighted by more sand-rich partings dipping at approximately 30° (Fig. 4D). The lapilli in this bed average 1.1 cm in diameter. Most lapilli have a clast-supported texture; however, areas of matrix-supported textures do exist. Portions of the bed also have common lithic lapilli, composed of amygdaloidal, devitrified glass, and pebbles of underlying lithologies, mainly black chert. This bed is overlain by 15 cm of interstratified upper-flow-regime, parallel-laminated accretionary lapilli beds and coarse-grained sandstone beds. These layers form lenses that taper out laterally, commonly over 1 to 2 m. The lapilli here average 0.8 cm in diameter. The uppermost unit in this outcrop is 20 cm thick and consists of 1- to 3-cm-thick beds of parallel-laminated, coarse- to medium-grained sandstone. Strata in the lower 8 cm of this unit contain stringers of accretionary lapilli associated with laminae of coarser sandstone with some scattered, isolated lapilli (Fig. 4F). No lapilli were observed in the upper portion of this sandstone succession.

The drill hole, though only 200 m away, intersected a different assemblage (Fig. 5). The impact deposit here lies on cherts forming the Gunflint bedrock. The lower two thirds of the deposit exhibit inverse, coarse-tail grading, going from very coarse-grained, granule-bearing sandstone at the base to pebbly, granule-rich, very coarse-grained sandstone with scattered accretionary lapilli. The upper third of the core is normally graded from the pebbly sandstone to a medium-grained sandstone. No sedimentary structures were observed, except for an indistinct parallel lamination in the upper 5 cm of the core.

MARTIAN SPHERULE-BEARING STRATA

For the first time in the history of the exploration of Mars, the observation by *Opportunity* of ancient sedimentary processes recorded in the strata preserved at Meridiani allowed a fresh approach to understanding the early history of surface processes on Mars. This approach implements standard techniques used in analysis of terrestrial strata. Reconstruction of depositional processes is based on the analysis of suites of physically produced sedimentary structures along with observations of other textural attributes such as grain size, sorting,

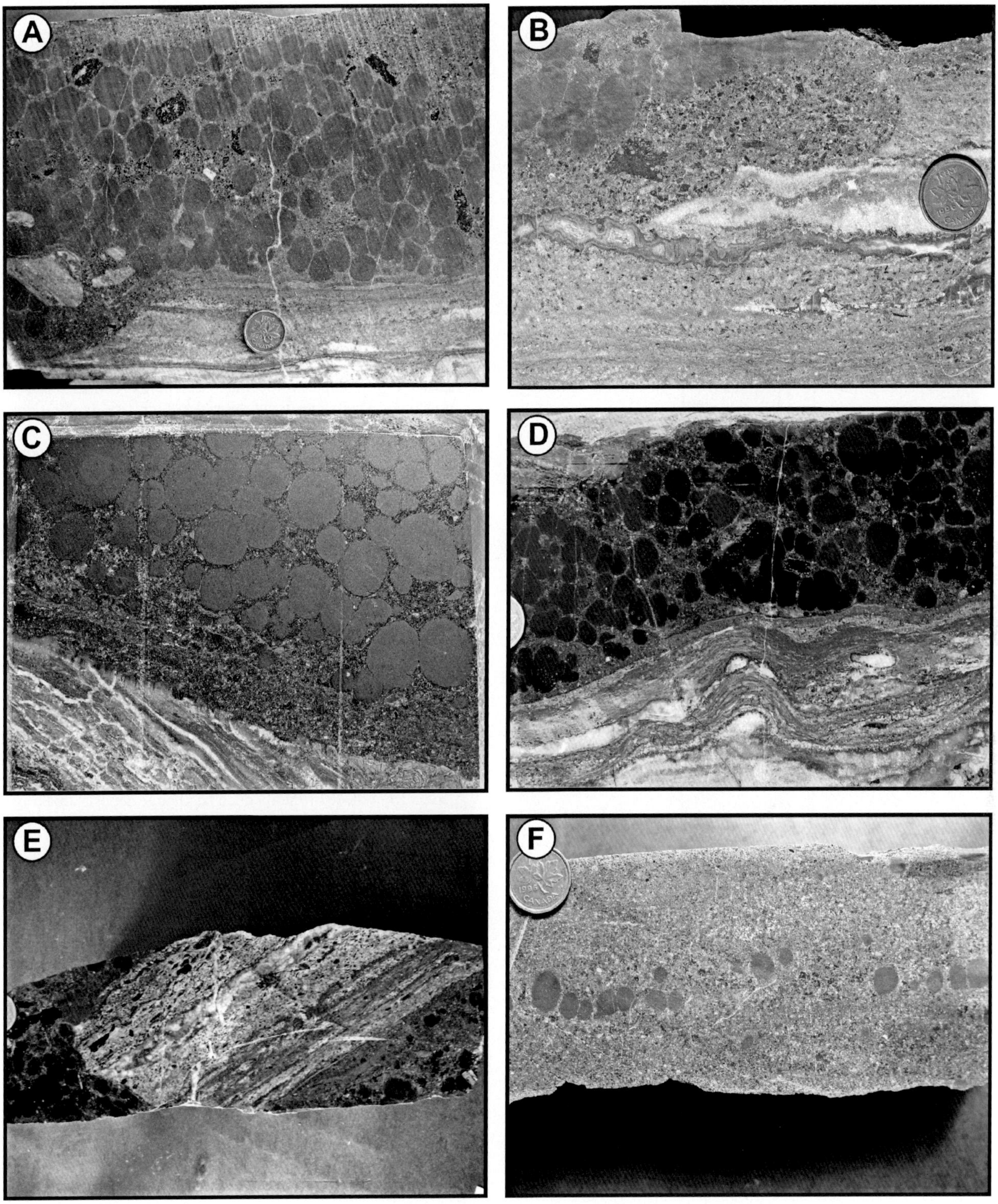

FIG. 4.—Photographs of cut slabs from site #3A. **A)** Erosively scoured bedrock composed of Gunflint carbonate and chert overlain by the impact deposit. Very coarse-grained sandstone with granules fills the scour. Coin is 19 mm. **B)** Another example of a scoured depression filled by a remnant of granule-rich sandstone. Note the very coarse-grained, granule-rich sandstone in the scour has a different grain size than the matrix of the overlying lapilli-rich bed. **C)** The truncated top of a hemispherical stromatolite overlain by a layer of very coarse-grained sandstone, which in turn is overlain by accretionary lapilli in clast-support. Slab courtesy of W. Addison. **D)** Lapilli directly overlying Gunflint carbonate bedrock. **E)** A large block of underlying bedrock encased in accretionary lapilli–rich ejecta. **F)** Coarse-grained sandstone, which lies above the more massive accretionary lapilli–rich bed (not present in this slab). Note the stringers of lapilli are associated with, and overlain by, coarser sand-sized material than that constituting the underlying layer.

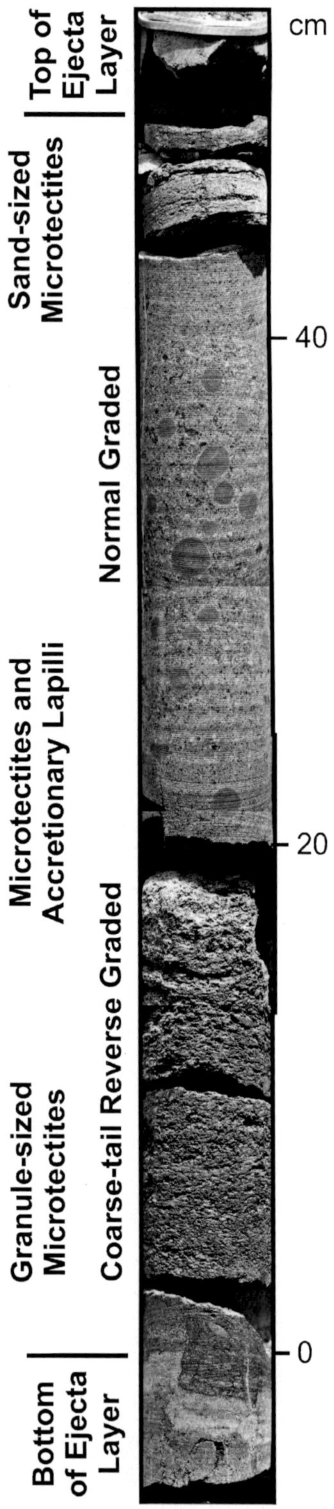

FIG. 5.—Drill-hole intersection with ejecta deposit from site #3B.

and associated diagenetic features (cf. Leeder 1982). Detailed observations and measurements related to the sedimentology, stratigraphy, mineralogy, provenance, and diagenesis at Meridiani, Mars, have been previously reported (Squyres et al. 2004a, 2006b; Clark et al. 2005; Grotzinger et al. 2005, 2006; McLennan et al. 2005; Squyres and Knoll 2005; Edgar et al. this volume). Here, we summarize the key observations that relate to the physical emplacement of these sedimentary rocks for the purpose of comparison with the Sudbury impact deposit.

The Burns formation is the informal name given to the succession of well-sorted, moderately indurated sandstones (Grotzinger et al. 2005) preserved immediately beneath the surface of the Meridiani plains and examined in most detail at exposures in the vicinity and within Eagle, Endurance, Erebus, and Victoria craters. Regional geology suggests that the deposit is of late Noachian age and thus likely in excess of 3.5 Ga. Detailed stratigraphic measurements (Fig. 6) and sedimentological observations (Fig. 7) generally support a model whereby eolian and locally subaqueous processes deposited these sediments (Fig. 8). The Burns formation has been divided into three stratigraphic units (lower, middle, and upper), and three major sedimentary facies associations have been identified (Grotzinger et al. 2005, Edgar et al., this volume):

1. Facies A is composed of large-scale cross-bedded well-sorted sandstones. This facies is interpreted to represent a migrating dune system of unknown extent that was deposited under dry conditions such that the sediment was noncohesive and thus transported in migrating dune fields. Bed set thicknesses suggest moderately large dunes in excess of 2 m.

2. Facies B is planar-laminated to low-angle cross-stratified, well-sorted sandstones. This type of stratification results from sediment limitation, forming low-relief bed forms produced by migrating impact ripples. In Endurance crater, the contact between the dune and overlying sand sheet facies marks the boundary between the lower and middle stratigraphic units. This important stratigraphic boundary, termed the Wellington contact, is interpreted to be a deflation surface formed where the groundwater capillary fringe limited erosion of previously deposited dune sediments. This facies is interpreted as an interdune sand sheet deposit. Smith and Katzman (1991) describe eolian deposits from the Jemez Mountains volcanic field, New Mexico, that are similar to facies A and B. They emphasize that the styles of stratification in the eolian deposits are distinct from those of interbedded base surge deposits, making their differentiation possible.

3. Facies C is centimeter-scale trough (or "festoon") ripple cross-lamination, diagnostic of subaqueous current transport in the lower flow regime. At Endurance and Eagle craters, this facies is found in the upper part of the upper unit of the Burns formation and marks where the groundwater table breached the surface and wind-driven subaqueous currents transported the sediment. Centimeter-scale trough cross-lamination is particularly well developed at Erebus crater, where additional features suggestive of desiccation also are present. Polygonal patterns of small cracks are interpreted to have formed during multiple wetting and drying events, and soft sediment deformation features are consistent with sediment liquefaction. This facies is interpreted as a wet to evaporitic interdune environment.

All facies present in all craters studied by *Opportunity* contain spherules. Comparison of their size and spacing has been presented by McLennan et al. (2005) and Edgar et al. (this volume). Spherules in the Burns formation have relatively uniform shape and size (Fig. 9). At Eagle and Endurance craters, they have a rather uniform size distribution of about 4 mm average and are almost perfectly spherical (McLennan et al. 2005). Spherules at Victoria crater are smaller, with an average diameter of about 1 mm (Edgar et al. this volume). At all outcrops observed to date, the spherules do not disrupt lamination and are highly dispersed relative to bedding, so they are not observed to concentrate along bedding planes or scour surfaces (Fig. 10). All factors considered, the Burns spherules are most likely to have formed

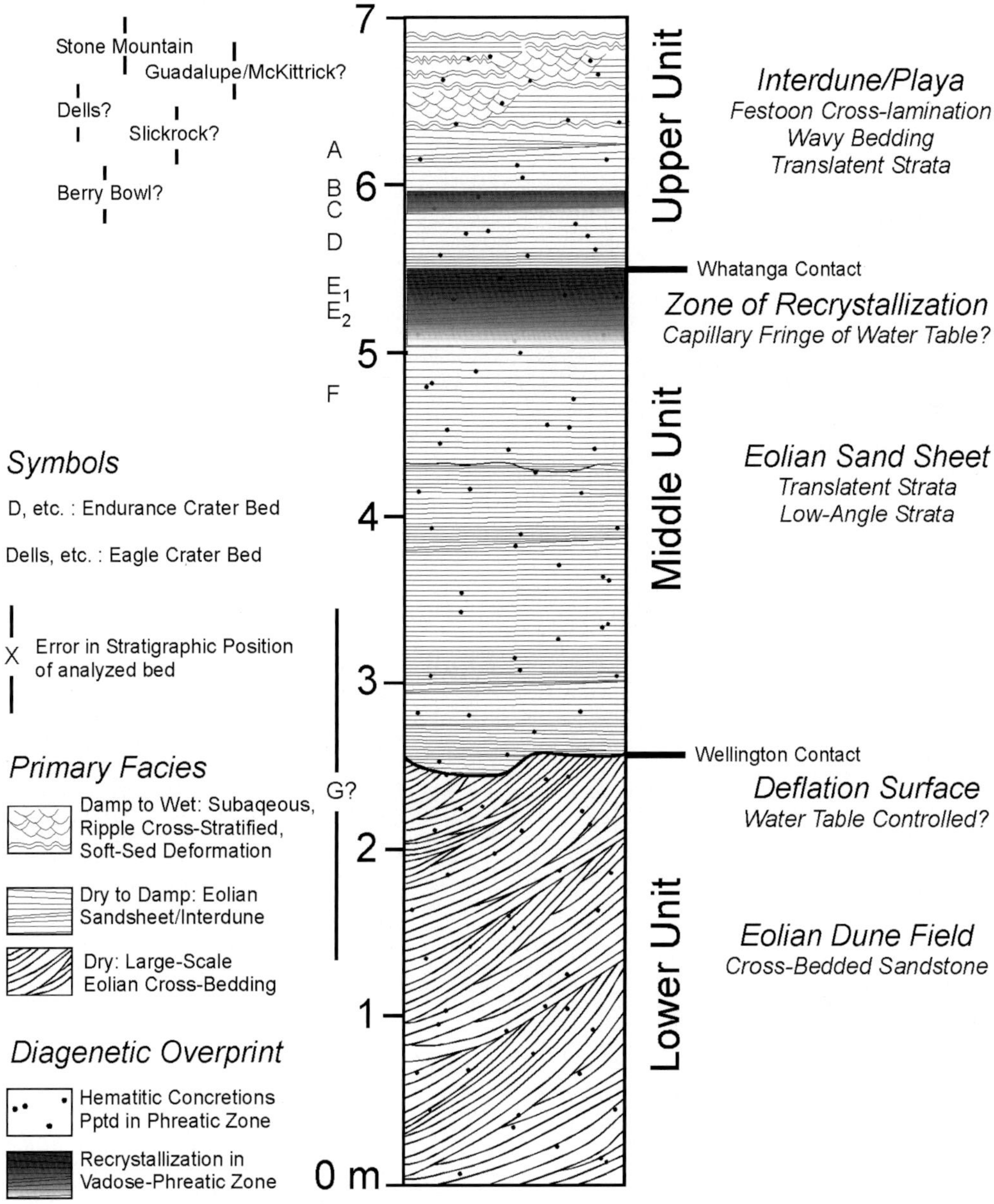

FIG. 6.—Stratigraphy of the Burns formation exposed at Burns cliff and within Eagle crater, Meridiani Planum. Names and letters on the left side denote the approximate stratigraphic locations of key rocks and targets. Sedimentological interpretations of the three stratigraphic units are shown on the right hand side of the column. See text for further discussion. From Grotzinger et al. (2005).

as diagenetic concretions (McLennan et al. 2005). However, see the following section for a discussion of alternatives.

DISCUSSION

Though direct observation of the events triggered by a large impact on a terrestrial planet has never been made, insight into the sediment dispersal dynamics can be gained from the literature on explosive volcanic eruptions and nuclear tests. Surges generated by subaerial, pyroclastic eruptions are low-concentration, high-velocity currents, which are dominated by high degrees of turbulence (Carey 1991, Orton 1996). The stratification and sorting characteristics of volcanic base surge deposits are largely determined by the particle concentration and velocity profiles across the flow-boundary zone from which the lithofacies aggrades (Branney and Kokelaar 2003). The particle concentration and velocity profiles follow clear trends based on proximity to the source and relative position within the flow, producing predictable patterns in stratification and sorting. Deposits formed from these flows show distinct proximal to distal and bottom to top changes consistent with decreases in flow velocity and sediment accumulation rate (Chough and Sohn 1990). In more vent-proximal areas and within valleys or channels, where sediment accumulation rates from suspension are high, thicker, massive, and coarser-grained deposits accumulate (Crowe and Fisher 1973, Chough and Sohn 1990, Druitt 1992, Orton 1996). As the cloud loses material and sediment concentration declines, deposition from the traction carpet dominates,

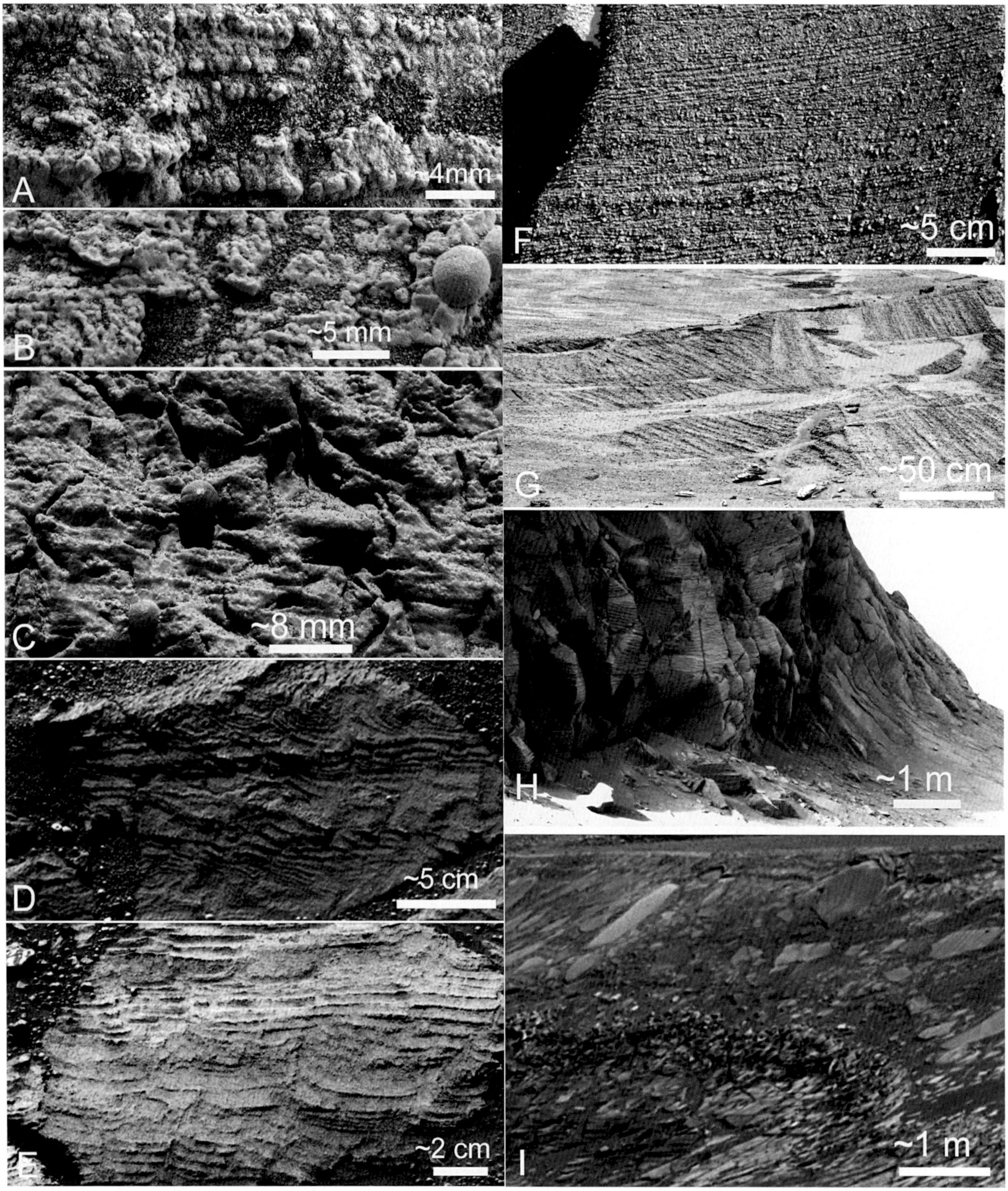

FIG. 7.—Sedimentologic observations that constrain the environment of deposition for the Burns formation. See Grotzinger et al. (2005), Grotzinger et al. (2006), McLennan et al. (2005), Metz et al. (2009), and Edgar et al. (this volume) for further descriptive and interpretive details. **A)** Microscopic Imager (MI) image showing distinct grains that form laminated sandstones ("Flatrock," Eagle crater) of the Burns formation. Grains range in size from 0.1 to 1.0 mm, are moderately well rounded, and are well sorted into discrete laminae. The excellent size sorting on the scale of individual laminae is strongly suggestive of migrating eolian impact ripples. The image (1M131912465) was acquired on sol 042. **B)** MI image showing effects of early cementation and recrystallization, visible in the center left ("Algerita," Eagle crater). Note the set of three or four laminae, expressed by fused grains. At center right, the effects of recrystallization on destruction of primary fabrics are well illustrated by

and stratified deposits form (Schmincke et al. 1973, Chough and Sohn 1990, Carey 1991, Orton 1996). Thus, deposits progress from massive to well stratified, both outward from the vent—though local topography complicates this (Schmincke et al. 1973, Wilson and Walker 1982, Druitt 1992, Brown et al. 2010)—and upward through the deposit. Grain size also generally decreases in these two directions, and sedimentary structures change from higher- to lower-velocity bed forms (Crowe and Fisher 1973, Chough and Sohn 1990).

In the deposits formed by the Sudbury impact event, basal, disorganized, pebble to boulder breccias, composed of underlying lithologies, commonly with little to no lithic or accretionary lapilli, overlie a bedrock surface that is, in places, severely jointed. The near-surface rock layers were fractured and rotated by the strong earthquake produced by the impact (Addison et al. 2010). The presence of abundant boulders over 1 m in diameter that were entrained by the flow indicates that the leading edge of the base surge was traveling at very high velocity. Additionally, the shock wave (see Wohletz et al. 1984) produced by the impact may have swept some of the pre-earthquake-arrival, surficial debris into depressions prior to entrainment by the base surge. The breccias commonly are coarse-tail graded upward (Addison et al. 2010), indicating decreasing flow velocities. They are overlain by upper-flow-regime, parallel-laminated sandstones composed of angular, aphanitic grains of various sericite- and chlorite-dominated lithologies (Cannon et al. 2010), probably representing devitrified glass. At site #1, where this sandstone is best exposed, all bed forms are upper flow regime with an upward progression in some sections from parallel lamination to undulatory layering. Undulatory layering associated with parallel lamination similar to that present in the Sudbury impact deposits has been described from base surge deposits in the Laacher See area (Schmincke et al. 1973), at Songaksan mount (Chough and Sohn 1990), and at Cora Maar volcano (Gencalioglu-Kuscu et al. 2007).

The absence of accretionary lapilli in the lower sandstones may reflect processes similar to those proposed to operate in ground-hugging pyroclastic density currents developed during eruptions on Tenerife (Brown et al. 2010). Here, the absence of accretionary lapilli in the basal portion of the deposits is attributed to a two-stage process. Ash pellets formed in the uppermost lofted, moist, and cool parts of the plume above the base surge. They then dropped into the high-velocity, ground-hugging turbulent current, accreted concentric layers, and were sedimented (Brown et al. 2010, Branney and Brown 2011). The leading front of the ground-hugging flow advanced in front of the overlying, billowing ash cloud and, thus, did not receive ash pellets to transform into accretionary lapilli.

The thickness of the lower sandstones is probably controlled by fluctuations in velocity and concentration of sediment in the highly turbulent flow. At site #3A, where these sandstones are mostly limited to low-velocity zones in depressions and behind obstacles, overall flow conditions did not fall within the depositional realm until the main upper ash plume had advanced over the area and accretionary lapilli began accumulating. At site #2, the flow conditions accommodated sand and granule deposition from the leading edge of the surge, but arrival of the upper ash plume over this area was accompanied by a change in conditions promoting erosion of an unknown thickness of the sand. Deposition of accretionary lapilli filled this broad scour surface and individual smaller scours superimposed upon it. Cross-stratification, consisting of alternating accretionary lapilli laminae and laminae of sand-sized material, developed during infilling of the scours and behind obstacles, such as boulders or small bedrock promontories. At site #1, no noticeable erosion occurred between deposition of the sandy succession and the overlying accretionary lapilli. Site #3B has a gradual transition as well, but in a massive assemblage of sediment.

The accretionary lapilli–rich units overlying the upper-flow-regime, parallel to undulatory beds of sand- to granule-sized material possess many of the attributes of the sandy succession. The common parallel-lamination and rarer development of cross-stratification dipping toward the impact site (similar to type III and V dunes of Schmincke et al. [1973], which were interpreted as antidunes, i.e., structures developed at higher velocities than parallel-lamination and in association with a hydraulic jump) indicate that the lapilli-bearing units were deposited under upper-flow-regime conditions. As in the underlying sand dominated beds, this type of layering developed when sediment in suspension rapidly dropped into the traction carpet and was deposited. Changes in velocity of the highly turbulent flow resulted in deposition of accretionary lapilli, with very coarse-grained sand, granules, and small pebbles forming the matrix, interbedded with

←

a halo, which envelops a hematitic spherule interpreted as a concretion. Note complete loss of primary fabric and development of blocky, interlocking crystals. The image was acquired on sol 028. **C)** MI mosaic showing both diagenetic concretions (positive relief) and molds after diagenetic minerals (negative relief). Note distinct geometric shapes of crystal molds, which crosscut primary lamination. The images were acquired on sol 029. **D)** Pancam image of trough (or "festoon") cross-lamination (upper "Overgaard," Erebus crater). Small-scale trough cross-lamination is interpreted to have formed in low-velocity subaqueous flows. This image was acquired on sol 716. **E)** Pancam image of probable desiccation cracks and soft-sediment deformation features ("Skull Valley," Erebus crater). Cracks crosscut lamination, some oblique to bedding. Note characteristic upward-deflected laminae along crack margins; termination of prominent crack in center of rock at a discrete bedding plane; and truncation of upward-deflected laminae along discrete bedding planes in center and upper parts of rock. This image was acquired on sol 713. **F)** Pancam image of low-angle cross-stratification featuring well-developed pinstripe texture, which has likely formed due to differential cementation of eolian wind ripple stratification ("Tipuna," Endurance crater). The Tipuna outcrop represents the uppermost 30–40 cm of the Burns middle unit. The image was obtained on sol 307, and a larger version is shown in Figure 10. **G)** Pancam image of the contact between the lower and upper units of the Burns formation. Bedding dips steeply to the right, and the trace of the contact runs nearly vertically through the image and is to the right of the center of the image. Below the contact, large-scale cross-bedding forms a bed set at least 2 m thick that was likely formed by migrating eolian dunes and consists of strata that are steeply inclined relative to the truncation surface. The contact—a truncation surface—is interpreted to represent deflation of the dune field down to the level of the groundwater table. The image was acquired on sol 288. **H)** Pancam image of large-scale cross-bedding in the cliffs of Victoria crater, along the west-facing wall of Cape St. Mary. This cross-stratification is also diagnostic of migrating eolian dunes, as indicated by the presence of meter-scale scours and wedge-shaped sets. The image is part of a mosaic acquired during sols 970–991. **I)** Pancam image of ejecta blanket deposited on Burns formation bedrock at Victoria crater, along the east face of Cabo Frio. The blanket consists of coarse block breccia and is capped by a stratified upper unit, which drapes protruding breccia blocks. It was formed by ejection of bedrock during the impact event that created Victoria crater. The image is part of a mosaic acquired on sol 952.

coarse-grained sandstone. A decrease in size of the accretionary lapilli up through this unit denotes a drop in flow competency, culminating in a sand-dominated succession with accretionary lapilli concentrated as stringers at the base of coarser sand-granule–dominated beds and, more rarely, isolated in massive beds. These massive beds with scattered accretionary lapilli probably formed as a result of an increase in the rainout of material from suspension, which will suppress the formation of parallel lamination (Arnott and Hand 1989). Further dropping of flow competency upward through the succession resulted in the development of lower-flow-regime bed forms (scour-fill festoon cross-stratification, dunes, and sand waves). Only rare accretionary lapilli are found at the base of this sandstone unit, i.e., scattered accretionary lapilli along the top surface of the basal erosive scour underlying the sigmoidal-shaped cross-stratification at site #1. Because of a large difference in mass between the lapilli and sand, the accretionary lapilli were probably remobilized from exposed accretionary lapilli layers and rolled in traction rather than transported into the area in suspension with the other material in the waning flow.

A general model for deposition of accretionary lapilli–bearing and related units can be developed using data collected from sections through the impact deposit (Fig. 11). This model does not reflect proximal to distal relationships, but rather decreasing energy level during sediment accumulation. Sites #3A and #3B are within 200 m of one another, yet their deposits are at opposite ends of the spectrum (Fig. 11). Work on volcanogenic base surges and ignimbrites has highlighted that local topography can exert an important control on the type and thickness of the sedimentary succession that accumulates (Schmincke et al. 1973, Wilson and Walker 1982, Druitt 1992, Brown et al. 2010, Branney and Brown 2011). It is envisaged that localized topographic differences also played an important role in defining the type of accretionary lapilli–bearing successions that developed in the Sudbury impact deposits. Examination of outcrops where basement topography is exposed confirms that thicker deposits of the basal breccia accumulated in bedrock lows, whereas adjacent higher levels were swept clean of debris and accumulated lithic lapilli–sized sediment. Although the basement topography at site #3 is unknown, a similar scenario is plausible there.

Coarse-tail normal grading in the basal breccia (Addison et al. 2010) combined with the overlying parallel-laminated sandstone to form a powering down succession. Erosive truncation of the top of this unit at some locations indicates that there was an increase in competency prior to the arrival of the lofted portion of the flow where accretionary lapilli could form. The lofted portion of the ash cloud provided a source for the accretionary lapilli, which were sedimented by traction processes in the highly turbulent, basal surging flow. The coarser accretionary lapilli in beds deposited on the, in places, eroded top of the sandstones form the base of a fining-upward succession to small accretionary lapilli and finally sands, forming another powering down series of beds. This decrease in energy is also reflected in the upper-flow-regime bed forms in the accretionary lapilli–rich section, which are replaced upward by lower-flow-regime bed forms in the overlying sand-rich sediments.

The ejecta-bearing sediments exposed in the drill hole at site #3B are massive, except for parallel lamination in the uppermost fine-grained sandstones. Their inverse to normal grading is similar to some layers deposited from base surges in the volcanogenic Songaksan tuff ring (Chough and Sohn 1990). There, the inverse grading is related to dispersive pressure within a suspension or a traction carpet with a dilute, turbulent overriding surge (Chough and Sohn 1990). The normally graded portion was probably deposited from suspension. This depositional scenario may be applicable to site #3B, but the lack of parallel lamination or other sedimentary structures also indicates that movement from suspension was not into a traction carpet, but rather resulted in individual particles being sedimented. This implies that sedimentation was not initiated at site #3B until the more diffuse, later stages of the flow were passing this area, and at this stage, the flow may

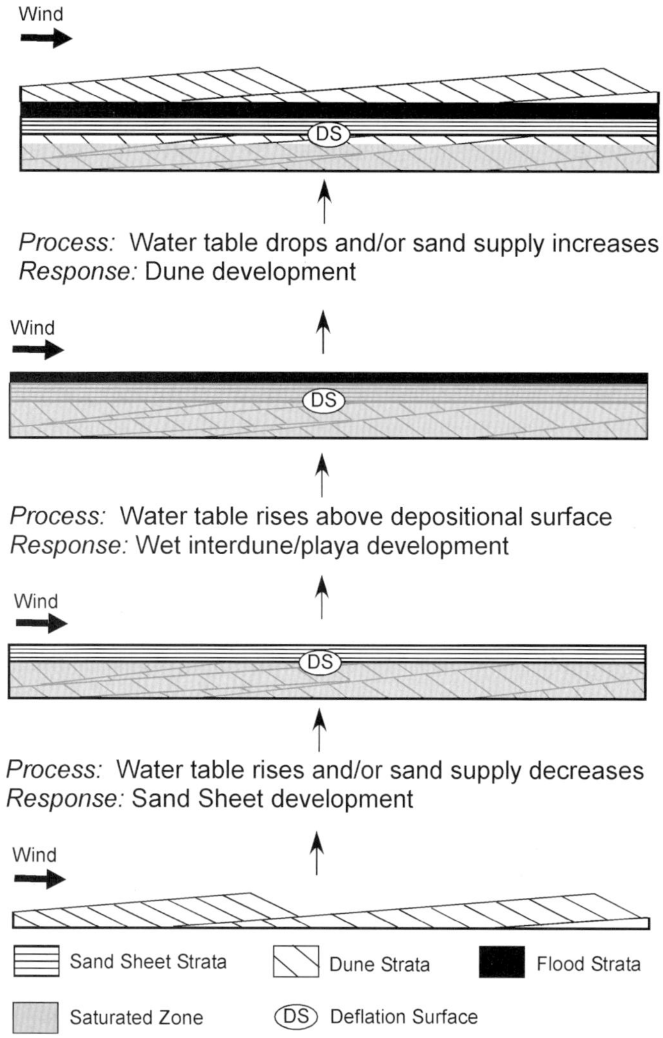

FIG. 8.—Depositional model for the Burns formation in the region of Eagle and Endurance craters (from Grotzinger et al. 2005) documenting a "wetting upward" eolian system. The lower unit accumulated by migration of eolian dunes. It is composed of sulfate-cemented altered basalt sand grains. Elevation of the water table (or decrease in sediment flux) resulted in the formation of a deflation surface (DS) down to the level of the capillary fringe of the water table. Sand sheet strata were deposited above the deflation surface. At some point, the groundwater table breached the surface, resulting in deposition of subaqueous interdune sandstones ("Flood Strata"). To date, only one such cycle has been confirmed by *Opportunity*, but such a model predicts the cycle would be repeated as the level of the groundwater table fluctuated.

have separated from the substrate so traction was not possible—as in the pyroclastic flow transforming into a buoyant plume in the Mount St. Helens eruption (Sparks et al. 1986).

Another type of massive impact deposit has been described by Branney and Brown (2011) from the Stac Fada impact layer in Scotland. This impact layer is up to 10 m thick and composed of reverse graded and then disorganized bedded, matrix-supported breccia without accretionary lapilli. This section fines upward (coarse tail grading) to massive, matrix-supported pebble breccia with accretionary

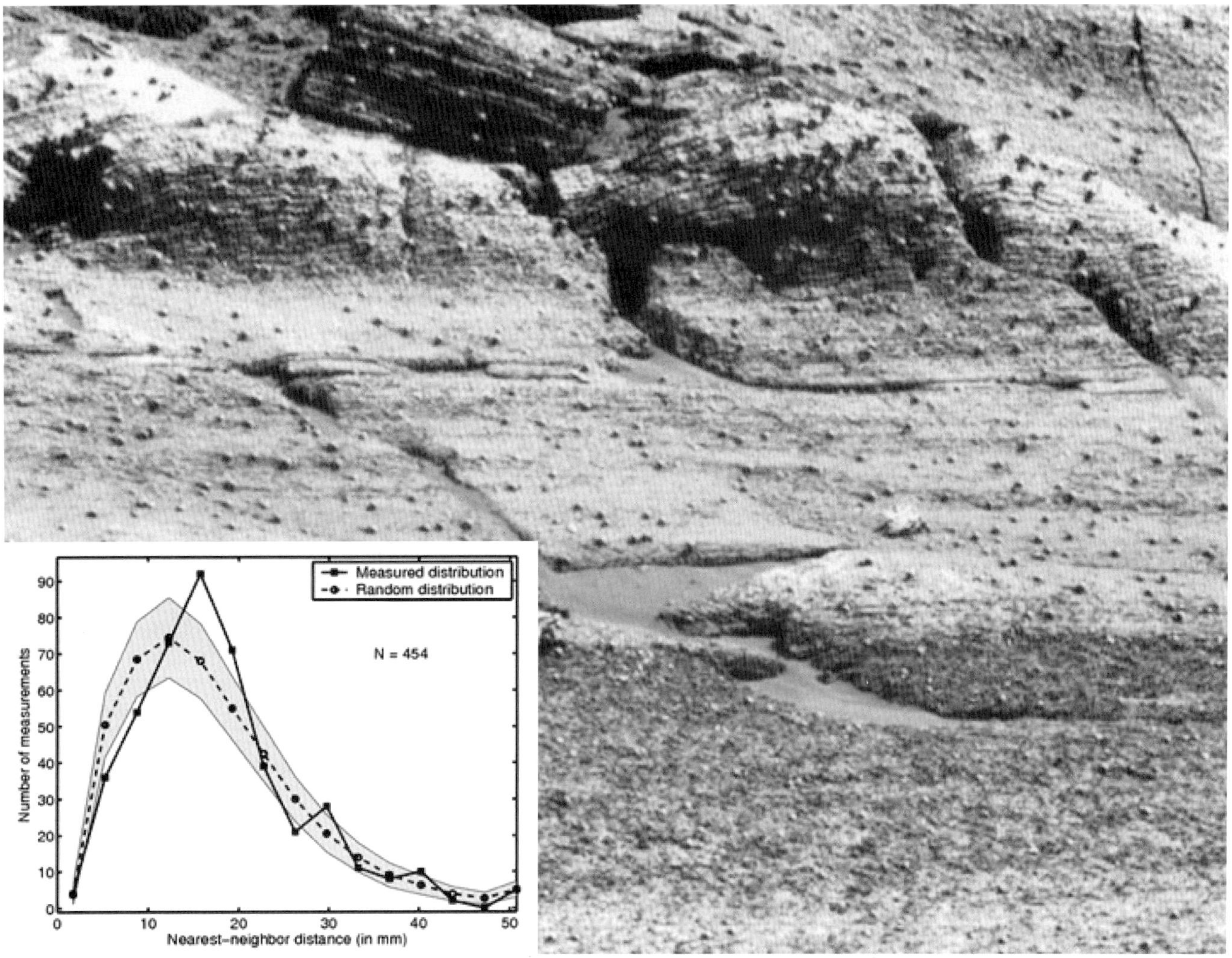

FIG. 9.—Image of upper part of Burns formation, Endurance crater, Meridiani Planum, Mars. Lower, dark-toned unit is overlain by upper, light-toned unit, which represents a diagenetic alteration front (see Figure 5c of Grotzinger et al. [2005] for image location and camera parameters). Note that spherules are highly dispersed throughout both facies and are not concentrated along stratification boundaries. Inset shows plot of frequency versus nearest-neighbor-distance for spherules (solid line) from four locations compared to a numerical random distribution (dashed line). This distribution shows quantitatively what can be observed qualitatively in the underlying image: that the spherules are highly dispersed, which generally supports of a model of formation by precipitation at point sources by iron-bearing pore fluids. After McLennan et al. (2005).

lapilli. Faint bedding developed immediately below a capping, thin layer of clast-supported rainout ash pellets. Branney and Brown (2011) hypothesize that this deposit formed from a high-concentration, steady, granular fluid-based density current. This is a higher concentration flow than that described for the Sudbury impact, possibly denoting a closer position to its source. As no layering is developed in the majority of the deposit, the accretionary lapilli are scattered throughout the upper portion of the layer.

Aside from thin, massive impact deposits, probably formed by the waning suspension cloud, or representing tsunami reworking (Pufahl et al. 2007), the accretionary lapilli–bearing Sudbury impact deposits are internally organized assemblages. They have well-developed stratification, and accretionary lapilli were sorted by traction processes according to their size and mass. Velocity variations in the surge, caused by turbulent eddies, separated the different medium-grained sand to pebble size fractions in transport into distinct beds. Only where sedimentation from suspension formed a massive layer, or where a limited number of accretionary lapilli were eroded from a clast-supported lapilli bed and rolled into a sand-dominated area, were isolated lapilli observed in thicker sandy units. In the latter example, the lapilli are commonly found at or near the base of the sandstone bed as continued sand deposition buried the source of the lapilli.

Relevance for the Burns Formation and Other Deposits on Mars

Two alternatives to the depositional model outlined previously have been suggested for the Burns formation. McCollom and Hynek (2005) argued that the sequence was a volcanic base surge deposit that experienced in situ isochemical acid-sulfate alteration in an acid-fog–like setting. They interpreted the spherules to represent diagenetic concretions formed during alteration. Knauth et al. (2005) suggested that the Meridiani strata were formed by an impact-induced base surge,

FIG. 10.—Well-stratified sandstone in Burns formation showing low-angle truncation surfaces and cross-stratification, typical of traction sedimentation processes in turbulent flows. Note dispersion of spherules throughout deposit and absence of concentration along bedding planes and scour surfaces. This outcrop ("Tipuna") represents the uppermost 30–40 cm of the Burns middle unit. The image was obtained on sol 307.

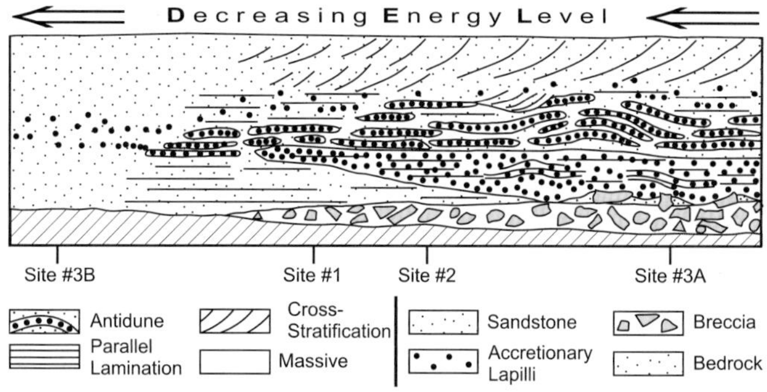

FIG. 11.—Depositional model for units associated with accretionary lapilli deposited by the Sudbury impact event. Decreasing energy of the flow does not necessarily reflect proximal to distal positioning as topography exerted a major control on the energy level that existed during deposition. The higher-energy deposits have a basal breccia zone with locally derived debris erosively overlain by upper-flow-regime, parallel-laminated and undulatory laminated (antidune) accretionary lapilli and sandstone. This fines upward, where the upper-flow-regime bed forms are replaced by lower-flow-regime trough cross-stratification. At this point, accretionary lapilli become rare. With decreasing energy levels, the basal breccia zone does not form, and accretionary lapilli become less numerous and confined to thinner beds. Upper-flow-regime parallel-lamination and some antidunes formed from hydraulic jumps are still present. At the lowest energy levels studied, the flow was probably not in contact with the bed, and a thinner, massive deposit developed.

whereby an iron meteorite struck an area with buried sulfate-rich evaporite deposits. Knauth et al. (2005) contended that the Burns spherules are iron condensation spherules—not diagenetic concretions. These alternative models have not been widely accepted in part because they are inconsistent with the regional geological context, which provides no evidence for contemporaneous sources of volcanism or age-appropriate impacts of sufficient size (see McLennan and Grotzinger 2008). In addition, there are other, more specific factors that are inconsistent with these models. Here, we specifically consider the possibility that the Burns formation spherules formed as impact-induced accretionary lapilli, or from condensation. In either case they would have been free particles interacting with the current that transported other sediments from the site of impact to the site of deposition, and thus the Sudbury impact deposit constitutes an important analog.

If the spherules are accretionary lapilli or condensation droplets, as might be predicted to exist in an impact deposit, they should show concentrations along bedding planes, as is the case for the Sudbury impact deposit. In contrast, in all outcrops observed to date by *Opportunity*, spherules are always highly dispersed relative to bedding (McLennan et al. 2005). Figure 12 shows three separate scenarios for the potential origin of spherules in the Burns formation. In the first case (Fig. 12, left), outsized particles (spherules) are entrained within a hyperconcentrated flow that causes them to be dispersed. The shear strength of such a fluid is high enough that viscous forces dominate over gravitational forces, and the larger, heavier particles do not preferentially settle out to form discrete beds (Nemec and Steel 1984). Note that ash plumes rising from the surface where a surge cloud loses contact with the ground, as sometimes happens (from the initial ash plume or from retrogressive collapse of loose material) (Branney and Kokelaar 2003) can deposit ash and accretionary lapilli directly from suspension with no traction (this likely occurred at site #3B). However, deposits produced directly from suspension result in massive bedding in which deposits are poorly sorted and lack well-defined stratification; this contrasts with the Burns formation, in which millimeter-scale lamination is pervasively developed (Grotzinger et al., 2005, 2006; Metz et al. 2009; Edgar et al. this volume).

In the second scenario (Fig. 12, center), the outsized particles (e.g., lapilli, condensation spherules) are mixed with other sediments in a more dilute flow that is fully turbulent. This turbulence results in velocity fluctuations during deposition that will produce beds of coarser- and finer-grained sediment. Movement of sediment in traction under turbulent flow can also lead to the development of bed forms, the migration of which may create scour surfaces, and multiple flow events may cause stacking of beds separated by hiatal surfaces. In this type of flow, gravitational forces may dominate, so that general behavior according to Stokes law is predicted. Particles will sort themselves given their respective settling velocities (w_s)

$$w_s = \frac{2gr^2(\rho_p - \rho_f)}{9\mu}$$

where r is the radius of the particle, ρ is the density of the particle or fluid, μ is the viscosity of the fluid, and g is the gravitational constant. Hydraulic segregation of particles with different settling velocities is why sorting occurs, and why outsized particles (for example, lapilli) are commonly concentrated along bedding planes in successions of flow-emplaced strata (see Figs. 2–5; Lowe et al. 2003). This is a distinctive textural attribute of the well-stratified Sudbury deposits, as well as other well-stratified volcanic surge deposits. In contrast, the Burns formation lacks evidence for hydrodynamic sorting of spherules despite being well stratified and showing clear evidence for scouring by currents, and also reworking above erosional surfaces. Equation 1 shows that the settling velocity goes as the square of the grain radius, so the outsized spherules—if sediment particles—are predicted to drop out very rapidly. In the Burns formation, they are on the order of 5–10 times larger than the grains that make up the stratified matrix in which the spherules are embedded (Grotzinger et al. 2005, McLennan et al. 2005), so they should show strong sorting, as in the Sudbury deposit. Furthermore, compositional data indicate that the density of the spherules is likely higher than the density of the grains in the stratified matrix; the former consist of a significant fraction of hematite, up to 50–60% (McLennan et al. 2005). Again, Eq. 1 shows that the density difference between the particles and the fluid in which they are makes a difference, so the hematitic spherules are again expected to have preferentially higher settling velocities, which will result in further enhanced separation of the spherule and sand fractions during deposition (Saxton et al. 2008). Finally, it is worth noting that the smaller g of Mars has no substantial effect on the *relative* settling velocities of the grains.

In the third scenario, spherules are dispersed across all strata and show no concentration, even at obvious erosional surfaces, such as the lower–middle unit boundary (Wellington contact) at Burns cliff (Grotzinger et al. 2005). Spherules are not concentrated along these or any other scour surfaces, despite clear evidence that the surfaces truncate spherule-bearing strata and represent erosion of spherule-bearing matrix sediments. This strongly indicates that the spherules must have developed in situ, after deposition of the sediments, favoring an origin as depicted in this third scenario. This interpretation is supported by the occurrence of "doublets" and even "triplets," where spherules touch and overgrow each other (see Figure 7g in McLennan et al. 2005).

In summary, the textural relationship between the outsized spherules in the Burns formation and the well-laminated sandstone matrix does not support their origin as sedimentary particles. This applies regardless of whether the spherules are interpreted to be accretionary lapilli or condensation droplets. In each case, these spherules should react in the flow as heavy particles that would preferentially travel in traction and be sedimented together at velocities that would segregate the majority of the sand to the saltation and suspension populations (Saxton et al. 2008). Concentrations of spherules should be particularly abundant along scour surfaces, where erosive lags would have developed. In addition, shielding from the current in scour troughs would have led to preferential deposition of material in traction transport, i.e., the much larger and denser spherules. Indeed, this is exactly what is observed in the spherule-bearing strata of the Sudbury impact deposits. These deposits highlight the improbability of the spherules in the strata observed at Meridiani Planum being related to an impact event. The Sudbury impact deposits also provide clues as to what a spherule deposit on Mars, which is related to an impact, may look like.

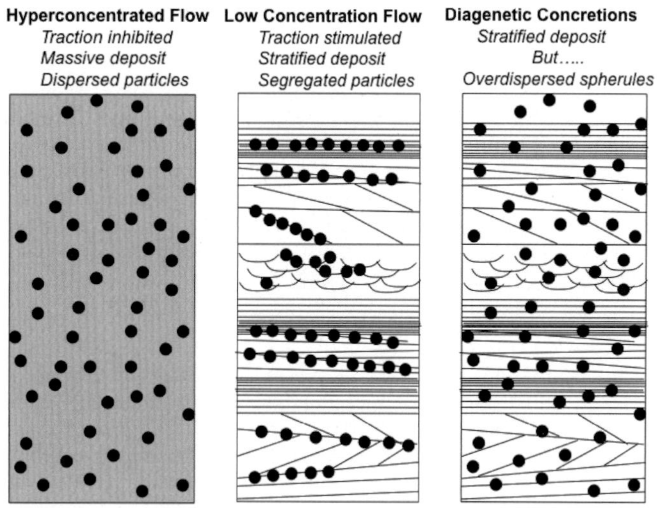

FIG. 12.—Models for relationships between spherule distribution and sedimentologic texture. **Left)** Spherules are dispersed within a massive matrix of poorly sorted, finer-grained sediments. This texture is commonly observed in debris flows, or even hyper-concentrated flows, where flows may be nonturbulent and suspended particle concentrations are very high (Jakob and Hungr 2005). Viscous fluid forces dominate the flow, and a combination of high fluid shear strength coupled with hindered settling of particles results in dispersion (poor sorting) of all grain sizes. The presence of excellent stratification excludes this option for the Burns formation, even though the spherules are highly dispersed. **Center)** Flows with low sediment concentrations are fully turbulent, and particles will become hydraulically sorted according to grain size. Heavier particles, such as spherules, will be systematically concentrated as erosive lags along scour surfaces, where smaller (less massive) particles are removed by turbulent eddies in the scour pits of bed forms. The Burns formation does not show evidence for concentration of the spherules as this scenario would predict. **Right)** spherules here are highly dispersed relative to primary bedding features, including scour-related truncation surfaces and other stratigraphic disconformities. The absence of concentrated spherule beds points to a different origin, perhaps as diagenetic concretions. In this case, the mineralizing pore fluids move through the previously deposited sediments, and precipitation occurs at discrete points. Concretions then form by growing radially in an outward direction from the point.

CONCLUSION

This study clearly outlines a means of distinguishing between impact-generated spherules and concretions. The differences produced by physical processes are of primary importance because the use of mineralogy or geochemistry can provide misleading results. The Sudbury impact accretionary lapilli present in the northern portion of their outcrop area have been massively replaced by ferroan dolomite. Thus, their mineralogy and geochemistry cause them to appear to be carbonate-rich concretions rather than lapilli generated by an impact on sialic crust. If impact-generated lapilli on Mars prove to be just as susceptible to replacement, their distribution within the strata will be of paramount importance in formulating the correct interpretation of their origin.

Differentiating between impact- and volcanic eruption–produced accretionary lapilli is difficult (Branney and Brown 2011). Here, the associated lithofacies become important. In an impact event, the initial burst of energy is directed down into the substrate, producing intense shock waves. In a volcanic eruption, the energy is released in an upward direction, with only minor resultant earthquake activity. Ground movement distal to the Sudbury impact event shattered the bedrock (Addison et al. 2010). This debris was entrained by the leading edge of the base surge and swept into topographic lows. Thus, at many outcrop occurrences of the Sudbury impact layer, the basal deposits are matrix-supported, cobble to boulder breccias composed of local material. They commonly do not have impact-derived sediment, i.e., devitrified glass or accretionary lapilli, mixed with the local debris. This type of basal unit, in conjunction with a generally broken and faulted underlying basement, is uniquely characteristic of impact-produced deposits. In addition, if the target rock included sedimentary strata, some far-traveled pieces of nonvolcanic debris may be present higher in the deposit. In contrast, large volcanic eruptions on Mars would be expected to have been relatively nonexplosive, due to the composition of the magma, and if a violent eruption did occur, it would have only rarely entrained nonvolcanic fragments.

The formation of ash pellets and accretionary lapilli requires moisture in the impact debris cloud. Some moisture can be released by the devolatilization of the target rocks. However, more favorable conditions for ash aggregates are produced by a wet sedimentary substrate, surface water, or atmospheric moisture incorporated into the debris cloud (Branney and Brown 2011). Thus, the presence or absence of ash pellets and accretionary lapilli in deposits of impact debris may provide information on surface moisture in the region at the time of impact.

ACKNOWLEDGMENTS

We would like to thank John Spray and Horton Newsom for critical reviews, in addition to comments by Editor Ralph Milliken, which led to improvements in the manuscript. Drafting of figures related to the Sudbury deposits was performed by Sam Spivik. Funding for this work was provided by Lakehead University and a Discovery Grant from the National Science and Engineering Research Council of Canada to P.W.F. The NASA Astrobiology Institute provided funding for J.P.G. and L.A.

REFERENCES

Addison WD, Brumpton GR, Davis DW, Fralick PW, Kissin SA. 2010. Debrisites from the Sudbury impact event in Ontario, north of Lake Superior, and a new age constraint: are they base-surge or tsunami deposits. *In* Gibson RL, Reimold WU (Editors). *Large Meteorite Impacts and Planetary Evolution IV.* Geological Society of America, Boulder, CO. Special Paper 465, p. 245–268.

Addison WD, Brumpton GR, Vallini DA, McNaughton NJ, Davis DW, Kissin SA, Fralick PW, Hammond AL. 2005. Discovery of distal ejecta from the 1850 Ma Sudbury impact event. *Geology* 33:193–196.

Arnott RWC, Hand BM. 1989. Bedforms, primary structures and grain fabrics in the presence of suspended sediment rain. *Journal of Sedimentary Petrology* 59:1062–1069.

Barlow NG. 2005. A review of Martian impact crater ejecta structures and their implications for target properties. *In* Kenkmann T, Horz F, Deutsch A, (Editors). *Large Meteorite Impacts III.* Geological Society of America, Boulder, CO. Special Paper 384, p. 433–442.

Branney MJ, Brown RJ. 2011. Impactoclastic density current emplacement of terrestrial meteorite-impact ejecta and the formation of dust pellets and accretionary lapilli: evidence from Stac Fada, Scotland. *The Journal of Geology* 119:275–292.

Branney MJ, Kokelaar P. 2003. Introduction and key concepts. *In* Branney MJ, Kokelaar P (Editors). *Pyroclastic Density Currents and the Sedimentation of Ignimbrites.* Geological Society of London. Memoir 27, p. 1–5.

Brown RJ, Branney MJ, Maher C, Davila-Harris P. 2010. Origin of accretionary lapilli within ground-hugging density currents: evidence from pyroclastic couplets on Tenerife. *Geological Society of America Bulletin* 122:305–320.

Burt DM, Knauth LP, Wohletz KH, Sheridan MF. 2008. Surge deposits misidentification at Spor Mountain, Utah and elsewhere: a cautionary message for Mars. *Journal of Volcanology and Geothermal Research* 177:755–759.

Cannon WF, Schulz KJ, Horton JW, Kring DA. 2010. The Sudbury impact layer in the Paleoproterozoic iron ranges of northern Michigan, USA. *Geological Society of America Bulletin* 122:50–75.

Carey SN. 1991. Transport and deposition of tephra by pyroclastic flows and surges. *In* Fisher RV and Smith GA (Editors). *Sedimentation in Volcanic Settings.* Society of Economic Paleontologists and Mineralogists, Tulsa, OK. Special Publication 45, p. 39–57.

Chough SK, Sohn YK. 1990. Depositional mechanics and sequences of base surges, Songaksan tuff ring, Cheju Island, Korea. *Sedimentology* 37:1115–1135.

Clark BC, 23 others. 2005. Chemistry and mineralogy of outcrops at Meridiani Planum. *Earth and Planetary Science Letters* 240:73–94.

Crowe BM, Fisher RV. 1973. Sedimentary structures in base-surge deposits with special reference to cross-bedding, Ubehebe Craters, Death Valley, California. *Geological Society of America Bulletin* 84:663–682.

Druitt TH. 1992. Emplacement of the 18 May 1980 lateral blast deposit ENE of Mount St Helens, Washington. *Bulletin of Volcanology* 54:554–572.

Earth Impact Database. 2004. Earth Impact Database. www.unb.ca/passe/ImpactDatabase. Accessed October 12, 2004.

Fralick PW, Burton J. 2008. Geochemistry of the Paleoproterozoic Gunflint Formation carbonate: implications for early hydrosphere-atmosphere evolution. *Geochimica et Cosmochimica Acta* 72:A280.

Fan C, Xie H, Schulze-Makuch D, Ackley S. 2010. A formation mechanism for hematite-rich spherules on Mars: *Planetary and Space Science* 58:401–410.

Gencalioglu-Kuscu G, Atilla C, Cas RAF, Kuscu L. 2007. Base surge deposits, eruption history, and depositional processes of a wet phreatomagmatic volcano in Central Anatolia (Cora Maar). *Journal of Volcanology and Geothermal Research* 159:198–209.

Golden DC, Ming DW, Morris RV, Graff TG. 2008. Hydrothermal synthesis of hematite and jarosite: implications for diagenesis and hematite spherule formation in sulfate outcrops at Meridiani Planum, Mars. *American Mineralogist* 93:1201–1214.

Grant JA, Rossman PI, Grotzinger JP, Milliken RE, Tornabene LL, McEwen AS, Weitz CM, Squyres SW, Glotch TD, Thomson BJ. 2008. HiRISE imaging of impact megabreccia and sub-meter aqueous strata in Holden Crater, Mars. *Geology* 36:195–198.

Graup G. 1981. Terrestrial chondrules, glass spherules and accretionary lapilli from the suevite, Ries Crater, Germany. *Earth and Planetary Science Letters* 55:407–418.

Grotzinger JP, Arvidson RE, Bell JF, Calvin WM, Clark BC, Fike DA, Golombek MP, Greeley R, Haldemann A, Herkenhoff KE, Jollif BL, Knoll AH, Malin MC, McLennan SM, Parker T, Soderblom L, Sohl-Dickstein JN, Squyres SW, Tosca NJ, Watters WA. 2005. Stratigraphy and sedimentology of a dry to wet eolian depositional system, Burns Formation, Meridiani, Mars. *Earth and Planetary Science Letters* 240:11–72.

Grotzinger JP, Bell J, Herkenhoff K, Johnson J, Knoll A, McCartney E, McLennan S, Metz J, Moore J, Squyres S, Sullivan S, Aharonson O, Jollif B, Golombek M, Lewis K, Parker T, and Soderblom J. 2006. Sedimentary textures formed by aqueous process, Erebus crater, Meridiani Planum, Mars. *Geology* 34:1085–1088.

Jakob M, Hungr O. 2005. *Debris-Flow Hazards and Related Phenomena.* Springer, Berlin. 781 p.

Jirsa MA. 2008. Scientists unearth ancient impact's secrets. *Astronomy* 36(12):32–37.

Kenkmann T, Schonian F. 2006. Ries and Chicxulub: impact craters on Earth provide insights for Martian ejecta blankets. *Meteoritics and Planetary Science* 41:1587–1603.

Knauth LP, Burt DM, Wohletz KH. 2005. Impact origin of sediments at the *Opportunity* landing site on Mars. *Nature* 438:1123–1128.

Koeberl C, Brandstatter F, Glass BP, Hecht L, Mader D, Reimold WU. 2007. Uppermost impact fallback layer in the Bosumtwi crater (Ghana): mineralogy, geochemistry, and comparison with Ivory Coast tektites. *Meteoritics and Planetary Science* 42:709–729.

Krogh TE, Davis DW, Corfu F. 1984. Precise U-b and zircon and baddeleyite ages for the Sudbury area. *In* Pye EG, et al. (Editors). *The Geology and Ore Deposits of the Sudbury Structure.* Ontario Geological Survey, Toronto, Canada. Special Volume 1, p. 431–446.

Leeder MR. 1982. *Sedimentology, Process and Product.* Unwin-Hyman, London. 344 p.

Lowe DR, Byerly GR, Kyte FT, Shukolyukov A, Krull A, Asaro F. 2003. Characteristics, origin and interpretation of Archean impact produced spherule beds, 3.47–3.22 Ga, in the Barberton Greenstone Belt, South Africa: keys to the role of large impacts on the evolution of early Earth. *Astrobiology* 3:7–48.

Malin MC, Edgett KS. 2000. Sedimentary rocks of early Mars. *Science* 290:1927–1937.

Malin MC, Edgett KS. 2001. Mars Global Surveyor Mars Orbiter Camera: interplanetary cruise through primary mission. *Journal of Geophysical Research* 106:23429–23570, DOI 10.1029/2000JE001455

Masaitis V. 2003. Obscure-bedded ejecta facies from the Popigai impact structure, Siberia; lithological features and mode of origin. *In* Koeberl C,

Martinez-Ruiz FC (Editors). *Impact Markers in the Stratigraphic Record.* Springer, Berlin, Impact Studies 26, p. 137–162.

McCollom TM, Hynek BM. 2005. A volcanic environment for bedrock diagenesis at Meridiani Planum on Mars. *Nature* 438:1129–1131.

McLennan SM, Grotzinger JP. 2008. The sedimentary rock cycle of Mars. *In* Bell JF (Editor). *The Martian Surface; Composition, Mineralogy and Physical Properties.* Cambridge University Press, New York. p. 541–577.

McLennan SM, 30 others. 2005. Provenance and diagenesis of the Burns Formation. Meridiani Planum, Mars. *Earth and Planetary Science Letters* 240:95–121.

Melosh HG. 1989. *Impact Cratering: A Geologic Process.* Oxford University Press, New York. 245 p.

Metz JM, Grotzinger JP, Rubin DM, Lewis KW, Squyres SW, Bell JF. 2009. Sulfate-rich eolian and wet interdune deposits, Erebus Crater, Meridiani Planum, Mars. *Journal of Sedimentary Research* 79:247–264.

Mouginis-Mark PJ, Garbeil H. 2007. Crater geometry and ejecta thickness of the Martian impact crater Tooting. *Meteoritics and Planetary Science* 42:1615–1625.

Nemec W, Steel RJ. 1984. Alluvial and coastal conglomerates: their significant features and some comments on gravelly mass-flow deposits. *In* Koster EH, Steel RH (Editors). *Sedimentology of Gravels and Conglomerates.* Canadian Society of Petroleum Geology, Calgary, Canada. Memoir 10, p. 1–31.

Newsom HE, Grauo G, Iseri D, Geissman JW, Keil K. 1990. The formation of the Reis Crater, West Germany: evidence of atmospheric interactions during a large cratering event. *In* Sharpton VL, Ward PD (Editors), *Global Catastrophies in Earth History: An Interdisciplinary Conference on Impacts, Volcanism and Mass Mortality.* Geological Society of America, Boulder, CO. Special Paper 247, p. 195–206.

Orton GJ. 1996. Volcanic environments. *In* Reading HG (Editor). *Sedimentary Environments: Processes, Facies and Stratigraphy.* Blackwell Science, London. p. 485–567.

Osinski G. 2006. Effects of volatiles and target lithology on the generation and emplacement of impact crater fill and ejecta deposits on Mars. *Meteoritics and Planetary Science* 41:1571–1586.

Pope KO, Ocampo AC, Fischer AG, Alvarez W, Fouke BW, Webster CL, Vega FJ, Smit J, Fritsche AE, Claeys Ph. 1999. Chicxulub impact ejecta from Albion Island, Belize. *Earth and Planetary Science Letters* 170:351–364.

Pope KO, Ocampo AC, Fischer AG, Vega FJ, Ames DE, King DT, Fouke BW, Wachtman RJ, Kletetschka G. 2005. Chicxulub impact ejecta deposits in southern Quintana Roo, Mexico, and central Belize. *In* Kenkmann T, Horz F, Deutsch A (Editors). *Large Meteorite Impacts III.* Geological Society of America, Boulder, CO. Special Paper 384, p. 171–190.

Pufahl PK, Hiatt EE, Stanley CR, Morrow JR, Nelson GJ, Edwards CT. 2007. Physical and chemical evidence for the 1850 Ma Sudbury impact event in the Baraga Group, Michigan. *Geology* 35:827–830.

Saxton J, Fralick PW, Panu U, Wallace K. 2008. Density segregation of minerals during high-velocity transport over a rough bed: implications for the formation of placers. *Economic Geology* 103:1657–1664.

Schmincke H.-U, Fisher RV, Waters AC. 1973. Antidune and chute and pool structures in base surge deposits of the Laacher see area, Germany. *Sedimentology* 20:553–574.

Smith GA, Katzman D. 1991. Discrimination of eolian and pyroclastic-surge processes in the generation of cross-bedded tuffs, Jemez Mountains volcanic field, New Mexico. *Geology* 19:465–468.

Sparks RSJ, Moore JC, Rice CJ. 1986. The initial giant umbrella cloud of the May 18, 1980, explosive eruption of Mount St Helens. *Journal of Volcanological and Geothermal Research* 28:257–274.

Spray JG, Butler HR, Thompson LM. 2004. Tectonic influences on the morphometry of the Sudbury impact structure: implications for terrestrial cratering and modeling. *Meteoritics and Planetary Science* 39:287–301.

Squyres SW, Aharonson O, Arvidson RE, Bell JF, Cristensen PR, Clark BC, Crisp JA, Farrand W, Glotch T, Golombek MP, Grant J, Grotzinger J, Herkenhoff KE, Johnson JR, Jollif BL, Knoll AH, McLennan SM, McSween HY, Moore JM, Rice JW, Tosca N. 2006a. Planetary science: bedrock formation at Meridiani Planum. *Nature* 443. DOI 10.1038/nature05212

Squyres SW, Arvidson RE, Bell JF, et al. 2004a. The *Opportunity* rover's Athena science investigation at Meridiani Planum, Mars. *Science* 306:1698–1703.

Squyres SW, Grotzinger JP, Arvidson RE, Bell JF, Calvin W, Christensen PR, Clark BC, Crisp JA, Farrand WH, Herkenhoff KE, Johnson JR, Klingelhofer G, Knoll AH, McLennan SM, McSween HY, Morris RV, Rice JW, Rieder R, Soderblom LA. 2004b. In situ evidence for an ancient aqueous environment at Meridiani Planum, Mars. *Science* 306:1709–1714.

Squyres SW, Knoll AH. 2005. Sedimentary rocks at Meridiani Planum: origin, diagenesis and implications for life on Mars. *Earth and Planetary Science Letters* 240:1–10.

Squyres SW, Knoll AH, Arvidson RE, Clark BC, Grotzinger JP, Jollif BL, McLennan SM, Tosca N, Bell JF, Calvin W, Farrand WH, Glotch TD, Golombek MP, Herkenhoff KE, Johnson JR, Klingelhofer G, McSween HY, Yen AS. 2006b. Two years at Meridiani Planum: results from the *Opportunity* rover. *Science* 313:1403–1407.

Warme JE, Morgan M, Kuehner H.-C. 2002. Impact generated carbonate accretionary lapilli in the Late Devonian Alamo Breccia. *In* Koeberl C, MacLeod KG (Editors). *Catastrophic Events and Mass Extinctions; Impacts and Beyond.* Geological Society of America, Boulder, CO. Special Paper 356, p. 489–504.

Wilson CJN, Walker GPL. 1982. Ignimbrite depositional facies: the anatomy of a pyroclastic flow. *Journal of the Geological Society of London* 139:581–592.

Wohletz KH, McGetchin TR, Sandford MT, Jones EM. 1984. Hydrodynamic aspects of caldera-forming eruptions: numerical models. *Journal of Geophysical Research* 89:8265–8269.

EARLY DIAGENESIS BY MODERN ACID BRINES IN WESTERN AUSTRALIA AND IMPLICATIONS FOR THE HISTORY OF SEDIMENTARY MODIFICATION ON MARS

BRENDA B. BOWEN

Department of Earth and Atmospheric Sciences, Purdue University, West Lafayette, Indiana 47907 USA
e-mail: bbowen@purdue.edu,

KATHLEEN C. BENISON

Department of Earth and Atmospheric Sciences, Central Michigan University, Mt. Pleasant, Michigan 48859 USA

AND

STACY STORY

Department of Earth and Atmospheric Sciences, Purdue University, West Lafayette, Indiana 47907 USA

ABSTRACT: Mineralogical and geochemical data collected from multiple sites on Mars suggest that acid saline surface waters and groundwater existed there in the past. The geologic context and sedimentology suggest that these acid saline waters were associated with groundwater-fed ephemeral lakes. Ephemeral acid saline lakes in southern Western Australia (WA) are some of the few known natural systems that have the same combination of extreme acid brine chemistry and lacustrine depositional setting as is observed on Mars. Thus, the WA acid saline environments provide a modern analog for understanding past depositional and diagenetic processes that may have occurred on Mars. Here, we examine surface sediments and sedimentary rocks that have been in contact with acid (pH down to ~1.5) and saline brines (total dissolved solids up to ~32%) in southern Western Australia. Through sedimentological, mineralogical, geochemical, and petrographic analyses, we identify the impacts of early diagenesis in and adjacent to eight acid saline lakes and evaluate the processes that have been important in creating these deposits. The combination of extreme chemistry, spatial variability, arid climate, and reworking by winds and floods contributes to make spatially complex depositional products that are a combination of siliciclastics and chemical sediments. Important syndepositional and very early diagenetic processes in these settings include the chemical precipitation of minerals from shallow groundwaters to form displacive crystals and cements, dissolution/partial dissolution of chemical sediments, replacement/partial replacement of some minerals, cracking due to repeated wetting and drying, and the formation of iron-oxide concretions. Minerals observed in these sediments include a variety of chlorides, sulfates, iron oxides, and phyllosilicates, many of which have textures and mineral associations that suggest authigenic formation. These observations are supported by the chemistry of the modern acid brines, which appear to be supersaturated with respect to these minerals. The range of early diagenetic products, compositions, and textures that are apparent in the WA acid saline lake sediments may provide insights into the processes that influenced the sediments on Mars and the timing of sedimentary formation processes on Mars.

KEY WORDS: Acid, brine, diagenesis, sedimentology, Mars

INTRODUCTION

Geologic Setting, Surface Processes, and Extreme Water Chemistry

Modern shallow saline lakes are abundant throughout the Archean Yilgarn Craton in southern Western Australia (WA). The lake waters and shallow groundwaters range in pH from 1.5 to ~9 and from nearly fresh to hypersaline (Benison et al. 2007; Fig. 1). They are Na-Cl to Na-Mg-Cl-SO_4 brines with variable yet locally high Ca, Br, Al, Fe, and Si (Bowen and Benison 2009; Fig. 2). Many of the lakes are seasonally dry, and even when they do have standing surface water, they are very shallow, with water depths typically less than 1 m deep. Surface waters cover areas as small as 1 km^2, although some lakes occupy basins up to ~700 km^2. Many of the lakes occupy closed depressions within buried fluvial inset valleys that were carved into the Archean basement (Fig. 3). Multiple mechanisms have been proposed for the initial formation of the valleys, including erosion from Permian glaciers (McArthur et al. 1989) and/or Jurassic to Middle Eocene fluvial incision (Woodward 1897, Gregory 1914, Clarke et al. 1996, de Broekert and Sandiford 2005). The Yilgarn Craton is characterized by subtle topography and highly weathered regolith, and it is dominantly an erosional vs. a depositional system (Anand and Paine 2002). Some of the lakes occur above paleovalley-hosted basins that locally contain marine, fluvial, and lacustrine sediment accumulations up to 100 m thick (McArthur et al. 1991, Clarke 1993), although other lakes occur directly on top of Archean basement outcrops and contain nearly no sediment (Fig. 3). Ephemeral lacustrine deposition may have existed in some of the basins since the mid-Miocene or earlier, and since that time, this region has experienced relative tectonic inactivity and increasing aridity (De Deckker 1983).

All of the WA lakes are ephemeral, saline, and fed by a combination of rainwater and diffuse influx of regional acid saline groundwaters. The range in pH in the lakes is due to varying contributions of acidity that originates in groundwaters from subsurface oxidation of pyrite and precipitation and dissolution of iron and aluminum hydroxides (e.g., Mann 1983, McArthur et al. 1991, Peiffer et al. 2009). Precipitation/evaporation ratios on the Yilgarn Craton are very low (P/E = 0.026 to 0.22), which is reflected

Sedimentary Geology of Mars
SEPM Special Publication No. 102, Copyright © 2012
SEPM (Society for Sedimentary Geology), Print ISBN 978-1-56576-312-8, CD/DVD ISBN 978-1-56576-313-5, p. 229–252.

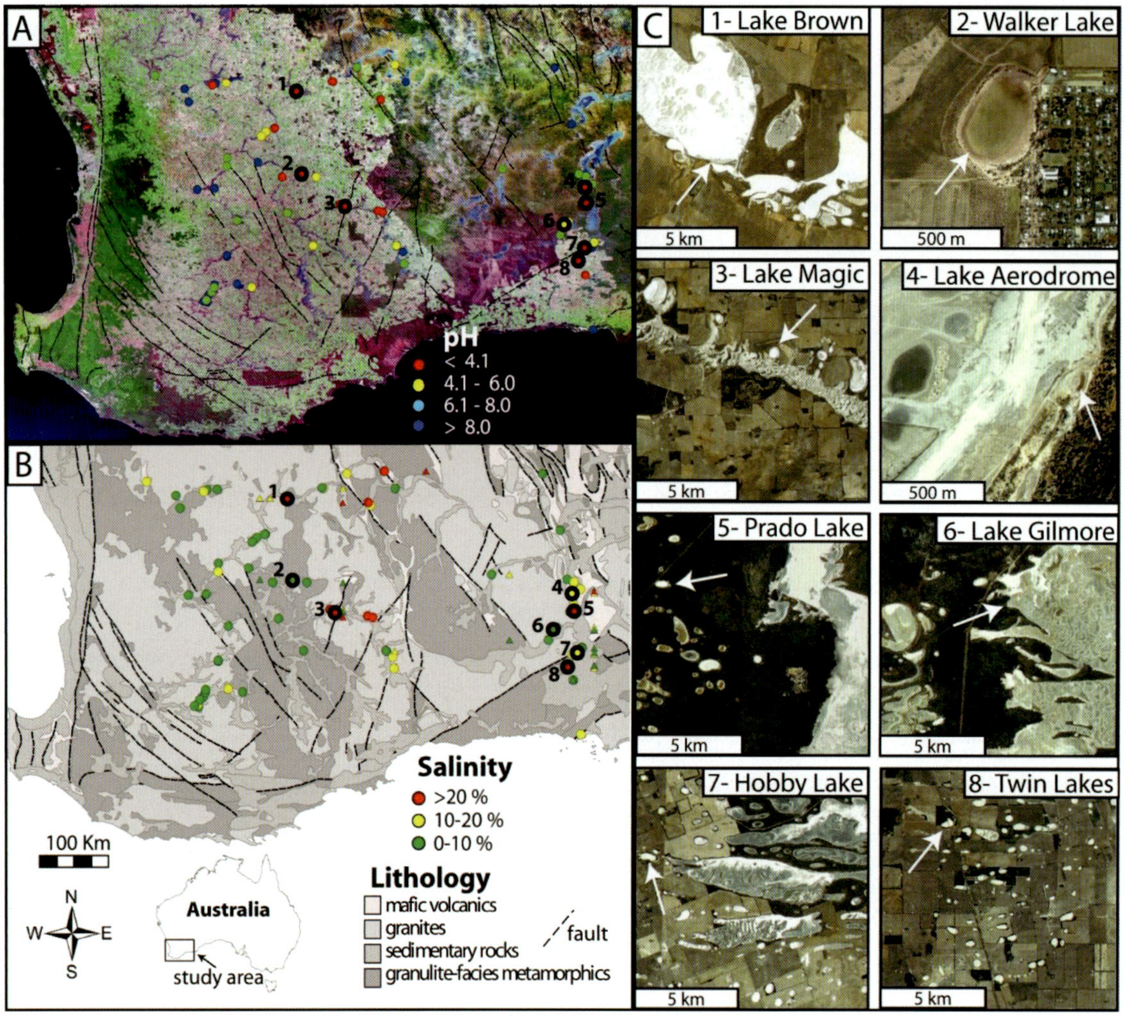

FIG. 1.—Southern Western Australia acid saline lake field area. **A)** False-color Landsat satellite image of field area with lake-water pH (minimum observed) and faults overlain. Numbers 1–8 show locations of lakes included in this study (see Table 1 for details of study sites). **B)** Simplified geologic map of same area showing salinity of lakes (circles) and groundwater (triangles) in percent total dissolved solid (maximum observed). **C)** Air photos of eight acid saline lakes described in this study (image credit: Google Earth™ mapping services).

in the high salinity of the lakes, although stable isotope data demonstrate that dissolution of evaporites with the influx of fresh meteoric water generates the highest salinity lake waters (Bowen and Benison 2009). The low-relief topography in the region leads to extremely low hydrologic gradients and the formation of the highly evolved acid groundwaters (Gray 2001).

The surface processes and modern depositional facies that exist in and around these lakes typically include lake, mud flat/sand flat, ephemeral channel, and dune facies (Benison et al. 2007; Fig. 4). In addition, many of the lakes have lithified and partially lithified sedimentary rocks exposed around the fringes and just below the surface. The detrital surface sediments appear to be locally derived, and eolian transport is one the dominant processes for moving sediments intra- and interbasinally. All of these sediments and rocks are exposed to extreme and fluctuating acid saline waters. The conditions that these environments experience change drastically through the year as the lakes go through multiple flooding, evapoconcentration, and desiccation cycles.

Importance of Western Australia Acid Brine Environments as a Mars Analog

Recent data from surface sediments and sedimentary rocks on Mars provide intriguing evidence of extraterrestrial weathering, depositional, and diagenetic processes. Orbital spectral data and ground-based rover data suggest that while water was certainly involved in the formation of many of these deposits, at least some of the fluids had an extreme acid saline chemistry (Grotzinger et al. 2005, McLennan et al. 2005, Squyres et al. 2009). Similar to Western Australia, the generation of acidity on Mars can be explained by surface oxidation of Fe^{2+}-rich groundwater (Hurowitz et al. 2010). The salinity may be a result of long-term weathering of basement rocks and subsequent weathering of the weathering products (e.g., Altheide et al. 2010). While the presence of water has long been considered a prerequisite for the existence of extraterrestrial life, the extreme chemistry of these waters has called into question the likelihood of life existing and/or being preserved in

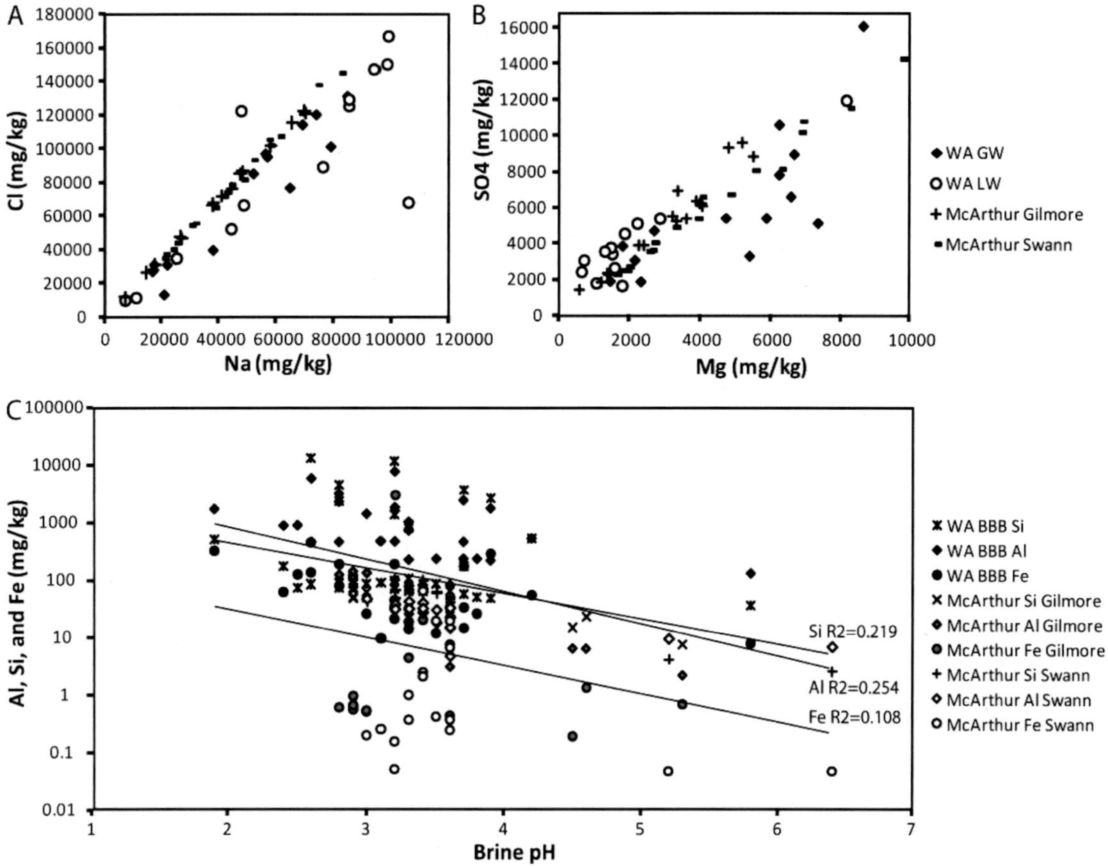

FIG. 2.—Water chemistry measured for the major ions Na, Cl, Mg, and SO_4 and Si, Al, and Fe in lake and groundwaters in southern Western Australia in association with sediments investigated in this study (data plotted in Table 2; after Bowen and Benison 2009). Plots also include acid brine chemistry from McArthur et al. (1989; shallow groundwater/pore fluids) from Lake Gilmore and Lake Swann. **A)** Cl vs. Na. **B)** SO_4 vs. Mg. **C)** pH vs. Si, Al, and Fe. Note log scale on *y*-axis. Best-fit exponential trends are shown for total Si, Al, and Fe.

environments exposed to these hostile conditions (Sumner 2004, Tosca et al. 2008).

Interpretation of sedimentary data from Mars requires knowledge of sedimentary processes on Earth. Terrestrial analogs provide ground truth to constrain interpretations of the type of processes that have been important on Mars. This type of comparative sedimentology (using modern to understand ancient, outcrop to understand subsurface, experimental to understand theory, or terrestrial to understand planetary) was achieved for many depositional environments decades ago, facilitating a uniformitarian approach for interpreting sedimentary records. However, acid saline lake systems are rare environments that have only recently been recognized in both the modern and ancient rock record (McArthur et al. 1991, Benison and Goldstein 2002, Risacher et al. 2002, Benison et al. 2007). Although many of the physical processes are analogous to those in other saline lake settings (e.g., influence of rock type, climate, and degree of evaporation), the acid brines and associated sedimentary deposits evolve via geochemical pathways distinctly different from other saline systems (e.g., Long et al. 2009). While lakes with varying geochemistry within the same geographic location are not uncommon (e.g., Last 1994), the range in salinity and pH in WA and the resulting variations in elemental concentrations are quite unusual and affect the resulting sedimentary records. The early diagenetic processes, that is, the mineralogical, textural, and geochemical modifications that occur within pore space and around detrital grains as they interact with the acid saline surface and pore fluids, result in an assemblage of Fe-, Al-, and Si-rich minerals not typically expected in a saline system.

As some of the only known naturally acidic and saline modern depositional and diagenetic systems on Earth, the abundant ephemeral lakes in WA provide important analogs for past acid saline environments on Mars (Benison and LaClair 2003, Benison and Bowen 2006, Bowen et al. 2008, Baldridge et al. 2009, West et al. 2009). In addition to the acid saline chemistry, the depositional environment in WA shares many characteristics with those that likely existed on Mars in the past. In WA, groundwater-fed interdune lakes desiccate and erode to supply the surrounding dunes with reworked evaporite grains; this scenario is also envisioned for the formation of many sedimentary deposits on Mars (Grotzinger et al. 2005, Squyres and Knoll 2005, Andrews-Hanna et al. 2007, Fishbaugh et al. 2007). Both systems contain a unique assemblage of chemical sedimentary minerals, including Ca- and Mg-sulfates (Baldridge et al. 2009), chlorides (Clark and Van Hart 1981, Bell et al. 2000, Osterloo et al. 2008), Fe-minerals including smectites, hematite, and jarosite, and Al/Si-precipitates such as alunite and kaolinite (Story et al. 2010).

Of course, there is no perfect terrestrial analog for the unique conditions on Mars, and it is important to understand the differences between the acid saline systems in WA and on Mars. One of the most significant differences between the WA analog and Mars is the nature of the basement material. While Mars is dominated by mafic igneous rocks, the WA acid lakes are hosted by a variety of rock types, most

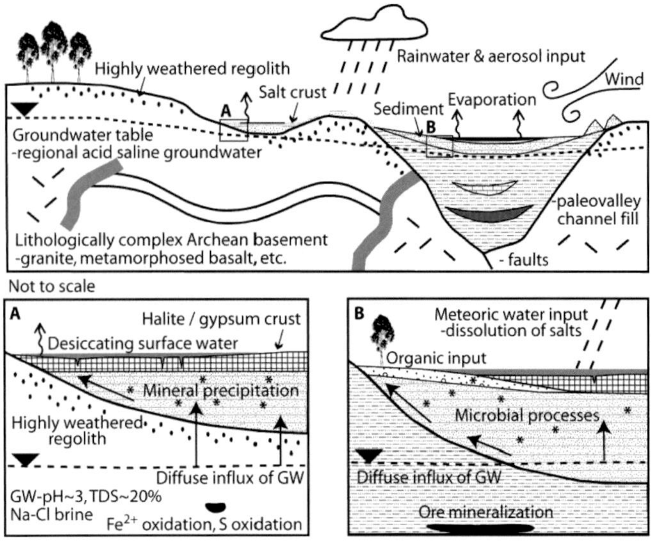

FIG. 3.—Schematic cartoon model of the geologic setting of ephemeral acid saline lakes in southern Western Australia. **A)** Basement-hosted shallow acid saline lake. **B)** Paleovalley sediment-hosted shallow acid saline lake. The boxes at each lake show approximate location of sediments characterized in association with acid saline lake waters and/or groundwaters in this study. All postdepositional processes that influence these sediments (e.g., in situ weathering reactions, mineral precipitation, mineral dissolution, etc.) are considered aspects of diagenesis. Zoomed-in views below depict some of the processes that are important in modifying the sediments, but they are not meant to be specific to a particular lake. Note that this schematic diagram is a simplified depiction of a complex system. For example, microbial processes likely occur throughout all waters, sediments, and weathered bedrock here. Temporal variations in the surface processes and products occur at various timescales and are described in detail in Benison et al. (2007) and Bowen and Benison (2009).

commonly felsic, but some rocks of intermediate and mafic composition as well (Table 1). This difference in starting compositions influences all later weathering products, precipitates, and fluid chemistry, and no doubt leads to significant differences in the specific geochemical reactions that occur. For example, in WA, Al-phyllosilicates tend to be the most common phyllosilicate mineral (Story et al. 2010) due to the dominance of highly weathered felsic bedrock. In contrast, some work has suggested that Mg-phyllosilicates are the most common phyllosilicate mineral on Mars due to the dominance of less weathered mafic volcanic bedrock (Baldridge et al. 2009). Other significant differences include the overwhelming long-term effects of tectonics, a comparatively warm climate, and the influence of biology in WA. Despite these differences, sediments exposed to extreme acid saline fluids in terrestrial ephemeral interdune lacustrine environments exhibit some characteristic early diagenetic features that may help to elucidate similar processes that have been important on Mars and highlight the potential for preservation of biological materials in these types of settings.

By studying the surface sediments in contact with acid brines in WA, we can investigate the complex relationships among detrital lithology, authigenic mineralogy, and water chemistry. Here, we can examine the spatial relationship between typical evaporite minerals such as halite and gypsum, and their relationship to the minerals specific to acid environments such as iron oxides and iron sulfates. It can be very difficult to definitively distinguish between detrital and authigenic components in highly weathered settings with complex mineralogy such as this (e.g., Story et al. 2010), and research in this setting can help to define characteristics that may help to define mineral provenance in sedimentary systems on Mars. In WA, some of the greatest clues for deciphering the relationships between mineral phases are the mineral associations or assemblages and the spatial relationships between minerals that can be observed petrographically.

Our previous work in WA included detailed studies of the variety of extreme geochemical conditions of ~60 lakes and their associated shallow groundwaters, depositional processes and products, and mineralogy, all of which demonstrate temporal and spatial complexity and diversity (Benison et al. 2007, Bowen and Benison 2009, Story et al. 2010). Previous work has also revealed the presence of diverse microbiological communities within these lakes (Mormile et al. 2009), and the potential for biological material to be entrapped and potentially preserved within the rapidly precipitating minerals in these systems (Benison et al. 2008). However, questions still remain as to whether these are chemolithotrophic organisms and the role that they play in modifying the chemistry of the brines and the sediments. Past work has also established these environments as important Mars analogs (Benison and LaClair 2003, Benison and Bowen 2006, Bowen et al. 2008), yet since these publications, many new discoveries have been made that further signify the potential importance of acid saline lacustrine systems in the geologic past on Mars.

Our objective is to identify the early diagenetic processes that are important in creating and modifying these deposits and to consider the way in which they may be specific to acid saline lacustrine settings. Understanding the formation of these sediments and rocks and variations in WA will help us to interpret deposits that have formed in potentially similar depositional systems and from similar fluids on Mars.

MATERIALS AND METHODS

Field Observations and Sample Collection

We observed a wide range in aqueous geochemistry, sedimentology, and surface processes at ~60 saline lakes and pans in WA during five field excursions from 2001 to 2009 that covered winter, summer, flooding, evapoconcentration, and desiccation conditions (Benison et al. 2007, Bowen and Benison 2009). In this study, we focus on new petrographic, mineralogic, and geochemical characterization of shallow sediments and sedimentary rocks from eight acid saline lake systems, which include some of the most acidic and well-characterized sites from our previous work (Table 1; Benison et al. 2007, Bowen and Benison 2009, Story et al. 2010). The three western sites (Lake Brown, Walker Lake, Lake Magic) occupy the paleo–Yilgarn River drainage catchment, while the five eastern sites (Lake Aerodrome, Prado Lake, Lake Gilmore, Hobby Lake, and the Twin Lakes) occupy the paleo–Lefroy River drainage catchment and depressions in the surrounding Archean bedrock (Fig. 1). Representative sediments and sedimentary rocks used in this study were collected from surface environments (e.g., lakes, sand flats/mud flats, ephemeral channels, dunes). In addition, shallow subsurface samples were collected by manually trenching pits down to 2 m with shovels and taking shallow manual cores down to 50 cm with polyvinyl chloride (PVC) pipes (Fig. 5). At select sites (Aerodrome, Brown, Prado, Twin Lake), samples were collected in detailed transects across lakes (Fig. 6). At sediment and sedimentary rock sample sites, we also measured lake water and groundwater geochemistry in the field (temperature, pH, salinity) and collected water samples for detailed isotopic, elemental, and compound analyses in the laboratory (Bowen and Benison 2009; Table 2; Fig. 2). Therefore, sediment and sedimentary rocks samples

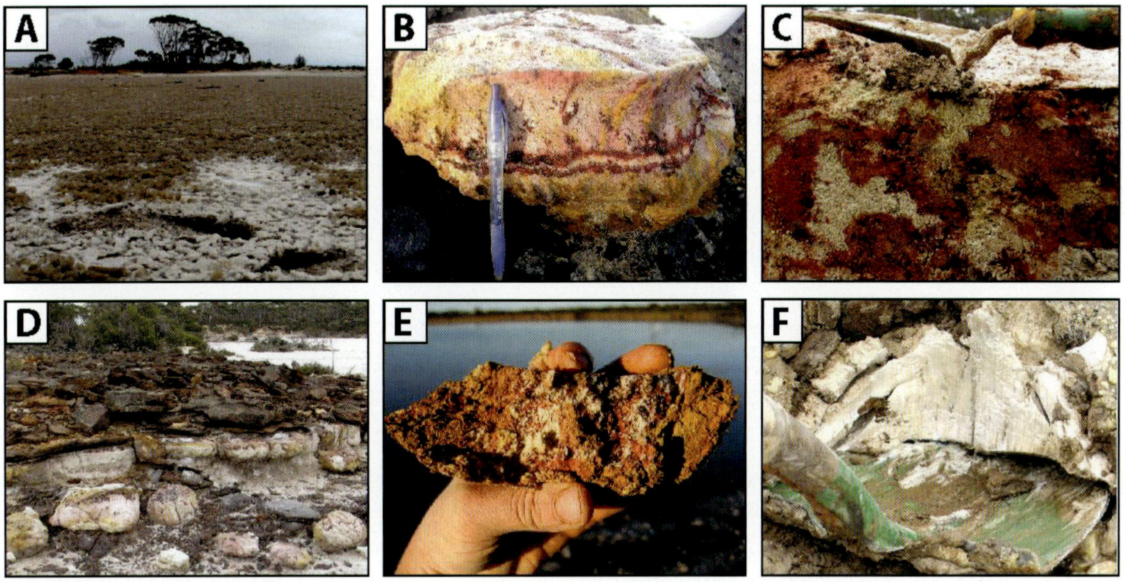

FIG. 4.—Field photos showing examples of acid saline lake settings and diagenetically modified sediments in the field. **A)** Exposed bottom-growth gypsum (reddish-brown) and halite (white) along desiccated lake bottom of Lake Aerodome. **B)** Iron-oxide banding in lithified sandstone along the shores of Prado Lake. **C)** Mottled sand near Prado Lake. **D)** Ferricrete bed along the shores of Twin Lake West. **E)** Red, yellow, and white mottling in mid-lake mud from Twin Lake West. **F)** White halloysite bed (confirmed via XRD and spectroscopy) in Twin Lake West mud flat.

can be paired with local water geochemistry (e.g., Fig. 6). Sediments and sedimentary rocks were sealed in air-tight sample bags and plastic containers and shipped back to the USA for subsequent analyses.

Mineralogy and Geochemistry

Representative bulk sediment and sedimentary rock samples were analyzed by X-ray diffraction and visible to near-infrared reflectance spectroscopy (350–2500 nm) to identify mineralogy. Additionally, the clay-size (<2 μm) fraction of select samples was analyzed for mineralogy via X-ray diffraction (XRD) (Story et al. 2010). The amount of Cl^- in the sediments was quantified with Mohr titrations (Story et al. 2010). Weight percent of major oxides, trace elements, rare earth elements, and some important compounds (e.g., SO_4) in 26 representative sediment samples were measured by a commercial laboratory (Activation Laboratories, Ltd.). For these analyses, major oxides and trace elements were measured via lithium metaborate/tetraborate fusion ICP (inductively coupled plasma mass spectrometry) and TD-ICP (total digestion inductively coupled plasma mass spectrometry) respectively, and SO_4 testing was performed via combustion/infrared analysis. The geochemistry of lake waters and shallow groundwaters spatially associated with the sediment samples was also characterized in detail (Bowen and Benison 2009). Optical petrography was also used to identify some minerals in thin section.

The geochemistry of waters associated with the sediments being analyzed was evaluated using Geochemists Workbench (Bethke and Yeakel 2009). These water data are a subset of previously published data (Bowen and Benison 2009). Aqueous species and mineral saturation indices were calculated using the thermo.dat thermodynamic database based on measured values of pH, total dissolved solids, and measured concentrations of Na^+, Mg^{2+}, Cl^-, SO_4^{2-}, K^+, Ca^{2+}, Br^-, Si, Al^{3+}, Fe^{3+}, HS^-, Sr^{2+}, Mn^{2+}, and Cu^{2+} from seven of the lakes (Table 2). Initial conditions were set to 1 kg of water, T = 25°C, surface pressure, assumed equilibrium between atmospheric oxygen and the oxygenated surface water ($fO_2 = 0.21$; Hem 1985), and measured pH and solute concentrations. Redox couples for Fe, Cu, and S were decoupled in SpecE8 (GWB), and the modeled reactions were charge balanced using Cl^- concentrations. Minerals saturation indices were calculated for each individual water sample and then compared to list only those minerals that were common to all the water samples at a specific lake (Table 3).

Petrography

Representative samples from 16 shallow PVC pipe cores and surface sedimentary rocks were prepared into thin sections for petrographic analyses (n = 84; Fig. 5). These samples included unconsolidated and semilithified surface sediments as well as lithified sedimentary rocks. The thin sections were prepared with techniques appropriate for samples sensitive to water and heat, so that the saline minerals and textures were preserved. The samples were vacuum-impregnated with blue epoxy (blue = porosity in micrographs). The thin sections were examined petrographically under both transmitted and reflected (and a combination of both) light at up to 400× magnification.

RESULTS

Geochemistry and Mineralogy

Geochemistry: The sediments that have been in contact with geochemically complex acid saline fluids in southern WA exhibit a wide range of compositions and textures that suggest high degrees of early syndepositional/diagenetic modification. Because of the extreme variability in water chemistry and sediment composition, it is difficult to describe a "typical" acid saline sediment (Fig. 6). The sediments contain a complex mixture of both detrital and authigenic components. The sediments contain up to ~20% Cl, but this may be misleading due to chloride salts that may have formed from surface or groundwater

TABLE 1.—Western Australia acid saline lake study sites focused on in this study. Site number correlates with locations shown on map in Figure 1. "PVC core" column indicates sites where surface sediments were collected with manual PVC core and prepared as thin sections. "SedArticle rocks" column indicates sites where lithified surface sedimentary rocks that have been in contact with acid brines were analyzed (thin section and spectroscopy). "Sed chem" column indicates sites where sediments were analyzed for major-oxide and trace-element geochemistry. GSWA (Geologic Survey of Western Australia) Map column indicates map where basement rock units associated with lakes were derived.

Site	Lake	S Lat	E Long	PVC core	Sed rocks	Sed chem	GSWA Map name	Primary basement rock units associated with lake
1	Brown	31.128	118.307	X	X	X	Kellerberrin	Agx: Archean seriate and porphyritic adamellite; contains abundant xenoliths of gneiss Age: Archean biotite granite and adamellite; biotite-rich, commonly foliated Agm: Archean mixed granitic rocks predominantly even-grained and seriate granite and adamellite
2	Walker	32.062	118.389	X		X	Corrigin	Agn: Archean adamellite and granodiorite; medium and coarse grained, strongly recrystallized Ans: Archean adamellitic, granodioritic, and tonalitic gneiss with biotite or hornblende folia; minor compositional banded phases Ahf: Archean mafic hornblende-plagioclase-hypersthene granulite
3	Magic	32.433	118.906	X		X	Hyden	Agl: Archean medium- and coarse-grained porphyritic granite and adamellite
4	Aerodrome	32.211	121.759	X		X	Norseman	Awb: Archean metamorphosed basaltic pillow lava, dolerite, and gabbro, minor graphitic slate
5	Prado	32.433	121.718	X	X		Norseman	Tog: Tertiary weathered granitic rocks Age: Archean biotite granite, equigranular Agm: Archean migmatite; marginal between Archean formations and granite
6	Gilmore	32.610	121.561		X		Norseman	Ttf: Tertiary residual and reworked deposits—ferruginous and siliceous; includes silcrete, ferricrete, yellow to buff sand, ironstone gravel limonite nodules, and ferruginous sandstone Ag: Archean granites; not subdivided, locally porphyritic, commonly foliated; unknown mutual relationships; abundant pegmatites in places
7	Hobby	32.924	121.746		X		Norseman	Tog, Ttf, and Ag (from above)
8	Twins	33.055	121.676	X	X	X	Esperance	Tep: Tertiary Pallinup Siltstone: yellow to gray claystone, siltstone, silty sandstone, with fossil sponges and molluscs pEn: Precambrian gneiss; banded, garnet-biotite gneiss, and granitic rocks of varying composition and texture

brines after collection. Bulk major oxide and elemental analyses show that SiO_2 is the dominant phase (average ~60 wt% for samples that have >1% SiO_2), and that Al_2O_3 accounts for up to ~20% of the sediment mass (Fig. 7).

SO_4 is also a significant phase in some of the sediments (up to ~65 wt%), with a distinct separation between sulfate-rich and sulfate-poor sediments. Comparisons of %S vs. %SO_4 show four populations of sediments: gypsum-dominated chemical sediments (with S/SO_4 ~ 0.15), SO_4- and S-poor halite-dominated chemical sediments, and two groups of clastics with varying amounts of S (Fig. 7B). Samples collected from the surface and in contact with lake waters have higher S amounts, while samples from the shallow subsurface (few to tens of centimeters deep) have S/SO_4 values more consistent with what would be expected if all of the S mass was in the form of sulfate.

Fe_2O_3 constitutes up to ~15% of the sediment mass and shows corresponding increases in the amount of many trace elements, including U, Ti, and Ni (Fig. 7C). Ni vs. Fe_2O_3 concentrations show two distinct populations that correspond to specific lake settings with

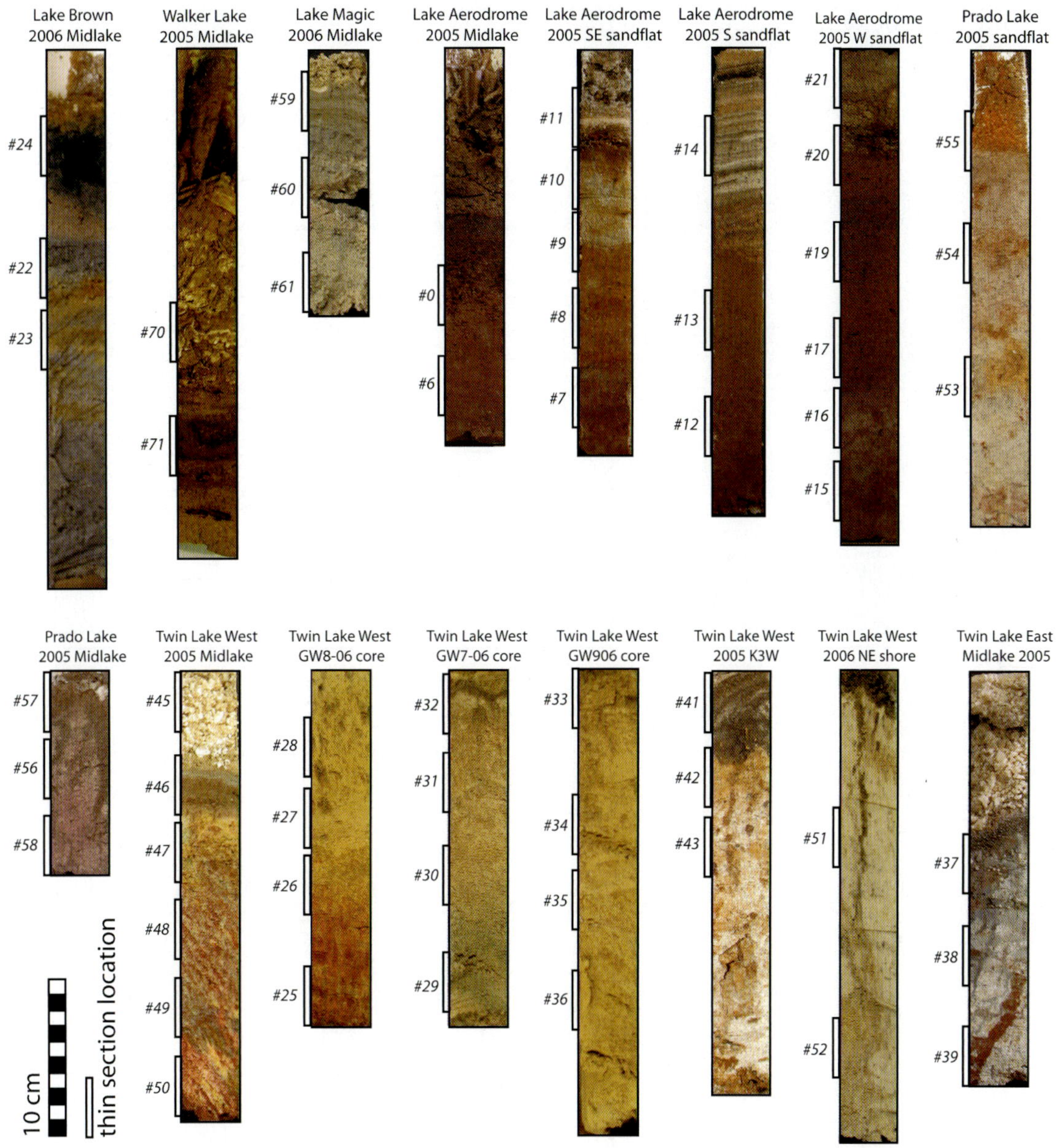

FIG. 5.—Photos of shallow PVC cores showing location of thin sections (and thin section sample numbers).

higher Ni in hematite and lower Ni in goethite based on spectral classification of the iron-oxide mineralogy. Detailed spectral data show that ferric iron exists in these sediments as a mixture of hematite, goethite, and jarosite (Fig. 8). Some sediment grain-size fractions are dominated by specific iron-oxide mineralogy. For example, the <2 μm fraction of sediments analyzed from Twin Lake (e.g., Story et al. 2010) has an absorption feature minima wavelength indicative of hematite, while the >62 μm fraction of these same samples is indicative of goethite (Fig. 8). The samples with the deepest ferric iron absorption features have minima indicative of more mixtures. These interpretations are supported by XRD and petrographic data as well.

Calculated mineral saturation indices suggest that lake waters and groundwaters are supersaturated with respect to many of the observed minerals, including alunite, amorphous silica, goethite, hematite, heulandite, illite, jarosite, kaolinite, smectite, and K-feldspar (Table 3).

Origin Classification: The minerals in the lakes and associated sand flats/mud flats can be classified by their four main origins: detrital grains, chemical sediments, reworked chemical sediments, and diagenetic phases. The detrital grains were transported into the environments by wind or sheetfloods. The chemical sediments precipitated directly from the lake waters, typically during evaporation. Some chemical sediments were first precipitated in lakes, and then they were entrained, transported, and redeposited, usually by wind. Finally, diagenetic minerals are those that either grew from or are the alteration products of shallow groundwater. While it can be very difficult to definitively distinguish between these different types, the textures observed petrographically support these interpretations, and our observations are supported by the associated geochemistry (e.g., mineral saturation indices, sediment geochemistry). The significant role of the extreme and varying salinity and acidity in the lakes and groundwaters accounts for a diverse range of chemical sediment and

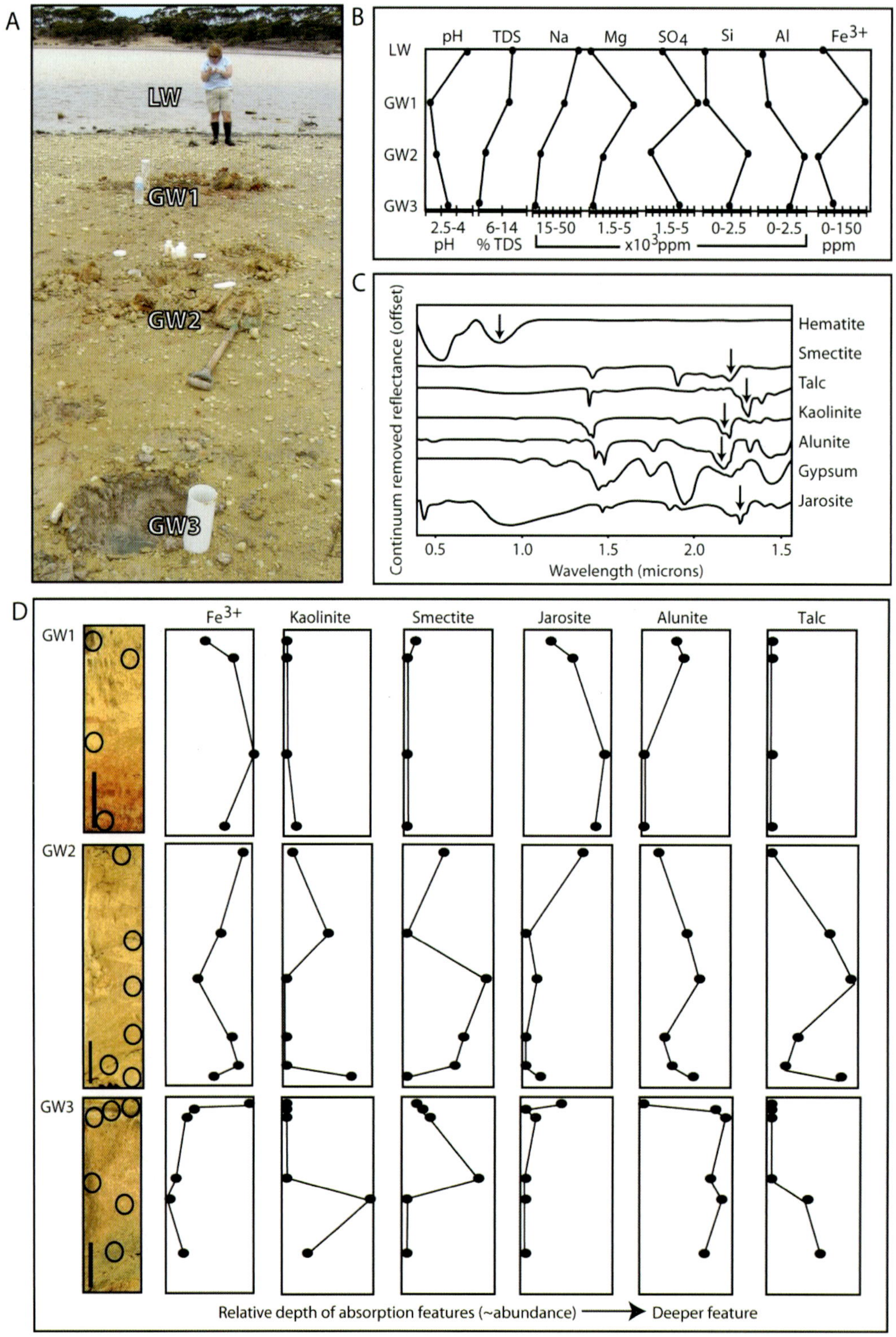

FIG. 6.—Transect along mud flat of Twin Lake, illustrating spatial heterogeneity in brine chemistry as well as associated mineralgy (interpreted from visible–near-infrared spectroscopy on sediment cores). **A)** Photograph of shallow pits along the shore of Twin Lake where subsurface pore fluids were sampled and associated sediment cores were obtained. **B)** Comparisons of brine chemistry along transect and into lake. **C)** Examples of minerals observed in spectra of sediment cores from this site. Arrows indicate specific absorption features that were analyzed to produce graphs in D. **D)** Three sediment cores collected along transect, locations of spectral measurements, and comparisons of relative depth of mineral-specific absorption features through the cores. X-axis scale is simply scaled to compare all of the samples analyzed, with the deepest absorption features all the way to the right, and the shallowest/absent features on the left. Depth of absorption feature is generally correlative with mineral amount, but it is also influenced by other factors such as grain size and mineral mixtures.

TABLE 2.—Basic acid brine water chemistry associated with sediments analyzed in this study.

Site	Lake	Sample ID	pH*	TDS (%)	Na^+ (mg/kg)	K^+ (mg/kg)	Ca^{2+} (mg/kg)	Mg^{2+} (mg/kg)	Cl^- (mg/kg)	SO_4^{2-} (mg/kg)	Si (mg/kg)	Al^{3+} (mg/kg)	Fe^{3+} (mg/kg)
1	Brown	4-G3–05	3.3	16	52,400	755	639	5420	85,000	3290	904	1057	32
1	Brown	4-G1–06	3.7	24	84,999	936	873	7390	131,000	5120	180	476	34
2	Walker	15-G1–05	3.1	6	17,600	221	846	1490	30,700	1890	91	492	10
3	Magic	20-G14–06	3.3	7	22,200	272	345	2180	30,600	3060	797	756	14
3	Magic	20-G1–05	3.2	15	56,600	599	1440	6260	96,800	10,600	11,700	8017	191
4	Aerodrome	36-G19–05	2.8	18	57,200	317	899	6260	94,900	7820	4580	3274	196
4	Aerodrome	36-G4–06	2.8	20	69,343	376	812	6690	114,000	8960	75	476	80
5	Prado	42-GC6–05	3.2	23	79,200	954	557	5900	101,000	5400	100	484	22
6	Gilmore	43-G11–06	2.6	19	65,000	543	447	6600	76,500	6600	13,300	6073	459
8	Twins	56-G2B-05	3.3	20	25,100	392	130	2720	33,800	4700	110	234	19
8	Twins	57-G7–06	3.2	6	17,000	349	41	1840	26,700	3850	1380	1645	46
8	Twins	57-G9–06	2.8	7	21,100	420	58	2350	12,800	1870	2320	2371	–
8	Twins	57-G8–06	2.6	11	38,300	834	97	4750	39,300	5400	86	476	141
8	Twins	57-G1F-05	2.4	24	74,100	1560	279	8670	120,000	16,100	180	911	62
1	Brown	4-L1–05	4.2	22	85,700	289	1210	1550	125,000	3380	544	548	55
1	Brown	4-L1–06	3.9	25	106,000	317	1330	1840	67,600	1650	2640	1815	291
2	Walker	15-L1–05	3.5	3	7100	61	1460	730	9600	3040	87	242	12
3	Magic	20-L17–06	1.9	28	48,000	4520	121	38,400	122,000	35,200	510	1774	331
3	Magic	20-L1–05	2.5	7	99,100	1030	818	8190	167,000	11,900	74	927	127
4	Aerodrome	36-L2D-05	3.7	9	25,300	130	1420	2250	34,300	5060	59	242	15
4	Aerodrome	36-L7–06	3.2	11	44,700	174	1780	2900	52,300	5340	–	105	32
5	Prado	42-L10–06	3.9	17	76,500	180	460	1100	88,800	1820	50	226	–
5	Prado	42-L2G-05	3.7	16	98,500	418	809	1520	150,000	3760	3730	2516	176
6	Gilmore	43-L14–06	5.8	5	11,100	87	1310	671	11,200	2420	37	135	8
8	Twins	56-L2D-05	3	16	85,500	307	406	1340	129,000	3550	88	1468	26
8	Twins	57-L12–06	3.8	12	48,800	276	151	1630	66,300	2630	51	242	26
8	Twins	57-L1E-05	3.2	20	94,400	387	370	1890	147,000	4510	72	1903	34

LW = surface lake water, GW = shallow groundwater, TDS = total dissolved solids. Ion and compound concentration data are after Bowen and Benison (2009).

*pH and salinity data compiled from field-collected data in 2001, 2005, 2006, 2008, and 2009.

diagenetic mineral types. The physical dynamics of the environments, including flooding, evaporation, and desiccation, and resulting variations in water depths, along with strong and variable-direction winds, cause the various mineral types to intermingle. Therefore, the diagenetic minerals in southern WA can be best understood in the context of the entire sedimentary system.

Detrital Grains: This siliciclastic component is commonly thinly bedded sand and contains grains that have experienced eolian reworking. These grains originated from weathering of basement material as well as reworked surface chemical sediments (such as gypsum and hematite). The most common detrital mineral is quartz. Other silicates, such as feldspar (both K- and Na/Ca-feldspar), amphibole (e.g., riebeckite $Na_2[Fe,Mg]_5Si_8O_{22}[OH]_2$), and micas (e.g., Fe-muscovite), are represented, but in lesser quantities. Other minerals that may also include a detrital component include Fe- and Al-phyllosilicates (Story et al. 2010).

Chemical Sediments: Field occurrence, textures, and mineralogical associations suggest that multiple mineral types precipitate as chemical sediments directly from the acid saline lake water. Here, chemical sediments are defined as those that precipitate from lake waters and form surface beds that can be millimeter to centimeter scale, vs. diagenetic phases that precipitate from pore fluids and form cements. Like diagenetic phases, these are authigenic precipitates that form in situ. Chemical sediments forming from the acid brine lake waters in WA include halite, gypsum, bassanite, iron oxides (hematite and goethite), and kaolinite/halloysite. At the extremely low pH values that are encountered at these lakes (pH commonly down to ~2), the solubility of Si is more than twice that of a fluid with a pH of 7–8 (Iler 1979). Similarly, Al solubility increases at these low pHs. The availability of these ions allows for supersaturation of phyllosilicates in these fluids, which is supported by equilibrium thermodynamic calculations (McArthur et al. 1991, Marion et al. 2009). When the pH fluctuates slightly with seasonal changes in meteoric water input and flooding-evaporation-desiccation cycles, all of these minerals appear to undergo episodes of both precipitation and dissolution.

Halite: Halite precipitates from lake water during evapoconcentration and forms cumulate crystals and chevron crystals. Cumulate crystals grow at the air–water interface or within the water column. Some cumulates grow together to form rafts on the lake surface. Cumulate crystals eventually sink to the lake bottom, where they form beds. Cumulate halite beds typically contain randomly oriented, small (commonly less than 1 cm) halite cubes. If halite growth continues once the cumulates have fallen to the lake floor, chevrons, larger (up to 4 cm) halite crystals pointing upward, form. In WA, these crystals form subaqueous halite beds up to at least 48 cm thick.

Petrographic observations show textures that are related to the

TABLE 3.—Mineral saturation indices results.

Mineral type	Mineral (s)	Observed?	Lakes where all water samples (GW & LW) yielded calculated saturation indices indicative of supersaturation of that mineral (or group of minerals)
SiO_2	Amorphous silica	X	Brown, Magic, Prado
SiO_2	Quartz, chalcedony, cristobalite, tridymite	X	Brown, Walker, Magic, Prado, Gilmore, Twins
Feldspar	Albite, K-feldspar, microcline	X	Brown, Prado, Gilmore
Fe oxide	Goethite, hematite	X	Brown, Walker, Aerodrome, Gilmore
MnO_2	Pyrolusite	–	Brown
Sulfate	Alunite	X	Brown, Walker, Prado, Gilmore
Sulfate	Jarosite	X	Brown, Walker, Aerodrome, Gilmore
Clay	Illite	X	Brown, Gilmore
Clay	Kaolinite	X	Brown, Prado, Gilmore
Smectite	Beidellite	?	Brown, Prado, Gilmore
Smectite	Nontronite	?	Brown, Walker, Magic, Gilmore
Mica	Muscovite, paragonite	X	Brown, Gilmore
Mica	Phengite	–	Gilmore
Mica	Pyrophyllite	X	Brown, Walker, Prado, Gilmore
Zeolite	Analcime, heulandite	?	Brown, Gilmore
Zeolite	Clinoptilolite	?	Brown, Prado, Gilmore, Walker, Magic, Twins
Zeolite	Mordenite	–	Brown, Walker, Magic, Prado, Gilmore, Twins
CuS	Covellite	–	Walker, Magic
S native element	Sulfur-rhombic form	X	Brown, Walker, Magic, Aerodrome, Prado, Gilmore, Twins

X = observed (via XRD, spectroscopy, geochemically, and/or petrographically); ? = observed mineral type in XRD but specific minerals not distinguished; – = not observed.

sensitivity of halite to dissolution and variability in the rate of mineral formation (Fig. 9). Rapid growth promotes the formation of many fluid inclusions along growth bands, and slower growth forms clear halite growth bands. Fluid inclusions in saline minerals from these lakes have been shown to be important in entrapping and potentially preserving organic remains in these environments (Benison et al. 2008), and they may have unique characteristics specific to the acid chemistry (Jagniecki and Benison 2010).

Gypsum: Gypsum precipitated from lakes takes three forms in the acid saline lakes in WA. There are bottom-growth gypsum crystals, many of which are swallow-tail twinned crystals. These crystals form at some lake bottoms at the water–sediment interface (e.g., Fig. 4A). They exhibit competitive crystal growth, widening upward. The largest bottom-growth gypsum crystal we measured is 12 cm long, and they make beds up to 15 cm thick. Each individual bottom-growth gypsum crystal contains alternating orange, white, and clear growth bands, representing periods in which the lakes precipitated hematite/goethite (orange) and kaolinite/halloysite (white) along with the gypsum, or precipitated gypsum alone (clear) (Benison et al. 2007). A less common type of lake-precipitated gypsum is in the form of clear, needle-shaped crystals that grow on and around solids in the lake. Common substrates are wood fragments from eucalyptus trees and salt bush. These crystals are rarely larger than 1 cm in length. During the latest stages of evapoconcentration, as a lake becomes desiccated, tiny (less than 1 mm long) gypsum needle-shaped crystals and cubic halite crystals grow on the desiccated lake bed. This forms an efflorescent crust that coats the underlying lake salts or muds. After desiccation of all the lake water, some groundwater may be pulled up to the surface by evaporation and precipitate more efflorescent salts. In this way, these efflorescent crusts are surface precipitates of both end-member lake water as well as shallow groundwater. Bassanite forms as yellow rounded crystals atop gypsum crystals during evapoconcentration in at least one of the acid saline lakes.

Iron oxides and phyllosilicates: Hematite, goethite, and kaolinite/halloysite precipitate from acid saline lake waters as mud-sized crystals. They are found in lake facies sediments interbedded with halite and gypsum beds, forming mud drapes over halite and gypsum crystals or forming a hematite/goethite or a kaolinite/halloysite mud bed up to approximately 6 cm thick (Fig. 4F).

Reworked Chemical Sediments: All of the chemical sediments that form in the shallow lakes are exposed during desiccation, eroded, and reworked by wind into sediments in the lakes, sand flats/mud flats, and surrounding sand dunes. Some reworked chemical sediments (rounded gypsum grains) have oolitic textures suggestive of reworking in shallow waters where they are coated by layers of iron oxide and clay (Fig. 10).

Diagenetic Minerals: Early diagenetic minerals are those that, by field and petrographic occurrence, show they were not deposited either as detrital grains or chemical sediments but were formed after those depositional minerals. Diagenetic minerals in this study have formed from acid saline groundwaters at or just below (centimeter to meter depths) the surface. Early diagenetic minerals observed include common saline minerals such as halite and gypsum, as well as hydrated sulfates such as hydrobasaluminite ($Al_4[SO_4][OH]_{10}$•12–36H_2O), and metal sulfate salts such as rozenite ($Fe^{2+}SO_4$•4H_2O), alunite ($KAl_3[SO_4]_2[OH]_6$), and jarosite ($KFe_3^{3+}[OH]_6[SO_4]_2$). Other diagenetic phases include hematite (Fe_2O_3), goethite (FeO[OH]), anatase (TiO_2), gibbsite ($Al[OH]_3$), and Al-phyllosilicates (e.g., kaolinite $Al_2Si_2O_5[OH]_4$, dickite, halloysite; see Story et al. 2010). For some phyllosilicate minerals that have been identified in these sediments, such as chlorite–smectite, it is very difficult to confirm

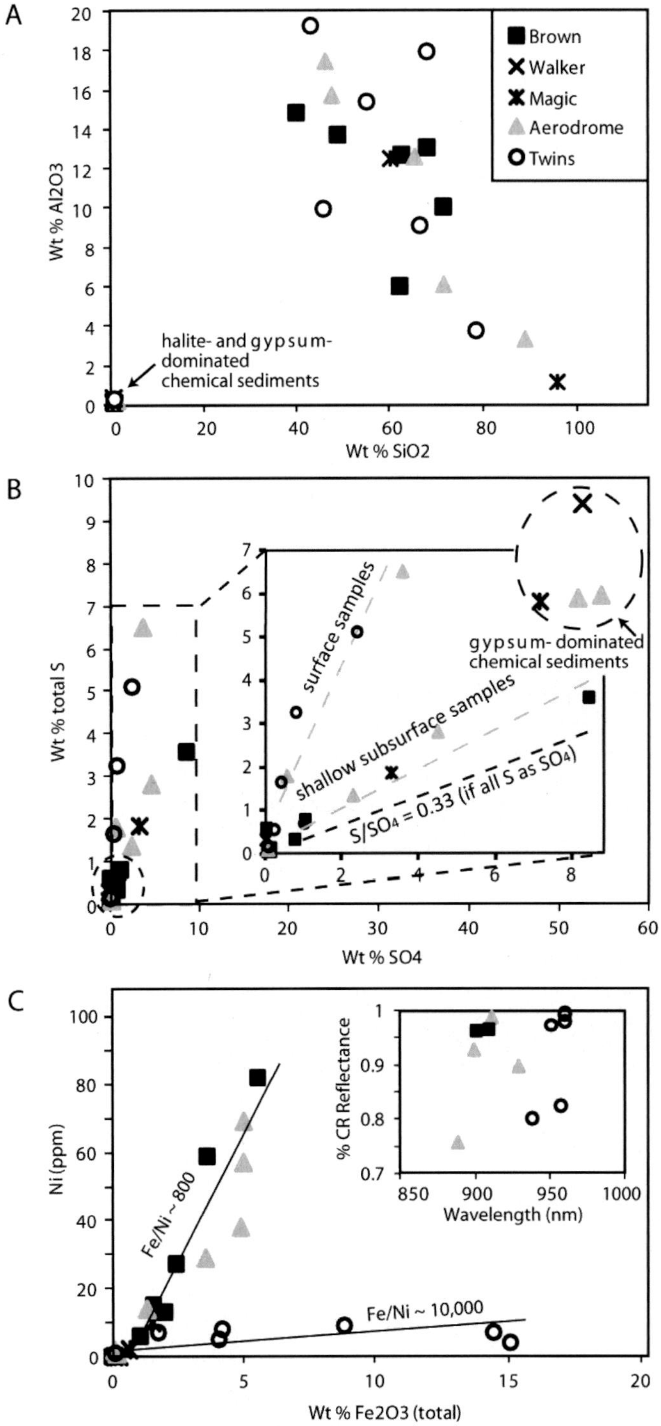

FIG. 7.—Geochemistry of sediments from Western Australia acid salt lake environments. Data are shown in Tables 4 and 5. **A)** Weight percent of Al_2O_3 vs. SiO_2. Note population of halite- and gypsum-dominated chemical sediments clustered near the origin. **B)** Weight percent total S vs. SO_4. Four populations of sediments are apparent: gypsum-dominated chemical sediments (highlighted in circle in upper right), halite-dominated chemical sediments (circle near origin), and two populations of clastic sediments. Zoom-in inset differentiates the two populations of sediments as "S-rich" surface ← sediments and "S-poor" sediments from the shallow subsurface. **C)** Ni concentrations vs. weight percent Fe_2O_3 (total). Patterns in Ni enrichment show two distinct populations, one with a Fe/Ni ratio of ~10,000 and ferric iron absorption features at higher wavlenghts indicative of hydrous ferric oxyhydroxides (inset) at the Twin Lakes, and one with a Fe/Ni ratio of ~800 and ferric iron absorption features more consistant with hematite.

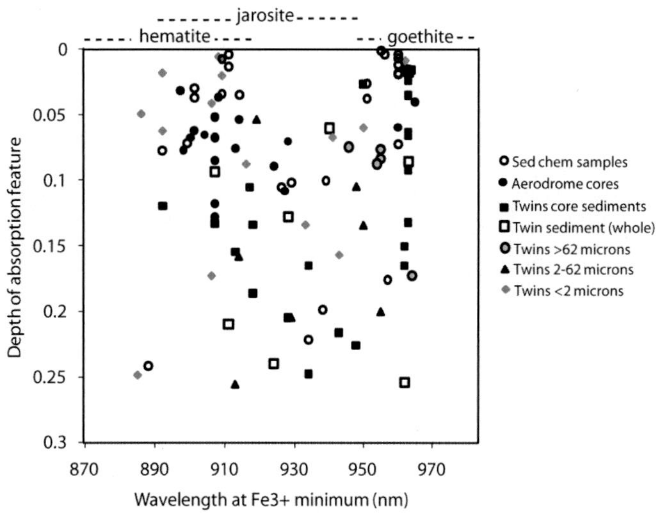

FIG. 8.—Wavelength and depth of near-infrared ferric iron absorption feature measured in sediment samples indicative of hematite, jarosite, and goethite as well as mixtures of these minerals. Wavelength fields are defined by minina observed in USGS spectral library end-member samples of hematite, jarosite, and goethite. The samples with smaller absorption features have distinctive hematite and goethite mineralogy, with specific sediment size fractions being dominated by specific minerals (e.g., <2 μm fraction from Twins dominated by hematite, whereas >62 μm fraction is dominated by goethite). Samples with deeper absroption features exhibit reflectance minima indicative of jarosite and iron mineral mixtures.

whether it is a detrital or a diagenetic constituent (or both). Past work has attempted to identify the authigenic component based on distinct differences in grain size and morphology, but the populations of the sediment fraction that was considered as detrital contain many of the same minerals as the presumed authigenic component (Story et al. 2010). Geochemical modeling supports the hypotheses that the minerals described here as showing textures and mineral associations suggestive of diagenetic/authigenic precipitation would be expected to be supersaturated in these extreme fluids (Table 3; McArthur et al. 1991, Marion et al. 2009).

Petrography

Because much of the mineralogy of the diagenetic minerals overlaps with that of the chemical sediments and reworked chemical sediments, fieldwork and petrographic examination of thin sections are especially important to distinguish between depositional and diagenetic textures

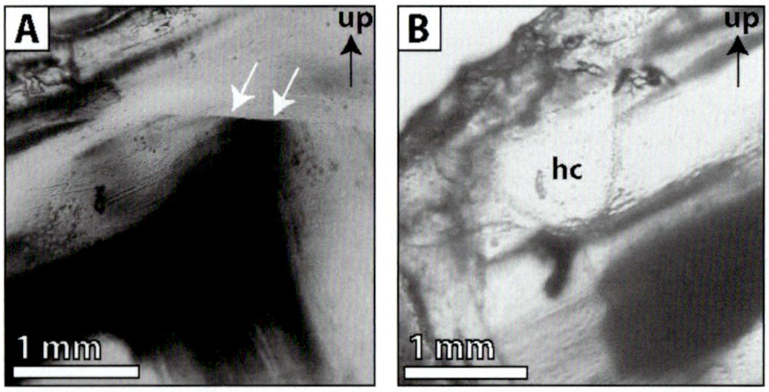

FIG. 9.—Micrographs of halite from an acid salt lake. **A)** Twin Lake West, plane transmitted light. Chevron halite crystal with horizontal dissolution surface (arrows). Dark bands are fluid-inclusion–rich growth bands (formed during fast growth). Note that top corner of dark growth band was truncated. Clear area is fluid-inclusion–poor halite (formed during slower growth). **B)** Twin Lake West, plane transmitted light. Chevron halite crystal with dissolution vug filled with clear halite cement (hc).

and to identify spatial relationships between minerals. The morphology and spatial context of the minerals relative to the grains can be used to determine the minerals and textures that are diagenetic (i.e., Chilingarian and Wolf 1988, McIlreath and Morrow 1990). Preliminary macroscopic observations of early diagenetic features, such as surface and near-surface cements, were made in the field, and they demonstrate the complex spatial associations between mineral types (e.g., Fig. 4B, C, E). Examination of thin sections allowed for more detailed study of diagenetic features and evaluation of the microscopic associations between detrital and diagenetic phases, the relative timing of these phases, and the nature of any biological preservation within the sediments (Figs. 9–14).

Cements: In the surface and shallow subsurface sediments in these acid salt lakes, there are zones with relatively simple pore-filling intergranular isopachous cements, intergranular meniscus cements, and complex multigenerational cements (Fig. 11). Cements are composed of multiple mineralogies, including soluble salts such as gypsum and halite, acid-sulfate minerals such as jarosite, alunite, and rozenite, phyllosilicates (e.g., kaolinite), and iron oxides (Fig. 11). The texture of the cement can reveal details about the conditions under which they formed. For example, cement crystals that are thicker on the bottom side of grains or that are concentrated near the grain-to-grain

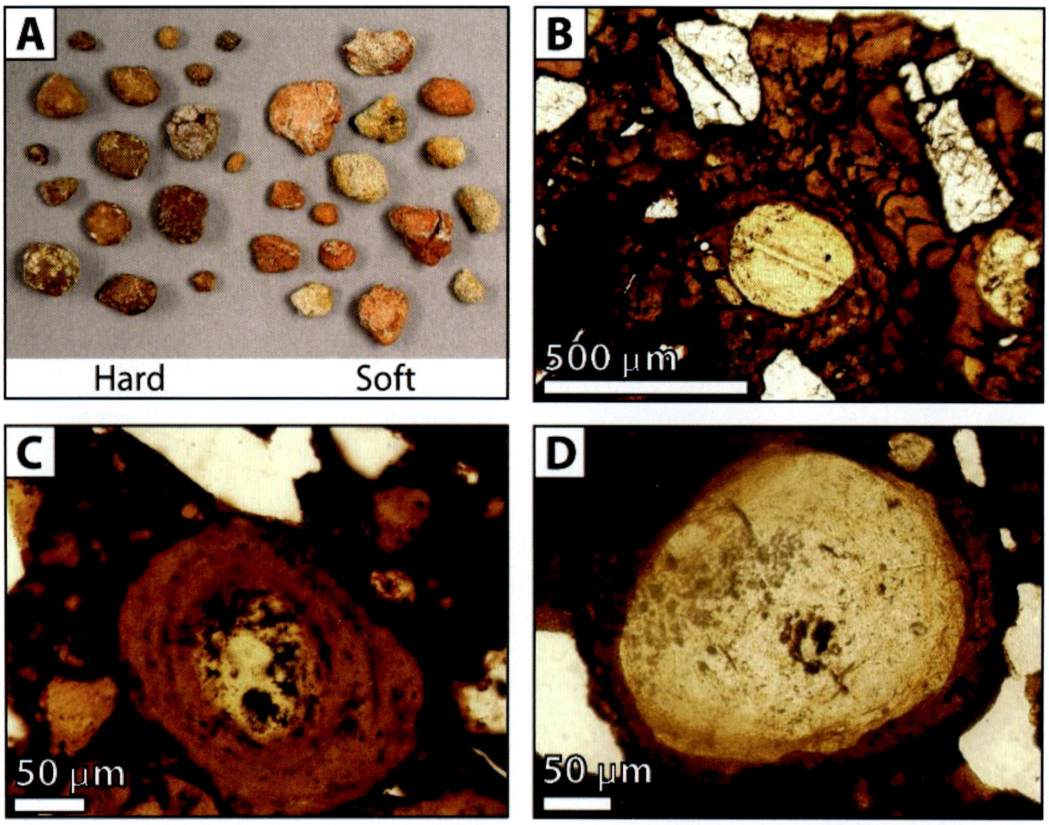

FIG. 10.—Examples of iron-oxide concretions in acid saline sediments, Lake Brown. **A)** Concretions extracted from unlithified sediments from subaqueous lake facies at Lake Brown, including both soft (semilithified) and hard (well-lithified) concretions. **B)** Concretion interiors showing multiple generations of iron-oxide cement precipitated around reworked detrital grains including quartz and reworked gypsum grains. **C)** Concretion interior with reworked gypsum grain (with organic material as solid inclusions in grain) with oolitic iron-oxide coatings encased in later generation of iron-oxide cement. **D)** Concretion interior with reworked gypsum grain with solid inclusions of organic material and clay and fluid inclusions. **B–D)** Combination of transmitted and reflected light.

TABLE 4.—Major components of sediment geochemistry for subset of surface sediments in contact with acid brines in Western Australia.

				Element							
				SiO_2 (%)	Al_2O_3 (%)	$Fe_2O_{3(T)}$ (%)	CaO (%)	Na_2O (%)	SO_4 (%)	S (%)	LOI (%)
			Detection limit	0.01	0.01	0.01	0.01	0.01	0.01	0.001	
			Reference method	FUS-ICP	FUS-ICP	FUS-ICP	FUS-ICP	FUS-ICP	IR	TD-ICP	FUS-ICP
	Lake	Sample type	Color								
1a	Lake Brown	Subaqueous halite	White	0.55	0.09	0	1.07	49.98	0.02	0.61	48.39
1b	Lake Brown	Subaqueous sand	Orangish-brown	40.12	14.88	5.59	33.73	0.16	0.06	0.08	2.728
1c	Lake Brown	Mud flat clay	White-tan	68.07	13.08	2.43	0.15	3.32	0.13	0.07	10.14
1d	Lake Brown	Subaqueous sand	Orangish-red	71.46	10.08	2.04	0.36	3.88	0.15	0.15	7.621
1e	Lake Brown	Subaqueous sand	Greenish-brown	48.94	13.78	3.65	0.47	8.32	1.06	0.81	21.43
1f	Lake Brown	Subaqueous sand	Reddish-brown	62.32	6.01	1.07	6.65	3.31	8.44	3.6	18.16
1g	Lake Brown	Subaqueous sand	Tan-brown	62.61	12.72	1.6	0.36	5.27	0.78	0.34	12.81
2a	Walker Lake	Subaqueous gypsum	Yellowish-white	0.41	0.31	0.24	34.02	0.13	52.5	9.41	21.98
3a	Magic	Surface gypsum	White	0.26	0.09	0.06	30.51	3.95	47.7	7.11	26.7
3b	Magic	Sand flat sand	White-tan	95.59	1.09	0.64	0.16	0.18	0.08	0.13	0.7545
3c	Magic	Sand flat sand	Tan	60.17	12.46	1.5	0.24	5.78	3.29	1.86	15.34
3d	Magic	Subaqueous halite	White	0.31	0.14	0.1	0.27	51.4	0.02	0.33	45.59
4a	Aerodrome	Subaqueous gypsum	Reddish-brown	0.64	0.23	0.16	34.1	0.23	52	7.17	21.88
4b	Aerodrome	Mud flat clay	White-tan	1.34	0.21	0.08	32.59	1.5	54.6	7.24	23.48
4c	Aerodrome	Dune sand	Orangish-red	88.87	3.32	3.55	0.3	0.35	0.07	0.12	1.813
4d	Aerodrome	Sand flat sand	Brown	71.52	6.09	5	1.8	2.65	2.3	1.37	10.66
4e	Aerodrome	Sand flat sand	White	46.39	17.45	1.3	0.28	1.86	3.59	6.5	26.8
4f	Aerodrome	Sand flat coarse sand	Brown	65.37	12.57	4.89	0.46	1.49	0.57	1.81	12.6
4g	Aerodrome	Sand flat sand	Pinkish white	47.77	15.72	4.99	1.27	5.49	4.5	2.82	19.78
8a	Twin Lake	Sand flat sand	Brick red	43.33	19.31	15.01	1.74	4.08	0.4	1.67	12.84
8b	Twin Lake	Channel sand	Dark yellow	66.24	9.07	8.79	0.08	2.6	0.22	0.56	10.2
8c	Twin Lake	Sand flat sand	Tan	45.88	10	14.4	0.41	1.94	2.4	5.13	22.75
8d	Twin Lake	Subaqueous halite crust	White	0.79	0.24	0.12	0.23	55.44	0.05	0.19	43.17
8e	Twin Lake	Sand flat sand	Brown	78.35	3.75	4.19	0.22	3.24	1.01	0.71	8.496
8f	Twin Lake	Mud flat very fine sand	Yellowish-white	55.28	15.41	4.09	0.11	2.59	0.81	3.26	18.91
8g	Twin Lake	Sand flat sand	Pinkish-peach	68.12	17.96	1.73	0.04	1.72	0.08	0.17	9.257

contacts are called pendant and meniscus cements, respectively, and are evidence of cementation in the vadose zone (Fig. 11). On the other hand, isopachous cements, having equal thickness around individual grains or pore spaces, form in the phreatic zone. In some samples, the cements predate syndepositional to early postdepositional features such as desiccation cracks, illustrating the very early nature of the cement formation (Fig. 12). Concretionary cements are common in these sediments, with massive spheroidal zones of iron-oxide mineralization. While these have been observed at several lakes, they are particularly abundant at Lake Brown (Bowen et al. 2008). The concretions encase detrital grains including reworked chemical sediments such as gypsum (Fig. 10).

Cracks: Desiccation cracks that display a vertical tapering "v" in cross section and a polygonal pattern on bedding planes are a temporary feature in the lake and mud flat/sand flat facies. Where thick halite crusts have remained on lake beds after desiccation, continued surface evaporation and wicking up of saline groundwaters produce buckled, positive-relief salt along the polygon patterns on salt crusts.

Cracking patterns typically associated with paleosols also form in the shallow subsurface. These cracks include horizontal cracks, vertical cracks, and crazed plane cracks. In addition, we have also observed autoclastic breccias and circumgranular cracks; i.e., rounded cracks that enclose individual large sand or gravel grains and clumps of finer sediment. Cracking strongly suggests repeated wetting and drying. In these sediments, the cracking both predates and postdates other diagenetic features (Fig. 12).

Dissolution: During flooding stages, the chemistry of the surface water changes considerably as freshwater is introduced into the environment. Initially, flooding serves to lower the salinity and slightly

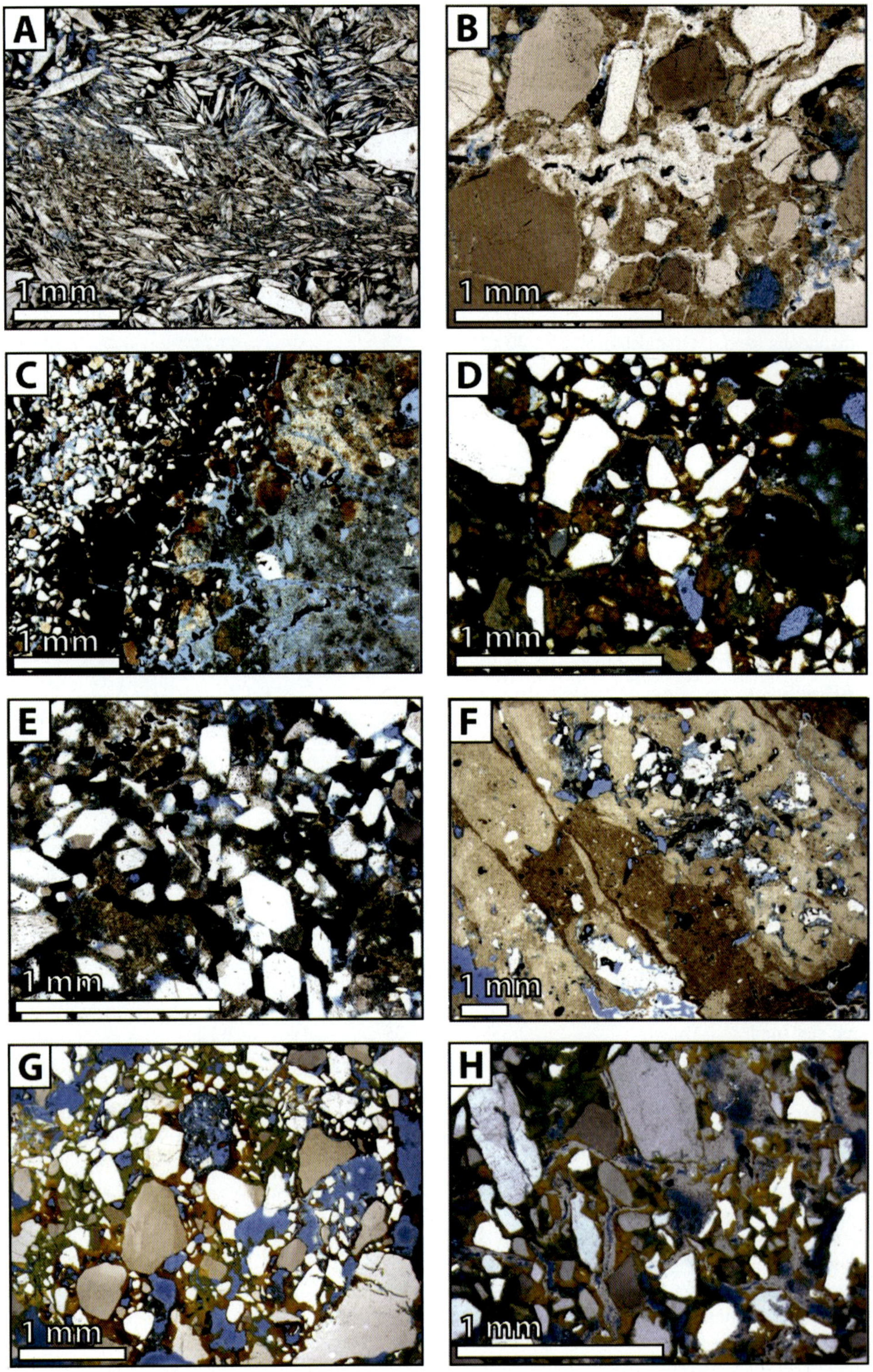

FIG. 11.—Photomicrograph examples of clastic-dominated acid saline sediments with cements that are diverse in mineralogy and morphology. Sample numbers correspond to thin section locations and depths shown in Figure 5. **A)** Lake Aerodrome 2005 sand flat core, sample #21, plane transmitted light; reworked gypsum cemented by iron oxide; note imbrication of gypsum clasts. **B)** Lake Magic 2006 midlake core, sample #61, polarized transmitted light; quartz grains cemented by at least three generations of intergranular isopachous clay, gypsum, and iron-oxide cement; clay cement formed before desiccation cracks in the sample; cracks formed by wetting and drying; gypsum and opaque iron-oxide cements formed after cracking and fill the cracks. **C)** Lake Aerodrome 2005 sand flat core, sample #15, polarized transmitted light; fine-grained laminated red iron-oxide layer that was mud cracked (note thin once-vertical cracks originating on left side), then pushed up and

raise the pH of the fluids. Halite chevron and cumulate crystals undergo dissolution; fresher waters dissolve the tops of upward-pointing chevron cubes and leave a flat dissolution surface (Fig. 9). These same floodwaters also dissolve "pipes"; i.e., vertically oriented, millimeter- to centimeter-scale holes in bedded halite. These vugs typically are filled with a coarse, clear halite cement during the next evapoconcentration stage (Fig. 9). All of this salt dissolution increases the salinity of the lake water significantly (Bowen and Benison 2009), which could then lead to dissolution of gypsum. Other minerals also show evidence of dissolution. For example, iron oxide–, clay-, and jarosite-rich samples have porosity with textures suggestive of dissolution of these phases (e.g., Fig. 11G). The flooding-evaporation-desiccation cycles not only influence the typical saline minerals, but they also change the water chemistry in ways that allow for both precipitation and dissolution of the other acid-related minerals precipitated by these fluids. Although the low pH values in these fluids are needed to concentrate the elements to precipitate the Fe, Al, and Si minerals observed here, when pHs are extremely low ($\sim<3$), some of these minerals would likely go through dissolution.

Displacive Crystals: Many of the diagenetic cements that are observed in these sediments exhibit textures suggestive of displacive precipitation (Fig. 13). The sediments commonly contain displacive halite and gypsum crystals. These are euhedral, millimeter- to centimeter-scale crystals that are randomly oriented in the siliciclastic sand or mud matrix and are found in the shallow subsurface below lakes and sand flats (e.g., Fig. 2J in Benison and Bowen 2006). Displacive gypsum can be distinguished from depositional gypsum crystals by their needle, lath, and rosette shapes, random orientation within host sediment of different composition, and their fairly uniform size. Displacive halite crystals have the same occurrence as gypsum displacive crystals, but they are common and smaller in size. Some sand beds contain both displacive halite and gypsum; some others contain only gypsum or halite. While commonly observed in the field, these features are not well represented in the core samples studied here, likely due to their ephemeral nature. In addition to displacive gypsum and halite, we also observe iron oxides and clays that precipitated in situ and exhibit relationships with clastic host grains suggestive of displacive growth (Fig. 13).

Fossils: Trace fossils and body fossils are observed in some of the sediments, encased within fine-grained sediments and associated with diagenetic cements (Fig. 14). They occur within very fine-grained phyllosilicate and iron-oxide–dominated sediments, as well as within sand beds. Biological materials are also observed encased within evaporite minerals (Benison et al. 2008).

DISCUSSION

Evolution of Acid Brines and Paragenetic Sequence of Events in WA

Multiple different physical and chemical processes occur in the acid saline sediments in WA very early after deposition of the siliciclastic grains (Fig. 15). These include (in a general paragenetic sequence based on field and petrographic observations): (1) syndepositional modification and reworking of detrital grains by winds and water, (2) precipitation of chemical sediments from lake waters, (3) solution/partial dissolution of soluble minerals, (4) displacive growth of cements from groundwater, (5) intergranular cement precipitation from groundwater (including iron-oxide concretions), (6) cracking due to drying (and repeated wetting and drying), and (7) precipitation of crack- and pore-filling cements. Many of these processes are concurrent and syndepositional.

Changing aqueous geochemical conditions cause these various processes to occur and are related to seasonal (or shorter timescale) surface processes such as evaporation and desiccation. Subtle changes in water geochemistry occur on these timescales, but major fluid fluxes that lead to wholesale changes in chemistry do not generally seem to have been necessary to cause a transition from one of these processes to the next. For example, flooding caused by local rainstorms is localized and temporally unpredictable. Flooding causes runoff, resulting in sheetfloods that carry meteoric water, sediment grains, and vegetative debris into the lakes. Lake waters are diluted and dissolve soluble chemical sediments and diagenetic minerals such as halite. Evaporation concentrates solutes in lake and shallow groundwater, resulting in the precipitation of halite, gypsum, iron oxides, and clays from lake waters. Some diagenetic mineral precipitation in the shallow subsurface is also driven by evaporation during arid weather. Desiccation causes cracking of sediments and further mineral precipitation by evapoconcentration. Finally, winds promote reworking of siliciclastic grains, chemical sediments, and diagenetic minerals. These surface conditions greatly affect the shallow subsurface and, in concert with the extreme geochemistry of the groundwaters, lead to an unusual assemblage of early diagenetic features.

Multiple different geochemical processes have influenced the chemistry of the brines and the complex mineral assemblage that occurs within these sediments. These processes are related to the composition of the bedrock, the history of weathering, the stratigraphy of the paleovalley fill sediments, the influence of seaspray aerosols, and multiple stages of fluid–sediment interaction. While a complete discussion of the geochemical history of this region is beyond the scope of this study, consideration of some of the most important

←

sideways (maybe as part of repeated desiccation and dissolution). **D)** Lake Prado 2005 sand flat core, sample #54, combination of plane transmitted and reflected light; multiple intergranular cement generations, iron oxide, probable jarosite, and rozenite. **E)** Lake Brown 2006 midlake core, sample #23; polarized transmitted light; poorly sorted sediment of reworked gypsum sand (large white grains) and silt-size quartz, cemented with intergranular cement of kaolinite followed by hematite (opaque) cement, and even later needle-like displacive gypsum crystal (in photo center). F) Twin Lake East 2005 midlake core, sample #38, polarized transmitted light; complex paragenetic relationship between zones of clay cement, iron oxide, and salt precipitation; silt partially cemented in patches with hematite (red) and clay (white) with dark organic matter, white displacive halite, and vugs where halite has been dissolved. **G)** Lake Gilmore 2005 sand flat (not pictured in Fig. 5), sample #75, combined polarized and reflected light; complex spatially heterogeneous mineralogy without any obvious paragenesis (mineralogy variations are transitional rather than overlapping); jarosite, hematite, and alunite meniscus cements around quartz grains. **H)** Lake Gilmore 2005 sand flat (not pictured in Fig. 5), sample #73, polarized and reflected light; similar cements to G, but followed by final phase of pore-lining/crack-lining gypsum cement, demonstrating the very early nature of the pre-gypsum cements.

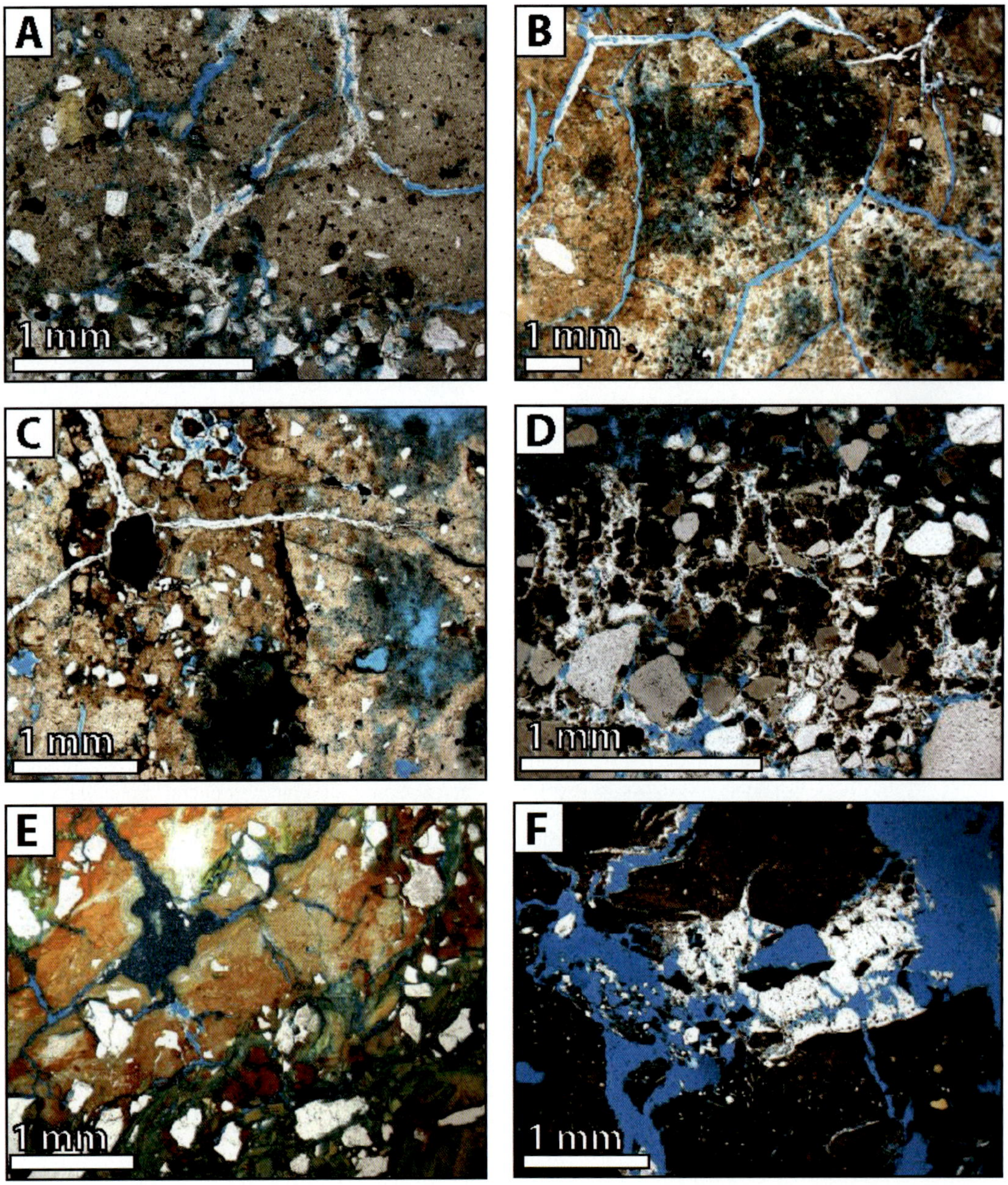

FIG. 12.—Photomicrograph examples of desiccation cracks in acid saline sediments. Crosscutting relationships between these syndepositional cracks and the authigenic minerals show the very early nature of mineral precipitation in these sediments. **A)** Lake Brown 2006 midlake core; sample #24, plane transmitted light; gypsum lining desiccation cracks; note iron-oxide spherules in host silt. **B)** Lake Aerodrome sand flat core; sample #15, polarized transmitted light; crazed plane cracks filled with halite in clay groundmass with spatially heterogeneous composition (kaolinite, gypsum, iron-oxide spherules, etc.). **C)** Lake Aerodrome 2005 sand flat core, sample #16, plane transmitted light; radial cracks around opaque hematite zone; upper center shows gypsum cement postdating hematite cement; possible paragenetic sequence for this sample: white clay precipitation/deposition, desiccation and cracking (so they now look like individual rounded coarse sand-pebble–sized grains of mud), precipitation of brown/red iron oxide within cracks, further cracking (large radial cracks), precipitation of white gypsum cement. **D)** Lake Aerodrome 2001 mud flat core (not pictured in Fig. 5), sample #2, polarized transmitted light; gypsum and iron-oxide cement around quartz grains with cracks forming autoclastic breccia appearance. **E)** Twin Lake West 2005 midlake core; sample #46, polarized and reflected light; multiple generations of goethite, alunite, and jarosite around detrital quartz grains; a later opaque iron oxide (hematite?) fills cracks. **F)** Twin Lake East 2005 midlake core, sample #37, combination of polarized transmitted and reflected light; halite cement partially filling cracks bridging zones of porosity in iron-oxide/clay cementation.

processes elucidates the significance of these events on diagenesis in these sediments.

The generation of acidity is likely related to a combination of processes, including oxidation of sulfides, iron, and precipitation of various minerals that release H^+ into the brines. Metal sulfides are abundant in the subsurface of this region, both in ores and as pyrite associated with Eocene lignites and coals. Weathering of these sulfides, due to interactions with water, oxygen, and potentially bacteria, leads to oxidation and the generation of acidity according to the reaction (Lottermoser 2007):

$$FeS_{2(s)} + 14Fe^{3+}_{(aq)} + 8H_2O_{(l)} \Rightarrow 15Fe^{2+} + 2SO_4^{2-} + 16H^+.$$

This reaction produces ferrous iron as a product, which is then also oxidized, precipitating iron hydroxides and generating additional acidity according to the reaction:

$$4Fe^{2+} + 10H_2O + O_2 = 4Fe(OH)_3 + 8H^+.$$

The complete oxidation (combination of the above reactions) of Fe-rich sulfides can be described as:

$$4FeS_{2(s)} + 15O_{2(g)} + 14H_2O_{(l)} \Rightarrow 4Fe(OH)_{3(s)} + 8SO^{2-}_{4(aq)} + 16H^+_{(aq)}.$$

In addition, the seasonal fluctuations in aqueous conditions and drying cycles likely result in dissolution of Fe- and Al-sulfate salt minerals such as jarosite and alunite, which can also generate acidity according to:

$$KFe_3(SO_4)_2(OH)_{6(s)} \Rightarrow K^+_{(aq)} + 3Fe(OH)_{3(s)} + 2SO^{2-}_{4(aq)} + 3H^+_{(aq)}$$

and

$$KAl_3(SO_4)_2(OH)_{6(s)} + 3H_2O_{(l)} \Rightarrow K^+_{(aq)} + 3Al(OH)_{3(s)} + 2SO^{2-}_{4(aq)} + 3H^+_{(aq)}.$$

The precipitation of ferric iron oxides/hydroxides and aluminum hydroxides also generates acidity:

$$Fe^{3+}_{(aq)} + 3H_2O_{(l)} \Leftrightarrow Fe(OH)_{3(s)} + 3H^+_{(s)}$$

and

$$Al^{3+}_{(aq)} + 3H_2O_{(l)} \Leftrightarrow Al(OH)_{3(s)} + 3H^+_{(aq)}.$$

In most terrestrial acid-generating systems, the hydrogen that is produced is then consumed through buffering reactions related to the weathering of silicates, carbonates, and hydroxides (Lottermoser 2007). In WA, lakes with higher pH are likely a result of acid groundwater neutralization by subsurface limestones. Alternatively, the higher-pH rainwater-fed lakes may remain separate from regionally acidic groundwater by localized impermeable sediments (e.g., clay-pans). Extreme and varied waters can coexist within meters of each other (e.g., Fig. 6). For example, a lake with a pH of 8 may have shallow groundwater with a pH of 3 in adjacent sand flats (Benison et al. 2007). The extremely acidic conditions in the lakes and shallow groundwaters cause dissolution of Fe, Al, and Si constituents through reactions that can consume hydrogen and help to neutralize the acid brines. For example, ferric oxides and hydroxides that have precipitated can redissolve in acid conditions, consuming hydrogen ions:

$$Fe(OH)_{3(s)} + 3H^+_{(aq)} \Leftrightarrow Fe^{3+}_{(aq)} + 3H_2O_{(l)}.$$

At extremely low pHs, kaolinite can dissolve via:

$$Al_2Si_2O_5(OH)_{4(s)} + 6H^+_{(aq)} \Rightarrow 2Al^{3+}_{(aq)} + 2H_4SiO_{4(aq)} + H_2O.$$

The dissolution of kaolinite consumes acidity and produces silicic acid, which may precipitate as opaline silica or chalcedony.

Geochemical modeling of the surface brines and shallow subsurface groundwater shows that these acid fluids are saturated with respect to multiple mineral groups, including silica, feldspar, iron oxides, manganese oxides, sulfates, clays, smectite, mica, zeolite, and sulfur and sulfides (Table 3). While there are many potential problems associated with modeling geochemical equilibrium in these complex acid brines, these results are encouraging in that they generally agree with the minerals we observe with XRD, spectroscopy, and petrography, and they suggest that many of these minerals may in fact be authigenic precipitates within these sediments. The thermo.dat database, which is used here, is known to be inaccurate when applied to high-ionic-strength solutions such as these; however, the thermodynamic databases that more accurately model high-ionic-strength solutions (e.g., PITZER, etc.) do not typically include Al, Si, and Fe^{3+}, which are important components in this system. Previous workers have also modeled saturation indices for acid brines from this area and have found comparable chemistry to the fluids observed in this study (e.g., Fig. 2) utilizing different modeling parameters, which may account for the high ionic strength of the solutions (e.g., PHREEQE [McArthur et al. 1991] and FREZCHEM modified to include Fe, Al, and Si [Marion et al. 2009]). These models also showed that the acid brines should be expected to precipitate multiple minerals, including iron oxides, jarosite, quartz, kaolinite, alunite, and gypsum (McArthur et al. 1991, Marion et al. 2009).

Comparisons with Mars and Lessons from an Analog

Our understanding of the history of sedimentary processes on Mars has changed dramatically in the last decade (Grotzinger et al. 2011). Significantly, multiple different types of evaporite minerals that record details about the physical and chemical characteristics of the environment where they are formed have been identified on Mars (Gendrin et al. 2005, Langevin et al. 2005, Tosca and McLennan 2006, Osterloo et al. 2008). Saline systems in general can evolve via multiple chemical pathways (e.g., Hardie and Eugster 1970). The addition of extreme acidity to these systems introduces an additional level of complexity, resulting in a unique suite of Fe, Al, and Si authigenic precipitates from the low-pH brines. Understanding the complex evolution of the acid brines and associated sediments in WA can help us to predict the processes that have been important in modifying acid saline sediments on Mars, and the potential that these types of sediments may have for preserving astrobiological signatures.

"Burns Formation": Sedimentary rocks in the Meridiani Planum region were extensively examined by the *Opportunity* rover and were dubbed the "Burns Formation" after Roger Burns, the exemplary planetary geologist who promoted the notion of acid fluids on the surface of the red planet (e.g., Burns 1987, 1989, 1993; Burns and Fisher 1990). The Burns formation shares many significant similarities with the sediments in WA. The Late Noachian or Early Hesperian (age ca. 3.5 Ga or older) Burns formation deposits are composed of an evaporitic sandstone that exhibits a transition from a dry dune facies to a sand sheet facies. These rocks contain wind-ripple laminations, evidence of water-table fluctuations, and an interdune facies that includes evidence of some subaqueous deposition, including festoon cross-bedding and tepee salt-ridge structures (Grotzinger et al. 2005). The grains are composed of reworked "dirty" evaporites that include a chemical precipitate from a fluid (either groundwater or surface water) around some weathered host basaltic volcaniclastic sediment (McLennan et al. 2005, Clark et al. 2005). While there is not clear evidence of clay minerals within the Burns formation, geochemical data at Eagle and Endurance crater do show Si in excess of Al, suggesting aluminosilicate minerals that are interpreted to possibly be phyllosilicates (e.g., kaolinite) in the fine-grained component of the outcrop matrix (Clark et al. 2005).

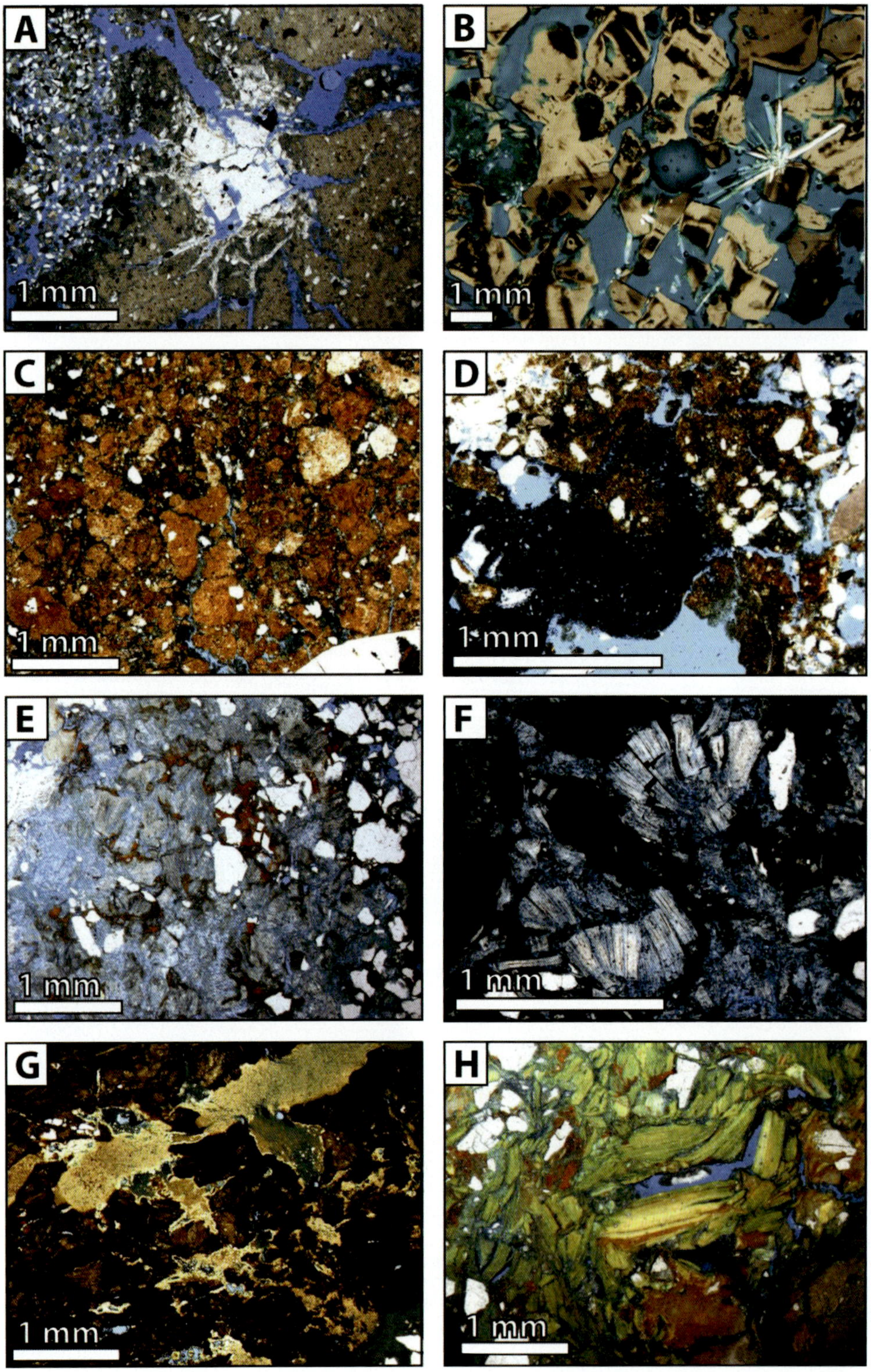

FIG. 13.—Examples of acid saline sediments that are composed of dominantly authigenic and potentially displacive cements. **A)** Lake Brown 2006 midlake core, sample #24, combined polarized transmitted and reflected light; displacive halite in cracked silt and sand matrix. **B)** Twin Lake West 2005 midlake core halite; polarized transmitted light; partially dissolved halite with displacive splay of gypsum. **C)** Lake Aerodrome 2005 midlake core, sample #6, plane transmitted light; typical spheroidal iron-oxide groundmass texture with autoclastic brecciation. **D)** Lake Aerodrome 2005 sand flat core, sample #18, combined transmitted plane and reflected light; detrital quartz and gypsum grains cemented by iron oxide (brown) and efflorescent rozenite (black, confirmed by XRD). **E)** Twin Lake West 2005 midlake core, sample #47, combined polarized transmitted and reflected light; iron-oxide grain coatings around individual displacive kaolinite books. **F)** Twin Lake

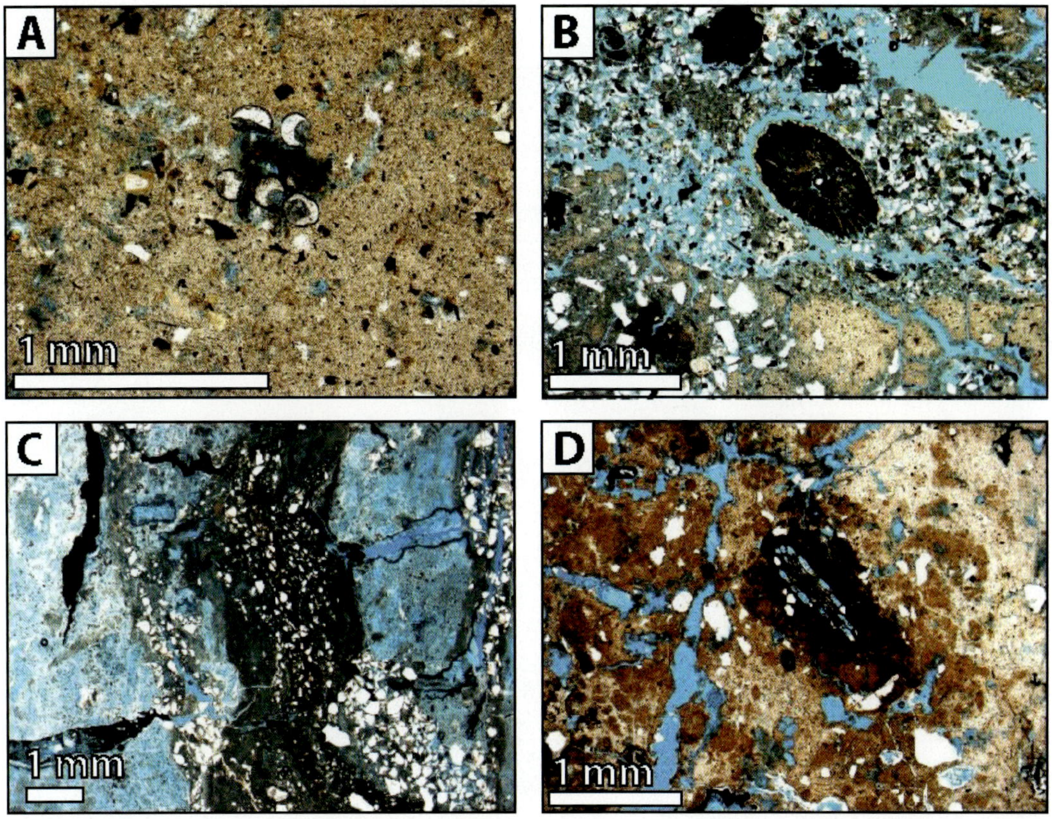

FIG. 14.—Photomicrograph examples of fossil material within acid saline sediments. **A)** Lake Brown midlake core, sample #24, plane transmitted light; remnants of microfossils (ostracodes?) trapped within clay and iron oxide groundmass. **B)** Lake Brown midlake core, sample #24, plane transmitted light; multichambered microfossil (dascyladacean algae?) with carbonate preserved within chambers) in sand bed. **C)** Twin Lake West mud flat core, sample #51, combination of polarized transmitted and reflected light; coarser detrital quartz grains filling in root feature and associated changes in mineralogy potentially related to gradients in redox chemistry around root. **D)** Lake Aerodrome, sand flat core, sample #16, plane transmitted light; texture of organic remains (plant material?) and cracking around groundmass of kaolinite with patchy nodular iron-oxide staining.

The Burns formation also contains evidence of groundwater diagenesis, such as diagenetic fronts represented by abrupt secondary color changes (e.g., the Whatanga contact; Grotzinger et al. 2005), elongate vugs, and hematite spherules interpreted as concretions within the sandstone. The presence of concretions in the sedimentary rocks on Mars suggests that there was not only episodic surface water in the past on Mars, but that liquid water also saturated the sediments or sedimentary rocks of the region (Chan et al. 2004). This implies a well-developed hydrologic groundwater system on the surface of Mars. The hematite mineralogy of these concretions suggests changing acid and/or redox conditions to facilitate iron mobility and precipitation.

Spheroidal hematite concretions are observed actively precipitating within sediments (<3000 years old) composed of siliciclastics and intrabasinally reworked evaporite grains in WA (Bowen et al. 2008). Nickel enrichment in hematite concretions and iron-rich sediments in WA has been observed to have Fe/Ni ratios that range from 800 to over 10,000, and it appears to be influenced by the type of iron-oxide precipitate, with the higher Ni concentrations in the hematite-bearing sediments and lower concentrations in goethite-bearing sediments (Fig. 7). The sediments with the highest amounts of Ni occur in lakes over paleovalleys with significant sedimentary fill, while the sediments with less Ni are all from a lake that occurs directly on Archean basement that has little sediment cover. Mars sedimentary rocks are also enriched in Ni. The Burns formation has 400–1000 ppm Ni, and it has been estimated (McClennan et al. 2005, Morris et al. 2005) that the concretions have up to 1800 ppm Ni. The significance of Ni enrichment on Mars is not yet understood, but it may relate to the

←

West 2006 sand flat core, sample #35, plane transmitted light; displacive kaolinite cement. **G)** Twin Lake West 2006 sand flat core, sample #33, combined plane transmitted and reflected light; zone of complex iron-oxide and jarosite cementation, white gypsum cements lining and partially filling crazed plane cracks. **H)** Twin Lake West 2005 midlake core, sample #46, combined polarized transmitted and reflected light; mixture of jarosite (large fibrous yellow grains) and hematite cement.

TABLE 5.—Select minor, trace, and rare earth element components of sediment geochemistry for subset of surface sediments in contact with acid brines.

		Element									
		MgO (%)	K_2O (%)	TiO_2 (%)	P_2O_5 (%)	Cu (ppm)	Ni (ppm)	Zn (ppm)	Pb (ppm)	V (ppm)	Ba (ppm)
	Detection limit	0.01	0.01	0.001	0.01	1	1	1	5	5	3
	Reference method	FUS-ICP	FUS-ICP	FUS-ICP	FUS-ICP	TD-ICP	TD-ICP	TD-ICP	TD-ICP	FUS-ICP	FUS-ICP
	Lake										
1a	Lake Brown	0.06	0.08	0.005	–	–	–	1	–	–	14
1b	Lake Brown	1.5	0.09	1.125	0.36	15	82	84	22	94	107
1c	Lake Brown	0.64	1.53	0.329	0.03	19	27	20	21	28	362
1d	Lake Brown	0.5	2.91	0.285	0.03	23	13	10	25	19	791
1e	Lake Brown	1.43	1.26	0.397	0.06	23	59	46	101	45	228
1f	Lake Brown	0.47	1.8	0.143	–	6	6	6	15	7	526
1g	Lake Brown	0.99	2.96	0.252	0.04	16	15	17	30	20	803
2a	Walker Lake	0.03	0.09	–	–	1	–	–	–	–	5
3a	Magic	0.33	0.11	0.004	–	–	–	–	–	–	4
3b	Magic	0.02	0.38	0.033	0.02	5	2	2	5	–	132
3c	Magic	0.4	2.45	0.376	0.02	11	10	7	17	41	380
3d	Magic	0.64	0.07	0.002	–	1	–	–	–	–	5
4a	Aerodrome	0.04	0.09	0.002	–	3	1	1	–	–	5
4b	Aerodrome	0.17	–	0.007	–	2	2	–	–	–	8
4c	Aerodrome	0.16	0.46	0.262	0.03	14	29	11	5	49	144
4d	Aerodrome	1.62	0.8	0.275	0.04	46	69	575	167	74	153
4e	Aerodrome	0.29	4.07	0.269	–	55	14	9	11	28	156
4f	Aerodrome	0.45	1.62	0.425	0.06	47	38	38	16	79	225
4g	Aerodrome	1.08	2.01	0.382	0.06	42	57	27	20	77	253
8a	Twin Lake	0.27	1.6	2.208	0.06	37	4	18	21	91	281
8b	Twin Lake	0.43	0.74	0.943	0.03	158	9	55	23	132	216
8c	Twin Lake	0.29	4.24	0.514	0.04	25	7	17	9	47	340
8d	Twin Lake	0.07	–	0.01	–	–	1	2	–	–	13
8e	Twin Lake	0.57	0.74	0.347	–	17	8	11	6	62	240
8f	Twin Lake	0.42	2.42	0.196	–	12	5	9	30	32	181
8g	Twin Lake	0.26	0.34	0.184	0.04	13	7	34	50	20	54

– = below detection limit.

associated iron-oxide mineralogy and the amount of iron cycling that has occurred within the sediments.

Hematite concretions on Mars postdate syndepositional to early diagenetic minerals now represented by crystal molds and pore-filling cements, but they appear to predate secondary vugs of moldic crystal porosity. The excellent preservation of the eolian sandstone sulfate grains shows that subsequent surface water (depositing the upper Burns formation) and groundwaters did not dissolve or significantly alter these grains, implying that the fluids were all saturated with respect to the sulfate minerals in the grains.

The mineral jarosite has been identified in the Burns formation with Mossbauer spectroscopy, alpha particle X-ray spectrometry, and infrared spectroscopy (Christensen et al. 2004, Klingelhöfer et al. 2004). On Earth, jarosite is stable under low-pH, oxidizing, and arid conditions, and it commonly forms during acid-sulfate alteration of volcanic rocks, or during sulfide oxidation in acid-mine drainage settings. In WA, jarosite forms from the low-pH, sulfate- and iron-rich brines as an *evaporite* mineral (e.g., Long et al. 1992). Acid-sulfate minerals such as jarosite occur in other settings such as acid bog lake sediments and hydrothermal acid-sulfate alteration systems (McCollom and Hynek 2005), but they do not coexist with iron oxides and evaporite minerals in these types of settings.

Comparisons between the Burns formation and modern acid saline sediments in WA show that both systems contain a unique mineral assemblage indicative of acid brines, in particular co-occurrence of halite, gypsum, hematite, jarosite, and phyllosilicates. In WA, we observe that these minerals are all intimately mixed at microscopic scales, and while they may not necessarily represent an equilibrium assemblage, they persist together in the acid saline conditions. Our petrographic observations illustrate how these minerals co-occur: as a complex combination of detrital grains, reworked chemical sediments, and diagenetic cements. As on Mars, early diagenetic features in WA include displacive sulfate crystals and hematite concretions as well as jarosite cement. Our observations in WA illustrate the short timescales (thousands of years and less) that are needed to form these diagenetic features and the complex spatial heterogeneity that can be expected in acid saline diagenetic systems (e.g., Fig. 6).

On Earth, modern dune and interdune environments are commonly

Table 5.—Extended.

Element										
Sr (ppm)	Zr (ppm)	Cr (ppm)	La (ppm)	Ce (ppm)	Nd (ppm)	Sm (ppm)	Eu (ppm)	Gd (ppm)	Hf (ppm)	U (ppm)
2	4	20	0.1	0.1	0.1	0.1	0.05	0.1	0.2	0.1
FUS-ICP	FUS-ICP	FUS-MS	FUS-MS	FUS-MS	FUS-MS	FUS-MS	FUS-MS	FUS-MS	FUS-MS	FUS-MS
107	–	30	1.5	4.8	1.3	0.2	–	0.2	–	0.1
248	39	80	6.7	19.5	4.5	1	0.29	1	4.3	12.9
72	125	140	25.9	26.8	6.4	1	0.22	0.8	3.4	5.1
171	281	30	19	25.2	6.5	1.2	0.38	1	6.9	5.5
145	140	50	93.4	285	91.8	16.4	2.93	14.7	3.8	18.9
191	106	–	10.9	22.8	7.4	1.4	0.35	1.1	2.5	7.4
288	165	30	27.6	35	6.8	1.2	0.4	1	4.3	13.2
720	–	–	13.5	41.8	12.4	2.1	0.44	1.2	–	–
3562	–	–	66	182	46.8	8.5	1.71	10.9	–	0.1
30	30	110	2	3.7	1.1	0.2	–	0.2	0.9	0.6
266	276	40	11.5	17.9	4.8	0.9	0.18	0.8	7.1	3
44	–	20	0.7	1.7	0.4	–	–	–	–	–
188	–	–	4.4	14.4	8.1	2	0.53	1.8	–	–
209	–	–	0.5	1.2	0.8	0.2	–	0.2	–	0.2
78	38	290	5.5	9.8	3.5	0.7	0.16	0.7	1.1	0.8
191	36	200	6.7	13.1	5.5	1.2	0.38	1.4	1.1	0.9
494	74	230	4.5	6.2	1.8	0.4	0.07	0.3	2	0.5
506	89	320	13.7	16.1	3.3	0.6	0.16	0.5	2.4	1.7
557	74	310	18.5	24.5	5.6	1	0.23	0.9	2.1	1.7
406	181	50	12.7	16.4	3.9	0.7	0.26	0.6	4.9	6.4
45	365	60	13	30.8	12.1	2.2	0.51	1.6	8.5	6.6
379	242	140	15.5	18.5	5.3	0.9	0.23	0.7	6	1.9
17	–	30	1.1	3.5	1.5	0.3	0.05	0.2	–	–
51	93	30	5	8.2	2.7	0.4	0.1	0.3	2.3	1.6
118	668	30	70.1	119	40.5	7	0.73	5.4	16	3.5
24	185	–	98.6	86.9	44.1	7.8	1	6.5	5.8	5.3

considered deflational vs. depositional, and they typically have a low preservation potential. In contrast, the Burns formation exhibits excellent preservation, which may be a result of the influence of the acid saline fluids. In WA, we observe very early or syndepositional precipitation of iron oxides and evaporitic cements that aid in preservation of these ephemeral and potentially soluble deposits. In addition, the rapid growth of chemical sediments such as iron oxides and evaporites suggests the possibility of microfossil entrapment and preservation within acid saline sediments (Benison et al. 2008, Fernandez-Remolar and Knoll 2008). In WA, early iron-oxide grain coatings armor the more soluble grains (Fig. 10) and prevent them from dissolving when waters freshen with meteoric input, allowing minerals with varying solubility to coexist in a single assemblage. The presence of soluble and insoluble mineral phases together and the presence of hematite concretions and evaporite crystal molds on Mars have led to diagenetic interpretations requiring multiple distinct fluids (McLennan et al. 2005). However, in WA, we have observed this same combination of diagenetic minerals and textures forming very early in shallow sediments from the same acid saline groundwater that undergoes minor changes in chemistry on seasonal scales as the result of flooding-evapoconcentration-desiccation cycles. Similar seasonal-scale diagenetic processes should be considered for Mars.

"Crater Lakes": With increasing spectral analysis of surface mineralogy, several more settings have been identified on Mars beyond the Burns formation that suggest a history of acid saline lacustrine systems analogous to the environments in WA. For example, intracrater clay-sulfate deposits were identified by Swayze et al. (2008), Wray et al. (2008), and Wray et al. (2009) in two closed-basin craters in Terra Sirenum. Of the proposed crater lakes, Columbus crater hosts the most diverse mineralogy and is thought to have been one of several groundwater-fed paleolakes in the Terra Sirenum region of the Martian southern highlands (Wray et al. 2011). Minerals identified in Columbus crater include gypsum, interbedded kaolinite, hydrated Fe/Mg-sulfates, montmorillonite, alunite, jarosite, and ferric oxide or hydroxide (e.g., Milliken et al. 2008, Weitz et al. 2010, Wray et al. 2011), which combine to form a similar suite of minerals as that observed in WA. Interestingly, the alkaline and/or high-temperature secondary minerals prevalent elsewhere on Mars (e.g. carbonates, chlorides, zeolites, and/or prehnite; Ehlmann et al. 2008, 2009;

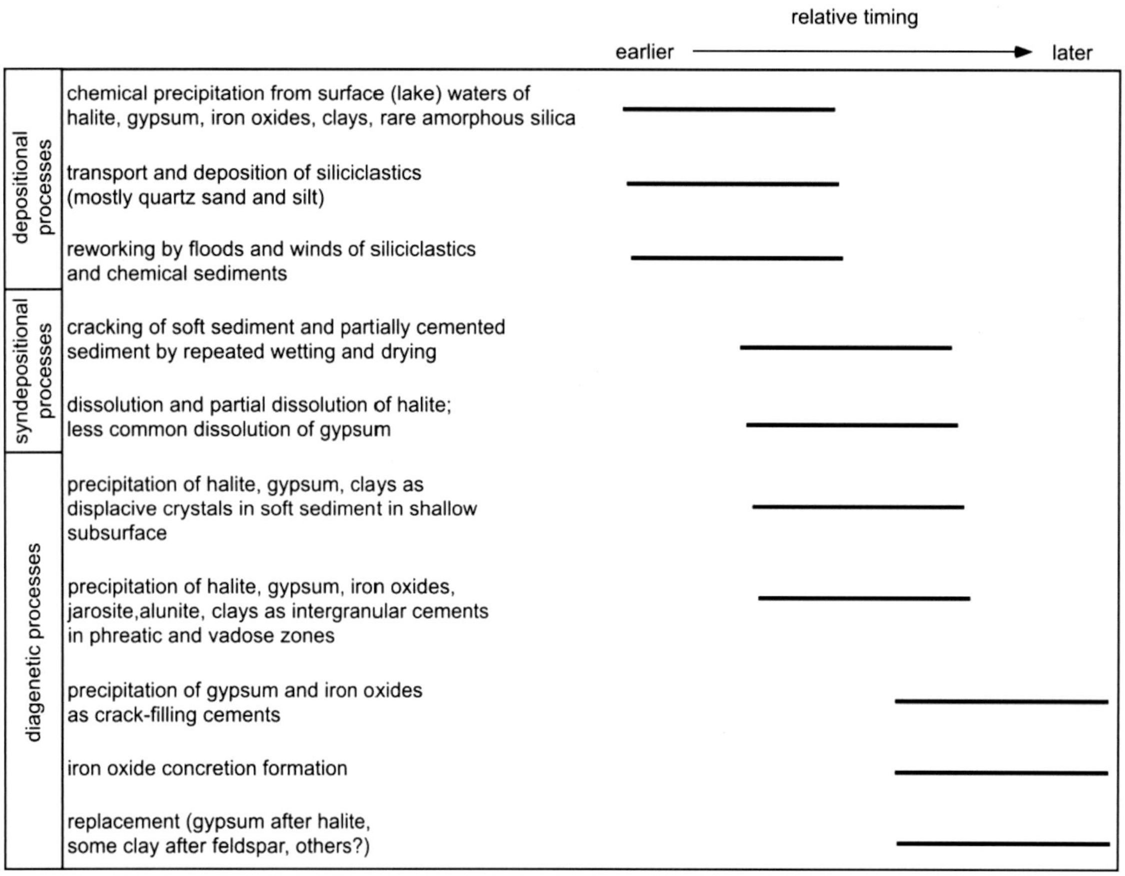

FIG. 15.—Paragenetic sequence for modern acid saline lakes and adjacent sand flat and mud flat facies in southern Western Australia. Please note that all diagenetic features form in shallow subsurface, only centimeters to tens of centimeters below the surface. Also note that depositional, syndepositional, and diagenetic processes overlap in timing and many are concurrently occuring.

Osterloo et al. 2008) have not yet been detected in the Terra Sirenum craters (Wray et al. 2011). The proposed formation mechanisms for these deposits on Mars include mineral precipitation from groundwater to form extensive cements, precipitation from surface springs, or chemical sedimentation from acid saline lakes (Baldridge et al. 2009; Wray et al. 2009, 2011). The types of acid saline groundwater-derived cements precipitated in the Terra Sirenum system may be analogous to those observed in WA. Modeling data suggest that the origin of the aqueous deposits within Columbus crater likely coincided with an inferred period of Late Noachian groundwater upwelling and evaporation; however, the proposed paleolake is hypothesized to have been ~900 m deep (Wray et al. 2011).

Similar to the sedimentary deposits on Mars, the sedimentary deposits in terrestrial acid saline systems contain a combination of siliciclastic and chemically precipitated sediments that can be difficult to differentiate. Acid saline mineral assemblages include jarosite, alunite, and hematite together with chloride and sulfate minerals such as halite and gypsum, as well as several types of phyllosilicates that exhibit complex spatial associations and heterogeneity on macro- and microscopic scales (e.g., Story et al. 2010). The occurrence of phyllosilicates is influenced by multiple factors, including the composition of the source rocks throughout the catchment, the hydrochemistry of fluids in contact with these rocks, and the climate, which determines the temperature and water availability (Fernandez-Remolar et al. 2010). Phyllosilicates found on the surface of Mars have been interpreted as remnants of a pH-neutral water-rich era with extreme weathering (Bibring et al. 2006), requiring a fluid and climate regime distinctly separate from that which would facilitate precipitation of sulfate- and iron-oxide–rich sediments. However, in WA, we observe sediments with all of these constituents, suggesting that an extreme change in climate and chemistry may not have been needed to produce these varied deposits on Mars. Other extreme acid systems, including Rio Tinto, also exhibit alternating phyllosilicate–acid-sulfate minerals due to seasonal changes (Fernandez-Remolar et al. 2010). All of these extreme acid terrestrial analogs suggest that seasonal variations in pH and fluid chemistry may be sufficient to explain the occurrence of these types of sulfate, phyllosilicate, and iron-oxide–rich acid saline sedimentary deposits on Mars.

CONCLUSIONS

Mineralogical, geochemical, and petrographic studies of sediments and sedimentary rocks from acid saline lakes and adjacent environments in Western Australia show that multiple different diagenetic processes occur very early after deposition. These processes include: (1) syndepositional modification and reworking of detrital grains by winds and water, (2) precipitation of chemical sediments from lake waters, (3) solution/partial dissolution of soluble minerals, (4) displacive growth of crystals from groundwater, (5) intergranular cement precipitation from groundwater (including iron-oxide concretions), (6) cracking due to drying (and repeated wetting and drying), and (7) precipitation of crack- and pore-filling cements. Many of these processes are concurrent and syndepositional. The acid saline systems in Western Australia provide a process analog for acid saline deposits on Mars, and while there are some important differences between the two systems, they share the same suite of minerals and likely have experienced many of the same processes. This work demonstrates the complex relationships among acid brine chemistry, mineral weathering, and authigenic precipitates within sedimentary deposits. Geochemical and mineralogical data can be best interpreted in conjunction with careful petrographic analysis, which illustrates spatial associations and crosscutting relationships between minerals that only co-occur in acid saline systems, such as halite, gypsum, iron oxide, jarosite, and phyllosilicates. Acid saline sediments in Western Australia do entrap and potentially preserve biological remains, suggesting that there is some likelihood that sediments associated with similar extreme fluids on Mars could host astrobiological materials.

ACKNOWLEDGMENTS

Partial funding for this study was provided by National Science Foundation grants EAR-0719822, EAR-0719838, 621, and EAR-0719892 (to F.E. Oboh-Ikuenobe, B.B. Bowen, and K.C. Benison), and EAR-0433044 622 and EAR-0433040 (to M. Mormile, F.E. Oboh-Ikuenobe, and K.C. Benison). Acknowledgment is made to the donors of the American Chemical Society–Petroleum Research Fund for partial support of this research. We thank F.E. Oboh-Ikuenobe, M. Mormile, B-Y. Hong, E. Jagniecki, J. Knapp, D. LaClair, and M. Hein for field assistance. Thanks also go to R. Bobick, M. Williams for laboratory assistance.

REFERENCES

Altheide TS, Chevrier VF, Dobrea EN. 2010. Mineralogical characterization of acid weathered phyllosilicates with implications for secondary Martian deposits. *Geochimica et Cosmochimica Acta* 74:6232–6248.

Anand R, Paine M. 2002. Regolith geology of the Yilgarn Craton. *Australian Journal of Earth Sciences* 49:3–162.

Andrews-Hanna JC, Phillips RJ, Zuber MT. 2007. Meridiani Planum and the global hydrology of Mars. *Nature* 446:163–166.

Baldridge AM, Hook SJ, Crowley JK, Marion GM, Kargel JS, Michalski JL, Thomson BJ, de Souza Filho CR, Bridges NT. 2009. Contemporaneous deposition of phyllosilicates and sulfates: Using Australian acidic saline lake deposits to describe geochemical variability on Mars. *Geophysical Research Letters* 36:L19201.

Bell JF III, McSween HY Jr, Crisp JA, Morris RV, Murchie SL, Bridges NT, Johnson JR, Britt DT, Golombek MP, Moore HJ, Ghosh A, Bishop JL, Anderson RC, Bruckner J, Economou T, Greenwood JP, Gunnlaugsson HP, Hargraves RM, Hviid S, Knudsen JM, Madsen MB, Reid R, Rieder R, Soderblom L. 2000. Mineralogic and compositional properties of Martian soil and dust: Results from *Mars Pathfinder. Journal of Geophysical Research–Planets* 105:1721–1755.

Benison KC, Bowen BB. 2006. Acid saline lake systems give clues about past environments and the search for life on Mars. *Icarus* 183:225–229.

Benison KC, Bowen BB, Oboh-Ikuenobe FE, Jagniecki EA, LaClair DA, Story SL, Mormile MR, Hong B. 2007. Sedimentology of acid saline lakes in southern Western Australia: Newly described processes and products of an extreme environment. *Journal of Sedimentary Research* 77:366–388.

Benison KC, Goldstein RH. 2002. Recognizing acid lakes and groundwaters in the rock record. *Sedimentary Geology* 151:177–185.

Benison KC, Jagniecki EA, Edwards TB, Mormile MR, Storrie-Lombardi MC. 2008. "Hairy blobs": Microbial suspects from modern and ancient ephemeral acid saline evaporates. *Astrobiology* 8(4):807–821.

Benison KC, LaClair DA. 2003. Modern and ancient extremely acid saline deposits: Terrestrial analogs for Martian environments? *Astrobiology* 3:609–618.

Bethke CM, Yeakel S. 2009. *The Geochemist's Workbench®, Version 8.0*: *GWB Essentials Guide*: Hydrogeology Program, University of Illinois, Urbana. 84 p.

Bibring J-P, Langevin Y, Mustard JF, Poulet F, Arvidson R, Gendrin A, Gondet B, Mangold N, Pinet P, Forget F, OMEGA Team. 2006. Global mineralogical and aqueous Mars history derived from *OMEGA/Mars Express* data. *Science* 312:400–404.

Bowen BB, Benison KC. 2009. Geochemical characteristics of naturally acid and alkaline saline lakes in southern Western Australia. *Applied Geochemistry* 24:268–284.

Bowen BB, Benison KC, Oboh-Ikuenobe FE, Story S, Mormile MR. 2008. Active hematite concretion formation in modern acid saline lake sediments, Lake Brown, Western Australia. *Earth and Planetary Science Letters* 288:52–63.

Burns RG. 1987. Ferric sulfates on Mars. *Journal of Geophysical Research* 92:E570–E574.

Burns RG. 1989. Terrestrial analogues of the surface rocks of Mars? *Nature* 320:55–56.

Burns RG. 1993. Rates and mechanisms of chemical weathering of ferromagnesian silicate minerals on Mars. *Geochimica et Cosmochimica Acta* 57:4555–4574.

Burns RG, Fisher DS. 1990. Iron-sulfur mineralogy of Mars: Magmatic evolution and chemical weathering products. *Journal of Geophysical Research* 95:14415–14421.

Chan MA, Beitler B, Parry WT, Ormö J, Komatsu G. 2004. A possible terrestrial analogue for hematite concretions on Mars. *Nature* 429:731–734.

Chilingarian GV, Wolf KH. 1988. *Analytical and Petrographic Methods Used in the Study of Diagenesis or Sandstones*. Elsevier, New York. Developments in Sedimentology 41, 591 p.

Christensen PR, and 26 colleagues. 2004. Mineralogy at Meridiani Planum from the Mini-TES experiment on the *Opportunity* rover. *Science* 306:1733–1739.

Clark BC, Van Hart DC. 1981. The salts of Mars. *Icarus* 45:370–378.

Clark BC, Morris RV, McLennan SM, Gelbert R, Jolliff B, Knoll AH, Squyres SW, Lowenstein TK, Ming DW, Tosca NJ, and 13 colleagues. 2005. Chemistry and mineralogy of outcrops at Meridiani Planum:. *Earth and Planetary Science Letters* 240:73–94.

Clarke JDA. 1993. Stratigraphy of the Lefroy and Cowan palaeodrainages, Western Australia. *Journal of the Royal Society of Western Australia* 76:13–23.

Clarke JDA, Bone Y, James NP. 1996. Cool-water carbonates in an Eocene paleoestuary, Norseman Formation, Western Australia. *Sedimentary Geology* 101:213–226.

de Broekert P, Sandiford M. 2005. Buried inset-valleys in the Eastern Yilgarn Craton, Western Australia: Geomorphology, age, and allogenic control. *The Journal of Geology* 113:471–493.

De Deckker P. 1983. Australian salt lakes: Their history, chemistry, and biota—a review. *Hydrobiologia* 105:231–244.

Ehlmann BL, Mustard JF, Murchie SL, Poulet F, Bishop JL, Brown AJ, Calvin WM, Clark RN, Des Marais DJ, Milliken RE, Roach LH, Roush TL, Swayze GA, Wray JJ. 2008. Orbital identification of carbonate-bearing rocks on Mars. *Science* 322:1828–1832.

Ehlmann, BL, Mustard JF, Swayze G, Clark RN, Bishop JL, Poulet F, Des Marais DJ, Roach LH, Milliken RE, Wray JJ, Barnounin-Jha O, Murchie S. 2009. Identification of hydrated silicate minerals on Mars using MRO-CRISM: Geologic context near Nili Fossae and implications for aqueous alteration. *Journal of Geophysical Research* 114:E00D08.

Fernandez-Remolar DC, Knoll AH. 2008. Fossilization potential of iron-bearing minerals in acidic environments of Rio Tinto, Spain: Implications for Mars exploration. *Icarus* 194:72–85.

Fernandez-Remolar DC, Prieto-Ballesteros O, Gomez-Ortiz D, Fernandez-Sampredo M, Sarrazin P, Gailhanou M, Amils R. 2010. Rio Tinto sedimentary mineral assemblages: A terrestrial perspective that suggests some formation pathways of phyllosilicates on Mars. *Icarus* 211:114–138.

Fishbaugh KE, Poulet F, Chevrier V, Langevin Y, Bibring J-P. 2007. On the origin of gypsum in the Mars north polar region. *Journal of Geophysical Research* 12:E07001, doi:10.1029/2006JE002862.

Gendrin A, Mangold N, Bibring JP, Langevin Y, Gondet B, Poulet F, Bonello G, Quantin C, Mustard J, Arvidson RE, LeMouelic S. 2005. Sulfates in Martian layered terrains: The *OMEGA/Mars Express* view. *Science* 307:1587–1591.

Gray DJ. 2001. Hydrogeochemistry in the Yilgarn Craton. *Geochemistry: Exploration, Environment, Analysis* 1:253–264.

Gregory JW. 1914. The lake system of Westralia. *The Geographical Journal* 43:656–664.

Grotzinger JP, Arvidson RE, Bell JF, Calvin W, Clark BC, Fike DA, Golombek M, Greeley R, Haldemann A, Herkenhoff KE, Jolliff BL, Knoll AH, Malin M, McLennan SM, Parker T, Soderblom L, Sohl-Kickstein JN, Squyres SW, Tosca NJ, Watters WA. 2005. Stratigraphy and sedimentology of a dry to wet eolian depositional system, Burns Formation, Meridiani Planum, Mars. *Earth and Planetary Science Letters* 240:11–72.

Grotzinger JP, Beaty D, Dromart G, Gupta S, Harris M, Hurowtiz J, Kocurek G, McLennan S, Milliken R, Ori GG, Sumner D. 2011. Mars sedimentary geology: Key concepts and outstanding questions. *Astrobiology* 11:77–87.

Hardie LA, Eugster HP. 1970. The evolution of closed-basin brines. *In* Morgan BA (Editor). *Mineralogy and petrology of the Upper Mantle; Sulfides; Mineralogy and geochemistry of non-marine evaporites,* Fiftieth anniversary symposia, Issue 3 of Special Paper: Mineralogical Society of America, Washington DC. 3:273–290.

Hem JD. 1985. Study and Interpretation of the Chemical Characteristics of Natural Water: U.S. Geological Survey, Water-Supply Paper 2254, Alexandria, VA. 263 p.

Hurowitz JA, Fischer WW, Tosca NJ, Milliken RE. 2010. Origin of acidic surface waters and the evolution of atmospheric chemistry on early Mars. *Nature Geoscience* 3:323–326.

Iler RK. 1979. The chemistry of silica: John Wiley and Sons, Inc., New York. 866 p.

Jagniecki EA, Benison KC. 2010. Criteria for the recognition of acid-precipitated halite. *Sedimentology* 57:273–292.

Klingelhöfer G, and 18 colleagues. 2004. Jarosite and hematite at Meridiani Planum from *Opportunity*'s Mössbauer spectrometer. *Science* 306:1740–1745.

Langevin Y, Poulet F, Bibring JP, Gondet B. 2005. Sulfates in the north polar region of Mars detected by *OMEGA/Mars Express. Science* 307:1584–1586.

Last WM. 1994. Paleohydrology of playas in the northern Great Plains: Perspectives from Pallister's Triangle. *In* Rosen MR (Editor), *Paleoclimate and Basin Evolution of Playa Systems*: Geological Society of America, Boulder, Colorado, Special Paper 289, p. 69–80.

Long DT, Fegan NE, McKee JD, Lyons WB, Hines ME, Macumber PG. 1992. Formation of alunite, jarosite and hydrous iron oxides in a hypersaline system: Lake Tyrrell, Victoria, Australia. *Chemical Geology* 96:183–202.

Long DT, Lyons WB, Hines ME. 2009. Influence of hydrogeology, microbiology and landscape history on the geochemistry of acid hypersaline waters, N.W. Victoria. *Applied Geochemistry* 24:285–296.

Lottermoser BB. 2007. *Mine Wastes, Characterization, Treatment and Environmental Impacts*, 2nd ed.: Springer-Verlag, Berlin. 304 p.

Mann AW. 1983. Hydrogeochemistry and weathering on the Yilgarn Block, Western Australia—ferrolysis and heavy metals in continental brines. *Geochimica et Cosmochimica Acta* 47:181–190.

Marion GM, Crowley JK, Thomson BJ, Kargel JS, Bridges NT, Hook SJ, Baldridge A, Brown AJ, Ribeiroda Luz B, de Souza Filho CR. 2009. Modeling aluminum-silicon chemistries and applications to Australian acidic playa lakes as analogues for Mars. *Geochimica et Cosmochimica Acta* 73:3493–3511.

McArthur JM, Turner J, Lyons WB, Osborn AO, Thirlwall MF. 1991. Hydrochemistry on the Yilgarn Block, Western Australia: Ferrolysis and mineralization in acidic brines. *Geochimica et Cosmochimica Acta* 55:1273–1288.

McArthur JM, Turner J, Lyons WB, Thirlwall MF. 1989. Salt sources and water–rock interactions on the Yilgarn Block, Australia: Isotopic and major element tracers. *Applied Geochemistry* 4:79–92.

McCollom TM, Hynek BM. 2005. A volcanic environment for bedrock diagenesis at Meridiani Planum on Mars. *Nature* 438, doi:10.1038/nature04390.

McIlreath I, Morrow D (Editors). 1990. *Diagenesis*: Geological Association of Canada, St John's, NL. Geoscience Canada Reprint Series 4. 338 p.

McLennan SM, Bell JF III, Calvin WM, Christensen PR, Clark BC, de Souza PA, Farmer J, Farrand WH, Fike DA, Gellert R, Ghosh A, Glotch TD, Grotzinger JP, Hahn B, Herkenhoff KE, Hurowitz JA, Johnson JR, Johnson SS, Jolliff B, Klingelhofer G, Knoll AH, Learner Z, Malin MC, McSween HY, Pocock J, Ruff SW, Soderblom LA, Squyres SW, Tosca NJ, Watters WA, Wyatt MB, Yen A. 2005. Provenance and diagenesis of the evaporite-bearing Burns formation, Meridiani Planum, Mars. *Earth and Planetary Science Letters* 240:95–121.

Milliken RE, Swayze GA, Arvidson RE, Bishop JL, Clark RN, Ehlmann BL, Green RO, Grotzinger JP, Morris RV, Murchie SL, Mustard JF, Weitz C. 2008. Opaline silica in young deposits on Mars. *Geology* 36:847–850.

Mormile MR, Hong B-Y, Benison KC. 2009. Molecular analysis of the microbial communities of Mars analog lakes in Western Australia. *Astrobiology* 9:919–930.

Morris RV, Ming DW, Graff TG, Arvidson RE, Bell JF, Squyres SW, Mertzman SA, Gruener JE, Golden DC, Le L, Robinson GA. 2005. Hematite spherules in basaltic tephra altered under aqueous, acid-sulfate conditions on Mauna Kea volcano, Hawaii: Possible clues for the occurrence of hematite-rich spherules in the Burns formation at Meridiani Planum, Mars. *Earth and Planetary Science Letters* 240:168–178.

Osterloo MM, Hamilton E, Bandfield JL, Glotch TD, Baldridge AM, Christensen PR, Tornabene LL, Anderson FS. 2008. Chloride-bearing materials in the Southern Highlands of Mars. *Science* 319:1651–1654.

Peiffer S, Oldham C, Salmon U, Lillicrap A, Kusel K. 2009. Does iron cycling trigger generation of acidity in groundwaters of Western Australia? *Environmental Science and Technology* 43:6548–6552.

Risacher F, Alonso H, Salazar C. 2002. Hydrochemistry of two adjacent acid saline lakes in the Andes of northern Chile. *Chemical Geology* 187:39–47.

Squyres SW, Knoll AH. 2005. Sedimentary rocks at Meridiani Planum: Origin, diagenesis, and implications for life on Mars. *Earth and Planetary Science Letters* 240:1–10.

Squyres SW, Knoll AH, Arvidson RE, Ashley JW, Bell JF, Calvin WM, Christensen PR, Clark BC, Cohen BA, de Souza PA, Edgar L, Farrand WH, Fleischer I, Gellert R, Golombek MP, Li R, McCoy TJ, McLennan SM, Ming DW, Mittlefehldt DW, Morris RV, Rice JW, Schroder C, Sullivan RJ, Yen A, Yingst RA. 2009. Exploration of Victoria Crater by the Mars Rover *Opportunity. Science* 324:1058–1061.

Story S, Bowen BB, Benison KC, Schulze D. 2010. Authigenic phyllosilicates in modern acid saline lake sediments and implications for Mars. *Journal of Geophysical Research* 115:E12012.

Sumner DY. 2004. Poor preservation potential of organics in Meridiani Planum hematite-bearing sedimentary rocks. *Journal of Geophysical Research* 109:E12007.

Swayze GA, Ehlmann BL, Milliken RE, Poulet F, Wray JJ, Rye RO, Clark RN, Desborough GA, Crowley JK, Gondet B, Mustard JF, Seelos KD, Murchie SL. 2008. Discovery of the acid-sulfate mineral alunite in Terra Sirenum, Mars, using MRO CRISM: Possible evidence for acid-saline lacustrine deposits? *Eos (Transactions AGU)* 89(53)(Fall Meet. Suppl.):abstract P44A-04.

Tosca NJ, Knoll AH, McLennan SM. 2008. Water activity and the challenge for life on early Mars. *Science* 320:1204–1207.

Tosca NJ, McLennan SM. 2006. Chemical divides and evaporite assemblages on Mars. *Earth and Planetary Science Letters* 241:21–31.

Weitz CM, Milliken RE, Grant JA, McEwen AS, Williams RME, Bishop JL, Thomson BJ. 2010. *Mars Reconnaissance Orbiter* observations of light-toned layered deposits and associated fluvial landforms on the plateaus adjacent to Valles Marineris. *Icarus* 205:73–102.

West MD, Clarke JDA, Thomas M, Pain CF, Walter MR. 2009. The geology of Australian Mars analogue sites. *Planetary and Space Science* 58:447–458.

Woodward HP. 1897. The dry lakes of Western Australia. *Geological Magazine* 4:363–366.

Wray JJ, Milliken, RE, Dundas CM, Swayze GA, Andrews-Hanna JC, Baldridge AM, Chojnacki M, Bishop JL, Ehlmann BL, Murchie SL, Clark RN, Seelos FP, Tornabene LL, Squyres SW. 2011. Columbus crater and other possible groundwater-fed paleolakes of Terra Sirenum, Mars. *Journal of Geophysical Research* 116:E01001.

Wray JJ, Milliken RE, Murchie SL, Swayze GA, Dundas CM, Seelos FP, Squyres SW. 2008. Clays and sulfates in a potential lacustrine evaporite sequence at Columbus Crater, Mars. *Eos (Transactions AGU)* 89(53)(Fall Meet. Suppl.):abstract P53B-1446.

Wray JJ, Murchie SL, Squyres SW, Seelos FP, Tornabene LL. 2009. Diverse aqueous environments on ancient Mars revealed in the southern highlands. *Geology* 37:1043–1046.

CHARACTERISTICS OF TERRESTRIAL FERRIC OXIDE CONCRETIONS AND IMPLICATIONS FOR MARS

MARJORIE A. CHAN AND SALLY L. POTTER
University of Utah, Department of Geology and Geophysics, 115 South 1460 E. Rm. 383 FASB, Salt Lake City, Utah 84112-0102 USA
e-mail: marjorie.chan@utah.edu

BRENDA B. BOWEN
Department of Earth and Atmospheric Sciences, Purdue University, 550 Stadium Mall Drive, West Lafayette, Indiana 47907 USA

W.T. PARRY
University of Utah, Department of Geology and Geophysics, 115 South 1460 E. Rm. 383 FASB, Salt Lake City, Utah 84112-0102 USA

LAURA M. BARGE
*University of Southern California, Department of Earth Sciences, 3651 Trousdale Parkway, Los Angeles, California 90089 USA *Present address: Jet Propulsion Laboratory, California Institute of Technology, 4800 Oak Grove Drive, Pasadena, California 91109 USA*

WINSTON SEILER
*University of Utah, Department of Geology and Geophysics, 115 South 1460 E. Rm. 383 FASB, Salt Lake City, Utah 84112-0102 USA *Present address: Chevron North America Exploration and Production, 9525 Camino Media, Bakersfield, California 93301 USA*

ERICH U. PETERSEN AND JOHN R. BOWMAN
University of Utah, Department of Geology and Geophysics, 115 South 1460 E. Rm. 383 FASB, Salt Lake City, Utah 84112-0102 USA

ABSTRACT: Concretions are diagenetic products of cementation that establish significant records of groundwater flow through porous sedimentary deposits. Common spheroidal ferric oxide concretions form by diffusive coupled with advective mass transfer and share similar physical characteristics with hematite spherules from Meridiani Planum (Mars "blueberries"), investigated by the Mars Exploration Rover *Opportunity*. Terrestrial concretions from the Jurassic Navajo Sandstone are not perfect analogs to Mars, particularly in terms of their geochemistry. However, the Navajo Sandstone contains exceptional examples that represent typical concretion characteristics from the geologic record. Both ancient and modern analogs provide information about concretion forming processes and their relationship to porosity and permeability, fluid flow events, subsequent weathering, and surficial reworking.

Concretions on Earth possess variable mineralogies and form in a variety of lithologies in formations of nearly all geologic ages. Despite the prevalence of concretions, many unknowns exist, including their absolute ages and their precise nucleation and growth mechanisms. Some opportunities for future concretion research lie in three approaches: (1) New analytical techniques may show geochemical gradients and important textures reflecting biotic (role of bacteria) or abiotic origins. (2) Concretion modeling can determine important formation mechanisms. Sensitivity tests and simulations for different parameters can help show the magnitude of influence for different input factors. (3) New age-dating methods that remove preservational bias and expand the supply of datable material may yield quantitative limits to the timing of diagenetic events beyond what relative cross-cutting relationships can show.

The discovery of hematite spherules on Mars has driven efforts to better understand both terrestrial examples of ferric oxide concretions and the competing mechanisms that produce spheroidal geometries. The integration of geologic and planetary sciences continues to encourage new findings in the quest to understand the role of water on Mars as well as the tantalizing possibility that extraterrestrial life is associated with mineral records of watery environments.

KEY WORDS: concretion, Mars, diagenesis, Navajo Sandstone, Burns formation

INTRODUCTION

Iron cycling is common in Earth surface systems. This cycling is often evidenced by the iron oxide minerals left behind in abundant concretions (cemented mineral masses) that can preserve the record of multiple generations of precipitation events and water–rock interactions. We herein use the broad term "ferric oxide," which includes hematite as well as any hydrous ferric oxide (HFO) or oxyhydroxide phases, such as goethite or ferrihydrite. Terrestrial studies of ferric oxide concretions have been encouraged by the Mars Exploration Rover (MER) *Opportunity* discovery of "blueberries" in the Burns formation (Squyres et al. 2004, Grotzinger et al. 2005). Concretion

Sedimentary Geology of Mars
SEPM Special Publication No. 102, Copyright © 2012
SEPM (Society for Sedimentary Geology), Print ISBN 978-1-56576-312-8, CD/DVD ISBN 978-1-56576-313-5, p. 253–270.

preservation in the ancient rock provides a comparative model for the Mars examples that are similarly embedded in host sediments. Although decades ago concretions were viewed simply as geologic "curiosities," it is now clear that the presence of concretions has important implications for understanding groundwater movement and chemistry, diagenesis, host rock properties, biogeochemical processes, and iron cycling (precipitation and mobilization) through time.

The purpose of this article is to review current knowledge of ferric oxide concretions and the implications for Mars and to discuss the current gaps in our understanding. The intent is not to argue why terrestrial concretions serve as an explanation for the hematite spherules on Mars, which has largely been discussed by others (e.g., Chan et al. 2004, Ormö et al. 2004, Grotzinger et al. 2005, McLennan et al. 2005, Potter et al. 2011). Additional types of hematitic concretions and concentrations (e.g., ferromanganese nodules, various pedogenic concretions, bog and marsh concretions, and others) are not addressed in this article as a result of space limitations. Here we focus on concretion occurrences from ancient terrestrial examples of the Jurassic Navajo Sandstone (Figs. 1, 2) (Chan et al. 2004, 2005) and modern examples from the acid lakes of Western Australia (Benison and Bowen 2006, Bowen et al. 2008); both of these have been cited as possible analogs to Meridiani Planum's hematite spherules.

The hematite spherules on Mars form from an abundant supply of iron sourced from basaltic components (Tosca et al. 2004). Under the acidic conditions existing on Mars, iron would be readily mobilized regardless of oxidation state (McLennan et al. 2005, Tosca et al. 2008a). Spherules would be precipitated by oxidation of ferrous iron or by rising pH to precipitate ferric iron (Klingelhöfer et al. 2004, Squyres and Knoll 2005, Tosca et al. 2008a). The pH conditions throughout the diagenetic history of the Burns formation are constrained to less than 4 to 5 because of the presence of jarosite (Madden et al. 2004, Tosca et al. 2005).

Mars is likely unique in its own particular setting of the reactive sulfate and basaltic sandstone host rock and extreme chemical solutions (by Earth standards), along with the acidic conditions for mobilizing iron. Thus, like many other terrestrial concretions, the quartz-rich Jurassic Navajo Sandstone, with relatively inert host grains, is clearly not a perfect analog to Mars, particularly because of its different host rock chemistry (e.g., Catling 2004). However, terrestrial analog studies are pivotal to understanding basic geologic processes and relationships that have application to the inference of fluid flow

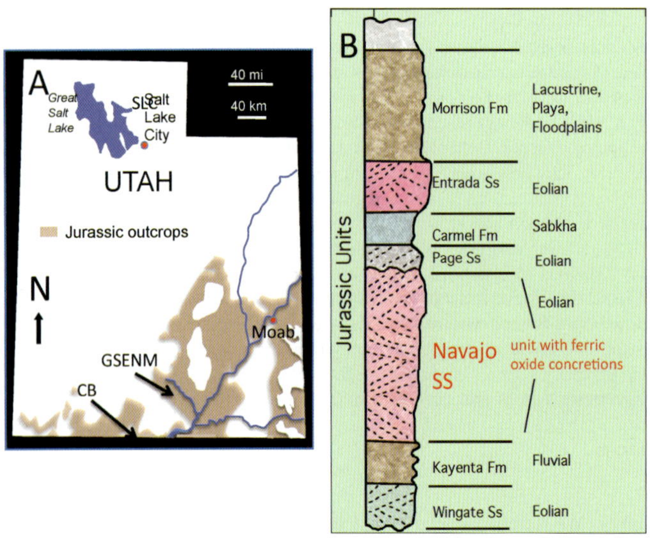

FIG. 1.—(A) Study area of Jurassic Navajo Sandstone terrestrial examples and (B) generalized stratigraphic section of Mesozoic units. Major study sites include localities near Moab, the Spencer Flat area within Grand Staircase Escalante National Monument (GSENM), and Coyote Buttes (CB) within the Vermillion Cliffs National Monument close to the Utah–Arizona border. Major urban center (Salt Lake City) is shown for reference.

FIG. 2.—Comparative concretion characteristics shared between the Jurassic Navajo Sandstone (top first and third rows), with comparative examples of Mars spherules (correlative second and fourth rows) from the Burns formation at Meridiani Planum. Comparative physical characteristics shared between the Jurassic Navajo Sandstone (top first and third rows: A, C, E, G, I, and K), with comparative examples of Mars spherules (correlative second and fourth rows: B, D, F, H, J, and L) from the Burns formation at Meridiani Planum. Mars photos: National Aeronautics and Space Administration/Jet Propulsion Laboratory/Cornell. (A, B) Loose concretions and spheroidal forms weathered out of host rock. These Utah concretions are ~4 cm in diameter, and Mars spherules are ~4.5 mm in diameter. (C, D) Two size populations present together in topographic low. Mars spherules are ~4 mm in diameter. (E, F) In situ concretions or spherules exhibiting ridges and furrows aligned with bedding (arrows). (G, H) Loose concretions with ridges and furrows (arrows). (I, J) Loose concretions with adhered host rock (arrows). (K, L) Conjoined concretion and spherule forms (doublets).

events and timing on Mars. These studies can provide input toward payload instrumentation to evaluate concretions on future Mars missions.

BACKGROUND AND PREVIOUS WORK

Concretions are composed of a localized mass of cement (e.g., ferric oxide) that precipitates in pore spaces during diagenesis in sedimentary rocks. Concretions are known to form in host rocks with diverse compositions; cements can include carbonates, iron oxides, sulfides, and silicates (Mozley 1989, Sellés-Martinez 1996, Seilacher 2001, Mozley and Davis 2005). Ferric oxide concretions form during early diagenesis (e.g., acid saline lake sediments; Benison and Bowen 2006, Bowen et al. 2008) or during burial diagenesis (e.g., Navajo Sandstone concretions; Chan et al. 2000, 2004, 2007; Beitler et al. 2005; Potter and Chan 2011). These are all products of low-temperature, near-surface, diagenetic reactions (Chan and Parry 2002; Chan et al. 2004, 2005; Beitler et al. 2005). They can form both at surface temperatures (Pantin 1958, Sellés-Martinez 1996, Bowen et al. 2008, Barge 2009) and at higher temperatures where fluids move along fault zones (Morrison and Parry 1986).

The widespread occurrence of terrestrial concretions in a variety of mineralogies indicates that concretion formation is a common geologic process in the near surface, porous sediments, and sedimentary rocks. Concretions (inclusively) occur in a wealth of sizes and mineralogies that are beyond the scope of this article; thus, herein we focus on summarizing ancient concretion characteristics from the rock record of the Jurassic Navajo Sandstone (unless otherwise noted) and from some modern examples that are representative of many terrestrial ferric oxide concretions.

TERRESTRIAL CONCRETIONS

Important localities of Navajo Sandstone occur in southern Utah and northern Arizona (Fig. 1). Of the many formations on the Colorado Plateau, the Navajo Sandstone unit is a thick, well-sorted, quartz arenite with high porosity and permeability that facilitated episodic fluid flow over geologic time. These host rock characteristics promoted the formation of a great variety of concretions (Fig. 2). The physical characteristics of concretions, such as size, shape/geometry, mineralogy, and internal structure, are all highly variable. These characteristics are likely to be controlled by factors such as reactant supply, precursor or earlier mineralogy, the time allowed for concretion growth, preferential permeability paths, and nucleation kinetics. The later section on mineralogy and geochemistry outlines a model for Navajo ferric oxide concretion formation that requires reducing conditions to mobilize iron and a change to oxidizing conditions to precipitate iron (Chan and Parry 2002; Chan et al. 2004, 2005). The features and models described here are relevant to the genesis of other ferric oxide concretions in general, although the specifics may vary under different environmental conditions.

Methodologies

There are a number of different approaches and data (Table 1) that are important to studying concretions. These approaches should be used in concert to fully understand complex diagenetic processes.

Physical Characteristics and Field Settings

A major objective of field characterization is to examine the geologic context of the concretions and to determine their distribution on a broad scale; it is also critical that one examine the characteristics of individual concretions. It is clear that the host rock characteristics are an important control on how fluids move through porous media that then affects the concretion forms and their growth. Many of the physical characteristics described here apply to most concretions, regardless of the mineralogical cement compositions.

TABLE 1.—Terrestrial methodologies and approaches for studying Jurassic Navajo Sandstone iron cycling and ferric oxide concretions.

Field characterizations
Host rock (texture, fabric, composition/mineralogy, porosity, permeability, role of clays) (e.g., Beitler et al. 2003, 2005; Bowen et al. 2008)
Concretions (external morphology geometries, size, shapes, in situ self-organized spacing) (Chan et al. 2000, Potter and Chan 2011, Potter et al. 2011)
Internal structure (solid, rinds, multiple layers) (Potter and Chan 2011)
Ancient examples (Jurassic Navajo Sandstone with much variety and other formation examples) (e.g., Chan et al. 2004, 2005; Ormö et al. 2004)
Modern examples (Western Australia) (Bowen et al. 2008)
Mineralogy and geochemistry
Visible near-infrared (VNIR) reflectance spectroscopy (Bowen et al. 2007, Potter et al. 2010)
Thermal infrared spectroscopy (TIR)
Whole rock analysis (Bowen 2005, Nielsen 2010, Potter and Chan 2011)
Trace element geochemistry (Beitler et al. 2005, Potter and Chan 2011)
X-ray diffraction (XRD)
Petrographic thin section
Iron, oxygen isotopes (Chan et al. 2006, Busigny and Dauphaus 2007)
QEMSCAN—quantitative evaluation via electron microscopy for minerals and element phases (Potter et al. 2011)
Cone beam X-ray microtomography (CT) (Bowen 2005)
Modeling
Laboratory bench chemical experiments (Chan et al. 2007, Barge 2009, Barge et al. 2011)
Diffusion and advection rates (Sefton-Nash and Catling 2008)
Numerical modeling and nucleation (Ortoleva 1984, 1994a; Chan et al. 2007)

For the ancient examples, the overall descriptions of ferric oxide terrestrial concretions, including dispersed in situ spacing, size, shape/geometry and conjoined forms, internal structure, and weathering and accumulation aspects (Chan et al. 2004, 2005; Potter et al. 2011), are all consistent with Mars "blueberry" characteristics (Calvin et al. 2008), which imply that the Martian spherules may have formed under diagenetic fluid conditions (Fig. 2). For each major characteristic, we briefly discuss the implications for Mars examples.

In Situ Spacing: In situ spheroidal concretions are distributed and dispersed throughout the host rock in a self-organized spacing pattern (Fig. 3). Numerical simulations indicate that the development of periodic self-organized nucleation centers occurs through Liesegang-type double-diffusion of iron and oxygen (Ortoleva 1984, 1994a, 1994b; Chan et al. 2007). Laboratory bench testing of diffusing chemicals (Barge 2009, Barge et al. 2011; also see later discussion) also shows self-organized precipitates. This nearest-neighbor spacing in ferric oxide concretions arises from self-organized nucleation of cement as reactants diffuse through the host rock in three-dimensional reaction fronts (Potter and Chan 2011, Potter et al. 2011). Spacing is related to concretion size, such that smaller diameter concretions are

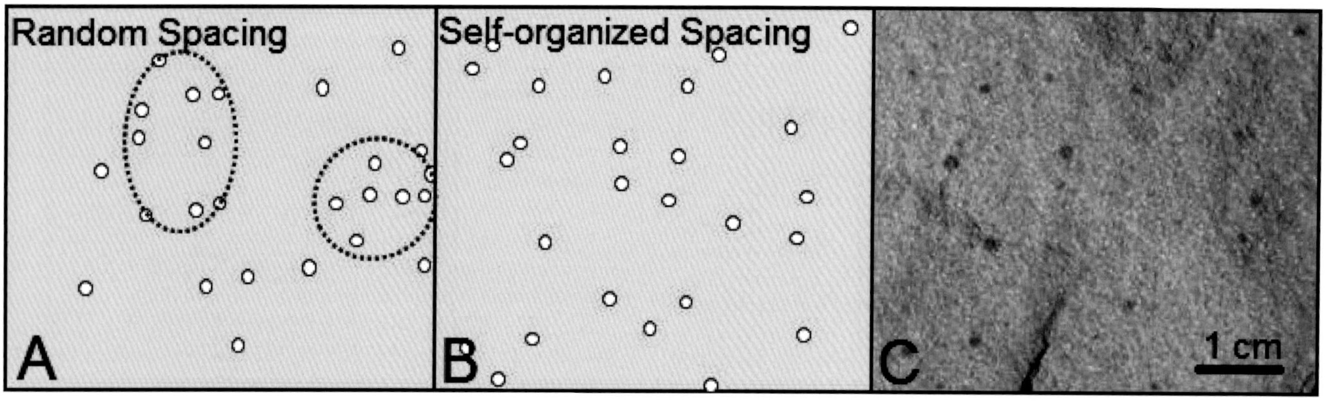

FIG. 3.—Possible concretion spacing distributions with consistent precipitation points (n = 26). (A) A random representation was generated using the Jmicrovision software random point generator. Note the close clusters of randomly spaced points (dotted circles) surrounded by open spaces. Schematic representation (B) replicates an actual photograph of Navajo Sandstone micro-concretions (C). Self-organized concretions (B, C) are more evenly spaced and dispersed throughout the rock. The comparison of χ^2 goodness-of-fit analysis in the measured nearest neighbor spacing (B) with the random spacing (A) returns a P value $\ll 0.001$ that the two sets are the same. Thus, it is statistically unlikely that the concretion spacing is random. Bottom edge of each figure is 6 cm.

more closely spaced than larger diameter concretions, which are more widely spaced (Chan et al. 2004). The χ^2 goodness-of-fit analysis (Fig. 3) shows that spheroidal concretions are not randomly spaced in one study locality of the Spencer Flat area of Grand Staircase Escalante National Monument (GSENM) (Fig. 1) but instead show a geochemically self-organized in situ spacing (Potter and Chan 2011).

In contrast to self-organized (nonrandom) deposition, accumulation of grains by primary hydrodynamic sedimentation processes would typically result in localized concentrations of grains along bedding planes. Of the many alternative explanations that have been given for the hematite spherules on Mars, this self-organized in situ spacing is the one characteristic that is typically overlooked and that is not adequately explained by other models that have spheroidal nodules or forms. Meridiani Planum hematite spherules exhibit a similar self-organized, nonrandom spacing (McLennan et al. 2005), consistent with a diagenetic concretion interpretation (Fig. 3).

Size: Sizes of Navajo Sandstone spheroidal concretions can range from 1 mm to >10 cm. Statistical analysis of size measurements in Spencer Flat (GSENM, Fig. 1) indicates that although there is a continuum of concretion diameters, a division between micro-concretions (1–5 mm) and macro-concretions (>5 mm) exists (Potter and Chan 2011). Macro-concretions are common in Spencer Flat, although they are present in other localized Navajo Sandstone exposures. Micro-concretions are ubiquitous throughout the region wherever the sandstone has undergone diagenetic iron mobilization and re-precipitation (Beitler et al. 2003, 2005; Seiler 2008; Potter et al. 2011).

Micro-concretions (typically ~1–2 mm) can coalesce to form macro-concretions, and these exhibit a bumpy, "avocado-skin" exterior texture (Fig. 4). This coalescence is similar to Ostwald ripening, in which smaller forms nucleate initially and are subsequently consumed to form larger crystals or structures (Steefel and Van Cappellen 1990, Potter et al. 2011).

Diffusive reaction fronts that move through the sandstone in three dimensions likely nucleate similar (consistent) size ranges of concretions. However, sizes of reaction fronts themselves are difficult to document because they may pass through the sandstone and may leave confined zones of fairly uniform diameters of concretions. Some localized reaction fronts produce populations of unique concretions, such as a 1-m^2 area with populations of micro-concretions between Liesegang bands or a 25-m^2 area of all doublets (Potter and Chan 2011). A widespread concretion population covering a large 120-km^2 area in the Spencer Flat locality (GSENM) is characterized by a mean concretion diameter/radius of 22.5 mm (median of 21.7 mm, for n = 1300). This common size indicates either consistent geochemical conditions and/or a large reaction front that covered this area (Potter et al. 2011).

The size of the concretions is likely a function of nucleation kinetics, diffusion rates, reactant supply, and the mobility of reactants. In Utah, the precipitation reactions were likely iron-limited with abundant water. However, in Meridiani Planum the reaction is inferred to be water-limited with abundant iron (Calvin et al. 2008, Potter et al. 2011). These differences probably account for the solid but small size range (1–5 mm) of Meridiani Planum spherules compared to Spencer Flat concretions, in which porosity and diffusion in a water-rich system allowed larger, more widely spaced cemented masses (Potter et al. 2011).

Shape/Geometry: Spheroidal shapes are the most common concretion geometries in the Navajo Sandstone; however, other shapes, such as pipes and concretionary plating of joint faces, are also present (Chan et al. 2000; Beitler et al. 2003, 2005; Seiler 2008; Potter and Chan 2011). Spheroidal shapes form where isotropic media regulate diffusion rates so that reactants can diffuse at an equal rate around the entire concretion (Clifton 1957; Berner 1968, 1980; Raiswell 1971; Johnson 1989; McBride et al. 1994, 2003; Ortoleva 1994b; Mozley and Goodwin 1995; Chan et al. 2004; Mozley and Davis 2005). Elongate shapes typically form in situations in which anisotropic advective mass transfer (due to fluid flow toward a hydraulic low) dominates over diffusion or where there is directional diffusion (due to anisotropies in porosity due to primary textures such as bedding; Schultz 1941; Johnson 1989; Cibin et al. 1993; McBride et al. 1994, 1995, 2003; Mozley and Goodwin 1995; Mozley 1996; Eichhubl et al. 2004; Mozley and Davis 2005; Potter and Chan 2011).

In GSENM, nearly all of the different concretionary forms are present, with many overlapping or showing multiple fluid-flow events that are distinguished by cross-cutting relationships. Conjoined forms are common (Figs. 2L, 4D), and some reaction fronts are composed almost entirely of twins (Potter et al. 2011). En echelon, northeast-striking joints are lined with associated concretionary cement as well as pipes and loops that preferentially extend from the joints to the

FIG. 4.—Micro-concretions commonly coalesce to form concretion rinds. (A) A natural collection of Navajo concretions of consistent large size (12–15 cm in diameter) and internal multiple layers from a chemical reaction front in Grand Staircase Escalante National Monument, Utah. (B) The bumpy "avocado" skin texture of these large concretions (from A) indicates coalescing micro-concretions at numerous nucleation sites. (C) Liesegang banding and micro-concretions near a Navajo concretion rind in cross section. (D) Micro-concretions comprising a larger concretion doublet/twin rind. Both (C) and (D) are from the Vermilion Cliffs National Monument, near the Utah–Arizona border (near CB in Fig. 1).

southeast. Documented relationships of the diagenetic events indicate that the spheroidal macro-concretions formed independently from the precipitation associated with the joints (e.g., Potter and Chan 2011).

Some concretion patterns appear to be regionally correlative. Similar northeast-striking joints and southeast precipitation are documented in the correlative Aztec Sandstone in the Valley of Fire (Nevada) (Eichhubl et al. 2004). This indicates a regional event whereby meteoric water infiltrated the formation via northeast striking, Miocene-aged joints. A hydraulic low was created by the downcutting of the Colorado River (~5 Ma) to the southeast of the region, and the advecting fluid precipitated elongate "flow lines" in a southeasterly direction (Eichhubl et al. 2004, Potter and Chan 2011).

On Mars, spheroidal geometries are the norm, with some conjoined forms, such as doublets (two conjoined spherules) or rare triplets (three attached and aligned spherules) (McLennan et al. 2005, Calvin et al. 2008). Rarely, the Mars spherules also show ridges parallel to the host rock laminae/bedding, which implies the influence of differential permeability along laminae (Potter and Chan 2011).

Internal Structure: Internal structures of concretions encompass three end-member types: rind, layered, and solid (Potter et al. 2011). Rind concretions exhibit a well-cemented exterior (~1–5 mm thick) surrounding an interior that contains very little cement. Layered concretions may have a rind, but they also exhibit well-cemented, interfingering layers in the interiors. These layers can persist throughout the entire interior of the concretion, or the concretion can have a small, depleted center. Solid concretions are cemented evenly throughout the entire concretion. Internal structure varies widely between these end members.

Navajo Sandstone ferric oxide concretions typically lack an obvious, physical nucleus, although field and laboratory studies (e.g., the microtomography discussed below) indicate that there could be a threshold chemical nucleus/nuclei that might be consumed in chemical reactions, instead of a visible "seed" nucleus. Nucleation centers that produce smaller spheroids are far more abundant and commonly aggregate and coalesce to make larger forms. Kinetic factors and iron supply, in addition to permeability/tortuosity paths, likely affect the concretion mineralogy, geometry, and spacing.

Carbonate concretions are some of the most common in the geologic record, and some can be quite large (up to the size of large boulders). Inside some carbonate concretions there appears to be a nucleus (e.g., calcite or organic material) that might have attracted the carbonate ions (Raiswell et al. 2000, McBride et al. 2003). Thus, these types of concretions are thought to grow from a nucleus outward to the edges with an increasing radius (Raiswell et al. 2000). Other carbonate concretions exhibit no nucleus, but rather multiple generations of cements within pore spaces; these concretions grow pervasively from a set radius (Raiswell et al. 2000). In contrast, cement textures in the Navajo ferric oxide concretions typically show a component of "inward" growth (similar to a geode), with small bulbous protrusions extending inward from concretion rims (Fig. 5).

If no obvious physical nucleus is present in ferric oxide concretions, it raises the question of whether there ever was a nucleus and/or whether a nucleus present initially was consumed later in some chemical reaction such that its presence is no longer detectable. Some concretions likely did have an early diagenetic nucleus, such as pyrite. Cretaceous sandstone examples from the Western Interior associated with coal-forming environments show pyrite precursors that typically alter to the ferric oxide mineralogies (Roberts and Chan 2010). However, in many concretions examples, such as those of the Jurassic Navajo Sandstone, there are no obvious or relict pyrite nuclei, indicating that they were not present initially and that the reactions and ferric oxide precipitation proceeded without any precursor pyrite.

The few interiors of Mars spherules ground by the MER Rock Abrasion Tool appear to be homogeneous with no obvious nuclei, although some broken spherules appear to show faint internal structure and rinds (Calvin et al. 2008). However, there is an intrinsic resolution issue of the MER microscopic imager whereby the spherule interiors are so small that nuclei or rinds could be undetectable.

Weathering and Accumulation: In most cases, concretions are heavier, denser, and more resistant to weathering than is the host rock because of preferential cementation. Over time, the formation weathers, and much of the host rock disaggregates and grains are removed by wind or water. Weakly cemented Navajo Sandstone is particularly susceptible to weathering. However, the well-cemented concretions are more resistant and commonly remain to accumulate in topographically low surfaces as a lag. These accumulated concretions are much more concentrated than when they are in situ, and they typically represent the vestiges of tens of meters of eroded host rock.

The accumulated Navajo concretions freed from the host rock show little difference in diameter size compared to the in situ concretions. This indicates that there has not been significant size reduction as a result of weathering. Similar features at Meridiani Planum show a similar size of accumulated spherules compared to those that are in situ within the Burns formation (Calvin et al. 2008).

Eolian Ripples in Eroded Micro-concretions—Analog Case Study

Jurassic Navajo Sandstone micro-concretions (Fig. 6) in the Coyote Buttes area of the Vermilion Cliffs National Monument, Arizona (Fig. 1), form complex ripples in areas with loose accumulations of concretions and strong wind gusts. We discuss these micro-concretions in some detail, as these have not been previously addressed in the literature, and because study of these terrestrial analog ripples can aid in the understanding of similar Mars micro-sized spherules, dubbed "microberries" (Squyres et al. 2006) (Fig. 6E), and their ripple formation at Meridiani Planum.

Micro-concretions in chemical reaction fronts occur as discrete spherules (Fig. 6) that are loosened by entrained sand that actively abrades sandstone walls. The micro-concretions average 1.5 mm in diameter (Fig. 6D) and 0.0056 g in grain mass. In contrast, the average grain mass of medium-grained quartz sand derived from the host rock is ~0.000075 g, resulting in a grain mass ratio of ~74:1. Relatively tight micro-concretion size population distribution allows wind to sort out very fine size differences and to rework the micro-concretions into trains of straight to sinuous, in-phase, coarse-grained ripples and catenary out-of-phase ripples (Fig. 6B, E). These local, complex, and transient wind ripples (typical ripple index of wavelength to height = 18) form in shallow depressions and alcoves where winds are strong or funneled, sometimes in pathways extending up to 14.5 m. Ripples have accumulations of micro-concretions up to 5 mm thick on the windward/stoss sides and ripple crests (Fig. 6B) and scattered micro-concretion on the leeward faces.

Reverse grading in the micro-concretions likely forms by two mechanisms, possibly in combination: (1) a common sieve or "Brazil nut" effect, such that the micro-concretions "float" on top of the finer grained quartz sand, or (2) motions typical of coarse-grained ripples, in which the larger grains (in this case, micro-concretions) move in creep and are overrun by the ripple itself (Sharp 1963, Anderson and Bunas 1993; Fig. 6B). Wind gusts of 6.6 m/s (measured at a height of ~5 cm from the air–sediment interface) move the micro-concretions by low saltation, although creep occurs if sufficient sand is in traction at sustained wind speeds of ~4.5 m/s. Although micro-concretions can be moved by eolian transport, larger concretion sizes (e.g., 1+ cm sizes too big to be moved) form deflation lags.

At Meridiani Planum, the mass ratio between the 1.5-mm Mars spherules and the basaltic, 100-μm sand likely ranges from 3000:1 to 6000:1, depending on the amount of hematite. This Mars ratio is higher than the terrestrial example, in which the micro-concretions contain finer grained hematite and have incorporated quartz grains. The limited local occurrence of terrestrial micro-concretion ripples is indicative of the number of environmental conditions required for their formation. These circumstances may be related and may be scaled to geologic and atmospheric conditions on Mars, where similar ripples are far more abundant and exhibit a wider variety of ripple types (e.g., asymmetrical straight, sinuous and catenary ripples) in a stronger wind regime (Jerolmack et al. 2006).

Modern Concretions

Ferric oxide concretions have been documented forming in situ in modern acid saline ephemeral lake sediments in Western Australia (Benison et al. 2007, Bowen et al. 2008). These concretions illustrate

FIG. 5.—Cross-sectional areas of Navajo concretions show small digitate inward projections (arrows) and indicate some inward growth.

the potential for early and rapid formation of hematite concretion-bearing evaporites and indicate the possibility of astrobiological preservation potential in such deposits. These modern interdune acid evaporite lakes in southern Western Australia have sedimentological, mineralogical, and diagenetic features that are somewhat analogous to those observed in the lithified strata in the Meridiani Planum region of Mars (Benison and LaClair 2003, Benison and Bowen 2006, Bowen et al. this volume). The sediments being deposited within these lakes

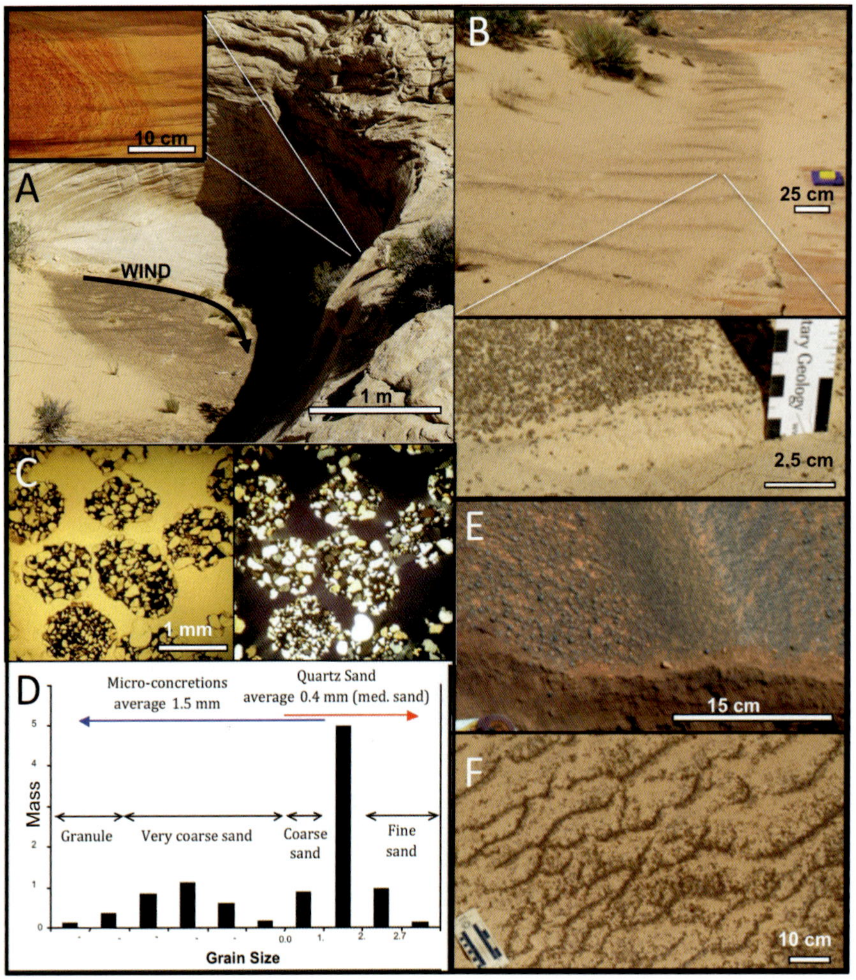

FIG. 6.—Micro-concretions that formed weathered accumulations and ripples. (A) Outcrop alcove of Navajo Sandstone showing wind action that loosens concretions from chemical reaction fronts in the host rock (Liesegang bands, upper left inset). (B) Ripple trains of Navajo micro-concretions within shallow depression where winds are funneled and accelerated (top is surface expression, lower is close-up cut-away section of sand ripples armored by a thin-veneer micro-concretion on the windward/stoss side of ripples). (C) Petrographic thin section of spheroidal Navajo micro-concretions in plane light (left) and crossed nicols (right) where dark hematite cement encases eolian quartz (light) grains. (D) Sieved data showing grain size distribution of quartz sand and micro-concretions sampled from area of photo (B). (Note: Grain size scale is nonlinear.) (E) Representative micro-concretion ripples at top from Meridiani Planum, Mars, imaged by the MER *Opportunity* (Pancam stretched false-color image, Sol 367B, Seq P2550; photo credit: National Aeronautics and Space Administration/Jet Propulsion Laboratory/Cornell). (F) Comparable terrestrial ripples of reworked Jurassic Navajo Sandstone and micro-concretions. Locale: All Navajo examples shown in A through D and F are from Coyote Buttes (Fig. 1).

contain a mixture of clastic and evaporite grains that have been modified by eolian and shallow water processes (ooids). Chemical precipitates from the acid saline surface waters and shallow groundwaters include evaporites (halite, gypsum, etc.), iron oxides (hematite, goethite), sulfates (jarosite, alunite), and clays (kaolinite, etc.). Early diagenetic spheroidal hematite concretions are actively precipitating in reworked evaporite and siliciclastic subsurface sediments: below shallow lake waters (total dissolved solids up to 23% and a pH of ~4) and salt crusts; in sediment ~30 cm deep saturated with shallow groundwater that is also hypersaline, with a pH measured down to 3.1. Concretions range in size from 2 to 4 cm in diameter and are composed of isopachous hematite cement that encases host grains composed of quartz and evaporites.

Observation of modern ferric oxide concretion growth in a natural sedimentary setting provides a rare glimpse into the conditions and the timescales of formation. The concretion host sediments are between 1400 and 2900 years old, based on radiocarbon dating of stratigraphically lower and higher organic-rich sediments. This indicates that concretions are less than 2900 years old (Bowen et al. 2008). These absolute age constraints and the observed field conditions (i.e., softness of concretions) indicate that these hematite concretions are actively forming in the modern acid saline environment and should therefore contain records of the existing geochemical and microbiological conditions. Diagenetic cements and evaporite minerals can serve as tombs of biological materials that are commonly involved in or simply trapped or preserved by rapid mineral precipitation processes (e.g., Lowenstein et al. 2011). The existence of minerals that indicate both acidic and extremely saline fluids on Mars has pointed to inhospitable conditions, even if liquid water was once present (Tosca et al. 2008b). Preservation of microfossils in oxidizing environments has been considered dubious (Sumner 2004). However, examination of fluids and sediments from terrestrial environments with hypersaline (low–water activity), acidic, and oxidizing conditions, such as are presumed to have existed in the ancient Meridiani environment, reveals that even in these extreme conditions, microbiological communities exist (Mormile et al. 2009) and can be preserved by rapidly precipitating evaporates and iron oxides (Benison et al. 2008, Fernandez-Remolar and Knoll 2008). If life existed in past aqueous environments on Mars, biosignatures may be preserved within authigenic mineral accumulations, such as hematite concretions.

Relative Timing and Age Dates of Concretions

Although the physical geometric characteristics and field settings of ferric oxide concretions have different controls, it is likely that the basic geologic process for concretion formation remains consistent. If the conditions are right, cementation can occur quickly. However, absolute timing of concretion formation is difficult to ascertain, and in some cases there may be multiple diagenetic events or additional cementation that cause "ripening" and growth of concretions over a longer period of time.

Modern concretions form in extreme acid and saline syndepositional to early diagenetic conditions over short timescales of a few thousands of years, as discussed above (Bowen et al. 2008). The timing of ferric oxide concretion formation in the geologic past is more difficult to decipher except in a relative sense, often using cross-cutting relationships at field and microscopic levels, and may be highly variable (e.g., Potter and Chan 2011). The concretions are diagenetic and form after deposition, but their formation could occur over a wide time span, that could range from very quickly after sediment deposition up to a few hundred million years later. A major difficulty is that there is little material in concretions to date with existing radiometric methods. In some rare cases where the K-bearing mineral cryptomelene is present and appears to be associated with ferric oxide mineralization, the cryptomelene can be selected for Ar–Ar dating (Chan et al. 2001). In Navajo Sandstone samples south of Moab (Chan et al. 2001) the cryptomelene cement date is ~22 million years, which is considerably later than the original Early Jurassic deposition of the host rock sediments. For the future, the development of new methods of (U-Th)/He and 4He/3He dating techniques applied to iron oxides appears to show promise for providing some absolute dates for diagenetic ferric oxide cements (Shuster and Farley 2004; Shuster et al. 2004, 2005; Heim et al. 2006). Eventually if this absolute dating technique works on terrestrial concretions, it could be applied to Mars "blueberries" from a sample return mission.

FERRIC OXIDE MINERALOGY AND GEOCHEMISTRY

The mineralogy and geochemistry of iron systems covers a wide range of spectrum of geochemical conditions and environments covered in the vast literature and summarized well in Cornell and Schwertmann (2003). We will not repeat this literature but instead focus on some of the important elements of mineralogy and geochemistry that may elucidate particular concretion processes. Much of the terrestrial concretion mineralogy can be distinguished by traditional laboratory methods (Table 1) of petrography, X-ray diffraction, whole rock analysis, spectroscopy, and electron microscopy. Highly detailed analyses of trace element geochemistry, isotopes, and tomography (discussed below) can provide additional insight.

Geochemical Relationships

There is a relationship between concretions and sandstone coloration that indicates that there can be significant cycling of iron over time and space. The formation of ferric oxide concretions in a quartz arenite is thought to proceed via a three-step process (Chan et al. 2000, 2004, 2007; see Fig. 7): (1) reduction and mobilization of Fe from hematite grain coatings; (2) transport of Fe^{2+} to the site of precipitation; and (3) precipitation of Fe to form iron sulfide or hydrous ferric oxides. Iron oxide concretions can precipitate via oxidation of ferrous iron in solution via the following hematite reaction,

$$2Fe^{2+}_{(aq)} + 0.5O_{2(g)} + 2H_2O_{(l)} = Fe_2O_{3(s)} + 4H^+_{(aq)},$$

or via the following goethite reaction:

$$2Fe^{2+}_{(aq)} + 0.5O_{2(g)} + 3H_2O_{(l)} = 2FeO(OH)_{(s)} + 4H^+_{(aq)}.$$

After initial precipitation, HFO can dehydrate to hematite, thus:

$$2FeO(OH)_{(s)} = Fe_2O_{3(s)} + H_2O_{(l)}.$$

We believe that the majority of the Navajo Sandstone concretions likely formed via these aforementioned reactions because of the lack of other

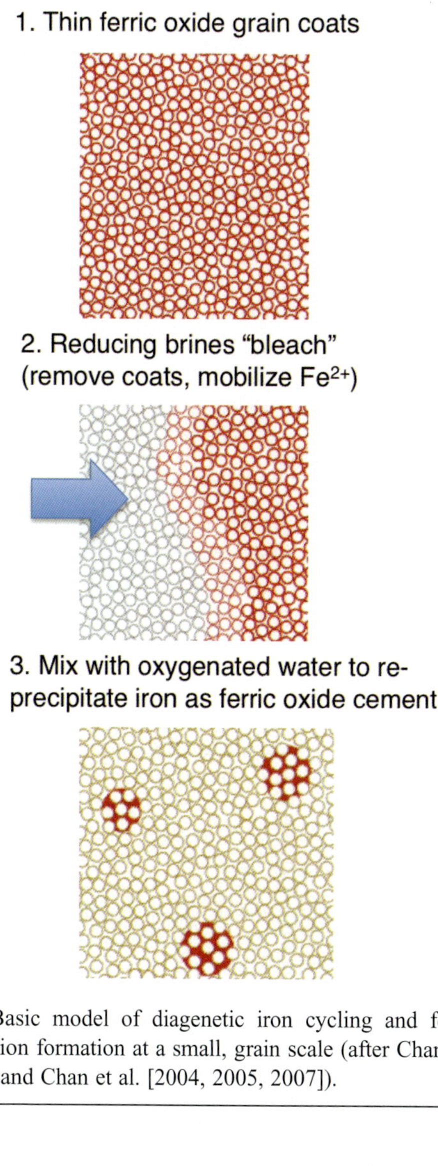

FIG. 7.—Basic model of diagenetic iron cycling and ferric oxide concretion formation at a small, grain scale (after Chan and Parry [2002] and Chan et al. [2004, 2005, 2007]).

TABLE 2.—There are multiple mineral pathways to ferric oxide end products, with various conditions and evidence that indicates past diagenetic history.

Original mineral	Intermediate phases	Products	Process[a]	Criteria or evidence
Goethite α-$Fe^{3+}O(OH)$		hematite	dehydration, thermal transform[c]	relict crystal habit
Ferrihydite $Fe_2^{3+}O_3 \bullet 0.5(H_2O)$	±goethite	hematite	dehydration/thermal tranform	possible relict microtexture
Amorphous HFO FeO_x	±goethite	hematite	dehydration/thermal tranform	possible relict microtexture
Lepidocrocite γ-$Fe^{3+}O(OH)$	maghemite	hematite	dehydroxylation[c]/thermal transform	relict crystal habit?
Akaganeite β-$Fe^{3+}O(OH)$		hematite	dehydration, thermal transform[c]	relict crystal habit?
Pyrite $Fe^{2+}S_2$	goethite sulfate[b]	hematite	oxidation–hydration/dehydration	relict crystal habit of pyrite (cubes)
Siderite $Fe^{2+}(CO_3)$	goethite CO_2[b]	hematite	decarbonation–hydration/ dehydration	relict morphology (spherules)
Schwertmannite $Fe_8^{3+}O_8(SO_4)(OH)_6$	goethite	hematite sulfate	thermal dehydration[c]	mass loss (develop sponge texture)
Jarosite $KFe_3^{3+}(SO_4)_2(OH)_6$	goethite sulfate, K^{+1}[b]	hematite	decompositon–dehydration/ dehydration	evidence of local low-pH environment
Hematite $Fe_2^{3+}O_3$		magnetite	reduction	relict hematite crystal habit (bladed or tabular), relict growth bands
Magnetite $Fe^{2+}Fe_2^{3+}O_4$		hematite	oxidation	relict magnetite crystal habit (cubes)
Magnetite $Fe^{2+}Fe_2^{3+}O_4$	maghemite	hematite	higher temperature oxidation[c]/ thermal transform	relict magnetite crystal habit (cubes)
Fayalite $Fe_2^{2+}SiO_4$	magnetite quartz	hematite quartz	high-temperature oxidation	metamorphic heritage

[a] Weathering or diagenetic changes to stable forms, thermal dehydration <120° C, and "/" separates intermediate product process from final process.
[b] Aqueous.
[c] Cornell and Schwertmann (2003).

iron-rich relict minerals, although several pathways to ferric oxide minerals exist and could be possible mechanisms for the resultant end products (Table 2).

In more reducing and sulfur-rich environments, ferrous iron in solution can precipitate with reduced sulfur (H_2S) to produce iron sulfide precipitates,

$$2H_2S_{(g)} + 0.5O_{2(g)} + Fe^{2+}_{(aq)} = FeS_{2(s)} + H_2O_{(l)} + 2H^{+}_{(aq)},$$

that can be altered to iron hydroxide/oxide under oxidizing conditions. Iron sulfide concretions in western US Mesozoic sandstone units have been the subject of previous studies (e.g., Thorson and MacIntyre 2005, Roberts and Chan 2010). These examples typically show relict pyrite with a clear association of reducing conditions from coal-forming environments (Roberts and Chan 2010). Other examples of siderite concretions (e.g., Gauthier 1982, Mozley 1989) also occur in reducing conditions.

Table 2 above outlines the multiple chemical pathways and successive transitions from initial different iron-rich mineralogies, ending at goethite or hematite end products. Although not all of these are documented from the Jurassic Navajo Sandstone examples cited here, they are all geochemically possible in natural settings and thus again show the complexities of diagenesis. Hematite concretions are relatively common in terrestrial systems partly because it is a stable phase. There are a number of precursor mineralogies that may have intermediate steps that can be stopped at any point in the history of its development, regardless of whether it gets to the stable end product (Table 2).

Trace Element Geochemistry

The trace element chemistry of concretions, compared to that of the host rock, may be an important indicator of diagenetic history. Certain trace elements in Navajo Sandstone ferric oxide concretions show elevated values relative to the host rock (e.g., As, V, U, Zn, Co, Ni, and U; Beitler et al. 2005, Potter et al. 2011). Some particular trace elements (e.g., As, Ni, and U) are likely adsorbed to the Fe precipitates (e.g., Langley and Colberg 2006). Furthermore, trace elements (present as either adsorbed species to the ferric oxide or in minor mineral phases) may record biogenic influence because microbial species uptake certain elements (e.g., Al, Cr, Mn, Cu, Pb, Ni, and Se) preferentially in minor amounts during their life cycles (Fein et al. 1997, Fowle and Fein 1999, Ferris et al. 2000, Liermann et al. 2000, Ferris 2005). Both Navajo Sandstone concretions and Mars spherules on Meridiani Planum are enriched in Ni relative to the host rock (Morris et al. 2006, Potter and Chan 2011). This may be the result of Ni mobilization and precipitation adsorbed to the more abundant iron species; however, this trace element phenomenon warrants further investigation with regard to its potential as a biogenic marker.

Isotopes

Iron isotopes have the potential to trace redox variations, bacteria-mediated reduction, and precipitation of iron oxide/hydroxides in surface and near-surface hydrological and diagenetic environments (Anbar 2004, Beard and Johnson 2004). Chan et al. (2006) and Busigny and Dauphas (2007) measured iron isotopes in goethite and hematite concretions in the Navajo Sandstone in the Colorado Plateau. Both studies document that the δ^{56}Fe values of these concretions are mostly negative, ranging from near zero to values as low as $-1.5‰$ (Chan et al. 2006), whereas iron in unbleached Navajo Sandstone samples (purported source of iron for concretions; Chan and Parry 2002; Chan et al. 2004, 2005) from the region has δ^{56}Fe values near 0‰. In general, iron in surface environments derived by abiotic, terrestrial weathering processes has a δ^{56}Fe value very near zero (Beard and Johnson 2004).

The two studies differ in interpretation of these negative δ^{56}Fe values. Chan et al. (2006) interpret the negative δ^{56}Fe values of the concretions to result from complete oxidation and precipitation of Fe from aqueous fluids (groundwaters) that had low δ^{56}Fe values of -0.5 to $-1.5‰$. They attribute the negative δ^{56}Fe values of the aqueous iron to isotopic fractionation accompanying bacteria-mediated reduction of ferric iron sand grain coatings in the source areas for the iron. This interpretation is supported by the fact that $Fe(II)_{aq}$ and reactive ferric oxides in modern marine sediments, in which iron-reducing bacteria are actively recycling Fe, have similar low δ^{56}Fe values (Johnson et al. 2004). In contrast, Busigny and Dauphas (2007) conclude that the negative δ^{56}Fe values of the concretions are best explained by progressive precipitation and/or adsorption of ^{56}Fe-enriched iron during flow of Fe-bearing groundwater, with initial δ^{56}Fe measuring $\sim$0‰ through the sandstone. This model implies that concretions form primarily in a single episode from a single, large-scale groundwater flow regime. Busigny and Dauphas (2007) document a regional trend of decreasing δ^{56}Fe toward the east within their sample suite, which they cite as evidence in support of their scenario. Chan et al. (2006) did not find any systematic trend of δ^{56}Fe with geographic location of the concretions in their sample suite, although in situ sampling with an ion microprobe in the future may be able to distinguish details that are not available with current methods.

Concretions may be zoned in Fe isotopes, depending on the nature of the processes involved in their growth. Eventually, if appropriate standards can be obtained, it should be possible to analyze iron isotopes in concretions in situ to look for isotopic gradients or shifts in δ^{56}Fe values in order to determine more about the growth origins of concretions.

Microtomography

Cone beam X-ray microtomography (three-dimensional density imaging) can be used to examine the textural heterogeneity in concretions (Beitler et al. 2004, Bowen 2005). Some of the density differences that can be detected may be related to mineralogy or mineral concentrations (Fig. 8). Examples presented here included ferric oxide concretions from the Navajo Sandstone (Fig. 8A) as well as copper carbonate concretions (e.g., azurite) from the Cretaceous Dakota Sandstone near the La Sal Mountains (south of Moab, Utah) (Fig. 8C). The sphericity of the concretions can be evaluated by comparing the concretion diameter in different dimensions (Fig. 8D). The imaging reveals subtle textural differences in the centers of the concretions that indicate a chemical reaction near the nucleus (Fig. 8). The copper carbonate concretions show very distinct "floating" detrital quartz grains (indicating some quartz dissolution) and a homogeneous pore-filling network of cement, while the ferric oxide concretions appear to have more heterogeneity of different minerals encased in the cement.

MODELING AND GENETIC PROCESSES

Experimental Laboratory Simulations

Laboratory simulations of self-organized mineral precipitation have been performed in various media, ranging from gels (Henisch 2005, and references therein) to sand-like media that more closely simulate natural sedimentary environments (Barge 2009, Barge et al. 2011). Typically this experimental design consists of a column of diffusion medium (gel or sand) saturated with a fluid containing one reactive ion, and a concentrated solution containing another reactive ion is introduced at the column interface (Fig. 9). The outer reactant diffuses through the column, reacting with the inner reactant to produce an insoluble precipitate that is held in place by the gel/sand medium so that a self-organized pattern can form. The initial precipitates form over a period of several days to a few weeks and can evolve via Ostwald ripening over even longer timescales. A recent study of self-organized silver chromate precipitates was performed in various media (Barge et al. 2011)—including gels; laboratory glass beads of specific sizes, similar to typical grain sizes (100–600 μm) in the Navajo Sandstone (Bowen 2005); and combinations of glass beads and gel (which allows use of large-grained sand while still maintaining a diffusion-controlled environment). Liesegang bands and crystals were formed in gels, and millimeter-size spheroidal precipitates formed in the sand-like media under a variety of conditions (Barge et al. 2011).

The significance of this study shows that in a diffusion-controlled system, where a reaction front passes through a porous sand-like medium, small spheroidal mineral precipitates (similar in size to the millimeter-size Navajo Sandstone "micro-concretions") can nucleate and grow in a self-organized fashion. These experimental simulations have implications for how micro-concretions without obvious eolia can form in natural sedimentary environments. Benchtop experiments can be very helpful for examining some physical and chemical parameters; however, there are limitations to experimental data because a laboratory setup can never replicate all the subtleties of the natural environment. A main problem with laboratory experiments is that these generally represent simple, closed systems designed to isolate specific parameters and produce a repeatable result, as opposed to the open complex natural systems. Furthermore, getting precipitation and other reactions to occur on a human timescale of weeks or months sometimes necessitates the use of very concentrated solutions compared to what would be found in natural systems. Laboratory bench tests cannot replicate the natural diagenetic settings or provide simulations of precipitation over the geologic timescales of hundreds, thousands, and even millions of years that may be necessary for concretions to form in natural environments. However, they can simulate concretion-like nucleation and growth with strong chemical solutions and varying combinations of reactants (Chan et al. 2007).

Genetic Models

The detailed aspects and processes involved in the origin of ferric oxide concretions are not fully understood. The concretion formation process appears to involve the self-organized diffusion of Fe, similar to the process proposed for the origin of Liesegang banding (Ortoleva 1984, 1994a, 1994b), and Liesegang banding is often associated with the diagenetic concretions (Chan et al. 2007). The characteristics of concretions in general can be quite complicated, as diagenetic systems can be open, with multiple precipitation and water infiltration events, and sometimes more than one type (chemically) of water (Chan et al. 2004, 2005; Potter et al. 2011). There are many factors to consider, including the supply of reactants, composition of the water(s), temperature, pH, water–rock interactions, diffusive rates, nucleation kinetics, biomediation, and fluctuating water table or reaction fronts. It is clear that geochemical conditions must be right in order to encourage

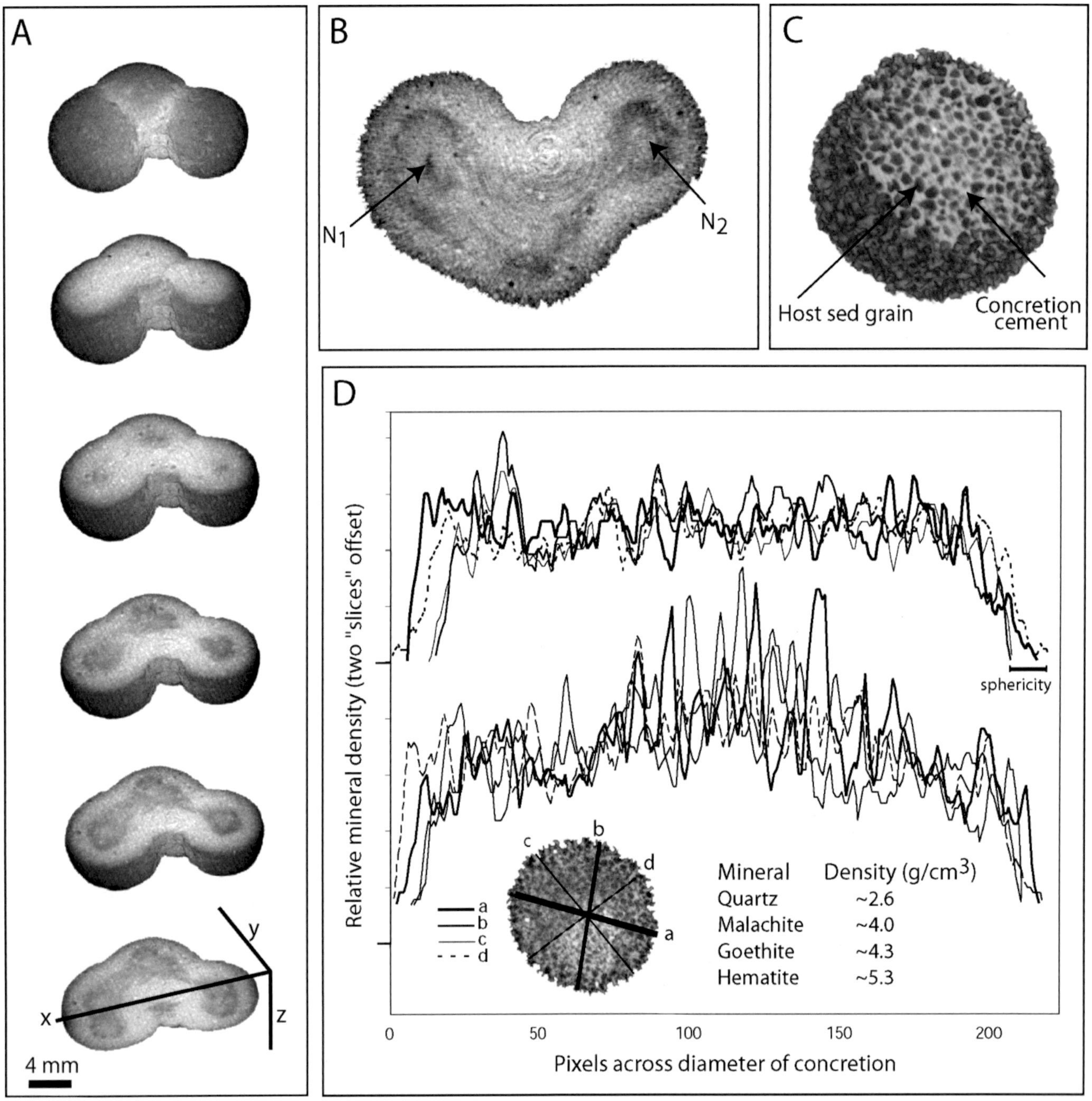

FIG. 8.—Cone beam X-ray microtomography (three-dimensional density imaging) of Utah concretions. Grayscale indicates relative density, with darker colors being less dense. (A, B) "Triplet" hematite concretion (2.5 cm across) from Navajo Sandstone in southern Utah. (A) Vertical series shows virtual "slices" down into the concretion and a final slice in from the front. The original spatial resolution of the imagery is 80 μm/pixel. Virtual slicing shows that where one chooses to cut open a concretion might lead to different interpretations about the interior homogeneity. (B) High-resolution imaging shows two possible nucleation scenarios: N1, where there is a void in the center of the left side sphere, and N2, where there is a dense zone, likely iron oxide, in the center of the right side sphere. The ferric oxide in this concretion has engulfed the detrital sandstone grains. (C) Three-dimensional volume of azurite concretion from the Cretaceous Dakota Sandstone, south of Moab (Fig. 1). Concretion has ~1-cm diameter, and image has 20 μm/pixel resolution. Grayscale illustrates density differences between host quartz grains and secondary concretion cements. (D) Density profiles for spherical hematite concretion from the Navajo Sandstone ~1 cm across. In this concretion, there are some "bright" spots that indicate a mineral with a higher density, perhaps hematite, in a primarily goethite concretion (mineral density data after Cornell and Schwertmann [2003]). Relative density profiles showing the apparent lack of physical nuclei and homogeneity of iron oxide cement. Upper graph is from slice 112, halfway through the concretion. Lower graph is from slice 168, three quarters of the way through the concretion. Difference in concretion diameters (A) through (D) illustrates the near-sphericity of the ferric oxide concretion. Vertical scale shows relative "brightness" of the image, which corresponds to material density. The brighter areas in the computed tomography scan are more dense, and the darker colors are less dense.

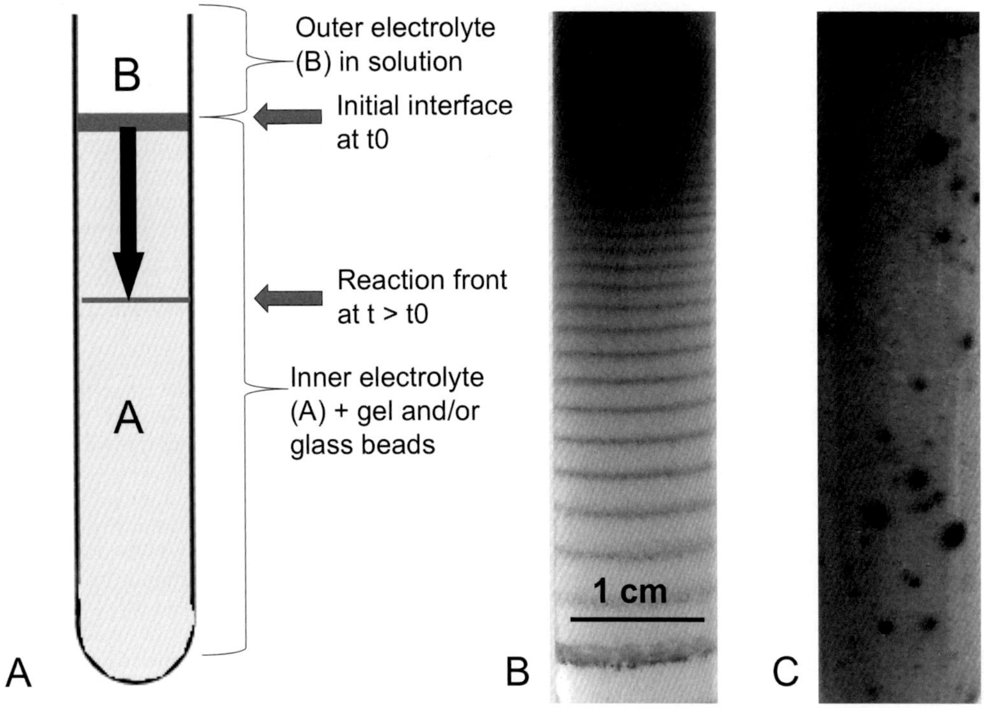

FIG. 9.—Self-organized mineral precipitates in diffusion experiments. (A) Experimental setup. The inner electrolyte (in this case chromate) is distributed throughout the diffusion medium, and a concentrated solution containing the outer electrolyte (in this case silver) is added at the column interface. The outer electrolyte diffuses through the column and reacts with the inner electrolyte to form an insoluble precipitate, and the type of self-organized precipitation morphology that results is dependent on the initial experimental conditions. (B) Silver chromate precipitates in gelatin gel, forming Liesegang bands. (C) Silver chromate precipitates in a mixture of 100-μm glass beads and silica gel, forming spheroidal precipitates. (Adapted from Barge et al. 2011.)

formation of mineral cements. Precipitation typically occurs once the concentrations of reactants reach some nucleation threshold, and mineral composition/structure may be altered after initial precipitation (Chan et al. 2007, Potter and Chan 2011).

The concretion formation process must additionally involve advection of Fe in solution to the site of concretion nucleation, as mass-balance calculations indicate that rocks contain far less Fe in situ than is required for the observed abundance and sizes of ferric oxide concretions in the rocks. An example from the Spencer Flat area illustrates the point. The average diameter of concretions and average spacing between concretions in this area are 5 cm and 40 cm, respectively (also see Chan et al. [2004]). The host rock surrounding the concretion averages 1.39% Fe, present as goethite and lepidocrocite nuclei on illite-coated quartz grains and K-feldspar grains. The volume of a 40-cm-diameter volume of porous rock surrounding each concretion is 33,000 cm^3. The porosity of the sandstone is approximately 15%, so the pore volume is 5000 cm^3. Chemical modeling of the reduction of hematite in the original sandstone by dissolved methane (Parry et al. 2004) indicates that the maximum dissolved iron in the pore space is near 15 mg/l, and measurement of Fe content in groundwater from nearby wells reaches a maximum of ~5 mg/l. Static pore fluid surrounding a concretion could contain a maximum of 0.075 g of Fe. A 5-cm-diameter concretion with a rind 2 mm thick has a rind volume of 14 cm^3 that is 20% goethite and contains 8 g of Fe, but 0.075 g of Fe is available in the surrounding pore fluid. Iron must therefore be supplied by pore fluid flow (approximate minimum of 100 pore volumes) or redistribution of Fe mineral within the surrounding rock. Navajo Sandstone with hematite grain coatings from early diagenesis originally contained on average 0.53 wt. % Fe_2O_3 (range 0.18–1.25%; Beitler et al. 2005), or 350 g of Fe, ample to supply the Fe in the concretion rind. However, no Fe-depleted zone is apparent surrounding the concretions, nor is a chemical reduction mechanism apparent in the rock surrounding a concretion that could mobilize iron locally. This mass-balance calculation shows that the original iron was not redistributed locally to form the concretion rind and that insufficient Fe is available in static pore fluid; therefore, iron must added to the rock by advection.

Precipitation of iron oxide cement–forming concretions is enhanced by oxidation and by rising pH. Precipitation takes place on eolia consisting of K-feldspar grains and illite rims on quartz grains. A small concretion may form around a single grain, but larger concretions incorporate many detrital grains of feldspar. Both feldspar and illite rims on quartz grains chemically react with solution to consume hydrogen ion and form kaolinite as a reaction product. The solution pH may rise if hydrogen ion is consumed by chemical reaction faster than it can be supplied by diffusion. The rise in pH initiates precipitation. Larger numbers of K-feldspar grains consume hydrogen faster, and so larger volumes of sandstone nucleus are more likely to initiate precipitation.

Rates and Timescales of Concretion Formation

Estimates of the amount of time necessary to form ferric oxide concretions of a given size can be made assuming equilibrium precipitation and using advection–diffusion models available from the literature. The equation 1 (modified from Berner [1968, 1980])

TABLE 3.—Defined parameters for growth time equation 1 after Parry (2011).

$$t = \frac{\left(R - \frac{D}{0.715U}\right)\left(1 + \frac{RU}{D}\right)^{0.715} + \frac{D}{0.715U}}{1.715Uv(c_\infty - c_R)} \quad (1)$$

t = time (years)

R = radius of concretion (m)

D = diffusion coefficient estimated as follows:

D_{aq}^{Fe2+} = diffusion coefficient of aqueous Fe^{2+}: 1.5×10^{-5} cm^2/s at 50° C (Oelkers and Helgeson 1988)

D_{eff}^{Fe2+} = effective diffusion coefficient of Fe in a porous medium, corrected for tortuosity, following Berner (1980) and Drever (1997). Correcting for tortuosity and converting to appropriate units of m^2/yr produces $D_{eff}^{Fe2+} = 7 \times 10^{-3}$ m^2/yr

U = advective flow velocity = 2 m/yr, based on an estimate of the paleo-hydraulic gradient in Navajo Sandstone on a regional scale (Parry 2011)

v = molar volume of precipitate (goethite) = 2.08×10^{-5} $m^3/mole$

c_∞ = concentration of Fe^{2+} far away from a concretion. We choose a conservative 5 mg/kg for these calculations = .0895 $moles/m^3$ based on regional analyses of groundwater from the Navajo Sandstone (Hood and Danielson 1979, 1981; Spangler et al. 1996; Parry et al. 2004; Chidsey et al. 2007; Parry 2011)

c_R = concentration of Fe^{2+} at the concretion surface. The solubility of goethite in groundwaters with pH between 6 and 10, and which have not been highly reduced, is very low, $<1 \times 10^{-3}$ mg/kg, and can be neglected in these calculations (Parry 2011)

calculates the growth time of a concretion of given radius by accounting for the supply of iron by both diffusion and flowing groundwater to a growing surface of a solid, spherical concretion (Parry 2011). This model of growth time (equation 1 and Table 3) assumes that transport of iron via diffusion and advection controls the rate of growth of the concretion; that is, oxidation of aqueous Fe^{2+} and nucleation and precipitation of ferric oxide (e.g., goethite) are rapid compared to the flux of Fe^{2+} to the concretion surface via diffusion and advection.

Diffusion and chemical reactions are temperature-dependent processes. Goethite in tens of micrometer particles is stable relative to hematite up to 60° C (Navrotsky et al. 2008). Mobilization of iron is associated with hydrocarbon migration that took place in the Late Cretaceous to Early Tertiary, and precipitation of iron oxide concretions must follow iron reduction and mobilization. Temperature in the Navajo Sandstone cooled to 50° C, a temperature at which goethite could precipitate, during exhumation in the Late Tertiary (Huntoon et al. 1999). The diffusion model has therefore been calculated using a temperature of 50° C. Other conditions of hydraulic gradient, hydraulic conductivity, and iron concentration in groundwater relevant to this model are presented in Parry (2011).

Growth times calculated with Eq. 1 are compiled in Table 4 for different sizes and types (solid vs. rind) of concretions. These calculations indicate that concretions grow quickly in a geological perspective, in timescales of thousands to tens of thousands of years. Doubling the Fe content of solution cuts the growth time by half. Increasing the temperature to 100° C doubles the diffusion coefficient and decreases the growth time by a factor of 2, but hematite is the more likely precipitate. Decreasing the flow velocity below the 2 m/yr used in the calculations produces growth times that approach the diffusion-only growth.

These calculated growth times (Table 4) assume that transport controls the rate of growth of the concretion. If the rate of growth is instead controlled by chemical reaction, then growth times are even shorter. Parry (2011) demonstrated that the rate of oxidation of aqueous Fe^{2+} by abiotic processes is rapid, being well less than a year timescale for oxidation levels of $P_{O_2} = 1 \times 10^{-5}$ bars or higher. Bacterial mediation of the oxidation reaction increases the rate by a factor of one million (Langmuir 1997). Recrystallization of $Fe(OH)_3$ at $T \geq 50°$ C and pH ≥ 6 is also very rapid (Parry 2011). Hence, concretion growth appears to be transport controlled because of the rapid oxidation of Fe^{2+} with P_{O_2} above 1×10^{-5}, the rapid conversion of $Fe(OH)_3$ to goethite, and the lack of a Fe-depleted zone surrounding concretions.

ANALOG CONCEPTS FOR THE FUTURE

It is difficult to find any terrestrial analog that is a perfect match to Mars (whether depositional environment, diagenetic, or other) because so many different fundamental conditions and parameters vary between Earth and Mars. Mars is likely a unique setting, with its basaltic, sulfate-rich sandstone and extreme chemical solutions (by Earth standards). Several important differences exist on Mars: the lack of plate tectonics, different current atmospheric and gravitational conditions, and the possibility of abiotic environments. Furthermore, Mars exploration techniques are currently conducted via remote evaluation, with a lack of absolute dating. However, the range of terrestrial examples does provide comparative context for understanding Mars processes and end products.

Characterization of both Mars "blueberries" (Calvin et al. 2008) and terrestrial analog examples (e.g., Chan et al. 2004, 2005; Bowen et al. 2008; Potter et al. 2011) provides a strong basis for interpreting broad Earth and Mars conditions. The common occurrence of terrestrial concretions in a wide range of mineralogies (from carbonates to iron oxides and iron sulfides) indicates that concretion formation is a common geologic process in the near surface, porous sediments, and sedimentary rocks.

Ancient terrestrial examples provide a wide range of distributions, geometries, and sizes that demonstrate the variability of both concretions and influences on their growth and development. Field relationships

TABLE 4.—Calculated growth times for concretions (based on equation 1 and Table 3 parameters). Note: Rind concretions are calculated by subtracting spherical growth of interior radius from the value of concretion growth from the exterior radius. NA = not applicable.

Concretion type	Radius (mm)	Rind thickness (mm)	Growth time (years)	Mass transfer conditions
Solid (center to edge)	25	NA	24,000	diffusion
Solid	25	NA	15,000	diffusion and advection
Rind	25 (outer)	2	3600	diffusion
Rind	25 (outer)	2	2000	diffusion and advection
Rind	25 (outer)	5	5000	diffusion and advection
Rind	25 (outer)	1	1000	diffusion and advection

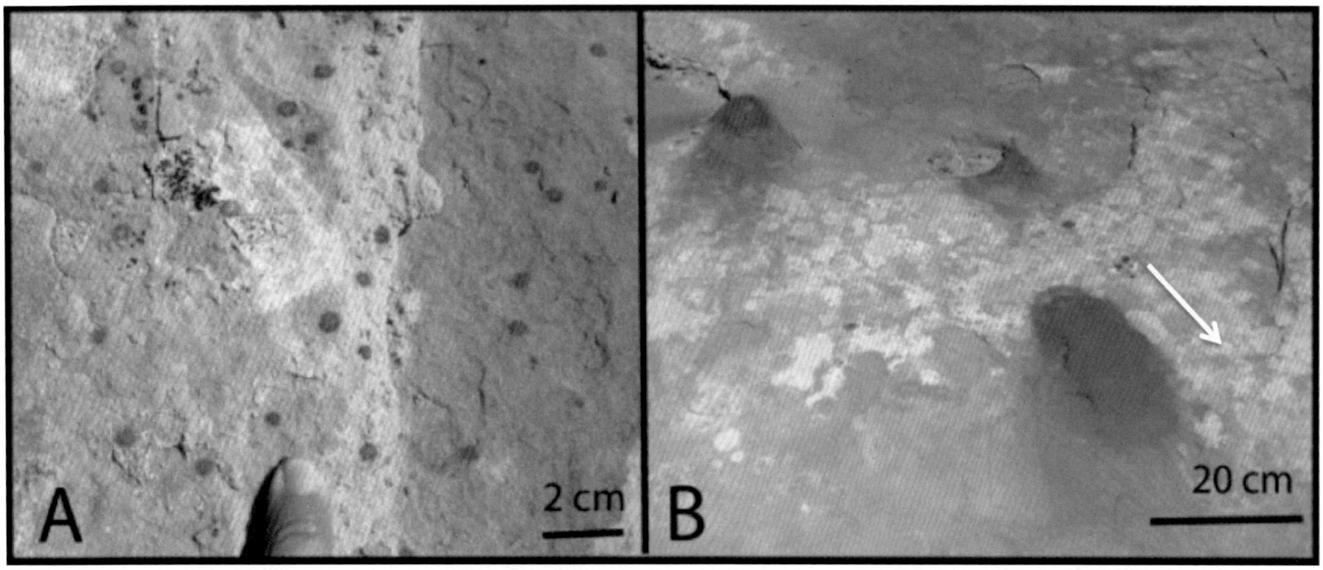

FIG. 10.—Types of mass transfer: (A) diffusion, which produces small spherical concretions (where proper reactants are present at a reaction front) and (B) flow anisotropies (possible advective component), which can produce large forms, such as these vertical pipes in plan view, with asymmetric flow "tails" (arrow showing directional flow). Examples are from the Jurassic Navajo Sandstone, Utah, Moab area.

indicate different types of mass transfer. Advective transport of iron is required for concretion formation based on mass-balance examples (Parry 2011), but diffusion-dominated mass transfer produces spherical concretions in isotropic media. Anisotropic geometries result either from inherent host rock properties that allow faster diffusion in preferential directions (e.g., along bedding or laminae) or from preferential cementation due to fluid flow or flow anisotropies (which may be later) that can produce elongate forms (Fig. 10).

Ferric oxide concretions typically lack a physical nucleus—an important observation in interpreting how concretions grow. Concretion structure and crystal growth examinations show a component of inward growth (much like a geode; Fig. 5), as opposed to the traditional idea of carbonate concretions that might be started from a nucleus and grow outward. Current studies indicate that several intermediate mineralogy stages exist from hydrous ferric oxide gels to goethite to hematite (in order of increasing dehydration over time; Cornell and Schwertmann 2003). Large terrestrial concretion examples commonly show aggregation of small nucleation centers that coalesce together. Small nucleation centers might be easier to form. It is likely that kinetic factors, iron supply, and even small chemical differences have potentially strong effects on concretion mineralogy or expression. Detailed studies of the Burns formation indicate that the Mars spherules could have formed under a range of temperatures (diagenetic to hydrothermal; Morris et al. 2005) and chemical conditions (e.g., Clark et al. 2005, McLennan et al. 2005, Tosca et al. 2008a).

Laboratory analyses provide clues to the complexities of diagenetic processes, with possible multiple compositions of waters and fluctuating water table effects developing over time (Potter and Chan 2011, Potter et al. 2011). Iron isotopes and trace element geochemistry warrant further exploration (Chan et al. 2006, Busigny and Dauphaus 2007). Modeling studies provide some constraints on ideas of nucleation, chemical reactions, and diffusion rates (Sefton-Nash and Catling 2008). It would not be surprising to find that with certain iron-oxidizing or iron-reducing bacteria, concretions could form within tens of years. Laboratory precipitation experiments in combined gel and sand-like media show that small concretionary precipitates with concentrated chemical solutions can begin quickly (in a matter of days; Barge 2009, Barge et al. 2011), although these conditions do not simulate the natural setting. Some modeling of diffusion- and diffusion plus advection-controlled growth of spherical concretions indicates timescales of a few thousand years (Sefton-Nash and Catling 2008, Parry 2011). Numerical models (Chan et al. 2007) help determine boundary conditions, although there are many assumptions that can be difficult to verify.

Major gaps in our knowledge of concretions still remain in several areas:

1. Although concretions may form quickly, natural settings are open (vs. closed) systems over potentially long geologic timescales, providing opportunities for multiple episodes of concretion formation and/or modification. Many variables must be evaluated, and even simple mass balancing of iron and the geochemistry are difficult to reconstruct.

2. Ancient concretions possess little material for traditional methods of absolute dating (e.g., radiometric dating) to pinpoint the timing of diagenetic events. However, more detailed studies of chemistry and careful study of textural and mineral relationships may help elucidate the relative timing.

3. The role of biomediation is still unclear, and while it seems very likely that bacteria played a role in terrestrial concretion formation, original organic matter is unlikely to remain in ancient, highly oxidized rocks such as the Navajo Sandstone. Due to the lack of a carbon signature, evidence of biogenicity would have to be inferred from a suite of diagnostic criteria, such as distinct textures, iron isotopes, and/or trace element compositions.

Despite these gaps and challenges, more studies of concretions will continue to shed light on these fascinating diagenetic artifacts, and we expect that other concretionary forms may be discovered as other sedimentary deposits and formations on Mars are further explored.

SUMMARY

Concretions are important records of groundwater flow through sedimentary deposits. Many terrestrial examples have likely undergone long and complex diagenesis as a result of the abundance of water and numerous biogenic influences on Earth. In addition, the active plate tectonic system on Earth can also encourage the development of multiple fluid pathways and superposition of different event and growth episodes. The inert quartz sandstone host rock mineralogy for the Jurassic Navajo Sandstone would maintain significant porosity for diagenesis even after burial, which may partly explain the diversity of concretion in this formation.

By comparison, in the water-limited Mars system, solid "blueberries" likely formed quickly in the reactive sulfate host rock of the Burns formation, with its abundant iron supply, under diffusive mass transfer, but limited water supply. The unusual chemistry (by Earth standards) might have affected sphericity, crystallinity, and size of the "blueberries."

Terrestrial ferric oxide concretion examples likely formed with limited iron supply and abundant water supply and have continued to be altered and affected by diagenesis for tens of millions of years since formation. Documented characteristics and models of Earth concretions will continue to provide valuable evidence to aid in the interpretation of the origins of spherules on Mars.

ACKNOWLEDGMENTS

This work is supported by the National Aeronautics and Space Administration (to M.A.C.) under grant NNG06GI10G issued through the Mars Fundamental Research Program. The reviews of Scott McLennan and Dorothy Oehler were very helpful in terms of improving the article. We thank various collaborators (on other aspects of concretion research), including Greg Nielsen (Weber State University); Clark Johnson, Brian Beard, and John Valley (all from the University of Wisconsin); Jens Ormö (Centro de Astrobiología in Madrid, Spain); Goro Komatsu (International Research School of Planetary Sciences in Pescara, Italy); and Tony Park (Sienna Geodynamics). We appreciate Rob Sullivan's input on "microberries" and eolian processes on Mars. The Bureau of Land Management–Vermillion Cliffs National Monument and Grand Staircase Escalante National Monument provided logistical support and sampling permission.

REFERENCES

Anbar AD. 2004. Iron stable isotopes: Beyond biosignatures. *Earth Planetary Science Letters (Frontiers)* 217:223–236.

Anderson RS, Bunas KL. 1993. Grain size segregation and stratigraphy in aeolian ripples modeled with a cellular automation. *Nature* 365:740–743.

Barge LM. 2009. Self-organized chemical precipitates: Laboratory and field studies [PhD thesis]: University of Southern California, Los Angeles.

Barge LM, Hammond DH, Chan MA, Potter S, Petruska J, Nealson K. 2011. Precipitation patterns formed by self-organizing processes in porous media. *Geofluids* 11:124–133. doi:10.1111/j.1468-8123.2010.00324.x.

Beard BL, Johnson CM. 2004. Inter-mineral Fe isotope variations in mantle-derived rocks and implications for the Fe geochemical cycle. *Geochimica et Cosmochimica Acta* 68:4727–4743.

Beitler B, Chan MA, Parry WT. 2003. Bleaching of Jurassic Navajo Sandstone on Colorado Plateau Laramide highs: Evidence of exhumed hydrocarbon supergiants? *Geology* 31:1041–1044.

Beitler B, Chan MA, Parry WT, Ormö J, Komatsu G. 2004. Diagenetic analogs to hematite regions on Mars: Examples from Jurassic sandstones of southern Utah, USA. *In* Hoover RB, Levin GV, Rozanov AY (Editors). *Instruments, Methods, and Missions for Astrobiology VIII*, Proceedings of SPIE, Vol. 5555: Society for Optics and Photonics, Bellingham, Washington. p. 162–169, doi:10.1117/12.581625.

Beitler B, Parry WT, Chan MA. 2005. Fingerprints of fluid flow: Chemical diagenetic history of the Jurassic Navajo Sandstone, southern Utah, USA. *Journal of Sedimentary Research* 75:547–561.

Berner R. 1968. Rate of concretion growth. *Geochimica et Cosmochimica Acta* 32:477–483.

Berner RA. 1980. *Early Diagenesis: A Theoretical Approach*: Princeton University Press, New Jersey. 241 p.

Benison KC, Bowen BB. 2006. Acid saline lake systems give clues about past environments and the search for life on Mars. *Icarus* 183:225–229.

Benison KC, Bowen BB, Oboh-Ikuenobe FE, Jagniecki EA, LaClair DA, Story SL, Mormile MR, Hong BY. 2007. Sedimentology of acid saline lakes in Southern Western Australia: Newly described processes and products of an extreme environment. *Journal of Sedimentary Research* 77:366–388.

Benison KC, Jagniecki EA, Edwards TB, Mormile MR, Storrie-Lombardi MC. 2008. "Hairy blobs": Microbial suspects from modern and ancient ephemeral acid saline evaporates. *Astrobiology* 8:807–821.

Benison KC, LaClair DA. 2003. Modern and ancient extremely acid saline deposits: Terrestrial analogs for martian environments? *Astrobiology* 3:609–618.

Bowen B, Martini BA, Chan MA, Parry WT. 2007. Reflectance spectroscopic mapping of diagenetic heterogeneities and fluid-flow pathways in the Jurassic Navajo Sandstone. *American Association of Petroleum Geologists Bulletin* 91:173–190.

Bowen BB. 2005. Sandstone bleaching and iron concretions: An index to fluid pathways and diagenetic history in Jurassic Navajo Sandstone, southern Utah [PhD dissertation]: University of Utah, Salt Lake City, 231 p.

Bowen BB, Benison KC, Oboh-Ikuenobe F, Story S, Mormile M. 2008. Active hematite concretion formation in modern acid saline lake sediments, Lake Brown, Western Australia. *Earth and Planetary Science Letters* 268:52–63.

Busigny V, Dauphas N. 2007. Tracing paleofluid circulations using iron isotopes: A study of hematite and goethite concretions from the Navajo Sandstone (Utah, USA). *Earth and Planetary Science Letters* 254:272–287.

Calvin WM, Shoffner JD, Johnson, JR, Knoll AH, Pocock JM, Squyres SW, Weitz CM, Arvidson, RE, Bell JF III, Christensen PR, de Souza, PA Jr., Farrand, WH, Glotch, TD, Herkenhoff KE, Jolliff BL, Knudson AT, McLennan SM, Rogers AD, Thompson SD. 2008. Hematite spherules at Meridiani: Results from MI, Mini-TES and Pancam. *Journal of Geophysical Research* 113:E12S37.

Catling DC. 2004. On Earth, as it is on Mars? *Nature* 429:707–708.

Chan MA, Beitler B, Parry WT, Ormö J, Komatsu G. 2004. A possible terrestrial analogue for hematite concretions on Mars. *Nature* 429:731–734.

Chan MA, Beitler B, Parry WT, Ormö J, Komatsu G. 2005. Red rock and red planet diagenesis: Comparisons of Earth and Mars concretions. *Geological Society of America Today* 15:4–10.

Chan MA, Johnson CM, Beard BL, Bowman JR, Parry WT. 2006. Iron isotopes constrain the pathways and formation mechanisms of terrestrial oxide concretions: A tool for tracing iron cycling on Mars? *Geosphere* 2:324–332.

Chan MA, Ormö J, Park AJ, Stich M, Souza-Egipsy V, Komatsu G. 2007. Models of iron oxide concretion formation: Field, laboratory, and numerical comparisons. *Geofluids* 7:1–13.

Chan MA, Parry WT. 2002. Rainbow of rocks: Mysteries of sandstone colors and concretions in Colorado Plateau Canyon Country. *Utah Geological Survey Public Information Series*, Salt Lake City, Utah. 77, 17 p.

Chan MA, Parry WT, Bowman JR. 2000. Diagenetic hematite and manganese oxides and fault-related fluid flow in Jurassic sandstones, southeastern Utah. *American Association of Petroleum Geologists Bulletin* 84:1281–1310.

Chan MA, Parry WT, Petersen EU, Hall CM. 2001. ^{40}Ar-^{39}Ar age and chemistry of manganese mineralization in the Moab to Lisbon fault systems, southeastern Utah. *Geology* 29:331–334.

Chidsey TC Jr, DeHamer JS, Hartwick EE, Johnson KR, Schelling DD, Sprinkel DA, Strickland DK, Vrona JP, Wavrek DA. 2007. Petroleum geology of Covenant oil field, central Utah thrust belt. *In* Willis GC, Hylland MD, Clark DL, Chidsey TC Jr (Editors). *Central Utah—Diverse Geology of a Dynamic Landscape*, Publication 36: Utah Geological Association, Salt Lake City. p. 273–296.

Cibin U, Cavazza W, Fontana D, Milliken KL, McBride EF. 1993. Comparison of composition and texture of calcite-cemented concretions and host sandstones, Northern Apennines, Italy. *Journal of Sedimentary Petrology* 63:945–954.

Clark BC, Morris RV, McLennan SM, Gellert R, Jolliff B, Knoll AH, Squyres WS, Lowenstein TK, Ming DW Tosca NJ, Yen A, Christensen PR, Gorevan S, Brukner J, Calvin W, Dreibus G, Farrand W, Klingelhoefer G, Waenke H, Zipfel J. Bell III JF, Grotzinger J, McSween HY, Rieder R. 2005. Chemistry and mineralogy of outcrops at Meridiani Planum. *Earth and Planetary Science Letters* 240:73–94, doi:10.1016/j.epsl.2005.09.040.

Clifton H. 1957. The carbonate concretions of the Ohio Shale. *The Ohio Journal of Science* 57:114–124.

Cornell RM, Schwertmann U. 2003. *The Iron Oxides: Structures, Properties, Reactions, Occurrences and Uses*: Wiley VCH, New York. 659 p.

Drever JI. 1997. *The Geochemistry of Natural Waters, Surface and Groundwater Environments*, 3rd ed.: Prentice Hall, Upper Saddle River, New Jersey. 436 p.

Eichhubl P, Taylor WL, Pollard DD, Aydin A. 2004. Paleo-fluid flow and deformation in the Aztec Sandstone at the Valley of Fire, Nevada: Evidence for the coupling of hydrogeologic, diagenetic, and tectonic processes. *Geological Society of America Bulletin* 116:1120–1136.

Fein JB, Daughney CJ, Yee N, Davis TA. 1997. A chemical equilibrium model for metal adsorption onto bacterial surfaces. *Geochimica et Cosmochimica Acta* 61:3319–3328.

Fernandez-Remolar DC, Knoll AH. 2008. Fossilization potential of iron-bearing minerals in acidic environments of Rio Tinto, Spain: Implications for Mars exploration. *Icarus* 194:72–85.

Ferris FG. 2005. Microbial mineral transformations in the Fe(II)-Fe(III)-H (sub 2) O system. *Geochimica et Cosmochimica Acta* 69 (Supplement):454 p.

Ferris FG, Hallberg RO, Lyven B, Pedersen K. 2000. Retention of strontium, cesium, lead and uranium by bacterial iron oxides from a subterranean environment. *Applied Geochemistry* 15:1035–1042.

Fowle DA, Fein JB. 1999. Competitive adsorption of metal cations onto two gram positive bacteria: Testing the chemical equilibrium model. *Geochimica et Cosmochimica Acta* 63:19–20.

Gauthier DL. 1982. Siderite concretions: Indicators of early diagenesis in the Gammon Shale (Cretaceous). *Journal of Sedimentary Petrology* 52:859–871.

Grotzinger JP, Arvidson RE, Bell JF III, Calvin W, Clark BC, Fike DA, Golombek M, Greeley R, Haldemann A, Herkenhoff KE, Jolliff BL, Knoll AH, Malin M, McLennan SM, Parker T, Soderblom L, Sohl-Dickstein JN, Squyres SW, Tosca NJ, Watters WA. 2005. Stratigraphy and sedimentology of a dry to wet eolian depositional system, Burns Formation, Meridiani Planum, Mars. *Earth and Planetary Science Letters* 240:11–72.

Heim JA, Vasconcelos PM, Shuster DL, Farley KA, Broadbent G. 2006. Dating palaeochannel iron ore by (U-Th)/He analysis of supergene goethite, Hamersley Province, Australia. *Geology* 34:173–176.

Henisch HK. 2005. *Crystals in Gels and Liesegang Rings*: Cambridge University Press, Cambridge, UK. 212 p.

Hood JW, Danielson TW. 1979. *Aquifer Tests of the Navajo Sandstone Near Caineville, Wayne County, Utah*: State of Utah Department of Natural Resources, Salt Lake City. Technical Publication 66, 69 p.

Hood JW, Danielson TW. 1981. *Bedrock Aquifers in the Lower Dirty Devil Basin Area, Utah with Special Emphasis on the Navajo Sandstone*: State of Utah Department of Natural Resources, Salt Lake City. Technical Publication 68, 143 p.

Huntoon JE, Hansley PL, Naeser ND. 1999. The search for a source rock for the giant tar sand triangle accumulation, southeastern Utah. *American Association of Petroleum Geologists Bulletin* 83:467–495.

Jerolmack DJ, Mohrig D, Grotzinger JP, Fike DA, Watters WA. 2006. Spatial grain size sorting in eolian ripples and estimation of wind conditions on planetary surfaces: Applications to Meridiani Planum, Mars. *Journal of Geophysical Research* 11:1–14.

Johnson CM, Beard B, Roden E, Newman D, Nealson K. 2004. Isotopic constraints on biogeochemical cycling of Fe. *In* Johnson C, Beard B, Albarède F (Editors). *Geochemistry of Non-Traditional Stable Isotopes, Reviews in Mineralogy and Geochemistry 55*: Mineralogical Society of America, Chantilly, Virginia. p. 359–408.

Johnson MR. 1989. Paleogeographic significance of oriented calcareous concretions in the Triassic Katberg Formation, South Africa. *Journal of Sedimentary Petrology* 59:1008–1010.

Klingelhoefer G, Morris RV, Bernhardt B, Schroeder C, Rodionov DS, de Souza PA, Yen A, Gellert R, Evlanov EN, Zubkov B, Foh J, Bonnes U, Kankeleit E, Guetlich P, Ming DW, Renz F. Wdowiak T, Squyres SW, Arvidson RE, Rowan L. 2004. Jarosite and hematite at Meridiani Planum from Opportunity's Moessbauer spectrometer. *Science* 306:1740–1745.

Langley T, Colberg MK. 2006. *Characterization of Cements in Moki Marbles as a Possible Source of Arsenic in Groundwater, Washington County, Utah*: Geological Society of America Rocky Mountain Section, Paper 19-1.

Langmuir D. 1997. *Aqueous Environmental Geochemistry*: Prentice Hall, Upper Saddle River, New Jersey. 600 p.

Liermann LJ, Kalinowski BE, Brantley SL, Ferry JG. 2000. Role of bacterial siderophores in dissolution of hornblende. *Geochimica et Cosmochimica Acta* 64:587–602.

Lowenstein TK, Schubert BA, Timofeeff MN. 2011. Microbial communities in fluid inclusions and long-term survival in halite. *Geological Society of America Today* 21:4–9.

Madden MEE, Bodnar RJ, Rimstidt JD. 2004. Jarosite as an indicator of water-limited chemical weathering on Mars. *Nature* 43:821–823.

McBride E, Picard MD, Milliken KL. 2003. Calcite-cemented concretions in Cretaceous sandstone, Wyoming and Utah, USA. *Journal of Sedimentary Research* 73:462–484.

McBride EF, Milliken KL, Cavazza W, Cibin U, Fontana D, Picard MD, Zuffa GG. 1995. Heterogeneous distribution of calcite cement at the outcrop scale in Tertiary sandstones, Northern Apennines, Italy. *American Association of Petroleum Geologists Bulletin* 79:1044–1063.

McBride EF, Picard MD, Folk RL. 1994. Oriented concretions, Ionian Coast, Italy: Evidence of groundwater flow direction. *Journal of Sedimentary Research Section A, Sedimentary Petrology and Processes* 64:535–540.

McLennan SM, Bell JF III, Calvin W, Christensen PR, Clark BC, de Souza PA, Farmer J, Farrand WH, Fike DA, Gellert R, Ghosh A, Glotch TD, Grotzinger JP, Hahn B, Herkenhoff KE, Hurowitz JA, Johnson JR, Johnson SS, Jolliff BL, Klingelhöfer G, Knoll AH, Learner ZA, Malin MC, McSween HY Jr, Pocock J, Ruff SW, Soderblom LA, Squyres SW, Tosca NJ, Watters WA, Wyatt MB, Yen A. 2005. Provenance and diagenesis of the evaporite-bearing Burns Formation, Meridiani Planum, Mars. *Earth and Planetary Science Letters* 240:95–121.

Mormile MR, Hong B-Y, Benison KC. 2009. Molecular analysis of the microbial communities of Mars analog lakes in Western Australia. *Astrobiology* 9:919–930.

Morris RV, Ming DW, Graff TG, Arvidson RE, Bell JF III, Squyres SW, Mertzman SA, Gruener JE, Golden DC, Le L, Robinson GA. 2005. Hematite spherules in basaltic tephra altered under aqueous, acid sulfate conditions on Mauna Kea volcano, Hawaii: Possible clues for the occurrence of hematite-rich spherules in the Burns Formation at Meridiani Planum, Mars. *Earth and Planetary Science Letters* 240:168–178.

Morris RV, Klingelhoefer G, Schroeder C, Rodionov DS, Yen A, Ming DW, deSouza Jr. PA, Wdowiak T, Fleischer I, Gellert R, Bernhardt B, Bonnes U, Cohen BA, Evlanov EN, Foh J, Guetlich, P, Kankeleit E, McCoy TJ, Mittlefehldt DW, Renz F, Schmidt ME, Zubkov B, Squyres, SW, Arvidson RE, Anonymous. 2006. Moessbauer mineralogy of rock, soil, and dust at Meridiani Planum, Mars: Opportunity's journey across sulfate-rich outcrop, basaltic sand and dust, and hematite lag deposits. *Journal of Geophysical Research* 111:E12.

Morrison SJ, Parry WT. 1986. Formation of carbonate-sulfate veins associated with copper ore deposits from saline basin brines, Lisbon Valley, Utah: Fluid inclusion and isotopic evidence. *Economic Geology* 81:1853–1866.

Mozley P. 1989. Complex compositional zonation in concretionary siderite: Implications for geochemical studies. *Journal of Sedimentary Petrology* 59:815–818.

Mozley P. 1996. The internal structure of carbonate concretions in mudrocks: A critical evaluation of the conventional concentric model of concretion growth. *Sedimentary Geology* 103:85–91.

Mozley P, Davis JM. 2005. Internal structure and mode of growth of elongate calcite concretions: Evidence for small-scale, microbially induced, chemical heterogeneity in groundwater. *Geological Society of America Bulletin* 117:1400–1412.

Mozley P, Goodwin LB. 1995. Patterns of cementation along a Cenozoic normal fault: A record of paleoflow orientations. *Geology* 23:539–542.

Navrotsky A, Mazeina L, Majzlan J. 2008. Size-driven structural and thermodynamic complexity in iron oxides. *Science* 319:1635–1638.

Nielsen G. 2010. Regional correlation of diagenetic coloration facies and analysis of iron oxide cementation processes, Jurassic Navajo Sandstone, southwestern Utah [PhD dissertation]: University of Utah, Salt Lake City, 268 p.

Oelkers EH, Helgeson HC. 1988. Calculation of the thermodynamic and transport properties of aqueous species at high pressures and temperatures: Aqueous tracer diffusion coefficients of ions to 1000°C and 5 kb. *Geochimica et Cosmochimica Acta* 52:63–85.

Ormö J, Komatsu G, Chan MA, Beitler B, Parry WT. 2004. Geological features indicative of processes related to the hematite formation in Meridiani Planum and Aram Chaos, Mars: A comparison with diagenetic hematite deposits in southern Utah, USA. *Icarus* 171:295–316.

Ortoleva P. 1994a. *Agates, Geodes, Concretions and Orbicules: Self-Organized Zoning and Morphology: Fractals and Dynamic Systems in Geosciences*: Springer-Verlag, Berlin. 283 p.

Ortoleva PJ. 1984. The self organization of Liesegang bands and other precipitate patterns. *In* Nicolis G, Baras F (Editors). *Chemical Instabilities, Applications in Chemistry, Engineering, Geology, and Materials Science*, NATO ASI Series C: D. Reidel Publishing Co., Boston. p. 289–297.

Ortoleva PJ. 1994b. *Geochemical Self-Organization*: Oxford University Press, New York. 411 p.

Pantin HM. 1958. Rate of formation of a diagenetic calcareous concretion. *Journal of Sedimentary Petrology* 28(3):366–371.

Parry WT. 2011. Composition, nucleation, and growth of iron oxide concretions in the Jurassic Navajo Sandstone, Utah. *Sedimentary Geology* 233:53–68.

Parry WT, Chan MA, Beitler B. 2004. Chemical bleaching indicates fluid flow in sandstone deformation bands. *American Association of Petroleum Geologists Bulletin* 88:175–191.

Potter SL, Chan M, Petersen E, Dyar MD, Sklute E. 2011. Characterization of Navajo Sandstone Concretions: Mars comparison and criteria for distinguishing diagenetic origins. *Earth and Planetary Science Letters* 301:444–456.

Potter SL, Chan MA. 2011. Iron mass transfer and fluid flow patterns in Jurassic Navajo Sandstone, southern Utah, USA. *Geofluids* 11:184–198.

Raiswell R. 1971. Cementation in some Cambrian concretions, South Wales. *Johns Hopkins University Studies in Geology* 19:196–197.

Raiswell R, Fisher QJ, Cope JCW, Curtis CD. 2000. Mudrock-hosted carbonate concretions: A review of growth mechanisms and their influence on chemical and isotopic composition. *Geological Society of London Journal* 157:239–251.

Roberts EM, Chan MA. 2010. Variations in iron oxide, iron sulfide, and carbonate concretions and their distributions in fluvio-deltaic and nearshore sandstones: Cretaceous examples from the Kaiparowits Plateau, Utah and San Juan Basin, New Mexico. *In* Carney SM, Tabet DE, Johnson CL (Editors). *Geology of South-Central Utah*, Publication 39: Utah Geological Association, Salt Lake City. 442 p.

Schultz CB. 1941. The pipy concretions of the Arikaree. *Nebraska University Bulletin* 2:69–82.

Sefton-Nash E, Catling DC. 2008. Hematitic concretions at Meridiani Planum, Mars: Their growth timescale and possible relationship with iron sulfates. *Earth and Planetary Science Letters* 269:366–376.

Seilacher A. 2001. Concretion morphologies reflecting diagenetic and epigenetic pathways. *Sedimentary Geology* 143:41–57.

Seiler WM. 2008. Jurassic Navajo Sandstone of Coyote Buttes, Utah/Arizona: Coloration and diagenetic history, preservation of a dinosaur trample surface, and a terrestrial analog to Mars [MS thesis]: University of Utah, Salt Lake City.

Sellés-Martinez J. 1996. Concretion morphology, classification and genesis. *Earth Science Reviews* 41:3–4.

Sharp RP. 1963. Wind ripples. *Journal of Geology* 71:617–636.

Shuster DL, Farley KA. 2004. 4He/3He thermochronometry. *Earth and Planetary Science Letters* 217:1–17.

Shuster DL, Farley KA, Sisterson JM, Burnett DS. 2004. Quantifying the diffusion kinetics and spatial distributions of radiogenic 4He in minerals containing proton-induced 3He. *Earth and Planetary Science Letters* 217:19–32.

Shuster DL, Vasconcelos PM, Heim JA, Farley KA. 2005. Weathering geochronology by (U-Th)/He dating of goethite. *Geochimica et Cosmochimica Acta* 69:659–673.

Spangler LE, Naftz DL, Peterman ZE. 1996. Hydrology, Chemical Quality, and Characterization of Salinity in the Navajo Aquifer in and Near the Greater Aneth Oil Field, San Juan County, Utah: US Geological Survey, Water-Resources Investigations Report 96–4155, 90 p.

Squyres SW, Grotzinger JP, Arvidson RE, Bell JF III, Calvin W, Christensen PR, Clark BC, Crisp JA, Farrand WH, Herkenhoff KE, Johnson JR, Klingelhöfer G, Knoll AH, McClennan SM, McSween HY, Morris RV, Rice JW, Rieder R, Soderblom LA. 2004. In situ evidence for an ancient aqueous environment at Meridiani Planum, Mars. *Science* 306:1709–1714.

Squyres SW, Knoll AH. 2005. Sedimentary rocks at Meridiani Planum: Origin, diagenesis, and implications for life on Mars. *Earth and Planetary Science Letters* 240:1–10.

Squyres SW, Knoll AH, Arvidson RE, Clark BC, Grotzinger JP, Jolliff BL, McLennan SM, Tosca N, Bell JF III, Calvin WM, Farrand WH, Glotch TD, Golombek MP, Herkenhoff KE, Johnson JR, Klingelhöfer G, McSween HY, Yen AS. 2006. Two years at Meridiani Planum: Results from the Opportunity Rover. *Science* 313:1403–1407.

Steefel CI, Van Cappellen P. 1990. A new kinetic approach to modeling water-rock interaction: The role of nucleation, and Ostwald ripening. *Geochimica et Cosmochimica Acta* 54:2657–2677.

Sumner DY. 2004. Poor preservation potential of organics in Meridiani Planum hematite-bearing sedimentary rocks. *Journal of Geophysical Research–Planets* 109:E12007.

Thorson JP, MacIntyre T. 2005. Geology of the Cashin Mine sandstone-hosted disseminated copper deposit, Montrose County, Colorado. *In Lisbon Valley Sediment-Hosted Copper Deposits and Paradox Basin Fluids*: Society of Economic Geologists, Guidebook Series, Vol. 37. p. 43–49.

Tosca NJ, Knoll AH, McLennan SM. 2008b. Water activity and the challenge for life on early Mars. *Science* 320:1204–1207, doi:10.1126/science.1155432.

Tosca NJ, McLennan SM, Clark BC, Grotzinger JP, Hurowitz JA, Knoll AH, Schroeder C, Squyres SW. 2005. Geochemical modeling of evaporation processes on Mars: Insight from the sedimentary record at Meridiani Planum. *Earth and Planetary Science Letters* 240:122–148.

Tosca NJ, McLennan SM, Dyar MD, Sklute EC, Michel FM. 2008a. Fe-oxidation processes at Meridiani Planum and implications for secondary Fe-mineralogy on Mars. *Journal of Geophysical Research–Planets* 113:E05005.

Tosca NJ, McLennan SM, Hindsley DH, Schoonen MAA. 2004. Acid-sulfate weathering of synthetic Martian basalt: The acid fog model revisited. *Journal of Geophysical Research* 109:E5.